T0183698

Lecture Notes in Artificial Intelligence 11726

Subseries of Lecture Notes in Computer Science

Series Editors

Randy Goebel
University of Alberta, Edmonton, Canada
Yuzuru Tanaka
Hokkaido University, Sapporo, Japan
Wolfgang Wahlster
DFKI and Saarland University, Saarbrücken, Germany

Founding Editor

Jörg Siekmann
DFKI and Saarland University, Saarbrücken, Germany

More information about this series at http://www.springer.com/series/1244

Gabriele Kern-Isberner · Zoran Ognjanović (Eds.)

Symbolic and Quantitative Approaches to Reasoning with Uncertainty

15th European Conference, ECSQARU 2019
Belgrade, Serbia, September 18–20, 2019
Proceedings

 Springer

Editors
Gabriele Kern-Isberner ⓘ
Technische Universität Dortmund
Dortmund, Germany

Zoran Ognjanović ⓘ
Mathematical Institute of the Serbian
Academy of Sciences and Arts
Belgrade, Serbia

ISSN 0302-9743   ISSN 1611-3349   (electronic)
Lecture Notes in Artificial Intelligence
ISBN 978-3-030-29764-0   ISBN 978-3-030-29765-7   (eBook)
https://doi.org/10.1007/978-3-030-29765-7

LNCS Sublibrary: SL7 – Artificial Intelligence

© Springer Nature Switzerland AG 2019
This work is subject to copyright. All rights are reserved by the Publisher, whether the whole or part of the material is concerned, specifically the rights of translation, reprinting, reuse of illustrations, recitation, broadcasting, reproduction on microfilms or in any other physical way, and transmission or information storage and retrieval, electronic adaptation, computer software, or by similar or dissimilar methodology now known or hereafter developed.
The use of general descriptive names, registered names, trademarks, service marks, etc. in this publication does not imply, even in the absence of a specific statement, that such names are exempt from the relevant protective laws and regulations and therefore free for general use.
The publisher, the authors and the editors are safe to assume that the advice and information in this book are believed to be true and accurate at the date of publication. Neither the publisher nor the authors or the editors give a warranty, expressed or implied, with respect to the material contained herein or for any errors or omissions that may have been made. The publisher remains neutral with regard to jurisdictional claims in published maps and institutional affiliations.

This Springer imprint is published by the registered company Springer Nature Switzerland AG
The registered company address is: Gewerbestrasse 11, 6330 Cham, Switzerland

Preface

The biennal ECSQARU conference is a major forum for advances in the theory and practice of reasoning under uncertainty. Contributions are provided by researchers in advancing the state of the art and practitioners using uncertainty techniques in applications. The scope of the ECSQARU conferences encompasses fundamental topics as well as practical issues, related to representation, inference, learning, and decision making both in qualitative and numeric uncertainty paradigms.

Previous ECSQARU events were held in Lugano (2017), Compiegne (2015), Utrecht (2013), Belfast (2011), Verona (2009), Hammamet (2007), Barcelona (2005), Aalborg (2003), Toulouse (2001), London (1999), Bonn (1997), Fribourg (1995), Granada (1993), and Marseille (1991).

The 15th European Conference on Symbolic and Quantitative Approaches to Reasoning with Uncertainty (ECSQARU 2019) was held in Belgrade, Serbia, during September 18–20, 2019. The 41 papers in this volume were selected from 62 submissions, after a rigorous peer-review process by the members of the Program Committee and some external reviewers. Each submission was reviewed by at least 2, and on the average 3.1, reviewers. ECSQARU 2019 also included invited talks by outstanding researchers in the field: Fabio Gagliardi Cozman (University of São Paulo), Lluís Godo (Artificial Intelligence Research Institute IIIA, Spanish National Research Council CSIC), and Francesca Toni (Imperial College London).

We would like to thank all those who submitted papers, the members of the Program Committee and the external reviewers for their valuable reviews, and the members of the local Organizing Committee for their contribution to the success of the conference. Financial support from the Ministry of Education, Science and Technological Development of the Republic of Serbia, as well as operational support from the Serbian Academy of Sciences and Arts Council was greatly appreciated. We are also grateful to Springer Nature for granting a Best Paper Award of the conference, and for the smooth collaboration when preparing the proceedings. Moreover, EasyChair proved to be a convenient platform for handling submissions, reviewing, and final papers for the proceedings of ECSQARU 2019, which was greatly appreciated.

July 2019

Gabriele Kern-Isberner
Zoran Ognjanović

Organization

Conference and PC Chairs

Gabriele Kern-Isberner   Technische Universität Dortmund, Germany
Zoran Ognjanović   Mathematical Institute of the Serbian Academy
     of Sciences and Arts, Serbia

Program Committee

Leila Amgoud	IRIT, CNRS, France
Alessandro Antonucci	IDSIA, Switzerland
Ringo Baumann	Leipzig University, Germany
Christoph Beierle	University of Hagen, Germany
Vaishak Belle	The University of Edinburgh, UK
Concha Bielza	Universidad Politécnica de Madrid, Spain
Isabelle Bloch	LTCI, Télécom Paris, France
Giovanni Casini	University of Luxembourg, Luxembourg
Laurence Cholvy	ONERA-Toulouse, France
Giulianella Coletti	University of Perugia, Italy
Inés Couso	University of Oviedo, Spain
Fabio G. Cozman	University of São Paulo, Brazil
Fabio Cuzzolin	Oxford Brookes University, UK
Luis M. De Campos	University of Granada, Spain
Thierry Denoeux	Université de Technologie de Compiègne, France
Sebastien Destercke	CNRS, UMR Heudiasyc, France
Dragan Doder	IRIT, Université Paul Sabatier, France
Florence Dupin de St-Cyr	IRIT, Université Paul Sabatier, France
Zied Elouedi	Institut Superieur de Gestion de Tunis, Tunisia
Patricia Everaere	CRIStAL - Université Lille, France
Alessandro Facchini	IDSIA, Switzerland
Eduardo Fermé	Universidade da Madeira, Portugal
Tommaso Flaminio	Artificial Intelligence Research Institute, IIIA-CSIC, Spain
Laurent Garcia	LERIA, Université d'Angers, France
Laura Giordano	DISIT, Università del Piemonte Orientale, Italy
Lluís Godo	Artificial Intelligence Research Institute, IIIA-CSIC, Spain
Anthony Hunter	University College London, UK
Nebojša Ikodinović	University of Belgrade, Serbia
Souhila Kaci	LIRMM, France
Sébastien Konieczny	CRIL, CNRS, France
Jérôme Lang	CNRS, LAMSADE, Università Paris-Dauphine, France

Florence Le Ber	Université de Strasbourg, ENGEES, France
Joao Leite	Universidade NOVA de Lisboa, Portugal
Philippe Leray	LS2N - DUKe, Nantes University, France
Peter Lucas	Leiden University, The Netherlands
Carsten Lutz	Universität Bremen, Germany
Francesca Mangili	IDSIA, USI-SUPSI, Switzerland
Pierre Marquis	CRIL, Université d'Artois, and CNRS - Institut Universitaire de France, France
Maria Vanina Martinez	CONICET, Universidad de Buenos Aires, Argentina
David Mercier	Université d'Artois, France
Thomas Meyer	University of Cape Town, and CAIR, South Africa
Enrique Miranda	University of Oviedo, Spain
Serafín Moral	University of Granada, Spain
Farid Nouioua	LIS CNRS UMR 7020, Aix-Marseille Université, France
Odile Papini	LIS CNRS UMR 7020, Aix-Marseille Université, France
Davide Petturiti	University of Perugia, Italy
Rafael Peñaloza	University of Milano-Bicocca, Italy
Nico Potyka	University of Osnabrueck, Germany
Henri Prade	IRIT, CNRS, France
Silja Renooij	Utrecht University, The Netherlands
Guillermo R. Simari	Universidad del Sur in Bahia Blanca, Argentina
Umberto Straccia	ISTI, CNR, Italy
Karim Tabia	Artois University, France
Choh Man Teng	Florida Institute for Human and Machine Cognition, USA
Rafael Testa	Unicamp, Brazil
Matthias Thimm	Universität Koblenz-Landau, Germany
Matthias Troffaes	Durham University, UK
Guy Van den Broeck	University of California, Los Angeles, USA
Linda van der Gaag	IDSIA, Switzerland
Leon van der Torre	University of Luxembourg, Luxembourg
Ivan Varzinczak	University of Artois, CNRS, France
Srdjan Vesic	CNRS, CRIL, Université d'Artois, France
Jirka Vomlel	Czech Academy of Sciences, Czech Republic
Renata Wassermann	University of São Paulo, Brazil
Stefan Woltran	Vienna University of Technology, Austria
Éric Würbel	LIS CNRS UMR 7020, Aix-Marseille Université, France

Additional Reviewers

Camille Bourgaux	École Normale Supérieure Paris, France
Arina Britz	CAIR, South Africa
Maximilian Heinrich	Leipzig University, Germany

Jean Christoph Jung	Universität Bremen, Germany
Václav Kratochvíl	Czech Academy of Sciences, Czech Republic
Aleksandar Perović	University of Belgrade, Serbia
Marco Wilhelm	Technische Universität Dortmund, Germany

Abstracts of Invited Talks

Knowledge Representation
in Knowledge-Enhanced Machine Learning:
How? Where?

Fabio Gagliardi Cozman ⓘ

Escola Politécnica, Universidade de São Paulo
fgcozman@usp.br

Amazing success has been attained by artificial intelligences that resort to data intensive machine learning, for instance in natural language processing and in recommendation systems. Can we build an artificial intelligence endowed with full logical and commonsense reasoning just out of pattern extraction from ever increasing datasets? Possibly. But it seems reasonable to assume that tasks at higher abstraction levels demand at least bits of knowledge representation mixed with machine learning. In any event, several questions must be answered before we can have *knowledge enhanced* machine learning at our fingertips.

How can we bring theoretical insights and practical tools from knowledge representation into machine learning tasks? Where is it worthwhile to add the power (and the cost) of knowledge representation to available datasets? How to evaluate the resulting combination of formalisms? This invited talk discusses these questions, necessarily focusing on a small subset of possible answers. Overall we emphasize the knowledge representation aspects of knowledge enhanced machine learning, optimistically assuming that optimization and estimation methods will be available whenever needed.

We start by examining languages that combine logical formulas/rules with probabilities, as such languages must be key tools in our intended mix. The combination of logic and probability has an old and rich history; connections have been rediscovered more than once in artificial intelligence research [7]. In particular, during the past two decades there has been steady interest in languages that mix probabilistic graphical models, such as Bayesian networks, and relational logic [4]. Another line of research under investigation for more than twenty years has focused on probabilistic logic programming [10]. There are now solid techniques, often imported from finite model theory, that support us in studying these languages; results discussed in the talk are extracted from Refs. [1–3]. We compare the various languages, arguing that several ideas behind probabilistic logic programming are particularly valuable.

However, given the often "unreasonable" effectiveness of data in producing ostensibly intelligent behavior [6], it seems that we should *not* try to force knowledge representation into any machine learning task. Rather, we should carefully look for those tasks where knowledge-enhanced techniques will really make a difference. In this talk we discuss the task of explaining a link prediction in a knowledge base. In such a

Partially supported by CNPq grant 312180/2018-7 and by FAPESP grant 2016/18841-0.

task we do have knowledge, and the state of art methods resort to embeddings that are very difficult to interpret (that is, all entities and relations are mapped into vectors, and relationships are then expressed by relatively simple mathematical operations such as addition) [9]. The difficulty with embeddings is that decisions depend on numerical values that are apparently disconnected from semantic meaning. We discuss how explanations for link predictions can be extracted from embeddings, explanations that for instance resort to Horn clauses and similar formalisms [5, 11].

But even explanations can be learned from data: one can learn how to explain the behavior of another learner… and so on. Thus one might just argue that we can keep improving our pattern extraction methods so that they learn both to decide and to explain decisions, leaving aside any need for knowledge representation. To investigate the limits of knowledge-free learning, we propose a test inspired by the Winograd challenge [8] that can exercise the connection between commonsense reasoning and data intensive language processing. We suggest that such a Winograd Explaining Challenge, where the goal is to explain the answer to a Winograd scheme, can help focus our attention on problems that can only be solved by a mix of machine learning and commonsense reasoning. We discuss how we might go about facing such a test, and which research directions it opens.

References

1. Cozman, F.G., Mauá, D.D.: On the semantics and complexity of probabilistic logic programs. J. Artif. Intell. Res. **60**, 221–262 (2017)
2. Cozman, F.G., Mauá, D.D.: The complexity of Bayesian networks specified by propositional and relational languages. Artif. Intell. **262**, 96–141 (2018)
3. Cozman, F.G., Mauá, D.D.: The finite model theory of Bayesian networks: descriptive complexity. In: International Joint Conferences on Artificial Intelligence, pp. 5229–5233 (2018)
4. Getoor, L., Taskar, B.: Introduction to Statistical Relational Learning. MIT Press (2007)
5. Gusmao, A.C., Correia, A.H.C., Bona, G.D., Cozman, F.G.: Interpreting embedding models of knowledge bases: a pedagogical approach. In: ICML Workshop on Human Interpretability in Machine Learning, pp. 79–86 (2018)
6. Halevy, A., Norvig, P., Pereira, F.: The unreasonable effectiveness of data. IEEE Intell. Syst. 8–12, March/April 2009
7. Halpern, J.Y.: Reasoning About Uncertainty. MIT Press (2003)
8. Levesque, H.: Common Sense, the Turing Test, and the Quest for Real AI. MIT Press (2019)
9. Nickel, M., Murphy, K., Tresp, V., Gabrilovich, E.: A review of relational machine learning for knowledge graphs. Proc. IEEE **104**(1), 11–33 (2016)
10. Riguzzi, F.: Foundations of probabilistic logic programming (2018)
11. Ruschel, A., Gusmão, A.C., Polleti, G.P., Cozman, F,G.: Explaining completions produced by embeddings of knowledge graphs. In: Kern-Isberner, G., Ognjanović, Z. (eds.) ECSQARU 2019. LNCS(LNAI), vol. 11726, pp. 324–335. Springer, Cham (2019). https://doi.org/10.1007/978-3-030-29765-7_27

Boolean Algebras of Conditionals, Probability and Logic

Lluis Godo

Artificial Intelligence Research Institute (IIIA - CSIC), Campus UAB,
Bellaterra 08193, Spain
godo@iiia.csic.es

Conditionals play a fundamental role both in qualitative and in quantitative uncertain reasoning, see e.g. [1, 2, 8, 12, 13]. In the former, conditionals constitute the core focus of non-monotonic reasoning [9–11]. In the latter, conditionals are central for the foundations of conditional uncertainty measures, in particular in connection to conditional probability [3, 6].

Conditionals have been investigated –largely independently– both in probability and in logic. Each has its own theory and deep questions arise if we consider combining the two settings as in the field of probability logic, which is of great interest to Artificial Intelligence.

In two previous ECSQARU conference papers [4, 5], we have preliminarily introduced and studied a new construction of a Boolean structure for conditionals motivated by the goal of "separating" the measure-theoretic from the logical properties of conditional probabilities. The question is well-posed: it is in fact well-known that if events a, b are to be taken as arbitrary elements of a Boolean algebra, the conditional probability $P(b \mid a)$ cannot be identified with the probability of the (material) implication $a \rightarrow b$. So the following questions about conditional probability become interesting: which of its properties depend on the properties of unconditional probability measures and not on the logical properties of conditional events, and which properties instead depend on the logic of conditional events. Motivated by these questions, our ultimate aims are:

(a) identify the desirable properties (axioms) which characterise the notion of a Boolean algebra of conditional events, and investigate the atomic structure of these algebras;
(b) show that the axioms of our Boolean algebras of conditional events give rise naturally to a logic of conditionals which satisfies widely accepted logical properties;
(c) investigate unconditional probabilistic measures on the algebra of conditional events;
(d) prove that classically defined conditional probability functions can be viewed as unconditional probability measures on the algebra of conditional events.

Joint work with T. Flaminio and H. Hosni.

Parts (a) and (b) have been mostly addressed in [4, 5], but (c) and (d) remained open.

In this talk we present an investigation on the structure of conditional events and on the probability measures which arise naturally in this context. In particular we introduce a construction which defines a (finite) Boolean algebra of conditionals from any (finite) Boolean algebra of events.

Moreover, as for (c) and (d) above, we provide positive and satisfying solutions. In particular, we have approached the following main problem, which is known in the literature as the *strong conditional event problem* [7]: given a measurable space $(\Omega, \mathbf{A})$ and a probability measure P over $(\Omega, \mathbf{A})$, find another measurable space $(\Omega^*, \mathbf{A}^*)$, of which the former is a subspace, and a probability measure P^* over $(\Omega^*, \mathbf{A}^*)$, satisfying the two following conditions:

1. Any conditional event of the form $(a \mid b)$ with $a, b \in \mathbf{A}$ is mapped to an element $(a \mid b)^*$ of $\mathbf{A}^*$.
2. For each conditional event $(a \mid b)$, $P^*((a \mid b)^*) = P(a \wedge b)/P(b)$ (whenever $P(b) > 0$).

A solution of the above was first proposed by Van Frassseen [14], and then reworked by Goodman and Nguyen [7] within the frame of *conditional event algebras*. They take Ω^* as the countably infinite Cartesian product space $\Omega^{\mathbb{N}}$, and A^* is always infinite, even if the original structure of (unconditional) events A is finite. Indeed, A^* has countably many atoms and conditional events in A^* are defined as countable unions of special cylinders sets. In contraposition, our approach provides a *finitary* solution to the strong conditional event problem in the setting of finite Boolean algebras of conditionals.

Acknowledgments. This research has been partially supported by the Spanish FEDER/MINECO project TIN2015-71799-C2-1-P.

References

1. Adams, E.W.: The Logic of Conditionals. Reidel, Dordrecht (1975)
2. Dubois, D., Prade, H.: Conditioning, non-monotonic logics and non-standard uncertainty models. In: Goodman, I.R., et al. (eds.) Conditional Logic in Expert Systems, North-Holland, pp. 115–158 (1991)
3. Coletti, G., Scozzafava, R.: Probabilistic Logic in a Coherent Setting. Trends Log. **15** (2002). Kluwer
4. Flaminio, T., Godo, L., Hosni, H.: On the algebraic structure of conditional events. In: Destercke, S., Denoeux, T. (eds.) ECSQARU 2015. LNCS, vol. 9161, pp. 106–116. Springer, Cham (2015). https://doi.org/10.1007/978-3-319-20807-7_10
5. Flaminio, T., Godo, L., Hosni, H.: On boolean algebras of conditionals and their logical counterpart. In: Antonucci, A., Cholvy, L., Papini, O. (eds.) ECSQARU 2017. LNCS, vol. 10369, pp. 246–256. Springer, Cham (2017). https://doi.org/10.1007/978-3-319-61581-3_23
6. Friedman, N., Halpern, J.Y.: Plausibility measures and default reasoning. J. ACM **48**(4), 648–685 (2001)

7. Goodman, I.R., Nguyen, H.T.: A theory of conditional information for probabilistic inference in intelligent systems: II. Product space approach. Inf. Sci. **76**, 13–42 (1994)

8. Goodman, I.R., Nguyen, H.T., Walker, E.A.: Conditional Inference and Logic for Intelligent Systems - A Theory of Measure-free Conditioning, North-Holland (1991)

9. Kern-Isberner, G. (eds.) Conditionals in Nonmonotonic Reasoning and Belief Revision. LNAI, vol. 2087, Springer, Heidelberg (2001). https://doi.org/10.1007/3-540-44600-1

10. Lehmann, D., Magidor, M.: What does a conditional knowledge base entail? Artif. Intell. **55** (1), 1–60 (1992)

11. Makinson, D.: Bridges From Classical to Non-monotonic Logic. College Publications, London (2005)

12. Makinson, D.: Conditional probability in the light of qualitative belief change. In: Hosni, H., Montagna, F. (eds.) Probability, Uncertainty and Rationality, Edizioni della Normale (2010)

13. Nguyen, H.T., Walker, E.A.: A history and introduction to the algebra of conditional events and probability logic. IEEE Trans. Syst. Man Cybern. **24**(12), 1671–1675, December 1994

14. van Fraassen, B.C.: Probabilities of conditionals. In: Harper, W.L., Stalnaker, R., Pearce, G. (eds.) Foundations of Probability Theory, Statistical Inference, and Statistical Theories of Science. The University of Western Ontario Series in Philosophy of Science, vol. 1, pp. 261–308. D. Reidel, Dordrecht (1976)

Dialectical Explanations

Francesca Toni [iD]

Department of Computing, Imperial College London, UK
ft@imperial.ac.uk

Abstract. The lack of transparency of AI techniques, e.g. prediction systems or recommender systems, is one of the most pressing issues in the field, especially given the ever-increasing integration of AI into everyday systems used by experts and non-experts alike, and the need to explain how and/or why these systems compute outputs, for any or for specific inputs. The need for explainability arises for a number of reasons: an expert may require more transparency to justify outputs of an AI system, especially in safety-critical situations, while a non-expert may place more trust in an AI system providing basic (rather than no) explanations, regarding, for example, items suggested by a recommender system. Explainability is also needed to fulfil the requirements of regulation, notably the General Data Protection Regulation (GDPR), effective from May 25, 2018. Furthermore, explainability is crucial to guarantee comprehensibility in human-machine interactions, to support collaboration and communication between human beings and machines.

In this talk I will overview recent efforts to use argumentative abstractions for data-centric methods in AI as a basis for generating dialectical explanations. These abstractions are formulated in the spirit of argumentation in AI, amounting to a (family of) symbolic formalism(s) where arguments are seen as nodes in a graph with relations between arguments, e.g. attack and support, as edges. Argumentation allows for conflicts to be managed effectively, an important capability in any AI system tasked with decision-making. It also allows for reasoning to be represented in a human-like manner, and can serve as a basis for a principled theory of explanation supporting human-machine dialectical exchanges and conversations.

Keywords: Explanation · Argumentation · Conversational AI

Contents

Conditional, Default and Analogical Reasoning

Learning and Decision Making

Precise and Imprecise Probabilities

Uncertain Reasoning for Applications

Argumentation

Similarity Measures Between Arguments Revisited

Leila Amgoud[1(✉)], Victor David[2(✉)], and Dragan Doder[2(✉)]

[1] CNRS – IRIT, Toulouse, France
leila.amgoud@irit.fr
[2] Paul Sabatier University – IRIT, Toulouse, France
{victor.david,dragan.doder}@irit.fr

Abstract. Recently, the notion of similarity between arguments, namely those built using propositional logic, has been investigated and several similarity measures have been defined. This paper shows that those measures may lead to inaccurate results when arguments are not *concise*, i.e., their supports contain information that is useless for inferring their conclusions. For circumventing this limitation, we start by refining arguments for making them concise. Then, we propose two families of similarity measures that extend existing ones and that deal with concise arguments.

Keywords: Logical arguments · Similarity

1 Introduction

Argumentation is a reasoning process based on the justification of claims by *arguments*. It has received great interest from the Artificial Intelligence community, which used it for solving various problems like decision making (eg., [1,2]), defeasible reasoning (eg., [3,4]), handling inconsistency in propositional knowledge bases (eg., [5,6]), etc.

In case of inconsistency handling, an argument is built from a knowledge base and contains two parts: a *conclusion*, which is a single propositional formula, and a *support*, which is a minimal (for set inclusion) and consistent subset of the base that infers logically the conclusion. Examples of arguments are $A = \langle \{p \wedge q\}, p \rangle$, $B = \langle \{p\}, p \rangle$ and $C = \langle \{p \wedge p\}, p \rangle$. Such arguments may be in conflict and thus an evaluation method, called also semantics in the literature, is used for evaluating their strengths. Some weighting semantics, like h-Categorizer [5], satisfy the Counting (or strict monotony) principle defined in [7]. This principle states that each attacker of an argument contributes to weakening the argument. For instance, if the argument $D = \langle \{\neg p\}, \neg p \rangle$ is attacked by A, B, C, then each of the three arguments will decrease the strength of D. However, the three attackers are somehow similar, thus D will loose more than necessary. Consequently, the authors in [8] have motivated the need for investigating the notion of similarity between pairs of such logical arguments. They introduced a set of principles that a reasonable similarity measure should satisfy, and provided several measures that satisfy them. In [9] the authors introduced three possible extensions of h-Categorizer that take into account similarities between arguments.

While the measures from [8] return reasonable results in most cases, they may lead to inaccurate assessments if arguments are not *concise*. An argument is concise

© Springer Nature Switzerland AG 2019
G. Kern-Isberner and Z. Ognjanović (Eds.): ECSQARU 2019, LNAI 11726, pp. 3–13, 2019.
https://doi.org/10.1007/978-3-030-29765-7_1

if its support contains only information that is useful for inferring its conclusion. For instance, the argument A is not concise since its support $\{p \wedge q\}$ contains q, which is useless for the conclusion p. Note that minimality of supports does not guarantee conciseness. For example, the support of A is minimal while A is not concise. The similarity measures from [8] declare the two arguments A and B as not fully similar while they support the same conclusion on the same grounds (p). Consequently, both A and B will have an impact on D using h-Categorizer. For circumventing this problem, we propose in this paper to clean up arguments from any useless information. This amounts to generating the concise versions of each argument. The basic idea is to weaken formulas of an argument's support. Then, we apply the measures from [8] on concise arguments in two ways, leading to two different families of measures.

The paper is organized as follows: Sect. 2 recalls the measures proposed in [8], Sect. 3 shows how to make arguments concise, Sect. 4 refines existing measures, and Sect. 5 concludes and presents some perspectives.

2   Background

We consider classical propositional logic $(\mathcal{L}, \vdash)$, where $\mathcal{L}$ is a propositional language built up from a finite set $\mathcal{P}$ of variables, called atoms, the two Boolean constants $\top$ (true) and $\bot$ (false), and the usual connectives ($\neg, \vee, \wedge, \rightarrow, \leftrightarrow$), and $\vdash$ is the consequence relation of the logic. A literal of $\mathcal{L}$ is either a variable of $\mathcal{P}$ or the negation of a variable of $\mathcal{P}$, the set of all literals is denoted by $\mathcal{P}^{\pm}$. A formula ϕ is in negation normal form (NNF) if and only if it does not contain implication or equivalence symbols, and every negation symbol occurs directly in front of an atom. $\text{NNF}(\phi)$ denotes the NNF of ϕ. For instance, $\text{NNF}(\neg((p \rightarrow q) \vee \neg t)) = p \wedge \neg q \wedge t$. $\text{Lit}(\phi)$ denotes the set of literals occurring in $\text{NNF}(\phi)$, hence $\text{Lit}(\neg((p \rightarrow q) \vee \neg t)) = \{p, \neg q, t\}$. Two formulas $\phi, \psi \in \mathcal{L}$ are *logically equivalent*, denoted by $\phi \equiv \psi$, iff $\phi \vdash \psi$ and $\psi \vdash \phi$. In [10], the authors defined the notion of *independence* of a formula from literals as follows.

Definition 1 (Literals Independence). *Let $\phi \in \mathcal{L}$ and $l \in \mathcal{P}^{\pm}$. The formula ϕ is* independent *from the literal l iff $\exists \psi \in \mathcal{L}$ such that $\phi \equiv \psi$ and $l \notin \text{Lit}(\psi)$. Otherwise, ϕ is dependent on l. $\text{DepLit}(\phi)$ denotes the set of all literals that ϕ is dependent on.*

For instance, $\text{DepLit}((\neg p \vee q) \wedge (\neg p \vee \neg q)) = \{\neg p\}$ while $\text{DepLit}(\neg p \wedge q) = \{\neg p, q\}$.

A finite subset Φ of $\mathcal{L}$, denoted by $\Phi \subseteq_f \mathcal{L}$, is *consistent* iff $\Phi \nvdash \bot$, it is *inconsistent* otherwise. Two subsets $\Phi, \Psi \subseteq_f \mathcal{L}$ are *equivalent*, denoted by $\Phi \cong \Psi$, iff $\forall \phi \in \Phi$, $\exists \psi \in \Psi$ such that $\phi \equiv \psi$ and $\forall \psi' \in \Psi$, $\exists \phi' \in \Phi$ such that $\phi' \equiv \psi'$. We write $\Phi \ncong \Psi$ otherwise. This definition is useful in the context of similarity where arguments are compared with respect to their *contents*. Assume, for instance, p and q that stand respectively for "bird" and "fly". Clearly, the two rules "birds fly" and "everything that flies is a bird" express different information. Thus, the two sets $\{p, p \rightarrow q\}$ and $\{q, q \rightarrow p\}$ should be considered as different. Note that $\{p, p \rightarrow q\} \ncong \{q, q \rightarrow p\}$ even if $\text{CN}(\{p, p \rightarrow q\}) = \text{CN}(\{q, q \rightarrow p\})$, where $\text{CN}(\Phi)$ denotes the set of all formulas that follow from the set Φ of formulas.

Let us now recall the backbone of our paper, the notion of logical *argument*.

Definition 2 (Argument). *An* argument *built under the logic* $(\mathcal{L}, \vdash)$ *is a pair* $\langle \Phi, \phi \rangle$, *where* $\Phi \subseteq_f \mathcal{L}$ *and* $\phi \in \mathcal{L}$, *such that:*

- Φ *is consistent,*   *(Consistency)*
- $\Phi \vdash \phi$,   *(Validity)*
- $\nexists \Phi' \subset \Phi$ *such that* $\Phi' \vdash \phi$.   *(Minimality)*

An argument $\langle \Phi, \phi \rangle$ *is* trivial *iff* $\Phi = \emptyset$.

It is worth noticing that trivial arguments support tautologies. It was shown in [11] that the set of arguments that can be built from a finite set of formulas is infinite.

Example 1. The following pairs are all arguments.

$A = \langle \{p \wedge q\}, p \rangle$	$B = \langle \{p\}, p \rangle$
$C = \langle \{p \wedge q \wedge r\}, r \rangle$	$D = \langle \{p \wedge q, p \wedge r\}, p \wedge q \wedge r \rangle$
$E = \langle \{p \wedge q, (p \vee q) \rightarrow r\}, r \rangle$	$F = \langle \{p \wedge q\}, p \vee q \rangle$

Notations: $\text{Arg}(\mathcal{L})$ denotes the set of all arguments that can be built under the logic $(\mathcal{L}, \vdash)$. For any $A = \langle \Phi, \phi \rangle \in \text{Arg}(\mathcal{L})$, the functions Supp and Conc return respectively the *support* ($\text{Supp}(A) = \Phi$) and the *conclusion* ($\text{Conc}(A) = \phi$) of A.

In [11], the notion of *equivalence of arguments* has been investigated, and different variants of equivalence have been proposed. The most general one states that two arguments are equivalent if their supports are equivalent (in the sense of $\cong$) and their conclusions are equivalent (in the sense of $\equiv$). For the purpose of our paper, we focus on the following one that requires equality of conclusions.

Definition 3 (Equivalent Arguments). *Two arguments* $A, B \in \text{Arg}(\mathcal{L})$ *are* equivalent, *denoted by* $A \approx B$, *iff* $(\text{Supp}(A) \cong \text{Supp}(B))$ *and* $(\text{Conc}(A) = \text{Conc}(B))$.

In [8], the authors have investigated the notion of similarity between pairs of arguments, and have introduced several measures which are based on the well-known Jaccard measure [12], Dice measure [13], Sorensen one [14], and those proposed in [15–18]. All these measures compare pairs of non-empty sets (X and Y) of objects. Table 1 shows how to adapt their definitions for comparing supports (respectively conclusions) of arguments, which are sets of propositional formulas. In that table, $\text{Co}(\Phi, \Psi)$ is a function that returns for all $\Phi, \Psi \subseteq_f \mathcal{L}$ a set of formulas such that:

$$\text{Co}(\Phi, \Psi) = \{\phi \in \Phi \mid \exists \psi \in \Psi \text{ such that } \phi \equiv \psi\}.$$

The definition of each similarity measure between sets of formulas follows the schema below that we illustrate with the Jaccard-based measure. For all $\Phi, \Psi \subseteq_f \mathcal{L}$,

$$s_j(\Phi, \Psi) = \begin{cases} \frac{|\text{Co}(\Phi, \Psi)|}{|\Phi| + |\Psi| - |\text{Co}(\Phi, \Psi)|} & \text{if } \Phi \neq \emptyset, \Psi \neq \emptyset \\ 1 & \text{if } \Phi = \Psi = \emptyset \\ 0 & \text{otherwise.} \end{cases}$$

In [8], a similarity measure between arguments is a function that assigns to every pair of arguments a value from the interval $[0, 1]$. The greater the value, the more similar are the arguments. Such measure should satisfy some properties including symmetry.

Table 1. Similarity Measures for Sets $\Phi, \Psi \subseteq_f \mathcal{L}$.

Extended Jaccard	$s_j(\Phi, \Psi) = \dfrac{|\mathsf{Co}(\Phi, \Psi)|}{|\Phi| + |\Psi| - |\mathsf{Co}(\Phi, \Psi)|}$
Extended Dice	$s_d(\Phi, \Psi) = \dfrac{2|\mathsf{Co}(\Phi, \Psi)|}{|\Phi| + |\Psi|}$
Extended Sorensen	$s_s(\Phi, \Psi) = \dfrac{4|\mathsf{Co}(\Phi, \Psi)|}{|\Phi| + |\Psi| + 2|\mathsf{Co}(\Phi, \Psi)|}$
Extended Symmetric Anderberg	$s_a(\Phi, \Psi) = \dfrac{8|\mathsf{Co}(\Phi, \Psi)|}{|\Phi| + |\Psi| + 6|\mathsf{Co}(\Phi, \Psi)|}$
Extended Sokal and Sneath 2	$s_{ss}(\Phi, \Psi) = \dfrac{|\mathsf{Co}(\Phi, \Psi)|}{2(|\Phi| + |\Psi|) - 3|\mathsf{Co}(\Phi, \Psi)|}$
Extended Ochiai	$s_o(\Phi, \Psi) = \dfrac{|\mathsf{Co}(\Phi, \Psi)|}{\sqrt{|\Phi|}\sqrt{|\Psi|}}$
Extended Kulczynski 2	$s_{ku}(\Phi, \Psi) = \dfrac{1}{2}\left(\dfrac{|\mathsf{Co}(\Phi, \Psi)|}{|\Phi|} + \dfrac{|\mathsf{Co}(\Phi, \Psi)|}{|\Psi|}\right)$

Definition 4 (Similarity Measure). *A* similarity measure *is a function* $\mathcal{S} : \mathtt{Arg}(\mathcal{L}) \times \mathtt{Arg}(\mathcal{L}) \to [0, 1]$ *such that:*

Symmetry: for all $a, b \in \mathtt{Arg}(\mathcal{L})$, $\mathcal{S}(a, b) = \mathcal{S}(b, a)$.
Maximality: for any $a \in \mathtt{Arg}(\mathcal{L})$, $\mathcal{S}(a, a) = 1$.
Substitution: for all $a, b, c \in \mathtt{Arg}(\mathcal{L})$, *if* $\mathcal{S}(a, b) = 1$ *then* $\mathcal{S}(a, c) = \mathcal{S}(b, c)$.

In [8], several similary measures have been defined. They apply any measure from Table 1 for assessing similarity of both arguments' supports and their conclusions. Furthermore, they use a parameter that allows a user to give different importance degrees to the two components of an argument. Those measures satisfy the three properties (Symmetry, Maximality, Substitution) and additional ones (see [8] for more details).

Definition 5 (Extended Measures). *Let* $0 < \sigma < 1$ *and* $x \in \{j, d, s, a, ss, o, ku\}$. *A similarity measure* $\mathcal{S}_x^\sigma$ *is a function assigning to any pair* $(A, B) \in \mathtt{Arg}(\mathcal{L}) \times \mathtt{Arg}(\mathcal{L})$ *a value* $\mathcal{S}_x^\sigma(A, B) = \sigma \cdot s_x(\mathtt{Supp}(A), \mathtt{Supp}(B)) + (1 - \sigma) \cdot s_x(\{\mathtt{Conc}(A)\}, \{\mathtt{Conc}(B)\})$.

Example 1 (Continued). Let $\sigma = 0.5$ and $x = j$.

- $\mathcal{S}_j^{0.5}(A, B) = 0.5 \cdot 0 + 0.5 \cdot 1 = 0.5$
- $\mathcal{S}_j^{0.5}(A, D) = 0.5 \cdot 0.5 + 0.5 \cdot 0 = 0.25$
- $\mathcal{S}_j^{0.5}(A, F) = 0.5 \cdot 1 + 0.5 \cdot 0 = 0.5$

3   Concise Arguments

The two arguments $A = \langle\{p \wedge q\}, p\rangle$ and $B = \langle\{p\}, p\rangle$ are not fully similar according to the existing measures from [8] while they support the same conclusion and on the same grounds. This inaccuracy is due to the non-conciseness of A, which contains the useless information q in its support. In what follows, we refine arguments by removing from their supports such information. The idea is to weaken formulas in supports.

Definition 6 (Refinement). *Let* $A, B \in \mathtt{Arg}(\mathcal{L})$ *such that* $A = \langle\{\phi_1, \ldots, \phi_n\}, \phi\rangle$ *and* $B = \langle\{\phi'_1, \ldots, \phi'_n\}, \phi'\rangle$. *$B$ is a refinement of A iff:*

1. $\phi = \phi'$,
2. *There exists a permutation* ρ *of the set* $\{1, \ldots, n\}$ *such that* $\forall k \in \{1, \ldots, n\}$, $\phi_k \vdash \phi'_{\rho(k)}$ *and* $\mathtt{Lit}(\phi'_{\rho(k)}) \subseteq \mathtt{DepLit}(\phi_k)$.

Let Ref *be a function that returns the set of all refinements of a given argument.*

The second condition states that each formula of an argument's support is weakened. Furthermore, novel literals are not allowed in the weakening step since such literals would negatively impact similarity between supports of arguments. Finally, literals from which a formula is independent should be removed since they are useless for inferring the conclusion of an argument. It is worth mentioning that an argument may have several refinements as shown in the following example.

Example 1 (Continued).

- $\{\langle\{p\}, p\rangle, \langle\{p \wedge p\}, p\rangle\} \subseteq \mathtt{Ref}(A)$
- $\{\langle\{p \wedge r\}, r\rangle, \langle\{q \wedge r\}, r\rangle, \langle\{r\}, r\rangle\} \subseteq \mathtt{Ref}(C)$
- $\{\langle\{p \wedge q, r\}, p \wedge q \wedge r\rangle, \langle\{q, p \wedge r\}, p \wedge q \wedge r\rangle\} \subseteq \mathtt{Ref}(D)$
- $\{\langle\{p \vee q, (p \vee q) \rightarrow r\}, r\rangle, \langle\{p, p \rightarrow r\}, r\rangle, \langle\{q, q \rightarrow r\}, r\rangle\} \subseteq \mathtt{Ref}(E)$
- $\{\langle\{p\}, p \vee q\rangle, \langle\{q\}, p \vee q\rangle, \langle\{p \vee q\}, p \vee q\rangle\} \subseteq \mathtt{Ref}(F)$

The following property shows that there exists a unique possible permutation ρ for each refinement of an argument.

Proposition 1. *For all* $A = \langle\{\phi_1, \ldots, \phi_n\}, \phi\rangle, B = \langle\{\phi'_1, \ldots, \phi'_n\}, \phi'\rangle \in \mathtt{Arg}(\mathcal{L})$ *such that* $B \in \mathtt{Ref}(A)$, *there exists a unique permutation* ρ *of the set* $\{1, \cdots, n\}$ *such that* $\forall k \in \{1, \ldots, n\}$, $\phi_k \vdash \phi'_{\rho(k)}$.

Obviously, a trivial argument is the only refinement of itself.

Proposition 2. *For any trivial argument* $A \in \mathtt{Arg}(\mathcal{L})$, $\mathtt{Ref}(A) = \{A\}$.

A non-trivial argument has a non-empty set of refinements. Moreover, such argument is a refinement of itself only if the formulas of its support do not contain literals from which they are independent.

Proposition 3. *Let* $A \in \mathtt{Arg}(\mathcal{L})$ *be a non-trivial argument. The following hold:*

- $\mathtt{Ref}(A) \neq \emptyset$,
- $A \in \mathtt{Ref}(A)$ *iff* $\forall \phi \in \mathtt{Supp}(A)$, $\mathtt{Lit}(\phi) = \mathtt{DepLit}(\phi)$.

We show next that the function Ref is somehow monotonic, and that equivalent arguments have the same refinements.

Proposition 4. *Let* $A, B \in \mathtt{Arg}(\mathcal{L})$. *The following hold:*

- *If* $B \in \mathtt{Ref}(A)$, *then* $\mathtt{Ref}(B) \subseteq \mathtt{Ref}(A)$.
- *If* $A \approx B$, *then* $\mathtt{Ref}(A) = \mathtt{Ref}(B)$.

We are now ready to define the backbone of the paper, the novel notion of concise argument. An argument is concise if it is equivalent to any of its refinements. This means that a concise argument cannot be further refined.

Definition 7 (Conciseness). *An argument $A \in \text{Arg}(\mathcal{L})$ is concise iff for all $B \in \text{Ref}(A)$, $A \approx B$.*

Example 1 (Continued). The two refinements $\langle \{p \wedge r\}, r \rangle$ and $\langle \{q \wedge r\}, r \rangle$ of the argument C are not concise. Indeed, $\langle \{r\}, r \rangle \in \text{Ref}(\langle \{p \wedge r\}, r \rangle)$, $\langle \{r\}, r \rangle \in \text{Ref}(\langle \{q \wedge r\}, r \rangle)$ while $\langle \{r\}, r \rangle \not\approx \langle \{p \wedge r\}, r \rangle$ and $\langle \{r\}, r \rangle \not\approx \langle \{q \wedge r\}, r \rangle$.

For any argument from $\text{Arg}(\mathcal{L})$, we generate its concise versions. The latter are simply its concise refinements.

Definition 8 (Concise Refinements). *A concise refinement of an argument $A \in \text{Arg}(\mathcal{L})$ is any concise argument B such that $B \in \text{Ref}(A)$. We denote the set of all concise refinements of A by $\text{CR}(A)$.*

Example 1 (Continued).

- $\langle \{p\}, p \rangle \in \text{CR}(A)$
- $\langle \{r\}, r \rangle \in \text{CR}(C)$
- $\{\langle \{p \wedge q, r\}, p \wedge q \wedge r \rangle, \langle \{q, p \wedge r\}, p \wedge q \wedge r \rangle\} \subseteq \text{CR}(D)$
- $\{\langle \{p \vee q, (p \vee q) \to r\}, r \rangle, \langle \{p, p \to r\}, r \rangle, \langle \{q, q \to r\}, r \rangle\} \subseteq \text{CR}(E)$
- $\{\langle \{p\}, p \vee q \rangle, \langle \{q\}, p \vee q \rangle, \langle \{p \vee q\}, p \vee q \rangle\} \subseteq \text{CR}(F)$

Next we state some properties of concise refinements.

Proposition 5. *Let $A \in \text{Arg}(\mathcal{L})$. The following hold:*

1. *For any $B \in \text{CR}(A)$ the following hold: $B \in \text{Ref}(B)$ and $\forall C \in \text{Ref}(B)$, $C \approx B$.*
2. *$\text{CR}(A) \neq \emptyset$.*
3. *If A is non-trivial, then $\text{CR}(A)$ is infinite.*
4. *If $A \approx B$, then $\text{CR}(A) = \text{CR}(B)$.*
5. *$\forall B \in \text{Ref}(A)$, $\text{CR}(B) \subseteq \text{CR}(A)$.*

The following result shows that any formula in the support of a concise argument cannot be further weakened without introducing additional literals.

Proposition 6. *Let $A, B \in \text{Arg}(\mathcal{L})$ such that $B \in \text{CR}(A)$. For any $\phi \in \text{Supp}(B)$, if $\exists \psi \in \mathcal{L}$ such that $\phi \vdash \psi$, $\psi \not\vdash \phi$, and $\langle (\text{Supp}(B) \setminus \{\phi\}) \cup \{\psi\}, \text{Conc}(B) \rangle \in \text{Arg}(\mathcal{L})$, then $\text{Lit}(\psi) \setminus \text{Lit}(\phi) \neq \emptyset$.*

4   Similarity Measures

As already said in previous sections, although the similarity measures from Definition 5 return reasonable results in most cases, they might lead to inaccurate assessments if the arguments are not concise. Indeed, as we illustrated in Sect. 2, the measures from Definition 5 declare the two arguments $A = \langle \{p \wedge q\}, p \rangle$ and $B = \langle \{p\}, p \rangle$ as not fully similar, while they support the same conclusion based on the same grounds (p).

In this section, we extend those measures in two ways, leading to two families of similarity measures, using concise refinements of arguments, and we show that they properly resolve the drawbacks of the existing measures. Note that by Proposition 5(3), every non-trivial argument A has infinitely many concise refinements. This is due to the fact that every formula α from a support of a concise refinement can be equivalently rewritten in infinitely many ways using the same set of literals (eg. $\alpha \equiv \alpha \wedge \alpha \equiv \alpha \wedge \alpha \wedge \alpha \equiv \cdots$). In the rest of the paper, we will consider only one argument from $\mathrm{CR}(A)$ per equivalence class. For that reason, we consider a fixed set $\overline{\mathcal{L}} \subset \mathcal{L}$ such that $\phi \in \mathcal{L}$ there exists a unique $\psi \in \overline{\mathcal{L}}$ such that $\psi \equiv \phi$. Furthermore, we assume that each $\psi \in \overline{\mathcal{L}}$ contains only dependent literals.

Definition 9. *Let $A \in \mathrm{Arg}(\mathcal{L})$. We define the set*

$$\mathrm{CR}(A) = \{B \in \mathrm{CR}(A) \mid \mathrm{Supp}(B) \subset \overline{\mathcal{L}}\}.$$

In this way, we obtain a finite set of non-equivalent concise refinements.

Proposition 7. *For every $A \in \mathrm{Arg}(\mathcal{L})$, the set $\overline{\mathrm{CR}}(A)$ is finite.*

Now we propose our first family of similarity measures. In the following definition, for $A \in \mathrm{Arg}(\mathcal{L})$, $\Sigma \subseteq_f \mathrm{Arg}(\mathcal{L})$ and a similarity measure S from Definition 5, we denote by $\mathrm{Max}(A, \Sigma, S)$ the maximal similarity value between A and an argument from Σ according to S, i.e.,

$$\mathrm{Max}(A, \Sigma, S) = \max_{B \in \Sigma} S(A, B).$$

The first family of measures compares the sets of concise refinements of the two arguments under study. Indeed, the similarity between A and B is the average of maximal similarities (using any existing measure from Definition 5) between any concise refinement of A and those of B.

Definition 10 (**A-CR Similarity Measures**). *Let $A, B \in \mathrm{Arg}(\mathcal{L})$, and let S be a similarity measure from Definition 5. We define* **A-CR** *similarity measure[1] by*

$$s_{\mathrm{CR}}^{\mathtt{A}}(A, B, S) = \frac{\displaystyle\sum_{A_i \in \overline{\mathrm{CR}}(A)} \mathrm{Max}(A_i, \overline{\mathrm{CR}}(B), S) + \sum_{B_j \in \overline{\mathrm{CR}}(B)} \mathrm{Max}(B_j, \overline{\mathrm{CR}}(A), S)}{|\overline{\mathrm{CR}}(A)| + |\overline{\mathrm{CR}}(B)|}.$$

The value of A-CR similarity measure always belongs to the unit interval.

Proposition 8. *Let $A, B \in \mathrm{Arg}(\mathcal{L})$, S_x^{σ} a similarity measure where $x \in \{\mathtt{j}, \mathtt{d}, \mathtt{s}, \mathtt{a}, \mathtt{ss},$ $\mathtt{o}, \mathtt{ku}\}$ and $0 < \sigma < 1$. Then $s_{\mathrm{CR}}^{\mathtt{A}}(A, B, S_x^{\sigma}) \in [0, 1]$.*

---

[1] The letter A in A-CR stands for "average".

Next we show that the new measure properly resolves the problem of non-conciseness of the argument $A = \langle \{p \land q\}, p \rangle$ from our running example. We illustrate that by considering Extended Jaccard measure with the parameter $\sigma = 0.5$.[2]

Example 1 (Continued). It is easy to check that $\overline{CR}(A) = \{\langle \{p\}, p \rangle\}$ and $\overline{CR}(B) = \{\langle \{p\}, p \rangle\}$. Then $s_{CR}^A(A, B, S_j^{0.5}) = 1$ while $S_j^{0.5}(A, B) = 0.5$.

Now we define our second family of similarity measures, which is based on comparison of sets obtained by merging supports of concise refinements of arguments. For an argument $A \in \text{Arg}(\mathcal{L})$, we denote that set by

$$\text{US}(A) = \bigcup_{A' \in \overline{CR}(A)} \text{Supp}(A').$$

Definition 11 (U-CR **Similarity Measures**). *Let $A, B \in \text{Arg}(\mathcal{L})$, $0 < \sigma < 1$, and s_x be a similarity measure from Table 1. We define* U-CR *similarity measure*[3] *by*

$$s_{CR}^U(A, B, s_x, \sigma) = \sigma \cdot s_x(\text{US}(A), \text{US}(B)) + (1 - \sigma) \cdot s_x(\{\text{Conc}(A)\}, \{\text{Conc}(B)\}).$$

Next example illustrates that U-CR also properly resolves the problem of non-conciseness of the argument $A = \langle \{p \land q\}, p \rangle$ from our running example.

Example 1 (Continued). Let $\sigma = 0.5$ and $x = j$. It is easy to check that $s_{CR}^U(A, B, s_j, 0.5) = 1$ while $S_j^{0.5}(A, B) = 0.5$.

Let us now consider another more complex example where existing similarity measures provide inaccurate values while the new ones perform well.

Example 2. Let us consider the following arguments:

- $A = \langle \{p \land q, (p \lor q) \rightarrow t, (p \lor t) \rightarrow r\}, t \land r \rangle$
- $B = \langle \{p, p \rightarrow t, p \rightarrow r\}, t \land r \rangle$

It is easy to check that $\overline{CR}(A) = \{A_1, A_2, A_3, A_4, A_5\}$ and $\overline{CR}(B) = \{B_1\}$, where:

- $A_1 = \langle \{p, p \rightarrow t, p \rightarrow r\}, t \land r \rangle$
- $A_2 = \langle \{p, p \rightarrow t, t \rightarrow r\}, t \land r \rangle$
- $A_3 = \langle \{q, q \rightarrow t, t \rightarrow r\}, t \land r \rangle$
- $A_4 = \langle \{p \lor q, (p \lor q) \rightarrow t, t \rightarrow r\}, t \land r \rangle$
- $A_5 = \langle \{p \land q, q \rightarrow t, p \rightarrow r\}, t \land r \rangle$
- $B_1 = \langle \{p, p \rightarrow t, p \rightarrow r\}, t \land r \rangle$

It is worth noticing that the Extended Jaccard measure could not detect any similarity between the supports of A and B while their concise arguments A_1 and B_1 are identical. Indeed, $s_j(\text{Supp}(A), \text{Supp}(B)) = 0$ and $S_j^{0.5}(A, B) = 0.5 \cdot 0 + 0.5 \cdot 1 = 0.5$ while $s_{CR}^U(A, B, s_j, 0.5) = 0.5 \cdot \frac{3}{9} + 0.5 \cdot 1 = \frac{2}{3} = 0.666$ and $s_{CR}^A(A, B, S_j^{0.5}) = 0.5 \cdot \frac{9}{20} + 0.5 \cdot 1 = \frac{29}{40} = 0.725$.

---

[2] In this section, we slightly relax the notation by simply assuming that $p \in \overline{\mathcal{L}}$. We will make similar assumptions throughout this section.

[3] U in U-CR stands for "union".

The following proposition characterizes the arguments which are fully similar according to the novel measures. It states that full similarity is obtained exactly in the case when two arguments have equivalent concise refinements.

Proposition 9. *Let* $A, B \in \text{Arg}(\mathcal{L})$, $0 < \sigma < 1$ *and* $x \in \{j, d, s, a, ss, o, ku\}$. *Then* $s_{\text{CR}}^{\text{A}}(A, B, S_x^{\sigma}) = s_{\text{CR}}^{\text{U}}(A, B, s_x, \sigma) = 1$ *iff:*

- $\forall A' \in \overline{\text{CR}}(A)$, $\exists B' \in \overline{\text{CR}}(B)$ *such that* $\text{Supp}(A') \cong \text{Supp}(B')$, $\text{Conc}(A') \equiv \text{Conc}(B')$ *and*
- $\forall B' \in \overline{\text{CR}}(B)$, $\exists A' \in \overline{\text{CR}}(A)$ *such that* $\text{Supp}(B') \cong \text{Supp}(A')$, $\text{Conc}(B') \equiv \text{Conc}(A')$.

In [8], the authors proposed a set of principles that a reasonable similarity measure should satisfy. Now we show that the new measures satisfy four of them but violate Monotony. The reason of violation is due to the definition itself of the principle. Indeed, it is based on the supports of arguments. The new measures do not handle those supports but those of the concise refinements of the initial arguments.

Proposition 10. *Let* $0 < \sigma < 1$ *and* $x \in \{j, d, s, a, ss, o, ku\}$. *The following hold:*

(Syntax Independence) *Let* π *be a permutation on the set of variables, and* $A, B, A', B' \in \text{Arg}(\mathcal{L})$ *such that*
- A' *is obtain by replacing each variable* p *in* A *with* $\pi(p)$,
- B' *is obtain by replacing each variable* p *in* B *with* $\pi(p)$.
Then $s_{\text{CR}}^{\text{A}}(A, B, S_x^{\sigma}) = s_{\text{CR}}^{\text{A}}(A', B', S_x^{\sigma})$ *and* $s_{\text{CR}}^{\text{U}}(A, B, s_x, \sigma) = s_{\text{CR}}^{\text{U}}(A', B', s_x, \sigma)$.
(Maximality) *For every* $A \in \text{Arg}(\mathcal{L})$, $s_{\text{CR}}^{\text{A}}(A, A, S_x^{\sigma}) = s_{\text{CR}}^{\text{U}}(A, A, s_x, \sigma) = 1$.
(Symmetry) *For all* $A, B \in \text{Arg}(\mathcal{L})$, $s_{\text{CR}}^{\text{A}}(A, B, S_x^{\sigma}) = s_{\text{CR}}^{\text{A}}(B, A, S_x^{\sigma})$ *and* $s_{\text{CR}}^{\text{U}}(A, B, s_x, \sigma) = s_{\text{CR}}^{\text{U}}(B, A, s_x, \sigma)$.
(Substitution) *For all* $A, B, C \in \text{Arg}(\mathcal{L})$,
- *if* $s_{\text{CR}}^{\text{A}}(A, B, S_x^{\sigma}) = 1$, *then* $s_{\text{CR}}^{\text{A}}(A, C, S_x^{\sigma}) = s_{\text{CR}}^{\text{A}}(B, C, S_x^{\sigma})$,
- *if* $s_{\text{CR}}^{\text{U}}(A, B, s_x, \sigma) = 1$, *then* $s_{\text{CR}}^{\text{U}}(A, C, s_x, \sigma) = s_{\text{CR}}^{\text{U}}(B, C, s_x, \sigma)$.

The next proposition shows that if we apply A-CR or U-CR to any similarity measure S_x^{σ} from Definition 5 (respectively s_x from Table 1), then both novel measures will coincide with S_x^{σ} on the class of concise arguments.

Proposition 11. *Let* $A, B \in \text{Arg}(\mathcal{L})$ *be two concise arguments. Then, for every* $0 < \sigma < 1$ *and* $x \in \{j, d, s, a, ss, o, ku\}$, *it holds*

$$s_{\text{CR}}^{\text{A}}(A, B, S_x^{\sigma}) = s_{\text{CR}}^{\text{U}}(A, B, s_x, \sigma) = S_x^{\sigma}(A, B). \tag{1}$$

Remark. Note that the Eq. (1) might also hold for some A and B that are not concise. For example, let $A = \langle \{p \wedge q, t \wedge s\}, p \wedge t \rangle$ and $B = \langle \{p, t \wedge s\}, p \wedge s \rangle$. Then $\overline{\text{CR}}(A) = \{\langle \{p, t\}, p \wedge t \rangle\}$ and $\overline{\text{CR}}(B) = \{\langle \{p, s\}, p \wedge s \rangle\}$, so $s_{\text{CR}}^{\text{A}}(A, B, S_j^{0.5}) = s_{\text{CR}}^{\text{U}}(A, B, s_j, 0.5) = S_j^{0.5}(A, B) = 0.25$.

The following example shows that A-CR and U-CR may return different results. Indeed, it is possible for three arguments A, B and C that A is more similar to B than to C according to one measure, but not according to the other one.

Example 3. Let $A = \langle \{p, p \rightarrow q_1 \wedge q_2\}, q_1 \vee q_2\rangle$, $B = \langle \{p, s\}, p \wedge s\rangle$ and $C = \langle \{p \rightarrow q_1\}, p \rightarrow q_1\rangle$. We have $\overline{CR}(A) = \{\langle \{p, p \rightarrow q_1\}, q_1 \vee q_2\rangle, \langle \{p, p \rightarrow q_2\}, q_1 \vee q_2\rangle, \langle \{p, p \rightarrow q_1 \vee q_2\}, q_1 \vee q_2\rangle\}$, $\overline{CR}(B) = \{\langle \{p, s\}, p \wedge s\rangle\}$, $\overline{CR}(C) = \{\langle \{p \rightarrow q_1\}, p \rightarrow q_1\rangle\}$. Consequently:

- $s_{CR}^A(A, B, S_j^{0.5}) = \frac{1}{6} > s_{CR}^A(A, C, S_j^{0.5}) = \frac{1}{8}$, but
- $s_{CR}^U(A, B, s_j, 0.5) = \frac{1}{10} < s_{CR}^U(A, C, s_j, 0.5) = \frac{1}{8}$.

The next example shows that none of the two novel measures dominates the other. Indeed, some pairs of arguments have greater similarity value according to A-CR, and other pairs have greater similarity value using U-CR.

Example 3 (Continued). Note that $s_{CR}^U(A, B, s_j, 0.5) < s_{CR}^A(A, B, S_j^{0.5})$. Let us consider $A' = \langle \{p \wedge q\}, p \vee q\rangle$, $B' = \langle \{p, q\}, p \wedge q\rangle \in \text{Arg}(\mathcal{L})$. $s_{CR}^U(A', B', s_j, 0.5) = 0.5 \cdot \frac{2}{3} + 0.5 \cdot 0 = \frac{1}{3} = 0.333$ and $s_{CR}^A(A', B', S_j^{0.5}) = 0.5 \cdot \frac{3}{8} + 0.5 \cdot 0 = \frac{3}{16} = 0.1875$, thus $s_{CR}^U(A', B', s_j, 0.5) > s_{CR}^A(A', B', S_j^{0.5})$.

5   Conclusion

The paper tackled the question of similarity between logical arguments. Starting from the observation that existing similarity measures may provide inaccurate assessments, the paper investigated the origin of this limitation and showed that it is due to the presence of useless information in the supports of arguments. It then introduced the novel notion of concise argument, and a procedure for generating the concise versions of any argument. These versions are then used together with existing similarity measures for extending the latter into more efficient measures.

This work can be extended in different ways. The first one consists of identifying a principle, or formal property for distinguishing the new measures. The second one consists of investigating other approaches for generating concise arguments, namely we plan to use the well-known forgetting operator for getting rid of useless literals in formulas. The Third one consists of using the new measures for refining argumentation systems that deal with inconsistent information. Finally, we plan to investigate the notion of similarity for other types of arguments, like analogical arguments.

Acknowledgment. Support from the ANR-3IA Artificial and Natural Intelligence Toulouse Institute is gratefully acknowledged.

References

1. Amgoud, L., Prade, H.: Using arguments for making and explaining decisions. Artif. Intell. **173**, 413–436 (2009)
2. Zhong, Q., Fan, X., Luo, X., Toni, F.: An explainable multi-attribute decision model based on argumentation. Expert Syst. Appl. **117**, 42–61 (2019)
3. García, A., Simari, G.: Defeasible logic programming: an argumentative approach. Theor. Pract. Logic Prog. **4**(1–2), 95–138 (2004)

4. Prakken, H., Sartor, G.: Argument-based extended logic programming with defeasible priorities. J. Appl. Non-Class. Logics **7**(1), 25–75 (1997)
5. Besnard, P., Hunter, A.: A logic-based theory of deductive arguments. Artif. Intell. **128**(1–2), 203–235 (2001)
6. Vesic, S.: Identifying the class of maxi-consistent operators in argumentation. J. Artif. Intell. Res. **47**, 71–93 (2013)
7. Amgoud, L., Ben-Naim, J.: Axiomatic foundations of acceptability semantics. In: Proceedings of the Fifteenth International Conference on Principles of Knowledge Representation and Reasoning KR, pp. 2–11 (2016)
8. Amgoud, L., David, V.: Measuring similarity between logical arguments. In: Proceedings of the Sixteenth International Conference on Principles of Knowledge Representation and Reasoning KR, pp. 98–107 (2018)
9. Amgoud, L., Bonzon, E., Delobelle, J., Doder, D., Konieczny, S., Maudet, N.: Gradual semantics accounting for similarity between arguments. In: Proceedings of the Sixteenth International Conference on Principles of Knowledge Representation and Reasoning KR, pp. 88–97 (2018)
10. Lang, J., Liberatore, P., Marquis, P.: Propositional independence-formula-variable independence and forgetting. J. Artif. Intell. Res. **18**, 391–443 (2003)
11. Amgoud, L., Besnard, P., Vesic, S.: Equivalence in logic-based argumentation. J. Appl. Non-Class. Logics **24**(3), 181–208 (2014)
12. Jaccard, P.: Nouvelles recherches sur la distributions florale, Bulletin do la Société Vaudoise des Sciences Naturelles **37**, 223–270 (1901)
13. Dice, L.R.: Measures of the amount of ecologic association between species. Ecology **26**(3), 297–302 (1945)
14. Sørensen, T.: A method of establishing groups of equal amplitude in plant sociology based on similarity of species and its application to analyses of the vegetation on Danish commons. Biol. Skr. **5**, 1–34 (1948)
15. Anderberg, M.R.: Cluster Analysis for Applications. Probability and Mathematical Statistics: A Series of Monographs and Textbooks. Academic Press Inc., New York (1973)
16. Sneath, P.H., Sokal, R.R., et al.: Numerical taxonomy. In: The Principles and Practice of Numerical Classification (1973)
17. Ochiai, A.: Zoogeographical studies on the soleoid fishes found in Japan and its neighbouring regions. Bull. Jpn. Soc. Sci. Fischeries **22**, 526–530 (1957)
18. Kulczynski, S.: Die pflanzenassoziationen der pieninen. Bulletin International de l'Académie Polonaise des Sciences et des Lettres, Classe des Sciences Mathématiques et Naturelles, Série B, pp. 57–203 (1927)

A New Generic Framework for Mediated Multilateral Argumentation-Based Negotiation Using Case-Based Reasoning

Rihab Bouslama[1(✉)], Raouia Ayachi[2], and Nahla Ben Amor[1]

[1] LARODEC, Institut Supérieur de Gestion de Tunis, Université de Tunis,
Tunis, Tunisia
rihabbouslama@yahoo.fr, nahla.benamor@gmx.fr
[2] LARODEC, École Supérieure des Sciences Économiques et commerciales de Tunis,
Université de Tunis, Tunis, Tunisia
raouia.ayachi@gmail.com

Abstract. Multilateral negotiation and its importance in today's society, makes it an interesting application in the Artificial Intelligence domain. This paper proposes a new generic framework for multilateral multi-issue Argumentation-Based Negotiation. Agents are first clustered based on their offers to reduce the negotiation complexity. These agents negotiate via a mediator who has two roles: (1) organize the dialogue between them and (2) step in when no mutual agreement is found to get them out of the bottleneck, we formulate this problem as a linear problem where the mediator maximizes both parties' utilities fairly. To highlight the performance of our framework, we tested it on the tourism domain using real data. We study the cases where agents are: not clustered, clustered using hard clustering techniques and clustered using soft clustering techniques. We also study the impact of the mediator in each variant by detecting agreements reached with and without his help.

Keywords: Argumentation-Based Negotiation · Multi-issue · Multilateral · Mediator · Multi-agent system · Clustering · CBR

1 Introduction

Automated negotiation is characterized by many factors influencing its complexity. These factors include the number of involved parties (bilateral or multilateral), the negotiation protocol, the number of issues and their interactions, time constraints, etc. In order to deal with this complexity, several researches incorporate advanced Artificial Intelligence technologies including predicting and learning methods, clustering techniques, Case-Based Reasoning etc. One of the prevailing approaches for automated negotiation is the Argumentation-based negotiation (ABN) where agents go beyond exchanging offers and have the possibility to exchange arguments that backup their positions. This enhances the negotiation process and its final outcome quality [1,2]. Moreover, Case-Based Reasoning

© Springer Nature Switzerland AG 2019
G. Kern-Isberner and Z. Ognjanović (Eds.): ECSQARU 2019, LNAI 11726, pp. 14–26, 2019.
https://doi.org/10.1007/978-3-030-29765-7_2

(CBR) may be used as mean of helping agents to generate offers and arguments in Argumentation-Based Negotiations. But contrarily to automated negotiation and even ABN where several researchers have proposed generic frameworks [3–5], most of proposed work gathering ABN and CBR is domain specific [6,7]. Recently, we have investigated this issue in [8] by proposing a generic framework for ABN (so-called, *GANC* for Generic Argumentation-Based Negotiation using CBR) gathering Case-Based Reasoning (CBR) to Argumentation-Based Negotiation. However, it is restricted to bilateral and one issue negotiations.

In this paper, we propose to extend *GANC* framework to multilateral and multi-issue setting. To this end, we propose to cluster agents starting from the idea that agents can form clusters where each cluster will have a representative agent who will negotiate for it. Then, all representative agents will negotiate with the intervention of a mediator which is a neutral and impartial agent that helps remaining agents at reaching a settlement. Grouping agents in automated negotiation has already been explored in the literature [9,10]. In the last decades, automated mediated negotiation has received a considerable attention [11,12]. Tasks ensured by the mediator differ depending on the attributed authority. Indeed, the mediator can: (i) organize the discussion between agents (*facilitator*) [13], (ii) impose a solution on different parties of the negotiation (*arbitrator*) [14], (iii) analyze the ongoing discussion like in [15] where the mediator's goal is to reduce the negative consequences of the conflict or (iv) give proposals to the conflicting parties [16].

Computational mediators have been the interest of many research works [11,16,17]. In ABN field, several computational mediators were proposed [6,15, 16,18]. Most of the research works concerning mediated Argumentation-Based Negotiation were focusing on the proposal of a computational mediator that can handle disputants' conflicts and find a solution that satisfies both of them. In this paper, we are focusing on the negotiation process as a whole where we handle the negotiation protocol, mediator's tasks, the exchange of arguments and the use of CBR.

The paper is organized as follows, Sect. 2 introduces the new proposed framework, discusses the clustering and the negotiation phases and highlights the role of the mediator. Section 3 discusses the experimental environment and the obtained results.

2   A New Framework for Multilateral Mediated ABN Using CBR

For multilateral and-issues settings, we propose *MGANC*: Multilateral Generic framework for Argumentation-Based Negotiation using CBR. *MGANC* is a generic framework that supports multi-argumentative negotiator agents. We opt for a mediated negotiation where the mediator helps agents to reach a mutually beneficial outcome. Moreover, agents will take advantage of their past experiences using their own CBR to generate arguments. In *MGANC*, the process of searching an agreement between many agents is based on two main phases: (i) the clustering phase and (ii) the negotiation phase detailed below.

2.1 Clustering Phase

The complexity of the negotiation process is highly influenced by the number of involved parties. In this work we explore the interesting idea of clustering agents. Generally, agents are grouped based on their goals (e.g., buyers and sellers) [19]. In our framework, we propose to cluster agents based on their offers. Several clustering algorithms are available in the literature (e.g., Hierarchical clustering, soft clustering, hard clustering) and some of them are used in automated negotiation such as Hierarchical clustering [10]. Since agents' offers can be close, we propose to apply clustering on the involved agents based on their offers. While clustering negotiating parties, they may switch clusters looking for better solutions, which is known as soft clustering. For this purpose, we propose to use Evidential C-Means (ECM) [20] and Fuzzy C-Means (FCM) [24] which are soft clustering algorithms that group parties in a soft way where one agent can belong to more than one cluster with different degrees. This flexibility allows to gain a deeper insight in the data. ECM and FCM showed better results than c-means since they better reflect real world situations.

From each cluster we choose one representative agent. This agent presents the center of the cluster and has the highest degree of belonging to it. Thus, we move from n agents entering the negotiation phase to k agents where $k < n$.

2.2 Negotiation Phase

In the proposed framework called $MGANC$, argumentative negotiator agents follow a state machine protocol that specifies the rules of interaction between them via a mediator. Figure 1 depicts the proposed protocol that explicitly expresses the rules that should be followed by different agents as well as the different locutions (i.e., messages) exchanged between them. Agents start from the "BEGIN" state and end at the "DIE" state. Once an agent enters the state "BEGIN", she will go to "OPEN" state waiting for all negotiation parties to start the negotiation at the same time. Next, according to the type of agents (i.e., opponent, proponent, mediator), they will follow the intermediary states (e.g., BEGIN, ENTER, ASSERT) where in each state they generate the corresponding messages. The most important messages are:

- *Propose:* with this locution, an agent sends her proposal to the commitment store where it can be checked by the mediator.
- *Assert:* with this locution, an agent asserts her offer by giving arguments explaining her choice.
- *Attack:* with this locution, agents attack each others' arguments and offers via the mediator.
- *Why:* with this locution, the opponent asks to the mediator for some explanations from the proponent.
- *SAY Why:* with this locution, the mediator asks the proponent to assert her offer.
- *Propose solution:* with this locution, at the end of each round, the mediator proposes a solution that may be accepted by both opponent and proponent.

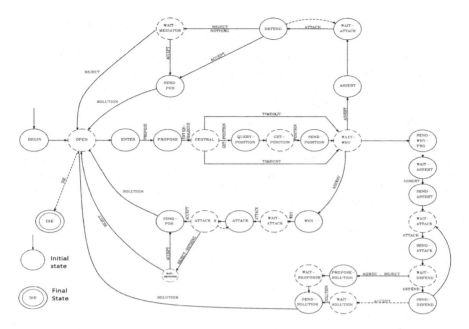

Fig. 1. Proposed multilateral argumentation-based negotiation protocol (dotted lines indicate wait states and solid lines are send states).

The end state is "DIE" and it can be reached through "OPEN" state triggered by the following locutions:

- *Agree:* with this locution, an agent agrees on the opponent's argument while she still believes that her offer is better and then a new round will start where she is the proponent agent.
- *Reject:* with this locution, an agent rejects the counter party's offer and indicates that she has no more arguments to present. Just like agreeing, agents exchange roles and the opponent becomes proponent and vice-versa.
- *Accept:* with this locution, agents accept an offer:
  1. If the offer is proposed by the counter party agent, it is accepted in two cases: (i) if the proposal is the agent's current preferred value. (ii) If the argument respects the agent's conditions such as: (a) the number of similar premises between premises characterizing her offer and premises sent by the opponent, (b) the number of the received distinguished premises that exists in her CBR (these premises were once a reason to choose a given offer). As an example, an agent can accept the other agent's proposal if their positions are characterized by two common premises with the same value. These conditions differ from agent to another based on their flexibility and how much they are open minded. Finally, in case where a counter-example argument is received, the agent will immediately accept the offer since this kind of argument has the highest power of convincing.

2. If the offer is proposed by the mediator, agents will compare the utility of the offer to their reservation point utility. If it exceeds than they will accept otherwise, they refuse the solution. This is explained by the fact that agents can't get better solution since the mediator proposes the last offer by trying to maximize both negotiation parties' utilities. If one of them rejects the mediator's offer, the negotiation ends with a disagreement.

The Role of the Mediator. Agents always negotiate via an autonomous agent: the mediator. The mediation process can be classified into two categories, *agreement centred* and *relationship centred* [21]. Our work belongs to the first category "agreement centred" where the goal of the mediator is to reach a solution accepted by the conflicting parties. Although the mediator in this framework is a *facilitator* that has no power on the negotiation parties, if after a series of arguments exchange (i.e., attacks and counter-attacks) still no agreement is reached, he proposes a solution that respects agents' reservation points and that he values beneficial for both parties. Indeed, the mediator tries to maximize the utility of both parties fairly. We propose to formulate the problem as a linear problem:

$$Maximize\ \alpha\ U_{Opp} + (1 - \alpha)\ U_{Prop} \tag{1}$$

s.t

$$U_{iOpp} = a_{iOpp}x_i + b_{iOpp} \tag{2}$$

$$U_{iProp} = a_{iProp}x_i + b_{iProp} \tag{3}$$

$$LB_i \leq x_i \leq UB_i \tag{4}$$

$$MinUtil_{opp} \leq U_{Opp} \leq MaxUtil_{opp} \tag{5}$$

$$MinUtil_{Prop} \leq U_{Prop} \leq MaxUtil_{Prop} \tag{6}$$

$$U_{Opp} = \sum_{i=1}^{n} w_{iopp}U_{iOpp} \tag{7}$$

$$U_{Prop} = \sum_{i=1}^{n} w_{iProp}U_{iProp} \tag{8}$$

$$i = 1, ..., n \tag{9}$$

Table 1 presents the nomenclature used to represent the linear problem.

Table 1. Nomenclature

α	Can take values from 0 to 1, used to calibrate the importance of each agent
U_{Opp} (resp. U_{Prop})	The opponent's (resp. proponent's) total utility
U_{iOpp} (resp. U_{iProp})	The opponent's (resp. proponent's) utility on issue i
a_i	The slope of the utility function calculated based on the utilities of two different values of an issue i
b_i	The intercept of the utility function calculated based on the utilities of two different values of an issue i
LB_i	The lower bound of issue i
UB_i	The upper bound of issue i
x_i	The value of issue i
n	Total number of issues
$MinUtil_{Prop}$ (resp. $MinUtil_{Opp}$)	The proponent (resp. the opponent) minimum utility under which he would never accept an offer
$MaxUtil_{Prop}$ (resp. $MaxUtil_{Opp}$)	The proponent (resp. the opponent) maximum utility, which is the utility of his first offer
w_{iopp} (resp. w_{iprop})	The weight attributed by the opponent (resp. proponent) to each utility

The Negotiation Process Description. This framework supports many agents $A = A_1...A_n$ where each agent has a starting offer $O = O_1...O_n$ about m issues. Thus, for each agent A_i, $O_i = O_{i1}...O_{im}$. Algorithm 1, outlines the main steps of the negotiation process. Agents are first clustered based on their initial offers using the function $Clustering(A, O)$. Then the center of each cluster C_i is detected using $SelectRepresentative(C_i)$. This agent A_i has the highest degree of belonging to her cluster. Then, each representative agent A_i will enter in a dialogue with the counter party agent Ag to negotiate the issues. During the negotiation, the representative agent and the counter party will negotiate by exchanging arguments in order to convince each other following the negotiation protocol depicted in Fig. 1 called by the function $FollowNegotiationProtocol(A_i, Ag)$. An argument takes the form of premises/claim where premises are the explanations and reasons to choose the given claim. For example, if agents are negotiating one issue, a holiday destination, an argument may take the form: *warm weather, beaches, 5 stars hotels: Bahamas.* The "Bahamas" represents the claim and the negotiation offer while "warm weather, beaches, 5 stars hotels" are the premises explaining the reasons to accept the claim. Arguments are selected using agents'

CBR. They represent agents' past experiences that are similar to the current negotiation case. At each round, agents start by discussing the proponent offer and then the opponent offer after exchanging roles. By the end of each round, if an agreement is found then the negotiation ends. Otherwise, the mediator steps in and proposes a solution using $MediatorPropose(Sol, A_i, Ag)$. If it is accepted by both agents, the negotiation ends successfully. In the other case a new round starts. This process is repeated until the number of maximum rounds fixed at the beginning of the negotiation is achieved or an agreement is reached. The acceptance is: (1) argument-based, an agent accepts an offer because she was convinced by the counter-party's argument, (2) utility-based, the agent accepts an offer only if the utility of this offer exceeds her reservation point utility. Finally, to determine the utility of each agent from the final outcome: (1) We compute the distance between the final outcome and each agent's initial offer. (2) We compute the similarity between them.

3 Experimental Study

3.1 Experimental Protocol

The framework is implemented on the multi-agent platform, *Magentix2* [22], in Java. Agents use their old argumentation and negotiation experiences (i.e., case bases) that are stored in files acceded by their own CBR.

Data. Experiments are conducted in the tourism domain. Real data was gathered from different sources: (1) the tourism ministry of Tunisia, (2) many websites such as Trivago[1], HolidayWeather[2], Bandsintown[3] and eDreams[4] and (3) the Travel and Tourism Competitiveness Report 2017 published by the World Economic Forum. The database contains 509 lines about 126 different destinations characterized by 29 features (e.g., safety rank, plane ticket's price etc.). In the tourism domain, many clients choose to travel through a travel agency to take advantage of their organized trips. Under this context, we study the case where many clients negotiate their trip. We suppose that agents are negotiating over 5 issues: the price of a trip, the destination, the number of stars of the stay in hotel, the board of the stay in hotel and the season while the rest of features are left to be used as arguments' premises in agents' case-bases. The price and the number of the stars of the hotel are numeric and ready to be clustered. As for the rest of the issues they are originally in a qualitative format. Thus, they were converted from qualitative to quantitative. The board of the stay can be: *housing*, *breakfast and bed*, *all inclusive*, *breakfast and dinner* and *breakfast and lunch and dinner*. They are represented in order from 1 to 5. The season represents the period of the trip: *June*, *July* or *August* and were represented respectively, 1, 2

---

[1] https://www.trivago.fr/.
[2] http://www.holiday-weather.com/.
[3] https://news.bandsintown.com/home.
[4] https://www.edreams.fr/.

Algorithm 1. Multilateral negotiation over multi-issue

Input: $A = A_1...A_n$ the set of agents, Ag the counter party, $I = I_1...I_m$ the set of issues, $O = O_{11}...O_{nm}$ the set of offers
Parameters: $MaxNbRounds$
Output: NegotiationOutcome

1: $agreement \leftarrow false$, $round \leftarrow 0$.
2: $clusters \leftarrow Clustering(A, O)$. {Clustering phase}
3: **for** each $C_i \in clusters$ **do**
4:     $A_i \leftarrow SelectRepresentative(C_i)$
5:     $EnterDialogue(A_i, Ag)$ {Negotiation phase}
6:     **repeat**
7:         $FollowNegotiationProtocol(A_i, Ag)$
8:         **if** $FindAgreement()$ **then**
9:             $agreement \leftarrow true$
10:             $NegotiationOutcome \leftarrow offer$
11:         **else**
12:             $ExchangeRoles(A_i, Ag)$
13:             $FollowNegotiationProtocol(A_i, Ag)$
14:         **end if**
15:         **if** $FindAgreement()$ **then**
16:             $agreement \leftarrow true$
17:             $NegotiationOutcome \leftarrow offer$
18:         **else**
19:             $MediatorPropose(Sol, A_i, Ag)$
20:             **if** $Accept(A_i, Sol)$ AND $Accept(Ag, Sol)$ **then**
21:                 $agreement \leftarrow true$
22:                 $NegotiationOutcome \leftarrow Sol$
23:             **else**
24:                 $round \leftarrow round + 1$
25:             **end if**
26:         **end if**
27:     **until** $(agreement = true)$ OR $(round = MaxNbRounds)$
28: **end for**
29: **return** $NegotiationOutcome$

and 3. For the issue destination that represents the country/city to visit were converted to numbers $(1, 2,...,n)$ with respect to their continent. This means, destinations having the closest numbers are the closest geographically and have many common characteristics (presented by the features).

Moreover, issues such as price and season varies in very different intervals. The price varies from 40 to 18000 euros and the season varies from 1 to 3. To overcome this gap, we normalized data using: $x' = (x - min)/(max - min)$ with x' is the new normalized value, x is the original value, min and max are the minimal and the maximal values of the interval, respectively.

Evaluation Metrics. The framework is evaluated based on: (1) the utility, which depicts how much an agent gained from the final outcome. More precisely, we compute the average utility of agents. (2) Time, which is a crucial criteria when it comes to negotiation. Thus, we compare the average needed time to reach an agreement in several scenarios. (3) Ratio of agreements, which is a ratio of the number of agreements reached with the help of the mediator by the total number of agreements.

Experiments Description. The experiments proceed following the steps outlined below:

Table 2. Example of clustered entities

Agent	Price	Destination	Stars	Accommodation	Season	Cluster 1	Cluster 2	Cluster 3
A1	774	1	5	1	1	0.008	**0.971**	0.020
A2	651	44	5	1	1	0.003	**0.88**	0.08
A3	132	9	4	1	3	0.18	0.05	**0.75**

In the first step, 50 agents (clients) are clustered to 3, 5 and then 10 clusters based on their offers. An example of agents' clustering results on 3 clusters using ECM is depicted in Table 2. Row 1 of Table 2 depicts agent $A1$'s offer where she proposes a price of 774 euros for the destination "Cancun" for a 5 stars hotel with "housing" as an accommodation and to travel in June. The cluster of this agent is cluster 2 and her second cluster is cluster 3. Once agents are clustered, the center element of each cluster is detected. The corresponding agent represents the cluster and negotiates on behalf of the whole group. Then, another agent represents the travel agency. Consequently, both agents start negotiating by exchanging offers and arguments. The negotiation is held over 3 rounds. By the end of each round, if no agreement is found, the mediator steps in and proposes a solution that can satisfy both of them. The mediator follows the linear problem discussed before to propose solutions to the conflicting parties. In order to guarantee the fairness between both agents, α is fixed to 0.5. w is fixed to 1 so that all issues will have the same importance. The agent representing the travel agency will negotiate with each cluster's representative to find an agreement. This agreement can be an already planned trip or it can be created. Actually, the travel agency's agent may accept to create a new trip only if it was proposed by the mediator and she conceives that this would be better for her.

At the end of the negotiation, we have 3 outcomes coming from the three negotiation scenarios between clusters' representative and the travel agency agent. To compute the distance between the final outcome and each agent's initial offer, we used the *Manhattan* distance. $Dis(X, Y) = |x1-y1|+|x2-y2|+\ldots+|xn-yn|$. Finally, to detect the similarity between the negotiation final outcome and the initial offer of each agent we computed: $Sim(X, Y) = 1 - Dis(X, Y)$.

Example 1. *We suppose that an agent A1 belongs to cluster c1 and her offer is:* $<price : 402, destination : Toulouse, France, stars : 4, board : housing(1), season : 3(August)>$. *If the final outcome is:* $<price : 249, destination : Porto, Portugal, stars : 1, board : housing(1), season : 1(June)>$ *then the Manhattan distances between the agent's first offer and the final solution on each issue are:* 0.138, 0.954, 0.66, 0 *and* 1 *for issue price, destination, number of the hotel stars, the board of the stay and the season in order.*

In order to evaluate our framework, we conducted the discussed experiments: (1) without any clustering among agents, (2) with a hard clustering algorithm namely, *K-means* [23] and (3) with a soft clustering namely, *ECM* and *FCM*.

3.2   Experimental Results

Computing agents' offers similarity to the final outcome depicts their utility from the solution found by their representative. In case of 3 clusters, the average utility of each cluster in order: 60%, 61% and 65%. These results are found using ECM, soft clustering where agents changed their original cluster to get solutions with bigger utilities from the other clusters. We also calculate the utility of the travel agency from all negotiations: 66% (with cluster 1), 40% (with cluster 2) and 88% (cluster 3). For the 2nd negotiation scenario, we can see that the travel agency made more concessions in order to convince her clients. Nevertheless, this is always with respect to the agent's reservation point. Table 3 depicts the results from different scenarios (i.e., different clustering techniques and no clustering situation) in terms of the average time needed for the negotiation phase, the average utility and the impact of the mediator. The time decreases remarkably when agents are clustered based on their offers. The highest utility is attended when agents were not clustered which is due to the fact that each agent negotiate by herself and thus, she maximizes her own utility without taking into consideration the rest of the agents. However, using soft clustering didn't cost agents in terms of utility since it still presents considerable utilities (i.e., column 3 and 4 of the Table 3) which is bigger than their utilities following a hard clustering. This is due to the fact that agents changed their clusters to get better solutions. In case where agents are clustered using ECM and FCM, agents may switch clusters. Thus, if no agreement is reached in their original cluster they switch to a cluster that presents a satisfying offer for them. However, when agents were not clustered, on 50 agents only 43 found agreements and 36 of them were reached with the help of the mediator. Similarly in hard clustering, agents can't switch clusters and thus, the average utility decreases.

Table 3. Results in terms of time, utility and mediator's impact

Nb clusters	Criteria	Clustering			
		Without clustering	Soft clustering (ECM)	Soft clustering (FCM)	Hard clustering (K-means)
0	Time	3.15 h	-	-	-
	Average utility	69%	-	-	-
	Ratio of agreements	36/43	-	-	-
3	Time	-	12.41 min	17.03 min	13.21 min
	Average utility	-	63%	63%	54%
	Ratio of agreements	-	3/3	3/3	3/3
5	Time	-	27.63 min	22.15 min	19.21 min
	Average utility	-	65%	68%	45%
	Ratio of agreements	-	2/2	3/4	3/4
10	Time	-	34.16 min	34.38 min	48.76 min
	Average utility	-	62%	66%	35%
	Ratio of agreements	-	5/6	6/9	5/6

4   Conclusion

This paper proposes a new framework for multi-agent negotiations over multi-issues. The framework combines three different fields: argumentation, negotiation and CBR. Agents negotiate via a mediator that has two main roles: (i) assist the negotiation process by sending messages to the negotiation parties and (ii) propose a solution that can satisfy both parties in case of no agreement. This will get them out of the bottleneck and help them to overcome their conflicts. To ensure the fairness for all negotiation parties, the mediator proposes solutions that maximizes all parties' utilities with respect to a set of conditions. In order to reduce the negotiation complexity, we propose soft clustering over agents. A representative agent is chosen from each cluster to lead the negotiation. The results showed that the time dedicated for negotiation decreases remarkably when we use the clustering technique while agents' utilities show that the clustering technique preserves their self-interested behaviour.

We assumed that if the agent client proposes a new trip that was unplanned by the travel agency it will be refused. However, it may be accepted if the proposal was coming from the mediator since the latter has a bigger view on the problem and ensures that both parties' utilities are maximized. Further experimental investigations are needed to test other negotiation strategies. Moreover, we assumed that all agents belonging to one cluster trust the representative agent and they all agreed on her. As perspective, we will include voting in the process of choosing a representative agent.

References

1. Dimopoulos, Y., Moraitis, P.: Advances in argumentation based negotiation. In: Negotiation and Argumentation in Multi-agent Systems: Fundamentals, Theories, Systems and Applications, pp. 82–125 (2014)
2. Rahwan, I., Ramchurn, S.D., Jennings, N.R., Mcburney, P., Parsons, S., Sonenberg, L.: Argumentation-based negotiation. Knowl. Eng. Rev. **18**(4), 343–375 (2003)
3. Kraus, S., Arkin, R.C.: Strategic Negotiation in Multiagent Environments. MIT Press, Cambridge (2001)
4. Lin, R., Kraus, S., Tykhonov, D., Hindriks, K., Jonker, C.M.: Supporting the design of general automated negotiators. In: Ito, T., Zhang, M., Robu, V., Fatima, S., Matsuo, T., Yamaki, H. (eds.) Innovations in Agent-Based Complex Automated Negotiations. SCI, vol. 319, pp. 69–87. Springer, Heidelberg (2010). https://doi.org/10.1007/978-3-642-15612-0_4
5. Amgoud, L., Dimopoulos, Y., Moraitis, P.: A general framework for argumentation-based negotiation. In: Rahwan, I., Parsons, S., Reed, C. (eds.) ArgMAS 2007. LNCS (LNAI), vol. 4946, pp. 1–17. Springer, Heidelberg (2008). https://doi.org/10.1007/978-3-540-78915-4_1
6. Sycara, K.P.: Persuasive argumentation in negotiation. Theory Decis. **28**(3), 203–242 (1990)
7. Soh, L.K., Tsatsoulis, C.: Agent-based argumentative negotiations with case-based reasoning. In: AAAI Fall Symposium Series on Negotiation Methods for Autonomous Cooperative Systems, North Falmouth, Massachusetts, pp. 16–25 (2001)
8. Bouslama, R., Ayachi, R., Amor, N.B.: A new generic framework for argumentation-based negotiation using case-based reasoning. In: Medina, J., Ojeda-Aciego, M., Verdegay, J.L., Pelta, D.A., Cabrera, I.P., Bouchon-Meunier, B., Yager, R.R. (eds.) IPMU 2018. CCIS, vol. 854, pp. 633–644. Springer, Cham (2018). https://doi.org/10.1007/978-3-319-91476-3_52
9. Breban, S., Vassileva, J.: A coalition formation mechanism based on inter-agent trust relationships. In: Proceedings of the First International Joint Conference on Autonomous Agents and Multiagent Systems: Part 1, pp. 306–307. ACM (2002)
10. Buccafurri, F., Rosaci, D., Sarnè, G.M.L., Ursino, D.: An agent-based hierarchical clustering approach for E-commerce environments. In: Bauknecht, K., Tjoa, A.M., Quirchmayr, G. (eds.) EC-Web 2002. LNCS, vol. 2455, pp. 109–118. Springer, Heidelberg (2002). https://doi.org/10.1007/3-540-45705-4_12
11. Aydoğan, R., Hindriks, K.V., Jonker, C.M.: Multilateral mediated negotiation protocols with feedback. In: Marsa-Maestre, I., Lopez-Carmona, M.A., Ito, T., Zhang, M., Bai, Q., Fujita, K. (eds.) Novel Insights in Agent-based Complex Automated Negotiation. SCI, vol. 535, pp. 43–59. Springer, Tokyo (2014). https://doi.org/10.1007/978-4-431-54758-7_3
12. Patrikar, M., Vij, S., Mukhopadhyay, D.: An approach on multilateral automated negotiation. Procedia Comput. Sci. **49**, 298–305 (2015)
13. Bratu, M., Andreoli, J.-M., Boissier, O., Castellani, S.: A software infrastructure for negotiation within inter-organisational alliances. In: Padget, J., Shehory, O., Parkes, D., Sadeh, N., Walsh, W.E. (eds.) AMEC 2002. LNCS (LNAI), vol. 2531, pp. 161–179. Springer, Heidelberg (2002). https://doi.org/10.1007/3-540-36378-5_10

14. Baydin, A.G., López de Mántaras, R., Simoff, S., Sierra, C.: CBR with common-sense reasoning and structure mapping: an application to mediation. In: Ram, A., Wiratunga, N. (eds.) ICCBR 2011. LNCS (LNAI), vol. 6880, pp. 378–392. Springer, Heidelberg (2011). https://doi.org/10.1007/978-3-642-23291-6_28

15. Tedesco, P.A.: MArCo: building an artificial conflict mediator to support group planning interactions. Int. J. Artif. Intell. Educ. **13**(1), 117–155 (2003)

16. Trescak, T., Sierra, C., Simoff, S., De Mantaras, R.L.: Dispute resolution using argumentation-based mediation. arXiv preprint arXiv:1409.4164 (2014)

17. Lang, F., Fink, A.: Learning from the metaheuristics: protocols for automated negotiations. Group Decis. Negot. **24**(2), 299–332 (2015)

18. Sierra, C., de Mantaras, R.L., Simoff, S.: The argumentative mediator. In: Criado Pacheco, N., Carrascosa, C., Osman, N., Julián Inglada, V. (eds.) EUMAS/AT -2016. LNCS (LNAI), vol. 10207, pp. 439–454. Springer, Cham (2017). https://doi.org/10.1007/978-3-319-59294-7_36

19. Garruzzo, S., Rosaci, D.: Agent clustering based on semantic negotiation. ACM Trans. Auton. Adapt. Syst. (TAAS) **3**(2), 7 (2008)

20. Masson, M.H., Denoeux, T.: ECM: an evidential version of the fuzzy c-means algorithm. Pattern Recognit. **41**(4), 1384–1397 (2008)

21. Silbey, S.S., Merry, S.E.: Mediator settlement strategies. Law Policy **8**(1), 7–32 (1986)

22. Pacheco, N.C., et al.: Magentix 2 user's manual (2015)

23. Kanungo, T., et al.: An efficient k-means clustering algorithm: analysis and implementation. IEEE Trans. Pattern Anal. Mach. Intell. **7**, 881–892 (2002)

24. Bezdek, J.C., Ehrlich, R., Full, W.: FCM: the fuzzy c-means clustering algorithm. Comput. Geosci. **10**(2–3), 191–203 (1984)

Interpretability of Gradual Semantics in Abstract Argumentation

Jérôme Delobelle[(✉)] and Serena Villata

Université Côte d'Azur, Inria, CNRS, I3S, Sophia-Antipolis, France
`jerome.delobelle@inria.fr, villata@i3s.unice.fr`

Abstract. Argumentation, in the field of Artificial Intelligence, is a formalism allowing to reason with contradictory information as well as to model an exchange of arguments between one or several agents. For this purpose, many semantics have been defined with, amongst them, gradual semantics aiming to assign an acceptability degree to each argument. Although the number of these semantics continues to increase, there is currently no method allowing to explain the results returned by these semantics. In this paper, we study the interpretability of these semantics by measuring, for each argument, the impact of the other arguments on its acceptability degree. We define a new property and show that the score of an argument returned by a gradual semantics which satisfies this property can also be computed by aggregating the impact of the other arguments on it. This result allows to provide, for each argument in an argumentation framework, a ranking between arguments from the most to the least impacting ones w.r.t. a given gradual semantics.

Keywords: Abstract argumentation · Gradual semantics · Interpretability

1 Introduction

The issue of interpreting the results obtained by Artificial Intelligence (AI) methods is receiving an increasing attention both in the AI community but also from a wider audience. In particular, the ability to interpret the rationale behind the results (e.g., classifications, decisions) returned by an artificial intelligent agent is of main importance to ensure the transparency of the interaction between the two entities in order to accomplish cooperative tasks. According to Miller [13], *interpretability* is the degree to which an observer can understand the cause(s) of a result. An algorithm, a program or a decision is said to be interpretable if it is possible to identify the elements or the features that have the greatest impact on (and thus lead to) the result. This term must not be confused with the term *explanation* which is the answer to a why-question or with the term *justification* which explains why a result is good, but does not necessarily aim to give an explanation of the process. Despite the numerous (formal and empirical) approaches [9,11,12,17] to tackle the problem of interpretability of

© Springer Nature Switzerland AG 2019
G. Kern-Isberner and Z. Ognjanović (Eds.): ECSQARU 2019, LNAI 11726, pp. 27–38, 2019.
https://doi.org/10.1007/978-3-030-29765-7_3

artificial intelligent systems, it is still an open research problem. As highlighted by Mittelstadt et al. [14], artificial argumentation [3] may play an important role in addressing this open issue, thanks to its inner feature of combining decision making with the pro and con arguments leading to a certain decision.

In this paper, we aim to study, from a formal point of view, how to cast the notion of interpretability in abstract argumentation so that the reasons leading to the acceptability of one or a set of arguments in a framework may be explicitly assessed. More precisely, this research question breaks down into the following sub-questions: *(i)* how to formally define and characterise the notion of *impact* of an argument with respect to the acceptability of the other arguments in the framework? and *(ii)* how does this impact play a role in the interpretation process of the acceptability of arguments in the framework?

To answer these questions, we start from the family of graded semantics [4,6], and we select two semantics which present different features so that we can show the generality of our approach to characterise the notion of impact. In particular, we select the h-categorizer semantics initially proposed by Besnard and Hunter [5] and the counting semantics from Pu et al. [16]. In both approaches, the acceptability of an argument, differently from standard Dung's semantics [10] where arguments are either (fully) *accepted* or *rejected*, is represented through an acceptability *degree* in the range $[0, 1]$. Roughly, we say that the impact of a certain argument (or a set of arguments) on the degree of acceptability of another argument can be measured by computing the difference between the current acceptability degree of the argument and its acceptability degree when the first argument is deleted. We study the formal properties of the notion of impact instantiated through these two graded semantics both for cyclic and acyclic abstract argumentation frameworks. Finally, we show that studying the impact of an argument on the other arguments allows us to answer to some main needs in terms of interpretability of argument-based decision maker's resolutions.

The remainder of the paper is as follows: in Sect. 2, we provide same basics about gradual semantics and more precisely, the h-categorizer [5] and the counting semantics [16], Sect. 3 discusses the notion of impact of an argument in an argumentation framework and its formal properties, Sect. 4 focuses on the balanced impact property, in Sect. 5 we highlight how the notion of impact and its properties play a role on the interpretability of abstract argumentation frameworks and the acceptability of the arguments. The discussion of the related literature and conclusions end the paper.

2   Preliminaries

An abstract argumentation framework (AF) is a set of abstract arguments connected by an attack relation.

Definition 1 (AF). *An (abstract) argumentation framework (AF) is a tuple* $F = \langle A, R \rangle$ *where* A *is a finite and non-empty set of (abstract) arguments, and* $R \subseteq A \times A$ *is a binary relation on* A, *called the attack relation. For two arguments* $x, y \in A$, *the notation* $(x, y) \in R$ *(or* xRy) *means that* x *attacks* y.

Definition 2 (Non-attacked set of arguments). *Let* $F = \langle \mathcal{A}, \mathcal{R} \rangle$ *be an AF. The set of arguments* $X \subseteq \mathcal{A}$ *is non-attacked if* $\forall x \in X, \nexists y \in \mathcal{A} \backslash X$ *s.t.* $(y, x) \in \mathcal{R}$.

Notation 1. *Let* $F = \langle \mathcal{A}, \mathcal{R} \rangle$ *be an AF and* $x, y \in \mathcal{A}$. *A* **path** P *from* y *to* x, *noted* $P(y, x)$, *is a sequence* $\langle x_0, \ldots, x_n \rangle$ *of arguments in* $\mathcal{A}$ *such that* $x_0 = x$, $x_n = y$ *and* $\forall i < n, (x_{i+1}, x_i) \in \mathcal{R}$. *The length of the path* P *is* n *(i.e., the number of attacks it is composed of) and is denoted by* $l_P = n$. *A* **cycle** *is a path from* x *to* x *and a* **loop** *is a cycle of length 1.*
Let $\mathcal{R}_n^-(x) = \{y \mid \exists P(y, x) \text{ with } l_P = n\}$ *be the multiset of arguments that are bound by a path of length* n *to the argument* x. *Thus, an argument* $y \in \mathcal{R}_n^-(x)$ *is a direct attacker (resp. defender) of* x *if* $n = 1$ *(resp.* $n = 2$*). More generally,* y *is an* **attacker** *(resp.* **defender***) of* x *if* n *is odd (resp. even).*

A gradual semantics assigns to each argument in an argumentation framework a score, called *acceptability degree*, depending on different criteria. This degree must be selected among the interval $[0, 1]$.

Definition 3 (gradual semantics). *A gradual semantics is a function* $\mathcal{S}$ *which associates to any argumentation framework* $F = \langle \mathcal{A}, \mathcal{R} \rangle$ *a function* $\mathrm{Deg}_F^{\mathcal{S}} : \mathcal{A} \to [0, 1]$. *Thus,* $\mathrm{Deg}_F^{\mathcal{S}}(x)$ *represents the acceptability degree of* $x \in \mathcal{A}$.

h-categorizer Semantics [5,15]. This gradual semantics uses a categorizer function to assign a value to each argument which captures the relative strength of an argument taking into account the strength of its attackers, which itself takes into account the strength of its attackers, and so on.

Definition 4. *Let* $F = \langle \mathcal{A}, \mathcal{R} \rangle$ *be an argumentation framework. The* **categorizer function** $\mathrm{Deg}_F^{Cat} : \mathcal{A} \to]0, 1]$ *is defined such that* $\forall x \in \mathcal{A}$,

$$\mathrm{Deg}_F^{Cat}(x) = \begin{cases} 1 & \text{if } \mathcal{R}_1^-(x) = \emptyset \\ \frac{1}{1 + \sum_{y \in \mathcal{R}_1^-(x)} \mathrm{Deg}_F^{Cat}(y)} & \text{otherwise} \end{cases}$$

Counting Semantics [16]. This gradual semantics allows to rank arguments by counting the number of their respective attackers and defenders. In order to assign a value to each argument, they consider an AF as a dialogue game between the proponents of a given argument x (i.e., the defenders of x) and the opponents of x (i.e., the attackers of x). The idea is that an argument is more acceptable if it has many arguments from proponents and few arguments from opponents. Formally, they first convert a given *AF* into a matrix $M_{n \times n}$ (where n is the number of arguments in *AF*) which corresponds to the adjacency matrix of *AF* (as an *AF* is a directed graph). The matrix product of k copies of M, denoted by M^k, represents, for all the arguments in *AF*, the number of defenders (if k is even) or attackers (if k is odd) situated at the beginning of a path of length k. Finally, a normalization factor N (e.g., the matrix infinite norm) is applied to M in order to guarantee the convergence, and a damping factor α is used to have a more refined treatment on different length of attacker and defenders (i.e., shorter attacker/defender lines are preferred).

Definition 5 (Counting model). *Let* $\mathsf{F} = \langle \mathcal{A}, \mathcal{R} \rangle$ *be an argumentation framework with* $\mathcal{A} = \{x_1, \ldots, x_n\}$, $\alpha \in {]}0,1{[}$ *be a damping factor and* $k \in \mathbb{N}$. *The n-dimensional column vector* v *over* $\mathcal{A}$ *at step* k *is defined by,*

$$v_\alpha^k = \sum_{i=0}^{k} (-1)^i \alpha^i \tilde{M}^i \mathcal{I}$$

where $\tilde{M}$ *is the normalized matrix such that* $\tilde{M} = M/N$ *with* N *as normalization factor and* $\mathcal{I}$ *the n-dimensional column vector containing only 1s.*
The counting model of F *is* $v_\alpha = \lim\limits_{k \to +\infty} v_\alpha^k$. *The strength value of* $x_i \in \mathcal{A}$ *is the* i^{th} *component of* v_α, *denoted by* $\mathrm{Deg}_F^{CS}(x_i)$.

3   Impact Measure

The impact of an argument on another argument can be measured by computing the difference when this argument exists and when it is deleted. To capture this notion of deletion, we need to define the complement operator which deletes a set of arguments from the initial argumentation framework w.r.t. a given argument (i.e., the targeted argument of the impact). These changes have also a direct impact on the set of attacks because the attacks directly related to the deleted arguments (attacking as well as attacked) are automatically deleted too.

Definition 6. *Let* $\mathsf{F} = \langle \mathcal{A}, \mathcal{R} \rangle$ *be an AF,* $X \subseteq \mathcal{A}$ *and* $y \in \mathcal{A}$. *The **complement operator** $\ominus$ is defined as* $\mathsf{F} \ominus_y X = \langle \mathcal{A}', \mathcal{R}' \rangle$, *where*

- $\mathcal{A}' = \mathcal{A} \backslash (X \backslash \{y\})$;
- $\mathcal{R}' = \{(x, z) \mid (x, z) \in \mathcal{R} \text{ and } x \in \mathcal{A} \backslash X, z \in \mathcal{A} \backslash X\}$.

Let us first formalise how to compute the impact of a non-attacked set of arguments on a given argument before generalising it for every set of arguments.

3.1   Impact of a Non-attacked Set of Arguments

The impact of a non-attacked set of arguments X on the degree of acceptability of an argument y can be measured by computing the difference between the current acceptability degree of y and its acceptability degree when X is deleted.

Definition 7 (Impact of a non-attacked set of arguments). *Let* $\mathsf{F} = \langle \mathcal{A}, \mathcal{R} \rangle$ *be an AF,* $y \in \mathcal{A}$ *and* $X \subseteq \mathcal{A}$ *be a non-attacked set of arguments. Let* $\mathcal{S}$ *be a gradual semantics. The impact of* X *on* y *is defined as follows:*

$$\mathrm{Imp}_F^{\mathcal{S}}(X, y) = \mathrm{Deg}_F^{\mathcal{S}}(y) - \mathrm{Deg}_{F \ominus_y X}^{\mathcal{S}}(y)$$

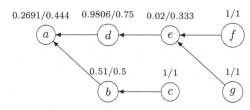

$\mathrm{Imp}_{\mathsf{F}}^{\mathcal{S}}(X,a)$	CS	Cat
$X = \mathcal{A}$	-0.7309	-0.556
$X = \{e,f,g\}$	0.0138	0.044
$X = \{b,c\}$	-0.2547	-0.127
$X = \{f,g\}$	-0.2215	-0.056
$X = \{c\}$	0.2353	0.0808

Fig. 1. On the left hand side, an AF with, above each argument, its scores returned by the counting semantics (with $\alpha = 0.98$) and the h-categorizer semantics [CS/Cat]. On the right hand side, the table contains the impact of some non-attacked sets of arguments on the degree of acceptability of argument a.

Globally, this definition is implicitly included in the formula of existing gradual semantics. A proof of this is that it is possible to compute the score of an argument by combining its basic score and the impact of each argument in the AF: $\mathrm{Deg}_{\mathsf{F}}^{\mathcal{S}}(y) = 1 + \mathrm{Imp}_{\mathsf{F}}^{\mathcal{S}}(\mathcal{A},y)$. Figure 1 illustrates this idea where $\mathrm{Deg}_{\mathsf{F}}^{\mathrm{CS}}(a) = 1 + \mathrm{Imp}_{\mathsf{F}}^{\mathrm{CS}}(\mathcal{A},a) = 1 + (\mathrm{Deg}_{\mathsf{F}}^{\mathcal{S}}(a) - \mathrm{Deg}_{\mathsf{F}\ominus_a\mathcal{A}}^{\mathcal{S}}(a)) = 1 - 0.7309 = 0.2691$.

Measuring the impact of these sets of arguments could be interesting for applications like the online debate platforms where people can argue on a given topic. A debate can be formalised with an AF which has, in many cases, a tree-shaped structure meaning that several sub-debates exist. For example, the arguments for/against the vegan diet can be divided into several categories like the environmental impact, health impact, psychological effects, etc. Checking the impact of these different categories (i.e., the sub-trees in the AF) on the topic implies to better know the influence of each part on the debate.

3.2   General Impact

As it stands, the formula of the impact (Definition 7) cannot be used for an attacked set of arguments. Indeed, calculating the impact of $\{e\}$ on a in Fig. 1 reverts to compute the impact of $\{e,f,g\}$ on a because, by deleting e, the path from f and g (the direct attackers of e) to a are also removed implying to indirectly take into account the impact of f and g on a too.

In order to compute the impact of any set of arguments X on an argument y, we propose to consider the degree of acceptability of y when the arguments in X are the strongest (i.e., when their direct attackers are deleted). The fact that these arguments are attacked will be taken into account during the computation of the impact of these attackers on y.

Definition 8 (Impact). *Let* $\mathsf{F} = \langle \mathcal{A}, \mathcal{R} \rangle$ *be an AF,* $y \in \mathcal{A}$ *and* $X \subseteq \mathcal{A}$. *Let* $\mathcal{S}$ *be a gradual semantics. The **impact** of* X *on* y *is:*

$$\mathrm{Imp}_{\mathsf{F}}^{\mathcal{S}}(X,y) = \mathrm{Deg}_{\mathsf{F}\ominus_y(\bigcup_{x \in X}\mathcal{R}_1^-(x))}^{\mathcal{S}}(y) - \mathrm{Deg}_{\mathsf{F}\ominus_y X}^{\mathcal{S}}(y)$$

This definition generalises Definition 7 because if $\bigcup_{x \in X} \mathcal{R}_1^-(x) = \emptyset$ (meaning that X is non-attacked) then the two formulae are equivalent.

As the acceptability degree of an argument is between 0 and 1 (see Definition 3), the impact of a set of arguments on an argument is in the interval $[-1, 1]$.

Proposition 1. *Let* $\mathsf{F} = \langle \mathcal{A}, \mathcal{R} \rangle$ *be an AF,* $y \in \mathcal{A}$ *and* $X \subseteq \mathcal{A}$. *Let* $\mathcal{S}$ *be a gradual semantics. We have* $\mathrm{Imp}_F^{\mathcal{S}}(X, y) \in [-1, 1]$.

Three categories of impact can be defined, i.e., *positive*, *negative* and *neutral*.

Definition 9. *Let* $\mathsf{F} = \langle \mathcal{A}, \mathcal{R} \rangle$ *be an AF,* $y \in \mathcal{A}$ *and* $X \subseteq \mathcal{A}$. *Let* $\mathcal{S}$ *be a gradual semantics. We say that* X *has a* **positive impact** *on* y *if* $\mathrm{Imp}_F^{\mathcal{S}}(X, y) > 0$, X *has a* **negative impact** *on* y *if* $\mathrm{Imp}_F^{\mathcal{S}}(X, y) < 0$, X *has a* **neutral impact** *on* y *if* $\mathrm{Imp}_F^{\mathcal{S}}(X, y) = 0$.

Note that the fact that a set of arguments has a specific impact (positive, negative or neutral) does not mean that all arguments belonging to this set also have this specific impact. For example, in Fig. 1, we can see that, when CS is used, the set $\{e, f, g\}$ has a positive impact whereas only e has a positive impact (f and g have a negative impact).

In order to be used for interpretability (Sect. 5), we define three notations to select the single arguments which have either a positive, negative or neutral impact on another argument.

Notation 2. *Let* $\mathsf{F} = \langle \mathcal{A}, \mathcal{R} \rangle$ *be an AF and* $y \in \mathcal{A}$. *Let* $\mathcal{S}$ *be a gradual semantics.*
$\mathrm{I}_{\mathcal{S}}^+(y) = \{x \in \mathcal{A} \mid \{x\} \text{ has a positive impact on } y\}$
$\mathrm{I}_{\mathcal{S}}^-(y) = \{x \in \mathcal{A} \mid \{x\} \text{ has a negative impact on } y\}$
$\mathrm{I}_{\mathcal{S}}^=(y) = \{x \in \mathcal{A} \mid \{x\} \text{ has a neutral impact on } y\}$.

Example 1. *Let us compute the impact of each single argument in the AF visualised in Fig. 1 on* a *when CS is used* ($\alpha = 0.98$). *Focusing on* e, *we have* $\mathrm{Imp}_F^{CS}(\{e\}, a) = \mathrm{Deg}_{F \ominus_a \{f,g\}}^{CS}(a) - \mathrm{Deg}_{F \ominus_a \{e\}}^{CS}(a) = 0.4906 - 0.25530 = 0.2353$. *For the other arguments, we have* $\mathrm{Imp}_F^{CS}(\{a\}, a) = 0$, $\mathrm{Imp}_F^{CS}(\{b\}, a) = \mathrm{Imp}_F^{CS}(\{d\}, a) = -0.49$, $\mathrm{Imp}_F^{CS}(\{c\}, a) = 0.2353$ *and* $\mathrm{Imp}_F^{CS}(\{f\}, a) = \mathrm{Imp}_F^{CS}(\{g\}, a) = -0.1108$.
Thus, we have $\mathrm{I}_{CS}^+(a) = \{c, e\}$, $\mathrm{I}_{CS}^-(a) = \{b, d, f, g\}$ *and* $\mathrm{I}_{CS}^=(a) = \{a\}$.

4   Balanced Impact Property

The definition of a new gradual semantics is often coupled with an axiomatic evaluation [1, 4]. Such axioms are mainly used to better understand the behaviour of gradual semantics in specific situations. The role and impact of an argument/attack are also discussed. Such axioms have the aim to answer questions like: Is an attack between two arguments killing (cf. Killing property [1]) or just weakening (cf. Weakening property [1]) the target of the attack? In addition, two semantics can both consider that an attack weakens its target (and then

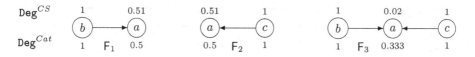

Fig. 2. Three argumentation frameworks F_1, F_2, F_3 showing the difference of impact among the counting semantics and the h-categorizer semantics.

both satisfy the Weakening property) but with different levels of weakening. Unfortunately, this distinction cannot be captured with such axioms.

For example, computing the impact of b and c on a in the three AFs visualised in Fig. 2 with the h-categorizer semantics shows that their impact on a is less important when they attack together ($\text{Imp}^{\text{Cat}}_{F_3}(\{b, c\}, a) = -0.667$) than when they attack it separately ($\text{Imp}^{\text{Cat}}_{F_1}(\{b\}, a) + \text{Imp}^{\text{Cat}}_{F_2}(\{c\}, a) = -0.5 + -0.5 = -1$). Conversely, for the counting semantics, both return the same result: $\text{Imp}^{\text{CS}}_{F_3}(\{b, c\}, a) = -0.98 = -0.49 + -0.49 = \text{Imp}^{\text{CS}}_{F_1}(\{b\}, a) + \text{Imp}^{\text{CS}}_{F_2}(\{c\}, a)$.

To capture this idea, we define a new property, called Balanced Impact (BI), which states that the sum of the impact of two arguments alone on an argument y should be equal to the impact of these two arguments together on y.

Property 1 (Balanced Impact (BI)). A gradual semantics S satisfies Balanced Impact if and only if for any $F = \langle \mathcal{A}, \mathcal{R} \rangle$ and $x, y, z \in \mathcal{A}$,

$$\text{Imp}^{S}_{F}(\{x\}, y) + \text{Imp}^{S}_{F}(\{z\}, y) = \text{Imp}^{S}_{F}(\{x, z\}, y)$$

Let us check which semantics (among CS and Cat) satisfies Balanced Impact.

Proposition 2. *The counting semantics satisfies Balanced Impact.*

Proposition 3. *The h-categorizer semantics does not satisfy Balanced Impact.*

Thus, this property allows to distinguish the semantics which distribute the impact of the arguments on another in a balanced way. Interestingly, this balance allows to go further because it is possible to compute the score of an argument w.r.t. a gradual semantics which satisfies BI from the impact of each single argument in the AF on this argument. Indeed, as explained in Sect. 3.1, the score of an argument y depends on the impact of all the arguments in the AF ($\text{Imp}^{S}_{F}(\mathcal{A}, y)$), but thanks to the balanced impact property, we can split $\text{Imp}^{S}_{F}(\mathcal{A}, y)$ into the impact of each individual argument in the AF. Let us first formally define it for the acyclic argumentation frameworks.

Definition 10. *Let $F = \langle \mathcal{A}, \mathcal{R} \rangle$ be an acyclic AF and $y \in \mathcal{A}$. Let S be a gradual semantics which satisfies BI. The score of y can be defined as follows:*

$$\text{Deg}^{S}_{F}(y) = 1 + \sum_{x \in \mathcal{A}} \text{Imp}^{S}_{F}(\{x\}, y)$$

Algorithm 1. Transformation function ACY

Data: $F = \langle \mathcal{A} = \{x_1, \ldots, x_n\}, \mathcal{R} \rangle$ and $x_1 \in \mathcal{A}$ the targeted argument.
Result: $F' = \langle \mathcal{A}', \mathcal{R}' \rangle$ the infinite acyclic AF of F
$C = \{x_1\}$; $\mathcal{A}' = \{x_1^0\}$; $\mathcal{R}' = \emptyset$ // x_1^0 is called the universal sink vertex of F'
for *every argument x_i in C* **do**
> $C = C \backslash \{x_i\}$
> $m_1 \leftarrow$ maximum value of m among $x_i^m \in \mathcal{A}'$
> **for** *every argument x_j in $\mathcal{R}_1^-(x_i)$* **do**
> > $C = C \cup \{x_j\}$
> > **if** $x_j^0 \notin \mathcal{A}'$ **then**
> > > $\mathcal{A}' = \mathcal{A}' \cup x_j^0$; $\mathcal{R}' = \mathcal{R}' \cup (x_j^0, x_i^{m_1})$
> >
> > **else**
> > > $m_2 \leftarrow$ (maximum value of m among $x_j^m \in \mathcal{A}'$) $+ 1$
> > > $\mathcal{A}' = \mathcal{A}' \cup x_j^{m_2}$; $\mathcal{R}' = \mathcal{R}' \cup (x_j^{m_2}, x_i^{m_1})$

Example 2. *Let us compute the score of a in the AF visualised in Fig. 1 using the impact of each single argument when CS is used.*

$$\text{Deg}_F^{CS}(a) = 1 + (\text{Imp}_F^{CS}(\{a\}, a) + \text{Imp}_F^{CS}(\{b\}, a) + \text{Imp}_F^{CS}(\{c\}, a) + \text{Imp}_F^{CS}(\{d\}, a)$$
$$+ \text{Imp}_F^{CS}(\{e\}, a) + \text{Imp}_F^{CS}(\{f\}, a) + \text{Imp}_F^{CS}(\{g\}, a))$$
$$= 1 + (0 - 0.49 + 0.2353 - 0.49 + 0.2353 - 0.1108 - 0.1108) = 0.2691$$

In order to generalise this definition for any AF, a preprocessing step is required. Indeed, deleting an argument in a cycle removes as well its impact as the ones of other arguments in the cycle. As the method works for acyclic AFs, we propose to transform a cyclic AF into an infinite acyclic AF[1] focused on a given argument a. Thus, as visualised in Fig. 3, we obtain a tree-shaped AF where the root node is a itself, its parent nodes are its direct attackers, the parent nodes of its parent nodes are its direct defenders, and so on. Algorithm 1 details the transformation mechanism called ACY.

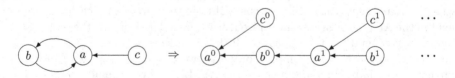

Fig. 3. Cyclic AF transformed into its infinite acyclic AF

---

[1] From a computational point of view, the scores of each argument are computed using a fixed-point approach. If the function used in the gradual semantics converges, the number of iterations needed for convergence can also be used to define the maximal depth of the tree-shaped AF.

We can now use the transformation of an AF, denoted by F, to define the impact of any argument x on a given argument y as the sum of the impact of all the sub-arguments of x $(x^0, x^1, \dots)$ on y^0 (the universal sink vertex) in $\mathsf{ACY}_y(\mathsf{F})$.

Definition 11. *Let* $\mathsf{F} = \langle \mathcal{A}, \mathcal{R} \rangle$ *be an AF with* $y \in \mathcal{A}$. *Let* $\mathsf{F}' = \mathsf{ACY}_y(\mathsf{F})$ *and* $\mathcal{X} = \{x^0, x^1, \dots\}$ *be the sub-arguments of* $x \in \mathcal{A}$ *in* F'. *Let* $\mathcal{S}$ *be a gradual semantics which satisfies BI. The impact of* x *on* y *is 0 if* $\mathcal{X} = \emptyset$, *otherwise it is defined as follows:*

$$\mathrm{Imp}_{\mathsf{F}}^{\mathcal{S}}(\{x\}, y) = \sum_{x^i \in \mathcal{X}} \mathrm{Imp}_{\mathsf{F}'}^{\mathcal{S}}(\{x^i\}, y^0)$$

This new definition of impact can then be used in Definition 10 to compute the score of a given argument.

Example 3. *By focusing on the AF visualised in Fig. 3, the impact of* b *on* a *is* $\mathrm{Imp}_{\mathsf{F}}^{CS}(\{b\}, a) = \mathrm{Imp}_{\mathsf{ACY}_a(\mathsf{F})}^{CS}(\{b^0\}, a^0) + \mathrm{Imp}_{\mathsf{ACY}_a(\mathsf{F})}^{CS}(\{b^1\}, a^0) + \cdots \simeq -0.63$. *We also have* $\mathrm{Imp}_{\mathsf{F}}^{CS}(\{c\}, a) \simeq -0.63$ *and* $\mathrm{Imp}_{\mathsf{F}}^{CS}(\{a\}, a) \simeq 0.3$. *We obtain* $\mathrm{Deg}_{\mathsf{F}}^{CS}(a) \simeq 0.04 = 1 + 0.3 - 0.63 - 0.63 = 1 + \sum_{x \in \{a,b,c\}} \mathrm{Imp}_{\mathsf{F}}^{CS}(\{x\}, a)$.

5   Interpretability of Gradual Semantics

One of the goals of interpretability for gradual semantics is to identify the elements which have an impact on the score assigned by the selected gradual semantics on each argument. Definition 9 allows to assess whether an argument has a positive, negative or neutral impact on the acceptability degree of an argument. It allows to answer questions about the impact of certain arguments on the others, like in the following example about the AF (F) in Fig. 1:

Q: *Which arguments have a positive impact on* a *in* F *when CS is used?*
A: c and e have a positive impact on a. $\mathrm{I}_{CS}^+(a) = \{c, e\}$

Through the impact values (see Definition 8), it is possible to provide, for each argument, a ranking between the arguments from the most positive to the most negative impacting ones w.r.t. a given gradual semantics.

Definition 12 (Impact ranking). *Let* $\mathsf{F} = \langle \mathcal{A}, \mathcal{R} \rangle$ *be an AF and* $\mathcal{S}$ *be a gradual semantics. The **impact ranking** $\succeq_y^{\mathcal{S}}$ on* $\mathcal{A}$ *with respect to* $y \in \mathcal{A}$ *is defined such that* $\forall x, z \in \mathcal{A}$, $x \succeq_y^{\mathcal{S}} z$ *iff* $\mathrm{Imp}_{\mathsf{F}}^{\mathcal{S}}(\{x\}, y) \geq \mathrm{Imp}_{\mathsf{F}}^{\mathcal{S}}(\{z\}, y)$.

This ranking allows us to select, for each argument, its most positive and negative impacting arguments, if they exist.

Definition 13. *Let* $\mathsf{F} = \langle \mathcal{A}, \mathcal{R} \rangle$ *be an AF and* $\mathcal{S}$ *be a gradual semantics. The **most positive** (resp. **negative**) **impacting arguments** on the acceptability degree of* $y \in \mathcal{A}$ *are defined as follows:*

$$PI_{\mathsf{F}}^{\mathcal{S}}(y) = argmax_{x \in \mathrm{I}_{\mathcal{S}}^+(y)} |\{z \in \mathrm{I}_{\mathcal{S}}^+(y) \mid x \succeq_y^{\mathcal{S}} z\}|$$
$$NI_{\mathsf{F}}^{\mathcal{S}}(y) = argmax_{x \in \mathrm{I}_{\mathcal{S}}^-(y)} |\{z \in \mathrm{I}_{\mathcal{S}}^-(y) \mid z \succeq_y^{\mathcal{S}} x\}|$$

Example 4. *Let us consider the AF depicted in Fig. 1. The impact ranking of argument* a, *when* CS *is used, is* $c \simeq_a^{CS} e \succ_a^{CS} a \succ_a^{CS} f \simeq_a^{CS} g \succ_a^{CS} b \simeq_a^{CS} d$. *Consequently, we have* $PI_F^{CS}(a) = \{c, e\}$ *and* $NI_F^{CS}(a) = \{b, d\}$.

In addition to providing a better understanding of the scores assigned to each argument, this information can also be used to develop strategies during a debate. For example, if someone wants to defend a point of view (i.e., increase the degree of acceptability of an argument in a debate), she can identify the argument(s) with the most negative impact and therefore look for solutions to attack them by introducing some counter-arguments.

6 Related Work

Interpretability has already been studied in the context of extension-based semantics in formal argumentation. Fan and Toni [11] first studied how to give explanations for arguments that are acceptable w.r.t. the admissible semantics in terms of arguments defending them, before formalising explanations for arguments that are not acceptable w.r.t. the admissible semantics by using a dispute tree [12]. Although the extension-based semantics and the gradual semantics share the same goal (i.e., evaluating the arguments), the two approaches are different (see the discussion in [7] for more details). Consequently, the investigation of the notion of interpretability for these two families of semantics also differs.

Concerning the gradual semantics, Amgoud et al. [2] have introduced the concept of contribution measure for evaluating the intensity of each attack in an argumentation graph. The Shapley value is used as contribution measure. However, only a specific family of gradual semantics is considered (i.e., the ones which satisfy the syntax-independent and monotonicity properties like the h-categorizer semantics). Moreover, unlike our method which checks the impact of all arguments in the framework, their method only measures the contribution of direct attacks on an argument which is coherent for the family of semantics studied in this work, but it is not necessarily the case for all existing semantics.

7 Conclusion

In this paper, we have presented a formal framework to interpret the results of gradual semantics in abstract argumentation. More precisely, we have considered the h-categorizer and the counting semantics, and we have formally studied the notion of impact of an argument with respect to the acceptability degree of another argument in the framework both for cyclic and acyclic frameworks. The impact of arguments on the acceptability degree of the other arguments is then employed to interpret the rationale behind the resulting ranking, and to provide a further understanding of the reasons why attacking one argument rather than another may be a strategically better choice.

Two main open issues will be considered as future work: first, in this paper we do not consider the *support* relation [8] between arguments but we aim to extend our formal framework to capture this relation too given its importance in many practical applications, and second, we plan to extend our analysis to the other gradual semantics proposed in the literature to provide a complete overview of the properties of the impact notion over such semantics.

Acknowledgements. This work benefited from the support of the project DGA RAPID CONFIRMA.

References

1. Amgoud, L., Ben-Naim, J.: Axiomatic foundations of acceptability semantics. In: Proceedings of the 15th International Conference on Principles of Knowledge Representation and Reasoning (KR 2016), pp. 2–11 (2016)
2. Amgoud, L., Ben-Naim, J., Vesic, S.: Measuring the intensity of attacks in argumentation graphs with Shapley value. In: Proceedings of the 26th International Joint Conference on Artificial Intelligence (IJCAI 2017), pp. 63–69 (2017)
3. Atkinson, K., et al.: Towards artificial argumentation. AI Magaz. **38**(3), 25–36 (2017). https://www.aaai.org/ojs/index.php/aimagazine/article/view/2704
4. Baroni, P., Rago, A., Toni, F.: From fine-grained properties to broad principles for gradual argumentation: a principled spectrum. Int. J. Approx. Reasoning **105**, 252–286 (2019). https://doi.org/10.1016/j.ijar.2018.11.019
5. Besnard, P., Hunter, A.: A logic-based theory of deductive arguments. Artif. Intell. **128**(1–2), 203–235 (2001)
6. Bonzon, E., Delobelle, J., Konieczny, S., Maudet, N.: A comparative study of ranking-based semantics for abstract argumentation. In: Proceedings of the 30th AAAI Conference on Artificial Intelligence (AAAI 2016), pp. 914–920 (2016)
7. Bonzon, E., Delobelle, J., Konieczny, S., Maudet, N.: Combining extension-based semantics and ranking-based semantics for abstract argumentation. In: Proceedings of the 16th International Conference on Principles of Knowledge Representation and Reasoning (KR 2018), pp. 118–127 (2018)
8. Cayrol, C., Lagasquie-Schiex, M.: Bipolarity in argumentation graphs: towards a better understanding. Int. J. Approx. Reasoning **54**(7), 876–899 (2013). https://doi.org/10.1016/j.ijar.2013.03.001
9. Cyras, K., et al.: Explanations by arbitrated argumentative dispute. Expert Syst. Appl. **127**, 141–156 (2019). https://doi.org/10.1016/j.eswa.2019.03.012
10. Dung, P.M.: On the acceptability of arguments and its fundamental role in nonmonotonic reasoning, logic programming and n-person games. Artif. Intell. **77**(2), 321–358 (1995)
11. Fan, X., Toni, F.: On computing explanations in argumentation. In: Proceedings of the Twenty-Ninth AAAI Conference on Artificial Intelligence, 25–30 January 2015, Austin, Texas, USA, pp. 1496–1502 (2015)
12. Fan, X., Toni, F.: On explanations for non-acceptable arguments. In: Black, E., Modgil, S., Oren, N. (eds.) TAFA 2015. LNCS (LNAI), vol. 9524, pp. 112–127. Springer, Cham (2015). https://doi.org/10.1007/978-3-319-28460-6_7
13. Miller, T.: Explanation in artificial intelligence: insights from the social sciences. Artif. Intell. **267**, 1–38 (2019)
14. Mittelstadt, B.D., Russell, C., Wachter, S.: Explaining explanations in AI. In: Proceedings of the Conference on Fairness, Accountability, and Transparency, FAT* 2019, Atlanta, GA, USA, 29–31 January 2019, pp. 279–288. ACM (2019). https://doi.org/10.1145/3287560.3287574

15. Pu, F., Luo, J., Zhang, Y., Luo, G.: Argument ranking with categoriser function. In: Buchmann, R., Kifor, C.V., Yu, J. (eds.) KSEM 2014. LNCS (LNAI), vol. 8793, pp. 290–301. Springer, Cham (2014). https://doi.org/10.1007/978-3-319-12096-6_26

16. Pu, F., Luo, J., Zhang, Y., Luo, G.: Attacker and defender counting approach for abstract argumentation. In: Proceedings of the 37th Annual Meeting of the Cognitive Science Society (CogSci 2015) (2015)

17. Rago, A., Cocarascu, O., Toni, F.: Argumentation-based recommendations: fantastic explanations and how to find them. In: Lang, J. (ed.) Proceedings of the Twenty-Seventh International Joint Conference on Artificial Intelligence, IJCAI 2018, Stockholm, Sweden, 13–19 July 2018, pp. 1949–1955. ijcai.org (2018). https://doi.org/10.24963/ijcai.2018/269

An Imprecise Probability Approach for Abstract Argumentation Based on Credal Sets

Mariela Morveli-Espinoza[1](✉), Juan Carlos Nieves[2], and Cesar Augusto Tacla[1]

[1] Program in Electrical and Computer Engineering (CPGEI),
Federal University of Technology of Paraná (UTFPR), Curitiba, Brazil
morveli.espinoza@gmail.com,tacla@utfpr.edu.br
[2] Department of Computing Science of Umeå University, Umeå, Sweden
jcnieves@cs.umu.se

Abstract. Some abstract argumentation approaches consider that arguments have a degree of uncertainty, which impacts on the degree of uncertainty of the extensions obtained from a abstract argumentation framework (AAF) under a semantics. In these approaches, both the uncertainty of the arguments and of the extensions are modeled by means of precise probability values. However, in many real life situations the exact probabilities values are unknown and sometimes there is a need for aggregating the probability values of different sources. In this paper, we tackle the problem of calculating the degree of uncertainty of the extensions considering that the probability values of the arguments are imprecise. We use credal sets to model the uncertainty values of arguments and from these credal sets, we calculate the lower and upper bounds of the extensions. We study some properties of the suggested approach and illustrate it with an scenario of decision making.

Keywords: Abstract argumentation · Imprecise probability ·
Uncertainty · Credal sets

1 Introduction

The AAF that was introduced in the seminal paper of Dung [3] is one of the most significant developments in the computational modelling of argumentation in recent years. The AAF is composed of a set of arguments and a binary relation encoding attacks between arguments. Some recent approaches on abstract argumentation assign uncertainty to the elements of the AAF to represent the degree of believe on arguments or attacks. Some of these works assign uncertainty to the arguments (e.g., [4,6–9,12–14]), others to the attacks (e.g., [9]), and others to both arguments and attacks (e.g., [11]). These works use precise probability approaches to model the uncertainty values. However, precise probability approaches have some limitations to quantify epistemic uncertainty,

© Springer Nature Switzerland AG 2019
G. Kern-Isberner and Z. Ognjanović (Eds.): ECSQARU 2019, LNAI 11726, pp. 39–49, 2019.
https://doi.org/10.1007/978-3-030-29765-7_4

for example, to represent group disagreeing opinions. These can be better represented by means of imprecise probabilities, which use lower and upper bounds instead of exact values to model the uncertainty values.

For a better illustration of the problem, consider a discussion between a group of medicine students (agents). The discussion is about the diagnose of a patient. In this context, arguments represent the student's opinions and the attacks represent the disagreements between such opinions. Figure 1 shows the argumentation graph where nodes represent arguments and edges the attacks between arguments. In the graph, two arguments represent two possible diagnoses namely measles and chickenpox, there is an argument against measles and two arguments against chickenpox, and there are three arguments that have no attack relations with the rest of arguments.

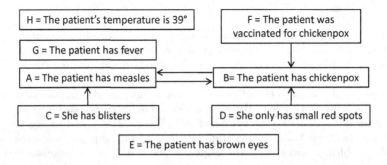

Fig. 1. Argumentation graph for the discussion about the diagnose of a patient.

Suppose that each opinion – i.e., argument – has a probability value between 0 and 1 that represents the degree of believe of each student. Since there is more than one opinion, this means that each argument has associated a set of probability values. Thus, we cannot model these degrees of believe by means of an unique probability value (precise probability value), what we need is to represent a range of the possible degrees of believe.

To the best of our knowledge, there is no work that models the uncertainty values of arguments by using an imprecise probability approach. Therefore, we aim to propose an approach for abstract argumentation in which the uncertainty of the arguments is modeled by an imprecise probability value. Thus, the research questions that are addressed in this paper are:

1. How to model the imprecise uncertainty values of arguments?
2. In abstract argumentation, several semantics have been proposed, which return sets of arguments – called extensions – whose basic characteristic is that these arguments do not attack each other, i.e. they are consistent. The fact that the arguments that belong to an extension are uncertain, causes that such extension also has a degree of uncertainty. How to calculate the lower and upper bounds of extensions?

In addressing the first question, we use credal sets to model the uncertainty values of arguments. Regarding the second question, we base on the credal sets of the arguments to calculate the uncertainty values of extensions obtained under a given semantics. These values are represented by lower and upper bounds. The way to aggregate the credal sets depends on a causal relation between the arguments.

The remainder of this paper is structured as follows. Next section gives a brief overview on credal sets and abstract argumentation. In Sect. 3, we present the AAF based on credal sets and the causality graph concept, which are the base for the calculation of the upper and lower bounds of extension. This calculation is tackled in Sect. 4. We study the main properties of our approach in Sect. 5. Related work is presented in Sect. 6. Finally, Sect. 7 is devoted to conclusions and future work.

2   Background

In this section, we revise concepts of credal sets and abstract argumentation.

2.1   Credal Sets

Assume that we have a finite set of events $\mathbb{E} = \{E_1, ..., E_n\}$ and a probability distribution p on this set, where p is a mapping $p : \mathbb{E} \rightarrow [0, 1]$. According to Levi [10], a closed convex set of probability distributions p is called a credal set. Given an event E, a credal set for E – denoted $K(E)$ – is a set of probability distributions about this event and $\mathbb{K} = \{K(E_1), ..., K(E_n)\}$ denotes a set of all credal sets. Every credal set has the same number of elements. In this work, we assume that the cardinality of the credal sets of $\mathbb{K}$ is the same (let us denote it by m); moreover, we assume that $p_i(E)$ denotes the suggested probability of the agent i w.r.t the event E such that $1 \leq i \leq m$ and $E \in \mathbb{E}$. Given a credal set $K(E)$, the lower and upper bounds for event E are determined as follows:

Lower probability: $\underline{P}(E) = inf\{p(E) : p(E) \in K(E)\}$

Upper probability: $\overline{P}(E) = sup\{p(E) : p(E) \in K(E)\}$ $\hspace{2cm}$ (1)

Given l events $\{E_1, ..., E_l\} \subseteq \mathbb{E}$ and their respective credal sets $K(E_1) = \{p_1(E_1), ..., p_m(E_1)\}, ..., K(E_l) = \{p_1(E_l), ..., p_m(E_l)\}$. If $\{E_1, ..., E_l\}$ are independent events, the lower and upper probabilities are defined as follows:

$$\underline{P}(\{E_1, ..., E_l\}) = min_{1 \leq j \leq m}\{\prod_{i=1}^{i \leq l} p_j(E_i)\} \text{ where } p_j \in K(E_i)$$

$$\overline{P}(\{E_1, ..., E_l\}) = max_{1 \leq j \leq m}\{\prod_{i=1}^{i \leq l} p_j(E_i)\} \hspace{1cm} (2)$$

On the other hand, when the independence relation is not assumed, the first step is to calculate a credal set for $\{E_1, ..., E_l\}$ as follows:

$$K(\{E_1, ..., E_l\}) = \{p_E | p_E = min_{1 \leq j \leq m}\{p_j(E_1), ...p_j(E_l)\}\} \text{ where}$$
$$p_j(E_i) \in K(E_i) \hspace{1cm} (3)$$

Based on $K(\{E_1, ..., E_l\})$, we obtain the lower and upper probabilities:

$$\underline{P}(\{E_1, ..., E_l\}) = min(K(\{E_1, ..., E_l\}))$$
$$\overline{P}(\{E_1, ..., E_l\}) = max(K(\{E_1, ..., E_l\})) \tag{4}$$

Example 1. Let $\{E_1, E_2, E_3\}$ be three events and $K(E_1) = \{p_1(E_1), p_2(E_1), p_3(E_1)\}, K(E_2) = \{p_1(E_2), p_2(E_2), p_3(E_2)\}$, and $K(E_3) = \{p_1(E_3), p_2(E_3), p_3(E_3)\}$ their respective credal sets. Next table shows the values of the probability distributions for each event.

	E_1	E_2	E_3
p_1	0.3	0.5	0.75
p_2	0.6	0.7	0.55
p_3	0.45	0.65	0.8

Assuming that E_1, E_2, and E_3 are independent, the lower and upper probabilities of (E_1, E_2, E_3) are calculated as follows: $\underline{P}(E_1, E_2, E_3) = min\{0.3 \times 0.5 \times 0.75, 0.6 \times 0.7 \times 0.55, 0.45 \times 0.65 \times 0.8\} = min\{0.1125, 0.231, 0.234\}$; hence $\underline{P}(E_1, E_2, E_3) = 0.1125$ and $\overline{P}(E_1, E_2, E_3) = max\{0.1125, 0.231, 0.234\} = 0.234$.

On the other hand, if we assume that E_1, E_2, and E_3 are not independent, then the lower and upper probabilities are calculated as follows: $K(E_1, E_2, E_3) = \{min\{0.3, 0.5, 0.75\}, min\{0.6, 0.7, 0.5\}, min\{0.45, 0.65, 0.8\}\} = \{0.3, 0.55, 0.45\}$. Thus, $\underline{P}(E_1, E_2, E_3) = 0.3$ and $\overline{P}(E_1, E_2, E_3) = 0.55$.

2.2   Abstract Argumentation

In this subsection, we will recall basic concepts related to the AAF defined by Dung [3], including the notion of acceptability and the main semantics.

Definition 1 *(Abstract AF).* *An abstract argumentation framework $\mathcal{AF}$ is a tuple $\mathcal{AF} = \langle ARG, \mathcal{R} \rangle$ where ARG is a finite set of arguments and $\mathcal{R}$ is a binary relation $\mathcal{R} \subseteq ARG \times ARG$ that represents the attack between two arguments of ARG, so that $(A, B) \in \mathcal{R}$ denotes that the argument A attacks the argument B.*

Next, we introduce the concepts of conflict-freeness, defense, admissibility and the four semantics proposed by Dung [3].

Definition 2 *(Argumentation Semantics).* *Given an argumentation framework $\mathcal{AF} = \langle ARG, \mathcal{R} \rangle$ and a set $\mathcal{E} \subseteq ARG$:*

- *$\mathcal{E}$ is conflict-free if $\forall A, B \in \mathcal{E}$, $(A, B) \notin \mathcal{R}$.*
- *$\mathcal{E}$ defends an argument A iff for each argument $B \in ARG$, if $(B, A) \in \mathcal{R}$, then there exist an argument $C \in \mathcal{E}$ such that $(C, B) \in \mathcal{R}$.*
- *$\mathcal{E}$ is admissible iff it is conflict-free and defends all its elements.*
- *A conflict-free $\mathcal{E}$ is a complete extension iff we have $\mathcal{E} = \{A | \mathcal{E} \, defends \, A\}$.*

- $\mathcal{E}$ is a preferred extension iff it is a maximal (w.r.t set inclusion) complete extension.
- $\mathcal{E}$ is a grounded extension iff it is the smallest (w.r.t set inclusion) complete extension.
- $\mathcal{E}$ is a stable extension iff $\mathcal{E}$ is conflict-free and $\forall A \in$ ARG and $A \notin \mathcal{E}$, $\exists B \in \mathcal{E}$ such that $(B, A) \in \mathcal{R}$.

In this article, there is a set of agents that give their opinions (degrees of belief) regarding each argument in ARG by means of probability distributions. The set of arguments can be compared with the events of set $\mathbb{E}$; hence, we can say that $\mathbb{E} =$ ARG. The number of agents that give their opinions determines the cardinality of credal sets. Thus, given m agents and an argument $A \in$ ARG, the credal set for A is represented by $K(A) = \{p_1(A), ..., p_m(A)\}$. Finally, $\mathbb{K}$ denotes all the credal sets of the arguments in ARG.

3   The Building Blocks

In this section, we present the definitions of AAF based on credal sets and causality graph. These concepts are important for the calculation of the lower and upper bounds of extensions.

We use credal sets to model the opinions (degrees of belief) of a set of agents about a set of arguments. Thus, each argument in an AAF has associated a credal set, which contains probability distributions that represent the opinions of the agents about it.

Definition 3 (Credal Abstract Argumentation Framework). An AAF based on credal sets is a tuple $\mathcal{AF}_{CS} = \langle$ARG$, \mathcal{R}, \mathbb{K}, f_{CS}\rangle$ where (i) ARG is a set of arguments, (ii) $\mathcal{R}$ is the attack relation presented in Definition 1, (iii) $\mathbb{K}$ is a set of credal sets, and (iv) f_{CS} : ARG $\to \mathbb{K}$ maps a credal set for each argument in ARG.

Recall that the cardinality of every credal set depends on the number of agents. Since all the agents give their opinions about all the arguments, all the credal sets have the same number of elements.

Definition 4 (Agent's Opinions). Let $\mathcal{AF}_{CS} = \langle$ARG$, \mathcal{R}, \mathbb{K}, f_{CS}\rangle$ be a Credal AAF and $\mathbb{AGT} = \{ag_1, ..., ag_m\}$ a set of agents. The opinion p_i of an agent ag_i (for $1 \leq i \leq m$) is ruled as follows:

1. If $A \in$ ARG, there is $p_i(A) \in K(A)$ where $K(A) \in \mathbb{K}$.
2. $\forall A \in$ ARG, $0 \leq p_i(A) \leq 1$.

Regarding the probability values given to the arguments, it is important to consider the notion of rational probability distribution given in [8]. According to Hunter [8], if the degree of belief in an argument is high, then the degree of belief in the arguments it attacks is low. Thus, a probability function p is rational for an $\mathcal{AF}_{CS}$ iff for each $(A, B) \in \mathcal{R}$, if $p(A) > 0.5$ then $p(B) \leq 0.5$ where $p(A) \in K(A)$ and $p(B) \in K(B)$.

Example 2. Consider that $\mathbb{AGT} = \{ag_1, ag_2, ag_3, ag_4\}$. The Credal AAF for the example given in Introduction is $\mathcal{AF}_{CS} = \langle \text{ARG}, \mathcal{R}, \mathbb{K}, f_{CS} \rangle$ where:
- $\text{ARG} = \{A, B, C, D, E, F, G.H\}$
- $\mathcal{R} = \{(A, B), (B, A), (F, B), (D, B), (C, A)\}$
- $\mathbb{K} = \{K(A), K(B), K(C), K(D), K(E), K(F), K(G), K(H)\}$. The table below shows the credal set of each argument
- $f_{CS}(A) = K(A), f_{CS}(B) = K(B), ..., f_{CS}(H) = K(H)$

	$K(A)$	$K(B)$	$K(C)$	$K(D)$	$K(E)$	$K(F)$	$K(G)$	$K(H)$
p_1	0.2	0.8	0.2	0.75	0.8	0.75	0.7	0.8
p_2	0.7	0.25	0.75	0.15	0.65	0.2	0.8	0.9
p_3	0.55	0.45	0.4	0.5	0.8	0.55	1	1
p_4	0.75	0.1	0.2	0.8	0.7	0.8	0.9	0.9

In a Credal AAF, besides the attack relation between the arguments, there may be a causality relation between them. To make this discussion more concrete, consider the following conflict-free sets:

- $\{G, E\}$: Having fever does not have to do with the eyes' color of the patient and vice-verse, so there is no relation between these arguments. This means that they are independent from each other.
- $\{A, G\}$ and $\{A, F\}$: In both cases the arguments are related in some way. In the first case, having fever (G) is a symptom of (causes) measles (A) and in the second case, the fact that the patient is vaccinated for chickenpox (F) causes that he may have measles and not chickenpox (A).

Definition 5 *(Causality Graph).* *Let* $\mathcal{AF}_{CS} = \langle \text{ARG}, \mathcal{R}, \mathbb{K}, f_{CS} \rangle$ *be a Credal AAF, a causality graph* $\mathbb{C}$ *is a tuple* $\mathbb{C} = \langle \text{ARG}, \mathcal{R}_{\text{CAU}} \rangle$ *such that:*
(i) $\text{ARG} = \text{ARG}_{\leftarrow} \cup \text{ARG}_{\rightarrow} \cup \text{ARG}_{\circ}$ *is a set of arguments,*
(ii) $\mathcal{R}_{\text{CAU}} \subseteq \text{ARG} \times \text{ARG}$ *represents a causal relation between two arguments of* ARG *(the existence of this relation depends on the domain knowledge), such that* $(A, B) \in \mathcal{R}_{\text{CAU}}$ *denotes that argument A causes argument B. It holds that if* $(A, B) \in \mathcal{R}$, *then* $(A, B) \notin \mathcal{R}_{\text{CAU}}$ *and* $(B, A) \notin \mathcal{R}_{\text{CAU}}$,
(iii) $\text{ARG}_{\leftarrow} = \{B | (A, B) \in \mathcal{R}_{\text{CAU}}\}$, $\text{ARG}_{\rightarrow} = \{A | (A, B) \in \mathcal{R}_{\text{CAU}}\}$, *and* $\text{ARG}_{\circ} = \{C | C \in \text{ARG} - (\text{ARG}_{\leftarrow} \cup \text{ARG}_{\rightarrow})\}$,
(iv) $\text{ARG}_{\leftarrow}$ *and* $\text{ARG}_{\rightarrow}$ *are not necessarily pairwise disjoint; however,* $(\text{ARG}_{\leftarrow} \cup \text{ARG}_{\rightarrow}) \cap \text{ARG}_{\circ} = \emptyset$.

Example 3. A causality graph for the Credal AAF of Example 2 is $\mathbb{C} = \langle \{A, B, C, D, E, F, G, H\}, \{(D, A), (F, A), (H, A), (G, A), (H, G), (G, B), (C, B)\} \rangle$ (see Fig. 2), where $\text{ARG}_{\leftarrow} = \{A, B, G\}$, $\text{ARG}_{\rightarrow} = \{D, F, H, G, C\}$, and $\text{ARG}_{\circ} = \{E\}$.

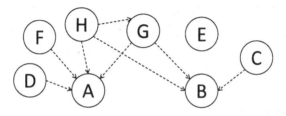

Fig. 2. Causality graph for Example 3. Traced edges represent the causality relation.

4 Lower and Upper Bounds of Extensions

Section 2 presented the definition of conflict-free (cf) and admissible (ad) sets and complete (co), preferred (pr), grounded (gr), and stable (st) semantics. Considering the causality graph, the arguments of an extension $\mathcal{E}_x$ (for x $\in$ {cf, ad, co, pr, gr, st}) may belong to ARG$_\rightarrow$, ARG$_\leftarrow$, or ARG$_\circ$. Depending on it, the calculation of the probabilistic lower and upper bounds of each extension is different. Thus, we can distinguish the following cases: (i) the extension is empty, (ii) the extension has only one argument, and (iii) the extension includes more than one argument.

Definition 6 (Upper and Lower Bounds of Extensions). *Let $\mathcal{AF}_{CS} = \langle ARG, \mathcal{R}, \mathbb{K}, f_{CS} \rangle$ be a Credal AAF, $\mathbb{C} = \langle ARG, \mathcal{R}_{CAU} \rangle$ a causality graph, and $\mathcal{E}_x \subseteq ARG$ (for x $\in$ {cf, ad, co, pr, gr, st}) an extension under semantics x. The lower and uppers bounds of $\mathcal{E}_x$ are obtained as follows:*
1. *If $\mathcal{E}_x = \{\}$, then $\underline{P}(\mathcal{E}_x) = 0$ and $\overline{P}(\mathcal{E}_x) = 1$, which denotes ignorance.*
2. *If $|\mathcal{E}_x| = 1$, then $\underline{P}(\mathcal{E}_x) = \underline{P}(A)$ and $\overline{P}(\mathcal{E}_x) = \overline{P}(A)$ s.t. $A \in \mathcal{E}_x$, where $\underline{P}(A)$ and $\overline{P}(A)$ are obtained by applying Eq. (1).*
3. *If $|\mathcal{E}_x| > 1$, then $(\underline{P}(\mathcal{E}_x), \overline{P}(\mathcal{E}_x)) = \text{UL\_BOUNDS}(\mathcal{E}_x)$ (see Algorithm 1). Consider the following functions:*
   - *$f_{CAU}(A) = \{B|(B,A) \in \mathcal{R}_{CAU} \cup f_{CAU}(B)\}$*
   - *TOP_CAU$(\mathcal{E}_x) = \{A|A \in ARG_\leftarrow \cap \mathcal{E}_x$ and $\forall B$ s.t. $A \in f_{CAU}(B), B \notin \mathcal{E}_x\}$*
   - *FREE_CAU$(\mathcal{E}_x) = \{A|A \in ARG_\rightarrow \cap \mathcal{E}_x$ and $\forall B \in f_{CAU}(A), B \notin \mathcal{E}_x\}$*

TOP_CAU and FREE_CAU consider only the arguments of $\mathcal{E}_x$ and their causal relations restricted to $\mathcal{E}_x$. The former returns the arguments that are caused by any of the other argument in $\mathcal{E}_x$ but do not cause other argument(s) in $\mathcal{E}_x$. If there is an argument that belongs to ARG$_\leftarrow$ and ARG$_\rightarrow$ in $\mathbb{C}$ but the argument(s) caused by it are not in $\mathcal{E}_x$, then it is returned by TOP_CAU. The latter returns the arguments that belong to ARG$_\rightarrow$ but whose caused arguments do not belong to extension $\mathcal{E}_x$.

Algorithm 1. Function UL_BOUNDS

Input: An extension $\mathcal{E}_x$ and a causality graph $\mathbb{C} = \langle \text{ARG}, \mathcal{R}_{\text{CAU}} \rangle$
Output: $(\underline{P}(\mathcal{E}_x), \overline{P}(\mathcal{E}_x))$
  1: **if** $(\mathcal{E}_x \cap \text{ARG}_{\leftarrow}) \neq \emptyset$ **then**
  2:    $\text{ARG}''_{\leftarrow} = \text{TOP\_CAU}(\mathcal{E}_x)$
  3:    **for** $i = 1$ to $|\text{ARG}''_{\leftarrow}|$ **do**
  4:       $E^i_A = A \cup (f_{\text{CAU}}(A) \cap \mathcal{E}_x)$
  5:       Calculate $K(E^i_A)$ //*Calculate the credal set for E^i_A by applying Equation (3)*
  6:    **end for**
  7: **end if**
  8: $\text{ARG}'_o = \mathcal{E}_x \cap \text{ARG}_o$
  9: **if** $(\mathcal{E}_x \cap \text{ARG}_{\rightarrow}) \neq \emptyset$ **then**
 10:    $\text{ARG}''_{\rightarrow} = \text{FREE\_CAU}(\mathcal{E}_x)$
 11: **end if**
 12: //* — $\mathcal{E}_x$ *contains only one set of related arguments* — *//
 13: **if** $|\text{ARG}''_{\leftarrow}| == 1$ && $\text{ARG}'_o == \emptyset$ && $\text{ARG}''_{\rightarrow} == \emptyset$ **then**
 14:    // *Apply Equation (4) for obtaining the lower and upper bounds of* $\mathcal{E}_x$
 15:    $\underline{P}(\mathcal{E}_x) = \underline{P}(\text{ARG}''_{\leftarrow}), \overline{P}(\mathcal{E}_x) = \overline{P}(\text{ARG}''_{\leftarrow})$
 16: **else**
 17:    //*Apply Equation (2) for obtaining the lower and upper bounds of* $\mathcal{E}_x$
 18:    $\underline{P}(\mathcal{E}_x) = \underline{P}(\bigcup_{i=1}^{i \leq |\text{ARG}''_{\leftarrow}|} E^i_A \cup \text{ARG}'_o \cup \text{ARG}''_{\rightarrow})$,
 19:    $\overline{P}(\mathcal{E}_x) = \overline{P}(\bigcup_{i=1}^{i \leq |\text{ARG}''_{\leftarrow}|} E^i_A \cup \text{ARG}'_o \cup \text{ARG}''_{\rightarrow})$
 20: **end if**
 21: **return** $(\underline{P}(\mathcal{E}_x), \overline{P}(\mathcal{E}_x)$

Example 4 (Cont. Example 2 considering the causality graph of Example 3). After applying the semantics presented in Definition 2, we obtain that $\mathcal{E}_{\text{CO}} = \mathcal{E}_{\text{PR}} = \mathcal{E}_{\text{GR}} = \mathcal{E}_{\text{ST}} = \{C, E, F, D, H, G\}$. Since this extension has more than one element, the Algorithm 1 has to be applied:

- We first evaluate the number of the caused arguments: $\mathcal{E}_y \cap \text{ARG}_{\leftarrow} = \{G\}$ (for $y \in \{\text{CO}, \text{PR}, \text{GR}, \text{ST}\}$), then we obtain $\text{TOP\_CAU}(\mathcal{E}_y) = \{G\}$ and $f_{\text{CAU}}(G) = \{H\}$; hence, $E_G = \{G, H\}$. At last, we calculate the credal set for E_G by applying Eq. (3): $K(E_G) = \{0.7, 0.8, 1, 0.9\}$.
- Next, we obtain those arguments that belong to the extension and that neither cause any other argument nor are caused by any other argument: $\text{ARG}'_o = \{E\}$.
- Then, we evaluate the number of causing arguments: $\mathcal{E}_y \cap \text{ARG}_{\leftarrow} = \{C, D, F\}$ and we obtain $\text{FREE\_CAU}(\mathcal{E}_y) = \{C, D, F\}$.
- Since $\mathcal{E}_y$ do not contains only related arguments, we apply Eq. (2) considering $K(E_G), K(E), K(C), K(D)$, and $K(F)$.
- Finally, we obtain: $(\underline{P}(\mathcal{E}_y), \overline{P}(\mathcal{E}_y) = [0.0117, 0.0806]$.

Let us also take some conflict-free sets: $\mathcal{E}^1_{\text{CF}} = \{A, F, H, D, E, G\}$, $\mathcal{E}^2_{\text{CF}} = \{A, F, H, D, G\}$, $\mathcal{E}^3_{\text{CF}} = \{B, C, G, H\}$, and $\mathcal{E}^4_{\text{CF}} = \{A\}$. The lower and upper bounds for these extensions are: $(\underline{P}(\mathcal{E}^1_{\text{CF}}), \overline{P}(\mathcal{E}^1_{\text{CF}})) = [0.13, 0.525]$, $(\underline{P}(\mathcal{E}^2_{\text{CF}}), \overline{P}(\mathcal{E}^2_{\text{CF}})) = [0.2, 0.75]$, $(\underline{P}(\mathcal{E}^3_{\text{CF}}), \overline{P}(\mathcal{E}^3_{\text{CF}})) = [0.02, 0.1875]$, and $(\underline{P}(\mathcal{E}^4_{\text{CF}}), \overline{P}(\mathcal{E}^4_{\text{CF}})) = [0.2, 0.75]$.

So far, we have calculated the lower and upper bounds of extensions obtained under a given semantics. The next step is to compare these bounds in order to determine an ordering over the extensions, which can be used to choose an extension that resolves the problem. In this case, the problem was making a decision about a possible diagnosis between two alternatives: measles or chickenpox. We are not going to tackle the problem of comparing and ordering the extensions because it is out of the scope of this article; however, we can do a brief analysis taking into account the result of the previous example. Arguments A and B represent each of the alternatives. The unique extension under any semantics y does not include any of the alternatives. On other hand, free-conflict sets $\mathcal{E}_{CF}^1$, $\mathcal{E}_{CF}^2$ and $\mathcal{E}_{CF}^4$ include argument A and conflict free set $\mathcal{E}_{CF}^3$ includes argument B. We can notice that there is a notorious difference between the lower and upper bounds of $\mathcal{E}_y$ and the lower and upper bounds of any of the other conflict-free sets. In fact, the lower and upper bounds of the conflict-free sets have a better location. This may indicate that lower and upper bounds of extensions that include one of the alternatives are better than others of extensions that do not include any of the alternatives. This in turn indicates that using uncertainty in AAF may improve the resolutions of some problems, which was demonstrated in [7] for precise uncertainty and it is showed in the example by using imprecise uncertainty.

5   Properties of the Approach

In this section, we study two properties of the proposed approach that guarantee (i) that the approach can be reduced to the AAF of Dung and (ii) that the values of both the lower and upper bounds of the extensions are between 0 and 1.

Given a Credal AAF $\mathcal{AF}_{CS} = \langle \text{ARG}, \mathcal{R}, \mathbb{K}, f_{CS} \rangle$, $\mathcal{AF}_{CS}$ is *maximal* if $\forall A \in \text{ARG}$ it holds that $p_i = 1$ $(1 \leq i \leq m)$ where $p_i \in K(A)$ and $K(A) = f_{CS}(A)$ and $\mathcal{AF}_{CS}$ is *uniform* if $0 \leq p_i \leq 1$. Be maximal transforms an $\mathcal{AF}_{CS}$ into a standard AAF of Dung, which means that every agent believes that every argument is believed without doubts. The next proposition shows that a $\mathcal{AF}_{CS}$ can be reduced to an AAF that follows Dung's definitions.

Proposition 1. *Given a credal AAF* $\mathcal{AF}_{CS} = \langle \text{ARG}, \mathcal{R}, \mathbb{K}, f_{CS} \rangle$ *and a extension* $\mathcal{E}_x$ *($x \in \{\text{cf}, \text{ad}, \text{co}, \text{pr}, \text{gr}, \text{st}\}$). If $\mathcal{AF}_{CS}$ is maximal, then $\forall \mathcal{E}_x \subseteq \text{ARG}$, $\underline{P}(\mathcal{E}_x) = \overline{P}(\mathcal{E}_x) = 1$.*

Proof. Since $\mathcal{AF}_{CS}$ is maximal, then $\forall A \in \text{ARG}$, $K(A) = \{1_1, ..., 1_m\}$. In order to obtain the $\underline{P}(\mathcal{E}_x)$ and $\overline{P}(\mathcal{E}_x)$, Eqs. (1), (2), or (4) have to be applied. For Eq. (1): the $inf\{1, ..., 1\} = sup\{1, ..., 1\} = 1$. For Eq. (2): $\forall A, \prod\{1, ..., 1\} = 1$, so the minimum and maximum of a set composed of 1s is always 1. The same happens with Eq. (4).

Proposition 2. *Given a credal AAF* $\mathcal{AF}_{CS} = \langle \text{ARG}, \mathcal{R}, \mathbb{K}, f_{CS} \rangle$ *and a extension* $\mathcal{E}_x$ *($x \in \{\text{cf}, \text{ad}, \text{co}, \text{pr}, \text{gr}, \text{st}\}$). If $\mathcal{AF}_{CS}$ is uniform, then $\forall \mathcal{E}_x \subseteq \text{ARG}$, $0 \leq \underline{P}(\mathcal{E}_x) \leq 1$ and $0 \leq \overline{P}(\mathcal{E}_x) \leq 1$.*

Proof. In order to obtain the $\underline{P}(\mathcal{E}_x)$ and $\overline{P}(\mathcal{E}_x)$, Eqs. (1), (2), or (4) have to be applied. Since $\mathcal{AF}_{CS}$ is uniform, we can say that the minimums (infimums) and maximums (supremums) are always between 0 and 1. Besides, the product of two numbers between 0 and 1 is always between 0 and 1.

6   Related Work

In this section, we present the most relevant works – to the best of our knowledge – that study probability and abstract argumentation. These works assign probability to the arguments, to the attacks, or to the extensions and all of them use precise probabilistic approaches. Thus, as far as we know, we are introducing the first abstract argumentation approach that employs imprecise probabilistic approaches.

Dung and Thang [4] propose an AF for jury-based dispute resolution, which is based on probabilistic spaces, from which are assigned probable weights – between zero and one – to arguments. In the same way, Li et al. [11] present and extension of Dung's original AF by assigning probabilities to both arguments and defeats. Hunter [7] bases on the two articles previously presented and focuses on studying the notion of probability independence in the argumentation context. The author also propose a set of postulates for the probability function regarding admissible sets and extensions like grounded and preferred. Following the idea of using probabilistic graphs, the author assigns a probability value to attacks in [9].

Thimm [13] focuses on studying probability and argumentation semantics. Thus, he proposes a probability semantics such that instead of extensions or labellings, probability functions are used to assign degrees of belief to arguments. An extension of this work was published in [14]. Gabbay and Rodrigues [6] also focus on studying the extensions obtained from an argumentation framework. Thus, they introduce a probabilistic semantics based on the equational approach to argumentation networks proposed in [5].

7   Conclusions and Future Work

This work presents an approach for abstract argumentation under imprecise probability. We defined a credal AAF, in which credal sets are used to model the uncertainty values of the arguments, which correspond to opinions of a set of agents about their degree of believe about each argument. We have considered that – besides the attack relation – there also exists a causality relation between the arguments of a credal AAF. Based on the credal sets and the causality relation, the lower and upper bounds of the extensions – obtained from a semantics – are calculated.

We have done a brief analysis about the problem of comparing and ordering the extensions based on their lower and upper bounds; however, a more complete analysis and study are necessary. In this sense, we plan to follow this direction in our future work. We also plan to further study the causality relations, more specifically in the context of credal networks [2]. Finally, we want to study the relation of this approach with bipolar argumentation frameworks [1].

References

1. Amgoud, L., Cayrol, C., Lagasquie-Schiex, M.C., Livet, P.: On bipolarity in argumentation frameworks. Int. J. Intell. Syst. **23**(10), 1062–1093 (2008)
2. Cozman, F.G.: Credal networks. Artif. Intell. **120**(2), 199–233 (2000)
3. Dung, P.M.: On the acceptability of arguments and its fundamental role in nonmonotonic reasoning, logic programming and n-person games. Artif. Intell. **77**(2), 321–357 (1995)
4. Dung, P.M., Thang, P.M.: Towards (probabilistic) argumentation for jury-based dispute resolution. In: International Conference on Computational Models of Argument, vol. 216, pp. 171–182 (2010)
5. Gabbay, D.M.: Equational approach to argumentation networks. Argument Comput. **3**(2–3), 87–142 (2012)
6. Gabbay, D.M., Rodrigues, O.: Probabilistic argumentation: an equational approach. Logica Universalis **9**(3), 345–382 (2015)
7. Hunter, A.: Some foundations for probabilistic abstract argumentation. In: International Conference on Computational Models of Argument, vol. 245, pp. 117–128 (2012)
8. Hunter, A.: A probabilistic approach to modelling uncertain logical arguments. Int. J. Approx. Reasoning **54**(1), 47–81 (2013)
9. Hunter, A.: Probabilistic qualification of attack in abstract argumentation. Int. J. Approx. Reasoning **55**(2), 607–638 (2014)
10. Levi, I.: The Enterprise of Knowledge: An Essay on Knowledge, Credal Probability, and Chance. MIT Press, Cambridge (1983)
11. Li, H., Oren, N., Norman, T.J.: Probabilistic argumentation frameworks. In: Modgil, S., Oren, N., Toni, F. (eds.) TAFA 2011. LNCS (LNAI), vol. 7132, pp. 1–16. Springer, Heidelberg (2012). https://doi.org/10.1007/978-3-642-29184-5_1
12. Riveret, R., Korkinof, D., Draief, M., Pitt, J.: Probabilistic abstract argumentation: an investigation with boltzmann machines. Argument Comput. **6**(2), 178–218 (2015)
13. Thimm, M.: A probabilistic semantics for abstract argumentation. In: European Conference on Artificial Intelligence, vol. 12, pp. 750–755 (2012)
14. Thimm, M., Baroni, P., Giacomin, M., Vicig, P.: Probabilities on extensions in abstract argumentation. In: Black, E., Modgil, S., Oren, N. (eds.) TAFA 2017. LNCS (LNAI), vol. 10757, pp. 102–119. Springer, Cham (2018). https://doi.org/10.1007/978-3-319-75553-3_7

A Model-Based Theorem Prover
for Epistemic Graphs for Argumentation

Anthony Hunter[1(✉)] and Sylwia Polberg[2(✉)]

[1] Department of Computer Science, University College London, London, UK
anthony.hunter@ucl.ac.uk
[2] School of Computer Science and Informatics, Cardiff University, Cardiff, UK
polbergs@cardiff.ac.uk

Abstract. Epistemic graphs are a recent proposal for probabilistic argumentation that allows for modelling an agent's degree of belief in an argument and how belief in one argument may influence the belief in other arguments. These beliefs are represented by probability distributions and how they affect each other is represented by logical constraints on these distributions. Within the full language of epistemic constraints, we distinguish a restricted class which offers computational benefits while still being powerful enough to allow for handling of many other argumentation formalisms and that can be used in applications that, for instance, rely on Likert scales. In this paper, we propose a model-based theorem prover for reasoning with the restricted epistemic language.

Keywords: Probabilistic argumentation · Epistemic argumentation · Abstract argumentation

1 Introduction

Both the constellations approach [4–6,9,11,14,16] and the epistemic approach [3, 10,13,18–20] to probabilistic argumentation offer a valuable way to represent and reason with various aspects of uncertainty arising in argumentation. The epistemic uncertainty is seen as the degree to which an argument is believed or disbelieved, thus providing a more fine–grained alternative to the standard Dung's approaches when it comes to determining the status of a given argument. Following the results of an empirical study with participants [17], epistemic graphs have been introduced as a generalization of the epistemic approach to probabilistic argumentation [8,12].

In this approach, the graph is augmented with a set of epistemic constraints that can restrict the belief we have in an argument with a varying degree of specificity and state how beliefs in arguments influence each other. This is illustrated in Example 1. The graphs can therefore model both attack and support as well as relations that are neither positive nor negative. The flexibility of this approach allows us to both model the rationale behind the existing semantics as well as completely deviate from them when required. The fact that we can specify the rules under which arguments should be evaluated and that we can include constraints between unrelated arguments permits the framework to be more context–sensitive. It also allows for better modelling of imperfect agents, which can be important in multi–agent applications.

© Springer Nature Switzerland AG 2019
G. Kern-Isberner and Z. Ognjanović (Eds.): ECSQARU 2019, LNAI 11726, pp. 50–61, 2019.
https://doi.org/10.1007/978-3-030-29765-7_5

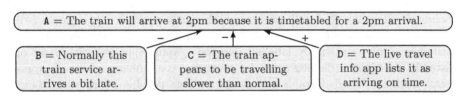

Fig. 1. Example of an epistemic graph. The + (resp. -) label denote support (resp. attack) relations. These are specified via the constraints given in Example 1.

Example 1. Consider the graph in Fig. 1, and let us assume that if D is strongly believed, and B or C is strongly disbelieved, then A is strongly believed, whereas if D is believed, and B or C is disbelieved, then A is believed. Furthermore, if B or C is believed, then A is disbelieved. These constraints could be reflected by the following formulae: $\varphi_1 : p(D) > 0.8 \wedge p(B \vee C) < 0.2 \Rightarrow p(A) > 0.8$; $\varphi_2 : p(D) > 0.5 \wedge p(B \vee C) \leq 0.5 \Rightarrow p(A) > 0.5$; and $\varphi_3 : p(B \wedge C) > 0.5 \Rightarrow p(A) < 0.5$.

Epistemic graphs are therefore a flexible and valuable tool for argumentation, and [12] has already provided methods for harnessing them in user modelling for persuasion dialogues. However, reasoning with the full epistemic language is non-trivial as the size of a probability distribution (i.e. the number of sets of arguments needing an assignment) is exponential w.r.t. the number of arguments, and there can potentially be infinitely many distributions satisfying a given set of constraints. As presented in [8], for certain applications a restricted form of logical constraint can be used, i.e. one where the probability values appearing in constraints and distributions come from a finite, restricted set of values. This may be appropriate if we want to represent beliefs in arguments as in a Likert scale [15], or we want to use epistemic graphs as a medium for existing extension-based or labeling-based methods. It also has the benefit of always producing a finite set of answers.

In order to reason with constraints based on a restricted set of values, we present a model-based theorem prover in this paper which can be used to check *(1)* whether constraints are consistent; *(2)* if one constraint entails another; and *(3)* find satisfying distributions. Our aim in this paper is to present a simple baseline system that can be implemented easily and used for small examples. This will help us understand some of the underlying issues in developing theorem provers for this formalism, and serve as a comparison for future systems.

We proceed as follows: Sect. 2 reviews epistemic graphs from [8]; Sect. 3 introduces a method for identifying the models for a constraint and Sect. 4 an algorithm for model-based reasoning (see [1] for proofs); and Sect. 5 discusses the contributions.

2 Epistemic Language

This section reviews the necessary basic definitions from [8]. We assume a directed graph $\mathcal{G} = (V, R)$, where each node in V denotes an argument (as illustrated by Fig. 1), an edge in R denotes a relation between arguments and a labeling $\mathcal{L}: R \to 2^{\{+,-,*\}} \setminus \{\varnothing\}$ tells us whether it is positive $(+)$, negative $(-)$, or neither $(*)$. We use $\mathsf{Nodes}(\mathcal{G})$ and $\mathsf{Arcs}(\mathcal{G})$ to denote V and R respectively. Epistemic graphs are simply labelled directed graphs equipped with a set of epistemic constraints (defined next) for capturing the influences between arguments. Both the labelled graph and the constraints provide information about the argumentation. In this paper, we focus on the constraints rather than on the full power of the graphs, and refer the readers to [8] for further details.

Like previously stated, the restricted epistemic language only allows values from a certain, finite set to appear in the formulae. However, in order for the approach to be coherent, this set should meet certain basic requirements. We thus use the notion of a (reasonable) restricted value set, which has to be closed under addition and subtraction (assuming the resulting value is still in the $[0, 1]$ interval) and contain value 1.

Definition 1. *A finite set of rational numbers from the unit interval Π is a* **reasonable restricted value set** *iff $1 \in \Pi$ and for any $x, y \in \Pi$ it holds that if $x + y \le 1$, then $x + y \in \Pi$, and if $x - y \ge 0$, then $x - y \in \Pi$.*

We can also create subsets of this set according to a given inequality and threshold value as well as sequences of values that can be seen as satisfying a given arithmetical formula, which will become useful in the next sections:

Definition 2. *With $\Pi_{\#}^{x} = \{y \in \Pi \mid y \# x\}$ we denote the subset of Π obtained according to the value x and relationship $\# \in \{=, \ne, \ge, \le, >, <\}$. The* **combination set** *for Π and a sequence of arithmetic operations $(*_1, \ldots, *_k)$ where $*_i \in \{+, -\}$ and $k \ge 0$ is defined as:*

$$
\Pi_{\#}^{x, (*_1, \ldots, *_k)} = \begin{cases} \{(v) \mid v \in \Pi_{\#}^{x}\} & k = 0 \\ \{(v_1, \ldots, v_{k+1}) \mid v_i \in \Pi, \, v_1 *_1 \ldots *_k v_{k+1} \# x\} & otherwise \end{cases}
$$

Example 2. Let $\Pi_1 = \{0, 0.5, 0.75, 1\}$. We can observe that it is not a restricted value set, since $0.75 - 0.5 = 0.25$ is missing from Π_1. Its modification, $\Pi_2 = \{0, 0.25, 0.5, 0.75, 1\}$, is a restricted value set. The subsets of Π_2 for $x = 0.25$ under various inequalities are as follows: $\Pi_{2>}^{x} = \{0.5, 0.75, 1\}$, $\Pi_{2<}^{x} = \{0\}$, $\Pi_{2\ge}^{x} = \{0.25, 0.5, 0.75, 1\}$, $\Pi_{2\le}^{x} = \{0, 0.25\}$, $\Pi_{2\ne}^{x} = \{0, 0.5, 0.75, 1\}$, and $\Pi_{2=}^{x} = \{0.25\}$.

Assume we have a reasonable restricted value set $\Pi_3 = \{0, 0.5, 1\}$, a sequence of operations $(+, -)$, an operator $=$ and a value $x = 1$. In order to find an appropriate combination set, we are simply looking for triples of values (τ_1, τ_2, τ_3) s.t. $x + y - z = 1$. This produces six possible value sequences, i.e. $\Pi_{2=}^{1, (+, -)} = \{(0, 1, 0),$ $(0.5, 0.5, 0), (0.5, 1, 0.5), (1, 0, 0), (1, 0.5, 0.5), (1, 1, 1)\}$.

2.1   Syntax and Semantics

Based on a given graph and restricted value set, we can now define the epistemic language. An epistemic formula can be seen as a propositional formula built out of components stating how the sums and/or subtractions of probabilities of argument terms should compare to values from Π.

Definition 3. *The **restricted epistemic language** based on $\mathcal{G}$ and a reasonable restricted value set Π is defined as follows:*

- *a **term** is a Boolean combination of arguments. We use $\vee$, $\wedge$ and $\neg$ as connectives and can derive secondary connectives, such as $\rightarrow$, as usual. $\mathsf{Terms}(\mathcal{G})$ denotes all the terms that can be formed from the arguments in $\mathcal{G}$.*
- *an **operational formula** is of the form $p(\alpha_i) *_1 \ldots *_{k-1} p(\alpha_k)$ where all $\alpha_i \in \mathsf{Terms}(\mathcal{G})$ and $*_j \in \{+, -\}$. $\mathsf{OForm}(\mathcal{G})$ denotes the set of all possible operational formulae of $\mathcal{G}$ and we read $p(\alpha)$ as the probability of α.*
- *an **epistemic atom** is of the form $\gamma \# x$ where $\# \in \{=, \neq, \geq, \leq, >, <\}$, $x \in \Pi$ and $\gamma \in \mathsf{OForm}(\mathcal{G})$.*
- *an **epistemic formula** is a Boolean combination of epistemic atoms. $\mathsf{EForm}(\mathcal{G})$ denotes the set of all possible epistemic formulae of $\mathcal{G}$.*

The full, unrestricted language simply permits x to be a rational value in the unit interval, hence we do not recall it here.

Example 3. Let $\Pi = \{0, 0.5, 1\}$. In the epistemic language restricted w.r.t. Π, we can only have atoms of the form $\beta \# 0$, $\beta \# 0.5$, and $\beta \# 1$, where $\beta \in \mathsf{OForm}(\mathcal{G})$ and $\# \in \{=, \neq, \geq, \leq, >, <\}$. From these atoms we compose epistemic formulae using the Boolean connectives, such as $p(\mathsf{A}) + p(\mathsf{B}) \leq 0.5 \wedge p(\mathsf{C}) = 0$.

The semantics for constraints come in the form of belief distributions, which assign probabilities to sets of arguments. Their restricted counterparts enforce that the assigned probabilities come from a restricted value set:

Definition 4. *A **belief distribution** on arguments is a function $P : 2^{\mathsf{Nodes}(\mathcal{G})} \rightarrow [0,1]$ s.t. $\sum_{\Gamma \subseteq \mathsf{Nodes}(\mathcal{G})} P(\Gamma) = 1$. With $\mathsf{Dist}(\mathcal{G})$ we denote the set of all belief distributions on $\mathsf{Nodes}(\mathcal{G})$. P is restricted w.r.t. Π iff for every $X \subseteq \mathsf{Nodes}(\mathcal{G})$, $P(X) \in \Pi^1$. With $\mathsf{Dist}(\mathcal{G}, \Pi)$ we denote the set of restricted distributions of $\mathcal{G}$.*

From the probability distribution, we can derive the probability of a term and therefore of an argument. Each $\Gamma \subseteq \mathsf{Nodes}(\mathcal{G})$ corresponds to an interpretation of arguments. We say that Γ *satisfies* an argument A and write $\Gamma \vDash \mathsf{A}$ iff $\mathsf{A} \in \Gamma$. Essentially $\vDash$ is a classical satisfaction relation and can be extended to complex terms as usual. For instance, $\Gamma \vDash \neg\alpha$ iff $\Gamma \nvDash \alpha$ and $\Gamma \vDash \alpha \wedge \beta$ iff $\Gamma \vDash \alpha$ and $\Gamma \vDash \beta$. With this, we can define the following:

Definition 5. *The **probability of a term** is defined as the sum of the probabilities (beliefs) of its models: $P(\alpha) = \sum_{\Gamma \subseteq \mathsf{Nodes}(\mathcal{G}) \ s.t. \ \Gamma \vDash \alpha} P(\Gamma)$.*

We say that an agent believes a term α to some degree if $P(\alpha) > 0.5$, disbelieves α to some degree if $P(\alpha) < 0.5$, and neither believes nor disbelieves α when $P(\alpha) = 0.5$. Please observe that in this notation, $P(\mathsf{A})$ stands for the probability of a simple term A (i.e. sum of probabilities of all sets containing A), which is different from $P(\{\mathsf{A}\})$, i.e. the probability assigned to set $\{\mathsf{A}\}$.

Using this, we can finally produce (restricted) satisfying distributions of a given atom, and therefore of a given formula:

---

[1] We note that this is a simpler, but still equivalent version of the notion in [8].

Definition 6. *Let* $\varphi : p(\alpha_i) \star_1 \ldots \star_{k-1} p(\alpha_k)\#b$ *be an epistemic atom. The* **satisfying distributions**, *or equivalently* **models**, *of* φ *are defined as* $\mathsf{Sat}(\varphi) = \{P' \in \mathsf{Dist}(\mathcal{G}) \mid P(\alpha_i) \star_1 \ldots \star_{k-1} P(\alpha_k)\#b\}$. *The* **restricted satisfying distribution** *of* φ *w.r.t.* Π *are defined as* $\mathsf{Sat}(\psi, \Pi) = \mathsf{Sat}(\psi) \cap \mathsf{Dist}(\mathcal{G}, \Pi)$.

The set of satisfying distributions for a given epistemic formula is as follows where ϕ and ψ are epistemic formulae: $\mathsf{Sat}(\phi \wedge \psi) = \mathsf{Sat}(\phi) \cap \mathsf{Sat}(\psi)$; $\mathsf{Sat}(\phi \vee \psi) = \mathsf{Sat}(\phi) \cup \mathsf{Sat}(\psi)$; and $\mathsf{Sat}(\neg \phi) = \mathsf{Sat}(\top) \setminus \mathsf{Sat}(\phi)$. For a set of epistemic formulae $\Phi = \{\phi_1, \ldots, \phi_n\}$, the set of satisfying distributions is $\mathsf{Sat}(\Phi) = \mathsf{Sat}(\phi_1) \cap \ldots \cap \mathsf{Sat}(\phi_n)$. The same holds for the restricted scenario.

Example 4. Let us assume we have a formula $\psi : p(\mathsf{A}) + p(\mathsf{B}) \leq 0.5$ on a graph s.t. $\{\mathsf{A}, \mathsf{B}\} = \mathsf{Nodes}(\mathcal{G})$. There can be infinitely many satisfying distributions of this formula, including P_1 s.t. $P_1(\varnothing) = 1$, P_2 s.t. $P_2(\varnothing) = P_2(\{\mathsf{A}\}) = 0.5$, P_3 s.t. $P_3(\varnothing) = P_3(\{\mathsf{B}\}) = 0.5$, or P_4 s.t. $P_4(\varnothing) = 0.68$, $P_4(\{\mathsf{A}\}) = 0.13$ and $P_4(\{\mathsf{B}\}) = 0.19$ (omitted sets are assigned 0). In contrast, a probability distribution P_5 s.t. $P_5(\{\mathsf{A}, \mathsf{B}\}) = 0.3$ and $P_5(\varnothing) = 0.7$ would not be satisfying. If we considered a restricted value set $\Pi = \{0, 0.5, 1\}$, then we could observe that P_1 to P_3 would be the all and only restricted satisfying distributions of ψ.

2.2 Epistemic Entailment Relation

In order to reason with the restricted epistemic language, we can use the consequence or the entailment relation. Given the focus of this paper, we will now recall the latter. From now on, unless stated otherwise, we will assume that the argumentation framework we are dealing with is finite and nonempty (i.e. the set of arguments in the graph is finite and nonempty).

Definition 7. *Let* Π *be a reasonable restricted value set,* $\psi \in \mathsf{EForm}(\mathcal{G}, \Pi)$ *an epistemic formula and* $\{\phi_1, \ldots, \phi_n\} \subseteq \mathsf{EForm}(\mathcal{G}, \Pi)$ *a set of epistemic formulae. The* **restricted epistemic entailment relation** *w.r.t.* Π, *denoted* $\Vdash_\Pi$, *is defined as follows.*

$$\{\phi_1, \ldots, \phi_n\} \Vdash_\Pi \psi \text{ iff } \mathsf{Sat}(\{\phi_1, \ldots, \phi_n\}, \Pi) \subseteq \mathsf{Sat}(\psi, \Pi)$$

Example 5. Consider $\Pi = \{0, 0.25, 0.5, 0.75, 1\}$ and restricted epistemic formulae $p(\mathsf{A}) + p(\neg \mathsf{B}) \leq 1$ and $p(\mathsf{A}) + p(\neg \mathsf{B}) \leq 0.75$. It holds that

$$\{p(\mathsf{A}) + p(\neg \mathsf{B}) \leq 0.75\} \Vdash_\Pi p(\mathsf{A}) + p(\neg \mathsf{B}) \leq 1$$

It is worth noting how changing the restricted valued set affects the entailment. We can observe that a less restricted entailment (i.e. one with Π permitting more values) implies a more restricted one, but not necessarily the other way around, as seen in Example 6.

Proposition 1. *(From [8]) Let* $\Pi_1 \subseteq \Pi_2$ *be reasonable restricted value sets. For a set of epistemic formulae* $\Phi \subseteq \mathsf{EForm}(\mathcal{G}, \Pi_1)$, *and an epistemic formula* $\psi \in \mathsf{EForm}(\mathcal{G})$, *if* $\Phi \Vdash_{\Pi_2} \psi$, *then* $\Phi \Vdash_{\Pi_1} \psi$.

Example 6. Consider two formulae $\varphi_1 : p(\mathtt{A}) \neq 0.5$ and $\varphi_2 : p(\mathtt{A}) = 0 \vee p(\mathtt{A}) = 1$ and a reasonable restricted set $\Pi = \{0, 0.5, 1\}$. We can observe that $\mathsf{Sat}(\varphi_1, \Pi) = \mathsf{Sat}(\varphi_2, \Pi)$ and therefore $\{\varphi_1\} \models_\Pi \varphi_2$. However, if we had set such as $\Pi' = \{0, 0.25, 0.5, 0.75, 1\}$, we could then consider a probability distribution P s.t. $P(\mathtt{A}) = 0.75$ in order to show that $\mathsf{Sat}(\varphi_1) \nsubseteq \mathsf{Sat}(\varphi_2)$.

3   Model-Based Reasoning

A simple route to theorem proving is to use the definition of entailment. This involves identifying the models of the formulae by decomposing them to find the models of their subformulae, and then composing these sets of models to identify the models of the original formulae. We first define decomposition rules to split the formulae (Definition 8). These rules are used to reduce an epistemic formula to epistemic atoms of the form $p(\alpha) = v$ (if possible), and then finally to a set of models that satisfy the epistemic atom. Once we have decomposed a formula, we use the model propagation function (Definition 10) to combine the models of the epistemic atoms into models of the original formula.

Definition 8. *The* **decomposition rules** *are as follows where for each rule, the condition is an epistemic formula, and where Π is a reasonable restricted value set and $\# \in \{=, \neq, \geq, \leq, >, <\}$.*

- *The* **propositional rules** *are as follows where $x|y$ denotes that x is the left child and y is the right child, and from left to right, they are the* **conjunction, disjunction, implication,** *and* **negation** *rules.*

$$\frac{\phi \wedge \psi}{\phi \mid \psi} \qquad \frac{\phi \vee \psi}{\phi \mid \psi} \qquad \frac{\phi \to \psi}{\neg \phi \mid \psi} \qquad \frac{\neg \phi}{\phi}$$

- *The* **operational rules** *are defined as follows, where either $n > 0$ or $\#$ is different from $=$.*

$$\frac{p(\alpha_1) *_1 \ldots *_n p(\alpha_{n+1}) \# x}{\bigvee_{(v_1, \ldots, v_{n+1}) \in \Pi_\#^{x,(*_1, \ldots, *_n)}} (p(\alpha_1) = v_1 \wedge \ldots \wedge p(\alpha_k) = v_{n+1})} \quad if \ \Pi_\#^x \neq \varnothing$$

$$\frac{p(\alpha_1) *_1 \ldots *_n p(\alpha_{n+1}) \# x}{\varnothing} \qquad\qquad otherwise$$

- *The* **term rule** *is defined as follows.*

$$\frac{p(\alpha) = v}{\{P \in \mathsf{Dist}(\mathcal{G}, \Pi) \mid (\sum_{X \subseteq \mathsf{Nodes}(\mathcal{G})} \ s.t. \ X \models \alpha \ P(X)) = v\}}$$

The decomposition of an epistemic formula using the above decomposition rules can be represented by a decomposition tree which we define next. For this, we assume that for a node n in a tree T, $\mathsf{Children}(n)$ is the set of children of n.

Definition 9. *A decomposition tree for an epistemic formula $\phi \in \mathsf{EForm}(\mathcal{G})$ is a tree where (1) the root is labelled with ϕ; (2) each non-leaf node is labelled with an epistemic formula $\psi \in \mathsf{EForm}(\mathcal{G})$; (3) each non-leaf node is associated with a decomposition rule such that the epistemic formula labelling the node satisfies the condition for the decomposition rule, and the child (or children in the case of the proposition rules) are obtained by the application of the decomposition rule; and (4) each leaf is a (possibly empty) set of models.* $\mathsf{Rule}(n)$ *denotes the decomposition rule that was applied to a non-leaf node n.*

Each decomposition tree is exhaustive, i.e. no further decomposition rules can be applied without violating the conditions of it being a decomposition tree. A possible decomposition tree can be seen in Fig. 2 paired with Table 1.

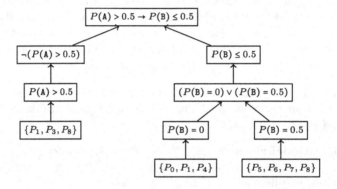

Fig. 2. A decomposition tree. Let P_0 to P_9 be defined as in Table 1. For the root, the set of models is $\{P_0, P_1, P_2, P_4, P_5, P_6, P_7, P_8, P_9\}$.

Table 1. Models for Fig. 2 where $\Pi = \{0, 0.5, 1\}$ and $\mathsf{Nodes}(\mathcal{G}) = \{\mathsf{A}, \mathsf{B}\}$

	P_0	P_1	P_2	P_3	P_4	P_5	P_6	P_7	P_8	P_9
$\varnothing$	1	0	0	0	0.5	0.5	0.5	0	0	0
$\{\mathsf{A}\}$	0	1	0	0	0.5	0	0	0.5	0.5	0
$\{\mathsf{B}\}$	0	0	1	0	0	0.5	0	0.5	0	0.5
$\{\mathsf{A}, \mathsf{B}\}$	0	0	0	1	0	0	0.5	0	0.5	0.5

Each leaf of a decomposition tree is a, possibly empty, set of models (i.e. a set of probability distributions) that satisfy the epistemic formula at its parent node. In other words, the models of the leaf can be used to determine the models of its parent. Furthermore, for each non-leaf node, its models can be used to determine the models of its parent. So in general, for any non-leaf node, its models are a function of the models of its children, as we specify next.

Definition 10. *For a decomposition tree T, the* **model propagation function** *for T, denoted* Models, *is defined a follows,*

1. *If* Rule(n) *is the conjunction rule, and* Children(n) $= \{n_1, n_2\}$, *then* Models(n) $=$ Models(n_1) $\cap$ Models(n_2).
2. *If* Rule(n) *is the disjunction or implication rule, and* Children(n) $= \{n_1, n_2\}$, *then* Models(n) $=$ Models(n_1) $\cup$ Models(n_2).
3. *If* Rule(n) *is the negation rule, and* Children(n) $= \{n_1\}$, *then* Models(n) $=$ Sat($\top, \Pi$) $\setminus$ Models(n_1).
4. *If* Rule(n) *is the term rule or the operational rule, and* Children(n) $= \{n_1\}$, *then* Models(n) $=$ Models(n_1).

For any given epistemic formula, the decomposition trees for the epistemic formula have the same set of leaves where Leaves(T) is the set of leaves in T.

Proposition 2. *If T_1 and T_2 are decomposition trees for ϕ, then* Leaves(T_1) $=$ Leaves(T_2).

Furthermore, the model propagation function ensures that the decompositions trees for an epistemic formula have the same set of models at the root.

Proposition 3. *If T_1 and T_2 are decomposition trees for ϕ, and the root of T_1 (respectively T_2) is n_1 (respectively n_2), then* Models(n_1) $=$ Models(n_2).

For each decomposition rule, the models of the epistemic formula in the condition of the rule are a function of the models in the consequent of the rule. For the conjunction (respectively disjunction) propositional decomposition rule, with condition ϕ, and consequent $\psi_1 \mid \psi_2$, $P \in$ Sat(ϕ, Π) iff $P \in$ Sat(ψ_1, Π) and (respectively or) $P \in$ Sat(ψ_2, Π). For the negation propositional decomposition rule, with condition ϕ, and consequent ψ, $P \in$ Sat(ϕ, Π) iff $P \notin$ Sat(ψ, Π). For the term rule or the operational rule, the models of the condition of the rule are the models of the consequent. Hence, given a decomposition tree for an epistemic formula, the models of that formula are the models returned by backwards induction.

Proposition 4. *If T is a decomposition tree for epistemic formula ϕ, and the root of the tree is node n, then* Sat(ϕ, Π) $=$ Models(n).

So constructing a decomposition tree is a method that is guaranteed to return exactly the models for the epistemic formula at the root.

4   Model-Based Theorem Proving

Our proposal for model based theorem proving is based on the `Entailment` method given in Algorithm 1 which is defined in terms of the `GetModels`. The advantage of the algorithm is that it is straightforward to implement.

Entailment(ϕ, ψ)
    return $\mathrm{GetModels}(\phi) \subseteq \mathrm{GetModels}(\psi)$

$\mathrm{GetModels}(\phi)$;
    if $\phi = \psi_1 \wedge \psi_2$ for some ψ_1, ψ_2
        return $\mathrm{GetModels}(\psi_1) \cap \mathrm{GetModels}(\psi_2)$
    else if $\phi = \psi_1 \vee \psi_2$ for some ψ_1, ψ_2
        return $\mathrm{GetModels}(\psi_1) \cup \mathrm{GetModels}(\psi_2)$
    else if $\phi = \psi_1 \rightarrow \psi_2$ for some ψ_1, ψ_2
        return $\mathrm{GetModels}(\neg\psi_1) \cup \mathrm{GetModels}(\psi_2)$
    else if $\phi = \neg\psi$ for some ψ
      return $\mathrm{Dist}(\mathcal{G}, \Pi) \setminus \mathrm{GetModels}(\psi)$
    else if $\phi = p(\alpha) = v$ for some α
      return $\{P \in \mathrm{Dist}(\mathcal{G}, \Pi) \mid (\sum_{X \subseteq \mathrm{Nodes}(\mathcal{G})} \text{ s.t. } X \vDash \alpha \, P(X)) = v\}$
    else if $\phi = p(\alpha_1) *_1 \ldots *_n p(\alpha_{n+1}) \# x$ for some $\alpha_1, \ldots, \alpha_{n+1}$
      if $\Pi_\#^{x,(*_1,\ldots,*_n)} = \varnothing$
        return $\varnothing$
      else return $\bigcup_{(v_1,\ldots,v_{n+1}) \in \Pi_\#^{x,(*_1,\ldots,*_n)}} (\bigcap_{1 \le i \le n+1} \mathrm{GetModels}(p(\alpha_i) = v_i))$

Algorithm 1: Entailment which if the entailment holds, returns true, otherwise returns false.

Proposition 5. *Algorithm 1 terminates.*

However, the disadvantage of this algorithm is that it is computationally naive, and does not scale well, because it considers the potentially large number of probability distributions. In order to investigate the algorithm in practice, we implemented it in Python (see [2] for code), and ran an evaluation on a Windows 10 HP Pavilion Laptop (with AMD A10 2 GHz processor and 8 GB RAM) on a number of examples taken from [8]. For instance, for the following formulae, we obtained the results in Table 2 for time taken for entailment.

(1) $p(\mathsf{A} + \mathsf{B}) \le 1$ (2) $p(\mathsf{A}) > 0.5 \rightarrow P(\mathsf{B}) \le 0.5$
(3) $p(\mathsf{A}) < 0.9 \wedge p(\mathsf{A}) > 0.7$ (4) $p(\mathsf{A}) > 0.7 \wedge \neg(p(\mathsf{A}) \ge 0.9)$
(5) $(p(\mathsf{B}) < 0.5 \wedge p(\mathsf{C}) > 0.5 \wedge p(\mathsf{D}) > 0.5) \rightarrow p(\mathsf{B}) > 0.5$
(6) $(p(\mathsf{B}) < 0.5 \wedge p(\mathsf{C} \wedge \mathsf{D}) > 0.5) \rightarrow p(\mathsf{B}) > 0.5$
(7) $p(\mathsf{A}) \vee p(\mathsf{B}) \vee p(\mathsf{C}) \vee p(\mathsf{D}) > 0.5$ (8) $p(\mathsf{A} \vee \mathsf{B} \vee \mathsf{C} \vee \mathsf{D}) > 0.5$

Since the implementation is based on generating and manipulating sets of models, the number of models is the dominant factor in the running time. To illustrate this, we focus on the method in the implementation for generating the methods. For example, for generating the models for $|\Pi| = 5$, the running time with $|\mathrm{Nodes}(\mathcal{G})| = 2$ (respectively 3, 4, and 5) is 0.001 (respectively 0.032, 1.927, and 59.12) s, and so the theoretical results (that are discussed below) are reflected in the running time. Essentially, the implementation takes a brute-force approach since it generates all the models for the given set of arguments in the graph and the restricted value set, before decomposing the formulae and finding the models of the subformulae.

Table 2. Average running time in seconds for implementation of entailment on examples of formulae for each column where $\Pi = \{0, 0.25, 0.5, 0.75, 1\}$. Time is average of 10 runs for each pair. For each pair (x, y), x is the assumption and y is the conclusion of the entailment. For all pairs, entailment holds, except for $(6, 5)$ and $(8, 7)$.

	(1, 2)	(2, 1)	(3, 4)	(4, 3)	(5, 6)	(6, 5)	(7, 8)	(8, 7)
Time (secs)	0.037	0.036	0.008	0.010	1.023	1.032	137.6	122.0

For comparison, we look at the number of models that are generated in general. Given Π and $\mathsf{Nodes}(\mathcal{G})$, we can calculate the number of probability distributions for any language for epistemic formulae. For this, we say that a set of rational numbers Ξ is **compatible** with an integer n iff there is a bijection $f : \Xi \to \{0, 1, \ldots, n\}$ and a value $k \in \mathbb{N}$ such that for each $x \in \Xi$, $f(x) = kx$. For example, $\Xi = \{0, 0.5, 1\}$ is compatible with 2 and $\Xi = \{0, 0.25, 0.5, 0.75, 1\}$ is compatible with 4.

Lemma 1. *If Π is a reasonable restricted value set, then there is an integer n s.t. Π is compatible with n.*

Proposition 6. *Let Π be compatible with integer n. The cardinality of the set of probability distributions for Π and $\mathcal{G}$ is given by the following binomial coefficient (using the stars and bars method [7]) where $k = 2^{|\mathsf{Nodes}(\mathcal{G})|}$*

$$\binom{n + k - 1}{n} = \frac{(n + k - 1)!}{(k - 1)! n!}$$

So, for a set $\Pi = \{0, 0.5, 1\}$ and $|\mathsf{Nodes}(\mathcal{G})| = 2$, we have $|\mathsf{Dist}(\mathcal{G}), \Pi| = 10$, for $\Pi = \{0, 0.25, 0.5, 0.75, 1\}$ and $|\mathsf{Nodes}(\mathcal{G})| = 2$, we have $|\mathsf{Dist}(\mathcal{G}, \Pi)| = 35$, and for $\Pi = \{0, 0.25, 0.5, 0.75, 1\}$ and $|\mathsf{Nodes}(\mathcal{G})| = 5$, we have $|\mathsf{Dist}(\mathcal{G}, \Pi)| = 52,360$.

5 Discussion

Epistemic graphs offer a rich formalism for modelling argumentation. There is some resemblance with variants of abstract argumentation such as ranking and weighted approaches, constrained argumentation frameworks, and weighted ADFs. However, the conceptual differences between epistemic probabilities and abstract weights lead to significant differences in modelling (see [8] for details). Also see [8] for a discussion of differences with Bayesian networks. In [8], a sound and complete proof theory is provided for constraints with restricted value sets but no algorithmic method is provided. In this paper, we have addressed this by giving a formal and transparent algorithmic method for reasoning with constraints. It is a practical alternative (for small examples) to the probabilistic optimization approach presented in [12]), and it can be used as a baseline system for which new algorithms can be compared. In future work, we will improve the efficiency of the algorithm (for example, by a lazy construction of models). We will also move beyond this baseline system by rewriting the constraints into a set of propositional clauses, and use a SAT solver.

References

1. http://www0.cs.ucl.ac.uk/staff/A.Hunter/papers/autoepigraphextra.pdf
2. http://www0.cs.ucl.ac.uk/staff/A.Hunter/papers/autoepigraph.py
3. Baroni, P., Giacomin, M., Vicig, P.: On rationality conditions for epistemic probabilities in abstract argumentation. In: Parsons, S., Oren, N., Reed, C., Cerutti, F. (eds.) COMMA 2014. FAIA, vol. 266, pp. 121–132. IOS Press (2014)
4. Bistarelli, S., Mantadelis, T., Santini, F., Taticchi, C.: Probabilistic argumentation frameworks with MetaProbLog and ConArg. In: Tsoukalas, L.H., Grégoire, É., Alamaniotis, M. (eds.) ICTAI 2018, pp. 675–679. IEEE (2018)
5. Doder, D., Woltran, S.: Probabilistic argumentation frameworks – a logical approach. In: Straccia, U., Calì, A. (eds.) SUM 2014. LNCS (LNAI), vol. 8720, pp. 134–147. Springer, Cham (2014). https://doi.org/10.1007/978-3-319-11508-5_12
6. Fazzinga, B., Flesca, S., Furfaro, F.: Complexity of fundamental problems in probabilistic abstract argumentation: beyond independence. Artif. Intell. **268**, 1–29 (2018)
7. Feller, W.: An Introduction to Probability Theory and Its Applications, vol. 1, 2nd edn. Wiley, London (1950)
8. Hunter, A., Polberg, S., Thimm, M.: Epistemic graphs for representing and reasoning with positive and negative influences of arguments. arXiv CoRR (2018). abs/1802.07489
9. Hunter, A.: Some foundations for probabilistic abstract argumentation. In: Verheij, B., Szeider, S., Woltran, S. (eds.) COMMA 2012. FAIA, vol. 245, pp. 117–128. IOS Press (2012)
10. Hunter, A.: A probabilistic approach to modelling uncertain logical arguments. Int. J. Approximate Reasoning **54**(1), 47–81 (2013)
11. Hunter, A.: Probabilistic qualification of attack in abstract argumentation. Int. J. Approximate Reasoning **55**, 607–638 (2014)
12. Hunter, A., Polberg, S., Potyka, N.: Updating belief in arguments in epistemic graphs. In: Thielscher, M., Toni, F., Wolter, F. (eds.) KR 2018, pp. 138–147. AAAI Press (2018)
13. Hunter, A., Thimm, M.: Probabilistic reasoning with abstract argumentation frameworks. J. Artif. Intell. Res. **59**, 565–611 (2017)
14. Li, H., Oren, N., Norman, T.J.: Probabilistic argumentation frameworks. In: Modgil, S., Oren, N., Toni, F. (eds.) TAFA 2011. LNCS (LNAI), vol. 7132, pp. 1–16. Springer, Heidelberg (2012). https://doi.org/10.1007/978-3-642-29184-5_1
15. Likert, R.: A technique for the measurement of attitudes. Arch. Psychol. **140**, 1–55 (1931)
16. Polberg, S., Doder, D.: Probabilistic abstract dialectical frameworks. In: Fermé, E., Leite, J. (eds.) JELIA 2014. LNCS (LNAI), vol. 8761, pp. 591–599. Springer, Cham (2014). https://doi.org/10.1007/978-3-319-11558-0_42
17. Polberg, S., Hunter, A.: Empirical evaluation of abstract argumentation: supporting the need for bipolar and probabilistic approaches. Int. J. Approximate Reasoning **93**, 487–543 (2018)
18. Polberg, S., Hunter, A., Thimm, M.: Belief in attacks in epistemic probabilistic argumentation. In: Moral, S., Pivert, O., Sánchez, D., Marín, N. (eds.) SUM 2017. LNCS (LNAI), vol. 10564, pp. 223–236. Springer, Cham (2017). https://doi.org/10.1007/978-3-319-67582-4_16
19. Thimm, M.: A probabilistic semantics for abstract argumentation. In: De Raedt, L., Bessiere, C., Dubois, D., Doherty, P., Frasconi, P., Heintz, F., Lucas, P. (eds.) ECAI 2012. FAIA, vol. 242, pp. 750–755. IOS Press (2012)

20. Thimm, M., Polberg, S., Hunter, A.: Epistemic attack semantics. In: Modgil, S., Budzynska, K., Lawrence, J. (eds.) COMMA 2018. FAIA, vol. 305, pp. 37–48. IOS Press (2018)

Discussion Games for Preferred Semantics of Abstract Dialectical Frameworks

Atefeh Keshavarzi Zafarghandi[✉], Rineke Verbrugge, and Bart Verheij

Department of Artificial Intelligence, Bernoulli Institute of Mathematics,
Computer Science and Artificial Intelligence, University of Groningen,
Groningen, The Netherlands
{A.Keshavarzi.Zafarghandi,L.C.Verbrugge,Bart.Verheij}@rug.nl

Abstract. Abstract dialectical frameworks (ADFs) are introduced as a general formalism for modeling and evaluating argumentation. However, the role of discussion in reasoning in ADFs has not been clarified well so far. The current work provides a discussion game as a proof method for preferred semantics of ADFs to cover this gap. We show that an argument is credulously acceptable (deniable) by an ADF under preferred semantics iff there exists a discussion game that can defend the acceptance (denial) of the argument in question.

Keywords: Argumentation · Abstract dialectical frameworks · Decision theory · Game theory · Structural discussion

1 Introduction

Abstract Dialectical frameworks (ADFs), first introduced in [7] and have been further refined in [5,6], are expressive generalizations of Dung's widely used argumentation frameworks (AFs) [15]. ADFs are formalisms that abstract away from the content of arguments but are expressive enough to model different types of relations among arguments. Applications of ADFs have been presented in legal reasoning [1,2] and text exploration [8].

Basically, the term 'dialectical method' refers to a discussion among two or more people who have different points of view about a subject but are willing to find out the truth by argumentation. That is, in classical philosophy, dialectic is a method of reasoning based on arguments and counter-arguments [20,22].

In ADFs, dialectical methods have a role in picking the truth-value of arguments under principles governed by several types of semantics, defined mainly based on three-valued interpretations, a form of labelings. Thus, in ADFs, beyond an argument being *acceptable* (the same as *defended* in AFs) there is a symmetric notion of *deniable*. One of the most common argumentation semantics are the *admissible* semantics, which in ADFs come in the form of interpretations that do not contain unjustifiable information. The other semantics

Supported by the Center of Data Science & Systems Complexity (DSSC) Doctoral Programme, at the University of Groningen.

© Springer Nature Switzerland AG 2019

G. Kern-Isberner and Z. Ognjanović (Eds.): ECSQARU 2019, LNAI 11726, pp. 62–73, 2019.
https://doi.org/10.1007/978-3-030-29765-7_6

of ADFs fulfil the admissibility property. Maximal admissible interpretations are called *preferred* interpretations. Preferred semantics have a higher computational complexity than other semantics in ADFs [25]. That is, answering the decision problems of preferred semantics is more complicated than answering the same problems of other semantics in a given ADF. Therefore, having a structural discussion to investigate whether a decision problem is fulfilled under preferred semantics in a given ADF has a crucial importance.

There exists a number of works in which the relation between semantics of AFs and structural discussions are studied [9,16,17,19,23,24]. As far as we know, the relation between semantics of ADFs and dialectical methods in the sense of discussion among agents has not been studied yet [3]. We aim to investigate whether semantics of ADFs are expressible in terms of discussion games.

In this paper we introduce the first existing discussion game for ADFs. We focus on preferred semantics and we show that for an argument being credulously accepted (denied) under preferred semantics in a given ADF there is a discussion game successfully defending the argument. Given the unique structure of ADFs, standard existing approaches known from the AFs setting could not be straightforwardly reused [11,12,27,28]. We thus propose a new approach based on interpretations that can be revised by evaluating the truth values of parents of the argument in question. The current methodology can be reused in other formalisms that can be represented in ADFs, such as AFs.

In the following, we first recall the relevant background of ADFs. Then, in Sect. 3, we present the *preferred discussion game*, which is a game with perfect information, that can capture the notion of preferred semantics. We show that there exists a proof strategy for arguments that are credulously acceptable (deniable) under preferred semantics in a given ADF and vice versa. Further, we show soundness and completeness of the method.

2   Background: Abstract Dialectical Frameworks

The basic definitions in this section are derived from those given in [5–7].

Definition 1. *An abstract dialectical framework (ADF) is a tuple $F = (A, L, C)$ where:*

- *A is a finite set of arguments (statements, positions);*
- *$L \subseteq A \times A$ is a set of links among arguments;*
- *$C = \{\varphi_a\}_{a \in A}$ is a collection of propositional formulas over arguments, called acceptance conditions.*

An ADF can be represented by a graph in which nodes indicate arguments and links show the relation among arguments. Each argument a in an ADF is attached by a propositional formula, called acceptance condition, φ_a over $par(a)$ such that, $par(a) = \{b \mid (b, a) \in R\}$. The acceptance condition of each argument clarifies under which condition the argument can be accepted [5–7]. Further, the acceptance conditions indicate the type of links. An *interpretation v (for F)* is a function $v : A \mapsto \{\mathbf{t}, \mathbf{f}, \mathbf{u}\}$, that maps arguments to one of the three truth values true ($\mathbf{t}$), false ($\mathbf{f}$), or undecided ($\mathbf{u}$). Truth values can be ordered via information ordering relation $<_i$ given by $\mathbf{u} <_i \mathbf{t}$ and $\mathbf{u} <_i \mathbf{f}$ and no other pair of truth values are related by $<_i$. Relation $\leqslant_i$ is the reflexive and transitive closure of $<_i$. Interpretations can be ordered via $\leqslant_i$ with respect to their information content. It is said that an interpretation v is an *extension* of another interpretation w, if $w(a) \leqslant_i v(a)$ for each $a \in A$, denoted by $w \leqslant_i v$. Interpretations v and w are incomparable if neither $w \not\leqslant_i v$ nor $v \not\leqslant_i w$, denoted by $w \not\sim v$.

Semantics for ADFs can be defined via the *characteristic operator* Γ_F which maps interpretations to interpretations. Given an interpretation v (for F), the partial valuation of φ_a by v, is $\varphi_a^v = \varphi_a[b/\top : v(b) = \mathbf{t}][b/\bot : v(b) = \mathbf{f}]$, for $b \in par(a)$. Applying Γ_F on v leads to v' such that for each $a \in A$, v' is as follows:

$$
v'(a) = \begin{cases} \mathbf{t} & \text{if } \varphi_a^v \text{ is irrefutable (i.e., a tautology),} \\ \mathbf{f} & \text{if } \varphi_a^v \text{ is unsatisfiable (i.e., } \varphi_a^v \text{ is a contradiction),} \\ \mathbf{u} & \text{otherwise.} \end{cases}
$$

From now on whenever there is no ambiguity, in order to make three-valued interpretations more readable, we rewrite them by the sequence of truth values, by choosing the lexicographic order on arguments. For instance, $v = \{a \mapsto \mathbf{t}, b \mapsto \mathbf{u}, c \mapsto \mathbf{f}\}$ can be represented by the sequence $\mathbf{tuf}$. The semantics of ADFs are defined via the characteristic operator as in Definition 2.

Definition 2. *Given an ADF F, an interpretation v is:*

- *admissible in F iff $v \leqslant_i \Gamma_F(v)$, denoted by adm;*
- *preferred in F iff v is $\leqslant_i$-maximal admissible, denoted by prf;*
- *a (two-valued) model of F iff v is two-valued and $\Gamma_F(v) = v$, denoted by mod.*

The notion of an argument being accepted and the symmetric notion of a argument being denied in an interpretation are as follows.

Definition 3. *Let $F = (A, L, C)$ be an ADF and let v be an interpretation of F.*

- *An argument $a \in A$ is called acceptable with respect to v if φ_a^v is irrefutable.*
- *An argument $a \in A$ is called deniable with respect to v if φ_a^v is unsatisfiable.*

One of the main decision problems of ADFs is whether an argument is credulously acceptable (deniable) under a particular semantics. Given an ADF $F = (A, L, C)$, an argument $a \in A$ and a semantics $\sigma \in \{adm, prf, mod\}$, argument a is *credulously acceptable (deniable)* under σ if there exists a σ interpretation v of F in which a is acceptable (a is deniable, respectively).

3   Discussion Game for Preferred Semantics

In this section, we present the structure of the discussion game for preferred semantics. The aim is to show that an argument is credulously accepted (denied) under preferred semantics in an ADF iff there exists a discussion game and a winning strategy for a player who starts the game.

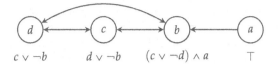

$$c \vee \neg b \qquad d \vee \neg b \qquad (c \vee \neg d) \wedge a \qquad \top$$

Fig. 1. ADF of Example 1

A *preferred discussion game*, which is similar to Socrates' form of reasoning [10,29], is a (non-deterministic) two-player game of perfect information between defender (proponent) and challenger (opponent). So, both agents know all acceptance conditions. The game starts with a belief of proponent (P) about credulous acceptance (denial) of an argument under preferred semantics in a given ADF. Then opponent (O) challenges the proponent by investigating the consequences of P's belief and demanding reasons for those consequences. The game continues alternately: P has to convince O why consequences of the claim can be held. Till the time that there is a new claim by P or there is a new challenge by O and there is no contradiction, the game will be continued.

Since each preferred interpretation is an admissible interpretation, if we want to investigate whether an argument is credulously acceptable (deniable) under preferred semantics, we study whether the argument is credulously acceptable (deniable) under admissible semantics. The key advantage of the current method is that the credulous acceptability (deniability) problem for preferred semantics in an ADF F can be solved without enumeration of all admissible interpretations of F. In the following, Examples 1 and 2 represent preferred discussion games, in which there are winning strategies for P's belief.

Example 1. Given an ADF $F = (\{a, b, c, d\}, \{\varphi_a : \top, \varphi_b : (c \vee \neg d) \wedge a, \varphi_c : d \vee \neg b, \varphi_d : c \vee \neg b\})$, depicted in Fig. 1.

- Assume that P claims that d is credulously acceptable under preferred semantics. The knowledge of P consists of information about the truth value of d, and there is no further information about the truth values of other arguments. This initial knowledge of P can be shown by the interpretation $v_0 = \mathbf{uuut}$.
- O checks the consequence of P's belief. O says that, based on the acceptance condition of d, argument d is acceptable in a preferred interpretation iff either c is accepted or b is denied in that interpretation. That is, O revises the information of v_0 to two interpretations; $v_1 = \mathbf{uutt}$ and $v_1' = \mathbf{ufut}$, and challenges P by asking, 'Why does either b have to be assigned to $\mathbf{f}$ or c have to be assigned to $\mathbf{t}$, if d is assigned to $\mathbf{t}$ in a preferred interpretation?'
- In both v_1 and v_1' there exists a new challenge, then the dialogue between players can be continued on any of them. P attempts to defeat the challenge by convincing O about the truth value of the arguments which are challenged by O in the preceding step.
  P chooses to work on v_1 in which the only new challenged argument is c. P checks under which condition c can be accepted in a preferred interpretation. Based on, $\varphi_c : d \vee \neg b$, c is assigned to $\mathbf{t}$ if and only if either d is assigned

to $\mathbf{t}$ or b is assigned to $\mathbf{f}$. That is, the new information of P about the truth values of arguments can be represented by $v_2 = \mathbf{uutt}$ and $v_2' = \mathbf{uftt}$. In the former one there is no new claim, that is, the dialogue v_0, v_1 and v_2 cannot be continued by O anymore. Further, in v_2 P answers the question of O (why is c assigned to $\mathbf{t}$), with no contradiction. Thus, P wins this dialogue. Since P can defend the initial claim via this dialogue, P wins the game and there is no need of continuing the game.

Definitions 4–6 are needed to define the systematic method of computation of moves of each player in Definition 8. In the following, w and v are interpretations such that $w \leqslant_i v$.

Definition 4. *An argument a is **recently presented** in interpretation v with respect to w if $w(a) = \mathbf{u}$ and $v(a) \neq \mathbf{u}$.*

In contrast with standard interpretations in ADFs, in Definition 5 we define so-called minimal interpretations that only give values to argument a and $par(a)$. In the following the notations of $v(b)$ and $w_a(b)$ are used to indicate the truth value of argument b in v and w_a, respectively.

Definition 5. *Let v be an interpretation of an ADF F, in which $a \mapsto \mathbf{t}/\mathbf{f}$ and $par(a) \neq \varnothing$. An interpretation w_a, which is defined over $(par(a) \cup \{a\})$, is called a **minimal interpretation around** a **in** F, if $\Gamma_F(w_a)(a) = v(a)$, and there exists no $w' <_i w_a$ such that $\Gamma_F(w')(a) = v(a)$. In contrast, when $par(a) = \varnothing$ then w_a assigns a to $\Gamma_F(v)(a)$.*

Since the acceptance condition of each argument is indicated by a propositional formula, argument a may have more than one minimal interpretation around a in F. The set of all minimal interpretations around a in F is denoted by W_a.

Definition 6. *Let $A' = \{a_1, \dots, a_n\}$ be the set of arguments recently presented in v w.r.t. w and choose $W_{A'} = \{w_{a_1}, \dots, w_{a_n}\}$ s.t. $w_{a_i} \in W_{a_i}$, for $1 \leqslant i \leqslant n$. The output of the binary function $\delta(v, W_{A'})$ is called an **evaluation of the parents of arguments in** A' **w.r.t.** v **and** $W_{A'}$ defined as follows:*

- *If $v(b) = \mathbf{t}/\mathbf{f}$ and $\nexists i$ s.t. $((w_{a_i}(b) = \mathbf{t}/\mathbf{f}) \vee (w_{a_i}(b) \neq v(b))) \wedge \nexists c$ s.t. $((w_b(c) \neq v(c)) \wedge (w_b(c) \neq w_{a_i}(c)))$ then $\delta(v, W_{A'})(b) = v(b)$.*
- *If $v(b) = \mathbf{u}$ and $\exists i$ s.t. $w_{a_i}(b) = \mathbf{t}/\mathbf{f} \wedge \nexists j$ s.t. $w_{a_i}(b) \neq w_{a_j}(b)$ then $\delta(v, W_{A'})(b) = w_{a_i}(b)$.*
- *If $(v(b) = \mathbf{t}/\mathbf{f}$ and $\exists i, c$ s.t. $(v(b) \neq w_{a_i}(b)) \vee (v(c) \neq w_b(c)) \vee (w_b(c) \neq w_{a_i}(c))) \vee (v(b) = \mathbf{u}$ and $(\exists i, j$ s.t. $w_{a_i}(b) \neq w_{a_j}(b)) \vee (\nexists i$ s.t. $w_{a_i}(b) = \mathbf{t}/\mathbf{f}))$ then $\delta(v, W_{A'})(b) = \mathbf{u}$.*

*The set of all possible evaluations of parents of arguments in A' is called **all evaluations of parents of** A', and denoted by $\delta_{A'}(v)$ such that:*

$$\delta_{A'}(v) = \{\delta(v, W_{A'}) \mid W_{A'} = \{w_{a_1}, \dots, w_{a_n}\} \text{ s.t. } w_{a_i} \in W_{a_i}, \text{ for } 1 \leqslant i \leqslant n\}$$

Note that when A' contains only one argument a, we address an evaluation of parents of a with $\delta(v, w_a)$, in which w_a is a minimal interpretation around a, and we denote the set of all evaluations of A' with $\delta_a(v)$.

In Example 1, it is assumed that d is credulously accepted, $v_0 = \mathbf{uuut}$. In comparison to interpretation $v_{\mathbf{u}} = \mathbf{uuuu}$, argument d is recently presented in v_0. Based on the acceptance condition of d, namely $\varphi_d : c \vee \neg b$, interpretations $w_d = \{b \mapsto \mathbf{u}, c \mapsto \mathbf{t}, d \mapsto \mathbf{t}\}$ and $w'_d = \{b \mapsto \mathbf{f}, c \mapsto \mathbf{u}, d \mapsto \mathbf{t}\}$ are minimal interpretations around d in F. As a consequence, the evaluation of the parents of the argument in question may lead to more than one interpretation. For instance, the evaluation of the parents of d with respect to v_0 and w_d is $\delta(v_0, w_d) = \mathbf{uutt}$, and with respect to v_0 and w'_d it is $\delta(v_0, w'_d) = \mathbf{ufut}$. Therefore, the set of evaluations of parents of d is $\delta_d(v_0) = \{\mathbf{uutt}, \mathbf{ufut}\}$.

Now we are going to define moves of each player based on the evaluation of the parents of the recently presented arguments, proposed in Definition 6. The information of each player in games can be represented by an interpretation. In the first claim of P there exists only information about the truth value of the argument which is claimed.

Definition 7. *The first claim of P about credulous acceptance (denial) of an argument is named **initial claim**, denoted by interpretation v_0, in which the argument in question is assigned to $\mathbf{t}$ ($\mathbf{f}$, respectively) and all other argument are assigned to $\mathbf{u}$.*

After each claim move of P, presented by interpretation v, O checks the conditions under which the claim of P can be valid. That is, O evaluates the truth values of the parents of arguments in A', recently presented by P in v with $\delta_{A'}(v)$. Then, O demands P to propose logical reasons for those results with the hope of leading to a contradiction. The game continues alternately: P has to convince O why at least one consequence of the claim can be held.

Definition 8. *Given interpretations v and w, such that $v \leqslant_i w$. Let A' be a set of arguments, recently presented in w. 1. If w is given by P, it is named that $a \in A'$ is **claimed** by P in w and $\delta_{A'}(w)$ is named **challenge move**. 2. If w is given by O, it is named that $a \in A'$ is **challenged** by O in w and $\delta_{A'}(w)$ is named **claim move**.*

Specifically, the initial claim is a claim move in comparison to the interpretation that assigns all arguments to $\mathbf{u}$. Actually, a preferred discussion game can be represented as a labeled rooted tree in which the root is labeled by the initial claim, v_0. The nodes of depth $i > 0$ are labeled by all $\delta(v, W_{A'})$ such that v is the label of the directly preceding node of the tree with depth $i - 1$, and $W_{A'} = \{w_a \mid \text{s.t. } a \in A'\}$ in which A' is a set of arguments that are recently presented in v with respect to the label of the directly preceding node of v. A part of the tree of Example 1, including a winning strategy for P, is depicted in Fig. 2.

Definition 9. *A **dialogue** is the sequence of labels of a branch of the tree corresponding to the game which is started by an initial claim, and continued by applying $\delta(v_i, W_{A'})$, for $i \geqslant 0$ s.t. $a \in A'$ is recently presented in v_i.*

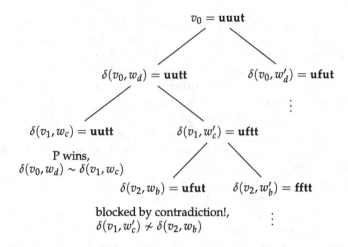

Fig. 2. Associated tree of the game in Example 1

We say that there is a *contradiction* in a dialogue if the dialogue consists of interpretations v_i and v_{i+1} that are incomparable. For instance, the dialogue $[v_0, \delta(v_0, w_d), \delta(v_1, w'_c), \delta(v_2, w_b)]$ in Fig. 2 leads to a contradiction. Definitions 10 and 11 explain under which conditions a dialogue can be continued or halted.

Definition 10. *Let* $[v_0, \ldots, v_n]$ *be a dialogue with no contradiction. **The dialogue is continued** on* v_n: *1. by O if an argument is claimed in* v_n *by P; or 2. by P if an argument is challenged in* v_n *by O.*

Definition 11. *Let* $[v_0, \ldots, v_n]$ *be a dialogue. It is said that **the dialogue is blocked** on* v_n *when: 1. a is challenged in* v_{n-1} *by O, and* $v_{n-1} \sim v_n$. *We say that the game is blocked by P in this step. Or, 2. a is claimed in* v_{n-1} *by P, and* $v_{n-1} \sim v_n$. *We say that the game is blocked by O in this step. Or 3. there is a contradiction, that is,* $v_{n-1} \not\sim v_n$.

In Example 1, dialogue $[v_0, \delta(v_0, w_{0d}), \delta(v_{1_0}, w_{0c})]$ is blocked by P. If a dialogue is blocked by P, it means that P could defeat a challenge of O without making a new claim. Thus, there is no further move for O. Therefore, P won the dialogue. Since P can defend the initial claim via this dialogue, P wins the game, as well. Thus, after this dialogue there is no need of continuing the game.

– *P wins the dialogue if the dialogue is blocked by P.*

Example 2 investigates the other condition under which P wins the dialogue.

Example 2. Let F be the ADF given in Example 1.

– P believes that d can be denied in a preferred interpretation in F, $v_0 = $ **uuuf**.
– The challenge move of O on d leads to $v_1 = \delta(v_0, w_d) = $ **utff**.
– The recently challenged arguments are b and c. The minimal interpretations around b are $w_b = \{a \mapsto \mathbf{t}, c \mapsto \mathbf{t}\}$ and $w'_b = \{a \mapsto \mathbf{t}, d \mapsto \mathbf{f}\}$, and the minimal interpretation around c is $w_c = \{b \mapsto \mathbf{t}, d \mapsto \mathbf{f}\}$. Thus, $v_2 = \delta(v_1, W_{bc}) = $ **ttuf** and $v'_2 = \delta(v_1, W'_{bc}) = $ **ttff**.

– Since $v_1 \not\sim v_2$, O cannot continue this dialogue. However, $v_1 <_i v_2'$ and the challenge move on v_2' is $\delta(v_2', w_a) = v_2'$. Thus, the game is blocked by O.

If a dialogue is blocked by O, it means that O cannot find a contradiction between P's claim and O's challenging, which is done by O in an element of the claim move, and O cannot make a new challenge for P. Thus, P wins the dialogue and the initial claim of P is proved via this dialogue.

– *P wins the dialogue if the dialogue is blocked by O.*

The ADF of Example 1 can also be used as an example in which there is a winning strategy for O, explained in Example 3.

Example 3. Given ADF F of Example 1.

– P believes that b can be denied in a preferred interpretation in F, $v_0 = $ **ufuu**.
– There are three different dialogues based on this initial claim; 1. [$v_0 = $ **ufuu**, $v_1 = $ **ufft**, $v_2 = $ **uufu**], 2. [$v_0 = $ **ufuu**, $v_1 = $ **ufft**, $v_2' = $ **uuuu**], 3. [$v_0 = $ **ufuu**, $v_1' = $ **ffuu**, $v_2'' = $ **ufuu**].

Each of the dialogues of this game is blocked by contradictions. That is, in each dialogue P cannot defeat the challenge of O. On the other hand, O defeats P in all the ways that P attempts to prove the initial claim, by finding contradictions. That is, P cannot make any reasonable discussion to defend the initial claim. Thus, O wins all dialogues and wins the game in consequence.

– *O wins the dialogue, when O can block the dialogue by contradiction.*

The examples which were studied above illustrate that each player only has to consider the arguments which are recently presented by the competitor in the directly preceding move. The discussion game that can decide the credulous acceptance (denial) problem in ADFs under preferred semantics is called, *preferred discussion game*, introduced in Definition 12.

Definition 12. *Given an ADF $F = (A, L, C)$. A preferred discussion game for credulous acceptance (denial) of an argument of A is a sequence $[\Delta_0, \ldots, \Delta_n](n \geqslant 0)$ such that all the following conditions hold:*

– Δ_0 *consists of an initial claim;*
– *for $i \geqslant 1$, $\Delta_i = \bigcup_v \delta_{A'}(v)$, for each $v \in \Delta_{i-1}$ such that set of arguments of A' are recently presented in v;*
– *each $[v_0, \ldots, v_m]$ such that $v_i \in \Delta_i$ is a dialogue of the game, for $1 \leqslant m \leqslant n$, when: $v_i = \delta(v_{i-1}, W_{A'})$, such that the set A' is recently presented in v_{i-1};*
– *the game is finished in Δ_n if at least a dialogue of the game is blocked by P or O, or if all the dialogues lead to contradictions.*

In Definition 12, 1. if i is odd, for each $v \in \Delta_{i-1}$, Δ_i consists of all challenge moves $\delta_{A'}(v)$ such that $a \in A'$ is claimed in v; and 2. if $i \geqslant 2$ is even, for each $v \in \Delta_{i-1}$, Δ_i consists of all claim moves $\delta_{A'}(v)$ such that $a \in A'$ is challenged in v. The winning strategy of each player is explained in Definition 13.

Definition 13. *Let F be a given ADF. Let $[\Delta_0, \ldots, \Delta_n]$ be a preferred discussion game for credulous acceptance (denial) of an argument.*

- *P has a winning strategy in the game if P wins a dialogue of the game.*
- *O has a winning strategy in the game if O wins all dialogues of the game.*

Let F be an ADF and let $[\Delta_0, \ldots, \Delta_n]$ be a preferred discussion game of an initial claim of F. The *length* of the preferred discussion game is the length of the sequence $[\Delta_0, \ldots, \Delta_n]$, which is the number of elements of the sequence.

Proposition 1. *Let $F = (A, L, C)$ be an ADF and $|A| = n$. The length of each preferred discussion game of F is at most $n + 1$.*

Proof. Toward a contradiction, assume that that there exists a preferred discussion game $[\Delta_0, \ldots, \Delta_m]$ of F such that $m > n$. On the other hand, each dialogue $[v_0, \ldots, v_i]$ of the game is continued in v_i if $v_{i-1} <_i v_i$. This can be done by indicating the truth value of an argument in v_i that is not indicated before. Since the number of arguments of F is n, the longest dialogue contains interpretations such that $v_0 < \cdots < v_{n-1}$, and in the next step, the parents of arguments of claimed or challenged items in v_{n-1} will be evaluated. That is, the longest dialogue can be a sequence of $n + 1$ interpretations. Thus, the length of each game cannot be more that $n + 1$.

Since we assumed in the definition of ADFs that each ADF is finite, the immediate result of Proposition 1 is that each preferred discussion game halts and there exists a winning strategy either for O or P.

Theorem 1. *Let an ADF $F = (A, L, C)$ be given.*

- *Soundness: if there exists a winning strategy in a preferred discussion game with initial claim of accepting (denying) an argument a, then a is credulously acceptable (deniable) under preferred semantics in F.*
- *Completeness: if an argument a is credulously acceptable (deniable) under preferred semantics in F, then there is a preferred discussion game with a winning strategy for the initial claim of accepting (denying) of a.*

Proof. Soundness: assume that there is winning strategy for P in a preferred discussion game $[\Delta_0, \ldots, \Delta_n]$, for accepting (denying) of an argument a. Therefore, there is a winning dialogue $[v_0, \ldots, v_m]$ for P, for $0 < m \leqslant n$. To show the soundness it is enough to investigate whether v_m is an admissible interpretation. Towards a contradiction, assume that v_m is not an admissible interpretation, that is, $v_m \not\leqslant_i \Gamma_F(v_m)$. Thus, there exists an argument b s.t. $b \mapsto \mathbf{t}/\mathbf{f} \in v_m$, however, the valuation of the acceptance condition of b under v_m is not the same as v_m; we prove the case that $b \mapsto \mathbf{t} \in v_m$. The proof method for the case in which $b \mapsto \mathbf{f} \in v_m$ is analogous.

$b \mapsto \mathbf{t} \in v_m$ means that either P claims this assignment in an interpretation v_i, $0 \leqslant i < m$, or O challenges it in an interpretation v_i, $0 < i < m$. Assume that this is claimed by P in v_i, $0 \leqslant i < m$. An element of the challenge move of O on

v_i is v_{i+1}. That is, O presents the truth values of $par(b)$ in v_{i+1}. Since there is a winning strategy for P in this dialogue, $v_{m-1} \sim v_m$. That is, $\varphi_b^{v_m} \equiv \top$, since v_m consists of the truth values of $par(b)$ presented in v_{i+1}. Thus, $\Gamma_F(v_m)(b) = \mathbf{t}$. Therefore, the assumption that v_m is not an admissible interpretation is rejected. The proof method for a challenge move is analogous.

Completeness: assume that an argument a is credulously accepted under preferred semantics in F (the proof method in case a is credulously denied is analogous). Then, there is a preferred interpretation v of F in which a is accepted. We construct the corresponding preferred discussion game as follows. Let v_0, the initial claim, be an interpretation in which a is assigned to $\mathbf{t}$ and all other arguments of A are assigned to $\mathbf{u}$. Extend v_0 to v_1 by changing the truth values of the parents of a in v_0 by their truth values in v. Continue this method and construct v_{i+1} by changing the truth value of the parents of arguments which are recently presented in v_i, by the ones which are in v, for $i > 0$. Since the number of arguments is finite, this procedure will end in some v_n. To construct v_{i+1} only the truth values of the arguments which are assigned to $\mathbf{u}$ in v_i can be changed, then $v_i < v_{i+1}$, for $0 \leqslant i < n$. Let $v_{n+1} = v_n$. The sequence $[v_0, \dots, v_{n+1}]$ is a dialogue of the preferred discussion game $[\Delta_0, \dots, \Delta_{n+1}]$ of F, in which $v_0 \in \Delta_0$. Further, this dialogue is a winning strategy for P in this game.

4   Conclusion and Future Work

In this paper, preferred discussion games between two agents, proponent and opponent, are considered as a proof method to investigate credulous acceptance (denial) of arguments in an ADF under preferred semantics. Some notable results of the current work are: 1. The method is sound and complete. 2. The presented methodology can be reused in AFs and generalizations of AFs that can be represented as subclasses of ADFs, namely set argumentation frameworks [21] and bipolar argumentation framework [13]. 3. Winning *one* dialogue of the game by P is sufficient to show that there exists a preferred interpretation in which the argument in question is assigned to the truth value which is claimed. In contrast, for AFs [23,26,27], P has a winning strategy if P can address *all* O's challenges. 4. In each move each player has to study the truth value of arguments that are recently presented in the directly preceding move. In contrast, in [9], O has to check all past moves of P to find a contradiction. 5. To investigate the credulous decision problem of ADFs under preferred semantics, there is no need to enumerate all preferred interpretations of an ADF. 6. Preferred semantics of an ADF corresponds to a preferred discussion game with winning strategy for P. 7. In [14] it is shown that in the class of acyclic ADFs all semantics coincide. Thus, in acyclic ADFs the presented game can be used to decide the credulous problem on other semantics. As future work, we could investigate structural discussion games for other semantics of ADFs. In addition, we could study discussion games for other decision problems of ADFs. Further, we could investigate whether the presented method is more effective than the methods used in current ADF-solvers, e.g. [4,18]. This study may lead to new ADF-solvers that work locally on an argument to answer decision problems.

References

1. Al-Abdulkarim, L., Atkinson, K., Bench-Capon, T.J.M.: Abstract dialectical frameworks for legal reasoning. In: Legal Knowledge and Information Systems JURIX. Frontiers in Artificial Intelligence and Applications, vol. 271, pp. 61–70. IOS Press (2014)
2. Al-Abdulkarim, L., Atkinson, K., Bench-Capon, T.J.M.: A methodology for designing systems to reason with legal cases using abstract dialectical frameworks. Artif. Intell. Law **24**(1), 1–49 (2016)
3. Barth, E.M., Krabbe, E.C.: From Axiom to Dialogue: A Philosophical Study of Logics and Argumentation. Walter de Gruyter, Berlin (1982)
4. Brewka, G., Diller, M., Heissenberger, G., Linsbichler, T., Woltran, S.: Solving advanced argumentation problems with answer-set programming. In: Conference on Artificial Intelligence, AAAI, pp. 1077–1083. AAAI Press (2017)
5. Brewka, G., Ellmauthaler, S., Strass, H., Wallner, J.P., Woltran, S.: Abstract dialectical frameworks. An overview. IFCoLog J. Logics Appl. (FLAP) **4**(8), 2263–2317 (2017)
6. Brewka, G., Strass, H., Ellmauthaler, S., Wallner, J.P., Woltran, S.: Abstract dialectical frameworks revisited. In: Proceedings of the Twenty-Third International Joint Conference on Artificial Intelligence (IJCAI 2013), pp. 803–809 (2013)
7. Brewka, G., Woltran, S.: Abstract dialectical frameworks. In: Proceedings of the Twelfth International Conference on the Principles of Knowledge Representation and Reasoning (KR 2010), pp. 102–111 (2010)
8. Cabrio, E., Villata, S.: Abstract dialectical frameworks for text exploration. In: Proceedings of the 8th International Conference on Agents and Artificial Intelligence (ICAART 2016), vol. 2, pp. 85–95. SciTePress (2016)
9. Caminada, M.: Argumentation semantics as formal discussion. In: Handbook of Formal Argumentation, vol. 1, pp. 487–518 (2017)
10. Caminada, M.W.: A formal account of Socratic-style argumentation. J. Appl. Logic **6**(1), 109–132 (2008)
11. Caminada, M.W., Dvořák, W., Vesic, S.: Preferred semantics as Socratic discussion. J. Logic Comput. **26**(4), 1257–1292 (2014)
12. Cayrol, C., Doutre, S., Mengin, J.: On decision problems related to the preferred semantics for argumentation frameworks. J. Logic Comput. **13**(3), 377–403 (2003)
13. Cayrol, C., Lagasquie-Schiex, M.: Bipolarity in argumentation graphs: towards a better understanding. Int. J. Approximate Reasoning **54**(7), 876–899 (2013)
14. Diller, M., Zafarghandi, A.K., Linsbichler, T., Woltran, S.: Investigating subclasses of abstract dialectical frameworks. In: Proceedings of Computational Models of Argument, COMMA 2018, Amsterdam, pp. 61–72 (2018)
15. Dung, P.M.: On the acceptability of arguments and its fundamental role in nonmonotonic reasoning, logic programming and n-person games. Artif. Intell. **77**, 321–357 (1995)
16. Dung, P.M., Thang, P.M.: A sound and complete dialectical proof procedure for sceptical preferred argumentation. In: Proceedings of the LPNMR-Workshop on Argumentation and Nonmonotonic Reasoning (ArgNMR 2007), pp. 49–63 (2007)
17. van Eemeren, F.H., Garssen, B., Krabbe, E.C.W., Snoeck Henkemans, A.F., Verheij, B., Wagemans, J.H.M.: Handbook of Argumentation Theory. Springer, Dordrecht (2014). https://doi.org/10.1007/978-90-481-9473-5
18. Ellmauthaler, S., Strass, H.: The DIAMOND system for computing with abstract dialectical frameworks. In: Proceedings of Computational Models of Argument, COMMA, vol. 266, pp. 233–240. IOS Press (2014)

19. Jakobovits, H., Vermeir, D.: Dialectic semantics for argumentation frameworks. In: Proceedings of the 7th International Conference on Artificial Intelligence and Law, pp. 53–62. ACM Press (1999)
20. Krabbe, E.C.: Dialogue logic. In: Gabbay, D., Woods, J. (eds.) Handbook of the History of Logic, pp. 665–704. Elsevier, Amsterdam (2006)
21. Linsbichler, T., Pührer, J., Strass, H.: A uniform account of realizability in abstract argumentation. In: 22nd European Conference on Artificial Intelligence, ECAI, vol. 285, pp. 252–260. IOS Press (2016)
22. Macoubrie, J.: Logical argument structures in decision-making. Argumentation **17**(3), 291–313 (2003)
23. Modgil, S., Caminada, M.: Proof theories and algorithms for abstract argumentation frameworks. In: Simari, G., Rahwan, I. (eds.) Argumentation in Artificial Intelligence, pp. 105–129. Springer, Boston (2009). https://doi.org/10.1007/978-0-387-98197-0_6
24. Prakken, H., Sartor, G.: Argument-based extended logic programming with defeasible priorities. J. Appl. Non-classical Logics **7**(1–2), 25–75 (1997)
25. Strass, H., Wallner, J.P.: Analyzing the computational complexity of abstract dialectical frameworks via approximation fixpoint theory. Artif. Intell. **226**, 34–74 (2015)
26. Thang, P.M., Dung, P.M., Hung, N.D.: Towards a common framework for dialectical proof procedures in abstract argumentation. J. Logic Comput. **19**(6), 1071–1109 (2009)
27. Verheij, B.: A labeling approach to the computation of credulous acceptance in argumentation. In: Proceedings of the 20th International Joint Conference on Artificial Intelligence, IJCAI, pp. 623–628. IJCAI (2007)
28. Vreeswik, G.A.W., Prakken, H.: Credulous and sceptical argument games for preferred semantics. In: Ojeda-Aciego, M., de Guzmán, I.P., Brewka, G., Moniz Pereira, L. (eds.) JELIA 2000. LNCS (LNAI), vol. 1919, pp. 239–253. Springer, Heidelberg (2000). https://doi.org/10.1007/3-540-40006-0_17
29. Walton, D., Krabbe, E.: Commitment in Dialogue: Basic Concepts of Interpersonal Reasoning. State University of New York Press, New York (1995)

Polynomial-Time Updates of Epistemic States in a Fragment of Probabilistic Epistemic Argumentation

Nico Potyka[1]([✉]), Sylwia Polberg[2], and Anthony Hunter[3]

[1] Institute of Cognitive Science, University of Osnabrück, Osnabrück, Germany
npotyka@uos.de
[2] School of Computer Science and Informatics, Cardiff University, Cardiff, UK
[3] Department of Computer Science, University College London, London, UK

Abstract. Probabilistic epistemic argumentation allows for reasoning about argumentation problems in a way that is well founded by probability theory. Epistemic states are represented by probability functions over possible worlds and can be adjusted to new beliefs using update operators. While the use of probability functions puts this approach on a solid foundational basis, it also causes computational challenges as the amount of data to process depends exponentially on the number of arguments. This leads to bottlenecks in applications such as modelling opponent's beliefs for persuasion dialogues. We show how update operators over probability functions can be related to update operators over much more compact representations that allow polynomial-time updates. We discuss the cognitive and probabilistic-logical plausibility of this approach and demonstrate its applicability in computational persuasion.

1 Introduction

Probabilistic epistemic argumentation [13,18,20,40] is an extension of Dung's classical argumentation framework [7]. While the original framework allows only for talking about attacks and accepting or rejecting arguments, probabilistic epistemic argumentation also allows more general relationships between arguments like support [4,5,29] and allows expressing more fine-grained beliefs by means of probabilities. Recent experiments give empirical evidence that these extensions are, in particular, beneficial when it comes to modelling human decision making [28]. One large application area of probabilistic epistemic argumentation is computational persuasion [15,16]. Computational persuasion aims at convincing the user of a persuasion goal such as giving up bad habits or living a healthier lifestyle. In order to derive persuasion strategies autonomously, we require a user model that represents the user's beliefs and simulates belief changes when new arguments are presented to the user. The user's epistemic state can be represented by a probability function and different update operators have been studied that can be used to adapt the current beliefs [15,17,19].

© Springer Nature Switzerland AG 2019
G. Kern-Isberner and Z. Ognjanović (Eds.): ECSQARU 2019, LNAI 11726, pp. 74–86, 2019.
https://doi.org/10.1007/978-3-030-29765-7_7

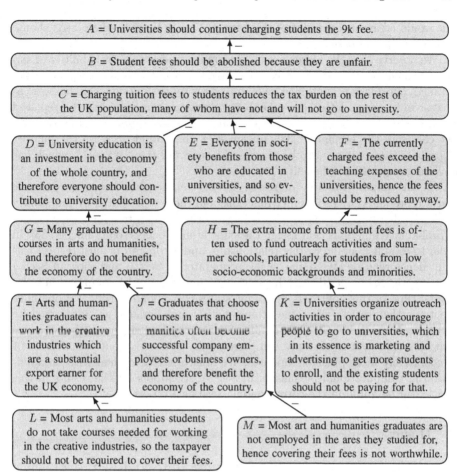

Fig. 1. Study fee dialogue.

Probability theory provides a strong foundational basis for probabilistic epistemic argumentation, but also comes with computational limitations. Without further assumptions, probability functions grow exponentially with the number of arguments. However, sometimes we are only interested in atomic beliefs in arguments, so that the full power of probability functions may not be required. For instance, we can consider the graph depicted in Fig. 1 induced by a dialogue between an automated dialogue system and a human participant that occurred in the empirical study considered in [11]. There are various constraints that could be attached to such a graph, as we will discuss further in Sect. 5. For instance, we could use postulates from the classical epistemic approach [13,40] such as *coherence*, which bounds the belief in an argument based on the belief of its attacker. Formally, in our scenario, for every argument-attacker pair X and Y

this would create a constraint of the form $\pi(X) + \pi(Y) \leq 1$, where $\pi(\alpha)$ should be read as the probability of α.

We observe that the aforementioned formulas operate on probabilities of single arguments rather than on complex logical expressions. Consequently, the detailed information contained in a full probability function can be seen as excessive. In such a situation, probability functions can sometimes be replaced by probability labellings that assign probabilities to arguments directly without changing the semantics [33]. In our case, this would decrease the number of probabilities that need to be processed from 8,192 (i.e. 2^{13}) to 13, which has obvious computational benefits.

In this paper, we are interested in the relationship between epistemic states represented by probability functions and those represented by probability labellings. Formally, probability labellings can be related to equivalence classes of probability functions that assign the same atomic beliefs to arguments [33]. In order to establish an interesting relationship, update operators must respect this equivalence relation. We define such an operator in Sect. 3 and show in Sect. 4 that it satisfies our desiderata. In particular, updates can be computed in polynomial time In this approach, epistemic states correspond to sets of probability functions that satisfy the same atomic beliefs and updates are performed by satisfying the new beliefs while minimizing the required changes. We will argue that this approach is not only computationally attractive, but can also result in cognitively more plausible updates. We illustrate our method with an application in computational persuasion in Sect. 5. All proofs for the results in this article can be found in the corresponding technical report [35].

2   Basics

We consider *bipolar argumentation frameworks (BAFs)* $(\mathcal{A}, \mathcal{R}, \mathcal{S})$ consisting of a set of arguments $\mathcal{A}$, an attack relation $\mathcal{R} \subseteq \mathcal{A} \times \mathcal{A}$ and a support relation $\mathcal{S} \subseteq \mathcal{A} \times \mathcal{A}$. $\Omega = \{w \mid w \subseteq \mathcal{A}\}$ denotes the set of *possible worlds*. Intuitively, each $w \in \Omega$ contains the arguments that are accepted in a particular state of the world. We represent beliefs by probability functions $P : \Omega \to [0, 1]$ such that $\sum_{w \in \Omega} P(w) = 1$. $\mathcal{P}_\mathcal{A}$ denotes the set of all probability functions over $\mathcal{A}$. The probability of an argument $A \in \mathcal{A}$ under P is defined by adding the probabilities of all worlds in which A is accepted, that is, $P(A) = \sum_{w \in \Omega, A \in w} P(w)$. $P(A)$ can be understood as a degree of belief, where $P(A) = 1$ means complete acceptance and $P(A) = 0$ means complete rejection[1].

The epistemic probabilistic argumentation approach developed in [13,18,20, 40] defines semantics of attack and support relations by means of constraints over probability functions. Some constraints can be automatically derived from the relations between arguments. For example, the *coherence constraint* demands that if A attacks B, we must have $P(B) \leq 1 - P(A)$, that is, the belief in

---

[1] Note that $P(A)$ denotes the probability of argument A (the sum of probabilities of all possible worlds that accept A), while $P(\{A\})$ denotes the probability of the possible world $\{A\}$.

an attacked argument B is bounded from above by the belief in an attacker A. However, it is also possible to design individual constraints manually. For example, if B is attacked by three related arguments A_1, A_2, A_3, we may want to bound the belief in B by the average belief in these attackers via $P(B) \leq 1 - \frac{1}{3} \sum_{i=1}^{3} P(A_i)$. To allow this flexibility, a general constraint language has been introduced in [17,18]. We will focus on the fragment of *linear atomic constraints* here because it is sufficiently expressive for most of the constraints considered in [13,20,40] and sometimes allows polynomial-time computations [33].

Formally, a *linear atomic constraint* over a set of arguments $\mathcal{A}$ is an expression of the form $\sum_{i=1}^{n} c_i \cdot \pi(A_i) \leq c_0$, where $A_i \in \mathcal{A}$ and $c_i \in \mathbb{Q}$. π is just a syntactic symbol that can be read as "the probability of". We let $\mathcal{C}_\mathcal{A}$ denote the set of all linear atomic constraints over $\mathcal{A}$. A probability function P *satisfies* such a linear atomic constraint iff $\sum_{i=1}^{n} c_i \cdot P(A_i) \leq c_0$. P satisfies a set of linear atomic constraints C, denoted as $P \models C$, iff it satisfies all $l \in C$. In this case, we call C *satisfiable*. We let $\mathrm{Sat}_\Pi(C) = \{P \in \mathcal{P}_\mathcal{A} \mid P \models C\}$ denote the set of all probability functions that satisfy C. We call sets of constraints C_1, C_2 equivalent and write $C_1 \equiv C_2$ iff they are satisfied by the same probability functions, that is, $\mathrm{Sat}_\Pi(C_1) = \mathrm{Sat}_\Pi(C_2)$.

Note that constraints with $\geq$ and $=$ can be expressed as well in our language. For $\geq$, just note that $\sum_{i=1}^{n} c_i \pi(A_i) \leq c_0$ is equivalent to $\sum_{i=1}^{n} -c_i \cdot \pi(A_i) \geq -c_0$. For $=$, note that $\sum_{i=1}^{n} c_i \cdot \pi(A_i) \leq c_0$ and $\sum_{i=1}^{n} c_i \cdot \pi(A_i) \geq c_0$ together are equivalent to $\sum_{i=1}^{n} c_i \cdot \pi(A_i) = c_0$. In particular, we can express probability assignments of the form $\pi(A) = p$ or probability bounds of the form $l \leq \pi(A) \leq u$.

Now assume that we are given an epistemic state represented as a probability function $P \in \mathcal{P}_\mathcal{A}$. Given some new evidence represented as a set of linear atomic constraints (and possibly some existing constraints that we want to preserve), we want to update P. To this end, different update operators have been studied in [15,17,19]. Here, we are interested in update operators of the following type.

Definition 1 (Epistemic Update Operator). *An* epistemic update operator *is a function* $\mathcal{U} : \mathcal{P}_\mathcal{A} \times \mathcal{C}_\mathcal{A} \to \mathcal{P}_\mathcal{A} \cup \{\bot\}$ *that satisfies the following properties:*

- **Success:** *If $C \subseteq \mathcal{C}_\mathcal{A}$ is satisfiable, then $\mathcal{U}(P, C) \in \mathrm{Sat}_\Pi(C)$.*
- **Failure:** *If $C \subseteq \mathcal{C}_\mathcal{A}$ is not satisfiable, then $\mathcal{U}(P, C) = \bot$.*
- **Representation Invariance:** *If $C_1 \equiv C_2$, then $\mathcal{U}(P, C_1) = \mathcal{U}(P, C_2)$.*
- **Idempotence:** *If $C \subseteq \mathcal{C}_\mathcal{A}$ is satisfiable, then $\mathcal{U}(\mathcal{U}(P, C), C) = \mathcal{U}(P, C)$.*

Success and failure guarantee a well-defined update. That is, if the constraints are satisfiable, the update operator will return a new epistemic state that satisfies the constraints. If the constraints are not satisfiable, $\bot$ will be returned to indicate an inconsistency. Representation invariance guarantees that the result is independent of the syntactic representation of the evidence. Finally idempotence guarantees that applying the same update twice does not change the outcome.

3 The Two-Stage Least-Squares Update Operator

Several update operators in [17,19] are based on the idea of satisfying new evidence by changing the current epistemic state in a minimal way. The distance between two probability functions is determined by looking at the probabilities that they assign to possible worlds. For example, one can use the least-squares distance $d_2(P, P') = \sum_{w \in \Omega}(P(w) - P'(w))^2$ or the KL-divergence $d_{KL}(P, P') = \sum_{w \in \Omega} P(w) \cdot \log \frac{P(w)}{P'(w)}$. While this makes perfect sense from a probability-theoretical point of view, the resulting belief changes may be intuitively implausible.

Table 1. Some probability functions over possible worlds used in Example 1.

w	P_1	P_2	P_3	P_4	P_5	P_6	P_7	P_8	P_9	P_{10}
$\emptyset$	0.1	0	0	0.4	0.45	0.26	0	0.3	0.15	0.35
$\{A\}$	0.2	0.4	0.33	0.05	0	0.19	0.3	0	0.15	0
$\{B\}$	0.3	0	0	0.15	0.1	0.29	0	0.1	0.35	0.15
$\{A, B\}$	0.4	0.6	0.67	0.4	0.45	0.26	0.7	0.6	0.35	0.5

Example 1. Consider a BAF $(\{A, B\}, \emptyset, \emptyset)$ with two unrelated arguments A, B. Suppose our current epistemic state is P_1 as defined in Table 1. Then we have $P_1(A) = 0.6$ and $P_1(B) = 0.7$. Now suppose that we want to update the belief in A to 1. A distance-minimizing update w.r.t. d_2 (i.e. update returning a probability distribution satisfying $\pi(A) = 1$ that is minimally different from P_1 w.r.t. d_2) yields the new epistemic state P_2 from Table 1. Now we have $P_2(A) = 1$ as desired. However, we also have $P_2(B) = 0.6 < 0.7 = P_1(B)$. Similarly, updating with respect to d_{KL} yields P_3 from Table 1 with $P_3(B) = \frac{2}{3} < 0.7 = P_1(B)$. This behaviour is rather counterintuitive in this context, since A and B are completely unrelated. Therefore, we should have $P_1(B) = P_2(B) = P_3(B)$.

In order to bring our model closer to humans' intuition, a two-stage minimization process has been proposed in [17]. In stage 1, we identify all probability distributions that minimize an atomic distance measure. Instead of comparing probability functions elementwise on possible worlds, atomic distance measures compare probability functions only based on the probabilities that they assign to arguments [19]. We consider a quadratic variant here that will allow us to compute some updates in polynomial time.

Definition 2 (Atomic Least-squares Distance (ALS)). *The **ALS** distance measure is defined as* $d_{At}^2(P, P') = \sum_{A \in \mathcal{A}}(P(A) - P'(A))^2$ *for all* $P, P' \in \mathcal{P}_\mathcal{A}$.

To begin with, we use the ALS distance to define a naive update operator which does not satisfy our desiderate from Definition 1 yet.

Definition 3 (Naive Least-squares Update Operator). *The* naive LS update operator u_{At} : $\mathcal{P}_A \times \mathcal{C}_A \rightarrow 2^{\mathcal{P}_A}$ *is defined by* $u_{At}(P, C) = \arg\min_{P' \in \text{Sat}_{\Pi}(C)} d^2_{At}(P, P')$.

u_{At} yields those probability functions that satisfy C and minimize the ALS distance to P. However, there is not necessarily a unique solution.

Example 2. Consider P_1 from Table 1. Suppose we recognize a conflict between A and B and want to update with the constraint $l_1 : \pi(A) + \pi(B) \leq 1$. We have $P_1(A) = 0.6$ and $P_1(B) = 0.7$. The cheapest way to satisfy the constraint with respect to the ALS distance is to decrease both probabilities by 0.15. That is, a solution P' must satisfy $P'(A) = P'(\{A\}) + P'(\{A, B\}) = 0.45$ and $P'(B) = P'(\{B\}) + P'(\{A, B\}) = 0.55$. P_4 and P_5 from Table 1 show two minimal solutions from the set $u_{At}(P_1, \{l_1\})$.

The second stage of the minimization process from [17] deals with the uniqueness problem. Among those probability functions that minimize the atomic distance, we pick the unique one that minimizes a sufficiently strong second distance measure. Here, we will consider again the least-squares distance for stage 2.

Definition 4 (Two-stage Least-squares Update Operator (2LS)) *The* 2LS update operator $\mathcal{U}^2_{At} : \mathcal{P}_A \times \mathcal{C}_A \rightarrow \mathcal{P}_A \cup \{\perp\}$ *is defined by*

$$\mathcal{U}^2_{At}(P, C) = \begin{cases} \arg\min_{P' \in u_{At}(P,C)} \sum_{w \in \Omega}(P(w) - P'(w))^2, & \text{if } u_{At}(P, C) \neq \emptyset \\ \perp & \text{otherwise.} \end{cases}$$

Before looking at an example, we note that $\mathcal{U}^2_{At}$ is an epistemic update operator as defined in Definition 1.

Proposition 1. *The 2LS update operator is an epistemic update operator.*

Example 3. Consider again P_1 and the constraint l_1 from Example 2. P_6, shown in Table 1, is the unique solution that minimizes the least-squares distance to P_1 among those distributions that minimize the ALS distance to P_1. That is, $\mathcal{U}^2_{At}(P_1, \{l_1\}) = P_6$.

Example 4. As another example, we consider again the scenario from Example 1 where a one-stage update changed the belief in B in an implausible way. We get $\mathcal{U}^2_{At}(P_1, \{\pi(A) = 1\}) = P_7$ shown in Table 1. In particular, we have $P_7(B) = 0.7 = P_1(B)$ as desired.

Intuitively, stage 1 determines which atomic beliefs in arguments have to be changed in order to satisfy the new constraints. This avoids the counterintuitive behaviour of elementwise minimization over the possible worlds, but does not yield a unique solution. Therefore, stage 2 performs an elementwise minimization over the possible worlds to pick a best solution among the ones that minimize the change in atomic beliefs.

4   Updates over Probability Labellings

The two-stage minimization process solves our semantical problems, but we are still left with a considerable computational problem. This is because we consider probability functions over possible worlds whose number grows exponentially with the number of arguments in our framework. However, as illustrated in our previous examples, human reasoning may be guided by atomic beliefs in arguments rather than by beliefs in possible worlds. Therefore, a natural question is, what changes semantically when considering belief functions over arguments rather than over possible worlds? As shown in [33], probability functions over possible worlds can sometimes just be replaced with *probability labellings* $L :$ $\mathcal{A} \to [0, 1]$ that assign beliefs to atomic arguments directly without changing the semantics. We let $\mathcal{L}_{\mathcal{A}}$ denote the set of all probability labellings.

Formally, probability functions can be related to probability labellings via an equivalence relation [33]. Two probability functions P_1, P_2 are called *atomically equivalent*, denoted as $P_1 \equiv P_2$, iff $P_1(A) = P_2(A)$ for all $A \in \mathcal{A}$. As usual, $[P] = \{P' \in \mathcal{P}_{\mathcal{A}} \mid P' \equiv P\}$ denotes the equivalence class of P and $\mathcal{P}_{\mathcal{A}}/\equiv =$ $\{[P] \mid P \in \mathcal{P}_{\mathcal{A}}\}$ denotes the set of all equivalence classes. As shown in [32], there is a one-to-one relationship between $\mathcal{P}_{\mathcal{A}}/\equiv$ and $\mathcal{L}_{\mathcal{A}}$.

Lemma 1 ([32])**.** *The function* $r : \mathcal{P}_{\mathcal{A}}/\equiv \to \mathcal{L}_{\mathcal{A}}$ *defined by* $r([P]) = L_P$, *where* $L_P(A) = P(A)$ *for all* $A \in \mathcal{A}$ *is a bijection.*

Intuitively, r determines a compact representation of the equivalence class $[P]$, namely the probability labelling $L_P = r([P])$. Since r is a bijection, every probability labelling can also be related to a set of probability functions $r^{-1}(L_P) = [P] = \{P' \in \mathcal{P}_{\mathcal{A}} \mid P' \equiv P\}$. Intuitively, $r^{-1}(L)$ is just the set of probability functions that satisfy the atomic beliefs encoded in L. We say that a probability labelling L satisfies a linear atomic constraint $\sum_{i=1}^{n} c_i \cdot \pi(A_i) \leq c_0$ iff $\sum_{i=1}^{n} c_i \cdot L(A_i) \leq c_0$. The set of probability labellings that satisfy a set of such constraints C is denoted by $\text{Sat}_{\Lambda}(C)$. The following observations from [32] are helpful to simplify computational problems by replacing probability functions with probability labellings.

Lemma 2 ([32])**.** *The following statements are equivalent: (1)* P *satisfies a linear atomic constraint* l; *(2) All* $P' \in [P]$ *satisfy* l; *(3)* $L_P = r([P])$ *satisfies* l.

For example, in order to decide whether a set of linear atomic constraints C is satisfiable by a probability function (of exponential size), we can just check whether it can be satisfied by a probability labelling (of linear size) [32]. If such a labelling L exists, all probability functions in $r^{-1}(L)$ satisfy C. Conversely, if some probability function P satisfies C, then $L = r([P])$ satisfies C as well.

In order to perform updates more efficiently, we could represent epistemic states by probability labellings. However, we should ask, what is the relationship between updates over probability functions and updates over probability labellings? We first note that update operators $\mathcal{U}_W$ that simply minimize the distance over possible worlds are not necessarily compatible with atomic equivalence. That is, given a set of linear atomic constraints C and two probability functions P_1 and P_2 such that $P_1 \equiv P_2$, we do not necessarily have $\mathcal{U}_W(P_1, C) \equiv \mathcal{U}_W(P_2, C)$.

Example 5. Consider P_1 and P_8 in Table 1. We have $P_1(A) = 0.6 = P_8(A)$ and $P_1(B) = 0.7 = P_2(B)$, that is, $P_1 \equiv P_8$. Suppose, we update with $C = \{\pi(A) = 0.5\}$ and update by just minimizing the least-squares distance to P_1. Then $\mathcal{U}_W(P_1, C) = P_9$ and $\mathcal{U}_W(P_8, C) = P_{10}$, where P_9, P_{10} are again shown in Table 1. We have $P_9(B) = 0.7 \neq 0.65 = P_{10}(B)$, that is, $P_9 \neq P_{10}$.

Update operators based on atomic distance measures give us compatibility guarantees that we explain in the following proposition.

Proposition 2. *Let $P_1, P_2 \in \mathcal{P}_A$ and let $C \subset C_A$ be a finite set of linear atomic constraints. If $P_1 \equiv P_2$, then*

1. *$d^2_{At}(P_1, P) = d^2_{At}(P_2, P)$ for all $P \in \mathcal{P}_A$,*
2. *$u_{At}(P_1, C) = u_{At}(P_2, C)$,*
3. *$P'_1 \equiv P'_2$ for all $P'_1, P'_2 \in u_{At}(P_1, C)$,*
4. *$\mathcal{U}^2_{At}(P_1, C) \equiv \mathcal{U}^2_{At}(P_2, C)$.*

Item 1 says that the ALS distance is invariant under atomically equivalent probability functions. This implies that the updates that minimize the ALS distance are invariant as well (item 2). As we demonstrated in Example 2, such updates do not necessarily yield a unique solution. However, when using the ALS distance, we can guarantee that all solutions are atomically equivalent (item 3). This implies that the 2LS update operator is invariant under atomically equivalent probability functions in the sense that it yields equivalent results when the prior probability functions are equivalent (item 4).

Hence, when updating with respect to linear atomic constraints, there is a well defined relationship between probability functions and probability labellings. If we start with an epistemic state represented by a probability labelling L, L can be understood as a compact representation of the set of probability functions $r^{-1}(L)$ that satisfy the atomic beliefs encoded in L. The 2LS update operator is compatible with this representation. That is, no matter which probability functions from $r^{-1}(L)$ we choose, an update with linear atomic constraints will always lead to the same equivalence class and therefore to a well defined next probability labelling L^*. We illustrate this in Fig. 2.

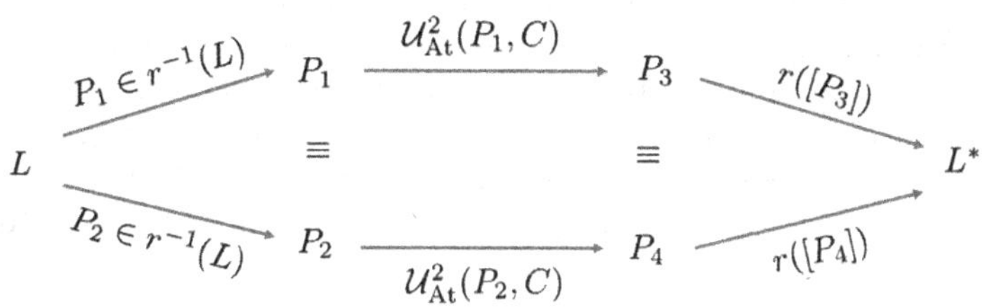

Fig. 2. The 2LS update operator $\mathcal{U}^2_{At}$ respects atomic equivalence.

If we are only interested in atomic beliefs, it would be convenient if we could move directly from L to L^* in Fig. 2 without generating (exponentially large) probability functions in the process. We can do this indeed in polynomial time for the 2LS update operator. In order to show this, we first define an update operator on labellings.

Definition 5 (Least-squares Labelling Update Operator (2LS)). *The* LS labelling update operator $LU_\lambda^2 : \mathcal{L}_\mathcal{A} \times \mathcal{C}_\mathcal{A} \to \mathcal{L}_\mathcal{A} \cup \{\bot\}$ *is defined by*

$$LU_\lambda^2(L, C) = \begin{cases} \arg\min_{L' \in \mathrm{Sat}_\mathcal{A}(C)} \sum_{A \in \mathcal{A}} (L(A) - L'(A))^2, & \text{if } \mathrm{Sat}_\mathcal{A}(C) \neq \emptyset \\ \bot & \text{otherwise.} \end{cases}$$

As we explain in the following theorem, LU_λ^2 provides us with a direct path from L to L^* and can be computed in polynomial time.

Theorem 1. *Let $C \subset \mathcal{C}_\mathcal{A}$ be a finite and satisfiable set of linear atomic constraints and let $L \in \mathcal{L}_\mathcal{A}$. Then $LU_\lambda^2(L, C) = L^*$ is well-defined and can be computed in polynomial time. Furthermore, $L^* = r([\mathcal{U}_{\mathrm{At}}^2(P, C)])$ for all $P \in r^{-1}(L)$.*

Hence, when we are only interested in atomic beliefs, we can use probability labellings to represent epistemic states and use the least-squares labelling update operator for updates. Semantically, this is equivalent to regarding epistemic states as sets of probability functions that satisfy the same atomic beliefs and updating with respect to the 2LS update operator. The benefit of the labelling representation is that we can perform updates in polynomial time.

5   Application Example

In this section we come back to the graph in Fig. 1 and analyze a scenario that, while being hypothetical, uses the data from an empirical study in [11]. In this study, the user's belief in argument A changed from 0 to 0.19 during the dialogue[2].

The graph in Fig. 1 is generated from an existing dialogue that involved an automated dialogue system and a human user. Arguments at even depth (starting from A) are system arguments (A, C, G, H, L and M), while the ones at odd depth are user arguments. The agents take turns in uttering their arguments (starting with A), and arguments at the same depth are uttered at the same point by a given party. We observe that not all user arguments are met with a system response (see arguments E and K). Despite this fact, the presented arguments have led to a positive change in belief in A, contrary to what would be the intuition from the classical Dungean approaches. It is possible that if all of the user's counterarguments were addressed, then the belief increase would be even more prominent.

We can try to provide an explanation for the belief change observed in [11] by modeling the reasoning process in our framework. Let us assume that the

---

[2] We note that the study data contained examples of dialogues that resulted in a bigger belief change, however, we have chosen this one due to its interesting structure.

Table 2. Probability labellings before and after the dialogue from Sect. 5.

L	A	B	C	D	E	F	G	H	I	J	K	L	M
L_0	0	1	0	1	1	1	0	0	1	1	1	0	0
$L_1 = LU_\lambda^2(L, C \cup \Phi)$	0.19	0.81	0.19	0.505	0.975	0.95	0.495	0.05	0.92	0.09	0.95	0.08	0.91

constraints representing the user's reasoning demand that the belief in an argument is dual to the belief in the average of its attackers. That is, we assume $P(X) = 1 - \frac{1}{|Att(X)|} \sum_{Y \in Att(X)} P(Y)$, where $Att(X) = \{Y \in \mathcal{A} \mid (Y, X) \in \mathcal{R}\}$). This assumption leads to the following set of constraints:

$$C = \{\pi(A) + \pi(B) = 1, \pi(B) + \pi(C) = 1, \pi(D) + \pi(G) = 1, \pi(F) + \pi(H) = 1,$$
$$\pi(C) + 0.33\pi(D) + 0.34\pi(E) + 0.33\pi(F) = 1, \pi(C) + 0.5\pi(I) + 0.5\pi(J) = 1,$$
$$\pi(H) + \pi(K) = 1, \pi(I) + \pi(L) = 1, \pi(J) + \pi(M) = 1\}$$

Let us further assume that the user initially completely accepts his or her own arguments and completely rejects the system's arguments. This belief state is represented by the labeling L_0 shown in Table 2. We now consider a possible persuasion system which, once a given dialogue branch is exhausted, asks the user about his or her beliefs in the unattacked arguments. In our case, the user states that he or she believes L, M, E and K with the degrees 0.08, 0.91, 0.975 and 0.95 respectively. This produces constraints $\Phi = \{\pi(L) = 0.08, \pi(M) = 0.91, \pi(E) = 0.975, \pi(K) = 0.95\}$. We can use this information along with C to update L_0 without asking the user his or her beliefs in all possible arguments. The resulting labeling $L_1 = LU_\lambda^2(L, C \cup \Phi)$ is shown in Table 2.

We observe that the belief in A in and L_0 and L_1 match the expected beliefs 0 and 0.19 based on the data in [11].

6   Related Work

There is a large variety of other probabilistic argumentation approaches [6,8,14, 24,25,27,37–39,41,42], which basically differ in the level of detail (e.g., structured or abstract argumentation), in the way how uncertainty is introduced (e.g. possible worlds correspond to argument interpretations or the graph structure) and in the nature of uncertainty (e.g., uncertainty about the acceptance state or uncertainty about the nature of a relation between arguments).

One limitation when restricting to probability labellings is that we cannot compute the probabilities of complex formulas over arguments anymore without adding further assumptions. However, as we demonstrated, we can sometimes do without complex formulas. In this context, probability labellings can be seen as an alternative to weighted argumentation frameworks that also assign a strength value between 0 and 1 to arguments [2,3,26,31,36]. What makes probability labellings an interesting alternative is their well-defined relationship to probability functions and probability theory.

The problem of adapting an epistemic state with respect to new knowledge has been studied extensively in the belief revision literature that evolved from the AGM theory developed in [1]. An up-to-date discussion of the main ideas can be found in [12]. Our postulates are inspired by AGM postulates. For example, *Success* and *Representation Invariance* can be seen as the counterparts of the *Closure* and *Extensionality* postulates in AGM theory. The closest relative to our setting is probably the probabilistic belief change framework from [21]. For a discussion of relationships between classical and probabilistic belief changes, see [21,22].

Other equivalence relations have been studied in order to improve the computational performance of probabilistic reasoning algorithms [9,10,23,30]. However, usually, these equivalence relations are introduced over possible worlds, not over probability functions. They can be applied to more expressive reasoning formalisms (they are not restricted to atomic beliefs), but identifying compact representatives for the corresponding equivalence classes remains intractable in general [34].

7 Conclusions

We demonstrated that, in the fragment of linear atomic constraints, it is possible to relate updates over probability labellings to equivalent updates over classes of probability functions. This is interesting from a cognitive, a probabilistic-logical and a computational perspective. Atomic beliefs are often easier to understand for humans. If we can relate these beliefs to probability functions, we get a strong foundational basis. Finally, they can be stored much more compactly and give us polynomial runtime guarantees. Our results can probably be generalized to other two-stage update operators. However, the building blocks for the two stages have to be chosen carefully in order to guarantee that the update operator respects atomic equivalence. For example, it may not be possible to relate the two-stage update process considered in [17], Sect. 5, to an update operator over probability labellings in a meaningful way. However, we may be able to construct similar relationships by replacing the least-squares distance with KL-divergence or more general classes of distance measures. An implementation of our update operator is available in the Java library *ProBabble*[3].

References

1. Alchourrón, C., Gärdenfors, P., Makinson, D.: On the logic of theory change: partial meet contraction and revision functions. J. Symbolic Logic **50**(2), 510–530 (1985)
2. Amgoud, L., Ben-Naim, J.: Evaluation of arguments in weighted bipolar graphs. In: Antonucci, A., Cholvy, L., Papini, O. (eds.) ECSQARU 2017. LNCS (LNAI), vol. 10369, pp. 25–35. Springer, Cham (2017). https://doi.org/10.1007/978-3-319-61581-3_3
3. Baroni, P., Romano, M., Toni, F., Aurisicchio, M., Bertanza, G.: Automatic evaluation of design alternatives with quantitative argumentation. Argument Comput. **6**(1), 24–49 (2015)

---

[3] https://sourceforge.net/projects/probabble/.

4. Cayrol, C., Lagasquie-Schiex, M.C.: Bipolarity in argumentation graphs: towards a better understanding. Int. J. Approximate Reasoning **54**(7), 876–899 (2013)
5. Cohen, A., Gottifredi, S., García, A.J., Simari, G.R.: A survey of different approaches to support in argumentation systems. Knowl. Eng. Rev. **29**(5), 513–550 (2014)
6. Doder, D., Woltran, S.: Probabilistic argumentation frameworks – a logical approach. In: Straccia, U., Calì, A. (eds.) SUM 2014. LNCS (LNAI), vol. 8720, pp. 134–147. Springer, Cham (2014). https://doi.org/10.1007/978-3-319-11508-5_12
7. Dung, P.M.: On the acceptability of arguments and its fundamental role in nonmonotonic reasoning, logic programming and n-person games. Artif. Intell. **77**(2), 321–357 (1995)
8. Dung, P.M., Thang, P.M.: Towards (probabilistic) argumentation for jury-based dispute resolution. In: Proceedings of COMMA 2010. FAIA, vol. 216, pp. 171–182. IOS Press (2010)
9. Finthammer, M., Beierle, C.: Using equivalences of worlds for aggregation semantics of relational conditionals. In: Glimm, B., Krüger, A. (eds.) KI 2012. LNCS (LNAI), vol. 7526, pp. 49–60. Springer, Heidelberg (2012). https://doi.org/10.1007/978-3-642-33347-7_5
10. Fischer, V.G., Schramm, M.: tabl-a tool for efficient compilation of probabilistic constraints. Technical report TUM-19636, Technische Universitaet Muenchen (1996)
11. Hadoux, E., Hunter, A., Polberg, S.: Strategic argumentation dialogues for persuasion: framework and experiments based on modelling the beliefs and concerns of the persuadee. Technical report. University College London (2019)
12. Hansson, S.: Logic of belief revision. In: The Stanford Encyclopedia of Philosophy. Metaphysics Research Lab, Stanford University, winter 2017 edn. (2017)
13. Hunter, A.: A probabilistic approach to modelling uncertain logical arguments. Int. J. Approximate Reasoning **54**(1), 47–81 (2013)
14. Hunter, A.: Probabilistic qualification of attack in abstract argumentation. Int. J. Approximate Reasoning **55**(2), 607–638 (2014)
15. Hunter, A.: Modelling the persuadee in asymmetric argumentation dialogues for persuasion. In: Proceedings of IJCAI 2015, pp. 3055–3061. AAAI Press (2015)
16. Hunter, A.: Computational persuasion with applications in behaviour change. In: Proceedings of COMMA 2016. FAIA, vol. 287, pp. 5–18. IOS Press (2016)
17. Hunter, A., Polberg, S., Potyka, N.: Updating belief in arguments in epistemic graphs. In: Proceedings of KR 2018, pp. 138–147. AAAI Press (2018)
18. Hunter, A., Polberg, S., Thimm, M.: Epistemic graphs for representing and reasoning with positive and negative influences of arguments. arXiv preprint arXiv:1802.07489v1 (2018)
19. Hunter, A., Potyka, N.: Updating probabilistic epistemic states in persuasion dialogues. In: Antonucci, A., Cholvy, L., Papini, O. (eds.) ECSQARU 2017. LNCS (LNAI), vol. 10369, pp. 46–56. Springer, Cham (2017). https://doi.org/10.1007/978-3-319-61581-3_5
20. Hunter, A., Thimm, M.: On partial information and contradictions in probabilistic abstract argumentation. In: Proceedings of KR 2016, pp. 53–62. AAAI Press (2016)
21. Kern-Isberner, G. (ed.): Conditionals in Nonmonotonic Reasoning and Belief Revision. LNCS (LNAI), vol. 2087. Springer, Heidelberg (2001). https://doi.org/10.1007/3-540-44600-1
22. Kern-Isberner, G.: Linking iterated belief change operations to nonmonotonic reasoning. In: Proceedings of KR 2008, pp. 166–176. AAAI Press, Menlo Park (2008)

23. Kern-Isberner, G., Lukasiewicz, T.: Combining probabilistic logic programming with the power of maximum entropy. Artif. Intell. **157**(1–2), 139–202 (2004)
24. Kido, H., Okamoto, K.: A Bayesian approach to argument-based reasoning for attack estimation. In: Proceedings of IJCAI 2017, pp. 249–255. AAAI Press (2017)
25. Li, H., Oren, N., Norman, T.J.: Probabilistic argumentation frameworks. In: Modgil, S., Oren, N., Toni, F. (eds.) TAFA 2011. LNCS (LNAI), vol. 7132, pp. 1–16. Springer, Heidelberg (2012). https://doi.org/10.1007/978-3-642-29184-5_1
26. Mossakowski, T., Neuhaus, F.: Modular semantics and characteristics for bipolar weighted argumentation graphs. arXiv preprint arXiv:1807.06685 (2018)
27. Polberg, S., Doder, D.: Probabilistic abstract dialectical frameworks. In: Fermé, E., Leite, J. (eds.) JELIA 2014. LNCS (LNAI), vol. 8761, pp. 591–599. Springer, Cham (2014). https://doi.org/10.1007/978-3-319-11558-0_42
28. Polberg, S., Hunter, A.: Empirical evaluation of abstract argumentation: supporting the need for bipolar and probabilistic approaches. Int. J. Approximate Reasoning **93**, 487–543 (2018)
29. Polberg, S., Oren, N.: Revisiting support in abstract argumentation systems. In: Proceedings of COMMA 2014. FAIA, vol. 266, pp. 369–376. IOS Press (2014)
30. Potyka, N.: Solving reasoning problems for probabilistic conditional logics with consistent and inconsistent information. Ph.D. thesis (2016)
31. Potyka, N.: Continuous dynamical systems for weighted bipolar argumentation. In: Proceedings of KR 2018, pp. 148–157. AAAI Press (2018)
32. Potyka, N.: A polynomial-time fragment of epistemic probabilistic argumentation (technical report). arXiv preprint arXiv:1807.06685 (2018)
33. Potyka, N.: A polynomial-time fragment of epistemic probabilistic argumentation (extended abstract). In: Proceedings of AAMAS 2019. IFAAMAS (2019, to appear)
34. Potyka, N., Beierle, C., Kern-Isberner, G.: A concept for the evolution of relational probabilistic belief states and the computation of their changes under optimum entropy semantics. J. Appl. Logic **13**(4), 414–440 (2015)
35. Potyka, N., Polberg, S., Hunter, A.: Polynomial-time updates of epistemic states in a fragment of probabilistic epistemic argumentation (technical report). arXiv preprint arXiv:1906.05066 (2019)
36. Rago, A., Toni, F., Aurisicchio, M., Baroni, P.: Discontinuity-free decision support with quantitative argumentation debates. In: Proceedings of KR 2016, pp. 63–73. AAAI Press (2016)
37. Rienstra, T.: Towards a probabilistic Dung-style argumentation system. In: Proceedings of AT 2012, pp. 138–152 (2012)
38. Rienstra, T., Thimm, M., Liao, B., van der Torre, L.: Probabilistic abstract argumentation based on SCC decomposability. In: Proceedings of KR 2018, pp. 168–177. AAAI Press (2018)
39. Riveret, R., Baroni, P., Gao, Y., Governatori, G., Rotolo, A., Sartor, G.: A labelling framework for probabilistic argumentation. Ann. Math. Artif. Intell. **83**(1), 21–71 (2018)
40. Thimm, M.: A probabilistic semantics for abstract argumentation. In: Proceedings of ECAI 2012. FAIA, vol. 242, pp. 750–755. IOS Press (2012)
41. Thimm, M., Baroni, P., Giacomin, M., Vicig, P.: Probabilities on extensions in abstract argumentation. In: Black, E., Modgil, S., Oren, N. (eds.) TAFA 2017. LNCS (LNAI), vol. 10757, pp. 102–119. Springer, Cham (2018). https://doi.org/10.1007/978-3-319-75553-3_7
42. Thimm, M., Cerutti, F., Rienstra, T.: Probabilistic graded semantics. In: Proceedings of COMMA 2018. FAIA, vol. 305, pp. 369–380. IOS Press (2018)

Ordering Argumentation Frameworks

Chiaki Sakama[1]([✉]) and Katsumi Inoue[2]

[1] Wakayama University, Wakayama, Japan
sakama@wakayama-u.ac.jp
[2] National Institute of Informatics, Tokyo, Japan
inoue@nii.ac.jp

Abstract. This paper introduces two orderings over abstract argumentation frameworks to compare justification status under argumentation semantics. Given two argumentation frameworks AF_1 and AF_2 and an argumentation semantics σ, AF_2 is *more ♯-general than* (or *equal to*) AF_1 (written $AF_1 \sqsubseteq_\sigma^\sharp AF_2$) if for any σ-extension F of AF_2 there is a σ-extension E of AF_1 such that $E \subseteq F$. In contrast, AF_2 is *more ♭-general than* (or *equal to*) AF_1 (written $AF_1 \sqsubseteq_\sigma^\flat AF_2$) if for any σ-extension E of AF_1 there is a σ-extension F of AF_2 such that $E \subseteq F$. We show that if $AF_1 \sqsubseteq_\sigma^\sharp AF_2$ then AF_2 skeptically accepts arguments more than AF_1 (under the σ-semantics) while if $AF_1 \sqsubseteq_\sigma^\flat AF_2$ then AF_2 credulously accepts arguments more than AF_1. Mathematically, these orders constitute pre-order sets over the set of all argumentation frameworks. Next we consider comparing two AFs under dynamic environments by observing the effect of incorporating new information into given AFs. We introduce two orderings in such dynamic environments and show its connection to strong equivalence between argumentation frameworks.

Keywords: Argumentation · Ordering · Strong equivalence

1 Introduction

There are several ways for comparing different theories. Given two first-order theories T_1 and T_2, if $T_1 \models T_2$ holds then every formula derived from T_2 is derived from T_1. In this case, T_1 is considered *more general* (or *informative*) than T_2. For instance, $p \models p \vee q$ means that p is more informative than $p \vee q$. In particular, T_1 is *equivalent* to T_2 $(T_1 \equiv T_2)$ if $T_1 \models T_2$ and $T_2 \models T_1$. Inoue and Sakama [7,8] argue that, in contrast to classical monotonic logic, there is difficulty in defining information ordering in *nonmonotonic logics*. A nonmonotonic theory generally has multiple extensions, and there are two kinds of consequences of a theory, i.e., *skeptical* and *credulous* consequences. This is contrasted to a first-order theory that has a unique extension as the logical consequences of the theory. Then, depending on types of consequences, there exist several definitions for determining that a theory is more informative than another theory. For instance, consider two (nonmonotonic) logic programs: $P_1 = \{\, p \leftarrow not\, q \,\}$ and $P_2 = \{\, p \leftarrow not\, q, \quad q \leftarrow not\, p \,\}$. Then P_1 has the single answer set (or stable model) $\{p\}$ and P_2 has two answer

© Springer Nature Switzerland AG 2019
G. Kern-Isberner and Z. Ognjanović (Eds.): ECSQARU 2019, LNAI 11726, pp. 87–98, 2019.
https://doi.org/10.1007/978-3-030-29765-7_8

sets $\{p\}$ and $\{q\}$. If we compare skeptical consequences, we can say that P_1 is more informative than P_2 because p is entailed from the former only. Instead, if we compare credulous consequences, P_2 is more informative than P_1 because q is derived from the latter only. As such, the result depends on the type of inference, and in this circumstance information ordering in classical logic cannot be applied. The study [7] then introduces two orderings to logic programs. Given two logic programs P_1 and P_2, $P_1 \models^\sharp P_2$ (P_1 is *more $\sharp$-general than P_2*) iff for any answer set S of P_1 there is an answer set T of P_2 such that $T \subseteq S$. Likewise, $P_1 \models^\flat P_2$ (P_1 is *more $\flat$-general than P_2*) iff for any answer set T of P_2 there is an answer set S of P_1 such that $T \subseteq S$. These two orderings are respectively called the *Smyth order* and the *Hoare order* in the *domain theory* [6]. The study [7] shows that if $P_1 \models^\sharp P_2$ (resp. $P_1 \models^\flat P_2$) then P_1 entails more skeptical (resp. credulous) consequences than P_2 under the answer set semantics [5]. These orderings are also applied to *default theories* [8] and *abductive theories* [9].

In this paper, we are interested in comparing justification status in *(abstract) argumentation frameworks* (AFs) [3]. Given an argumentation framework AF, an argument x is *skeptically accepted* (or *justified*) under the σ semantics if it is included in every σ-extension of AF, while x is *credulously accepted* if it is included in some σ-extension of AF. The notion of skeptical/credulous justification is of interest in the field of argumentation because "skepticism is related with making more or less committed evaluations about the justification state of arguments in a given situation: more skeptical attitude corresponds to less committed (i.e. more cautious) evaluations" [1]. Baroni and Giacomin [1] then provide systematic comparison of argumentation semantics with respect to their skepticism. They compare skeptical/credulous consequences of different argumentation semantics on a single argumentation framework. In contrast, the current study aims at comparing skeptical/credulous consequences of different argumentation frameworks under the same semantics. Suppose agents (or groups) who have their own argumentation frameworks in which each AF represents an agent's private view of arguments and attack relations. Then it is meaningful to compare those AFs to see which party is more skeptical/credulous in reasoning about arguments. We apply two orderings of [7,8] to argumentation frameworks and show that those orderings are useful for comparing skeptical/credulous acceptance among different argumentation theories. We also compare AFs under dynamic environments and provide a connection to *strong equivalence* of AFs. The rest of this paper is organized as follows. Section 2 reviews notions used in this paper. Section 3 introduces two orderings between AFs. Section 4 introduces orderings in dynamic environments, and Sect. 5 addresses final remarks.

2   Preliminaries

2.1   Argumentation Framework

Let $\mathcal{U}$ be the universe of all *arguments*. An *argumentation framework* (AF) [3] is a pair (A, R) where $A \subseteq \mathcal{U}$ is a finite set of arguments and $R \subseteq A \times A$ is the attack relation. The collection of all AFs (induced by $\mathcal{U}$) is denoted by $\mathcal{AF}$. We write

$a \rightarrow b$ (a *attacks* b) iff $(a, b) \in R$. A set S of arguments *attacks* an argument a (written $S \rightarrow a$) iff there is an argument $b \in S$ that attacks a. A set S of arguments is *conflict-free* if there are no arguments $a, b \in S$ such that a attacks b. A set S of arguments *defends* an argument a if S attacks every argument that attacks a. We write $D(S) = \{ a \mid S$ defends $a \}$. Given $AF = (A, R)$, a conflict-free set of arguments $S \subseteq A$ is:

- an *admissible set* iff $S \subseteq D(S)$;
- a *complete extension* iff $S = D(S)$;
- a *stable extension* iff S attacks each argument in $A \setminus S$;
- a *preferred extension* iff S is a maximal complete extension of AF (wrt $\subseteq$);
- a *grounded extension* iff S is the minimal complete extension of AF (wrt $\subseteq$).

Let $\mathcal{E}_{AF}^{adm}$, $\mathcal{E}_{AF}^{com}$, $\mathcal{E}_{AF}^{stb}$, $\mathcal{E}_{AF}^{prf}$, and $\mathcal{E}_{AF}^{grd}$ be the sets of admissible sets, complete extensions, stable extensions, preferred extensions, and the grounded extension of an AF, respectively. Then the following relations hold:

$$\mathcal{E}_{AF}^{stb} \subseteq \mathcal{E}_{AF}^{prf} \subseteq \mathcal{E}_{AF}^{com} \subseteq \mathcal{E}_{AF}^{adm} \quad \text{and} \quad \mathcal{E}_{AF}^{grd} \subseteq \mathcal{E}_{AF}^{com} .$$

$\mathcal{E}_{AF}^{stb}$ is possibly empty, while others are not. In particular, $\mathcal{E}_{AF}^{grd}$ is a singleton set. We often write $\mathcal{C}_{AF}^{\tau}$ where σ means either *adm*, *com*, *prf*, *stb* or *grd*. We say that two argumentation frameworks AF_1 and AF_2 are σ-*equivalent* (written $AF_1 \equiv_\sigma AF_2$) if $\mathcal{E}_{AF_1}^\sigma = \mathcal{E}_{AF_2}^\sigma$. An argument $a \in A$ is *credulously* (resp. *skeptically*) *accepted* under the σ semantics of $AF = (A, R)$ iff $a \in E$ for some (resp. every) $E \in \mathcal{E}_{AF}^\sigma$. The set of all credulously (resp. skeptically) accepted arguments under the σ semantics of AF is denoted by $crd^\sigma(AF)$ (resp. $skp^\sigma(AF)$). When $\mathcal{E}_{AF}^{stb} = \emptyset$, we define $crd^{stb}(AF) = \emptyset$ and $skp^{stb}(AF) = \mathcal{U}$.

2.2 Ordering on Powersets

We recall some mathematical definitions about domains [6]. A *pre-order* (or *quasi-order*) $\preccurlyeq$ is a binary relation which is reflexive and transitive. A pre-order $\preccurlyeq$ is a *partial order* if it is also anti-symmetric. A *pre-ordered set* (resp. *partially ordered set*; *poset*) is a set D with a pre-order (resp. partial order) $\preccurlyeq$ on D. For a pre-ordered set $\langle D, \preccurlyeq \rangle$ and $x, y \in D$, we write $x \prec y$ if $x \preccurlyeq y$ and $y \not\preccurlyeq x$. For a poset $\langle D, \preccurlyeq \rangle$, two elements $x, y \in D$ are *comparable* if $x \preccurlyeq y$ or $y \preccurlyeq x$; otherwise, they are *incomparable*. A *chain* in $\langle D, \preccurlyeq \rangle$ is a subset C of D in which each pair of elements is comparable. An *antichain* in $\langle D, \preccurlyeq \rangle$ is a subset A of D in which each pair of different elements is incomparable, i.e., there is no order relation between any two different elements in A. For a pre-ordered set $\langle D, \preccurlyeq \rangle$ and any set $X \subseteq D$, we denote the maximal and minimal elements of X as follows.

$$min_{\preccurlyeq}(X) = \{ x \in X \mid \neg \exists y \in X \text{ s.t. } y \prec x \},$$
$$max_{\preccurlyeq}(X) = \{ x \in X \mid \neg \exists y \in X \text{ s.t. } x \prec y \}.$$

We often denote these as $min(X)$ and $max(X)$ by omitting $\preceq$. We also assume that the relation $\preceq$ is well-founded (resp. upwards well-founded) on D^1 whenever $min_{\preceq}(X)$ (resp. $max_{\preceq}(X)$) is concerned in order to guarantee the existence of a minimal (resp. maximal) element of any $X \subseteq D$. Note that, when D is finite, any pre-order is both well-founded and upwards well-founded on D.

For any set D, let $\mathcal{P}(D)$ be the powerset of D. Given a poset $\langle D, \preceq \rangle$ and $X, Y \in \mathcal{P}(D)$, two orders are defined as follows:

$$X \preceq^{\sharp} Y \quad \text{iff} \quad \forall y \in Y \, \exists x \in X \ s.t. \ x \preceq y,$$
$$X \preceq^{b} Y \quad \text{iff} \quad \forall x \in X \, \exists y \in Y \ s.t. \ x \preceq y.$$

The relations $\preceq^{\sharp}$ and $\preceq^{b}$ are respectively called the *Smyth order* and the *Hoare order*, and both $\langle \mathcal{P}(D), \preceq^{\sharp} \rangle$ and $\langle \mathcal{P}(D), \preceq^{b} \rangle$ are pre-ordered sets.

Example 1. Consider the poset $\langle \mathcal{P}(\{p, q\}), \subseteq \rangle$. It holds that $\{\{p\}, \{q\}\} \preceq^{\sharp}$ $\{\{p\}\} \preceq^{\sharp} \{\{p, q\}\}$ and $\{\{p\}\} \preceq^{b} \{\{p\}, \{q\}\} \preceq^{b} \{\{p, q\}\}$. Since $\{\emptyset, \{p\}\} \preceq^{\sharp}$ $\{\emptyset, \{q\}\} \preceq^{\sharp} \{\emptyset, \{p\}\}$ and $\{\{p\}, \{p, q\}\} \preceq^{b} \{\{q\}, \{p, q\}\} \preceq^{b} \{\{p\}, \{p, q\}\}$ hold, both $\preceq^{\sharp}$ and $\preceq^{b}$ are not partial orders.

For notational convenience, we often denote two orderings as $\preceq^{\sharp/b}$ when distinction between them is unimportant.

3 Ordering Argumentation Frameworks

3.1 Ordering AFs

In this section, we consider a pre-ordered set $\langle D, \preceq \rangle$ in which the domain D is $\mathcal{P}(\mathcal{U})$, i.e., the class of sets of arguments in $\mathcal{U}$, and the pre-order $\preceq$ is the inclusion relation $\subseteq$ over $\mathcal{P}(\mathcal{U})$. In this case $\langle \mathcal{P}(\mathcal{U}), \subseteq \rangle$ becomes a poset. The Smyth and Hoare orderings on $\mathcal{P}(\mathcal{P}(\mathcal{U}))$ are then defined, which enables us to order classes of sets of arguments.

Definition 1 (orderings over sets of arguments). Let $\langle \mathcal{P}(\mathcal{U}), \subseteq \rangle$ be a poset. For any Σ_1 and Σ_2 in $\mathcal{P}(\mathcal{P}(\mathcal{U}))$,

$$\Sigma_1 \preceq^{\sharp} \Sigma_2 \quad \text{iff} \quad \forall T \in \Sigma_2 \, \exists S \in \Sigma_1 \ s.t. \ S \subseteq T,$$
$$\Sigma_1 \preceq^{b} \Sigma_2 \quad \text{iff} \quad \forall S \in \Sigma_1 \, \exists T \in \Sigma_2 \ s.t. \ S \subseteq T.$$

Definition 2 (ordering AFs). Let AF_1 and AF_2 be two argumentation frameworks.

$$AF_1 \sqsubseteq_{\sigma}^{\sharp} AF_2 \quad \text{iff} \quad \mathcal{E}_{AF_1}^{\sigma} \preceq^{\sharp} \mathcal{E}_{AF_2}^{\sigma},$$
$$AF_1 \sqsubseteq_{\sigma}^{b} AF_2 \quad \text{iff} \quad \mathcal{E}_{AF_1}^{\sigma} \preceq^{b} \mathcal{E}_{AF_2}^{\sigma}$$

---

[1] A relation R is *well-founded* on a class D iff every non-empty subset of D has a minimal element with respect to R. A relation R is *upwards well-founded* on D iff the inverse relation R^{-1} is well-founded on D.

where $\sigma \in \{adm, com, prf, stb, grd\}$. We say that AF_2 is *more* (or *equally*) $\natural$-*general* (resp. $\flat$-*general*) than AF_1 (under the σ-semantics) if $AF_1 \sqsubseteq_\sigma^\natural AF_2$ (resp. $AF_1 \sqsubseteq_\sigma^\flat AF_2$).

We write $AF_1 \equiv_\sigma^\natural AF_2$ (resp. $AF_1 \equiv_\sigma^\flat AF_2$) iff $AF_1 \sqsubseteq_\sigma^\natural AF_2$ and $AF_2 \sqsubseteq_\sigma^\natural AF_1$ (resp. $AF_1 \sqsubseteq_\sigma^\flat AF_2$ and $AF_2 \sqsubseteq_\sigma^\flat AF_1$).

For notational convenience, we often denote two orderings as $\sqsubseteq_\sigma^{\natural/\flat}$ when distinction between them is unimportant.

Proposition 1. *Let $\mathcal{AF}$ be the collection of all AFs. Then $\langle \mathcal{AF}, \sqsubseteq_\sigma^{\natural/\flat} \rangle$ is a pre-ordered set where $\sigma \in \{adm, com, prf, stb, grd\}$.*

Example 2. Consider $AF_1 = (\{a, b, c\}, \{(a, b), (b, a), (b, c), (c, c)\})$ and $AF_2 = (\{a, b, c, d\}, \{(a, d), (d, a), (b, d), (d, b)\})$.

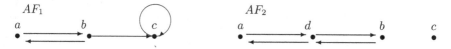

Then, $\mathcal{E}_{AF_1}^{adm} = \mathcal{E}_{AF_1}^{com} = \{\emptyset, \{a\}, \{b\}\}$, $\mathcal{E}_{AF_1}^{prf} = \{\{a\}, \{b\}\}$, $\mathcal{E}_{AF_1}^{stb} = \{\{b\}\}$, $\mathcal{E}_{AF_1}^{grd} = \{\emptyset\}$; and $\mathcal{E}_{AF_2}^{adm} = \{\emptyset, \{u\}, \{b\}, \{c\}, \{a, b\}, \{c, d\}, \{a, b, c\}\}$, $\mathcal{E}_{AF_2}^{com} = \{\{c\}, \{c, d\}, \{a, b, c\}\}$, $\mathcal{E}_{AF_2}^{prf} = \mathcal{E}_{AF_2}^{stb} = \{\{c, d\}, \{a, b, c\}\}$, $\mathcal{E}_{AF_2}^{grd} = \{\{c\}\}$. In this case, it holds that $AF_1 \sqsubseteq_\sigma^\natural AF_2$ for $\sigma \in \{adm, com, grd\}$; and $AF_1 \sqsubseteq_\sigma^\flat AF_2$ for $\sigma \in \{adm, com, prf, stb, grd\}$.

In what follows, some formal properties are addressed.

Proposition 2. *Let AF_1 and AF_2 be two argumentation frameworks. It holds that (i) $AF_1 \sqsubseteq_{adm}^\natural AF_2$, and (ii) $AF_1 \sqsubseteq_{grd}^\natural AF_2$ iff $AF_1 \sqsubseteq_{grd}^\flat AF_2$.*

Proof. For any AF, $\emptyset \in \mathcal{E}_{AF}^{adm}$, and $\mathcal{E}_{AF}^{grd}$ is a singleton set. Hence, the results hold. $\square$

Two relations $\preceq^\natural$ and $\preceq^\flat$ are monotonic with respect to the increase of extensions.

Proposition 3. *For any set Σ_1 and Σ_2 in $\mathcal{P}(\mathcal{P}(\mathcal{U}))$, $\Sigma_1 \subseteq \Sigma_2$ implies $\Sigma_1 \preceq^\flat \Sigma_2$ and $\Sigma_2 \preceq^\natural \Sigma_1$.*

Proof. If $\Sigma_1 \subseteq \Sigma_2$, then $\forall S \in \Sigma_1$, $S \in \Sigma_2$ thereby $\Sigma_1 \preceq^\flat \Sigma_2$ and $\Sigma_2 \preceq^\natural \Sigma_1$. $\square$

Proposition 4. *Let AF_1 and AF_2 be two argumentation frameworks. If $\mathcal{E}_{AF_1}^\sigma \subseteq \mathcal{E}_{AF_2}^\sigma$ then $AF_1 \sqsubseteq_\sigma^\flat AF_2$ and $AF_2 \sqsubseteq_\sigma^\natural AF_1$ hold for $\sigma \in \{adm, com, prf, stb, grd\}$.*

Proof. The result follows from Proposition 3. $\square$

Proposition 5. *Let AF_1 and AF_2 be two argumentation frameworks. Then the following results hold for $\sigma \in \{adm, com, prf, stb, grd\}$.*

(1) $AF_1 \equiv_\sigma^\sharp AF_2$ iff $min_\subseteq(\mathcal{E}_{AF_1}^\sigma) = min_\subseteq(\mathcal{E}_{AF_2}^\sigma)$.

(2) $AF_1 \equiv_\sigma^\flat AF_2$ iff $max_\subseteq(\mathcal{E}_{AF_1}^\sigma) = max_\subseteq(\mathcal{E}_{AF_2}^\sigma)$.

Proof. In what follows, $min_\subseteq$ is written as min. (1) If $AF_1 \sqsubseteq_\sigma^\sharp AF_2$, then $\forall S \in min(\mathcal{E}_{AF_2}^\sigma) \exists T \in \mathcal{E}_{AF_1}^\sigma$ s.t. $T \subseteq S$, and then $\exists U \in min(\mathcal{E}_{AF_1}^\sigma)$ s.t. $U \subseteq T$. Thus, $min(\mathcal{E}_{AF_1}^\sigma) \preceq^\sharp min(\mathcal{E}_{AF_2}^\sigma)$. Likewise, $AF_2 \sqsubseteq_\sigma^\sharp AF_1$ implies $min(\mathcal{E}_{AF_2}^\sigma) \preceq^\sharp min(\mathcal{E}_{AF_1}^\sigma)$. Assume $min(\mathcal{E}_{AF_1}^\sigma) \neq min(\mathcal{E}_{AF_2}^\sigma)$. Then, (i) $\exists U \in min(\mathcal{E}_{AF_1}^\sigma) \setminus min(\mathcal{E}_{AF_2}^\sigma)$ or (ii) $\exists V \in min(\mathcal{E}_{AF_2}^\sigma) \setminus min(\mathcal{E}_{AF_1}^\sigma)$. In case of (i), $\exists U' \in min(\mathcal{E}_{AF_2}^\sigma)$ s.t. $U' \subseteq U$ by $min(\mathcal{E}_{AF_2}^\sigma) \preceq^\sharp min(\mathcal{E}_{AF_1}^\sigma)$. Also, $\exists U'' \in min(\mathcal{E}_{AF_1}^\sigma)$ s.t. $U'' \subseteq U'$ by $min(\mathcal{E}_{AF_1}^\sigma) \preceq^\sharp min(\mathcal{E}_{AF_2}^\sigma)$. Thus, $U'' \subseteq U$. Since both U and U'' are in $min(\mathcal{E}_{AF_1}^\sigma)$, $U = U''$ thereby $U' = U$. This contradicts the assumption $U \notin min(\mathcal{E}_{AF_2}^\sigma)$. Similarly, (ii) also leads to contradiction. Hence, $min(\mathcal{E}_{AF_1}^\sigma) = min(\mathcal{E}_{AF_2}^\sigma)$. (2) is shown in a similar manner. $\square$

Proposition 6. *Let AF_1 and AF_2 be two argumentation frameworks. Then the following three are equivalent for $\sigma \in \{prf, stb, grd\}$: (1) $AF_1 \equiv_\sigma^\sharp AF_2$, (2) $AF_1 \equiv_\sigma^\flat AF_2$, (3) $AF_1 \equiv_\sigma AF_2$.*

Proof. Consider a poset $\langle \mathcal{P}(\mathcal{U}), \subseteq \rangle$. Since $\mathcal{E}_{AF}^\sigma$ is an antichain set for $\sigma \in \{prf, stb, grd\}$, $max_\subseteq(\mathcal{E}_{AF}^\sigma) = min_\subseteq(\mathcal{E}_{AF}^\sigma) = \mathcal{E}_{AF}^\sigma$. Hence, the result holds by Proposition 5. $\square$

Example 3. Consider $AF_1 = (\{p, q\}, \{(p, q), (q, p), (q, q)\})$ and $AF_2 = (\{p, q\}, \{(p, q), (q, p), (p, p)\})$ where $\mathcal{E}_{AF_1}^{com} = \{\emptyset, \{p\}\}$ and $\mathcal{E}_{AF_2}^{com} = \{\emptyset, \{q\}\}$. Then, $AF_1 \equiv_{com}^\sharp AF_2$ but $AF_1 \not\equiv_{com} AF_2$.

Two orderings are related to credulous/skeptical acceptance of arguments.

Proposition 7. *Let AF_1 and AF_2 be two argumentation frameworks. Then the following relations hold for $\sigma \in \{adm, com, prf, stb, grd\}$.*

1. If $AF_1 \sqsubseteq_\sigma^\flat AF_2$ then $crd^\sigma(AF_1) \subseteq crd^\sigma(AF_2)$.

2. If $AF_1 \sqsubseteq_\sigma^\sharp AF_2$ then $skp^\sigma(AF_1) \subseteq skp^\sigma(AF_2)$.

Proof. (1) Assume $AF_1 \sqsubseteq_\sigma^\flat AF_2$. If $\mathcal{E}_{AF_1}^\sigma = \emptyset$ then $crd^\sigma(AF_1) = \emptyset$ by definition, and the result holds immediately. Suppose that $\mathcal{E}_{AF_1}^\sigma \neq \emptyset$ and $\psi \in crd^\sigma(AF_1)$. Then $\psi \in E$ for some $E \in \mathcal{E}_{AF_1}^\sigma$. By $AF_1 \sqsubseteq_\sigma^\flat AF_2$, for any $E \in \mathcal{E}_{AF_1}^\sigma$ there is $F \in \mathcal{E}_{AF_2}^\sigma$ such that $E \subseteq F$. Then $\psi \in E$ implies $\psi \in F$, thereby $\psi \in crd^\sigma(AF_2)$. Hence, $crd^\sigma(AF_1) \subseteq crd^\sigma(AF_2)$.

(2) Assume $AF_1 \sqsubseteq_\sigma^\sharp AF_2$. If $\mathcal{E}_{AF_2}^\sigma = \emptyset$ then $skp^\sigma(AF_2) = \mathcal{U}$ by definition, and the result holds immediately. Suppose that $\mathcal{E}_{AF_2}^\sigma \neq \emptyset$. In this case, $\mathcal{E}_{AF_1}^\sigma \neq \emptyset$ by $AF_1 \sqsubseteq_\sigma^\sharp AF_2$. If $\psi \in skp^\sigma(AF_1)$ then $\psi \in E$ for every $E \in \mathcal{E}_{AF_1}^\sigma$. By $AF_1 \sqsubseteq_\sigma^\sharp AF_2$, for any $F \in \mathcal{E}_{AF_2}^\sigma$ there is $E \in \mathcal{E}_{AF_1}^\sigma$ such that $E \subseteq F$. Then $\psi \in E$ implies $\psi \in F$, thereby $\psi \in skp^\sigma(AF_2)$. Hence, $skp^\sigma(AF_1) \subseteq skp^\sigma(AF_2)$. $\square$

Example 4. Consider AFs in Example 2. By $AF_1 \sqsubseteq_{prf}^\flat AF_2$, $crd^{prf}(AF_1) = \{a, b\}$ is a subset of $crd^{prf}(AF_2) = \{a, b, c, d\}$. By $AF_1 \sqsubseteq_{com}^\sharp AF_2$, $skp^{com}(AF_1) = \emptyset$ is a subset of $skp^{com}(AF_2) = \{c\}$.

By Proposition 7, when $AF_1 \sqsubseteq_\sigma^b AF_2$, AF_2 has more (or equally) credulously accepted arguments than AF_1. In contrast, when $AF_1 \sqsubseteq_\sigma^\sharp AF_2$, AF_2 has more (or equally) skeptically accepted arguments than AF_1. As such, two orderings over AFs characterize the amount of acceptable arguments in two different modes of reasoning.

3.2 Comparing Different Semantics

In this section, we compare different semantics of a single AF under two orderings. By Proposition 3 and the relations $\mathcal{E}_{AF}^{stb} \subseteq \mathcal{E}_{AF}^{prf} \subseteq \mathcal{E}_{AF}^{com} \subseteq \mathcal{E}_{AF}^{adm}$ and $\mathcal{E}_{AF}^{grd} \subseteq \mathcal{E}_{AF}^{com}$, we have: $\mathcal{E}_{AF}^{stb} \preceq^b \mathcal{E}_{AF}^{prf} \preceq^b \mathcal{E}_{AF}^{com} \preceq^b \mathcal{E}_{AF}^{adm}$, $\mathcal{E}_{AF}^{grd} \preceq^b \mathcal{E}_{AF}^{com}$, $\mathcal{E}_{AF}^{adm} \preceq^\sharp \mathcal{E}_{AF}^{com} \preceq^\sharp \mathcal{E}_{AF}^{prf} \preceq^\sharp \mathcal{E}_{AF}^{stb}$, and $\mathcal{E}_{AF}^{com} \preceq^\sharp \mathcal{E}_{AF}^{grd}$. Moreover, we have the next results.

Proposition 8. *Let AF be an argumentation framework. Then, (1) $\mathcal{E}_{AF}^{grd} \preceq^\sharp \mathcal{E}_{AF}^\lambda$ for $\lambda \in \{com, prf, stb, grd\}$, and (2) $\mathcal{E}_{AF}^\sigma \preceq^b \mathcal{E}_{AF}^{com}$ for $\sigma \in \{adm, com, prf, stb, grd\}$.*

Proof. (1) Since a grounded extension is the least element of $\mathcal{E}_{AF}^{com}$, $\forall E \in \mathcal{E}_{AF}^\lambda$, $F \in \mathcal{E}_{AF}^{grd}$ and $F \sqsubseteq E$, thereby $\mathcal{E}_{AF}^{grd} \preceq^\sharp \mathcal{E}_{AF}^\lambda$. (2) The results $\mathcal{E}_{AF}^\sigma \preceq^b \mathcal{E}_{AF}^{com}$ for $\sigma \in \{com, prf, stb, grd\}$ is already known. If $E \in \mathcal{E}_{AF}^{adm}$ then $\exists F \in \mathcal{E}_{AF}^{com}$ such that $E \subseteq F$. Hence, $\mathcal{E}_{AF}^{adm} \preceq^b \mathcal{E}_{AF}^{com}$. □

The above results are combined with the ordering of different AFs. For instance, suppose that $AF_1 \sqsubseteq_{stb}^\sharp AF_2$ holds. By $\mathcal{E}_{AF_1}^\sigma \preceq^\sharp \mathcal{E}_{AF_1}^{stb}$ for $\sigma = \{adm, com, prf, stb, grd\}$, for any stable extension F of AF_1 there is a σ-extension E of AF_1 such that $E \subseteq F$. This means that if AF_2 employs the stable semantics, then AF_2 is more $\sharp$-general than AF_1 that employs any semantics. Suppose, on the other hand, that $AF_1 \sqsubseteq_\sigma^b AF_2$ holds. By $\mathcal{E}_{AF_2}^\sigma \preceq^b \mathcal{E}_{AF_2}^{com}$ (Proposition 8(2)), for any σ-extension E of AF_2 there is a complete extension F of AF_2 such that $E \subseteq F$. This means that if AF_2 employs the complete semantics, then AF_2 is more b-general than AF_1 that employs any semantics.

3.3 Minimal Upper and Maximal Lower Bounds

In this section, we consider a *minimal upper bound* and a *maximal lower bound* of the sets of extensions with respect to two orderings $\preceq^\sharp$ and $\preceq^b$.

Definition 3. (mub, mlb). *Let $\langle \mathcal{P}(\mathcal{P}(\mathcal{U})), \preceq^{\sharp/b} \rangle$ be a pre-ordered set. For any Σ_1 and Σ_2 in $\mathcal{P}(\mathcal{P}(\mathcal{U}))$, a set $\Sigma \in \mathcal{P}(\mathcal{P}(\mathcal{U}))$ is an upper bound of Σ_1 and Σ_2 if $\Sigma_1 \preceq^{\sharp/b} \Sigma$ and $\Sigma_2 \preceq^{\sharp/b} \Sigma$. An upper bound Σ is a minimal upper bound (mub) of Σ_1 and Σ_2 if for any upper bound Σ' of Σ_1 and Σ_2, $\Sigma' \preceq^{\sharp/b} \Sigma$ implies $\Sigma \preceq^{\sharp/b} \Sigma'$.*

On the other hand, a set $\Sigma \in \mathcal{P}(\mathcal{P}(\mathcal{U}))$ is a lower bound of Σ_1 and Σ_2 if $\Sigma \preceq^{\sharp/b} \Sigma_1$ and $\Sigma \preceq^{\sharp/b} \Sigma_2$. A lower bound Σ is a maximal lower bound (mlb) of Σ_1 and Σ_2 if for any lower bound Σ' of Σ_1 and Σ_2, $\Sigma \preceq^{\sharp/b} \Sigma'$ implies $\Sigma' \preceq^{\sharp/b} \Sigma$.

Proposition 9. *Let Σ_1 and Σ_2 be two antichain sets in $\langle \mathcal{P}(\mathcal{U}), \subseteq \rangle$.*

1. *$\Sigma \in \mathcal{P}(\mathcal{P}(\mathcal{U}))$ is an mub of Σ_1 and Σ_2 in $\langle \mathcal{P}(\mathcal{P}(\mathcal{U})), \preceq^\sharp \rangle$ iff $\Sigma = min_\subseteq(X)$ where $X = \{ S \cup T \mid S \in \Sigma_1 \text{ and } T \in \Sigma_2 \}$.*
2. *$\Sigma \in \mathcal{P}(\mathcal{P}(\mathcal{U}))$ is an mub of Σ_1 and Σ_2 in $\langle \mathcal{P}(\mathcal{P}(\mathcal{U})), \preceq^\flat \rangle$ iff $\Sigma = max_\subseteq(X)$ where $X = \{ S \cap T \mid S \in \Sigma_1 \text{ and } T \in \Sigma_2 \}$.*
3. *$\Sigma \in \mathcal{P}(\mathcal{P}(\mathcal{U}))$ is an mlb of Σ_1 and Σ_2 in $\langle \mathcal{P}(\mathcal{P}(\mathcal{U})), \preceq^\sharp \rangle$ iff $\Sigma = min_\subseteq(\Sigma_1 \cup \Sigma_2)$.*
4. *$\Sigma \in \mathcal{P}(\mathcal{P}(\mathcal{U}))$ is an mlb of Σ_1 and Σ_2 in $\langle \mathcal{P}(\mathcal{P}(\mathcal{U})), \preceq^\flat \rangle$ iff $\Sigma = max_\subseteq(\Sigma_1 \cup \Sigma_2)$.*

Proof. We show (1) and (3). The results of (2) and (4) are shown in similar ways.

(1) Σ is an upper bound of Σ_1 and Σ_2 in $\langle \mathcal{P}(\mathcal{P}(\mathcal{U})), \preceq^\sharp \rangle$ iff $\Sigma_1 \preceq^\sharp \Sigma$ and $\Sigma_2 \preceq^\sharp \Sigma$

iff $\forall S \in \Sigma \, \exists T_1 \in \Sigma_1$ s.t. $T_1 \subseteq S$ and $\forall S \in \Sigma \, \exists T_2 \in \Sigma_2$ s.t. $T_2 \subseteq S$

iff $\forall S \in \Sigma \, \exists T_1 \in \Sigma_1 \, \exists T_2 \in \Sigma_2$ s.t. $T_1 \cup T_2 \subseteq S$ (*).

Now suppose that Σ is given as $min_\subseteq(\{ S \cup T \mid S \in \Sigma_1 \text{ and } T \in \Sigma_2 \})$. Σ is an antichain set. Then Σ is an upper bound of Σ_1 and Σ_2 because (*) is satisfied. Assume that Σ is not an mub. Then there is an antichain set[2] $\Gamma \in \mathcal{P}(\mathcal{P}(\mathcal{U}))$ s.t. (i) Γ is an upper bound of Σ_1 and Σ_2, and (ii) $\Gamma \preceq^\sharp \Sigma$ and (iii) $\Sigma \npreceq^\sharp \Gamma$. Thus, $\Gamma \neq \Sigma$. For any $U \in \Sigma$, there are $S_1 \in \Sigma_1$ and $T_1 \in \Sigma_2$ s.t. $U = S_1 \cup T_1$ by the definition of Σ. For this U, there is a set $V \in \Gamma$ such that $V \subseteq U$ by (ii) and that $S_2 \cup T_2 \subseteq V$ for some $S_2 \in \Sigma_1$ and $T_2 \in \Sigma_2$ by (i) and (*). So $S_2 \cup T_2 \subseteq S_1 \cup T_1$. Since Σ is the collection of minimal sets, $S_2 \cup T_2 = S_1 \cup T_1$. Thus, $U = V$. Hence, $\Sigma \subseteq \Gamma$. By $\Sigma \neq \Gamma$, there is $W \in \Gamma \setminus \Sigma$. Again $S_3 \cup T_3 \subseteq W$ for some $S_3 \in \Sigma_1$ and $T_3 \in \Sigma_2$ by (i) and (*). However, there must be some $X \in \Sigma$ such that $X \subseteq W$ by the construction of Σ and the minimality of Σ. Because $W \notin \Sigma$, $X \subset W$ holds. However, by $\Gamma \preceq^\sharp \Sigma$ there is $Y \in \Gamma$ such that $Y \subseteq X$ and hence $Y \subset W$. This contradicts the fact that Γ is an antichain set.

(3) Σ is a lower bound of Σ_1 and Σ_2 in $\langle \mathcal{P}(\mathcal{P}(\mathcal{U})), \preceq^\sharp \rangle$ iff $\Sigma \preceq^\sharp \Sigma_1$ and $\Sigma \preceq^\sharp \Sigma_2$

iff $\forall S_1 \in \Sigma_1 \, \exists T \in \Sigma$ s.t. $T \subseteq S_1$ and $\forall S_2 \in \Sigma_2 \, \exists T \in \Sigma$ s.t. $T \subseteq S_2$

iff $\forall S \in \Sigma_1 \cup \Sigma_2 \, \exists T \in \Sigma$ s.t. $T \subseteq S$ (†).

Now suppose that $\Sigma = min_\subseteq(\Sigma_1 \cup \Sigma_2)$. Then Σ is a lower bound of Σ_1 and Σ_2 because (†) is satisfied. Assume that Σ is not an mlb. Then there is an antichain set $\Gamma \in \mathcal{P}(\mathcal{P}(\mathcal{U}))$ s.t. (i) Γ is a lower bound of Σ_1 and Σ_2, and (ii) $\Sigma \preceq^\sharp \Gamma$ and (iii) $\Gamma \npreceq^\sharp \Sigma$. Thus, $\Sigma \neq \Gamma$. By (ii), for any $V \in \Gamma$, there is $U \in \Sigma$ such that $U \subseteq V$. By this and the fact that Γ is a lower bound of Σ_1 and Σ_2, we have that $\forall W \in \Sigma_1 \cup \Sigma_2, \exists V \in \Gamma \, \exists U \in \Sigma$ such that $U \subseteq V \subseteq W$. As $U \in \Sigma_1 \cup \Sigma_2$, it must be $U = V$ by the minimality of Σ, and thus $\Gamma \subseteq \Sigma$. By $\Sigma \neq \Gamma$, there is $X \in \Sigma \setminus \Gamma$. Since $X \in \Sigma_1 \cup \Sigma_2$ by the construction of

---

[2] Without loss of generality, Γ is assumed to be an antichain set. If Γ is not an antichain set, there is $S, T \in \Gamma$ s.t. $S \subseteq T$. Put $\Gamma' = \Gamma \setminus \{T\}$. Then Γ' is an upper bound of Σ_1 and Σ_2 (because if Γ satisfies (*) then Γ' satisfies (*)) and also satisfies (ii) and (iii).

Σ, there must be some $Y \in \Gamma$ such that $Y \subseteq X$ by (†). As $X \notin \Gamma$, $Y \subset X$ holds. However, by (ii) there is $Z \in \Sigma$ such that $Z \subseteq Y$ and thus $Z \subset X$. This contradicts the fact that Σ is an antichain set. Therefore, Σ is a mlb of Σ_1 and Σ_2 in $\langle \mathcal{P}(\mathcal{P}(\mathcal{U})), \preceq^\sharp \rangle$.   □

Proposition 9 states that an mub or mlb of two antichain sets in $\langle \mathcal{P}(\mathcal{P}(\mathcal{U})), \preceq^{\sharp/\flat} \rangle$ is constructed by the operations min or max. Suppose two argumentation frameworks AF_1 and AF_2 having the sets of σ-extensions $\mathcal{E}^\sigma_{AF_1}$ and $\mathcal{E}^\sigma_{AF_2}$, respectively. Then, a question is whether there is $AF \in \mathcal{AF}$ such that $\mathcal{E}^\sigma_{AF}$ is obtained as an mub (or mlb) of $\mathcal{E}^\sigma_{AF_1}$ and $\mathcal{E}^\sigma_{AF_2}$. If $\sigma = grd$, there is an AF that has the extension obtained as the mub of Proposition 9(1) or (2). This is because if $\mathcal{E}^{grd}_{AF_1} = \{E\}$ and $\mathcal{E}^{grd}_{AF_2} = \{F\}$ then we can construct an AF s.t. $\mathcal{E}^{grd}_{AF} = \{E \cup F\}$ or $\mathcal{E}^{grd}_{AF} = \{E \cap F\}$ as $AF = (E \cup F, \emptyset)$ or $AF = (E \cap F, \emptyset)$. On the other hand, an AF having the grounded extension as the mlb of Proposition 9(3) or (4) does not always exist. This is because $min_\subseteq(\mathcal{E}^{grd}_{AF_1} \cup \mathcal{E}^{grd}_{AF_2})$ or $max_\subseteq(\mathcal{E}^{grd}_{AF_1} \cup \mathcal{E}^{grd}_{AF_2})$ is not a singleton set in general. When an AF has multiple extensions, the answer is also negative in general.

Example 5. Consider AF_1 and AF_2 such that $\mathcal{E}^{stb}_{AF_1} = \{\{a,b\},\{a,c\}\}$ and $\mathcal{E}^{stb}_{AF_2} = \{\{b,c\}\}$. Then, $min_\subseteq(\mathcal{E}^{stb}_{AF_1} \cup \mathcal{E}^{stb}_{AF_2}) = max_\subseteq(\mathcal{E}^{stb}_{AF_1} \cup \mathcal{E}^{stb}_{AF_2}) = \{\{a,b\},\{a,c\},\{b,c\}\}$, but there is no AF such that $\mathcal{E}^{stb}_{AF} = \{\{a,b\},\{a,c\},\{b,c\}\}$.

Any stable extension must be incomparable and *tight*, and the set $\{\{a,b\},\{a,c\},\{b,c\}\}$ does not satisfy this condition [2,4]. As such, the existence of an mub or mlb as a set of extensions as in Proposition 9 does *not* imply that it is *realizable* under a particular semantics [2,4], that is, it is not necessarily the case that there is an AF having the set of σ-extensions that coincide with an mub or mlb of two sets of extensions of two AFs. Investigating necessary and/or sufficient conditions for the existence of an mub/mlb of two AFs under σ-semantics is left for future study.

4   Strong Ordering

This section considers comparing two AFs under dynamic environments by observing the effect of incorporating new information into given argumentation frameworks. In this section we consider $AF = (A, R)$ where $A \subseteq \mathcal{U}$ and $R \subseteq \mathcal{U} \times \mathcal{U}$.[3] Given $AF_1 = (A_1, R_1)$ and $AF_2 = (A_2, R_2)$, define $AF_1 \sqcup AF_2 = (A_1 \cup A_2, R_1 \cup R_2)$.

Definition 4. Let AF_1 and AF_2 be two argumentation frameworks. Then,

$$AF_1 \trianglelefteq^\sharp_\sigma AF_2 \text{ iff } (AF_1 \sqcup AF) \sqsubseteq^\sharp_\sigma (AF_2 \sqcup AF) \text{ for any } AF \in \mathcal{AF},$$
$$AF_1 \trianglelefteq^\flat_\sigma AF_2 \text{ iff } (AF_1 \sqcup AF) \sqsubseteq^\flat_\sigma (AF_2 \sqcup AF) \text{ for any } AF \in \mathcal{AF}$$

where $\sigma \in \{adm, com, prf, stb, grd\}$.

---

[3] We relax the condition by technical reasons but it does not affect the results of previous sections. This is because attack relations in $(\mathcal{U} \times \mathcal{U}) \setminus (A \times A)$ do not change extensions of AF.

We write $\trianglelefteq_\sigma^{\sharp/\flat}$ to represent both $\trianglelefteq_\sigma^\sharp$ and $\trianglelefteq_\sigma^\flat$ together. The relation $AF_1 \trianglelefteq_\sigma^{\sharp/\flat}$ AF_2 implies $AF_1 \sqsubseteq_\sigma^{\sharp/\flat} AF_2$ by putting $AF = (\emptyset, \emptyset)$.

Proposition 10. *Let AF_1 and AF_2 be two argumentation frameworks. If $AF_1 \trianglelefteq_\sigma^{\sharp/\flat} AF_2$ then $AF_1 \sqsubseteq_\sigma^{\sharp/\flat} AF_2$ where $\sigma \in \{adm, com, prf, stb, grd\}$.*

By Proposition 2, the next result holds.

Proposition 11. *Let AF_1 and AF_2 be two argumentation frameworks. Then, (i) $AF_1 \trianglelefteq_{adm}^\sharp AF_2$, and (ii) $AF_1 \trianglelefteq_{grd}^\sharp AF_2$ iff $AF_1 \trianglelefteq_{grd}^\flat AF_2$.*

Two argumentation frameworks AF_1 and AF_2 are *strongly equivalent* (wrt σ semantics) if $AF_1 \sqcup AF \equiv_\sigma AF_2 \sqcup AF$ for any $AF \in \mathcal{AF}$ [10]. The notion of strong equivalence is related to the orderings $\trianglelefteq_\sigma^{\sharp/\flat}$ as follows.

Proposition 12. *Let AF_1 and AF_2 be two argumentation frameworks. Then the following three are equivalent for $\sigma \in \{prf, stb, grd\}$: (1) $AF_1 \trianglelefteq_\sigma^\sharp AF_2 \trianglelefteq_\sigma^\sharp AF_1$, (2) $AF_1 \trianglelefteq_\sigma^\flat AF_2 \trianglelefteq_\sigma^\flat AF_1$, (3) AF_1 and AF_2 are strongly equivalent.*

Proof. $AF_1 \trianglelefteq_\sigma^{\sharp/\flat} AF_2 \trianglelefteq_\sigma^{\sharp/\flat} AF_1$

iff $(AF_1 \sqcup AF) \sqsubseteq_\sigma^{\sharp/\flat} (AF_2 \sqcup AF) \sqsubseteq_\sigma^{\sharp/\flat} (AF_1 \sqcup AF)$ for any $AF \in \mathcal{AF}$

iff $(AF_1 \sqcup AF) \equiv_\sigma (AF_2 \sqcup AF)$ for any $AF \in \mathcal{AF}$ (Proposition 6)

iff AF_1 and AF_2 are strongly equivalent. $\qquad\qquad\square$

Example 6. ([10]) Two argumentation frameworks $AF_1 = (\{a, b, c\}, \{(a, b), (b, c), (c, a)\})$ and $AF_2 = (\{a, b, c\}, \{(a, c), (c, b), (b, a)\})$ have the same preferred extension $\emptyset$, but they are not strongly equivalent. This is explained by the fact that for $AF = (\{a, b\}, \{(a, b)\})$, $AF_1 \sqcup AF$ has the preferred extension $\emptyset$, while $AF_2 \sqcup AF$ has the preferred extension $\{a\}$, thereby $(AF_2 \sqcup AF) \not\sqsubseteq_{prf}^\sharp (AF_1 \sqcup AF)$.

Proposition 13. *Let AF_1 and AF_2 be two argumentation frameworks. Then the following results hold for $\sigma \in \{prf, stb, grd\}$.*

1. *If $AF_1 \trianglelefteq_\sigma^\flat AF_2$ then $\mathcal{E}_{AF_1}^\sigma \subseteq \mathcal{E}_{AF_2}^\sigma$.*
2. *If $AF_1 \trianglelefteq_\sigma^\sharp AF_2$ then $\mathcal{E}_{AF_2}^\sigma \subseteq \mathcal{E}_{AF_1}^\sigma$.*

Proof. (1) Let $AF_1 = (A_1, R_1)$ and $AF_2 = (A_2, R_2)$. If $AF_1 \trianglelefteq_\sigma^\flat AF_2$, then $AF_1 \sqsubseteq_\sigma^\flat$ AF_2 (Proposition 10). Assume $\mathcal{E}_{AF_1}^\sigma \not\subseteq \mathcal{E}_{AF_2}^\sigma$. Then there is an extension $E \in$ $\mathcal{E}_{AF_1}^\sigma \setminus \mathcal{E}_{AF_2}^\sigma$. By $AF_1 \sqsubseteq_\sigma^\flat AF_2$, there is $F \in \mathcal{E}_{AF_2}^\sigma$ such that $E \subset F$. For any F satisfying $E \subset F$, there is an argument $a \in F \setminus E$. Since F is conflict-free, $E \not\rightarrow a$. Suppose that $a \in A_1$. The fact $a \notin E$ implies $a \notin D(E)$. Then there is $(b, a) \in R_1$ s.t. $b \in A_1$ and $E \not\rightarrow b$. Since $(E \not\rightarrow a), b \notin E$ thereby $b \notin D(E)$. Then there is $(c, b) \in R_1$ s.t. $c \in A_1$, $c \neq a$ and $E \not\rightarrow c$. (If $c = a$ then $E' = E \cup \{a\}$ defends every element in E'. So $E' \in \mathcal{E}_{AF_1}^\sigma$ which contradicts the antichain property of $\mathcal{E}_{AF_1}^\sigma$.) Since $(E \not\rightarrow b), c \notin E$ thereby $c \notin D(E)$. Repeating the above argument, A_1 becomes an infinite set. This contradicts the assumption that A_1 is finite. Hence, there is an argument $a \in F \setminus E$ s.t. $a \notin A_1$. Consider $AF = (\{d\}, \{(a, d)\})$ where $d \notin A_1 \cup A_2$. Then $AF_1 \sqcup AF$ has an extension $E' = E \cup \{d\}$, while F is an extension of $AF_2 \sqcup AF$. So $E' \not\subseteq F$. Moreover, for any $G \in \mathcal{E}^\sigma_{AF_2}$ such that $E \not\subseteq G$, $E' = E \cup \{d\} \not\subseteq G$. Thus, for any extension G' of $AF_2 \sqcup AF$, $E' \not\subseteq G'$. Hence, $(AF_1 \sqcup AF) \not\sqsubseteq^b_\sigma (AF_2 \sqcup AF)$, thereby $AF_1 \not\trianglelefteq^b_\sigma AF_2$. Contradiction. (2) is shown in a similar manner. $\qquad\square$

Proposition 13 shows that $\trianglelefteq^{\sharp/b}_\sigma$ provides a sufficient condition for inclusion between the sets of extensions, while $\sqsubseteq^{\sharp/b}_\sigma$ provides a necessary condition for it (Proposition 4).

Proposition 14. *Let AF_1 and AF_2 be two argumentation frameworks. Then the following three are equivalent for $\sigma \in \{\,prf, stb, grd\,\}$: (1) $AF_1 \trianglelefteq^b_\sigma AF_2$, (2) $AF_2 \trianglelefteq^\sharp_\sigma AF_1$, (3) $\mathcal{E}^\sigma_{AF_1 \sqcup AF} \subseteq \mathcal{E}^\sigma_{AF_2 \sqcup AF}$ for any $AF \in \mathcal{AF}$.*

Proof. We show (1)$\Leftrightarrow$(3). The relation (2)$\Leftrightarrow$(3) is shown in a similar way. Suppose $AF_1 \trianglelefteq^b_\sigma AF_2$. By definition, $(AF_1 \sqcup AF) \sqsubseteq^b_\sigma (AF_2 \sqcup AF)$ for any $AF \in \mathcal{AF}$. Then $(AF_1 \sqcup AF) \sqcup AF' \sqsubseteq^b_\sigma (AF_2 \sqcup AF) \sqcup AF'$ for any AF and AF' in $\mathcal{AF}$. So $AF_1 \sqcup AF \trianglelefteq^b_\sigma AF_2 \sqcup AF$ for any $AF \in \mathcal{AF}$. By Proposition 13(1), $\mathcal{E}^\sigma_{AF_1 \sqcup AF} \subseteq \mathcal{E}^\sigma_{AF_2 \sqcup AF}$. Conversely, suppose $\mathcal{E}^\sigma_{AF_1 \sqcup AF} \subseteq \mathcal{E}^\sigma_{AF_2 \sqcup AF}$ for any $AF \in \mathcal{AF}$. By Proposition 4, $AF_1 \sqcup AF \sqsubseteq^b_\sigma AF_2 \sqcup AF$ for any $AF \in \mathcal{AF}$. Hence, $AF_1 \trianglelefteq^b_\sigma AF_2$. $\qquad\square$

As such, two relations $\trianglelefteq^b_\sigma$ and $\trianglelefteq^\sharp_\sigma$ are symmetric for $\sigma \in \{\,prf, stb, grd\,\}$.

5 Concluding Remarks

We introduced several orderings for comparing sets of extensions in argumentation frameworks. We showed that two orderings $\sqsubseteq^\sharp_\sigma$ and $\sqsubseteq^b_\sigma$ are used for comparing skeptical/credulous acceptance of arguments in different argumentation frameworks. Moreover, those relations have connections to inclusion/equivalence relations between sets of extensions. Since argumentation theories are nonmonotonic, some formal properties addressed in this paper have their counterpart in [7–9]. On the other hand, we show that those orderings are used for comparing different semantics of argumentation, which is not considered in the context of default theories or logic programming. The existence of an AF that has a set of extensions as an mub or mlb of given two sets of extensions is not always guaranteed, which is in contrast with the cases of default theories and logic programming where the existence of an mub or mlb is guaranteed. We considered five semantics of AFs in this paper, but the most results obtained in this paper are independent of particular semantics and applied to other semantics as well.

References

1. Baroni, P., Giacomin, M.: Skepticism relations for comparing argumentation semantics. J. Approximate Reasoning **50**(6), 854–866 (2009)

2. Baumann, R.: On the nature of argumentation semantics: existence and uniqueness, expressibility, and replaceability. In: Baroni, P., Gabbay, D., Giacomin, M, van der Torre, L. (eds.) Handbook of Formal Argumentation, chapter 17, pp. 839–936. College Publication (2018)

3. Dung, P.M.: On the acceptability of arguments and its fundamental role in nonmonotonic reasoning, logic programming and n-person games. Artif. Intell. **77**, 321–357 (1995)

4. Dunne, P.E., Dvořák, W., Linsbichlerc, T., Woltran, S.: Characteristics of multiple viewpoints in abstract argumentation. Artif. Intell. **228**, 153–178 (2015)

5. Gelfond, M., Lifschitz, V.: Classical negation in logic programs and disjunctive databases. New Gener. Comput. **9**, 365–385 (1991)

6. Gunter, C. A., Scott, D. S.: Semantic domains. In: van Leeuwen, J. (ed.) Handbook of Theoretical Computer Science, vol. B, North-Holland, pp. 633–674 (1990)

7. Inoue, K., Sakama, C.: Generality relations in answer set programming. In: Etalle, S., Truszczyński, M. (eds.) ICLP 2006. LNCS, vol. 4079, pp. 211–225. Springer, Heidelberg (2006). https://doi.org/10.1007/11799573_17

8. Inoue, K., Sakama, C.: Generality and equivalence relations in default logic. In: Proceedings AAAI-07, pp. 434–439 (2007)

9. Inoue, K., Sakama, C.: Exploring relations between answer set programs. In: Balduccini, M., Son, T.C. (eds.) Logic Programming, Knowledge Representation, and Nonmonotonic Reasoning. LNCS (LNAI), vol. 6565, pp. 91–110. Springer, Heidelberg (2011). https://doi.org/10.1007/978-3-642-20832-4_7

10. Oikarinen, E., Woltran, S.: Characterizing strong equivalence for argumentation frameworks. Artif. Intell. **175**, 1985–2009 (2011)

Constructing Bayesian Network Graphs from Labeled Arguments

Remi Wieten[1(✉)], Floris Bex[1,2], Henry Prakken[1,3], and Silja Renooij[1]

[1] Information and Computing Sciences, Utrecht University, Utrecht, The Netherlands
{g.m.wieten,f.j.bex,h.prakken,s.renooij}@uu.nl
[2] Institute for Law, Technology and Society, Tilburg University,
Tilburg, The Netherlands
[3] Faculty of Law, University of Groningen, Groningen, The Netherlands

Abstract. Bayesian networks (BNs) are powerful tools that are well-suited for reasoning about the uncertain consequences that can be inferred from evidence. Domain experts, however, typically do not have the expertise to construct BNs and instead resort to using other tools such as argument diagrams and mind maps. Recently, we proposed a structured approach to construct a BN graph from arguments annotated with causality information. As argumentative inferences may not be causal, we generalize this approach to include other types of inferences in this paper. Moreover, we prove a number of formal properties of the generalized approach and identify assumptions under which the construction of an initial BN graph can be fully automated.

Keywords: Bayesian networks · Argumentation · Inference · Reasoning

1 Introduction

Bayesian networks (BNs) [11] are compact graphical models of joint probability distributions that have found applications in many different fields where uncertainty plays a role, including medicine, forensics and law [6]. BNs are well-suited for reasoning about the uncertain consequences that can be inferred from evidence. However, especially in data-poor domains, their construction needs to be done mostly manually, which is a difficult, time-consuming and error-prone process [7], and domain experts typically resort to using other tools such as argument diagrams, mind maps and ontologies [4,8]. Hence, we believe BN construction can be facilitated by automatically extracting information relevant for a BN from such tools. More specifically, in this paper we study how information expressed as structured arguments [2] about the domain can inform the design of a BN *graph*, a directed acyclic graph (*DAG*) which captures the independence relation among variables.

In previous research, Bex and Renooij [3] identified constraints on BNs given structured arguments, but these only suffice for constructing an undirected skeleton of a BN graph. Recently, we were able to derive a directed graph [18], but only

© Springer Nature Switzerland AG 2019
G. Kern-Isberner and Z. Ognjanović (Eds.): ECSQARU 2019, LNAI 11726, pp. 99–110, 2019.
https://doi.org/10.1007/978-3-030-29765-7_9

by assuming that all inferences in the initial structured arguments are explicitly labeled with causality information [1,14]. Arcs in the BN graph are then set in the causal direction, following the heuristic typically used in the manual construction of BN graphs [11]. However, in [18] it is assumed that *all* inferences are labeled with causality information, which precludes the use of other types of inferences, such as mere statistical correlations and definitions. Furthermore, formal properties of the proposed proposals were not studied in [3,18].

Accordingly, in this paper we present an approach that generalizes our previously proposed construction approach [18] to other types of inference. In addition, we formally prove that BN graphs constructed by our approach allow reasoning patterns similar to the inferences represented in the original structured arguments. Moreover, we identify assumptions under which the fully automatically constructed initial graph is guaranteed to be a DAG, and we identify bounds on the complexity of inference in BNs constructed by our approach.

The paper is structured as follows. Section 2 provides preliminaries on argumentation and BNs. In Sect. 3, we present our generalized approach for constructing BN graphs from inferences. In Sect. 4, we prove a number of formal properties of the approach. In Sect. 5, we discuss related research and conclude.

2   Preliminaries

2.1   Argumentation

Throughout this paper, we assume that the domain experts' analysis is captured in an *argument graph* (AG), in which claims are substantiated by chaining inferences from the observed evidence; an example is depicted in Fig. 1a. AGs are closely related to argument diagrams and mind maps [4], familiar to many domain experts. Formally, an AG is a directed graph $G_A = (\mathbf{P}, \mathbf{A}_A)$, where $\mathbf{P}$ is a set of nodes representing propositions from a literal language with ordinary negation symbol $\neg$, and $\mathbf{A}_A$ is a set of directed (hyper)arcs. We write $p = -q$ in case $p = \neg q$ or $q = \neg p$. Nodes $\mathbf{E_p} \subseteq \mathbf{P}$ corresponding to the (observed) *evidence* are root nodes in G_A. We assume that for every $p \in \mathbf{E_p}$ it holds that $\neg p \notin \mathbf{P}$. $\mathbf{A}_A$ is comprised of three pairwise disjoint sets $\mathbf{S}$, $\mathbf{R}$ and $\mathbf{U}$, which are sets of support arcs, rebuttal arcs and undercutter arcs, respectively. A *support arc* is a (hyper)arc $s: \{p_1, \ldots, p_n\} \to p \in \mathbf{S}$, indicating an inference step from $\{p_1, \ldots, p_n\} \subseteq \mathbf{P}$ (called the *tails* of s, denoted by $\mathbf{Tails}(s)$) to a single proposition $p \in \mathbf{P}$ (called the *head* of s, denoted by $head(s)$). Here, curly brackets are omitted in case $|\mathbf{Tails}(s)| = 1$. Support arcs $s_1, \ldots, s_m$ form a *support chain* $(s_1, \ldots, s_m)$ iff $head(s_i) \in \mathbf{Tails}(s_{i+1})$ for $1 \le i < m$.

There are two types of attack arcs. A *rebuttal arc* $r \in \mathbf{R}$ is a bidirectional arc $r: p \longleftrightarrow \neg p$ in G_A that exists for every pair $p, \neg p \in \mathbf{P}$. An *undercutter arc* $u \in \mathbf{U}$ is a hyperarc $u: p \to (s)$, where $p \in \mathbf{P}$ undercuts $s \in \mathbf{S}$. Informally, a rebuttal is an attack on a proposition, while an undercutter attacks an inference by providing exceptional circumstances under which the inference may not be applicable. In figures in this paper, nodes in G_A corresponding to elements of

$\mathbf{E_p}$ are shaded. Support arcs are denoted by solid (hyper)arcs and rebuttal arcs and undercutter arcs are denoted by dashed (hyper)arcs.

In reasoning about evidence, a distinction can be made between causal and evidential inferences [1,14]. Causal inferences are of the form "*c is a cause for e*" (e.g. fire causes smoke), whereas evidential inferences are of the form "*e is caused by c*" (e.g. smoke is caused by fire). Inferences may also be neither causal nor evidential. For instance, definitions, or abstractions [5], allow for reasoning at different levels of abstraction, such as stating that guns can generally be considered deadly weapons. Another example of a different type of inference is an inference representing a mere statistical correlation, such as a correlation between homelessness and criminality. While there may be one or more confounding factors that cause both homelessness and criminality (e.g. unemployment), a domain expert may be unaware of these factors or may wish to refrain from capturing them in the AG. For our current purposes, we assume that support arcs in $\mathbf{S}$ are either annotated with a causal "c" label, an evidential "e" label, or are labeled "o" for all other types of inferences. $\mathbf{S}$ then divides into three disjoint sets $\mathbf{S^c}$, $\mathbf{S^e}$ and $\mathbf{S^o}$ of causal, evidential and other types of support arcs, respectively. In figures in this paper, "o" labels are omitted.

In this paper, some further assumptions are made. We assume that support chains are *non-repetitive* in that there does not exist a support chain $(s, \ldots, s)$ in AG. We assume that for every support chain $(s_1, \ldots, s_n)$ the heads of $s_1, \ldots, s_n$ are *consistent* in that $\nexists i, j \in \{1, \ldots, n\}$, $i \neq j$ such that $head(s_i) = -head(s_j)$. Furthermore, we assume that AGs do not include *causal cycles* in that there do not exist two support chains $(s_1, \ldots, s_n)$ and $(s'_1, \ldots, s'_m)$ in AG with $s_1, \ldots, s_n \in \mathbf{S^c}$, $s'_1, \ldots, s'_m \in \mathbf{S^e}$, $\mathbf{Tails}(s_1) \cap \mathbf{Tails}(s'_1) \neq \emptyset$ and $head(s_n) = head(s'_m)$ or $head(s_n) = -head(s'_m)$. Informally, this assumption says that for every $p, q \in \mathbf{P}$, if p is a cause of q, then q (or $-q$) cannot be a cause of p (see also [1]).

As noted by Pearl [14], the chaining of a causal inference and an evidential inference can lead to undesirable results. Consider the example in which a causal inference states that a smoke machine causes smoke and an evidential inference states that smoke is evidence for fire. Chaining these inferences would make us conclude there is a fire when seeing a smoke machine, which is clearly undesirable. We therefore assume that an AG does not include a support chain (s_1, s_2) where $s_1 \in \mathbf{S^c}$, $s_2 \in \mathbf{S^e}$, and refer to this assumption as *Pearl's C-E constraint*.

For those familiar with argumentation, we note that, although we use the term "argument graph", the graph only represents inferences and attacks between propositions by means of arcs; actual arguments are not represented in the graph. Preferences over arguments, as well as their status, are thus not taken into account in our formalism, since they are not needed for our current purposes. Our formalism can be straightforwardly mapped to ASPIC$^+$ (cf. [2]) if all inferences are considered to be defeasible.

2.2   Bayesian Networks

A BN [11] compactly represents a joint probability distribution $\Pr(\mathbf{V})$ over a finite set of discrete random variables $\mathbf{V}$; in this paper we assume all variables to be Boolean. The variables are represented as nodes in a DAG $G_\mathcal{B} = (\mathbf{V}, \mathbf{A}_\mathcal{B})$, where $\mathbf{A}_\mathcal{B} \subseteq \mathbf{V} \times \mathbf{V}$ is a set of directed arcs $V_i \rightarrow V_j$ from parent V_i to child V_j. The BN further includes, for each node, a conditional probability table (CPT) specifying the probabilities of the values of the node conditioned on the possible joint value combinations of its parents. A node is called instantiated iff it is set to a specific value. Given a set of instantiations, or evidence, for nodes $\mathbf{E_V} \subseteq \mathbf{V}$, the probability distributions over the other nodes in the network can be updated through *probabilistic inference* [11]. An example of a BN graph is depicted in Fig. 1b, where ovals represent nodes and instantiated nodes are shaded.

The BN graph $G_\mathcal{B}$ captures the independence relation among its variables. Let a *chain* be defined as a sequence of distinct nodes and arcs in the BN graph. A node V is called a *head-to-head node* on a chain c if it has two incoming arcs on c. A chain c between nodes V_1 and V_2 is *blocked* iff it includes a node $V \notin \{V_1, V_2\}$ such that (1) V is an uninstantiated head-to-head node on c without instantiated descendants; or (2) V is instantiated and has at most one incoming arc on c. A chain that is not blocked is called *active*. If no active chains exist between V_1 and V_2 given instantiations of $\mathbf{Z} \subseteq \mathbf{V}$, then they are considered conditionally independent given $\mathbf{Z}$. In case a head-to-head node or one of its descendants is instantiated, an active chain is *induced* between its parents, allowing for interparental interactions. If one of the parents is now true, then the probability of another parent being true as well may change, depending on the specific synergistic effect modeled in the CPT for the head-to-head node.

BN construction is typically an iterative process. After constructing an initial BN graph, we should verify that this graph is acyclic and that it correctly captures the (conditional) independencies. If the graph does not yet exhibit these properties, arcs should be reversed, added or removed by the BN modeler in consultation with the domain expert. We call this the *"graph validation step"*.

3   Constructing BN Graphs from Argument Graphs

To facilitate the BN construction process, we previously proposed a stepwise approach for constructing an initial BN graph from domain knowledge represented in AGs with support arcs in $\mathbf{S}^c \cup \mathbf{S}^e$ only [18]. In this section, we generalize this approach to include inferences in $\mathbf{S}^o$.

Upon using an AG to inform BN construction, we have to consider their difference in semantics. An AG, by means of its support chains, describes the iterative inference steps that can be made from the observed evidence towards the conclusions. In comparison, a BN describes a joint probability distribution which does not model such directionality. Only when probabilistic inference is performed is available evidence propagated through the network using the existing active chains. To mimic the inferences described by an AG in a BN, we will focus on ensuring that the (chains of) support arcs in the AG, originating from evidence $\mathbf{E_p} \subseteq \mathbf{P}$, are captured in the BN graph by means of active chains for propagating instantiations of $\mathbf{E_V} \subseteq \mathbf{V}$ (see also [18]). Note that since the notion of an active chain is a symmetrical concept, a BN graph will also capture reasoning patterns in the direction opposite of the support chains present in the

AG. In Sect. 4, we formally prove that all support chains in an AG indeed have corresponding active chains in the BN when following our generalized approach.

In the manual construction of BN graphs, arcs are typically directed using the *notion of causality* as a guiding principle [11]. By following this heuristic, two competing causes form a head-to-head connection in the node corresponding to the common effect, allowing synergistic effects between the causes to be directly captured in the CPT for this node. Hence, we propose to use the same heuristic in automatically directing arcs, where we exploit causality information explicitly expressed in an AG by means of "c" and "e" labels.

Undercutters attack inferences in support chains by providing exceptions to the inference. For instance, if an inference is in the evidential direction, then an undercutter suggests an alternative cause for the same effect. Accordingly, we propose to enable capturing such interactions between an undercutter and a support arc in the CPT of a head-to-head node formed in the BN graph.

3.1   The Generalized Approach

In this subsection, we present and explain the steps of the generalized approach. Let $var \colon \mathbf{P} \to \mathbf{V}$ be an operator mapping every proposition p or $\neg p \in \mathbf{P}$ in an AG to a BN variable $var(p) = var(\neg p) \in \mathbf{V}$ describing values p and $\neg p$. For an AG $G_\mathcal{A} = (\mathbf{P}, \mathbf{A}_\mathcal{A})$, a BN graph $G_\mathcal{B} = (\mathbf{V}, \mathbf{A}_\mathcal{B})$ is constructed as follows:

(1) $\forall p, \neg p \in \mathbf{P}$, include $var(p)$ in $\mathbf{V}$; if p or $\neg p \in \mathbf{E_p}$, also include $var(p)$ in $\mathbf{E_V}$.
(2) For every support arc $s \colon \{p_1, \ldots, p_n\} \to p$:
  (2a) If $s \in \mathbf{S^e}$, include $var(p) \to var(p_i)$, $i = 1, \ldots, n$ in $\mathbf{A}_\mathcal{B}$.
  (2b) If $s \in \mathbf{S^c}$, include $var(p_i) \to var(p)$, $i = 1, \ldots, n$ in $\mathbf{A}_\mathcal{B}$.
  (2c) If $s \subset \mathbf{S^o}$ and $\nexists s_1 \in \mathbf{S^e}$ such that (s, s_1) form a support chain, include $var(p_i) \to var(p)$, $i = 1, \ldots, n$ in $\mathbf{A}_\mathcal{B}$.
  (2d) If $s \in \mathbf{S^o}$ and $\exists s_1, \ldots, s_m \in \mathbf{S^e}$ such that $(s_1, \ldots, s_m)$ is a maximal chain of evidential support arcs in AG following s, include $var(p_i) \to var(head(s_m))$, $i = 1, \ldots, n$ in $\mathbf{A}_\mathcal{B}$.
(3) For every undercutter arc $u \colon p \to (s) \in \mathbf{U}$ with $s \colon \{q_1, \ldots, q_n\} \to q$:
  (3a) If $s \in \mathbf{S^e}$, include $var(p) \to var(q_i)$, $i = 1, \ldots, n$ in $\mathbf{A}_\mathcal{B}$.
  (3b) If $s \notin \mathbf{S^e}$, include $var(p) \to var(q)$ in $\mathbf{A}_\mathcal{B}$.
(4) Verify the properties of the constructed graph $G_\mathcal{B}$:
  (4a) Break cycles in $G_\mathcal{B}$ introduced by so-called *evidential shortcuts* resulting from the combination of steps 2a and 2d (see Sect. 3.3 for further details).
  (4b) Apply the standard graph validation step (see Sect. 2.2).

While our approach exploits the domain knowledge captured in the AG in constructing a BN graph, the AG may lack information needed to prevent cycles and unwarranted (in)dependencies in the obtained BN graph; hence the manual validation step (step 4b above), which is standard in BN construction.

The first step is to capture every proposition in $G_\mathcal{A}$ and its negation as two values of a random variable in $G_\mathcal{B}$. By the same step, two propositions involved in a rebuttal are captured as two mutually exclusive values of the same node.

The steps pertaining to $s \in \mathbf{S^c} \cup \mathbf{S^e}$ are analogous to those proposed previously in [18]. These steps formalize the approach of setting arcs using the notion of causality as a guiding principle [11]. For further details, the reader is referred to [18]. In Sects. 3.2 and 3.3, we motivate and explain the steps pertaining to $s \in \mathbf{S^o}$ with several examples.

3.2 Explanation and Motivation of Steps 2c and 3b

Consider Fig. 1, illustrating steps 1−2c and 3b of the generalized approach for a forensic example. A dead body was found and we are interested in the cause of death of this person. According to witness testimony (*tes1*), the person was hit with a hammer (*hammer*); however, according to another testimony (*tes2*), the person was hit with a stone (*stone*). We conclude that the person was hit with an angular object (*angular*), as hammers and stones can generally be considered to be angular. Note that the relation between *hammer* (*stone*) and *angular* is neither causal nor evidential; instead, the support arcs between these propositions express that, at a higher level of abstraction, both hammers and stones can generally be considered angular objects. A mallet was found at the crime scene (*mallet*), which undercuts inference *hammer* → *angular* since a mallet is an exceptional type of hammer that is not angular but instead has a large cylindrical head. Finally, an autopsy report (*autopsy*) further supports the claim that the person was hit with an angular object. By following steps 1−3 of the generalized approach, the BN graph of Fig. 1b is constructed from the AG in Fig. 1a. By steps 2c and 3b, variables Hammer and Stone and variables Mallet and Hammer respectively form head-to-head connections in Angular.

In general, by step 2c head-to-head nodes are formed in the nodes corresponding to the heads of support arcs in $\mathbf{S^o}$. Specifically, let $p_1, \ldots, p_n$ be tails of one or more $s_i \in \mathbf{S^o}$ with $head(s_i) = p$. Then $\mathbf{A}_\mathcal{B}$ includes arcs $var(p_j) \rightarrow var(p)$, $j = 1, \ldots, n$ by step 2c; head-to-head nodes are, therefore, formed in $var(p)$. By setting arcs as per step 2c, we thus allow for including synergistic effects, if any, of the tails on the probability of p in the CPT for the head-to-head node.

Similarly, by step 3b head-to-head nodes are formed in the nodes corresponding to the heads of undercut support arcs in $\mathbf{S^c} \cup \mathbf{S^o}$. Specifically, let $u: p \rightarrow (s) \in \mathbf{U}$ be an undercutter of $s: \{q_1, \ldots, q_n\} \rightarrow q \in \mathbf{S^c} \cup \mathbf{S^o}$. Then by step

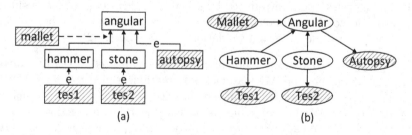

(a)   (b)

Fig. 1. An AG including support arcs in $\mathbf{S^o}$ (a); the corresponding BN graph constructed by steps 1−2c and 3b of the generalized approach (b).

$3b$, head-to-head nodes are formed in $var(q)$ as $\mathbf{A}_\mathcal{B}$ includes arc $var(p) \rightarrow var(q)$. Again, this allows for modeling possible interactions between p and q_i, and hence between $var(p)$ and $var(q_i)$, directly in the CPT for $var(q)$. Bex and Renooij [3] previously noted that the presence of an undercutter should decrease the probability that the conclusion of the undercut inference is true. By setting arcs as per step $3b$, this interaction can be directly captured by the following constraints on the CPT for $var(q)$: $\Pr(q \mid p, q_i) < \Pr(q \mid \neg p, q_i)$ for $i = 1, \ldots, n$.

3.3   Explanation and Motivation of Steps $2d$ and $4a$

Next, consider Figs. 2a and b, illustrating step $2d$ of the generalized approach for a medical example (taken from [7]). After performing a CT scan (*scan*) on a patient who has severe difficulty swallowing, it is established that a tumor is present in the lower (*distal*) part of his esophagus. Clinical studies indicate a strong correlation between the location of an esophageal tumor and its cell type; however, neither can be considered a cause of the other. Distal tumors generally consist of cylindrical cells (*cylindrical*), often formed as a result of frequent gastric reflux (*reflux*). The BN graph constructed by steps $1-2a$ and $2d$ of the generalized approach from the AG in Fig. 2a is depicted in Fig. 2b. As arcs Distal $\rightarrow$ Reflux and Reflux $\rightarrow$ Cylindrical are included in $\mathbf{A}_\mathcal{B}$ and the involved nodes are not instantiated, active chains exist between Distal and Reflux and Distal and Cylindrical. Note that we do not wish to set arcs as per step $2c$, as in this case a head-to-head node would instead be formed in Cylindrical which would block the chain between Distal and Reflux.

Under specific conditions, cycles are introduced in step $2d$ of the generalized approach, namely when a so-called *evidential shortcut* exists in the AG, i.e. if in addition to the conditions of step $2d$, also $\exists s'_1, \ldots, s'_k \in \mathbf{S^e}$ such that $(s'_1, \ldots, s'_k)$ form a support chain, $\mathbf{Tails}(s) \cap \mathbf{Tails}(s'_1) \neq \emptyset$ and $head(s'_k) = head(s_j)$ or $head(s'_k) = -head(s_j)$ for a $j \in \{1, \ldots, m\}$. An example is depicted in Fig. 2c. In this example, $s: p \rightarrow q_1 \in \mathbf{S^o}$ is followed by a chain of support arcs $s_1: q_1 \rightarrow r, s_2: r \rightarrow s \in \mathbf{S^e}$, where there also exists a chain of support arcs $s'_1: p \rightarrow$

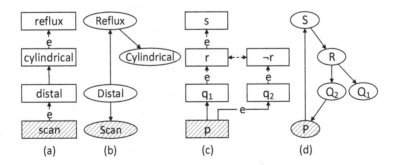

(a)   (b)   (c)   (d)

Fig. 2. An AG (a) and the corresponding BN graph (b) illustrating step $2d$ of the generalized approach; an AG (c) and the corresponding BN graph (d), illustrating the conditions under which a cycle is introduced in step $2d$.

q_2, s_2': $q_2 \rightarrow \neg r \in \mathbf{S^e}$. By step 2a, arcs $S \rightarrow R, R \rightarrow Q_2$ and $Q_2 \rightarrow P$ are included in $\mathbf{A}_{\mathcal{B}}$. By step 2d, arc $P \rightarrow S$ is also included, introducing a cycle in $G_{\mathcal{B}}$. We note that this arc can safely be removed, as an active chain already exists between P and S via Q_2 and R. In general, cycles are broken in step 4a by removing arcs $var(p_l) \rightarrow var(head(s_m))$ from $\mathbf{A}_{\mathcal{B}}$ $\forall p_l \in \mathbf{Tails}(s) \cap \mathbf{Tails}(s_1')$.

4 Properties of the Generalized Approach

In this section, we prove a number of formal properties of the generalized approach. The first property states that for every support chain in a given AG there indeed exists a corresponding active chain in the BN graph.

Proposition 1. *Let $G_{\mathcal{A}} = (\mathbf{P}, \mathbf{A}_{\mathcal{A}})$ be an AG with root nodes $\mathbf{E_p}$, and let $G_{\mathcal{B}} = (\mathbf{V}, \mathbf{A}_{\mathcal{B}})$ be the corresponding BN graph constructed according to steps 1–4a of the generalized approach. Let $(s_1, \ldots, s_n)$ be any support chain in $G_{\mathcal{A}}$, where $\mathbf{Tails}(s_1) = \{p_1, \ldots, p_m\}$ and $head(s_n) = q$. Then there exist active chains between $var(p_i)$ and $var(q)$ in $G_{\mathcal{B}}$ given $\mathbf{E_V}$ for every $i \in \{1, \ldots, m\}$.*

Proof (Sketch). The following cases are distinguished:

- If $s_k \in \mathbf{S^c} \cup \mathbf{S^e}$ $\forall k \in \{1, \ldots, n\}$, then when following steps 2a and 2b a head-to-head node can only be formed in $var(head(s_j))$ for an arbitrary s_j, $j \in \{1, \ldots, n-1\}$ if $s_j \in \mathbf{S^c}$, $s_{j+1} \in \mathbf{S^e}$; however, this construction is prohibited as it violates Pearl's C-E constraint (see Sect. 2.1). Furthermore, since heads of support arcs are not propositions in $\mathbf{E_p}$, corresponding nodes in $G_{\mathcal{B}}$ are not instantiated. Chains between $var(p_i)$ and $var(q)$ are thus never blocked.
- If $(s_1, \ldots, s_n)$ includes support arcs in $\mathbf{S^o}$ and none of these arcs is followed by an $s \in \mathbf{S^e}$, then arcs in $\mathbf{A}_{\mathcal{B}}$ are set similarly as for $s \in \mathbf{S^c}$ by step 2c. As per the above proof, chains are not blocked.
- Let an $s_j \in \mathbf{S^o}$, $1 \leq j < n$ be followed by a chain of support arcs in $\mathbf{S^e}$, and let $(s_{j+1}, \ldots, s_{j+l})$ be a maximal such chain. If $j + l \leq n$, then step 2d introduces direct arcs, and therefore active chains, between nodes in $\{var(p) \mid p \in \mathbf{Tails}(s_j)\}$ and $var(head(s_{j+l}))$. If $j + l > n$, then $\mathbf{A}_{\mathcal{B}}$ in addition includes a directed path from $var(head(s_{j+l}))$ to $var(head(s_n))$ by step 2a; therefore, chains between nodes in $\{var(p) \mid p \in \mathbf{Tails}(s_j)\}$ and $var(head(s_n))$ via $var(head(s_{j+l}))$ are active, as $var(head(s_{j+l}))$ is not a head-to-head node. In step 4a, a subset of the arcs introduced in step 2d is removed (see Sect. 3.3) iff an evidential shortcut and a corresponding active chain already exist.

Finally, $\mathbf{A}_{\mathcal{B}}$ is only extended for undercutter arcs in step 3; active chains formed between $var(p_i)$ and $var(q)$ in step 2 are, therefore, not affected by this step. □

In Proposition 2, we prove that under specific conditions on AGs an acyclic graph is automatically obtained when following steps 1–4a of the approach, which simplifies the manual verification involved in step 4b. Conditions (a) and (b) concern the existence of undercutter arcs within and between connected subgraphs of AGs. Condition (c) is a generalization of our assumption that no causal cycles exist in AGs (see Sect. 2.1) to support arcs in $\mathbf{S^o}$.

Proposition 2. *Let $G_\mathcal{A} = (\mathbf{P}, \mathbf{A}_\mathcal{A})$, and let $G_\mathcal{A}^* = (\mathbf{P}, \mathbf{A}_\mathcal{A}^*)$ be the subgraph of $G_\mathcal{A}$ with $\mathbf{A}_\mathcal{A}^* = \mathbf{A}_\mathcal{A} \setminus \mathbf{U}$. Let an AG component of $G_\mathcal{A}$ be defined as a connected component of $G_\mathcal{A}^*$. Assume the following conditions are satisfied:*

(a) For any AG component $C = (\mathbf{P}', \mathbf{A}_\mathcal{A}')$ of $G_\mathcal{A}$ with $\mathbf{P}' \subseteq \mathbf{P}$, $\mathbf{A}_\mathcal{A}' \subseteq \mathbf{A}_\mathcal{A}^*$, there does not exist a $u: p \to (s) \in \mathbf{U}$ with $p \in \mathbf{P}'$, $s \in \mathbf{A}_\mathcal{A}'$.

(b) For every pair of AG components $C_1 = (\mathbf{P}', \mathbf{A}_\mathcal{A}')$ and $C_2 = (\mathbf{P}'', \mathbf{A}_\mathcal{A}'')$ of $G_\mathcal{A}$ with $\mathbf{P}', \mathbf{P}'' \subseteq \mathbf{P}$, $\mathbf{A}_\mathcal{A}', \mathbf{A}_\mathcal{A}'' \subseteq \mathbf{A}_\mathcal{A}^*$, there does not exist both a $u_1: p_1 \to (s_1) \in \mathbf{U}$ with $p_1 \in \mathbf{P}'$, $s_1 \in \mathbf{A}_\mathcal{A}''$ and a $u_2: p_2 \to (s_2) \in \mathbf{U}$ with $p_2 \in \mathbf{P}''$, $s_2 \in \mathbf{A}_\mathcal{A}'$.

(c) There do not exist two support chains $(s_1, \ldots, s_n)$ and $(s_1', \ldots, s_m')$ with $s_1, \ldots, s_n \in \mathbf{S}^c \cup \mathbf{S}^o$, $s_1', \ldots, s_m' \in \mathbf{S}^e$, $\mathrm{Tails}(s_1) \cap \mathrm{Tails}(s_1') \neq \emptyset$, and $head(s_n) = head(s_m')$ or $head(s_n) = -head(s_m')$.

Let $G_\mathcal{B} = (\mathbf{V}, \mathbf{A}_\mathcal{B})$ be the graph constructed from $G_\mathcal{A}$ according to steps 1–4a of the generalized approach. Then $G_\mathcal{B}$ is a DAG.

Proof (Sketch). The following cases are distinguished:

- In steps 2a and 2b, no cycles are introduced. Specifically, our non-repetitiveness assumption and our consistency assumption (see Sect. 2.1) jointly assume that for every $p \subset \mathbf{P}$, p or $-p$ cannot be inferred via a chain of support arcs. Therefore, no chain of arcs exists in $\mathbf{A}_\mathcal{B}$ from a node P to itself. The only other case in which cycles can be introduced is when a causal cycle exists in $G_\mathcal{A}$, which is also prohibited by assumption (see Sect. 2.1).
- No cycles are introduced in step 2c if condition (c) is satisfied. Cycles are only introduced in step 2d if an evidential shortcut exists; however, these cycles are broken again in step 4a as described in Sect. 3.3.
- After step 2, there is a correspondence between AG components and the connected components of the underlying undirected graph S of the thus far constructed BN graph. Under condition (a), no cycles are introduced within a connected component of S when including additional arcs in $\mathbf{A}_\mathcal{B}$ for every $u \in \mathbf{U}$ in step 3. Furthermore, for every pair of AG components C_1 and C_2 of $G_\mathcal{A}$ with corresponding connected components C_1' and C_2' of S, no cycles are introduced between components C_1' and C_2' in step 3 under condition (b). $\square$

Figures 3a and c depict examples of AGs that do not satisfy conditions (a) and (b) of Proposition 2, respectively. In the validation step that follows the initial construction of these BN graphs, arcs can be reversed or removed to make these graphs acyclic. The choice of arc to reverse or remove will depend on its effect on active chains, including those between nodes not directly incident on the arc. We note that this type of manual verification is standard in BN construction, especially in data-poor domains. While the domain knowledge expressed in the original AG has been exploited to construct an initial BN graph, additional domain knowledge may need to be elicited to obtain a valid graph.

Proposition 3 gives an upper-bound on the number of parents introduced by the approach for each node $var(p)$ in a BN graph, which bounds both the size of the CPTs and the complexity of inference in the BN. This bound captures the

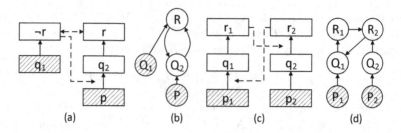

Fig. 3. Examples of AGs (a, c) for which a cyclic graph is constructed by steps 1–4a of the generalized approach (b, d).

number of support arcs and undercutters that involve either proposition p or $\neg p$. The proof of this result is straightforward and omitted due to space limitations.

Proposition 3. *Let $G_A = (\mathbf{P}, \mathbf{A}_A)$ be an AG, and let $G_B = (\mathbf{V}, \mathbf{A}_B)$ be the BN graph constructed according to steps 1–4a of the generalized approach. For every $p \in \mathbf{P}$, let $\mathbf{Par}_p = \{p_i \mid p_i \in \mathbf{Tails}(s), s \in \mathbf{S^c} \cup \mathbf{S^o}, head(s) = p\}$ and let $\mathbf{Par}'_p = \{p_i \mid p_i \in \mathbf{Tails}(s), s \in \mathbf{S^o}, s \text{ is followed by maximal chain } s_1, \ldots, s_m \in \mathbf{S^e} \text{ with } head(s_m) = p \text{ or } head(s_m) = \neg p\}$. Let $\mathbf{S}^e_p$ be a subset of $\mathbf{S^e}$, where $s \in \mathbf{S}^e_p$ iff $p \in \mathbf{Tails}(s)$. Let $\mathbf{U}^e_p \subseteq \mathbf{U}$ be the subset of undercutter arcs directed to an $s \in \mathbf{S}^e_p$ or $s \in \mathbf{S}^e_{\neg p}$. Similarly, let $\mathbf{U}^c_p, \mathbf{U}^o_p \subseteq \mathbf{U}$ be the subsets of undercutter arcs directed to an $s \in \mathbf{S^c}$ respectively $\mathbf{S^o}$ for which $head(s) = p$ or $head(s) = \neg p$. Then an upper-bound for the number of parents of $var(p)$ is:*

(1) $|\mathbf{Par}_p| + |\mathbf{Par}_{\neg p}| + |\mathbf{Par}'_p| + |\mathbf{U}^c_p| + |\mathbf{U}^o_p|$ *if* $\mathbf{S}^e_p = \mathbf{S}^e_{\neg p} = \emptyset$;
(2) $|\mathbf{Par}_p| + |\mathbf{S}^e_{\neg p}| + |\mathbf{Par}'_p| + |\mathbf{U}^e_p| + |\mathbf{U}^c_p| + |\mathbf{U}^o_p|$ *if* $\mathbf{S}^e_p = \emptyset$ *and* $\mathbf{S}^e_{\neg p} \neq \emptyset$;
(3) $|\mathbf{Par}_{\neg p}| + |\mathbf{S}^e_p| + |\mathbf{Par}'_p| + |\mathbf{U}^e_p| + |\mathbf{U}^c_p| + |\mathbf{U}^o_p|$ *if* $\mathbf{S}^e_p \neq \emptyset$ *and* $\mathbf{S}^e_{\neg p} = \emptyset$;
(4) $|\mathbf{S}^e_p| + |\mathbf{S}^e_{\neg p}| + |\mathbf{U}^e_p| + |\mathbf{U}^o_p|$ *if* $\mathbf{S}^e_p \neq \emptyset$ *and* $\mathbf{S}^e_{\neg p} \neq \emptyset$.

5   Conclusion

In this paper, we have studied how domain knowledge expressed as labeled arguments can be exploited to construct a BN graph. Firstly, we have generalized our previously proposed approach [18] by allowing inference types that are neither causal nor evidential. Moreover, we have formally proven that, as intended, our approach captures all support chains in an AG in the form of active chains in the BN graph. We have also identified conditions on AGs under which a DAG is automatically constructed by the approach, simplifying the manual verification step. Lastly, we have identified bounds on the size of the CPTs and the complexity of inference in BNs constructed by our approach. All properties also hold for the limited case considered in [18] but were not proven in that paper.

The generalized approach allows us to construct an initial BN graph from a domain expert's initial argument-based analysis, capturing similar reasoning patterns as their original AG; it thereby simplifies the BN elicitation process.

We note that BN construction is an iterative process in which both the domain expert and BN modeler should stay involved; this also holds when applying our approach, as the provided AG may be incomplete or incorrect. To aid in this iterative process, approaches were proposed in related work which allow experts to use argumentation to argue about the BN under construction instead of about the domain [13,19]. In other related work, approaches for explaining the reasoning patterns captured in BNs in terms of argumentation were proposed [12,17], which allow domain experts more accustomed to argumentation to understand the probabilistic reasoning captured in a BN. Compared to the present paper, this work is in the reverse direction, namely from BNs to arguments.

Recently, there has been much other work on probabilistic argumentation. However, most approaches concern *abstract* argumentation (see e.g. [10] for an overview) while we need structured arguments. Rienstra [16] considers probabilistic structured argumentation; however, he takes what Hunter [9] calls the constellations approach to probabilistic argumentation by considering uncertainty in the existence of arguments. Instead, we take what Hunter calls the epistemic approach to probabilistic argumentation by considering probabilities to express uncertainty concerning the reliability of an argument's inferences. There is some work on the epistemic approach to probabilistic structured argumentation (e.g. [9,15]). In future work, this may become relevant for deriving probabilistic constraints on BNs.

References

1. Bex, F.: An integrated theory of causal stories and evidential arguments. In: Proceedings of the Fifteenth International Conference on Artificial Intelligence and Law, pp. 13–22. ACM Press, New York (2015)
2. Bex, F., Modgil, S., Prakken, H., Reed, C.A.: On logical specifications of the argument interchange format. J. Logic Comput. **23**(5), 951–989 (2013)
3. Bex, F., Renooij, S.: From arguments to constraints on a Bayesian network. In: Baroni, P., Gordon, T.F., Scheffler, T., Stede, M. (eds.) Computational Models of Argument: Proceedings of COMMA 2016, vol. 287, pp. 95–106. IOS Press, Amsterdam (2016)
4. Buckingham Shum, S.J.: The roots of computer supported argument visualization. In: Kirschner, P.A., Buckingham Shum, S.J., Carr, C.S. (eds.) Visualizing Argumentation: Software Tools for Collaborative and Educational Sense-Making, pp. 3–24. Springer, London (2003). https://doi.org/10.1007/978-1-4471-0037-9_1
5. Console, L., Dupré, D.T.: Abductive reasoning with abstraction axioms. In: Lakemeyer, G., Nebel, B. (eds.) Foundations of Knowledge Representation and Reasoning. LNCS (LNAI), vol. 810, pp. 98–112. Springer, Heidelberg (1994). https://doi.org/10.1007/3-540-58107-3_6
6. Fenton, N., Neil, M.: Risk Assessment and Decision Analysis with Bayesian Networks. CRC Press, Boca Raton (2012)
7. van der Gaag, L.C., Helsper, E.M.: Experiences with modelling issues in building probabilistic networks. In: Gómez-Pérez, A., Benjamins, V.R. (eds.) EKAW 2002. LNCS (LNAI), vol. 2473, pp. 21–26. Springer, Heidelberg (2002). https://doi.org/10.1007/3-540-45810-7_4

8. Helsper, E.M., van der Gaag, L.C.: Building Bayesian networks through ontologies. In: van Harmelen, F. (ed.) Proceedings of the Fifteenth European Conference on Artificial Intelligence, vol. 77, pp. 680–684. IOS Press, Amsterdam (2002)

9. Hunter, A.: A probabilistic approach to modelling uncertain logical arguments. Int. J. Approx. Reason. **54**(1), 47–81 (2013)

10. Hunter, A., Thimm, M.: On partial information and contradictions in probabilistic abstract argumentation. In: Baral, C., Delgrande, J., Wolter, F. (eds.) Principles of Knowledge Representation and Reasoning: Proceedings of the Fifteenth International Conference (KR-2016), pp. 53–62. AAAI Press, Palo Alto (2016)

11. Jensen, F.V., Nielsen, T.D.: Bayesian Networks and Decision Graphs, 2nd edn. Springer, New York (2007). https://doi.org/10.1007/978-0-387-68282-2

12. Keppens, J.: Argument diagram extraction from evidential Bayesian networks. Artif. Intell. Law **20**(2), 109–143 (2012)

13. Keppens, J.: On modelling non-probabilistic uncertainty in the likelihood ratio approach to evidential reasoning. Artif. Intell. Law **22**(3), 239–290 (2014)

14. Pearl, J.: Embracing causality in default reasoning. Artif. Intell. **35**(2), 259–271 (1988)

15. Prakken, H.: Probabilistic strength of arguments with structure. In: Thielscher, M., Toni, F., Wolter, F. (eds.) Principles of Knowledge Representation and Reasoning: Proceedings of the Sixteenth International Conference (KR-2018), pp. 158–167. AAAI Press, Palo Alto (2018)

16. Rienstra, T.: Towards a probabilistic Dung-style argumentation system. In: Ossowski, S., Toni, F., Vouros, G.A. (eds.) Proceedings of the First International Conference on Agreement Technologies, pp. 138–152. CEUR, Aachen (2012)

17. Timmer, S.T., Meyer, J.-J.C., Prakken, H., Renooij, S., Verheij, B.: Explaining Bayesian networks using argumentation. In: Destercke, S., Denoeux, T. (eds.) ECSQARU 2015. LNCS (LNAI), vol. 9161, pp. 83–92. Springer, Cham (2015). https://doi.org/10.1007/978-3-319-20807-7_8

18. Wieten, R., Bex, F., Prakken, H., Renooij, S.: Exploiting causality in constructing Bayesian networks from legal arguments. In: Palmirani, M. (ed.) Legal Knowledge and Information Systems: JURIX 2018: The Thirty-First Annual Conference, vol. 313, pp. 151–160. IOS Press, Amsterdam (2018)

19. Wieten, R., Bex, F., Prakken, H., Renooij, S.: Supporting discussions about forensic Bayesian networks using argumentation. In: Proceedings of the Seventeenth International Conference on Artificial Intelligence and Law, pp. 143–152. ACM Press, New York (2019)

Belief Functions

Toward the Evaluation of Case Base Maintenance Policies Under the Belief Function Theory

Safa Ben Ayed[1,2]([⊠]), Zied Elouedi[1], and Eric Lefevre[2]

[1] Institut Supérieur de Gestion de Tunis, LARODEC,
Université de Tunis, Tunis, Tunisia
`safa.ben.ayed@hotmail.fr, zied.elouedi@gmx.fr`
[2] Univ. Artois, EA 3926, Laboratoire de Génie Informatique et d'Automatique
de l'Artois (LGI2A), 62400 Béthune, France
`eric.lefevre@univ-artois.fr`

Abstract. The life cycle of Case-Based Reasoning (CBR) systems implies the maintenance of their knowledge containers for reasons of efficiency and competence. However, two main issues occur. First, knowledge within such systems is full of uncertainty and imprecision since they involve real-world experiences. Second, it is not obvious to choose from the wealth of maintenance policies, available in the literature, the most adequate one to preserve the competence towards problems' solving. In fact, this competence is so difficult to be actually estimated due to the diversity of influencing factors within CBR systems. For that reasons, we propose, in this work, an entire evaluating process that allows to assess Case Base Maintenance (CBM) policies using information coming from both a statistical measure and a competence model under the belief function theory.

Keywords: Case-Based Reasoning · Case-Base Maintenance ·
Competence evaluation · Uncertainty · Belief function theory ·
Combination

1 Introduction

Case-Based Reasoning (CBR) is a methodology of problem solving that reuses past experiences to solve new problems according to their similarities [1]. Every new solved problem by a CBR system is retained in a memory structure called a Case Base (CB) to serve for future problems resolution. Although the incremental learning of CBR systems presents a strong point, it is not free of drawbacks. In fact, this evolution can be uncontrollable, caused by the retention of redundant and noisy cases which conduct to the degradation of systems' problem-solving competence and performance. For those reasons, the Case Base Maintenance (CBM) field presents the key factor's success of CBR systems. As has been

© Springer Nature Switzerland AG 2019
G. Kern-Isberner and Z. Ognjanović (Eds.): ECSQARU 2019, LNAI 11726, pp. 113–124, 2019.
https://doi.org/10.1007/978-3-030-29765-7_10

defined in [2], *"Case-base maintenance implements policies for revising the orga-nization or contents (representation, domain contents, accounting information, or implementation) of the case base in order to facilitate future reasoning for a particular set of performance objectives"*. During the last five decades [4], a wide range of CBM policies have been proposed, even in Machine Learning or CBR communities, that aim to update CBs content in such a way to be per-former and more competent to make high quality decisions. Different attempts to classify them have been proposed in different papers [5,6,9,10]. One of the simplest categorizations consists at regrouping CBM policies by their ability for uncertainty management (*hard and soft*). *Condensed Nearest Neighbor* (CNN) [11] and *Reduced Nearest Neighbor* (RNN) [12] present the baseline of the CB maintenance task. For the *soft* CBM policies, less work have been proposed, where two are implemented within the framework of the belief function theory which are *Evidential Clustering and case Types Detection for CBM* (ECTD) [6] and *Dynamic policy for CBM* (DETD) [13].

After performing a maintenance task, the question that arises is whether the original CBR is better, tantamount, or worse than the maintained one. The intuitive answer to this question is to measure the competence of the CB before and after maintenance. Therefore, this allows us to estimate the support degree of the CBM policy as well as its adequacy to be applied. However, estimating the real competence of a given CBR system in problem-solving is a very complex task since this competence depends on many affecting factors, such as statistical and problem solving properties [3]. To deal with these problems, available research are even measuring the accuracy of the CBR system using a statistical measure [6,7] or estimating their competence using a competence model [3,8]. Some of them are aware of the great importance of managing uncertainty within such knowledge since they reflect real-world situations. Consequently, we aim, in this work, to evaluate CBM policies by offering a support/adequacy degree through combining information coming from an accuracy measure and a competence model. To offer high quality aggregation with managing conflict within both sources' information, and to deal with uncertainty within case knowledge, we use one among the most powerful tools for uncertainty management called the belief function theory.

The rest of the paper is organized as follows. In the next section, we overview the key factors that affect CBs competence and the two used ways for CBR eval-uation. Section 3 presents, then, the basics of the belief function theory, as well as the used tools. Throughout Sect. 4, our CBM evaluating process is detailed to indicate the adequacy of the used CBM policy and estimate its support degree. In Sect. 5, we elaborate the experimental study on different CBM policies and using different CBs. Finally, Sect. 6 concludes the paper and proposes some future work.

2   Case Base Competence Evaluation

The competence (or coverage) of a CBR system presents the range of problems that it can successfully solve [3]. Actually, this criterion cannot be well estimated when we use a simple metric due to the diversity of influencing factors (Subsect. 2.1). In the literature, this competence is even estimated using statistical measure such as the accuracy (Subsect. 2.2), or using some competence model such as CEC-Model [14] (Subsect. 2.3).

2.1   Key Factors Affecting CBs Competence

Estimating the competence of a CBR system needs an awareness regarding the set of elements that may affect it. Actually, we note that the statistical properties of cases within a CB is highly influencing its ability in covering the problem space. Besides, problem-solving properties are an intuitive influencing factor of CBR systems competence. As done in [3,14], we can enumerate these factors as follows: CB size*, cases distribution*, density of cases*, cases vocabulary*, Similarity*, and adaptation knowledge[1].

2.2   Statistical Measures for CBR Evaluation

Some works mention that the precision or the accuracy presents a kind of true competence [3] with some limitations. Actually, the competence of a CBR system can be recognized as input problem solving capability with the right solutions. The most common and straightforward practice consists at using a test set from the original CB and applying a classification algorithm[2] to solve problems. By this way, we can estimate the competence of the CBR system using statistical measures such as the accuracy as the percentage of correct classifications, the specificity as the true negative rate, and others [15].

Actually, this kind of measures has to be taken into account when measuring the competence of a CB. However, it is not sufficient since it does not cover several affected factors. Hence, competence models are also used for this matter.

2.3   Competence Models for CBR Evaluation

Various competence models have been proposed to take into account different influencing factors. For instance, we find Case Competence Categories Model [16] which consists at dividing cases into four types so as to fix a maintenance strategy to be followed. However, it is not able to tangibly and mathematically quantify the global competence of the entire CB. Besides, we find Coverage model based on Mahalanobis Distance and clustering [17], that uses a density-based clustering method to distinguish three types of cases on which the overall CB competence depends. However, we cannot well estimate this competence without deeply studying the relation between cases. Although Smyth & McKenna model [3] is able to deal with different influencing factors, it suffers from its disability to manage the uncertainty within the real stored situations. Hence, the Coverage & Evidential Clustering based Model (CEC-Model) [14] has been proposed in a preliminary work to tackle the problem of uncertainty management while regrouping cases and measuring similarities. Its entire cycle is described in Fig. 1.

---

[1] The factors identified with a star (*) are taken into account in the current work.

[2] The (k-NN) classifier is the most used within the CBR community.

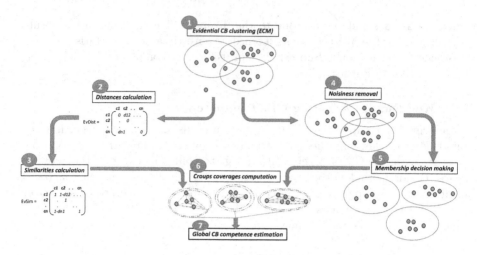

Fig. 1. CEC-Model's process

By this way, we use the latter mentioned CEC-Model [14], and the accuracy measure, during our proposed CBM evaluation process, where both of knowledge uncertainty and information fusion are taken into account under the belief function theory framework.

3   Belief Function Theory: Basic Concepts

The belief function theory [18,19], called also Evidence theory, is a mathematical framework for reasoning under partial and unreliable knowledge. This model is basically defined by a frame of discernment Ω which represents a set of a finite elementary events. The major strength of this theory is its ability to model all levels of uncertainty, from the complete ignorance to the total certainty, on a power set 2^{Ω} which contains all the possible subsets of Ω.

The key point of this theory is the *basic belief assignment (bba)* m which is defined as follows:

$$m : 2^{\Omega} \rightarrow [0, 1]$$
$$A \mapsto m(A) \tag{1}$$

with m is satisfying the following constraint:

$$\sum_{A \subseteq \Omega} m(A) = 1 \tag{2}$$

It aims at allocating to every set $A \in 2^{\Omega}$ a degree of belief to represent the partial knowledge about the actual value of y defined on Ω. A mass function is normalized if it assigns to the empty set partition null degree of belief ($m(\emptyset) = 0$). Contrariwise, the assigned amount of belief to the empty set reflects the flexibility to consider that the value of y may not belong to Ω. The latter situation has usually been used during the evidential clustering to identify noisy instances [20,21], where the frame of discernment Ω defines the set of clusters.

Actually, we often need to calculate the distance between two mass functions defined in the same frame of discernment. To do so, Jousselme Distance [22] presents one among the most used tools to measure distances between two pieces of evidence. It is defined as follows:

$$d(m_1, m_2) = \sqrt{\frac{1}{2}(\overrightarrow{m_1} - \overrightarrow{m_2})^T \underline{\underline{D}} \, (\overrightarrow{m_1} - \overrightarrow{m_2})} \tag{3}$$

where $\underline{\underline{D}}$ is a square matrix of size 2^K ($K = |\Omega|$), and its elements are calculated such that:

$$D(A, B) = \begin{cases} 1 & \text{if } A = B = \emptyset \\ \frac{|A \cap B|}{|A \cup B|} & \text{otherwise} \end{cases} \tag{4}$$

In the framework of belief function theory, various combination rules of evidence have been proposed. The conjunctive rule of combination [23] is one of the most used ones to combine two pieces of evidence induced from two independent and reliable sources of information. When the normality constraint ($m(\emptyset) = 0$) is imposed, we may use the Dempster rule of combination [18].

Ultimately, to make decision under the belief function theory, we may use the pignistic probability transformation, denoted $BetP$, which is considered as one of the best ways for decision making. If the mass function is normalized, then $BetP$ is defined as follows:

$$BetP(y) = \frac{1}{1 - m(\emptyset)} \sum_{y \in A, A \subseteq \Omega} \frac{m(A)}{|A|} \tag{5}$$

where $y \in \Omega$ and $|A|$ is the cardinality of the subset $A \subseteq \Omega$.

4 Evidential CBM Evaluating Process

In this section, we propose an evaluation method for Case Base Maintenance policies that aims to estimate their support/adequacy degree for a given CBR system. Its main idea consists at combining two mass functions reflecting their adequacy. These mass functions are deduced from the improvement degree of competence, extracted respectively from the CEC-Model [14] and the accuracy criterion before and after applying the CBM policy. For the sake of clarity, a general depict of the proposed evaluating method is shown in Fig. 2.

4.1 Two-Level Original CBR Evaluation

First of all, we aim at measuring the competence of the original non-maintained CBR system using both the evidential competence model CEC-Model [14] to provide $Comp_O$ and the accuracy criterion to provide Acc_O.

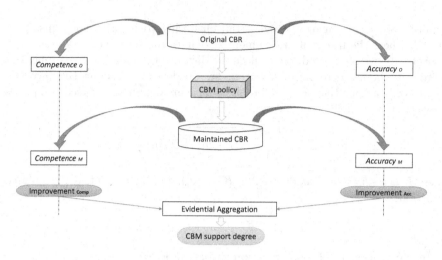

Fig. 2. The proposed CBM evaluating process

Comp$_O$ Estimation: The original CB denoted CBR_O presents the input of CEC-Model. As shown in Fig. 1, it applies the Evidential C-Means (ECM) [20] for cases clustering, Jousselme Distance [22] (Eqs. 3 and 4) for similarity calculation, and pignistic probability transformation (Eq. 5) for cases membership decision. Finally, groups coverage and CB competence are estimated through the density and size properties. The output result is bounded in $[0, 1]$, where the more it is near to 1 the more CBR_O is considered as competent in solving problems.

Acc$_O$ Estimation: The accuracy criterion is studied using 10-fold cross validation and the k-NN as a classifier (we chose to take $k = 1$). To be measured, the original CB is divided into training set ($Tr = 0.8 \times CB$) and test set ($Ts = 0.2 \times CB$), where Tr plays the role of the entire CB and Ts contains the set of input problems to be solved from Tr. Acc_O is therefore calculated as follows:

$$Acc_O = \frac{\#Correct\ Classifications\ on\ Ts}{Size\ of\ Ts} \tag{6}$$

4.2   Case Base Maintenance Application

After measuring the Original CB competence through the two previous identified sources (competence model and accuracy criterion), we perform on CBR_O the CBM policy to be evaluated. Actually, the main purpose of CBM policies is to detect the subset of cases that let a high problem-solving capability. In this step, we may consider the applied policy as a black box and we only focus on its input (CBR_O) and output, which is the maintained CB (CBR_M). By this way, any CBM policy, in the literature, may be applied at the aim to be evaluated, thereafter, by our evaluating process.

4.3   Two-Level Maintained CBR Evaluation

Once the CBM policy completes its execution, our next step consists at measuring the edited CB competence using the same tools and settings as the first step to generate $Comp_M$ and Acc_M values.

$Comp_M$ **Estimation:** As previously done, we evaluate the CB using the CEC-Model, whereas that time it is applied to assess the maintained CB (CBR_M) and provide $Comp_M$ between 0 and 1.

Acc_M **Estimation:** The testing strategy of the accuracy after the maintenance task consists at dividing CBR_O into training set Tr and test set Ts. Then, the used CBM policy is applied on Tr to generate an edited training set Tr_M. Using 1-NN, the accuracy is measured through classifying Ts using Tr_M. Finally, Acc_M is obtained by averaging ten trials values using 10-fold cross validation.

4.4   Extracting CBM Adequacy Knowledge from Statistical Measure and Competence Model Independently

Up to now, we have four different competence estimation values (in $[0, 1]$) that come from two sources: CEC-Model and Accuracy measure. The first is measuring the competence of CBR_O, and the second assesses the quality of the applied CBM task through CBR_M. During this step, we highlight the improvement of CBR_O against CBR_M, in terms of both competence and accuracy. Therefore, we define these two improvements $(Imp_{Comp}$ and $Imp_{Acc})$ as follows:

$$Imp_{Comp} = Comp_M - Comp_O \qquad (7)$$

and

$$Imp_{Acc} = Acc_M - Acc_O \qquad (8)$$

Knowing that their offered values are in $[-1, 1]$, three distinguished situations arise regarding Imp_x, where x replaces even $Comp$ or Acc terms:

- If $Imp_x \simeq 1$, then a high degree of adequacy is assigned to the applied CBM policy for the CBR system.
- If $Imp_x \simeq -1$, then the used CBM policy is not adequate at all for the CBR system.
- If $Imp_x \simeq 0$, then we have no preference regarding the maintenance task.

4.5   Knowledge Combination Under the Belief Function Theory

Based on the situations mentioned above, we build two mass functions on the same frame of discernment which contains two events. The first consists at indicating that the CBM policy is adequate to be applied on a given CBR system, and the second presents its complementary event. Hence, this frame is defined as follows:

$$\Omega = \{Adequate, \overline{Adequate}\} \qquad (9)$$

By this way, the mass functions, defined on Ω, indicates the evaluation of the CBM policy. The first m_{Comp} describes the knowledge coming from the improvement in terms of competence, and the second m_{Acc} informs the knowledge originated by the improvement in terms of accuracy. Consequently, we similarly define them as follows:

$$m_{Comp}\begin{cases} m_{Comp}(\emptyset) = 0 \\ m_{Comp}(Adequate) = \begin{cases} Imp_{Comp} & If\ Imp_{Comp} \geq 0 \\ 0 & Otherwise \end{cases} \\ m_{Comp}(\overline{Adequate}) = \begin{cases} |Imp_{Comp}| & If\ Imp_{Comp} < 0 \\ 0 & Otherwise \end{cases} \\ m_{Comp}(\Omega) = 1 - |Imp_{Comp}| \end{cases} \qquad (10)$$

and

$$m_{Acc}\begin{cases} m_{Acc}(\emptyset) = 0 \\ m_{Acc}(Adequate) = \begin{cases} Imp_{Acc} & If\ Imp_{Acc} \geq 0 \\ 0 & Otherwise \end{cases} \\ m_{Acc}(\overline{Adequate}) = \begin{cases} |Imp_{Acc}| & If\ Imp_{Acc} < 0 \\ 0 & Otherwise \end{cases} \\ m_{Acc}(\Omega) = 1 - |Imp_{Acc}| \end{cases} \qquad (11)$$

Obviously, knowledge obtained from each source is not perfect. Hence, their aggregation presents an interesting solution to reach more relevant information. For that reason, we opt to synthesize the knowledge obtained in m_{Comp} and m_{Acc} by combining them using tools offered within the evidence theory. Since m_{Comp} and m_{Acc} present normalized mass functions that are defined in the same frame of discernment Ω and induced from two distinct information sources, which are considered to be reliable, we use the conjunctive rule of combination defined in [23] as follows:

$$(m_{Comp} \bigcirc m_{Acc})(C) = \sum_{A \cap B = C} m_{Comp}(A) m_{Acc}(B), \quad \forall C \subseteq \Omega \qquad (12)$$

In the current work, we are not interested in making decision regarding whether the applied CBM policy is adequate or not, but we aim to estimate the adequacy support degree for the applied maintenance task. To do, we interpret this rate as the pignistic probability of the event "Adequate". Consequently, we measure this probability using Eq. 5 in such a way that:

$$CBM\ support\ degree = BetP(Adequate) \qquad (13)$$

5   Experimentation

The following experiments aim at projecting our proposal on the maintenance field within CBR systems and use it to evaluate this CBM policies adequacy. In this section, we present used data and the followed settings during implementation and tests (Subsect. 5.1). Offered results and discussion are then provided in Subsect. 5.2.

5.1   Experimental Setup

Our proposed evaluating process of the current work have been tested on five case bases from UCI Machine Learning Repository[3] to assess CBM policies available in the literature. These datasets are described in Table 1 in term of size, number of problems attributes, and number of classes or solutions.

Table 1. Case bases description

	Case base	# instances	# attributes	# solutions
1	Breast Cancer	569	32	2
2	Glass	214	9	6
3	Ionosphere	351	34	2
4	Indian	583	10	2
5	Sonar	208	60	2

For every CB, we estimate the support maintenance degree of four CBM policies. We have chosen CNN [11] and RNN [12] as the most widely used CBM algorithms, as well as ECTD [6] and DETD [13] as the two existing CBM policies under the belief function theory. These methods have been developed according to their default settings as described in their referenced papers.

5.2   Results and Discussion

As regards to the study of results offered, in Table 2, by our proposed evaluating process, some particular situations should be pointed out. If the offered support degree is equal to 50%, then the applied CBM method was able to retain exactly the initial competence of the CBR system. The amount above 50 represents the capability rate of the CBM policy to improve that competence. Therefore, the higher this value, the more the CBM policy is adequate to be applied. On the contrary, the amount below 50 reflects the amount of competence degradation after maintenance. In Table 2, we note that almost all the offered CBM support degrees are in [40, 60], which means that performed CBM policies slightly reduce or improve the CBR competence in problem-solving. Nevertheless, we remark that CNN and RNN algorithms are not adequate to be applied on some CBs such as "Ionosphere" and "Glass" datasets (25.74% and 29.98% with CNN, and 17.95% and 29.98% with RNN). In our sense, we may tolerate values in [45, 50] if other evaluation criteria are improved such as CBR performance and response time[4]. Ultimately, we note that the ECTD policy is the most supported CBM method to be applied on the different tested CBs, where it offers support values equal to 57.27% with "Breast Cancer", 50.84% with "Glass", 47.66% with

---

[3] https://archive.ics.uci.edu/ml/.

[4] Forthcoming research work will carry out with other evaluation criteria.

"Ionosphere", 50.68% with "Indian", and 50.78% with "Sonar". These values indicate that maintenance task applied by ECTD improves the performance of almost all the original tested CBs.

Table 2. Support degree results of some CBM policies applied on some CBs

CB	CBM	$Comp_O$ (%)	$Comp_M$ (%)	Acc_O (%)	Acc_M (%)	CBM support degree (%)
Cancer	CNN	83.44	81.52	59.39	71.45	55.08
	RNN	83.44	82.11	59.39	71.45	55.37
	ECTD	83.44	83.12	59.39	74.25	57.27
	DETD	83.44	83.06	59.39	70.12	55.18
Glass	CNN	55.86	54.24	87.38	48.33	29.98
	RNN	55.86	54.24	87.38	48.33	29.98
	ECTD	55.86	55.18	87.38	89.75	50.84
	DETD	55.86	54.26	87.38	73.81	42.52
Ionosphere	CNN	93.86	69.74	86.61	54.46	25.74
	RNN	93.86	68.88	86.61	34.46	17.95
	ECTD	93.86	89.17	86.61	86.61	47.66
	DETD	93.86	88.72	86.61	77.53	43.12
Indian	CNN	74.22	72.12	65.26	61.88	47.30
	RNN	74.22	71.03	65.26	61.75	46.71
	ECTD	74.22	73.68	65.26	67.15	50.68
	DETD	74.22	70.13	65.26	59.87	47.05
Sonar	CNN	78.11	73.87	81.28	64.22	39.71
	RNN	78.11	72.96	81.28	62.85	39.63
	ECTD	78.11	76.32	81.28	84.62	50.78
	DETD	78.11	76.01	81.28	78.55	47.31

6   Conclusion

In this paper, a process for evaluating Case Base Maintenance policies is proposed. Its main idea consists at applying a given CBM policy and measuring the CB competence before and after maintenance using both of an evidential competence model and the statistical accuracy measure. The output of these two sources are modeled and aggregated within the belief function framework to offer a high-quality CBM support degree estimation. During the experimentation, this process has been performed on different CBM policies and using different datasets. As future work, we opt to intervene on the opposite sense by setting parameters of some CBM policies at the aim of maximizing the support degree offered by the proposed evaluation process.

References

1. Aamodt, A., Plaza, E.: Case-based reasoning: foundational issues, methodological variations, and system approaches. In: Artificial Intelligence Communications, pp. 39–52 (1994)

2. Leake, D.B., Wilson, D.C.: Categorizing case-base maintenance: dimensions and directions. In: Smyth, B., Cunningham, P. (eds.) EWCBR 1998. LNCS, vol. 1488, pp. 196–207. Springer, Heidelberg (1998). https://doi.org/10.1007/BFb0056333

3. Smyth, B., McKenna, E.: Modelling the competence of case-bases. In: Smyth, B., Cunningham, P. (eds.) EWCBR 1998. LNCS, vol. 1488, pp. 208–220. Springer, Heidelberg (1998). https://doi.org/10.1007/BFb0056334

4. Juarez, J.M., Craw, S., Lopez-Delgado, J.R., Campos, M.: Maintenance of case bases: current algorithms after fifty years. In: proceedings of the International Joint Conferences on Artificial Intelligence, pp. 5458–5463 (2018)

5. Smiti, A., Elouedi, Z.: Overview of maintenance for case based reasoning systems. Int. J. Comput. Appl. 32, 49–56 (2011)

6. Ben Ayed, S., Elouedi, Z., Lefevre, E.: ECTD: evidential clustering and case types detection for case base maintenance. In: Proceedings of the 14th International Conference on Computer Systems and Applications (AICCSA), pp. 1462–1469. IEEE (2017)

7. Ben Ayed, S., Elouedi, Z., Lefevre, E.: Exploiting domain-experts knowledge within an evidential process for case base maintenance. In: Destercke, S., Denoeux, T., Cuzzolin, F., Martin, A. (eds.) BELIEF 2018. LNCS (LNAI), vol. 11069, pp. 22–30. Springer, Cham (2018). https://doi.org/10.1007/978-3-319-99383-6_4

8. Smiti, A., Elouedi, Z.: SCBM: soft case base maintenance method based on competence model. Int. J. Comput. Sci. 25, 221–227 (2018)

9. Lupiani, E., Juarez, J.M., Palma, J.: Evaluating case-base maintenance algorithms. Knowl.-Based Syst. 67, 180–194 (2014)

10. Chebel-Morello, B., Haouchine, M.K., Zerhouni, N.: Case-based maintenance: structuring and incrementing the case base. Knowl.-Based Syst. 88, 165–183 (2015)

11. Hart, P.: The condensed nearest neighbor rule. IEEE Trans. Inf. Theory 14(3), 515–516 (1968)

12. Gates, G.: The reduced nearest neighbor rule. IEEE Trans. Inf. Theory 18(3), 431–433 (1972)

13. Ben Ayed, S., Elouedi, Z., Lefevre, E.: DETD: dynamic policy for case base maintenance based on EK-NNclus algorithm and case types detection. In: Medina, J., Ojeda-Aciego, M., Verdegay, J.L., Pelta, D.A., Cabrera, I.P., Bouchon-Meunier, B., Yager, R.R. (eds.) IPMU 2018. CCIS, vol. 853, pp. 370–382. Springer, Cham (2018). https://doi.org/10.1007/978-3-319-91473-2_32

14. Ben Ayed, S., Elouedi, Z., Lefèvre, E.: CEC-model: a new competence model for CBR systems based on the belief function theory. In: Cox, M.T., Funk, P., Begum, S. (eds.) ICCBR 2018. LNCS (LNAI), vol. 11156, pp. 28–44. Springer, Cham (2018). https://doi.org/10.1007/978-3-030-01081-2_3

15. Mosqueira-Rey, E., Moret-Bonillo, V.: Validation of intelligent systems: a critical study and a tool. Expert Syst. Appl. 18(1), 1–16 (2000)

16. Smyth, B., Keane, M.T.: Remembering to forget: a competence-preserving deletion policy for CBR systems. In: The Thirteenth International Joint Conference on Artificial Intelligence, pp. 377–382 (1995)

17. Smiti, A., Elouedi, Z.: Modeling competence for case based reasoning systems using clustering. In: Proceedings of the 26th International FLAIRS Conference, the Florida Artificial Intelligence Research Society, pp. 399–404 (2013)

18. Dempster, A.P.: Upper and lower probabilities induced by a multivalued mapping. Ann. Math. Stat. 38, 325–339 (1967)

19. Shafer, G.: A Mathematical Theory of Evidence. Princeton University Press, Princeton (1976)

20. Masson, M.H., Denœux, T.: ECM: an evidential version of the fuzzy c-means algorithm. Pattern Recognit. **41**(4), 1384–1397 (2008)
21. Antoine, V., Quost, B., Masson, H.M., Denœux, T.: CECM: constrained evidential c-means algorithm. Comput. Stat. Data Anal. **56**, 894–914 (2012)
22. Jousselme, A.L., Grenier, D., Bossé, E.: A new distance between two bodies of evidence. Inf. Fusion **2**(2), 91–101 (2001)
23. Smets, P.: Application of the transferable belief model to diagnostic problems. Int. J. Intell. Syst. **13**(2–3), 127–157 (1998)

Belief Functions and Degrees of Non-conflictness

Milan Daniel[1(✉)] [iD] and Václav Kratochvíl[2] [iD]

[1] Jan Becher - Karlovarská Becherovka, a.s., Pernod Ricard Group,
Přemyslovská 43, 130 00 Prague 3, Czech Republic
milan.daniel@pernod-ricard.com
[2] Institute of Information Theory and Automation, Czech Academy of Sciences,
Pod Vodárenskou veží 4, 182 08 Prague 8, Czech Republic
velorex@utia.cas.cz

Abstract. A hidden conflict of belief functions in the case where the sum of all multiples of conflicting belief masses being equal to zero was observed. To handle that, degrees of non-conflictness and full non-conflictness are defined. The family of these degrees of non-conflictness is analyzed, including its relation to full non-conflictness. Further, mutual non-conflictness between two belief functions accepting internal conflicts of individual belief functions are distinguished from global non-conflictness excluding both mutual conflict between belief functions and also all internal conflicts of individual belief functions. Finally, both theoretical and computational issues are presented.

Keywords: Belief functions · Dempster-Shafer theory · Uncertainty · Conflicting belief masses · Internal conflict · Conflict between belief functions · Hidden conflict · Degree of non-conflictness · Full non-conflictness

1 Introduction

When combining belief functions (BFs) by the conjunctive rules of combination, some conflicts often appear (they are assigned either to $\emptyset$ by non-normalised conjunctive rule $\odot$ or distributed among other belief masses by normalization in Dempster's rule of combination $\oplus$). Combination of conflicting BFs and interpretation of their conflicts are often questionable in real applications.

Sum of all multiples of conflicting belief masses (denoted by $m_{\odot}(\emptyset)$) was interpreted as a conflict between BFs in the classic Shafer's approach [19]. Nevertheless, non-conflicting BFs with high $m_{\odot}(\emptyset)$ have been observed already in 90's examples. Classification of a conflict is very important in the combination of BFs from different belief sources. Thus a series of papers related to conflicts of BFs was published, e.g. [1, 6, 7, 10, 11, 13–15, 18, 21].

—————————————————
Supported by grant GAČR no. 19-04579S.

© Springer Nature Switzerland AG 2019
G. Kern-Isberner and Z. Ognjanović (Eds.): ECSQARU 2019, LNAI 11726, pp. 125–136, 2019.
https://doi.org/10.1007/978-3-030-29765-7_11

A new interpretation of conflicts of belief functions was introduced in [4]: an important distinction of an internal conflict of individual BF (due to its inconsistency) from a conflict between two BFs (due to conflict/contradiction of evidence represented by the BFs). Note that zero-sum of all multiples of conflicting belief masses $m_\ominus(\emptyset)$ is usually considered as non-conflictness of the belief functions in all the above mentioned approaches.

On the other hand, when analyzing the conflict between BFs based on their non-conflicting parts[1] [7] a positive value of conflict was observed even in a situation when the sum of all multiples of conflicting belief masses equals to zero. The observed conflicts—hidden conflicts [9]—are against the generally accepted classification of BFs, i.e. to be either mutually conflicting or mutually non-conflicting. Above that, different "degrees" of non-conflictness were observed. This also arose a question of what is a sufficient condition for full non-conflictness of BFs.

Section 5 presents the entire family of "non-conflictness" of different degrees between $m_\ominus(\emptyset) = 0$ and a full non-conflictness. Results for both general BFs and special classes of BFs are included. Relations to other approaches to non-conflictness are analysed in Sect. 6. Further computational complexity and other computational aspects are presented in Sect. 7.

2   Preliminaries

We assume classic definitions of basic notions from theory of *belief functions* [19] on finite exhaustive frames of discernment $\Omega_n = \{\omega_1, \omega_2, ..., \omega_n\}$. $\mathcal{P}(\Omega) = \{X | X \subseteq \Omega\}$ is a *power-set* of Ω.

A *basic belief assignment (bba)* is a mapping $m : \mathcal{P}(\Omega) \longrightarrow [0, 1]$ such that $\sum_{A \subseteq \Omega} m(A) = 1$; the values of the bba are called *basic belief masses (bbm)*. $m(\emptyset) = 0$ is usually assumed.

A *belief function (BF)* is a mapping $Bel : \mathcal{P}(\Omega) \longrightarrow [0, 1]$, such that $Bel(A) = \sum_{\emptyset \neq X \subseteq A} m(X)$. A *plausibility function* $Pl : \mathcal{P}(\Omega) \longrightarrow [0, 1]$, $Pl(A) = \sum_{\emptyset \neq A \cap X} m(X)$. Because there is a unique correspondence among m and corresponding Bel and Pl, we often speak about m as of a belief function.

A *focal element* is a subset of the frame of discernment $X \subseteq \Omega$, such that $m(X) > 0$; if $X \subsetneq \Omega$ then it is a *proper focal element*. If all focal elements are *singletons* (i.e. one-element subsets of Ω), then we speak about a *Bayesian belief function*; in fact, it is a probability distribution on Ω. If there are only focal elements such that $|X| = 1$ or $|X| = n$ we speak about *quasi-Bayesian BF*. In the case of $m(\Omega) = 1$ we speak about *vacuous BF* and about *a non-vacuous BF* otherwise. In the case of $m(X) = 1$ for $X \subset \Omega$ we speak about *categorical BF*. If all focal elements have a non-empty intersection, we speak about *a consistent BF*; and if all of them are nested, about a *consonant BF*.

Dempster's (normalized conjunctive) rule of combination $\oplus$: $(m_1 \oplus m_2)(A) = \sum_{X \cap Y = A} K m_1(X) \, m_2(Y)$ for $A \neq \emptyset$, where $K = \frac{1}{1-\kappa}$, $\kappa =$

---

[1] Conflicting and non-conflicting parts of belief functions originally come from [5].

$\sum_{X \cap Y = \emptyset} m_1(X) m_2(Y)$, and $(m_1 \oplus m_2)(\emptyset) = 0$, see [19]. Putting $K = 1$ and $(m_1 \odot m_2)(\emptyset) = \kappa = m_\odot(\emptyset)$ we obtain the *non-normalized conjunctive rule of combination* $\odot$, see e.g. [20].

Smets' *pignistic probability* is given by $BetP(\omega_i) = \sum_{\omega_i \in X \subseteq \Omega} \frac{1}{|X|} \frac{m(X)}{1-m(\emptyset)}$, see e.g. [20]. *Normalized plausibility of singletons*[2] of Bel is a probability distribution Pl_P such that $Pl\_P(\omega_i) = \frac{Pl(\{\omega_i\})}{\sum_{\omega \in \Omega} Pl(\{\omega\})}$ [2,3]. Sometimes we speak about *pignistic* and *plausibility transform* of respective BF.

3   Conflicts of Belief Functions

Original Shafer's definition of the conflict measure between two belief functions [19] is the following: $\kappa = \sum_{X \cap Y = \emptyset} m_1(X) m_2(Y) = (m' \odot m'')(\emptyset) = m_\odot(\emptyset)$, more precisely its transformation $log(1/(1 - \kappa))$.

After several counter-examples, W. Liu's approach [14] appeared in 2006 followed by a series of other approaches and their modifications. W. Liu suggested a two-dimensional conflict measure composed from $m_\odot(\emptyset)$ and $DifBetP_{m_j}^{m_i}$—a maximal difference of $BetP(\omega)$ for m_i, m_j over singletons $\omega \in \Omega$ (as kind of a distance); as it was shown, neither $m_\odot(\emptyset)$ nor any distance of BFs alone may be used as a convenient measure of conflict of BFs.

Further, we have to mention two axiomatic approaches to conflict of BFs by Desterke and Burger [11] and by Martin [15]. In 2010, Daniel distinguished internal conflict inside an individual BF from the conflict between them [4] and defined three new approaches to conflict; the most prospective of them - *plausibility conflict* - was further elaborated in [6,10]. Finally, Daniel's *conflict based on non-conflicting parts of BFs* was introduced in [7]. This last-mentioned measure motivated our research of hidden conflict [9], hidden auto-conflict [8] and also current research of degrees of non-conflictness.

Among the other approaches, we can mention e.g. Burger's geometric approach [1].

A conflict of BFs Bel', Bel'' based on their non-conflicting parts Bel'_0, Bel''_0 is defined by the expression $Conf(Bel', Bel'') = (m'_0 \odot m''_0)(\emptyset)$, where non-conflicting part Bel_0 (of a BF Bel) is unique consonant BF such that $Pl\_P_0 = Pl\_P$ (normalized plausibility of singletons corresponding to Bel_0 is the same as that corresponding to Bel); m_0 is a bba related to Bel_0. For an algorithm to compute Bel_0 see [7].

This measure of conflict analogously to Daniel's approaches from [4] does not include internal conflict of individual BFs in conflict between them. Similarly to plausibility conflict, it respects plausibilities equivalent to the BFs; and it better generalises the original idea to general frame of discernment.

---

[2] Plausibility of singletons is called *contour function* by Shafer in [19], thus $Pl\_P(Bel)$ is a normalization of contour function in fact.

4   Hidden Conflict

Example 1. **Introductory example:** Let us assume two simple consistent belief functions Bel' and Bel'' on $\Omega_3 = \{\omega_1, \omega_2, \omega_3\}$ given by the bbas $m'(\{\omega_1, \omega_2\}) = 0.6$, $m'(\{\omega_1, \omega_3\}) = 0.4$, and $m''(\{\omega_2, \omega_3\}) = 1.0$.

For the better understanding of the problem, see Fig. 1: The only focal element of m'' has a non-empty intersection with both focal elements of m', thus $\sum_{(X \cap Y) = \emptyset} m'(X) m''(Y) = (m' \odot m'')(\emptyset)$ is an empty sum. Considering the conflict based on non-conflicting parts, respective consonant BFs with the same plausibility transform has to be found. Because Bel'' is consonant then $Bel_0'' = Bel''$, $m_0'' = m''$. In case of m' we can easily calculate that $Pl'(\{\omega_1\}) = 1$, $Pl'(\{\omega_2\}) = 0.6$, $Pl'(\{\omega_3\}) = 0.4$, thus $m_0'(\{\omega_1\}) = 0.4$, $m_0'(\{\omega_1, \omega_2\}) = 0.2$, $m_0'(\{\omega_1, \omega_2, \omega_3\}) = 0.4$, hence $Conf(Bel', Bel'') = (m_0' \odot m_0'')(\emptyset) = m_0'(\{\omega_1\}) \cdot m_0''(\{\omega_2, \omega_3\}) = 0.4 \cdot 1 = 0.4$. Let us recall that the computational algorithm has been published in [7]—we are not putting it here because of the lack of space.

Fig. 1. Introductory Example: focal elements of m', m'', and of $m' \odot m''$.

Then $(m' \odot m'')(\emptyset) = 0$. This seems—and it is usually considered—to be a proof of non-conflictness of m' and m''. Nevertheless, the conflict based on non-conflicting parts $Conf(Bel', Bel'') = (m_0' \odot m_0'')(\emptyset) = 0.4 > 0$ (which holds true despite of Theorem 4 from [7] which should be revised in future).

Observation of a Hidden Conflict in Example 1

The following questions arise: Does $(m' \odot m'')(\emptyset) = 0$ represent non-conflictness of respective BFs as it is usually assumed? Is the definition of conflict based on non-conflicting parts correct? Are m' and m'' conflicting? What does $(m' \odot m'')(\emptyset) = 0$ mean?

For the moment, suppose that Bel' and Bel'' are non-conflicting. Thus both of them should be non-conflicting with the result of their combination as well. Does it hold for BFs from Example 1? It does if one combines $m' \odot m''$ with m'' one more time (assuming two instances of m'' coming from two independent belief sources). It follows from the idempotency of categorical m'': $m' \odot m'' \odot m'' = m' \odot m''$ and therefore $(m' \odot m'' \odot m'')(\emptyset) = 0$ again. On the other hand, we obtain positive $(m' \odot m'' \odot m')(\emptyset) = (m' \odot m' \odot m'')(\emptyset) = 0.48$ (assuming m' coming from two independent belief sources again). See Table 1 and Fig. 2. When m'' and m' are combined once, then we observe $m_\odot(\emptyset) = 0$. When combining m'' with m' twice then $m_\odot(\emptyset) = 0.48$. We observe some kind of *a hidden conflict*. Moreover, because both individual BFs are consistent, there are no internal conflicts. Thus our hidden conflict is a *hidden conflict between the BFs* and we have an argument for correctness of positive value of $Conf(Bel', Bel'')$.

Table 1. Hidden conflict in the introductory example

X	$\{\omega_1\}$	$\{\omega_2\}$	$\{\omega_3\}$	$\{\omega_1, \omega_2\}$	$\{\omega_1, \omega_3\}$	$\{\omega_2, \omega_3\}$	$\{\omega_1, \omega_2, \omega_3\}$	$\emptyset$
$m'(X)$	0.0	0.0	0.0	0.60	0.40	0.00	0.00	–
$m''(X)$	0.0	0.0	0.0	0.00	0.00	1.00	0.00	–
$(m' \odot m'')(X)$	0.00	0.60	0.40	0.00	0.00	0.00	0.00	0.00
$(m' \odot m'' \odot m'')(X)$	0.00	0.60	0.40	0.00	0.00	0.00	0.00	0.00
$(m' \odot m'' \odot m')(X)$	0.00	0.36	0.16	0.00	0.00	0.00	0.00	0.48
$(m' \odot m'' \odot m' \odot m'')(X)$	0.00	0.36	0.16	0.00	0.00	0.00	0.00	0.48

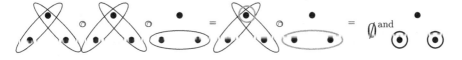

Fig. 2. Arising of a hidden conflict between BFs in the Introductory Example: focal elements of m', m', m''—$m' \odot m', m''$ and of $(m' \odot m') \odot m''$.

What is a decisional interpretation of our BFs? Since *contours* (plausibilities of singletons) are $Pl' = (1.0, 0.6, 0.4)$ and $Pl'' = (0.0, 1.0, 1.0)$, then by normalization we obtain $Pl\_P' = (0.5, 0.3, 0.2)$ and $Pl\_P'' = (0.0, 0.5, 0.5)$. This can be interpreted in a way that ω_1 is significantly preferred by Bel', while it is the opposite in case of Bel''. This is also an argument for a positive value of mutual conflict of the BFs.

Note that in this special case, Smets' pignistic transform and plausibility transform lead to the same result. We obtain $BetP' = (0.5, 0.3, 0.2)$ and $BetP'' = (0.0, 0.5, 0.5)$. Both the probabilistic approximations $BetP$ and Pl_P (in general different) give the highest value to a different singleton for Bel' and Bel''. Thus the argument for mutual conflictness of the BFs is strengthened and we obtain the same pair of incompatible decisions based on the BFs in both frequent decisional approaches: using either normalized contour (which is compatible with the conjunctive combination of BFs) or pignistic probability (designed for betting).

Hence $(m' \odot m'')(\emptyset)$ does not mean real non-conflictness of the BFs. It means simple or partial compatibility of their focal elements only. Or we can accept it as some weak version of non-conflictness.

5   Degrees of Non-conflictness

A case of a hidden conflict could be seen in the introductory example: Note that the example describes a situation when $(m'\odot m'')(\emptyset) = 0$ while $(m'\odot m'\odot m''\odot m'')(\emptyset) > 0$. I.e. there is some type of non-conflictness, but weak as both $Conf(m', m'') > 0$ and $(m'\odot m'\odot m''\odot m'')(\emptyset) > 0$.

Thus the following question arises now: Is $(m'\odot m'\odot m''\odot m'')(\emptyset) = 0$ sufficient for full non-conflictness of belief functions? The answer is of course "no".

Example 2. **Little Angel example:** Assume for example the following bbas defined over $\Omega_5 = \{\omega_1, \ldots, \omega_5\}$—as described in Table 2 (the example and its title comes from [9], the title is inspired by graphical visualization of respective focal elements structure).

Table 2. Little Angel Example

X	$A = \{\omega_1, \omega_2, \omega_5\}$	$B = \{\omega_1, \omega_2, \omega_3, \omega_4\}$	$C = \{\omega_1, \omega_3, \omega_4, \omega_5\}$	$D = \{\omega_2, \omega_3, \omega_4, \omega_5\}$
$m'(X)$	0.10	0.30	0.60	0.00
$m''(X)$	0.00	0.00	0.00	1.00

Indeed, while we can observe both $(m'\odot m'')(\emptyset) = 0$ and $(m'\odot m''\odot m'\odot m'')(\emptyset) = 0$ here, note that $(m'\odot m'\odot m'\odot m''\odot m''\odot m'')(\emptyset) = 0.108 > 0$, which witnesses some kind of a hidden conflict again. Nevertheless, one can feel that the *degree* of the non-conflictness is higher than in the case described by Example 1.

To make our findings more formal, note that due to associativity and commutativity of conjunctive combination rule $\odot$ we can write $(m'\odot m'\odot m'\odot m''\odot m''\odot m'')(\emptyset) = ((m'\odot m'')\odot(m'\odot m'')\odot(m'\odot m''))(\emptyset) = (\odot_{i=1}^{3}(m'\odot m''))(\emptyset)$. Thus, in case of Example 2, one can say that while $m_{\odot}(\emptyset) = (\odot_1^1(m'\odot m''))(\emptyset) = (\odot_1^2(m'\odot m''))(\emptyset) \doteq 0$, there is $(\odot_1^3(m'\odot m''))(\emptyset) = 0.108 > 0$. See Table 3.

Table 3. Hidden conflict in the Little Angel Example—Example 2

X	$A \cap D$	$B \cap D$	$C \cap D$	$A \cap B \cap D$	$A \cap C \cap D$	$B \cap C \cap D$	$\emptyset$
$(m'\odot m'')(X)$	0.10	0.30	0.60	0.00	0.00	0.00	0.00
$(\odot_1^2(m'\odot m''))(X)$	0.01	0.09	0.36	0.06	0.12	0.36	0.00
$(\odot_1^3(m'\odot m''))(X)$	0.001	0.027	0.216	0.036	0.126	0.486	0.108

Definition 1. *(i) Let Bel' and Bel'' be BFs defined by bbms m' and m''. We say that the BFs are non-conflicting in k-th degree if $(\odot_1^k(m'\odot m''))(\emptyset) = 0$.*
(ii) BFs Bel' and Bel'' are fully non-conflicting if they are non-conflicting in any degree.

Thus we can say that BFs from Table 2 are non-conflicting in the second degree, nevertheless, they are still conflicting in the third degree due to the observed hidden conflict.

Utilizing our results on hidden conflicts we obtain the following theorem.

Theorem 1. *Any two BFs on n-element frame of discernment Ω_n non-conflicting in the n-th degree are fully non-conflicting.*

Idea of the Proof: When combining two conflicting BFs defined over Ω_n repeatedly then, because of set intersection operator properties, we either obtain the least focal element of a cardinality lower than in the previous step, or a stable structure of focal elements as the least focal element is already contained in all others. Hence the empty set will appear as a focal element either in n steps or it will not appear at all.

The theorem offers an upper bound for a number of different degrees of non-conflictness of BFs. If a pair of BFs is non-conflicting in n-th degree then it is non-conflicting in any degree. Note that it is possible to find a pair of BFs non-conflicting in $(n-2)$-th degree but conflicting in $(n-1)$-th degree, as it is shown in the general example below.

Example 3. Assume n element Ω_n and BFs m^i and m^{ii} are given by:

$$m^i(\{\omega_1, \omega_2, ..., \omega_{n-1}\}) = \tfrac{1}{n-1},$$
$$m^i(\{\omega_1, \omega_2, ..., \omega_{n-2}, \omega_n\}) = \tfrac{1}{n-1},$$
$$m^i(\{\omega_1, \omega_2, ..., \omega_{n-3}, \omega_{n-1}, \omega_n\}) = \tfrac{1}{n-1},$$
$$...,$$
$$m^i(\{\omega_1, \omega_3, \omega_4, ..., \omega_n\}) = \tfrac{1}{n-1}, \text{ and}$$
$$m^{ii}(\{\omega_2, \omega_3, ..., \omega_n\}) = 1.$$

There is $(\bigcirc_1^k(m^i \textcircled{\tiny\bigcirc} m^{ii}))(\emptyset) = 0$ for $k \leq n-2$, $(\bigcirc_1^2(m^i \textcircled{\tiny\bigcirc} m^{ii}))(\emptyset) = 0.5$ on Ω_3 and e.g. $(\bigcirc_1^{15}(m^i \textcircled{\tiny\bigcirc} m^{ii}))(\emptyset) = 2.98 \cdot 10^{-6}$ on Ω_{16}.

Following the proof of Theorem 1, we can go further in the utilization of results on hidden conflicts and obtain the following theorem, which decreases the number of different degrees of BFs.

Theorem 2. *Any two non-vacuous BFs on any finite frame of discernment non-conflicting in degree c are fully non-conflicting for $c = min(c', c'') + |sgn(c' - c'')|$, where c', c'' are maximal cardinalities of proper focal elements of BFs Bel', Bel'' and $sgn()$ stands for signum.*

Idea of Proof. The smaller is the maximal cardinality of a proper focal element the faster an empty set—as a result of repeated combination of the BFs—may appear.

Corollary 1. *(i) There is only one degree of non-conflictness of any BFs on any two-element frame of discernment Ω_2. In the other words, all degrees of non-conflictness of BFs are equivalent on any two-element frame Ω_2.*

(ii) There is only one degree of non-conflictness of any quasi-Bayesian BFs on any finite frame of discernment Ω_n.

(iii) There are at most two different degrees of non-conflictness of a quasi-Bayesian BF an any other BF on any finite frame of discernment Ω_n.

6   Relation to Other Approaches to Non-conflictness

6.1   Degrees of Non-conflictness and $Conf = 0$.

We have described that there are $n - 1$ different degrees of non-conflictness on Ω_n in the previous section. Besides that, we can observe also different types of non-conflictness. Note that $(m' \odot m'')(\emptyset) = 0$ and $Conf(m', m'') > 0$ in both Examples 1 and 2. On the other hand, the opposite situation can be found—as follows:

Example 4. Let us recall W. Liu's Example 2 from [14] on Ω_5 where $m_i(\{\omega_j\}) = 0.2$ for $i = 1, 2$ and $j = 1, 2, \ldots, 5$ and $m_i(X) = 0$ otherwise (i.e. Bayesian bbas corresponding to uniform probability distributions). Note that while $Conf(Bel_1, Bel_2) = 0$, then $(m_1 \odot m_2)(\emptyset) = 0.8$ and $(\odot_1^k(m_1 \odot m_2))(\emptyset) > 0.8$ for any $k > 1$. Specifically, $0.9922, 0.99968, \ldots$.

Example 5. Similarly, we can present more general example on frame Ω_n for an arbitrary $n \geq 3$ – see Table 4.

Table 4. BFs from Example 5

X	$\{\omega_1\}$	$\{\omega_2\}$	$\{\omega_1, \omega_2\}$	Ω	$\emptyset$
$m^i(X)$	0.4	0.2	0.2	0.2	–
$m^{ii}(X)$	0.3	0.1	0.1	0.5	–
$(m^i \odot m^{ii})(X)$	0.48	0.18	0.14	0.10	0.10
$(\odot_1^2(m^i \odot m^{ii}))(X)$	0.4608	0.1188	0.0476	0.0100	0.3628

Our $n - 1$ degrees of non-conflictness are related to conjunctive combination of BFs, it covers general/global non-conflictness. If $(\odot_1^k(m' \odot m''))(\emptyset) = 0$ hold for any $k < n$ then there is neither internal conflict of any of individual BFs nor a mutual conflict between the two BFs. On the other hand, $Conf(m', m'') = 0$ is related only to mutual conflict between the BFs. Indeed, both the BFs in Example 4 are identical. There is no mutual conflict between them, but both of them are highly internally conflicting. Therefore there is also high conflict $(\odot_1^k(m_1 \odot m_2))(\emptyset)$ for any k.

In Example 5 (Table 4), there are two different BFs with the same order of bbms of proper focal elements. Their $\odot$ combination has the same order of bbms as well. Thus, there is no mutual conflict between them, but, there is an internal conflict inside both of them. We can obtain analogous results also in the case when the internal conflict is hidden in only one of the BFs.

6.2   A Comparison of the Approaches

From the above examples, we can simply see that the 1-st degree of non-conflictness is not comparable with $Conf(m', m'') = 0$.

A relation of the other degrees of non-conflictness to $Conf(m', m'') = 0$ is an open issue for further investigation. We can only see that full non-conflictness is stronger than $Conf(m', m'') = 0$. This is nicely illustrated by the following theorem. We can also see the full non-conflictness is equivalent to strong non-conflictness and that the 1-st degree of non-conflictness is equivalent to non-conflictness both from Destercke & Burger approach [11]. A relation of $Conf(m', m'') = 0$ to Destercke & Burger approach is also an open problem for future.

Theorem 3. (i) Non-conflictness of the 1-st degree is equivalent to Destercke-Burger non-conflictness ($(m_1 \odot m_2)(\emptyset) = 0$, see [11]).
(ii) Full non-conflictness is equivalent to Destercke-Burger strong non-conflictness (non-empty intersection of all focal elements of both BFs, see [11]).
(iii) If BFs m' and m'' are fully non-conflicting then $Conf(m', m'') = 0$ as well.

Idea of Proof:

(i) The first statement just follows the definition of the of the 1-st degree of non-conflictness.
(ii) Computing $\odot_1^n(m' \odot m'')$, the intersection of all focal elements of both the BFs appears among the resulting focal elements.
(iii) The intersection of all focal elements of both the BFs is non-empty in the case of full non conflictness. Thus the intersection of sets of elements with maximal plausibility is non-empty.

7   Computational Complexity and Computational Aspects

When looking for maximal degree of non-conflictness m of two BFs Bel^i and Bel^{ii} on general frame of discernment Ω_n we need to compute $\odot_1^m(m^i \odot m^{ii})$. Following Theorem 1, we know that $m \leq n$. Based on this we obtain complexity $O(n)$ of $\odot$ operations. Analogously to the case of complexity of looking for hidden conflict [9] we can reduce the complexity to $O(log_2(n))$ of $\odot$ operations utilizing a simplification of computation based on $\odot_{j=1}^{2k}(m^i \odot m^{ii}) = \odot_{j=1}^k(m^i \odot m^{ii})$ $\odot \odot_{j=1}^k(m^i \odot m^{ii})$. Note that the complexity of $\odot$ operation depends on the number and the structure of focal elements. Utilizing Theorem 3 we can go further in reduction of computational complexity to $O(n)$ of intersection operations $\cap$.

Beside theoretical research of properties degrees of non-conflictness we have also performed a series of example computations on frames of discernment of cardinality from 5 to 16. A number of focal elements rapidly grows up to $|\mathcal{P}(\Omega)| = 2^{|\Omega|} - 1$ when conjunctive combination $\odot$ is repeated. Note that there are 32.766 and 32.767 focal elements on Ω_{16} in Example 3. Because the conflictness/non-conflictness of BFs depends on the number and the structure of their focal elements not on their bbms, we have frequently used same bbms for all focal elements of BFs in our computations on frames of cardinality greater than 10.

All our experiments were performed in Language R [16] using R Studio [17]. We are currently developing an R package for dealing with belief functions on various frames of discernment. It is based on a relational database approach - nicely implemented in R, in a package called data.table [12].

8   An Important Remark

Repeated applications of the conjunctive combination $\odot$ of a BF with itself is used here to simulate situations where different independent believers have numerically the same bbm. Thus this has nothing to do with idempotent belief combination (where, of course, no conflict between two BFs is possible).

Our study was motivated by the investigation of conflict $Conf$ of BFs based on their non-conflicting parts [7], thus we were interested in independent BFs when a hidden conflict was observed. But we have to note that conflictness/non-conflictness of BFs has nothing to do with dependence/independence of the BFs. Repeated computation of several (up to n) numerically identical BFs, when looking for hidden conflict is just a technical tool for computation of $m(\emptyset)$ or more precisely say for computation of $\kappa = \sum_{X \cap Y = \emptyset} m_j(X) m_j(Y)$. We are not interested in entire result of repeated application of $\odot$, we are interested only in $m_\odot(\emptyset)$ or, more precisely, in $\kappa = \sum_{X_1 \cap X_2 \cap ... \cap X_k = \emptyset} m_j(X_1)\ m_j(X_2)...m_j(X_k)$. Thus our computation has nothing to do with any idempotent combination of BFs. We can look for non-conflictness of higher degrees using $\odot_1^k$ (or κ) in the same way for both dependent and independent BFs. It is also not necessary to include any independence assumption in Definition 1.

9   Summary and Conclusion

Based on existence and observation of hidden conflicts (when the sum of all multiples of conflicting belief masses is zero) a family of degrees of non-conflictness has been observed. Number of non-equivalent/different degrees of non-conflictness depends on the size of the corresponding frame of discernment.

Maximal size of degrees of non-conflictness is $n - 1$ for belief functions on a general finite frame of discernment Ω_n. Nevertheless, for special types of BFs or for particular BFs, a size of the family may be reduced in accordance to the sizes of the focal elements of the BFs in question. The highest degree of non-conflictness (different from lower ones) is equivalent to full non-conflictness and also to strong non-conflictness defined by Destescke and Burger [11]. The family of non-conflictness is further compared with non-conflictness given by Daniel's $Conf(Bel^i, Bel^{ii}) = 0$ [7].

The presented approach to non-conflictness includes both the internal non-conflictness of individual BFs and also mutual non-conflictness between them.

Presented theoretical results move us to a better understanding of the nature of belief functions in general. Due to the important role of conflictness/non-conflictness of BFs within their combination, the presented results may consequently serve as a basis for a better combination of conflicting belief functions and better interpretation of the results of belief combination whenever conflicting belief functions appear in real applications.

References

1. Burger, T.: Geometric views on conflicting mass functions: from distances to angles. Int. J. Approximate Reasoning **70**, 36–50 (2016)
2. Cobb, B.R., Shenoy, P.P.: On the plausibility transformation method for translating belief function models to probability models. Int. J. Approximate Reasoning **41**(3), 314–330 (2006)
3. Daniel, M.: Probabilistic transformations of belief functions. In: Godo, L. (ed.) ECSQARU 2005. LNCS (LNAI), vol. 3571, pp. 539–551. Springer, Heidelberg (2005). https://doi.org/10.1007/11518655_46
4. Daniel, M.: Conflicts within and between belief functions. In: Hüllermeier, E., Kruse, R., Hoffmann, F. (eds.) IPMU 2010. LNCS (LNAI), vol. 6178, pp. 696–705. Springer, Heidelberg (2010). https://doi.org/10.1007/978-3-642-14049-5_71
5. Daniel, M.: Non-conflicting and conflicting parts of belief functions. In: 7th International Symposium on Imprecise Probability: Theories and Applications (ISIPTA 2011), pp. 149–158. SIPTA, Innsbruck (2011)
6. Daniel, M.: Properties of plausibility conflict of belief functions. In: Rutkowski, L., Korytkowski, M., Scherer, R., Tadeusiewicz, R., Zadeh, L.A., Zurada, J.M. (eds.) ICAISC 2013. LNCS (LNAI), vol. 7894, pp. 235–246. Springer, Heidelberg (2013). https://doi.org/10.1007/978-3-642-38658-9_22
7. Daniel, M.: Conflict between belief functions: a new measure based on their non-conflicting parts. In: Cuzzolin, F. (ed.) BELIEF 2014. LNCS (LNAI), vol. 8764, pp. 321–330. Springer, Cham (2014). https://doi.org/10.1007/978-3-319-11191-9_35
8. Daniel, M., Kratochvíl, V.: Hidden auto-conflict in the theory of belief functions. In: Proceedings of the 20th Czech-Japan Seminar on Data Analysis and Decision Making Under Uncertainty, pp. 34–45 (2017)
9. Daniel, M., Kratochvíl, V.: On hidden conflict of belief functions. In: Proceedings of EUSFLAT 2019 (2019, in print)
10. Daniel, M., Ma, J.: Conflicts of belief functions: continuity and frame resizement. In: Straccia, U., Calì, A. (eds.) SUM 2014. LNCS (LNAI), vol. 8720, pp. 106–119. Springer, Cham (2014). https://doi.org/10.1007/978-3-319-11508-5_10
11. Destercke, S., Burger, T.: Toward an axiomatic definition of conflict between belief functions. IEEE Trans. Cybern. **43**(2), 585–596 (2013)
12. Dowle, M., Srinivasan, A.: data.table: extension of 'data.frame' (2016). https://CRAN.R-project.org/package=data.table, r package version 1.10.0
13. Lefèvre, E., Elouedi, Z.: How to preserve the conflict as an alarm in the combination of belief functions? Decis. Support Syst. **56**, 326–333 (2013)
14. Liu, W.: Analyzing the degree of conflict among belief functions. Artif. Intell. **170**(11), 909–924 (2006)

15. Martin, A.: About conflict in the theory of belief functions. In: Denoeux, T., Masson, M.H. (eds.) Belief Functions: Theory and Applications, pp. 161–168. Springer, Heidelberg (2012). https://doi.org/10.1007/978-3-642-29461-7_19

16. R Core Team: R: A Language and Environment for Statistical Computing. R Foundation for Statistical Computing, Vienna (2016). https://www.R-project.org/

17. RStudio Team: RStudio: Integrated Development Environment for R. RStudio Inc., Boston (2015). http://www.rstudio.com/

18. Schubert, J.: The internal conflict of a belief function. In: Denoeux, T., Masson, M.H. (eds.) Belief Functions: Theory and Applications, pp. 169–177. Springer, Heidelberg (2012)

19. Shafer, G.: A Mathematical Theory of Evidence, vol. 1. Princeton University Press, Princeton (1976)

20. Smets, P.: Decision making in the TBM: the necessity of the pignistic transformation. Int. J. Approximate Reasoning **38**(2), 133–147 (2005)

21. Smets, P.: Analyzing the combination of conflicting belief functions. Inf. Fusion **8**(4), 387–412 (2007)

On Expected Utility Under Ambiguity

Radim Jiroušek[1,2]([✉])[ID] and Václav Kratochvíl[1,2][ID]

[1] Faculty of Management, University of Economics, Prague, Czech Republic
[2] Institute of Information Theory and Automation, CAS, Prague, Czech Republic
{radim,velorex}@utia.cas.cz

Abstract. The paper introduces a new approach to constructing models exhibiting the ambiguity aversion. The level of ambiguity aversion is described by a subjective parameter from the unit interval with the semantics: the higher the aversion, the higher the coefficient. On three examples, we illustrate the approach is consistent with the experimental results observed by Ellsberg and other authors.

Keywords: Belief function · Credal set · Probability transform · Decision-making · Vagueness

1 Introduction

It is well known, and it has also been confirmed by our experiments that people prefer lotteries, in which they know the content of a drawing drum to situations when the constitution of the drum's content is unknown. In our experiments, the participants were asked to choose one from six predetermined colors and they got the prize when the color of a randomly drawn ball coincided with their choice. It appeared that the participants were willing to pay in average by 90% more to take part in games when they knew that the urn contained the same number of balls of all six colors in comparison with the situation when they knew only that the urn contained balls of the specified colors and their proportion was unknown. This well known, seemingly paradoxical phenomenon, can hardly be explained by different subjective utility functions or by different subjective probability distributions. To explain this fact, we accepted a hypothesis that humans do not use their personal probability distributions but just *capacity functions* that do not sum up to one [13]. Roughly speaking, the subjective probability of drawing a red ball is $\frac{1}{6}$ in the case that the person knows that all colors are in the same amount in the drum. However, the respective "subjective probability" in the case of lack of knowledge is $\varepsilon < \frac{1}{6}$. *The lack of knowledge psychologically decreases the subjective chance of drawing the selected color – it decreases the subjective chance of success.*

This paper is one of many studying the so-called *ambiguity aversion*, which is used to model the fact that human behavior violates Savage's expected utility

This work is supported by funds from grant GAČR 19-06569S.

© Springer Nature Switzerland AG 2019

G. Kern-Isberner and Z. Ognjanović (Eds.): ECSQARU 2019, LNAI 11726, pp. 137–147, 2019.
https://doi.org/10.1007/978-3-030-29765-7_12

theory [17]. We present one possible way how to find a personal weight function (the above-mentioned capacity) that can be used, similarly to probability function, to compute the personal subjective expected value of a reward in case that the description of the situation is ambiguous. It is clear from the literature [6–8,15] that it cannot be a probability function. It cannot be normalized because our experiments show that people usually expect smaller reward under total ignorance than in case they know that all alternatives are of equal probabilities. As we will see later (when discussing the Ellsberg's experiments), this function is neither additive. Thus, the considered function will belong to the class of *superadditive capacities*.

To find a way, how to compute this personal weight function we will take advantage of the fact that situations with ambiguity are well described by tools of a theory of belief functions. This theory distinguishes between two types of uncertainty: the uncertainty connected with the fact that we do not know the result of a random experiment (a result of a random lottery) and the ignorance arising when we do not know the content of a drawing urn. In this paper, we start with describing the situation by belief functions that can be interpreted as *generalized probability* [9], i.e., each belief function corresponds to a set of probability functions, which is called a *credal set* [9]. Then, we adopt a decision-theoretic framework used also by other authors based on the transformation of the belief function into a probability function. However, we do not use the achieved probabilistic representative directly to decision, we add one additional step. Before computing the expected reward, we reduce the probabilities to account for ambiguity aversion. This is the only point in which our approach differs from Smets' decision-making framework [20], which is based on the Dempster-Shafer theory of belief functions [5,18].

Before describing the process in more details, let us stress that our aim is not as ambitious as developing a mathematical theory describing the ambiguity aversion within the theory of belief functions. In fact, it was already done by Jaffray [12], who shows how to compute generalized expected utility for belief function. We do not even consider all elements from a credal set with all the preference relations as, for example, in [3]. The ambition of our approach is to provide tools making it possible to assign a personal coefficient of ambiguity to experimental persons. Then, we will have a possibility to study its stability with respect to different decision tasks and/or its stability in time. Such a coefficient of ambiguity is considered also by Srivastava [22] and the suggested approach repeats some of his basic ideas. For example, we use almost the same idea to identify the amount of ambiguity connected with individual states of the considered state space.

2   Belief Functions

The basic concepts and notations are taken over from [13], where the described approach was introduced for the first time. We consider only a finite *state space* Ω. In the examples described below, Ω is the set of six considered colors: $\Omega = \{red, black, white, yellow, green, azure\}$ ($\Omega = \{r, b, w, y, g, a\}$ for short in the sequel). Similar to probability theory, where a probability measure is a set function defined on some *algebra* of the considered events, belief functions are represented by functions defined on the set of all nonempty subsets of Ω [5,18]. Let 2^{Ω} denote the set of all subsets of Ω.

The fundamental notion is that of a *basic probability assignment* (bpa), which describes all the information we have about the considered situation. It is a function $m : 2^{\Omega} \to [0, 1]$, such that $\sum_{\mathbf{a} \in 2^{\Omega}} m(\mathbf{a}) = 1$ and $m(\emptyset) = 0$.

For bpa m, $\mathbf{a} \in 2^{\Omega}$ is said to be a *focal element* of m if $m(\mathbf{a}) > 0$. This enables us to distinguish the following two special classes of bpa's representing the extreme situations:

(1) m is said to be *vacuous* if $m(\Omega) = 1$, i.e., it has only one focal element, Ω. A vacuous bpa is denoted by m_ι. It represents total ignorance. In our examples, m_ι represents situations when we do not have any information as for the proportion of colors in the drawing urn.

(2) m is said to be *Bayesian*, if all its focal elements are singletons, i.e., for Bayesian bpa m, $m(\mathbf{a}) > 0$ implies $|\mathbf{a}| = 1$. Bayesian bpa's represents exactly the same knowledge as probability functions. As all focal elements of a Bayesian bpa m are singletons, we can define probability distribution P_m for Ω such that

$$P_m(x) = m(\{x\}) \tag{1}$$

for all $x \in \Omega$. Thus, Bayesian bpa's represent in our examples situations when the proportion of colors in a drawing ball is known.

The same knowledge that is expressed by a bpa m can also be expressed by a *belief function*, and by *plausibility function*.

$$Bel_m(\mathbf{a}) = \sum_{\mathbf{b} \in 2^{\Omega}: \mathbf{b} \subseteq \mathbf{a}} m(\mathbf{b}). \tag{2}$$

$$Pl_m(\mathbf{a}) = \sum_{\mathbf{b} \in 2^{\Omega}: \mathbf{b} \cap \mathbf{a} \neq \emptyset} m(\mathbf{b}). \tag{3}$$

We have already mentioned that we interpret the belief function theory as a generalization of the probability theory. It means that for each bpa we consider its *credal set*, which is a convex set of probability distributions P on Ω defined as follows ($\mathcal{P}$ denote the set of all probability distributions on Ω):

$$\mathcal{P}(m) = \left\{ P \in \mathcal{P} : \sum_{x \in \mathbf{a}} P(x) \geq Bel_m(\mathbf{a}) \text{ for } \forall \mathbf{a} \in 2^{\Omega} \right\}.$$

Notice that P_m defined by Eq. (1) for a Bayesian bpa m is such that $\mathcal{P}(m) = \{P_m\}$, and that $\mathcal{P}(m_\iota) = \mathcal{P}$. It is also easy to show that for all $P \in \mathcal{P}(m)$

$$Bel_m(\mathbf{a}) \leq P(\mathbf{a}) \leq Pl_m(\mathbf{a}),$$

for all $\mathbf{a} \in 2^{\Omega}$. Thus, if $Bel(\mathbf{a}) = Pl(\mathbf{a})$ then we are sure that the probability of $\mathbf{a}$ equals $Bel(\mathbf{a})$. Otherwise, the larger the difference $Pl(\mathbf{a}) - Bel(\mathbf{a})$, the more uncertain we are about the value of the probability of $\mathbf{a}$.

In this paper, we use belief functions only to represent the knowledge regarding the content of a drawing drum. How can we model the computation of a subjective expected gain if we know that in situation $x \in \Omega$ our reward will be $g(x)$? Since we want to reduce the expected value on the account of ambiguity we do not apply any direct formula (e.g., Choquet integral [2], Shenoy expectation [19]). We propose to use some of the probability transforms suggested to find a probabilistic representation of a belief function [4]. In this paper, we take advantage of the fact that for the examples presented in the next section it was shown in [14] that several probabilistic transforms yield the same results. Therefore we choose the simplest of them, the famous *pignistic transform*, which was for this purpose strongly advocated by Smets [20,21]):

$$Bet\_P_m(x) = \sum_{a \in 2^{\Omega}:x \in a} \frac{m(a)}{|a|}. \tag{4}$$

3   Experimental Lotteries

In our experiments, we considered 12 simple lotteries described below. For each lottery, the subjects were asked how much they are maximally willing to pay to be allowed to take part in the specified lottery. The considered lotteries should reveal the behavior of subjects in the following three situations.

Ellsberg's Example. First, we wanted to verify whether the behavior of our subjects corresponds to what was observed by many other authors. Therefore we included a simple modification of the original Ellsberg's example ([6], pp. 653–654) with an urn containing 30 red balls and 60 black or yellow balls, the latter in an unknown proportion. With this urn, Ellsberg considered two experiments. The first experiment (Ellsberg's Actions I and II) studied whether people prefer betting on the red or black ball, in which case they get the reward ($100) if the ball of the respective color is drawn at random. In the second experiment (Ellsberg's Actions III and IV), a person has a possibility to bet on red and yellow, or, alternatively, on black and yellow. Again, the participant gets the reward ($100) in case that the randomly drawn ball is of one of the selected colors.

Following the Ellsberg's idea we included two lotteries:

E1 The drawing urn contains 15 red, black and yellow balls, you know that exactly 5 of them are red, you do not know the proportion of the remaining black and yellow balls. How much you are maximally willing to pay to take part in the lottery in which you choose a color and get 100 CZK if the randomly drawn ball has the color of your choice?

E2 The drawing urn contains 15 red, black and yellow balls, you know that exactly 5 of them are red, you do not know the proportion of the remaining black and yellow balls. How much you are maximally willing to pay to take part at the lottery in which you choose a color and get 100 CZK if the randomly drawn ball is either yellow or of the color of your choice?

One Red Ball Example. This example is designed to test the decrease of a subjective "probability" in comparison with the combinatorial probability. For this, we included eight lotteries, which differ from each other just in the total number of balls in the drawing urn: the number n. We included lotteries with $n = 5, 6, 7, 8, 9, 10, 11, 12$:

> **Rn** The drawing urn contains n balls, each of which is either red, or black, or yellow, or white, or green, or azure. You know that one and only one of them is red, nothing more. You even do not know how many colors are present in the urn. How much you are maximally willing to pay to take part in the lottery in which you choose a color and get 100 CZK if the randomly drawn ball is of the color of your choice?

6-Color Example. This example concerns situations, in which six colors are considered and we do not have any reason to prefer one of them to others. Such situations occur in two completely different setting: *fair distribution of colors* and *total ignorance*. Thus, the following two lotteries considered:

> **F1** The drawing urn contains 30 balls, five of each of the following colors: red, black, yellow, white, green, and azure. How much you are maximally willing to pay to take part in the lottery in which you choose a color and get 100 CZK if the randomly drawn ball is of the color of your choice?

> **F2** The drawing urn contains 30 balls, they may be of the following colors: red, black, yellow, white, green, and azure. You know nothing more, you even do not know how much colors are present in the urn. How much you are maximally willing to pay to take part in the lottery in which you choose a color and get 100 CZK if the randomly drawn ball is of the color of your choice?

4 Decision Models

As said in the introduction, to describe the considered situations we define the respective bba's, and belief and plausibility functions. These belief function models are further transformed into probabilistic ones. As we have already mentioned in Sect. 2, for the specified simple situations we consider only the pignistic transform Bet_P_m defined by Eq. (4). However, the resulting probability distribution is not directly used to compute an expected reward. Before computing the subjective expected reward, the considered probabilities are reduced using a coefficient of ambiguity α, and the subjective expected reward is computed using the resulting capacity function $r_{m,\alpha}$. Let us stress again that $r_{m,\alpha}$ is not a probability distribution because it does not sum up to one. Now, we describe this process in more details.

Denote m the bpa describing the situation under consideration. Let Bet_P_m be the corresponding probability distribution obtained by the pignistic transform. Denote by Bel_m and Pl_m belief and plausibility functions corresponding to bpa m. Let us recall that the higher $Pl_m(\{x\}) - Bel_m(\{x\})$, the higher ambiguity about the probability of state $x \in \Omega$. Our intuition says, the higher ambiguity about the probability of a state x, the greater reduction of the respective probability should be done. Therefore we define a *reduced capacity function* $r_{m,\alpha}$ for all $x \in \Omega$ as follows:

$$r_{m,\alpha}(x) = (1 - \alpha)Bet\_P_m(x) + \alpha Bel_m(\{x\}), \tag{5}$$

where $\alpha \in [0, 1]$ denote a subjective *coefficient of ambiguity aversion* $\alpha \in [0, 1]$. Its introduction is inspired by the Hurwicz's *optimism-pessimism* coefficient [10,11]. In contrary to Hurwicz, who suggests that everybody *can choose* a personal coefficient expressing her optimism, we assume that each person *has* a personal coefficient of ambiguity aversion. The higher the aversion the higher the coefficient α. The detection of this coefficient for experimental persons is one of the goals why do we propose the described approach.

Notice that the amount of reduction realized in Formula (5) depends on the ambiguity aversion coefficient α, and the amount of ignorance associated with the state x. If we are certain about the probability of state x, it means that $Bet\_P_m(x) = Bel_m(\{x\})$, then the corresponding probability is not reduced: $r_{m,\alpha}(x) = Bet\_P_m(x)$. On the other hand, the maximum reduction is achieved for the states connected with maximal ambiguity, i.e., for the states for which $Bel_m(\{x\}) = 0$.

Some trivial properties of function $r_{m,\alpha}$ (we will call it *r-weight function*, or simply *r*-weight, in the sequel) are as follows:

1. $\sum_{x \in \Omega} r_{m,\alpha}(x) \le 1$; and
2. m is Bayesian if and only if $m(\{x\}) = Bet\_P_m(x) = r_{m,\alpha}(x)$ for all $x \in \Omega$, and $\alpha \in [0, 1]$.

This *r*-weight function is then used to compute *expected subjective reward*, which is computed similarly to expected value, but the probabilities are substituted by the respective *r*-weights.

$$R_{m,\alpha} = \sum_{x \in \Omega} r_{m,\alpha}(x)g(x), \tag{6}$$

where $g(x)$ denote the reward (gain) one expects in case $x \in \Omega$ occurs. Thus, $R_{m,\alpha}$ does not express a mathematical expected reward, but a *subjectively reduced expectation* of a decision maker, whose subjectivity, i.e., level of ambiguity aversion, is described by α. Let us note that for $\alpha > 0$, betting the amount $R_{m,\alpha}$ guarantees a sure gain [1,15].

Let us now apply this computational process to the situations considered in the preceding section. To proceed from simpler models to more complex ones, let us consider the respective examples in reverse order.

6-Color Example. For this example, $\Omega = \{r, b, y, w, g, a\}$. The knowledge about the content of the drawing urn differs; in case of lottery F1, the situation is described by a Bayesian bpa defined $m_\phi(\{x\}) = \frac{1}{6}$ for all $x \in \Omega$; in case of lottery F2, the situation is described by the vacuous bpa m_ι.

For both the lotteries, the pignistic transforms coincide: $Bet\_P_{m_\phi}(x) = Bet\_P_{m_\iota}(x) = \frac{1}{6}$ for all colors $x \in \Omega$. However, the respective subjective r-weight functions differ because the respective belief functions differ: $Bel_{m_\phi}(\{x\}) = \frac{1}{6}$ for all $x \in \Omega$, whilst $Bel_{m_\iota}(\{x\}) = 0$ for all $x \in \Omega$. Therefore, using Formula (5), $r_{m_\phi,\alpha}(x) = \frac{1}{6}$, and $r_{m_\iota,\alpha}(x) = \frac{1-\alpha}{6}$ for all $x \in \Omega$.

Consider that a player chose, let us say, red color. Let $g(x)$ denote the gain received in case when color x is drawn, i.e., $g(r) = 100$, and for $x \neq r$, $g(x) = 0$. The expected subjective rewards are as follows:

$$R_{m_\phi,\alpha} = \sum_{x \in \Omega} r_{m_\phi,\alpha}(x)g(x) = \sum_{x \in \Omega} \frac{1}{6}\, g(x) = \frac{100}{6},$$

$$R_{m_\iota,\alpha} = \sum_{x \in \Omega} r_{m_\iota,\alpha}(x)g(x) = \sum_{x \in \Omega} \frac{1-\alpha}{6}g(x) = \frac{100 \cdot (1-\alpha)}{6},$$

for F1 and F2, respectively. This can be interpreted as follows. If there were not for the subjective utility functions and for a different subjective risk attitude, a person should be willing to pay a maximum amount of $\frac{100}{6}$ CZK and $\frac{100 \cdot (1-\alpha)}{6}$ CZK for taking part at lottery F1 and F2, respectively. The fact that in case of lottery F1 the person is willing to pay maximally $b \neq \frac{100}{6}$ CZK is explained by her personal risk attitude and utility functions. Nevertheless, the difference between the amounts the person is willing to pay for F1 and F2 can be explained only by her ambiguity aversion measured by the coefficient α. Assuming a linear dependence, it gives us a possibility to estimate the value of a personal coefficient of aversion. If a person is willing to pay a CZK for taking part at lotteries F1/F2 and b CZK for taking part at I1/I2 one can assume that her personal coefficient of ambiguity is about

$$\alpha = \frac{a-b}{a}. \tag{7}$$

One Red Ball Example. For this example, again $\Omega = \{r, b, y, w, g, a\}$, and the uncertainty is described by the bpa m_ϱ as follows:

$$m_\varrho(\mathbf{a}) = \begin{cases} \frac{1}{n}, & \text{if } \mathbf{a} = \{r\}; \\ \frac{n-1}{n}, & \text{if } \mathbf{a} = \{b, g, o, y, w\}; \\ 0, & \text{otherwise.} \end{cases}$$

Using the pignistic transform, we get:

$$Bet\_P_{m_\varrho}(x) = \begin{cases} \frac{1}{n}, & \text{if } x = r; \\ \frac{n-1}{5n}, & \text{for } x \in \{b, g, o, y, w\}. \end{cases}$$

Since $Bel_{m_\varrho}(\{x\}) = 0$ for all $x \in \{b, g, o, y, w\}$, and $Bel_{m_\varrho}(\{r\}) = \frac{1}{n}$ we get the following reduced weights:

$$r_{m_\varrho,\alpha}(x) = \begin{cases} \frac{1}{n}, & \text{if } x = r; \\ (1-\alpha) \cdot \frac{n-1}{5n}, & \text{for } x \in \{b, g, o, y, w\}. \end{cases}$$

Considering (for the sake of simplicity just two) gain functions $g^r(x)$, and $g^w(x)$, the total subjective rewards are as follows. When betting on red it equals

$$R_{m_\varrho,\alpha}(r) = \frac{1}{n}g^r(r) + \sum_{x\in\Omega:x\neq r}\frac{(1-\alpha)(n-1)}{5n}g^r(x) = \frac{100}{n},$$

and analogously, for betting on white

$$R_{m_\varrho,\alpha}(w) = \frac{1}{n}g^w(r) + \sum_{x\in\Omega:x\neq r}\frac{(1-\alpha)(n-1)}{5n}g^w(x) = \frac{100(1-\alpha)(n-1)}{5n}.$$

Table 1. One Red Ball Example: Total subjective reward as a function of the coefficient of ambiguity aversion α, and the number of balls n.

n	$R_{m_\varrho,\alpha}(r)$	$R_{m_\varrho,\alpha}(w)$						
		$\alpha = 0$	$\alpha = 0.1$	$\alpha = 0.2$	$\alpha = 0.28$	$\alpha = 0.3$	$\alpha = 0.4$	$\alpha = 0.5$
5	20.00	16.00	14.40	12.80	11.52	11.20	9.60	8.00
6	16.67	16.67	15.00	13.33	12.00	11.67	10.00	8.33
7	14.29	17.14	15.43	13.71	12.34	12.00	10.29	8.57
8	12.50	17.50	15.75	14.00	12.60	12.25	10.50	8.75
9	11.11	17.78	16.00	14.22	12.80	12.44	10.67	8.89
10	10.00	18.00	16.20	14.40	12.96	12.60	10.80	9.00

Some of the values of these functions are tabulated in Table 1. From this table we see that, for example, a person with $\alpha = 0.28$ should bet on red color for $n \leq 7$, because for these $R_{m_\varrho}(r) > R_{m_\varrho,\alpha}(x)$ $(x \neq r)$, and bet on any other color for $n \geq 8$, because for these n, $R_{m_\varrho,\alpha}(r) \leq R_{m_\varrho,\alpha}(x)$ $(x \neq r)$. This means that for $n \leq 7$, it is subjectively more advantageous to bet on the red color.

Ellsberg's Example. Before showing how the idea of reduced weights is applied to Ellsberg's experiment, let us confess that to clear the main idea to the reader, we have purposely simplified the exposition. The computation of a r-weight function by Formula (5) and its application to computation of a total subjective reward by Formula (6) can be used only in simple situations when the gain function $g : \Omega \to \mathbb{R}$ does not assign the same positive value to two different states from Ω, i.e.,

$$x_1, x_2 \in \Omega, x_1 \neq x_2, g(x_1) > 0 \implies g(x_1) \neq g(x_2). \tag{8}$$

This condition was obviously met by the gain functions considered above because the gain function was positive just for one state from Ω. Let us now introduce a proper general belief function approach that can be used for any gain function.

Generally, we have to consider distribution Bet_P_m that is got from bpa m by the pignistic transform as a set function, and, analogously, also the r-weight function must be defined for all nonempty subsets a of Ω

$$r_{m,\alpha}(a) = (1 - \alpha)P_m(a) + \alpha Bel_m(a), \tag{9}$$

with the same subjective coefficient of ambiguity aversion α. The reader can easily show that this r-weight is monotonous and superadditive

1. for $a \subseteq b$, $r_{m,\alpha}(a) \leq r_{m,\alpha}(b)$;
2. for $a \cap b = \emptyset$, $r_{m,\alpha}(a \cup b) \geq r_{m,\alpha}(a) + r_{m,\alpha}(b)$.

Realize also that we can use the same symbol to denote it, because for singletons it coincide with Formula (5).

As it can be expected, this r-weight set function is used to compute the expected subjective reward. For this, denote $\Gamma = \{g(x) : x \in \Omega\} \setminus \{0\}$, then

$$R_{m,\alpha} = \sum_{\gamma \in \Gamma} \gamma\, r_{m,\alpha}(g^{-1}(\gamma)), \tag{10}$$

where $g^{-1}(\gamma) = \{x \in \Omega : g(x) = \gamma\}$. Notice that most of authors use for this purpose Choquet integral [3, 16], which is not, in our opinion, as intuitive as the proposed formula, and which can be shown to be always less or equal to the introduced $R_{m,\alpha}$.

Now, let us apply this general approach to the belief function model corresponding to E1 and E2 lotteries. For this, $\Omega = \{r, b, y\}$ and the bpa m_ε is as follows:

$$m_\varepsilon(a) = \begin{cases} \frac{1}{3}, & \text{if } a = \{r\}; \\ \frac{2}{3}, & \text{if } a = \{b, y\}; \\ 0, & \text{otherwise.} \end{cases}$$

Its pignistic transform yields a uniform distribution $Bet\_P_{m_\varepsilon}(x) = \frac{1}{3}$ for all $x \in \Omega$. The corresponding belief function is $Bel_{m_\varepsilon}(\{r\}) = \frac{1}{3}$, and $Bel_{m_\varepsilon}(\{b\}) = Bel_{m_\varepsilon}(\{y\}) = 0$, $Bel_{m_\varepsilon}(\{r, b\}) = Bel_{m_\varepsilon}(\{r, y\}) = \frac{1}{3}$, $Bel_{m_\varepsilon}(\{b, y\}) = \frac{2}{3}$, and $Bel_{m_\varepsilon}(\Omega) = 1$. Therefore,

$$r_{m_\varepsilon,\alpha}(a) = \begin{cases} \frac{1}{3}, & \text{if } a = \{r\}; \\ \frac{(1-\alpha)}{3}, & \text{for } a = \{b\}, \{y\}; \\ \frac{(2-\alpha)}{3}, & \text{for } a = \{r, b\}, \{r, y\}; \\ \frac{2}{3}, & \text{if } a = \{b, y\}. \end{cases}$$

For E1, we have to consider two gain functions: $g^r(x)$, and $g^b(x)$ for betting on red and black balls, respectively. These functions are as follows:

$$g^r(r) = 100, g^r(b) = g^r(y) = 0,$$
$$g^b(b) = 100, g^b(r) = g^b(y) = 0.$$

Using Formula (10), the total subjective reward for betting on red ball is

$$R_{m_\varepsilon,\alpha}(r) = 100 \; r_{m_\varepsilon,\alpha}((g^r)^{-1}(100)) = 100 \; r_{m_\varepsilon,\alpha}(\{r\}) = \frac{100}{3},$$

and analogously, for betting on black ball is as follows:

$$R_{m_\varepsilon,\alpha}(b) = 100 \; r_{m_\varepsilon,\alpha}((g^b)^{-1}(100)) = 100 \; r_{m_\varepsilon,\alpha}(\{b\}) = \frac{100(1-\alpha)}{3}.$$

Thus, for positive α, we get $R_{m_\varepsilon,\alpha}(r) > R_{m_\varepsilon,\alpha}(b)$, which is consistent with the Ellsberg's observation that "very frequent pattern of response is that betting on red is preferred to betting on black."

Let us consider the lottery E2, which involves betting on a couple of colors. In comparison with the first experiment, the situation changes only in the respective gain functions; denote them $g^{ry}(x)$ and $g^{by}(x)$ for betting on red and yellow, and for betting on black and yellow balls, respectively.

$$g^{ry}(r) = g^{ry}(y) = 100, g^{ry}(b) = 0,$$
$$g^{by}(b) = g^{by}(y) = 100, g^{by}(r) = 0.$$

Thus, the expected subjective rewards are as follows:

$$R_{m_\varepsilon,\alpha}(ry) = 100 \; r_{m_\varepsilon,\alpha}((g^{ry})^{-1}(100)) = 100 \; r_{m_\varepsilon,\alpha}(\{ry\}) = 100 \; \frac{(2-\alpha)}{3},$$

$$R_{m_\varepsilon,\alpha}(by) = 100 \; r_{m_\varepsilon,\alpha}((g^{by})^{-1}(100)) = 100 \; r_{m_\varepsilon,\alpha}(\{by\}) = 100 \; \frac{2}{3}.$$

Thus, we observe that, for positive α, $R_{m_\varepsilon,\alpha}(by) > R_{m_\varepsilon,\alpha}(ry)$, which is consistent with Ellsberg's observations that "betting on black and yellow is preferred to betting on red and yellow balls."

5   Conclusions

In the paper, we have introduced a belief function model manifesting a similar ambiguity aversion as human decision-makers. The intensity of this aversion is expressed by the subjective coefficient $\alpha \in [0,1]$ with the semantics: the higher the aversion, the higher the coefficient. In the time of submitting the paper for the conference, we have data about the behavior of 32 experimental subjects (university and high school students), who were offered a possibility to take part at the lotteries described in Sect. 3. Thus, one can hardly make serious conclusions. Nevertheless, it appears that computing the ambiguity aversion coefficient as suggested in Formula (7), the experimental subjects show a great variety of the intensity of ambiguity aversion; in fact, the individual coefficients are from the whole interval $[0,1]$, including both extreme values. The average value of this coefficient is about 0.36.

References

1. Camerer, C., Weber, M.: Recent developments in modeling preferences: uncertainty and ambiguity. J. Risk Uncertainty **5**(4), 325–370 (1992)
2. Choquet, G.: Theory of capacities. Ann. Inst. Fourier **5**, 131–295 (1955)
3. Coletti, G., Petturiti, D., Vantaggi, B.: Rationality principles for preferences on belief functions. Kybernetika **51**(3), 486–507 (2015)
4. Cuzzolin, F.: On the relative belief transform. Int. J. Approximate Reasoning **53**(5), 786–804 (2012)
5. Dempster, A.P.: Upper and lower probabilities induced by a multivalued mapping. Ann. Math. Stat. **38**(2), 325–339 (1967)
6. Ellsberg, D.: Risk, ambiguity and the Savage axioms. Q. J. Econ. **75**(2), 643–669 (1961)
7. Ellsberg, D.: Risk, Ambiguity and Decision. Garland Publishing, New York (2001)
8. Halevy, Y.: Ellsberg revisited: an experimental study. Econometrica **75**(2), 503–536 (2007)
9. Halpern, J.Y., Fagin, R.: Two views of belief: belief as generalized probability and belief as evidence. Artif. Intell. **54**(3), 275–317 (1992)
10. Hurwicz, L.: The generalized Bayes minimax principle: a criterion for decision making under uncertainty. Discussion Paper Statistics 335, Cowles Commission (1951)
11. Hurwicz, L.. A criterion for decision making under uncertainty. Technical report 355, Cowles Commission (1952)
12. Jaffray, J.Y.: Linear utility theory for belief functions. Oper. Res. Lett. **8**(2), 107–112 (1989)
13. Jiroušek, R., Shenoy, P.P.: Ambiguity aversion and a decision-theoretic framework using belief functions. In: 2017 SSCI, Proceedings of the 2017 IEEE Symposium Series on Computational Intelligence, Honolulu, pp. 326–332. IEEE, Piscataway (2017)
14. Jiroušek, R., Kratochvíl, V., Rauh, J.: A note on approximation of Shenoy's expectation operator using probabilistic transforms. Int. J. Gen. Syst. (submitted)
15. Kahn, B.E., Sarin, R.K.: Modeling ambiguity in decisions under uncertainty. J. Consum. Res. **15**(2), 265–272 (1988)
16. Klement, E.P., Mesiar, R., Pap, E.: A universal integral as common frame for Choquet and Sugeno integral. IEEE Trans. Fuzzy Syst. **18**(1), 178–187 (2009)
17. Savage, L.J.: Foundations of Statistics. Wiley, New York (1954)
18. Shafer, G.: A Mathematical Theory of Evidence. Princeton University Press, Princeton (1976)
19. Shenoy, P.P.: An expectation operator for belief functions in the Dempster-Shafer theory. In: Kratochvíl, V., Vejnarová, J. (eds.) Proceedings of the 11th Workshop on Uncertainty Processing, pp. 165–176, MatfyzPress, Praha (2018)
20. Smets, P.: Decision making in a context where uncertainty is represented by belief functions. In: Srivastava, R.P., Mock, T.J. (eds.) Belief Functions in Business Decisions. STUDFUZZ, vol. 88, pp. 316–332. Physica, Heidelberg (2002). https://doi.org/10.1007/978-3-7908-1798-0_2
21. Smets, P.: Decision making in the TBM: the necessity of the pignistic transformation. Int. J. Approximate Reasoning **38**(2), 133–147 (2005)
22. Srivastava, R.P.: Decision making under ambiguity: a belief-function perspective. Arch. Control Sci. **6**(1–2), 5–27 (1997)

Combination in Dempster-Shafer Theory Based on a Disagreement Factor Between Evidences

Joaquín Abellán[1], Serafín Moral-García[1($\boxtimes$)], and María Dolores Benítez[2]

[1] Department of Computer Science and Artificial Intelligence,
University of Granada, Granada, Spain
{jabellan,seramoral}@decsai.ugr.es
[2] BANKIA Central Office, Granada, Spain
mbenitez@bankia.com

Abstract. There exist many rules to combine the available pieces of information in Dempster-Shafer theory of Evidence (DST). The first one of them was the Dempster's rule of combination (DRC), which has some known drawbacks. In the literature, many rules have tried to solve the problems founds on DRC but normally they have other non-desirable behaviors too. In this paper, it is proposed a set of mathematical properties that a rule of that type should verify; it is analyzed some of the most used alternatives to the DRC including some of the last hybrid rules, via their properties and behaviors; and it is presented a new hybrid rule that satisfies an important set of properties and does not suffer from the counterintuitive behaviors of other rules.

Keywords: Theory of evidence · Combination rules ·
Dempster's rule · Conflict · Disagreement factor · Hybrid rules

1 Introduction

Dempster-Shafer theory (DST) [4,10] is based on an extension of the probability distributions in probability theory (PT), called *basic probability assignment* (or *evidence*). When we want to measure the degree of disagreement between 2 evidences, a measure called *conflict* is normally used. It is must be distinguished from the known concept of conflict as a measure of uncertainty-based-information in DST for a evidence [1,3,8]. Hence, here we will use the concept of conflict-based-combination (cbc) when we refer to the disagreement degree between two evidences when they are considered for combining or fusing in DST.

Dempster's rule of combination (DRC) was the first rule to combine information from different sources in DST. It can be considered as a natural extension of the Bayes rule in PT. Normally DRC gives us intuitive results but it has been shown that it can give us counterintuitive results when it is used to combined evidences with a high degree of cbc.

© Springer Nature Switzerland AG 2019
G. Kern-Isberner and Z. Ognjanović (Eds.): ECSQARU 2019, LNAI 11726, pp. 148–159, 2019.
https://doi.org/10.1007/978-3-030-29765-7_13

Due to the problems found in the application of DRC, many alternatives to this rule have appeared in the last years, using different ways to manage the cbc. The last alternatives to the DRC that seem to be the better ones are the called hybrid rules. Examples of this issue can be found in [7]. Normally, they use a mixture of 2 rules to obtain a final one that pretends to have a good performance in all the situations.

A combination rule (CR) must verify a set of mathematical properties coherent with the operation that it represents. A set of desirable properties for a CR in DST will be proposed in this work by analyzing the importance of most of the mathematical properties used in the literature, under our point of view. It is obvious that not only a CR must verify a set of properties, but we also hope that a rule for such an aim has good behavior, i.e. a CR should give us coherent results in very different situations.

In this paper, we analyze some of the most used alternatives to the DRC and some of the last hybrid rules. Our first aim is to compare them (i) under a theoretical point of view, analyzing the properties that a rule of combination in DST must verify; and (ii) under a practical point of view, studying behaviors in different situations to show if they give us non-counterintuitive results. Our second aim is to analyze the way to quantify the cbc between two bpas with the aim of presenting a new alternative of CR that can satisfy the majority of the needed properties without counterintuitive results.

The paper is arranged as follows: Sect. 2 presents a resume of the known Dempster-Shafer theory of Evidence. Section 3 describes some of the classic rules for the combination of information in DST; analyzes the properties that such a type of rule must verify, and shows some of the more problematic situations that can appear for these rules. Section 4 exposes known alternative rules to the classic ones, taking into account their properties and behaviors. Section 5 presents a new factor to quantify the maximum degree of disagreement between evidences; and used it to present a new hybrid rule. Conclusions are given in Sect. 6.

2   Dempster-Shafer Theory

Let X be a finite set considered as a set of possible situations, $|X| = k$, $\wp(X)$ the power set of X and x any element in X.

Dempster-Shafer theory [4, 10] is based on the concept of *mass assignment*, *basic probability assignment* (bpa) or *evidence*. A *mass assignment* or bpa is a mapping $m : \wp(X) \to [0, 1]$, such that $m(\emptyset) = 0$ and $\sum_{A \subseteq X} m(A) = 1$.

The value $m(A)$ represents the degree of belief that a specific element of X belongs to set A, but not to any particular subset of A.

The subsets A of X for which $m(A) \neq 0$ are called focal elements.

There are two functions associated with each bpa: a belief function, Bel, and a plausibility function, Pl:

$$Bel(A) = \sum_{B \subseteq A} m(B), \; Pl(A) = \sum_{A \cap B \neq \emptyset} m(B)$$

We may note that belief and plausibility are connected for all $A \in \wp(X)$

$$Pl(A) = 1 - Bel(\overline{A}), \tag{1}$$

where $\overline{A}$ denotes the complement of A. Furthermore, $Bel(A) \leq Pl(A)$.

3   Combination Rules in DST

A combination rule (CR) in DST can be seen as a procedure to combine different sources of information, i.e., unify the information from 2 different evidences in one bpa[1].

In the literature, there exist many procedures for such an aim and they can give us different results when they are applied to the same evidences. The problem is that every rule is based on a different approach and normally every one works well in certain situations but has problems in other ones.

The first rule exposed in DST was the Dempster's rule of combination (DRC). It is considered as a generalization of the Bayes's rule in probability theory. DRC is based on the orthogonal sum of two bpa that is expressed as follows, considering m^1 and m^2 two bpas on the finite set X:

$$m^1 \oplus m^2(A) = \sum_{B \cap C = A} m^1(B)m^2(C), \quad \forall A \subseteq X \tag{2}$$

Hence, the DRC is defined as follows:

$$m_{12}^D(A) = \frac{m^1 \oplus m^2(A)}{1 - K}, \quad \forall A \neq \emptyset, \quad m_{12}^D(\emptyset) = 0, \quad K = \sum_{B \cap C = \emptyset} m^1(B)m^2(C)$$

Here, K represents the mass assigned to the "conflict" between two sources of evidence. When $K = 1$ it is not possible to use this expression to combine the information in DST, it is the case of maximum conflict. In this point we must remark that the "conflict" concept here is different to the one used in uncertainty-based-information measures [3,8].

The origin of the conflict between two bpas for combination is the K value of Dempster. We will call the concept related to the meaning of K as *conflict-based-combination* (cbc). The K value has been analyzed in the literature and it has been shown that it has been considered as not a good way for measuring the cbc because its use in the DRC expression can give us counterintuitive results in situations where a high grade of cbc appears. This has been the principal drawback found on the application of the DRC.

---

[1] In this work we consider that the sources of information are independent, which is not always true in the reality.

Based on the drawbacks found on DRC, Yager [11] exposed a rule of combination where the mass of the empty set is assigned to the whole set of alternatives. Considering m^1 and m^2 two bpas on the finite set X, the Yager's rule can be expressed as follows:

$$m_{12}^Y(A) = m^1 \oplus m^2(A), \quad \forall A \neq X, \emptyset \tag{3}$$

$$m_{12}^Y(X) = m^1 \oplus m^2(X) + m^1 \oplus m^2(\emptyset) \quad m_{12}^Y(\emptyset) = 0 \tag{4}$$

3.1 Properties

A CR should verify a set of desired properties. To determinate the goodness of a rule, we cannot only focus on a set of the mathematical properties that a rule should verify because a rule can have bad behavior in some situations although it verifies a large set of properties.

About the set of desired properties expressed in the literature, we remark the following ones:

- **Idempotency**: When two similar sources of information are combined, the rule must give us the same information.
- **Commutativity**: When a rule of combination is used on two bpas the resulting bpa must do not depend on the order used on the bpas.
- **Associativity**: When a rule is used on several bpas, the resulting one must not vary with the order on the bpas used.
- **Continuity**: If we have two very similar information and we use a rule to combine each one with a third one, then both bpas obtained must be very similar too.
- **Absorption**: When a bpa is combined with the total ignorance the resulting bpa must be the original one.

About these properties, we must express the following comments:

- **Idempotency**: It makes sense that when the same information is repeated we have only that information. If a CR does not verify this property and we combine similar information many times via that CR, the final bpa could be quite different from the original, producing important incoherent results. Hence, this property can be considered as an essential one.
- **Commutativity**: When two bpas are combined it makes sense that the resulting one must not depend on the order used.
- **Associativity**: When we combine two informations we use a one-to-one function to obtain a final information. When a new information appears to be combined with that last one, the resulting bpa obtained depends only on the two bpas used in the procedure of combination. Hence, it makes sense that some information of the first two bpas were lost, because they are used in a combination rule with less strength than the third one. If we reorder the 3 bpas to combine, it is logical that we obtain a different final bpa because the one-to-one procedure used to combine.

- **Continuity**: It makes sense that few variations in information must produce few variations on the information obtained by combination with other information.
- **Absorption**: We do not agree with this property. If b.p.a is combined with the total ignorance, that absence of information must be taken into account in the combination, in such a way that it must imply a decreasing of the mass values for the focal sets of the first bpa that are different from the whole set of alternatives.

To be coherent with the above comments, under our point of view, only the following basic properties must be verified by a rule of combination in DST[2]:

1.- **Idempotency**
2.- **Commutativity**
3.- **Continuity**

Example 1. Let $X = \{x_1, x_2, x_3\}$ be a finite set and the following set of bpas $\{m^k, k = 1, .., 4\}$ on X:

$$m_1^1 = 0.80 \quad m_2^1 = 0.10 \quad m_3^1 = 0 \quad m_{23}^1 = 0.10$$
$$m_1^2 = 0.70 \quad m_2^2 = 0.10 \quad m_3^2 = 0.10 \quad m_{23}^2 = 0.10$$
$$m_1^3 = 0.80 \quad m_2^3 = 0 \quad m_3^3 = 0 \quad m_{23}^3 = 0.20$$
$$m_1^4 = 0.80 \quad m_2^4 = 0.10 \quad m_3^4 = 0 \quad m_{23}^4 = 0.10$$

where m_i^k expresses the value $m^k(x_i)$, m_{ij}^k the one of $m^k(\{x_i, x_j\})$, and so on (Table 1).

Table 1. Results of the combination by the rules of Dempster and Yager. The expressions m^{ijk} in columns indicate, in their superscript, the order of the bpas in the combination. Also m_i expresses the value $m(x_i)$, m_{ij} the one of $m(\{x_i, x_j\})$, and so on for each order of combination.

Rule	m^{12}	m^{123}	m^{1234}
m^D	$m_1^D = 0.918$	$m_1^D = 0.9782$	$m_1^D = 0.995$
	$m_2^D = 0.0492$	$m_2^D = 0.0131$	$m_2^D = 0.004$
	$m_3^D = 0.0164$	$m_3^D = 0.00435$	$m_3^D = 0.0005$
	$m_{23}^D = 0.0164$	$m_{23}^D = 0.00435$	$m_{23}^D = 0.0005$
	$m_{123}^D = 0$	$m_{123}^D = 0$	$m_{123}^D = 0$
m^Y	$m_1^Y = 0.56$	$m_1^Y = 0.048$	$m_1^Y = 0.0384$
	$m_2^Y = 0.03$	$m_2^Y = 0.006$	$m_2^Y = 0.0012$
	$m_3^Y = 0.01$	$m_3^Y = 0.002$	$m_3^Y = 0.0002$
	$m_{23}^Y = 0.01$	$m_{23}^Y = 0.002$	$m_{23}^Y = 0.0002$
	$m_{123}^Y = 0.39$	$m_{123}^Y = 0.942$	$m_{123}^Y = 0.96$

---

[2] If we have the aim to use these rules in applications, other property could be added about the complexity of the rule that allows us to use it easily.

The results obtained by both rules are very counterintuitive due to the no idempotency of these rules. The DRC produces a very high value for x_1 and values close to 0 for the rest ones. On the other hand, the rule of Yager produces a very high value for the total set, being the rest of values close to 0. Consequently, the result provided by the Yager rule is close to total ignorance.

With respect to the behaviors of a rule, there exist different situations that give us counterintuitive results for a determinate rule. Almost every rule gives us debatable results when it is used in a certain situation. By the large set of different situations, it is not possible to give a list of each possible situation. The most known one is the one represented by a high degree of cbc via the K value between 2 bpas that represents the principal drawback found on the DRC, and can be seen in the following example of Zadeh [12]:

Example 2. A patient is analyzed by two doctors. The first one founds that the patient has Meningitis with a mass of assignment of 0.99 or a Cerebral Tumor with a mass of 0.01. The second doctor founds that the patient has a Concussion with a mass of assignment of 0.99 or a Cerebral Tumor with a mass of 0.01. Using the DRC we have that the final combined information says us that the patient has a Cerebral Tumor with a mass of assignment of 1, which was very unlikely for each doctor.

In the above example, we have a high degree of cbc expressed by the K value, in this case $K = 0.99 \cdot 0.99 + 2 \cdot 0.01 \cdot 0.99 = 0.9999$. More discussion about this example can be found in [5].

It is known that the DRC satisfies Property 2. About Property 3, we show that also it is not verified by DRC in the following example:

Example 3. Let X be the finite set $X = \{x_1, x_2, x_3\}$ and m' and m'' the following bpas on X: $m'_{12} = 0.98$, $m'_2 = 0.01$, $m'_3 = 0.01$; $m''_{12} = 0.99$, $m'_2 = 0.01$, where m'_i expresses the value $m'(x_i)$, m_{ij} the one of $m'(\{x_i, x_j\})$; and similar for m''. We note that m and m' are very close bpas. We combine each one with the following bpa m via DRC: $m_3 = 1$.

Noting as m'^D and m''^D to the combination of m' and m'' with m via DRC respectively, we have the following values: $m'^D_3 = \frac{0.01 \cdot 1}{1 - 0.99} = 1$, whereas m''^D cannot be obtained because the K value is 1.

Yager's rule has not the problem expressed by Example 2 and verifies the Property 2, but its main drawback is that it does not verify the essential idempotency property.

4   Some Alternative Rules to the DRC

There exist many mathematical expressions to combine two evidences in the DST. Most of them have the same problem than the one of Yager and DRC: they do not verify the essential Property 1.

The Averaging rule [9] is expressed as the mean values for each focal set. The expression to combine two bpa m and m' on a finite set X is: $m^{Av}(A) = \frac{m(A)+m'(A)}{2}$, $\forall A \subseteq X$.

This is a rule that verifies all the needed properties, as it can be easily checked, but it cannot be considered as a good rule because in some situations its application has little sense. We show two cases about this issue:

(c1) We suppose 2 bpas m^1, m^2 on a finite set $X = \{x_1, .. x_{10}\}$, such that they have many and different focal sets but with $K = 0$ for these bpas. It makes little sense that the resulting bpa has many focal elements that are not in each one of the original bpas.

(c2) Suppose the case that all the focal sets are different but the majority of them share only the element x_t. Then the shared element must have a high mass of evidence in the resulting bpa, but if the set $\{x_t\}$ is not a focal set of one of the bpas, this does not occur.

To avoid the problem caused by the managing of the cbc, Dubois and Prade [6] presented the following rule based on the orthogonal sum and on a disjunction sum, using same notation than in the other expressions:

$$m^{DP}(A) = m^1 \oplus m^2(A) + \sum_{A_1 \cup A_2 = A,\, A_1 \cap A_2 = \emptyset} m^1(A_1)m^2(A_2) \qquad (5)$$

It is considered as a hybrid rule. This rule gives coherent results when it is applied in cases where the K value is high and the DRC gives counterintuitive results. But it does not verify the Property 1, producing counterintuitive results in many situations, though it verifies Properties 2 and 3.

The K value can be arguable as a measure of the cbc between two bpas. In Dymova et al. [7], this value is analyzed and it is showed that it has incorrect behaviors in some situations. Also, in that work, the authors presented another expression to measure the cbc and use it in a new hybrid combination rule. The expressions are the following ones:

– The new value to measure the cbc between two bpas m^1 and m^2 on a finite set X is expressed as follows:

$$Mc(m^1, m^2) = \frac{1}{Nc} \sum_{A \subseteq X} |m^1(A) - m^2(A)|, \qquad (6)$$

with Nc the number of focal sets A where $|m^1(A) - m^2(A)| > 0$.

– Based on the above measure of cbc, the following combination rule is presented in Dymova et al. [7]:

$$m^{Dy}(A) = Mc(m^{Av}) + (1 - Mc)m^D(A) \qquad (7)$$

Authors expressed that this rule gives better results than the DRC and the one of Dubois and Prade. Their rule is based on a new way to measure the cbc via the Mc value.

The Dymova's rule is not idempotent because in that case $Mc = 0$ and then the rule coincides with the DRC rule. It is obviously commutative. However, it has some problems with Property 3. Via the following example, we can see that the use of the Mc value can be problematic because for very close bpas we can obtain very different Mc values.

Example 4. We consider the following bpas on the finite set $X = \{x_1, x_2, x_3, x_4\}$

$$m_4^1 = 0.1, \ m_{12}^1 = 0.9; \quad m_4^2 = 0.9, \ m_{12}^2 = 0.1$$

and m'^1 the following one built very close to m^1

$$m_1'^1 = 0.001, \ m_2'^1 = 0.001, \ m_3'^1 = 0.001, \ m_4'^1 = 0.1, \ m_{12}'^1 = 0.897$$

In this case, we obtain the following Mc values:

$$Mc(m^1, m^2) = 0.8; \quad Mc(m'^1, m^2) = \frac{(0.003 + 0.8 + 0.797)}{5} = 0.32$$

The difference between these two Mc values can imply very different values of masses for the resulting bpa obtained via the combination rule. Hence, we can say that a rule using the Mc value does not verify the Property 3. It has been shown that MC is not a good measure of cbc.

We have that the Dymova's rule only verifies the Property 2, and has some problems of behavior motivated by the way to quantify the cbc and its use in the expression of the hybrid rule used.

These two rules, DRC and Av, can be used to obtain a hybrid rule with good performance in all the conflictive situations exposed in this paper.

5   A Proposal of a New Rule

With the above notation, we propose the use of the following value that can be considered as a factor related to the cbc concept between bpas:

$$Mx(m^1, m^2) = \max_{A \subseteq X} |m^1(A) - m^2(A)| \tag{8}$$

The Mx value can be considered as a factor to quantify the maximum disagreement between two bpas. Perhaps it is not an excellent measure of cbc but it makes more sense that the Mc value.

In this point, we could use the Mx and K values to define new hybrid rules with sense. But if we want to correct the problems of the Dymova's rule, it is not a good way to change only the expression to quantify the conflict, i.e. it is not a good alternative the following variation:

$$m^{ADy}(A) = Mx(m^{Av}) + (1 - Mx)m^D(A), \quad \forall A \subseteq X$$

This new rule corrects some problems with Dymova's rule. The Mx value has not the problem of the Mc expressed in Example 4. Also, this rule has not the problem expressed in Example 3 because in that case $K = M_x = 1$ and the m^{Av} is only applied. However, in the case of Property 1, it still has problems because $Mx = 0$ and the rule coincides with the DRC that is not idempotent.

We propose the following hybrid rule:

$$m^N(A) = \left((1 - Mx|K - Mx|)\, m^{Av} + Mx|K - Mx|\, m^D\right)(A), \ \forall A \subseteq X \quad (9)$$

The new hybrid rule has the following characteristics expressed in the following 3 points:

(1) It has no problem when the maximum K value appears (see Example 2). Here, $Mx = K = 1$ and the new rule coincides with the m^{Av} rule.
(2) It has not the problem expressed in Example 4. When we produce little changes in a bpa, the variation of the values obtained for the maximum value of differences Mx is little too, and then the final bpas obtained by combination with a third bpa, are similar too.
(3) It has not the problem expressed in Example 3. In this case, $K = 1$ and the rule coincides with the m^{Av} rule.

As a consequence of some of the comments expressed above, the new hybrid rule verifies the 3 properties:

Property 1 : It is Idempotent because when $Mx = 0$ the hybrid rule coincides with the Idempotent rule m^{Av}.

Property 2 : It is a convex combination of 2 commutative rules, then it is commutative.

Property 3 : It is a consequence of the points above commented. By point (2) little variations in the values of a bpa produce little variations in the resulting bpas. The problematic case that appears when we are very close to the maximum K value is solved too, taking into account the above point (3). In that case, from Example 3, the resulting bpa is very close to the one obtained only by m^{Av} rule.

5.1   Applications of the New Rule

Finally, we want to apply the new proposed rule on the examples shown in this paper to see its performance. We will use it on the examples where some of the rules showed here have counterintuitive behavior. Each item "EXi" corresponds with the *Example i* in this paper.

Ex1: In each step of the process of combination, we obtain the following mass values with the new combination rule:
The results are very coherent and different from the ones obtained by DRC and Yager's rule.

Rule m^{12}		m^{123}	m^{1234}
m^N	$m_1^N = 0.7549$	$m_1^N = 0.7811$	$m_1^N = 0.7927$
	$m_2^N = 0.0985$	$m_2^N = 0.0488$	$m_2^N = 0.0738$
	$m_3^N = 0.0490$	$m_3^N = 0.0243$	$m_3^N = 0.0120$
	$m_{23}^N = 0.0976$	$m_{23}^N = 0.1458$	$m_{23}^N = 0.1214$
	$m_{123}^N = 0$	$m_{123}^N = 0$	$m_{123}^N = 0$

Ex2: In this case, $K = 0.9999$ and $Mx = 0.99$. Then $Mx|K - Mx| = 0.0099$ and the new rule produces similar values than the ones of the Av rule, that have a correct sense: $m^N(Meningitis) = m^N(Concussion) = 0.4901$ and $m^N(CT) = 0.0198$

Ex3: With the same notation than in that example, we have the following values

$$m_2'^N = 0.005, \ m_3'^N = 0.505, \ m_{12}'^N = 0.49,$$

$$m_2''^N = 0.005, \ m_3''^N = 0.5, \ m_{12}''^N = 0.495,$$

For m''^N we have a coefficient of 0 for the DRC, then we can combine m'' with m; and in both cases we obtain very similar coherent results.

Ex4: The new rule obtains similar results when m^1 is combined with m^2 than when m'^1 is combined with m^2, because their K and Mx values are very similar in both cases, and m^1 is very close to m'^1. Here we have that $K(m^1, m^2) = 0.82$, $K(m'^1, m^2) = 0.8203$; and $Mx(m^1, m^2) = 0.8 = Mx(m'^1, m^2)$. Hence, we have the following values when m^1 is combined with m^2:

$$m_4^N = 0.5, \ m_{12}^N = 0.5$$

and the following ones when m'^1 is combined with m^2:

$$m_1^N = 0.0005, \ m_2^N = 0.0005, \ m_3^N = 0.00049, \ m_4^N = 0.5, \ m_{12}^N = 0.4985$$

In the situation expressed by the item (c2) about the performing of the Av rule, this new rule obtains the same values than the DRC, i.e. it is coherent with the Bayesian updating reasoning. In this case, $Mx = 1$ and $K = 0$, producing a coefficient of 0 for the Av rule in the expression of m^N; and 1 for the DRC.

6   Conclusions

In this paper we have analyzed the properties that a CR must verify, presenting a set of desirable properties. The idempotency property has been considered here as an essential one. We have shown that no verification of this property can give us important not logical results.

As the behavior in certain situations of a CR is as important as the verification of mathematical properties, we have done a short analysis of some of the most known CRs in the literature under those two ways. We have studied a recent hybrid rule presented as a good alternative, and we found that it has some important drawbacks.

We have presented a factor to quantify the maximum disagreement between 2 bpas and we have used it to define a new hybrid rule based on the Dempster's rule of combination and on the Averaging rule. We have shown that in some situations the application of DRC gives counterintuitive results; also we have shown situations where the Av gives little coherent results. With the new hybrid rule, one can fill the problems of the other one. The new CR has shown to verify all the proposed properties and not to suffer from the bad behaviors that the other CRs exposed here have.

The development of functions to information fusion has important applications in many areas. Our nest goal will be to apply these type of functions to combine information from diverse methods that extract knowledge from financial data, as we use in [2].

Acknowledgments. This work has been supported by the Spanish "Ministerio de Economía y Competitividad" and by "Fondo Europeo de Desarrollo Regional" (FEDER) under Project TEC2015-69496-R.

References

1. Abellán, J.: Combining nonspecificity measures in Dempster-Shafer theory of evidence. Int. J. Gen. Syst. **40**(6), 611–622 (2011)
2. Abellán, J., Castellano, J.G., Moral-García, S., Mantas, C.J., Benítez, M.D.: A decision support tool for credit domains: Bayesian network with a variable selector based on imprecise probabilities. Int. J. Inf. Technol. Decis. Making (2017, submitted)
3. Abellán, J., Klir, G., Moral, S.: Disaggregated total uncertainty measure for credal sets. Int. J. Gen. Syst. **35**(1), 29–44 (2006)
4. Dempster, A.P.: Upper and lower probabilities induced by a multivalued mapping. Ann. Math. Stat. **38**(2), 325–339 (1967)
5. Dubois, D., Prade, H.: Combination and propagation of uncertainty with belief functions: a reexamination. In: Proceedings of the 9th International Joint Conference on Artificial Intelligence - Volume 1, IJCAI 1985, pp. 111–113. Morgan Kaufmann Publishers Inc., San Francisco (1985)
6. Dubois, D., Prade, H.: On the combination of evidence in various mathematical frameworks. In: Flamm, J., Luisi, T. (eds.) Reliability Data Collection and Analysis. EURR, vol. 3, pp. 213–241. Springer, Dordrecht (1992). https://doi.org/10.1007/978-94-011-2438-6_13
7. Dymova, L., Sevastjanov, P., Tkacz, K., Cheherava, T.: A new measure of conflict and hybrid combination rules in the evidence theory. In: Rutkowski, L., Korytkowski, M., Scherer, R., Tadeusiewicz, R., Zadeh, L.A., Zurada, J.M. (eds.) ICAISC 2014. LNCS (LNAI), vol. 8468, pp. 411–422. Springer, Cham (2014). https://doi.org/10.1007/978-3-319-07176-3_36
8. Klir, G.J.: Uncertainty and Information: Foundations of Generalized Information Theory. Wiley, Hoboken (2005)

9. Murphy, C.K.: Combining belief functions when evidence conflicts. Decis. Support Syst. **29**(1), 1–9 (2000)
10. Shafer, G.: A Mathematical Theory of Evidence. Princeton University Press, Princeton (1976)
11. Yager, R.R.: On the Dempster-Shafer framework and new combination rules. Inf. Sci. **41**(2), 93–137 (1987)
12. Zadeh, L.A.: Review of a mathematical theory of evidence. AI Mag. **5**(3), 81–83 (1984)

Conditional, Default and Analogical Reasoning

A New Perspective on Analogical Proportions

Nelly Barbot[1], Laurent Miclet[1], Henri Prade[2(✉)], and Gilles Richard[2]

[1] ENSSAT, 22300 Lannion, France
nelly.barbot@irisa.fr, laurent.miclet@gmail.com
[2] IRIT - CNRS, 118, route de Narbonne, 31062 Toulouse Cedex 09, France
{prade,richard}@irit.fr

Abstract. Analogical proportions are statements of the form "a is to b as c is to d". They have known a revival of interest after they have been formalized and used in analogical inference. In particular their meaning has been made clear through a logical modeling. The paper shows that they are closely related to the heart of the reasoning process, since dichotomic trees built from pairs of mutually exclusive properties have also a reading in terms of Boolean analogical proportions. This provides a link between analogy and logically expressed taxonomies. Moreover, this gives birth to noticeable opposition structures, and can be also related to formal concept analysis.

1 Introduction

Analogical proportions, which are statements of the form "a is to b as c is to d", are closely related to analogical reasoning that puts in parallel two situations regarded as similar, one including a and b, and the other c and d. The proportion states that the relationship between a and b is the same as the one between c and d. Analogical reasoning amounts to infer that something may be true in situation 2 since an homologous statement is known to be true in situation 1 (considered as similar enough to situation 2 in other respects). As such, it has been regarded for a long time as a useful, but brittle, mode of reasoning that yields plausible conclusions (which may turn to be wrong).

Clearly analogical reasoning does not at all offer the guarantees of deductive reasoning. Maybe for this reason, it has not attracted the interest of logicians up to few exceptions [12,32]. It is only quite recently that a Boolean modeling of analogical proportions have been proposed [23,27]. It exactly expresses that "a differs from b as c differs from d and b differs from a as d differs from c". This modeling (and its gradual extension for handling numerical features) has proved to be of interest in classification tasks [4,20] or in solving IQ tests [6].

In this paper, we enlarge the logical perspective on analogical proportions by showing that they are naturally associated with dichotomic trees reflecting taxonomic deduction. The paper is organized as follows. After a brief reminder of their Boolean modeling, which is proved to be compatible with a function-based view of analogical proportion in Sect. 2, Sect. 3 shows that pairs of mutually exclusive Boolean properties induce both a dichotomic tree and an equivalent set of analogical proportions; moreover these proportions can be organized in remarkable opposition structures. Section 4 exhibits another noticeable linkage between analogical proportions and semi-products of formal contexts in formal concept analysis [11].

© Springer Nature Switzerland AG 2019
G. Kern-Isberner and Z. Ognjanović (Eds.): ECSQARU 2019, LNAI 11726, pp. 163–174, 2019.
https://doi.org/10.1007/978-3-030-29765-7_14

2   Analogical Proportions: Logical and Functional Views

As numerical proportions, an analogical proportion is a quaternary relation, denoted $a : b :: c : d$ between items a, b, c, d, supposed to obey the three following postulates (e.g., [9, 16, 21]):

1. $\forall a, b,\ a : b :: a : b$ (*reflexivity*);
2. $\forall a, b, c, d,\ a : b :: c : d \rightarrow c : d :: a : b$ (*symmetry*);
3. $\forall a, b, c, d,\ a : b :: c : d \rightarrow a : c :: b : d$ (*central permutation*).

These postulates entail that an analogical proportion $a : b :: c : d$ has eight equivalent forms: $a : b :: c : d = c : d :: a : b = c : a :: d : b = d : b :: c : a = d : c :: b : a = b : a :: d : c = b : d :: a : c = a : c :: b : d$.

Boolean Definition. From now on, a, b, c, d denote Boolean variables. This may be thought as encoding the fact that a given property is true or false for the considered item. Since items are usually described in terms of several properties, the Boolean modeling of analogical proportions is then extended to vectors in a component-wise manner. As shown in [28], the minimal Boolean model obeying the analogical proportion postulates makes $a : b :: c : d$ true only for the 6 patterns exhibited in Table 1 $a : b :: c : d$ is false for the 10 other patterns of values for a, b, c, d.

Table 1. Boolean patterns making $a : b :: c : d$ true

a	b	c	d
0	0	0	0
1	1	1	1
0	0	1	1
1	1	0	0
0	1	0	1
1	0	1	0

Thus, it can be checked that the analogical proportion "a is to b as c is to d" more formally states that "a differs from b as c differs from d and b differs from a as d differs from c", which means $a = b \Leftrightarrow c = d$, and $a \neq b \Leftrightarrow c \neq d$ (with the further requirement that both changes are in the same direction (either from 1 to 0, or from 0 to 1). This is logically expressed as a quaternary connective [23] by

$$a : b :: c : d = ((a \wedge \neg b) \equiv (c \wedge \neg d)) \wedge ((\neg a \wedge b) \equiv (\neg c \wedge d)) \tag{1}$$

Besides, it has been noticed [23] that $a : b :: c : d$ can be equivalently written as

$$a : b :: c : d = ((a \wedge d) \equiv (b \wedge c)) \wedge ((\neg a \wedge \neg d) \equiv (\neg b \wedge \neg c)) \tag{2}$$

or still equivalently

$$a : b :: c : d = ((a \wedge d) \equiv (b \wedge c)) \wedge ((a \vee d) \equiv (b \vee c)) \tag{3}$$

Expression (3) can be viewed as the logical counterpart of a well-known property of geometrical proportions: the product of the means is equal to the product of the extremes.

Boolean analogical proportions are transitive, namely $a : b :: c : d$ and $c : d :: e : f$ entails that $a : b :: e : f$ holds as well. They are also code independent, namely $a : b :: c : d = \neg a : \neg b :: \neg c : \neg d$. This latter property means that the Boolean variable of the considered attribute pertaining to items underlying a, b, c, d can be encoded positively or negatively without changing anything.

Representing objects with a single Boolean value is not generally sufficient and when items are represented by *vectors* of Boolean values, each component being the value of a binary attribute, a simple extension of the previous definition to Boolean vectors in $\mathbb{B}^n$ of the form $\vec{a} = (a_1, ..., a_n)$ can be defined as follows:

$$\vec{a} : \vec{b} :: \vec{c} : \vec{d} \text{ iff } \forall i \in [1, n], \ a_i : b_i :: c_i : d_i$$

It has been recently pointed out that it is easy to build analogical proportions as soon as we compare two items that differ at least on two attribute values [7].

Function-Based View. As often mentioned (see, e.g., [26]), $x : f(x) :: y : f(y)$ looks like a good prototype of analogical proportion. Indeed a statement of the form "x is to $f(x)$ as y is to $f(y)$" sounds as a statement making sense, namely one applies the same function f for obtaining $f(x)$ and $f(y)$ from x and y respectively. However, note that such a view differs from the view of a numerical proportion, since $a : a^2 :: b : b^2$, for some integers a and b, makes sense with $f(x) = x^2$, but not in terms of differences or of ratios. If we accept $x : f(x) :: y : f(y)$, some derived form (according to postulates) such as "x is to y as $f(x)$ is to $f(y)$" suggests that f should be injective (one-to-one) for making sure that $f(x) \neq f(y)$ as soon as $x \neq y$.

The above remark suggests that if we consider 4 items a, b, c, d, and we are wondering if $a : b :: c : d$ can be stated, one may think in terms of the change from a to b (and c to d), hypothesizing that b is obtained by the application of some unknown function f, i.e., $b = f(a)$. Such intuition is implicitly underlying the approach developed in COPYCAT [13] for completing a, b, c with a plausible d. It can be also found in formal models such as the ones of [8] in terms of a mapping between algebras (with an algorithm that computes a fourth pattern such that an identical relation holds between the items of the two pairs making the analogical proportion), or of [19] based on category-based view (advocated much earlier in formal anthropology). So d should be obtained as $f(c)$, when $b = f(a)$ (assuming that $a : b :: c : x$ has a unique solution). Thus, one may consider that there is no harm to assume that f is onto. Thus f is bijective and can be inverted.

Still, it is also natural, especially when trying to complete a, b, c, to look at the change from a to c and to hypothesize that c is obtained from a by the application of some unknown function g, i.e., $c = g(a)$. This leads to $a : f(a) :: g(a) : f(g(a))$[1], which indeed sounds right. However, due to central permutation postulates we have $a : g(a) :: f(a) : f(g(a))$, and thus we should also have $a : g(a) :: f(a) : g(f(a))$. This means that $f(g(a)) = g(f(a))$, i.e., f and g commute. Moreover g, as f, is bijective and can be inverted.

[1] It can be noticed that there is no analogical proportion equivalent to $a : b :: c : d$ of the form $b : c :: x : y$, or $c : b :: x : y$, or $b : x :: c : y$, or $c : x :: b : y$, which suggests that there is no need for considering a function $h(b) = c$, or $h'(c) = b$.

As we are going to see this view is compatible with the Boolean modeling. When $a : b :: c : d$ holds true, we can state $a \wedge \neg b = c \wedge \neg d = \alpha$ and $b \wedge \neg a = d \wedge \neg c = \beta$, where $\alpha \wedge \beta = \bot$. Similarly, since $a : c :: b : d$ also holds true, let $a \wedge \neg c = b \wedge \neg d = \varphi$ and $c \wedge \neg a = d \wedge \neg b = \psi$, where $\varphi \wedge \psi = \bot$. Thus, we can introduce the two Boolean functions

$$f(x) = (x \wedge \neg \alpha) \vee \beta; \quad g(x) = (x \wedge \neg \varphi) \vee \psi$$

and check that we indeed have

$$a : c :: b : d = a : f(a) :: g(a) : f(g(a)) = a : f(a) :: g(a) : g(f(a))$$

Proof 1. 1. $f(a) = (a \wedge \neg \alpha) \vee \beta = (a \wedge \neg(a \wedge \neg b)) \vee (b \wedge \neg a) = (a \wedge b) \vee (b \wedge \neg a) = b$
2. Similarly $g(a) = (a \wedge \neg(a \wedge \neg c)) \vee (c \wedge \neg a) = c$.
3. $g(f(a)) = g(b) = (b \wedge \neg(b \wedge \neg d)) \vee (d \wedge \neg b) = (b \wedge d) \vee (d \wedge \neg b) = d$.   □

This shows the agreement of the Boolean view with function-based view.

3   Analogical Structures of Opposition and Binary Trees

In this section, we investigate another aspect of the pervasiveness of analogical proportions by building them from mutually exclusive properties that also give birth to binary classification trees. Indeed a binary tree is a way of cataloguing objects by means of a set of relevant attributes. As such, it is a taxonomic structure, where a node represents a sub-class of any other node on the path from the root to the former node. This has been observed for a long time. In that respect, Johann Christian Lange (1669–1756) deserves a particular mention since he invented a tree-like diagram for solving syllogisms and finding the valid ones [14]; see [15] for details.

The analogical proportions associated to binary classification trees lead to a particular type of square, of cube, and more generally of hypercube of opposition. A preliminary sketch of these ideas can be found in an extended abstract [1].

3.1   Oppositions Underlying an Analogical Proportion Organized into a Square

An analogical proportion can be obtained by taking pairs of mutually exclusive Boolean properties, (p, p'), (q, q'), (r, r') (i.e., such that $p \wedge p' = \bot$, $q \wedge q' = \bot$, $r \wedge r' = \bot$), and then by considering the four items $\mathbf{a}, \mathbf{a'}, \mathbf{b}, \mathbf{b'}$ respectively described on the six properties (p, q, r, r', q', p') according to the following Table 2.

Table 2. Analogical proportion obtained from pairs of mutually exclusive properties

	p	q	r	r'	q'	p'
a	1	1	1	0	0	0
a'	1	1	0	1	0	0
b	1	0	1	0	1	0
b'	1	0	0	1	1	0

It can be seen that for any vector component,

$$(a_i \wedge \neg a'_i \equiv b_i \wedge \neg b'_i) \wedge (\neg a_i \wedge a'_i \equiv \neg b_i \wedge b'_i)$$

holds true, where $\mathbf{a} = (a_1, a_2, a_3, a_4, a_5, a_6)$, $\mathbf{a'}$, $\mathbf{b}$, $\mathbf{b'}$ being similarly defined, and each component a_i ($i = 1, \cdots, 6$) refers to the truth-value of $\mathbf{a}$ for properties p, q, r, r', q', p' respectively. We can observe that (a_i, a'_i, b_i, b'_i) for $i = 1, \cdots, 6$ takes respectively the values $(1, 1, 1, 1)$, $(1, 1, 0, 0)$, $(1, 0, 1, 0)$, $(0, 1, 0, 1)$, $(0, 0, 1, 1)$, and $(0, 0, 0, 0)$ which are the six 4-tuples that make an analogical proportion true. Thus the proportion $\mathbf{a}$: $\mathbf{a'} :: \mathbf{b} : \mathbf{b'}$ holds true.

It is worth noticing that $\mathbf{a}, \mathbf{a'}, \mathbf{b}, \mathbf{b'}$ make a kind of square of opposition (not to be confused with the traditional one [3, 25]), as pictured below:

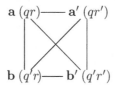

Indeed $\mathbf{a}, \mathbf{a'}, \mathbf{b}, \mathbf{b'}$ can be organized as a square exhibiting oppositions in the following sense: $\mathbf{a}, \mathbf{a'}$ satisfy q while $\mathbf{b}, \mathbf{b'}$ satisfy q', and $\mathbf{a}, \mathbf{b}$ satisfy r while $\mathbf{a'}, \mathbf{b'}$ satisfy r'. Moreover, diagonals $\mathbf{ab'}$ and $\mathbf{a'b}$ link items that are opposite with respect to properties q, q', r, r'.

Note that an equivalent binary (classification) tree can be built from the properties p, q, q', r, r', where we have indicated in bold the item associated to each path going from the root to the corresponding leave.

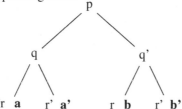

Although the above tree could be closely related to a decision tree, it does not follow the usual convention of associating each node with a property that is true in the left branch and false in the right branch, below the node. It is rather a classification tree organizing the items into classes and sub-classes. Indeed, the root node is associated with the property which is true for the whole class (p in the above example), then the nodes below are associated with the property that specializes the two subclasses (q and q' above), and so on. Following the path from the root to a leaf, we read the collection of properties that are true for the item in bold, near the leaf (e.g., $pq'r$ for $\mathbf{b}$).

In this paper, we only consider Boolean analogical proportions defined for Boolean variables. However, the definition can be straightforwardly extended to n-valued variables [29]. Assuming that the domain of a considered variable may be now $D = \{d^1, \cdots, d^n\}$ (instead of $\{0, 1\}$), the patterns $d^i : d^j :: d^i : d^j$ and $d^i : d^i :: d^j : d^j$, for all $d_i, d_j \in D$, would be the only ones that make an analogical proportion true, which would be false otherwise. This would enable us to extend the above approach to non binary trees where we deal with n-tuples of mutually exclusive properties. For instance, suppose we have objects whose size can be big or $small$ and whose color is a value in $D = \{blue, red, yellow\}$; then we would be led to consider analogical proportions such that $(big, red) : (small, red) :: (big, blue) : (small, blue)$.

3.2   Analogical Cube of Opposition

The above process can be obviously iterated. Let us introduce one more pair of mutually exclusive properties, say o, o', and thus one more level in the tree, as below:

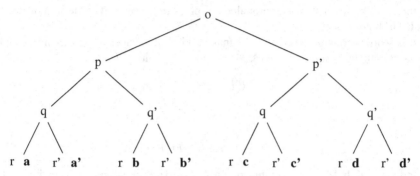

This corresponds to a table (see Table 3) which has now eight items $\mathbf{a}, \mathbf{a'}, \mathbf{b}, \mathbf{b'}$, $\mathbf{c}, \mathbf{c'}, \mathbf{d}, \mathbf{d'}$. Note that the table is also associated with an example of practical interpretation, involving the four oppositions: animal/plant, canid/suidae, tame/wild and young/adult.

Table 3. Example of a super analogical proportion obtained from pairs of mutually exclusive properties

	o (animal)	p (canid)	q (tame)	r (young)	r' (adult)	q' (wild)	p' (suidae)	o' (plant)
a *puppy*	1	1	1	1	0	0	0	0
a' *dog*	1	1	1	0	1	0	0	0
b *wolf cub*	1	1	0	1	0	1	0	0
b' *wolf*	1	1	0	0	1	1	0	0
c *piglet*	1	0	1	1	0	0	1	0
c' *pig*	1	0	1	0	1	0	1	0
d *yg.wd.boar*	1	0	0	1	0	1	1	0
d' *wild boar*	1	0	0	0	1	1	1	0

This gives birth to the cube[2] in Fig. 1 where parallel facets are in opposition on one property (e.g., facet $\mathbf{a}, \mathbf{b}, \mathbf{c}, \mathbf{d}$ corresponds to young animals (or), while facet $\mathbf{a'}, \mathbf{b'}, \mathbf{c'}, \mathbf{d'}$) corresponds to adult animals (or'). As can be seen, the edges of the cube link items in opposition on only one property over three (e.g., $\mathbf{a'}$ (pqr') and $\mathbf{c'}$ ($p'qr'$)). Diagonals in a facet link items in opposition on two properties over three (e.g., $\mathbf{a'}$ (pqr') and $\mathbf{c}$ ($p'qr$)), while diagonals in the cube link items in complete opposition on three properties (e.g., $\mathbf{a'}$ (pqr') and $\mathbf{d}$ ($p'q'r$)).

3.3   Super Analogical Proportion

Looking at Table 3, we may see it as the truth table of a connective with 8 *Boolean variables* $a, a', b, b', c, c', d, d'$ (we no longer use here the bold notation since we are

---

[2] This cube is distinct from the cube of opposition obtained as an extension of the traditional square of opposition; see [5, 10, 30].

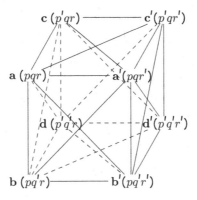

Fig. 1. Analogical cube

no longer dealing with vectors), which would be true only for the eight Boolean
8-tuples corresponding to the eight columns of Table 3, and false for any other Boolean
8-tuples (among the $2^8 = 256$ possible ones). This defines a connective with 8 entries
which would be denoted $a/a'/b/b'/c/c'/d/d'$. Since the 8 patterns that make it true are
obtained by concatenation of 4-tuples defining the analogical proportion semantically,
we call it *super proportion*. It can be shown that we have the following result:[3]

$$a/a'/b/b'/c/c'/d/d'$$
$$= (a : a' :: b : b') \wedge (a : a' :: c : c') \wedge (a : b :: c : d) \wedge (a' : b' :: c' : d') \quad (4)$$

Proof of Equation (4). Let us show that $(a : a' :: b : b') \wedge (a : a' :: c : c') \wedge (a :
b :: c : d) \wedge (a' : b' :: c' : d')$ is true only for the 8 patterns exhibited in the columns of
Table 3 and false for any other 8-tuple (among 2^8). Since $a : a' :: b : b'$ should be true,
it is impossible to have either $a = a'$ and $b \neq b'$, or $a \neq a'$ and $b = b'$. We have 4 cases

- $(a, a') = (0, 1)$. Thus $(b, b') = (0, 1) = (c, c')$ since $a : a' :: c : c'$ should hold. Then
  we should have $(a, a', b : b', c : c', d, d') = (0, 1, 0, 1, 0, 1, 0, 1)$ since $(a : b :: c :
  d) \wedge (a' : b' :: c' : d')$ should also hold.
- $(a, a') = (1, 0)$. Then similarly $(b, b') = (1, 0) = (c, c')$.
  It leads to $(a, a', b, b', c, c', d, d') = (1, 0, 1, 0, 1, 0, 1, 0)$.
- $(a, a') = (1, 1)$. Due to the two first conjuncts in (4), there are four sub-cases
  - $(b, b') = (1, 1)$ and $(c, c') = (1, 1)$. The two last conjuncts in (4) lead to
    $(a, a', b, b', c, c', d, d') = (1, 1, 1, 1, 1, 1, 1, 1)$.
  - $(b, b') = (1, 1)$ and $(c, c') = (0, 0)$. The two last conjuncts in (4) lead to
    $(a, a', b, b', c, c', d, d') = (1, 1, 1, 1, 0, 0, 0, 0)$.
  - $(b, b') = (0, 0)$ and $(c, c') = (1, 1)$. The two last conjuncts in (4) lead to
    $(a, a', b, b', c, c', d, d') = (1, 1, 0, 0, 1, 1, 0, 0)$.
  - $(b, b') = (0, 0)$ and $(c, c') = (0, 0)$. Then there is no solution for the equations
    $a : b :: c : x$ and $a' : b' :: c' : x'$.

---

[3] The third conjunct is unfortunately missing in [1].

- $(a, a') = (0, 0)$. There are the same four sub-cases again
  - $(b, b') = (1, 1)$ and $(c, c') = (1, 1)$ Then there is no solution for the equations $a : b :: c : x$ and $a' : b' :: c' : x'$.
  - $(b, b') = (1, 1)$ and $(c, c') = (0, 0)$. The two last conjuncts in (4) lead to $(a, a', b, b', c, c', d, d') = (0, 0, 1, 1, 0, 0, 1, 1)$.
  - $(b, b') = (0, 0)$ and $(c, c') = (1, 1)$. The two last conjuncts in (4) lead to $(a, a', b, b', c, c', d, d') = (0, 0, 0, 0, 1, 1, 1, 1)$.
  - $(b, b') = (0, 0)$ and $(c, c') = (0, 0)$. The two last conjuncts in (4) lead to $(a, a', b : b', c : c', d, d') = (0, 0, 0, 0, 0, 0, 0, 0)$.

As can be seen, the only patterns that make (4) true are precisely the eight 8-tuples appearing in Table 3.   □

The four conjuncts of expression (4) correspond to four facets of the cube. The two other facets of the cube are also associated with analogical proportions, as well as its six diagonal plans. Indeed the truth of (4) entails that the following eight analogical proportions hold as well (as can be checked in Table 3):

- $c : c' :: d : d'$,
- $b : b' :: d : d'$,
  and for the six diagonal plans:
- $b : b' :: c : c'$,
- $a : a' :: d : d'$,
- $a : b :: c' : d'$,
- $a' : b' :: c : d$,
- $a : b' :: c : d'$,
- $a' : b :: c' : d$.

Thus, all facets and all diagonal plans correspond to analogical proportions in the *analogical cube*. The cube is associated to 128 *syntactically* distinct analogical proportions, since each of the 12 above proportions have 8 different syntactic forms that are semantically equivalent due to symmetry and central permutation properties. Moreover there are 32 degenerated proportions with only one, or two distinct items (having two syntactic forms each in this latter case) corresponding respectively to the eight vertices and twelve edges. This makes a total of $12 \times 8 + 32 = 128$ analogical proportions.

It is clear that the procedure that has led us from a square to a cube can be iterated by considering more properties, and thus generalized to hypercubes, involving still more analogical proportions. This iteration process can be also understood in the setting of formal concept analysis, as we are now going to see briefly.

4   A Linkage with Formal Concept Analysis

The construction can be related to semi-products[4] in formal concept analysis (FCA) [11]. FCA defines a formal concept as a pair made of a subset of objects and a subset of Boolean attributes describing the extension and the intension of the concept respectively. It starts from a so-called formal context, which is a relation defined on a Cartesian

---

[4] This subsection follows from an idea suggested by Bernhard Ganter to the second author.

$$
\begin{array}{c|cc}
 & a_s & a_t \\
\hline
 & \times & \\
 & & \times
\end{array}
$$

product of a referential of objects and a referential of attributes, which records the information that each object is described by a subset of attributes that it possesses. The reader is referred to [11] for details. In the following, we only make use of the notion of semi-product of formal contexts. The formal contexts considered are basic ones, all have the same form, as sketched above with two attributes a_s and a_t and two objects (corresponding to the two lines of the above table, where the symbol $\times$ means that the object has the attribute of the column where the symbol is). As can be seen there is a form of opposition between the two objects, where an object has one of the two attributes and not the other, while it is the converse for the other one. The semi-product, denoted ⊠ , of two formal contexts yields another formal context whose number of columns is the sum of the number of columns of the formal contexts involved in the semi-product, and the number of lines is the product of their number of lines. Let us first consider the semi-product of two basic formal contexts of the above form. As can be seen, the first line of the result is obtained by inserting the first line of the second context in the first line of the first context, then the second line of the result is obtained by inserting the second line of the second context in the first line of the first context, the third line of the result is obtained by inserting the first line of the second context in the second line of the first context, and the fourth line of the result is obtained by inserting the second line of the second context in the second line of the first context. In FCA, the ordering of the attributes (and the objects) is a matter of convenience; it is here chosen for the purpose of making clear a link with analogical proportions. Indeed, we recognize in the columns of the result four of the patterns making an analogical proportion true (replacing each $\times$ by 1, and blanks by 0). The only missing patterns are $(1, 1, 1, 1)$ and $(0, 0, 0, 0)$, but lines or columns full of $\times$, as well as blank lines or columns, are not considered in FCA for mathematical convenience and because it corresponds to extreme situations where attributes are not discriminating. The above result is called an analogical complex [22] which plays an important role for finding out analogical proportions between formal concepts in a formal context [2].

$$
\begin{array}{c|cc}
 & a_1 & a_6 \\
\hline
 & \times & \\
 & & \times
\end{array}
\;\;⊠\;\;
\begin{array}{c|cc}
 & a_2 & a_5 \\
\hline
 & \times & \\
 & & \times
\end{array}
\;=\;
\begin{array}{c|cccc}
 & a_1 & a_2 & a_5 & a_6 \\
\hline
 & \times & \times & & \\
 & \times & & & \times \\
 & & \times & \times & \\
 & & & \times & \times
\end{array}
$$

Let us iterate the application of the semi-product by "multiplying" the result of first semi-product again by a basic formal context. The result has now six columns and eight lines. It can be checked that the six columns correspond to six of the eight 8-tuples making a super analogical proportion (see Table 3). The two missing patterns are $(1, 1, 1, 1, 1, 1, 1, 1)$ and $(0, 0, 0, 0, 0, 0, 0, 0)$ corresponding again to a column full of $\times$ and to a blank column.

$$
\begin{array}{c|cccc}
 & a_1\ a_2\ a_5\ a_6 \\
\hline
 & \times\ \times \\
 & \times\qquad \times \\
 & \qquad \times\ \times \\
 & \qquad\ \ \times\ \times
\end{array}
\ \boxtimes\
\begin{array}{c|cc}
 & a_3\ a_4 \\
\hline
 & \times \\
 & \quad \times
\end{array}
\ =\
\begin{array}{c|cccccc}
 & a_1\ a_2\ a_3\ a_4\ a_5\ a_6 \\
\hline
 & \times\ \times\ \times \\
 & \times\ \times\qquad\ \times \\
 & \times\qquad\ \times\qquad\ \times \\
 & \times\qquad\qquad\ \times\ \times \\
 & \qquad\times\ \times\qquad\qquad\ \times \\
 & \qquad\times\qquad\ \times\qquad\ \times \\
 & \qquad\qquad\ \times\qquad\ \times\ \times \\
 & \qquad\qquad\ \times\ \times\ \times
\end{array}
$$

A new "multiplication" of the result by a basic formal context, would lead to a formal context with $6 + 2 = 8$ columns and $2^3 \times 2 = 16$ lines, corresponding to a "super super" analogical proportion true for $8 + 2 = 10$ 16-tuples, and false for all other 16-tuples. This would have an hypercube with 4 dimensions, and a binary tree of depth 4, as counterparts. More generally, this construction process shows that a "super^n" analogical proportion is true for $2 \cdot (n + 3)$ 2^{n+2}-tuples only (a "super^0" analogical proportion is just an analogical proportion).

5   Conclusion

In this paper, we have seen that a binary classification tree not only supports an understanding in terms of taxonomic (deductive) reasoning, but also can be seen as equivalent to a collection of analogical proportions! This suggests an unexpected interplay between two modes of reasoning, usually regarded as very different in nature. Both have their roots in the differentiation process of the items in terms of their characteristic attributes. The (hyper)cube representation reflects the levels of opposition between two items in such a structure. It may help explaining results in classification by lying bare what items have in common and how they differ. Formal concept analysis provides another view of the relation linking an object to its properties. This paper has only identified hidden relationships between classification trees, analogical proportions, and formal concept analysis. How to take advantage of these relationships is a still a largely open question: for doing that they have also to be better understood and developed (for instance for handling n-ary trees).

Besides, the example leading to the cube of opposition suggests that analogical proportions do not exist in complete isolation. For instance, we have "a puppy is to a dog as a piglet is to a pig", "a puppy is to a dog as a wolf cub is to a wolf" and "a piglet is to a pig as a wolf cub is to a wolf". This might be useful for trying to solve analogical proportion equations between words in natural language. Indeed, following the pioneering work of Rumelhart and Abrahamson [31] in computational linguistics, the parallelogram-based modeling of analogical proportions for solving word analogies has become a standard evaluation tool for vector-space semantic models where words are represented as numerical vectors (e.g., [17, 18, 24]). However computing $\vec{d}$ as the solution of $\vec{a} - \vec{b} = \vec{c} - \vec{d}$, yields only a family of potential solutions (the neighbors of $\vec{d}$ corresponding to a word), and the closest neighbor is rarely the right solution. Taking advantage that the equation to be solved is related to other equations might help find the right solutions. This would contribute to bridge this computational view of analogical proportions with the work presented here.

Acknowledgements. This work was partially supported by ANR-11-LABX-0040-CIMI (Centre International de Mathématiques et d'Informatique) within the program ANR-11-IDEX-0002-02, project ISIPA.

References

1. Barbot, N., Miclet, L., Prade, H.: The analogical cube of opposition. In: Béziau, J.Y., Buchsbaum, A., Vandoulakis, I. (eds.) Handbook of Abstracts of the 6th World Congress on the Square of Opposition, Crete, 1–5 November 2018, pp. 9–10 (2018)
2. Barbot, N., Miclet, L., Prade, H.: Relational proportions between objects and attributes. In: Kuznetsov, S.O., Napoli, A., Rudolph, S. (eds.) Proceedings of 6th International Workshop "What can FCA do for Artificial Intelligence?", Co-located with (IJCAI/ECAI 2018). CEUR Workshop Proceedings, Stockholm, Sweden, 13 July 2019, vol. 2149, pp. 33–44 (2018)
3. Béziau, J.-Y.: New light on the square of oppositions and its nameless corner. Log. Invest. **10**, 218–233 (2003)
4. Bounhas, M., Prade, H., Richard, G.: Analogy-based classifiers for nominal or numerical data. Int. J. Approx. Reasoning **91**, 36–55 (2017)
5. Ciucci, D., Dubois, D., Prade, H.: Structures of opposition induced by relations - the Boolean and the gradual cases. Ann. Math. Artif. Intell. **76**(3–4), 351–373 (2016)
6. Correa Beltran, W., Prade, H., Richard, G.: Constructive solving of Raven's IQ tests with analogical proportions. Int. J. Intell. Syst. **31**(11), 1072–1103 (2016)
7. Couceiro, M., Hug, N., Prade, H., Richard, G.: Behavior of analogical inference w.r.t. Boolean functions. In: Lang, J. (ed.) Proceedings of the 27th International Joint Conference on Artificial Intelligence (IJCAI 2018), Stockholm, 13–19 July 2018, pp. 2057–2063 (2018)
8. Dastani, M., Indurkhya, B., Scha, R.: Analogical projection in pattern perception. J. Exp. Theor. Artif. Intell. **15**(4), 489–511 (2003)
9. Dorolle, M.: Le Raisonnement par Analogie. PUF, Paris (1949)
10. Dubois, D., Prade, H.: From Blanché's hexagonal organization of concepts to formal concept analysis and possibility theory. Log. Univers. **6**(1–2), 149–169 (2012)
11. Ganter, B., Wille, R.: Formal Concept Analysis. Springer, Heidelberg (1999). https://doi.org/10.1007/978-3-642-59830-2
12. Gust, H., Kühnberger, K., Schmid, U.: Metaphors and heuristic-driven theory projection (HDTP). Theoret. Comput. Sci. **354**(1), 98–117 (2006)
13. Hofstadter, D., Mitchell, M.: The Copycat project: a model of mental fluidity and analogy-making. In: Hofstadter, D., The Fluid Analogies Research Group (eds.) Fluid Concepts and Creative Analogies: Computer Models of the Fundamental Mechanisms of Thought, pp. 205–267. Basic Books Inc., New York (1995)
14. Langius, I.C.: Inventvm Novvm Quadrati Logici Vniversalis [...]. Henningius Millerus, Gissa Hassorum (1714)
15. Lemanski, J.: Logic diagrams in the Weigel and Weise circles. Hist. Philos. Log. **39**(1), 3–28 (2018)
16. Lepage, Y.: Analogy and formal languages. Electr. Not. Theor. Comp. Sci. **53**, 180–191 (2002). Moss, L.S., Oehrle, R.T. (eds.) Proceedings of Joint Meeting of the 6th Conference on Formal Grammar and the 7th Conference on Mathematics of Language
17. Levy, O., Goldberg, Y., Dagan, I.: Improving distributional similarity with lessons learned from word embeddings. Trans. Ass. Comput. Ling. **3**, 211–225 (2015)
18. Linzen, T.: Issues in evaluating semantic spaces using word analogies. CoRR, abs/1606.07736 (2016)
19. Lorrain, F.: Réseaux Sociaux et Classifications Sociales. Essai sur l'Algèbre et la Géométrie des Structures Sociales. Hermann, Paris (1975)

20. Miclet, L., Bayoudh, S., Delhay, A.: Analogical dissimilarity: definition, algorithms and two experiments in machine learning. J. Artif. Intell. Res. (JAIR) **32**, 793–824 (2008)
21. Miclet, L., Delhay, A.: Relation d'analogie et distance sur un alphabet défini par des traits. Technical report 1632, IRISA, July 2004
22. Miclet, L., Nicolas, J.: From formal concepts to analogical complexes. In: Ben Yahia, S., Konecny, J. (eds.) Proceedings of 12th International Conference on Concept Lattices and Their Applications. CEUR Workshop Proceedings, Clermont-Ferrand, 13–16 October 2015, vol. 1466, pp. 159–170 (2015)
23. Miclet, L., Prade, H.: Handling analogical proportions in classical logic and fuzzy logics settings. In: Sossai, C., Chemello, G. (eds.) ECSQARU 2009. LNCS (LNAI), vol. 5590, pp. 638–650. Springer, Heidelberg (2009). https://doi.org/10.1007/978-3-642-02906-6_55
24. Mikolov, T., Sutskever, I., Chen, K., Corrado, G.S., Dean, J.: Distributed representations of words and phrases and their compositionality. In: Burges, C.J.C., Bottou, L., Ghahramani, Z., Weinberger, K.Q. (eds.) Advances in Neural Information Processing Systems 26: 27th Annual Conference on Neural Information Processing Systems 2013. Proceedings of NIPS 2013, Lake Tahoe, 5–8 December 2013, pp. 3111–3119 (2013)
25. Parsons, T.: The traditional square of opposition. In: Zalta, E.N. (ed.) The Stanford Encyclopedia of Philosophy. Metaphysics Research Lab, Stanford University, Stanford (2017)
26. Prade, H., Richard, G.: Cataloguing/analogizing: a non monotonic view. Int. J. Intell. Syst. **26**(12), 1176–1195 (2011)
27. Prade, H., Richard, G.: From analogical proportion to logical proportions. Log. Univers. **7**(4), 441–505 (2013)
28. Prade, H., Richard, G.: Analogical proportions: from equality to inequality. Int. J. Approx. Reasoning **101**, 234–254 (2018)
29. Prade, H., Richard, G.: Multiple-valued logic interpretations of analogical, reverse analogical, and paralogical proportions. In: Proceedings of 40th IEEE International Symposium on Multiple-Valued Logic (ISMVL 2010), Barcelona, pp. 258–263 (2010)
30. Reichenbach, H.: The syllogism revised. Philos. Sci. **19**(1), 1–16 (1952)
31. Rumelhart, D.E., Abrahamson, A.A.: A model for analogical reasoning. Cogn. Psychol. **5**, 1–28 (2005)
32. Russell, S.J.: The Use of Knowledge in Analogy and Induction. Pitman, London (1989)

On the Antecedent Normal Form
of Conditional Knowledge Bases

Christoph Beierle$^{(\boxtimes)}$ and Steven Kutsch

FernUniversität in Hagen, 58084 Hagen, Germany
{christoph.beierle,steven.kutsch}@fernuni-hagen.de

Abstract. Desirable properties of a normal form for conditional knowledge are, for instance, simplicity, minimality, uniqueness, and the respecting of adequate equivalences. In this paper, we propose the notion of antecedentwise equivalence of knowledge bases. It identifies more knowledge bases as being equivalent and allows for a simpler and more compact normal form than previous proposals. We develop a set of transformation rules mapping every knowledge base into an equivalent knowledge base that is in antecedent normal form (ANF). Furthermore, we present an algorithm for systematically generating conditional knowledge bases in ANF over a given signature. The approach is complete in the sense that, taking renamings and equivalences into account, every consistent knowledge base is generated. Moreover, it is also minimal in the sense that no two knowledge bases are generated that are antecedentwise equivalent or that are isomorphic to antecedentwise equivalent knowledge bases.

Keywords: Conditional · Knowledge base · Equivalence ·
Antecedentwise equivalence · Antecedent normal form · ANF ·
Renaming · Knowledge base generation

1 Introduction

A core question in knowledge representation and reasoning is what a knowledge base consisting of a set of conditionals like "If A then usually B", formally denoted by $(B|A)$, entails [20]. For investigating this question and corresponding properties of a knowledge base, for comparing the inference relations induced by different knowledge bases, for implementing systems realizing reasoning with conditional knowledge bases, and for many related tasks a notion of normal form for knowledge bases is advantageous. Desirable properties of a normal form for conditional knowledge bases are, for instance, simplicity, minimality, uniqueness, and the respecting of adequate equivalences of knowledge bases. Normal forms of conditional knowledge bases have been investigated in e.g. [3,4]. In this paper, we propose the new notion of antecedentwise equivalence of conditional knowledge bases and the concept of antecedent normal form (ANF) of a knowledge base. Antecedentwise equivalence identifies more knowledge bases as being equivalent and allows for a simpler and more compact normal form

© Springer Nature Switzerland AG 2019
G. Kern-Isberner and Z. Ognjanović (Eds.): ECSQARU 2019, LNAI 11726, pp. 175–186, 2019.
https://doi.org/10.1007/978-3-030-29765-7_15

than previous proposals. As an effective way of transforming every knowledge base $\mathcal{R}$ into an equivalent knowledge base being in ANF, we develop a set of transformation rules Θ achieving this goal. Furthermore, we present an algorithm KB_{gen}^{ae} enumerating conditional knowledge bases over a given signature. The algorithm is complete in the sense that every consistent knowledge base is generated when taking renamings and antecedentwise equivalences into account. Moreover, KB_{gen}^{ae} is also minimal: It will not generate any two different knowledge bases $\mathcal{R}$, $\mathcal{R}'$ such that $\mathcal{R}$ and $\mathcal{R}'$ or any isomorphic images of $\mathcal{R}$ and $\mathcal{R}'$ are antecedentwise equivalent. This algorithm is a major improvement over the approach given in [9] because it generates significantly fewer knowledge bases, while still being complete and minimal. Systematic generation of knowledge bases as achieved by KB_{gen}^{ae} is fruitful for various purposes, for instance for the empirical comparison and evaluation of different nonmonotonic inference relations induced by a knowledge base (e.g. [5,17,20,22]) with the help of implemented reasoning systems like InfOCF [6].

For illustrating purposes, we will use ranking functions, also called ordinal conditional functions (OCF) [23,24], as semantics for conditionals. However, it should be noted that all notions and concepts developed in this paper are independent of the semantics of ranking functions we use in this paper. They also apply to every semantics satisfying system P [1,17], e.g., Lewis' system of spheres [21], conditional objects evaluated using Boolean intervals [12], possibility distributions [10], or special classes of ranking functions like c-representations [15]. A common feature of these semantics is that a conditional $(B|A)$ is accepted if its verification $A \wedge B$ is considered more plausible, more possible, less surprising, etc. than its falsification $A \wedge \neg B$.

After recalling required basics in Sect. 2, antecedentwise equivalence and ANF is introduced in Sect. 3. The system Θ transforming a knowledge base into ANF is presented in Sect. 4. Orderings and renamings developed in Sect. 5 are exploited in knowledge base generation by KB_{gen}^{ae} in Sect. 6, before concluding in Sect. 7.

2   Background: Conditional Logic

Let $\mathcal{L}$ be a propositional language over a finite signature Σ of atoms $a, b, c, \ldots$. The formulas of $\mathcal{L}$ will be denoted by letters $A, B, C, \ldots$. We write AB for $A \wedge B$ and $\overline{A}$ for $\neg A$. We identify the set of all complete conjunctions over Σ with the set Ω of possible worlds over $\mathcal{L}$. For $\omega \in \Omega$, $\omega \models A$ means that $A \in \mathcal{L}$ holds in ω, and the set of worlds satisfying A is $\Omega_A = \{\omega \mid \omega \models A\}$. By introducing a new binary operator $|$, we obtain the set $(\mathcal{L} \mid \mathcal{L}) = \{(B|A) \mid A, B \in \mathcal{L}\}$ of conditionals over $\mathcal{L}$. For a conditional $r = (B|A)$, $ant(r) = A$ is the antecedent of r, and $cons(r) = B$ is its consequent. The counter conditional of $r = (B|A)$ is $\overline{r} = (\overline{B}|A)$. As semantics for conditionals, we use ordinal conditional functions (OCF) [24]. An OCF is a function $\kappa : \Omega \to \mathbb{N}$ expressing degrees of plausibility of possible worlds where a lower degree denotes "less surprising". At least one world must be regarded as being normal; therefore, $\kappa(\omega) = 0$ for at least one $\omega \in \Omega$. Each κ uniquely extends to a function mapping sentences to $\mathbb{N} \cup \{\infty\}$ given by

$\kappa(A) = \min\{\kappa(\omega) \mid \omega \models A\}$ where $\min \emptyset = \infty$. An OCF κ *accepts* a conditional $(B|A)$, written $\kappa \models (B|A)$, if the verification of the conditional is less surprising than its falsification, i.e., if $\kappa(AB) < \kappa(A\overline{B})$; equivalently, $\kappa \models (B|A)$ iff for every $\omega' \in \Omega_{A\overline{B}}$ there is $\omega \in \Omega_{AB}$ with $\kappa(\omega) < \kappa(\omega')$. A conditional $(B|A)$ is *trivial* if it is *self-fulfilling* $(A \models B)$ or *contradictory* $(A \models \overline{B})$; a set of conditionals is self-fulfilling if every conditional in it is self-fulfilling. A finite set $\mathcal{R} \subseteq (\mathcal{L}|\mathcal{L})$ of conditionals is called a *knowledge base*. An OCF κ accepts $\mathcal{R}$ if κ accepts all conditionals in $\mathcal{R}$, and $\mathcal{R}$ is *consistent* if an OCF accepting $\mathcal{R}$ exists [14]. We use $\diamond$ to denote an inconsistent knowledge base. $Mod(\mathcal{R})$ denotes the set of all OCFs κ accepting $\mathcal{R}$. Two knowledge bases $\mathcal{R}, \mathcal{R}'$ are *model equivalent*, denoted by $\mathcal{R} \equiv_{mod} \mathcal{R}'$, if $Mod(\mathcal{R}) = Mod(\mathcal{R}')$. We say $(B|A) \equiv (B'|A')$ if $A \equiv A'$ and $AB \equiv A'B'$. Example 1 presents a knowledge base we will use for illustration.

Example 1 ($\mathcal{R}_{car}$ [4]). Let $\Sigma_{car} = \{c, e, f\}$ where c indicates whether something is a car, e indicates whether something is an e-car, and f indicates whether something needs fossil fuel. The knowledge base $\mathcal{R}_{car}$ contains seven conditionals:

q_1: $(f|c)$ *"Usually cars need fossil fuel."*
q_2: $(\overline{f}|e)$ *"Usually e-cars do not need fossil fuel."*
q_3: $(c|e)$ *"E-cars usually are cars."*
q_4: $(e|e\overline{f})$ *"E-cars that do not need fossil fuel usually are e-cars."*
q_5: $(e\overline{f}|e)$ *"E-cars usually are e-cars that do not need fossil fuel."*
q_6: $(\overline{e}|\top)$ *"Usually things are no e-cars."*
q_7: $(cf \vee \overline{c}f|ce \vee c\overline{e})$ *"Things that are cars and e-cars or cars but not e-cars are cars that need fossil fuel or are no cars but need fossil fuel."*

3   Antecedentwise Equivalence of Knowledge Bases

For comparing or generating knowledge bases, it is useful to abstract from merely syntactic variants. In particular, it is desirable to have minimal versions and normal forms of knowledge bases at hand. The following notion of equivalence presented in [4] employs the idea that each piece of knowledge in one knowledge base directly corresponds to a piece of knowledge in the other knowledge base.

Definition 1 (equivalence $\equiv_{ee}$ [4]). *Let $\mathcal{R}, \mathcal{R}'$ be knowledge bases.*

- *$\mathcal{R}$ is an elementwise equivalent sub-knowledge base of $\mathcal{R}'$, denoted by $\mathcal{R} \ll_{ee} \mathcal{R}'$, if for every conditional $(B|A) \in \mathcal{R}$ that is not self-fulfilling there is a conditional $(B'|A') \in \mathcal{R}'$ such that $(B|A) \equiv (B'|A')$.*
- *$\mathcal{R}$ and $\mathcal{R}'$ are strictly elementwise equivalent if $\mathcal{R} \ll_{ee} \mathcal{R}'$ and $\mathcal{R}' \ll_{ee} \mathcal{R}$.*
- *$\mathcal{R}$ and $\mathcal{R}'$ are elementwise equivalent, denoted by $\mathcal{R} \equiv_{ee} \mathcal{R}'$, if either both are inconsistent, or both are consistent and strictly elementwise equivalent.*

Elementwise equivalence is a stricter notion than model equivalence. In [3], as a simple example the knowledge bases $\mathcal{R}_1 = \{(a|\top), (b|\top), (ab|\top)\}$ and $\mathcal{R}_2 = \{(a|\top), (b|\top)\}$ are given which are model equivalent, but not elementwise equivalent since for $(ab|\top) \in \mathcal{R}_1$ there is no corresponding conditional in $\mathcal{R}_2$.

The idea of the notion of antecedentwise equivalence we will introduce here is to take into account the set of conditionals having the same (or propositionally equivalent) antecedent when comparing to knowledge bases.

Definition 2 ($Ant(\mathcal{R})$, $\mathcal{R}_{|A}$, **ANF**). *Let $\mathcal{R}$ be a knowledge base.*

- $Ant(\mathcal{R}) = \{A \mid (B|A) \in \mathcal{R}\}$ *is the set of* antecedents *of* $\mathcal{R}$.
- *For $A \in Ant(\mathcal{R})$, the set $\mathcal{R}_{|A} = \{(B'|A') \mid (B'|A') \in \mathcal{R}$ and $A \equiv A'\}$ is the set of* A-conditionals *in* $\mathcal{R}$.
- $\mathcal{R}$ *is in* antecedent normal form *(ANF) if either $\mathcal{R}$ is inconsistent and $\mathcal{R} = \diamond$, or $\mathcal{R}$ is consistent, does not contain any self-fulfilling conditional, contains only conditionals of the form $(AB|A)$, and $|\mathcal{R}_{|A}| = 1$ for all $A \in Ant(\mathcal{R})$.*

Definition 3 ($\ll_{ae}$, **equivalence** $\equiv_{ae}$). *Let $\mathcal{R}$, $\mathcal{R}'$ be knowledge bases.*

- $\mathcal{R}$ *is an* antecedentwise equivalent sub-knowledge base *of $\mathcal{R}'$, denoted by $\mathcal{R} \ll_{ae} \mathcal{R}'$, if for every $A \in Ant(\mathcal{R})$ such that $\mathcal{R}_{|A}$ is not self-fulfilling there is an $A' \in Ant(\mathcal{R}')$ with $\mathcal{R}_{|A} \equiv_{mod} \mathcal{R}'_{|A'}$.*
- $\mathcal{R}$ *and $\mathcal{R}'$ are* strictly antecedentwise equivalent *if $\mathcal{R} \ll_{ae} \mathcal{R}'$ and $\mathcal{R}' \ll_{ae} \mathcal{R}$.*
- $\mathcal{R}$ *and $\mathcal{R}'$ are* antecedentwise equivalent, *denoted by $\mathcal{R} \equiv_{ae} \mathcal{R}'$, if either both are inconsistent, or both are consistent and strictly antecedentwise equivalent.*

Note that any two inconsistent knowledge bases are also antecedentwise equivalent according to Definition 3, e.g., $\{(b|a), (\bar{b}|b)\} \equiv_{ae} \{(b|b), (a\bar{a}|\top)\}$, enabling us to avoid cumbersome case distinctions when dealing with consistent and inconsistent knowledge bases. In general, we have:

Proposition 1 ($\equiv_{ae}$). *Let $\mathcal{R}, \mathcal{R}'$ be consistent knowledge bases.*

1. *If $\mathcal{R} \ll_{ae} \mathcal{R}'$ then $Mod(\mathcal{R}') \subseteq Mod(\mathcal{R})$.*
2. *If $\mathcal{R} \equiv_{ae} \mathcal{R}'$ then $\mathcal{R} \equiv_{mod} \mathcal{R}'$.*
3. *If $\mathcal{R} \ll_{ee} \mathcal{R}'$ then $\mathcal{R} \ll_{ae} \mathcal{R}'$.*
4. *If $\mathcal{R} \equiv_{ee} \mathcal{R}'$ then $\mathcal{R} \equiv_{ae} \mathcal{R}'$.*
5. *None of the implications (1.)–(4.) holds in general in the reverse direction.*

Proof. (1.) If $\mathcal{R} \ll_{ae} \mathcal{R}'$, Definition 3 implies that there is a function $f : Ant(\mathcal{R}) \to Ant(\mathcal{R}')$ with $\mathcal{R}_{|A} \equiv_{mod} \mathcal{R}'_{|f(A)}$ for each $A \in Ant(\mathcal{R})$. Thus, $\mathcal{R} = \bigcup_{A \in Ant(\mathcal{R})} \mathcal{R}_{|A} \equiv_{mod} \bigcup_{A \in Ant(\mathcal{R})} \mathcal{R}'_{|f(A)} \subseteq \mathcal{R}'$ implies $Mod(\mathcal{R}') \subseteq Mod(\mathcal{R})$. Employing (1.) in both directions, we get (2.).

(3.) If $\mathcal{R} \ll_{ee} \mathcal{R}'$, Definition 1 ensures a function $f : \mathcal{R} \to \mathcal{R}'$ with $\{(B|A)\} \equiv_{mod} \{f((B|A))\}$ for each $(B|A) \in \mathcal{R}$. Hence, $A \equiv A'$ must hold if $(B'|A') = f((B|A))$. Thus, $\{(B|A) \mid (B|A) \in \mathcal{R}_{|A}\} \equiv_{mod} \{f((B|A)) \mid (B|A) \in \mathcal{R}_{|A}\}$ for each $A \in Ant(\mathcal{R})$. Together with $\mathcal{R} = \bigcup_{A \in Ant(\mathcal{R})} \mathcal{R}_{|A}$ and $\{f((B|A)) \mid (B|A) \in \mathcal{R}_{|A}\} \subseteq \mathcal{R}'$ this implies $\mathcal{R} \ll_{ae} \mathcal{R}'$. Employing (3.) in both directions yields (4.).

For proving (5.) w.r.t. both (1.) and (2.), consider $\mathcal{R}_3 = \{(c|a), (c|b)\}$ and $\mathcal{R}_4 = \{(c|a), (c|b), (c|a \vee b)\}$. Then $\mathcal{R}_3 \equiv_{mod} \mathcal{R}_4$ and $\mathcal{R}_3 \ll_{ae} \mathcal{R}_4$, but $\mathcal{R}_4 \not\ll_{ae} \mathcal{R}_3$ and therefore $\mathcal{R}_3 \not\equiv_{ae} \mathcal{R}_4$. For (5.) w.r.t. both (3.) and (4.), consider again $\mathcal{R}_1 = \{(a|\top), (b|\top), (ab|\top)\}$ and $\mathcal{R}_2 = \{(a|\top), (b|\top)\}$. We have $\mathcal{R}_1 \equiv_{ae} \mathcal{R}_2$ because $\mathcal{R}_{1|\top} \equiv_{mod} \mathcal{R}_{2|\top}$, but $\mathcal{R}_1 \not\ll_{ee} \mathcal{R}_2$ and therefore $\mathcal{R}_1 \not\equiv_{ee} \mathcal{R}_2$. $\square$

In the proof of Proposition 1 $\mathcal{R}_1 \not\equiv_{ee} \mathcal{R}_2$ and $\mathcal{R}_1 \equiv_{ae} \mathcal{R}_2$ holds, but also $\mathcal{R}_2 \ll_{ee} \mathcal{R}_1$. The following example shows that two knowledge bases may be antecedentwise equivalent even if they are not comparable with respect to $\ll_{ee}$.

Example 2 ($\equiv_{ae}$). Let $\mathcal{R}_5 = \{(bc|a), (cd|a)\}$ and $\mathcal{R}_6 = \{(bd|a), (bcd|a)\}$. Then $\mathcal{R}_5 \equiv_{ae} \mathcal{R}_6$, but $\mathcal{R}_5 \not\equiv_{ee} \mathcal{R}_6$, $\mathcal{R}_5 \not\ll_{ee} \mathcal{R}_6$, and $\mathcal{R}_6 \not\ll_{ee} \mathcal{R}_5$.

4 Transforming Knowledge Bases into ANF

In order to be able to deal with normal forms of formulas in $\mathcal{L}$ without having to select a specific representation, we assume a function ν mapping a propositional formula A to a unique normal form $\nu(A)$ such that $A \equiv A'$ iff $\nu(A) = \nu(A')$. We also use a function Π with $\Pi(\mathcal{R}) = \diamond$ iff $\mathcal{R}$ is inconsistent; Π can easily be implemented by the tolerance test for conditional knowledge bases [14]. Using Π and the propositional normalization function ν, the system Θ given in Fig. 1 contains four transformation rules:

(SF) removes a self-fulling conditional $(B|A)$ with $A \not\equiv \bot$.

(AE) merges two conditionals $(B|A)$ and $(B'|A')$ with propositionally equivalent antecedents to a conditional having this antecedent and the conjunction of the consequents.

(NO) transforms a conditional $(B|A)$ by sharpening its consequent to the conjunction with its antecedent and propositionally normalizes the antecedent and the resulting consequent.

(IC) transforms an inconsistent knowledge base into $\diamond$.

Example 3 ($\mathcal{N}(\mathcal{R}_{car})$). Consider the knowledge base $\mathcal{R}_{car}$ from Example 1.

(SF) As $ef \models e$, q_4 is self-fulfilling, and the application of (SF) removes q_4.

(AE) Applying this rule to q_3 and q_5 yields $q_8 : (ce\overline{f}|e)$.

(SF) *self-fulfilling* :	$\dfrac{\mathcal{R} \cup \{(B|A)\}}{\mathcal{R}}$	$A \models B,\ A \not\equiv \bot$
(AE) *antecedence equivalence* :	$\dfrac{\mathcal{R} \cup \{(B|A), (B'|A')\}}{\mathcal{R} \cup \{(BB'|A)\}}$	$A \equiv A'$
(NO) *normalization* :	$\dfrac{\mathcal{R} \cup \{(B|A)\}}{\mathcal{R} \cup \{(\nu(AB)|\nu(A))\}}$	$A \neq \nu(A)$ or $B \neq \nu(AB)$
(IC) *inconsistency* :	$\dfrac{\mathcal{R}}{\diamond}$	$\mathcal{R} \neq \diamond, \Pi(\mathcal{R}) = \diamond$

Fig. 1. Transformation rules Θ and their applicability conditions for the normalization of knowledge bases respecting antecedence equivalence; Π is a consistency test, e.g. the tolerance criterion [14], and ν a normalization function for propositional formulas.

(NO) Applying this rule to q_1 or to q_7 yields $\widetilde{q_1} : (\nu(cf)|\nu(c))$ in both cases, applying it to q_2 or to q_5 yields $\widetilde{q_2} : (\nu(e\overline{f})|\nu(e))$, applying it to q_3 yields $\widetilde{q_3} : (\nu(ce)|\nu(e))$, and applying it to q_6 yields $\widetilde{q_6} : (\nu(\overline{e})|\nu(\top))$. Applying (NO) to $q_8 : (ce\overline{f}|e)$ yields $\widetilde{q_8} : (\nu(ce\overline{f})|\nu(e))$; note that first applying (AE) to $\widetilde{q_2}$ and $\widetilde{q_3}$ and then (NO) to the result also yields exactly $\widetilde{q_8}$.
(IC) As $\mathcal{R}_{car}$ is consistent, (IC) can not be applied to $\mathcal{R}_{car}$.

Thus, applying Θ exhaustively and in arbitrary sequence to $\mathcal{R}_{car}$ gives us the knowledge base $\Theta(\mathcal{R}_{car}) = \{\widetilde{q_1}, \widetilde{q_6}, \widetilde{q_8}\}$. In contrast, the transformation system $\mathcal{T}$ given in [4] would yield $\mathcal{T}(\mathcal{R}_{car}) = \{\widetilde{q_1}, \widetilde{q_2}, \widetilde{q_3}, \widetilde{q_6}\}$ containing more conditionals.

Proposition 2 (properties of Θ). *Let $\mathcal{R}$ be a knowledge base.*

1. **(termination)** Θ *is terminating.*
2. **(confluence)** Θ *is confluent.*
3. **($\equiv_{mod}$ correctness)** $\mathcal{R} \equiv_{mod} \Theta(\mathcal{R})$.
4. **($\equiv_{ae}$ correctness)** $\mathcal{R} \equiv_{ae} \Theta(\mathcal{R})$.
5. **($\equiv_{ae}$ minimizing)**   *If $\mathcal{R}$ is inconsistent then $\Theta(\mathcal{R}) = \diamond$. If $\mathcal{R}$ is consistent, then for all knowledge bases $\mathcal{R}'$ it holds that $\mathcal{R}' \subsetneq \Theta(\mathcal{R})$ implies $\mathcal{R}' \not\equiv_{ae} \mathcal{R}$.*
6. **(ANF)** $\Theta(\mathcal{R})$ *is in antecedent normal form.*

Proof. (1.) (SF), (AE), and (IC) remove at least one conditional, and (NO) can be applied at most once to any conditional. Hence, Θ is terminating.

(2.) Since Θ is terminating, local confluence of Θ implies confluence of Θ; local confluence of Θ in turn can be shown by ensuring that for every critical pair obtained form superpositioning two left hand sides of rules in Θ reduces to the same knowledge base [2,16]: Any critical pair obtained from (IC) and another rule in Θ reduces to $\diamond$ since all rules preserve the consistency status of a knowledge base. Any critical pair obtained from (SF) with (NO) reduces to the same knowledge base since applying (NO) to a self-fulfilling conditional yields again a self-fulfilling conditional. Regarding critical pairs with respect to (NO), we observe that if $\mathcal{R}$ contains two distinct conditionals $(B|A)$ and $(B'|A')$ with $(\nu(AB)|\nu(A)) = (\nu(A'B')|\nu(A'))$, then applying (NO) first to either of the conditionals and second to the other one yields the same result. Critical pairs between (AE) and (NO) reduce to the same result because propositional normalization commutes with (AE). For a critical pair of (SF) and (AE) consider $\mathcal{R}_0 = \mathcal{R} \cup \{(B|A), (B'|A')\}$ with $A \equiv A'$ and $A' \models B'$. Applying (SF) yields $\mathcal{R}_1 = \mathcal{R} \cup \{(B|A)\}$, and applying (AE) yields $\mathcal{R}_2 = \mathcal{R} \cup \{(BB'|A)\}$. Applying (NO) to both $\mathcal{R}_1$ and $\mathcal{R}_2$ yields the same result because $A \equiv A'$, $A' \models B'$ and therefore $AB \equiv ABB'$. Thus, we are left with critical pairs obtained from (AE) which arise from $\mathcal{R} \cup \{(B|A), (B'|A'), (B''|A'')\}$ with $A \equiv A' \equiv A''$ so that (AE) could be applied to $\{(B|A), (B'|A')\}$ and to $\{(B'|A'), (B''|A'')\}$. Applying (AE) to the result followed by (NO) yields $\mathcal{R} \cup \{(\nu(BB'B'')|\nu(A))\}$ in both cases.

(3.) By Proposition 1, (3.) will follow from the proof of (4.).

(4.) We will show that $\equiv_{ae}$-equivalence is preserved by every rule in Θ.

(*IC*) Since Π is a consistency test, $\mathcal{R} \equiv_{ae} \diamond$ because all inconsistent knowledge bases are $\equiv_{ae}$-equivalent. Because all other rules preserve the consistency status of $\mathcal{R}$, we assume that $\mathcal{R}$ is consistent when dealing with the other rules in Θ. (*SF*) By Definition 3 we get $\mathcal{R} \cup \{(B|A)\} \equiv_{ae} \mathcal{R}$. (*AE*) This rule preserves $\equiv_{ae}$-equivalence because $A \equiv A'$ implies $\{(B|A), (B'|A')\} \subseteq (\mathcal{R} \cup \{(B|A), (B'|A')\})_{|A}$, $(BB'|A) \in (\mathcal{R} \cup \{(BB'|A)\})_{|A}$, and $Mod(\{(B|A), (B'|A')\}) = Mod(\{(BB'|A)\})$. (*NO*) This rule preserves $\equiv_{ae}$-equivalence because $(B|A) \in (\mathcal{R} \cup \{(B|A)\})_{|A}$, $(\nu(AB)|\nu(A)) \in (\mathcal{R} \cup \{(\nu(AB)|\nu(A))\})_{|A}$, and $Mod(\{(B|A)\}) = Mod(\{(\nu(AB)|\nu(A))\})$.

(5.) The $\equiv_{ae}$-minimizing property will follow from the proof of (6.).

(6.) From (1.) and (2.) we conclude that $\Theta(\mathcal{R})$ is well defined. If $\Theta(\mathcal{R})$ was not in ANF then at least one of the rules in Θ would be applicable to $\Theta(\mathcal{R})$, contradicting that Θ has been applied exhaustively.   □

Proposition 2 ensures that applying Θ to a knowledge base $\mathcal{R}$ always yields the unique normal form $\Theta(\mathcal{R})$ that is in ANF. This provides a convenient decision procedure for antecedentwise equivalence and thus also for model equivalence.

Proposition 3 (antecedentwise equivalence). *Let $\mathcal{R}$, $\mathcal{R}'$ be knowledge bases. Then $\mathcal{R} \equiv_{ae} \mathcal{R}'$ iff $\Theta(\mathcal{R}) = \Theta(\mathcal{R}')$.*

5   Orderings and Renamings for Conditionals

For developing a method for the systematic generation of knowledge bases in ANF, we will represent each formula $A \in \mathcal{L}$ uniquely by its set Ω_A of satisfying worlds. The two conditions $B \subsetneq A$ and $B \neq \emptyset$ then ensure the falsifiability and the verifiability of a conditional $(B|A)$, thereby excluding any trivial conditional [8]. This yields a propositional normalization function ν, giving us:

Proposition 4 (*NFC(Σ)* [9]). *For $NFC(\Sigma) = \{(B|A) \mid A \subseteq \Omega_\Sigma, B \subsetneq A, B \neq \emptyset\}$, the set of normal form conditionals over a signature Σ, the following holds:*

(**nontrivial**) *$NFC(\Sigma)$ does not contain any trivial conditional.*
(**complete**) *For every nontrivial conditional over Σ there is an equivalent conditional in $NFC(\Sigma)$.*
(**minimal**) *All conditionals in $NFC(\Sigma)$ are pairwise non-equivalent.*

For instance, for $\Sigma_{ab} = \{a, b\}$ we have $(\{ab, a\bar{b}\}|\{ab, \bar{a}b\}) \equiv (\{ab\}|\{ab, \bar{a}b\})$ where the latter is in $NFC(\Sigma_{ab})$. Out of the different 256 conditionals over Σ_{ab} obtained when using sets of worlds as formulas, only 50 are in $NFC(\Sigma_{ab})$ [9].

For defining a linear order on $NFC(\Sigma)$, we use the following notation. For an ordering relation $\leqslant$ on a set M, its lexicographic extension to strings over M is denoted by $\leqslant_{lex}$. For ordered sets $S, S' \subseteq M$ with $S = \{e_1, \ldots, e_n\}$ and $S' = \{e'_1, \ldots, e'_{n'}\}$ where $e_i \leqslant e_{i+1}$ and $e'_j \leqslant e'_{j+1}$ its extension $\leqslant_{set}$ to sets is:

$$S \leqslant_{set} S' \text{ iff } n < n', \text{ or } n = n' \text{ and } e_1 \ldots e_n \leqslant_{lex} e'_1 \ldots e'_{n'} \tag{1}$$

For Σ with ordering $<$, $[\![\omega]\!]_<$ is the usual interpretation of a world ω as a binary number; e.g., for Σ_{ab} with $a < b$, $[\![ab]\!]_< = 3$, $[\![a\bar{b}]\!]_< = 2$, $[\![\bar{a}b]\!]_< = 1$, and $[\![\bar{a}\bar{b}]\!]_< = 0$.

Definition 4 (induced ordering on formulas and conditionals). *Let Σ be a signature with linear ordering $<$. The orderings induced by $<$ on worlds ω, ω' and conditionals $(B|A), (B'|A')$ over Σ are given by:*

$$\omega \overset{w}{\leqslant} \omega' \text{ iff } [\![\omega]\!]_< \geqslant [\![\omega']\!]_< \tag{2}$$

$$(B|A) \overset{c}{\leqslant} (B'|A') \text{ iff } \Omega_A \overset{w}{<}_{set} \Omega_{A'}, \text{ or } \Omega_A \overset{w}{=} \Omega_{A'} \text{ and } \Omega_B \overset{w}{\leqslant}_{set} \Omega_{B'} \tag{3}$$

In order to ease our notation, we will omit the upper symbol in $\overset{w}{<}$ and $\overset{c}{<}$, and write just $<$ instead, and analogously $\leqslant$ for the non-strict variants. For instance, for Σ_{ab} with $a < b$ we have $ab < a\overline{b} < \overline{a}b < \overline{a}\overline{b}$ for worlds, and $(ab|ab \vee a\overline{b}) < (ab|ab \vee \overline{a}\overline{b})$ and $(ab \vee \overline{a}\overline{b}|ab \vee a\overline{b} \vee \overline{a}\overline{b}) < (\overline{a}\overline{b}|ab \vee a\overline{b} \vee \overline{a}b \vee \overline{a}\overline{b})$ for conditionals.

Proposition 5 *($NFC(\Sigma)$, $<$ [9]). For a linear ordering $<$ on a signature Σ, the induced ordering $<$ according to Definition 4 is a linear ordering on $NFC(\Sigma)$.*

Given the ordering $<$ on $NFC(\Sigma)$ from Proposition 5, we will now define a new ordering $\prec$ on these conditionals that takes isomorphisms (or *renamings*) $\rho : \Sigma \to \Sigma$ into account and prioritizes the $<$-minimal elements in each isomorphism induced equivalence class. As usual, ρ is extended canonically to worlds, formulas, conditionals, knowledge bases, and to sets thereof. We say that X and X' are *isomorphic*, denoted by $X \simeq X'$, if there exists a renaming ρ such that $\rho(X) = X'$. For a set M, $m \in M$, and an equivalence relation $\equiv$ on M, the set of equivalence classes induced by $\equiv$ is denoted by $[M]_{/\equiv}$, and the unique equivalence class containing m is denoted by $[m]_\equiv$. For instance, for Σ_{ab} the only non-identity renaming is the function ρ_{ab} with $\rho_{ab}(a) = b$ and $\rho_{ab}(b) = a$, $[\Omega_{\Sigma_{ab}}]_{/\simeq} = \{[ab], [a\overline{b}, \overline{a}b], [\overline{a}\overline{b}]\}$ are the three equivalence classes of worlds over Σ_{ab}, and we have $[(ab|ab \vee a\overline{b})]_\simeq = [(ab|ab \vee \overline{a}b)]_\simeq$.

Definition 5 *($cNFC(\Sigma)$, $\prec$ [9]). Given a signature Σ with linear ordering $<$, let $[NFC(\Sigma)]_{/\simeq} = \{[r_1]_\simeq, \ldots, [r_m]_\simeq\}$ be the equivalence classes of $NFC(\Sigma)$ induced by isomorphisms such that for each $i \in \{1, \ldots, m\}$, the conditional r_i is the minimal element in $[r_i]_\simeq$ with respect to $<$, and $r_1 < \ldots < r_m$. The canonical normal form conditionals over Σ are $cNFC(\Sigma) = \{r_1, \ldots, r_m\}$. The canonical ordering on $NFC(\Sigma)$, denoted by $\prec$, is given by the schema*

$$r_1 \prec \ldots \prec r_m \prec [r_1]_\simeq \setminus \{r_1\} \prec \ldots \prec [r_m]_\simeq \setminus \{r_m\}$$

where $r \prec r'$ iff $r < r'$ for all $i \in \{1, \ldots, m\}$ and all $r, r' \in [r_i]_\simeq \setminus \{r_i\}$.

Proposition 6 *($NFC(\Sigma)$, $\prec$ [9]). For a linear ordering $<$ on a signature Σ, the induced ordering $\prec$ according to Definition 5 is a linear ordering on $NFC(\Sigma)$.*

While $NFC(\Sigma_{ab})$ contains 50 conditionals, there are 31 equivalence classes in $[NFC(\Sigma_{ab})]_{/\simeq}$; hence $cNFC(\Sigma_{ab})$ has 31 elements [9]. The three smallest elements in $NFC(\Sigma_{ab})$ w.r.t. $\prec$ are $(\{ab\}|\{ab, a\overline{b}\})$, $(\{a\overline{b}\}|\{ab, a\overline{b}\})$, $(\{ab\}|\{ab, \overline{a}b\})$, and their corresponding equivalence classes are $[(\{ab\}|\{ab, a\overline{b}\}), (\{ab\}|\{ab, \overline{a}b\})]$, $[(\{a\overline{b}\}|\{ab, a\overline{b}\}), (\{\overline{a}b\}|\{ab, \overline{a}b\})]$, and $[(\{ab\}|\{ab, \overline{a}\overline{b}\})]$.

6 Generating Knowledge Bases in ANF

The algorithm KB_{gen}^{ae} (Algorithm 1) generates all consistent knowledge bases up to antecedentwise equivalence and up to isomorphisms. It uses pairs $\langle \mathcal{R}, C \rangle$ where $\mathcal{R}$ is a knowledge base and C is a set of conditionals that are candidates for extending $\mathcal{R}$ to obtain a new knowledge base. For extending $\mathcal{R}$, conditionals are considered sequentially according to their $\prec$ ordering. Note that in Line 3, only the *canonical* conditionals (which are minimal with respect $\prec$) are used for initializing the set of one-element knowledge bases. In Line 3 (and in Line 11, respectively), a conditional r is selected for initializing (or extending, respectively) a knowledge base. In Lines 4–6 (and in lines 13–15, respectively), in the set D conditionals are collected that do not have to be considered as candidates for further extending the current knowledge base: D_1 contains all conditionals that are smaller than r w.r.t. $\prec$, D_2 contains all conditionals having the same antecedent as r (since R should be ANF), and $\bar{r}$ would make $\mathcal{R}$ inconsistent. The consistency test used in Line 12 can easily be implemented by the well-known tolerance test for conditional knowledge bases [14].

Proposition 7 (KB_{gen}^{ae}). *Let Σ be a signature with linear ordering $<$. Then applying KB_{gen}^{ae} to it terminates and returns $\mathcal{KB}$ for which the following holds:*

1. (correctness)   *If $\mathcal{R} \in \mathcal{KB}$ then $\mathcal{R}$ is a knowledge base over Σ.*
2. (ANF)   *If $\mathcal{R} \in \mathcal{KB}$ then $\mathcal{R}$ is in ANF.*

Algorithm 1. KB_{gen}^{ae} – Generate knowledge bases over Σ up to $\equiv_{ae}$

Input: signature Σ with linear ordering $<$
Output: set $\mathcal{KB}$ of knowledge bases in ANF of over Σ that are consistent, pairwise antecedentwise non-equivalent and pairwise non-isomorphic

```
 1: L₁ ← ∅
 2: k ← 1
 3: for r ∈ cNFC(Σ) do              ▷ only canonical conditionals for initialization
 4:    D₁ ← {d | d ∈ NFC(Σ), d ⋠ r}          ▷ conditional d can not extend {r}
 5:    D₂ ← {(B|A) | (B|A) ∈ NFC(Σ), A = ant(r)}      ▷ (B|A) can not extend {r}
 6:    D ← D₁ ∪ D₂ ∪ {r̄}                      ▷ r̄ can not extend {r}
 7:    L₁ ← L₁ ∪ {⟨{r}, NFC(Σ) \ D⟩}
 8: while Lₖ ≠ ∅ do
 9:    Lₖ₊₁ ← ∅
10:    for ⟨R, C⟩ ∈ Lₖ do          ▷ R knowledge base, C candidates for extending R
11:       for r ∈ C do
12:          if R ∪ {r} is consistent then              ▷ extend R with conditional r
13:             D₁ ← {d | d ∈ C, d ⋠ r}          ▷ conditional d can not extend R ∪ {r}
14:             D₂ ← {(B|A) | (B|A) ∈ C, A = ant(r)}      ▷ (B|A) can not extend R ∪ {r}
15:             D ← D₁ ∪ D₂ ∪ {r̄}                  ▷ r̄ can not extend R ∪ {r}
16:             Lₖ₊₁ ← Lₖ₊₁ ∪ {⟨R ∪ {r}, C \ D⟩}
17:    k ← k + 1
18: return KB = {R | ⟨R, C⟩ ∈ Lᵢ, i ∈ {1, . . . , k}}
```

---

3. ($\equiv_{ae}$ minimality) If $\mathcal{R}, \mathcal{R}' \in \mathcal{KB}$ and $\mathcal{R} \neq \mathcal{R}'$ then $\mathcal{R} \not\equiv_{ae} \mathcal{R}'$.
4. ($\simeq$ minimality)   If $\mathcal{R}, \mathcal{R}' \in \mathcal{KB}$ and $\mathcal{R} \neq \mathcal{R}'$ then $\mathcal{R} \not\simeq \mathcal{R}'$.
5. (consistency)   If $\mathcal{R} \in \mathcal{KB}$ then $\mathcal{R}$ is consistent.
6. (completeness)   If $\mathcal{R}$ is a consistent knowledge base over Σ then there is $\mathcal{R}' \in \mathcal{KB}$ and an isomorphism ρ such that $\mathcal{R} \equiv_{ae} \rho(\mathcal{R}')$.

Proof. The proof is obtained by formalizing the description of KB_{gen}^{ae} given above and the following observations. Note that KB_{gen}^{ae} exploits the fact that every subset of a consistent knowledge base is again a consistent knowledge base. Thus building up knowledge bases by systematically adding remaining conditionals according to their linear ordering $\prec$ ensures completeness; the removal of candidates in Lines 5 and 14 does not jeopardize completeness since Proposition 2 ensures that for each knowledge base an antecedentwise equivalent knowledge base exists that for any propositional formula A contains at most one conditional with antecedent A. Checking consistency when adding a new conditional ensures consistency of the resulting knowledge base. ANF is ensured because all conditionals in $NFC(\Sigma)$ are of the form $(AB|A)$. Because for all A, each generated $\mathcal{R}$ contains at most one conditional with antecedent A, $\equiv_{ae}$-minimality is guaranteed, and $\simeq$-minimality can be shown by induction on the number of conditionals in a knowledge base.   □

Note that KB_{gen}^{ae} generates significantly fewer knowledge bases than the algorithm *GenKB* given in [9]. For each formula A, each $\mathcal{R} \in GenKB(\Sigma)$ may contain up to half of all conditionals in $NFC(\Sigma)$ with antecedent A,[1] while $\mathcal{R} \in KB_{gen}^{ae}(\Sigma)$ may contain at most one conditional with antecedent A.

For instance, $KB_{gen}^{ae}(\Sigma_{ab})$ will generate the knowledge base $\mathcal{R}_7 = \{(\{\overline{a}\overline{b}\}|\{a\overline{b}, \overline{a}\overline{b}\}), (\{a\overline{b}\}|\{ab, a\overline{b}, \overline{a}\overline{b}\})\}$, but it will not generate the knowledge base $\mathcal{R}_8 = \{(\{\overline{a}\overline{b}\}|\{a\overline{b}, \overline{a}\overline{b}\}), (\{ab, a\overline{b}\}|\{ab, a\overline{b}, \overline{a}\overline{b}\}), (\{a\overline{b}, \overline{a}\overline{b}\}|\{ab, a\overline{b}, \overline{a}\overline{b}\})\}$ which is antecedentwise equivalent to $\mathcal{R}_7$, i.e., $\mathcal{R}_8 \equiv_{ae} \mathcal{R}_7$. Furthermore, $KB_{gen}^{ae}(\Sigma_{ab})$ will also not generate, e.g., the knowledge bases $\mathcal{R}_9 = \{(\{a\overline{b}, \overline{a}b\}|\{ab, a\overline{b}, \overline{a}b\}), (\{\overline{a}\overline{b}\}|\{\overline{a}b, \overline{a}\overline{b}\}), (\{ab, \overline{a}b\}|\{ab, a\overline{b}, \overline{a}b\})\}$ or $\mathcal{R}_{10} = \{(\{\overline{a}\overline{b}\}|\{\overline{a}b, \overline{a}\overline{b}\}), (\{\overline{a}b\}|\{ab, a\overline{b}, \overline{a}b\})\}$ which are both antecedentwise equivalent to $\mathcal{R}_7$ when taking isomorphisms into account; specifically, we have $\rho_{ab}(\mathcal{R}_{10}) = \mathcal{R}_7$, and $\rho_{ab}(\mathcal{R}_9) = \mathcal{R}_8$ and hence also $\rho_{ab}(\mathcal{R}_9) \equiv_{ae} \mathcal{R}_7$.

7   Conclusions and Further Work

Aiming at a compact and unique normal form of conditional knowledge bases, we introduced the new notion of antecedentwise equivalence. We developed a system Θ transforming every knowledge base into its unique antecedent normal form. The algorithm KB_{gen}^{ae} is complete in the sense that it generates, for any signature Σ, knowledge bases in ANF such that all knowledge bases over Σ are

---

[1] Note that it can not be more than half of these conditionals with the same antecedent because otherwise there would be a conditional together with its counter conditional, leading to inconsistency of the knowledge base.

covered up to isomorphisms and antecedentwise equivalence. Furthermore, the set of knowledge bases returned by KB_{gen}^{ae} is minimal because no two different knowledge bases are generated such that they or any isomorphic images of them are antecedentwise equivalent. Currently, we are working with KB_{gen}^{ae} and the reasoning system InfOCF [6] for empirically evaluating different nonmonotonic inference relations induced by a conditional knowledge base and for computing the full closures of such inference relations [18]. Another part of our future work is the investigation of inferential equivalence of ANF (for another normal form see [3,7]) with respect to semantics that are not syntax independent like rational closure (cf. [11,13]), but that are syntax dependent like lexicographic closure [19].

References

1. Adams, E.W.: The Logic of Conditionals: An Application of Probability to Deductive Logic. Synthese Library. Springer, Dordrecht (1975). https://doi.org/10.1007/978-94-015-7622-2
2. Baader, F., Nipkow, T.: Term Rewriting and All That. Cambridge University Press, Cambridge (1998)
3. Beierle, C.: Inferential equivalence, normal forms, and isomorphisms of knowledge bases in institutions of conditional logics. In: Hung, C., Papadopoulos, G.A. (eds.) The 34th ACM/SIGAPP Symposium on Applied Computing (SAC 2019), Limassol, Cyprus, 8–12 April 2019, pp. 1131–1138. ACM, New York (2019)
4. Beierle, C., Eichhorn, C., Kern-Isberner, G.: A transformation system for unique minimal normal forms of conditional knowledge bases. In: Antonucci, A., Cholvy, L., Papini, O. (eds.) ECSQARU 2017. LNCS (LNAI), vol. 10369, pp. 236–245. Springer, Cham (2017). https://doi.org/10.1007/978-3-319-61581-3_22
5. Beierle, C., Eichhorn, C., Kern-Isberner, G., Kutsch, S.: Skeptical, weakly skeptical, and credulous inference based on preferred ranking functions. In: Kaminka, G.A., et al. (eds.) Proceedings 22nd European Conference on Artificial Intelligence, ECAI-2016. Frontiers in Artificial Intelligence and Applications, vol. 285, pp. 1149–1157. IOS Press, Amsterdam (2016)
6. Beierle, C., Eichhorn, C., Kutsch, S.: A practical comparison of qualitative inferences with preferred ranking models. KI Künstliche Intelligenz **31**(1), 41–52 (2017)
7. Beierle, C., Kern-Isberner, G.: Semantical investigations into nonmonotonic and probabilistic logics. Ann. Math. Artif. Intell. **65**(2–3), 123–158 (2012)
8. Beierle, C., Kutsch, S.: Computation and comparison of nonmonotonic skeptical inference relations induced by sets of ranking models for the realization of intelligent agents. Appl. Intell. **49**(1), 28–43 (2019)
9. Beierle, C., Kutsch, S.: Systematic generation of conditional knowledge bases up to renaming and equivalence. In: Calimeri, F., Leone, N., Manna, M. (eds.) JELIA 2019. LNCS (LNAI), vol. 11468, pp. 279–286. Springer, Cham (2019). https://doi.org/10.1007/978-3-030-19570-0_18
10. Benferhat, S., Dubois, D., Prade, H.: Possibilistic and standard probabilistic semantics of conditional knowledge bases. J. Log. Comput. **9**(6), 873–895 (1999)
11. Booth, R., Paris, J.B.: A note on the rational closure of knowledge bases with both positive and negative knowledge. J. Log. Lang. Comput. **7**(2), 165–190 (1998)
12. Dubois, D., Prade, H.: Conditional objects as nonmonotonic consequence relationships. Spec. Issue Conditional Event Algebra IEEE Trans. Syst. Man Cybern. **24**(12), 1724–1740 (1994)

13. Giordano, L., Gliozzi, V., Olivetti, N., Pozzato, G.L.: Semantic characterization of rational closure: from propositional logic to description logics. Artif. Intell. **226**, 1–33 (2015)
14. Goldszmidt, M., Pearl, J.: Qualitative probabilities for default reasoning, belief revision, and causal modeling. Artif. Intell. **84**, 57–112 (1996)
15. Kern-Isberner, G. (ed.): Conditionals in Nonmonotonic Reasoning and Belief Revision. LNCS (LNAI), vol. 2087. Springer, Heidelberg (2001). https://doi.org/10.1007/3-540-44600-1
16. Knuth, D.E., Bendix, P.B.: Simple word problems in universal algebra. In: Leech, J. (ed.) Computational Problems in Abstract Algebra, pp. 263–297. Pergamon Press, Oxford (1970)
17. Kraus, S., Lehmann, D., Magidor, M.: Nonmonotonic reasoning, preferential models and cumulative logics. Artif. Intell. **44**, 167–207 (1990)
18. Kutsch, S., Beierle, C.: Computation of closures of nonmonotonic inference relations induced by conditional knowledge bases. In: Kern-Isberner, G., Ognjanović, Z. (eds.) ECSQARU 2019. LNAI, vol. 11726, pp. 226–237. Springer, Cham (2019)
19. Lehmann, D.: Another perspective on default reasoning. Ann. Math. Artif. Intell. **15**(1), 61–82 (1995)
20. Lehmann, D.J., Magidor, M.: What does a conditional knowledge base entail? Artif. Intell. **55**(1), 1–60 (1992)
21. Lewis, D.: Counterfactuals. Harvard University Press, Cambridge (1973)
22. Paris, J.: The Uncertain Reasoner's Companion - A Mathematical Perspective. Cambridge University Press, Cambridge (1994)
23. Spohn, W.: Ordinal conditional functions: a dynamic theory of epistemic states. In: Harper, W., Skyrms, B. (eds.) Causation in Decision, Belief Change, and Statistics, II, pp. 105–134. Kluwer Academic Publishers, Dordrecht (1988)
24. Spohn, W.: The Laws of Belief: Ranking Theory and Its Philosophical Applications. Oxford University Press, Oxford (2012)

Revisiting Conditional Preferences: From Defaults to Graphical Representations

Nahla Ben Amor[1]([⊠]), Didier Dubois[2]([⊠]), Henri Prade[2]([⊠]),
and Syrine Saidi[1]([⊠])

[1] LARODEC Laboratory, ISG de Tunis, 41 rue de la Liberté, 2000 Le Bardo, Tunisia
nahla.benamor@gmx.fr, syrine.saidi@irit.fr
[2] IRIT - CNRS, 118, route de Narbonne, 31062 Toulouse Cedex 09, France
{dubois,prade}@irit.fr

Abstract. A conditional preference statement takes the form "in context c, a is preferred to *not a*". It is quite similar to the piece of knowledge "if c is true, a is more plausible than *not a*", which is a standard way of understanding the default rule "if c then generally a". A set of such defaults translates into a set of constraints that can be represented in the setting of possibility theory. The application of a minimum specificity principle, natural when handling knowledge, enables us to compute a priority ranking between possible worlds. The paper investigates if a similar approach could be applied to preferences as well. Still in this case, the use of a maximum specificity principle is as natural as the converse principle, depending on the decision maker attitude in terms of pessimism or optimism. The paper studies the differences between this approach and qualitative graphical approaches to preference modeling such as π-pref-nets (based on possibility theory) and CP-nets (relying on ceteris paribus principle). While preferences in a conditional preference network can always be expressed as "default-like" constraints, there are cases where "non monotonic" preferences cannot be associated with a preference network structure, but can still be dealt with as constraints. When both approaches can be applied, they may lead to different orderings of solutions. The paper discusses this discrepancy and how to remedy it.

1 Introduction

Possibilistic preference networks (π-pref-nets for short) [1] have been recently proposed as a model, whose graphical structure is identical to the one of conditional preference networks (CP-nets for short), but where each node is associated with a conditional possibility table (with symbolic weights) that represents the conditional preferences corresponding to the node. Then a chain rule enables us to attach a symbolic expression to any configuration or solution (i.e., to any complete instantiation of the variables). This method induces a partial ordering of the solutions, which has been shown to coincide with the inclusion-based ordering of the sets of violated preferences characterizing the solutions (when no further constraints is added between the symbolic weights) [4]. This partial order agrees with the CP-net partial ordering, but is more cautious. Adding appropriate constraints between the symbolic weights enables us to approach the CP-net partial order. More generally, the addition of meaningful constraints guarantees a better control of the representation of the preferences really expressed by the user, without the

© Springer Nature Switzerland AG 2019
G. Kern-Isberner and Z. Ognjanović (Eds.): ECSQARU 2019, LNAI 11726, pp. 187–198, 2019.
https://doi.org/10.1007/978-3-030-29765-7_16

blind introduction of extra preferences (as, e.g., in the case of CP nets, where the violation of preferences associated to father nodes is made more important than the violation of preferences associated to children nodes).

Conditional preferences are statements of the form "in context p, q is preferred to $\bar{q}$", where $\bar{q}$ denotes "not q". This sounds a bit similar to a default rule "if p then generally q", understood as "if p is true, q is more plausible than $\bar{q}$", although it is a piece of knowledge rather than the expression of some preference. Possibility theory is a framework that can be used for representing either knowledge or preferences. In the first case the degree of possibility is understood as a degree of plausibility, and the dual necessity means certainty; in the second case the degree of possibility is a degree of satisfaction and the degree of necessity is a priority level. In both cases, the conditional statement translates into the constraint $\Pi(p \wedge q) > \Pi(p \wedge \bar{q})$. Thus, a set of such default rules is turned into a set of possibilistic constraints. Then the application of a minimum specificity principle, natural when handling knowledge, ensures that each interpretation remains as much possible as allowed by the constraints. This method enables us to compute a priority ranking on interpretations among defaults then represented by possibilistic formulas. This leads to an approach to default reasoning with a simple semantics, which is in agreement with Lehmann et al. postulates [11, 14]. The minimum specificity principle makes sense for knowledge, while for preferences, the maximum specificity principle is natural as well [10]. So it is tempting to investigate a "default-like" treatment of conditional preferences and to compare it with π-pref-nets.

The paper is organized as follows. Sections 2 and 3 provide the necessary background on possibility theory and preference modeling respectively. Section 4 first studies how the topology of a preference graph influences the number of layers in the well-ordered partition obtained with the minimum specificity approach. Then we show that there are preference statements not representable by neither π-pref-nets nor CP-nets that can still be handled by the "default-like" method. Finally, we discuss the effects of the minimum and maximum specificity principles, and compare them to what is obtained with the π-pref-nets ordering when they are both applicable.

2   Background

Finding an order or ranking between configurations describing complete instantiations of preference variables can be achieved in several manners. One way of proceeding is to use default statements and non-monotonic logic interpreted in terms of possibility distributions. In a first subsection, we recall the background on possibility theory understood in terms of preference. The second subsection explains how to induce a well-ordered partition of configurations by means of some information principle.

2.1   Possibility Theory

Let us assume a finite set of configurations $\Omega = \{\omega_1, ..., \omega_n\}$ composed of all possible interpretations of a set of Boolean decision variables $X = \{X_1, ..., X_m\}$, where $n = 2^m$

possible combinations. Each interpretation ω_i is a vector which is a complete instantiation of variables in $\mathcal{X}$. In order to rank order these alternatives, a possibility distribution π is used. It is a mapping from Ω to a totally ordered scale S = [0,1]. Based on possibility degrees $\pi(\omega_i)$, this distribution provides a complete pre-order between interpretations. Originally, $\pi(\omega)$ was only used to evaluate to what extent ω is possible\plausible. However, further studies [7] showed that this encoding is also convenient for expressing preferences over a set of choices. Indeed, possibility values are adapted to satisfaction degrees in view of finding a rank ordering between interpretations of Ω. A constraint of the form $\pi(\omega) > \pi(\omega\prime)$ stipulates that the largest π is, the more satisfactory ω is.

A distribution π is said to be normalized if $\exists\ \omega \in \Omega$, such that, $\pi(\omega) = 1$, meaning that there is at least one configuration which is totally satisfactory. By contrast, $\pi(\omega) = 0$ amounts to saying that ω is rejected. Based on a possibility distribution π, the possibility measure Π of the event P, s.t. $\forall P \subseteq \Omega$ is defined by,

$$\Pi(P) = \max_{\omega \in P} \pi(\omega)\ \forall P \subseteq \Omega \tag{1}$$

$\Pi(P)$ estimates to what extent at least one configuration in P is satisfactory. Another measure using the minimum operator called *guaranteed possibility measure* can be defined,

$$\Delta(P) = \min_{\omega \in P} \pi(\omega)\ \forall P \subseteq \Omega \tag{2}$$

It estimates the extent to which the least preferred model of P is satisfactory. So $\Delta(P)$ represents a guaranteed satisfaction level when taking a configuration in P.

Let p be a proposition that models P ($\omega_i \in P$ if and only if $\omega_i \models p$). Using the two measures, a preference specification $p > \bar{p}$ can be interpreted in different ways [12,13]. Given a subset of interpretations P the statement "I prefer the best case in which p is true to the best case in which $\bar{p}$ is true" is seen as an optimistic modeling of preference and is formally expressed by $\Pi(P) > \Pi(\bar{P})$. In contrast, a pessimistic approach for expressing $p > \bar{p}$ would be "I prefer the worst case in which p is true to the worst case in which $\bar{p}$ is true", namely $\Delta(P) > \Delta(\bar{P})$. The claim "I prefer the best configurations in which p is true to the worst configurations in which $\bar{p}$ is true" expresses an opportunistic approach and is encoded by $\Pi(P) > \Delta(\bar{P})$. Finally a cautious, stronger statement "I prefer the worst configurations in which p is true to the best configurations in which $\bar{p}$ is true" is expressed by $\Delta(P) > \Pi(\bar{P})$.

2.2 Possibilistic Approach to Default Preferences

A conditional preference $p \rightsquigarrow q$ may be encoded by the constraint $\Pi(p \wedge q) > \Pi(p \wedge \bar{q})$. It explicitly means that "In the context defined by p, the best situation that models q is preferred to the best situation that models $\bar{q}$". A possibility distribution on configurations of Ω can be deduced from such constraints, based on some informational principle. Then, a well-ordered partition of configurations can be generated [8]. Considering possibility distributions π_1 and π_2, π_1 is said to be less specific than π_2 in the wide sense if $\forall\omega \in \Omega, \pi_1(\omega) \geq \pi_2(\omega)$. Then agent 1 is considered less demanding than agent 2.

Definition 1 *Minimum specificity principle. Assuming a set of possibilistic constraints* $\Pi(p_i \wedge q_i) > \Pi(p_i \wedge \bar{q}_i)$, *the least specific distribution* π^* *in accordance with these statements is the one that maximizes possibility degrees of configurations.*

The meaning of this principle is that if a configuration ω has not been explicitly rejected, it is considered as preferred ($\pi(\omega) = 1$). Given a conditional preference base $S = \{p_i \rightsquigarrow q_i : i = 1, \ldots, k\}$, C_Π denotes the set of constraints derived from these statements. Using an optimistic modeling of preference, C_Π is formally denoted by

$$C_\Pi = \{\Pi(p_i \wedge q_i) > \Pi(p_i \wedge \bar{q}_i) : p_i \rightsquigarrow q_i \in S\} \tag{3}$$

Maximizing possibility degrees of configurations based on the minimum specificity principle is achieved via the following algorithm which outputs a well-ordered partition composed of sets E_j of configurations [8].

Algorithm 1. Algorithm of partitioning of Ω

Input: a set of possibilistic constraints
Begin
1. $E_0 = \{\emptyset\}$;
2. **While** $\Omega \neq \emptyset$, repeat
    2.1 $E_j = \{\omega_i, i = 1, \cdots, n\}$ s.t. ω_i does not appear on the right-hand side of any constraint (ω_i is never dominated)
    2.2 Remove the added configurations to E_j from Ω
    2.3 Remove from C_Π all satisfied constraints (their left-hand side are consistent with configurations of E_j)
  End while
End

Given a set of constraints, the first step consists of finding configurations that are never dominated. They can be derived from calculating the negation of the disjunction of formulas that appear on the right side of constraints of C_Π. In accordance with the minimum specificity principle and using an optimistic interpretation approach, the resulting configurations are then associated to the highest possibility degree (e.g. $\pi(\omega_i) = 1$) and are assigned to the first partition E_0. Constraints that are satisfied are then deleted from C_Π. The same process is repeated until no constraints are left. In a final step, the remaining configurations of Ω are assigned to a final last level.

Note that this algorithm allows to rank configurations in terms of preference in the most compact way in accordance to the given constraints.

In contrast, $p \rightsquigarrow q$ may express the conditional constraint $\Delta(p \wedge q) > \Delta(p \wedge \bar{q})$ under a pessimistic view, evaluating $p \wedge q$ by its worst configuration. They can be similarly handled by the opposite principle.

Definition 2 *Maximum specificity principle. Assuming a set of possibilistic constraints* $\Delta(p_i \wedge q_i) > \Delta(p_i \wedge \bar{q}_i)$, *the maximum specific distribution* π^* *in accordance with these statements is the one that minimizes possibility degrees of configurations.*

3 Background on Graphical Preferences

Graphical representations can be used to express conditional preferences. In this section, we will closely examine two such structures: CP-nets [9] and π-pref nets [6]. Both represent qualitative counterparts of Bayesian networks and are based on the same type of preferential specification. They happen to share the same graphical structure, which is a directed acyclic graph (shortly *DAG*) between variables. Each node of the graph is associated to conditional data tables representing local preferences of variable values in the context of values assigned to their parents. Each model uses a specific independence property between variables, which enables to construct preference orderings between configurations [2,5].

3.1 CP-nets

CP-nets rely on the *Ceteris Paribus* preferential independence. It states that the preference of a partial configuration over another (as stated in the conditional preference tables attached to each variable in the acyclic graph) holds everything else being equal. Since this assumption leads to compare configurations that only differ by a single flip, a directed a-cyclic worsening flip graph of configurations can be built. It is a partial order. A configuration ω is said to be preferred to $\omega\prime$, if there exists a chain of worsening flips that links ω to $\omega\prime$. Finding the optimal solution amounts to sweeping through conditional preferences from top to bottom of the preference graph and to assigning its preferred value to each variable in the context of its parents.

3.2 π-pref Nets

The data component of a π-pref net involves conditional symbolic possibility distributions over the domains of each variable and its parents. The assignment $\pi(x_i|Par(X_i)) = \alpha$ is interpreted as "In the context of $Par(X_i)$, I prefer x_i to $\bar{x}_i$ with a satisfaction degree of α" (an unspecified value in $[0, 1]$). The degree of satisfaction of full configurations is computed using the chain rule associated to the product-based conditioning, namely, the symbolic expression $\pi(X_i, ..., X_n) = \prod_{i=1,...,n} \pi(X_i|Par(X_i))$. Comparing two outcomes comes down to comparing such symbolic expressions. Yet, finding the optimal solution can proceed in the same way as for CP-nets, by choosing for each node the best instantiation in the context of parents. π-pref nets are constructed based on the Markovian independence property which stipulates that each variable is independent from its non-descendants (N) in the context of its parents (Par). It is noticeable that, assuming the same specifications in both representations, comparisons generated by the Ceteris Paribus assumption can be expressed by adding constraints between products of symbolic weights of the π-pref net. The preference orderings induced by the two model kinds are consistent [3].

4 Application of Default Reasoning to Conditional Preferences

Conditional statements in a preference network can always be translated into "default-like" rules of the form "one value of a variable is generally preferred to another in some instantiation of its parents". This section studies this approach in more details.

4.1 Well-Ordered Partition Induced by a Conditional Preference Graph

A preference statement of the form $x_1 x_2 : x > \bar{x}$ is now expressed by the default preference rule $x_1 x_2 \rightsquigarrow x$, which translates into the constraint $\Pi(x_1 x_2 x) > \Pi(x_1 x_2 \bar{x})$ (in case of two parent variables X_1 and X_2). A conditional preference network can be expressed as a collection of such constraints. Generating a ranking of configurations can then be achieved by Algorithm 1 that uses the minimum specificity assumption. As seen now, for some preference networks having specific graph structures, the procedure outputs a well-ordered partition of exactly 3 level sets whatever the size of the preference network. First we consider the case of a path graph, where each variable has exactly one variable as a parent (except for the root one) and the graph forms a single path (as on Fig. 1). Hence conditional preference constraints are of the form $x_i : x_{i+1} > \bar{x}_{i+1}$.

Fig. 1. A path preference graph   **Fig. 2.** Example of a preference network

Proposition 1. *Interpreting conditional preference statements as possibilistic constraints under the minimum specificity principle, any conditional preference path graph results into a well-ordered partition of solutions with exactly 3 elements.*

Proof. Let us assume a path graph G of n vertices. The root node holds a preference constraint of the form $x_1 > \bar{x}_1$, whereas, for $i = 2, n$, the remaining nodes hold conditional preferences of the form $x_{i-1} \wedge x_i > x_{i-1} \wedge \overline{x_i}$ for the preferred instantiation of the parent X_i and $\overline{x_{i-1}} \wedge \overline{x_i} > \overline{x_{i-1}} \wedge x_i$ for its negation. The non-dominated solution is unique and is defined by

$$\bar{x}_1 \vee \bigvee_{i=2}^{n}(x_{i-1} \wedge \bar{x}_i) \vee \bigvee_{i=2}^{n}(\overline{x_{i-1}} \wedge x_i) = x_1 \wedge \bigwedge_{i=2}^{n}(\overline{x_{i-1}} \vee x_i) \wedge \bigwedge_{i=2}^{n}(x_{i-1} \vee \bar{x}_i) = \wedge_{i=1}^{n} x_i$$

At the end of this iteration, the root constraint and the children constraints in the context of preferred parents configurations are satisfied by this best solution and can be deleted. The remaining constraints are $\overline{x_{i-1}} \wedge \overline{x_i} > \overline{x_{i-1}} \wedge x_i, i = 1, \ldots n$. The dominated solutions are the models of $\bigvee_{i=2}^{n} \overline{x_{i-1}} \wedge x_i$. The non-dominated ones are thus of the form $\bigwedge_{i=2}^{n} x_{i-1} \vee \overline{x_i}$. This formula is consistent with the left-hand sides of the constraints $\overline{x_{i-1}} \wedge \overline{x_i} > \overline{x_{i-1}} \wedge x_i$ since they have in common the solution $\wedge_{i=1}^{n} \overline{x_i}$. Hence the solutions can be ranked in three levels: $\wedge_{i=1}^{n} x_i$ at the top forming E_0, and $\bigvee_{i=2}^{n} \overline{x_{i-1}} \wedge x_i$ at the bottom forming E_2, the rest being of the form $E_1 = (\bigvee_{i=2}^{n} \overline{x_{i-1}} \wedge \overline{x_i}) \wedge \bigvee_{i=1}^{n} \overline{x_i}$.

Actually, the number of layers for ordering preferences using the constraint based algorithm increases by adding edges between the grandparent nodes and those of children nodes. The graph on Fig. 2 differs from graph on Fig. 1 by an additional edge going from node A to node C. Applying the algorithm yields 4 preference levels.

Example 1. *Adding the edge $A \rightarrow C$ to the preference network of Fig. 7, changes statements and constraints of the node C. The set of constraints is of the form: $a > \bar{a}, ab > a\bar{b}, ab > \bar{a}b, \bar{a}\bar{b} > \bar{a}b, abc > ab\bar{c}, \bar{a}bc > \bar{a}b\bar{c}, a\bar{b}c > a\bar{b}\bar{c}, \bar{a}\bar{b}\bar{c} > \bar{a}\bar{b}c$. Again $\omega = abc$ is preferred. Only remain constraints $\bar{a}\bar{b} > \bar{a}b, \bar{a}bc > \bar{a}b\bar{c}, a\bar{b}c > a\bar{b}\bar{c}, \bar{a}\bar{b}\bar{c} > \bar{a}\bar{b}c$, which puts models of $(a \vee \bar{b}) \wedge (a \vee \bar{b} \vee c) \wedge (\bar{a} \vee b \vee c) \wedge (\bar{a} \vee b \vee c)$ at the second level or higher. This enforces $a\bar{b}\bar{c}$ down to a fourth level.*

Whatever the topology of the graph, if the network does not hold edges from the grandparents nodes to children nodes, the number of elements forming the well-ordered partition remains constant and equal to 3. The following propositions confirm this claim for topologies of Fig. 3 and Fig. 4, respectively.

Proposition 2. *Given any conditional preference network with one parent node and $n-1$ children, the well-ordered partition of configurations output by the minimum specificity principle based algorithm has exactly 3 levels.*

Proof. Assume the graph $\mathcal{G}$ of Fig. 3 with one parent and $n-1$ children node. The root has a preference statement $x_1 > \bar{x}_1$. For $i = 2 \cdots, n$, each child node bears conditional constraints of the form $x_1 x_i > x_1 \bar{x}_i$ and $\bar{x}_1 \bar{x}_i > \bar{x}_1 x_i$. The un-dominated set is the complement of propositions on the left of constraints, namely $E_0 = x_1 \wedge \bigwedge_{i=2}^{n} x_i$. Constraints $x_1 > \bar{x}_1$ and $x_1 x_i > x_1 \bar{x}_i$ are satisfied by this solution and are then deleted. The second level set E_1 contains models of $\overline{\bar{x}_1 \wedge \bigvee_{i=2}^{n} x_i} = x_1 \vee \bigwedge_{i=2}^{n} \bar{x}_i$. Note that all the left-hand side propositions $\bar{x}_1 \bar{x}_i$ of the remaining constraints are consistent with $x_1 \vee \bigwedge_{i=2}^{n} \bar{x}_i$. Hence again 3 levels are obtained.

Proposition 3. *Given any conditional preference network with $n-1$ independent parent nodes and one child variable, the well-ordered partition of configurations output by the minimum specificity principle based algorithm has exactly 3 levels.*

Proof. Assume the graph $\mathcal{G}$ of Fig. 4. In the same vein as Propositions 1 and 2, parent nodes bear constraints $x_i > \bar{x}_i$ for $i = 1, \cdots, n-1$. Denote by u the disjunction of parents configurations such that x_n is preferred to $\bar{x}_n$, where it is supposed that u is satisfied by $\bigwedge_{i=1}^{n-1} x_i$ and $\bar{u}$ is satisfied by $\bigwedge_{i=1}^{n-1} \bar{x}_i$. The remaining conditional constraints at step 2 reduce to $u x_n > u \bar{x}_n$ and $\bar{u} \bar{x}_n > \bar{u} x_n$. Obviously we get 3 levels again.

The last result considers a more general structure (see Fig. 5) we call quasi-linear and subsumes the preceding results.

Proposition 4. *Consider a conditional preference network $\mathcal{G} = \{\mathcal{V}, \mathcal{E}\}$, where the set $\mathcal{V}$ of variables is partitioned in $\mathcal{V}_1, \cdots, \mathcal{V}_n$. Suppose $\forall j \in [1, m]$, each variable $X \in \mathcal{V}_i$ has its parents only at the previous level $i - 1$, i.e., $Par(X) \subseteq \mathcal{V}_{i-1} \forall X \in \mathcal{V}_i$. The minimum specificity principle results in a well-ordered 3-partition of solutions.*

Proof. $\forall i = 2, \cdots, n$ all nodes $X_i \in \mathcal{V}_i$ are associated to the conditional constraints $u_i x_i > u_i \bar{x}_i$ and $\overline{u_i x_i} > \overline{u_i} x_i$, where u_i is the disjunction of configurations of $Par(X_i)$ such that x_i is preferred to $\bar{x}_i$, plus $x_1 > \overline{x_1}$ for nodes $X_1 \in \mathcal{V}_1$. Assuming $\bigwedge_{X_{i-1} \in Par(X_i)} x_i \models u_i$ and $\bigwedge_{X_{i-1} \in Par(X_i)} \bar{x}_i \models \bar{u}_i$, the non-dominated set E_0 reduces to $\left(\bigwedge_{X_1 \in \mathcal{V}_1} x_1 \right) \wedge \bigwedge_{i=2}^{n} \bigwedge_{X_i \in \mathcal{V}_i} \left[(\bar{u}_i \vee x_i) \wedge (u_i \vee \bar{x}_{ij}) \right] = \bigwedge_{X \in \mathcal{V}} x$. After deleting the satisfied constraints, the remaining ones are $\forall X_i \in \mathcal{V}_i$, $\bar{u}_i \bar{x}_i > \bar{u}_i x_i$, $\forall i = 2, \cdots, n$. The un-dominated set $E_1 \cup E_0$ forms the models of $\bigwedge_{i=2}^{n} \bigwedge_{X_i \in \mathcal{V}_i} (u_i \vee \bar{x}_i)$. We can easily check that $\bar{u}_i \bar{x}_i$ is consistent with E_1 since they share $\bar{x}_i$, $\forall i = 2, \cdots, n$ and $\forall X_i \in \mathcal{V}_i$. By consequence the third element of the well-ordered partition E_2 equals $\bigvee_{i=2}^{n} \bigvee_{j=1}^{m} \bigvee_{X_{ij} \in \mathcal{V}_i} \bar{u}_i \wedge x_i$.

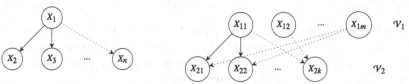

Fig. 3. A graph with one parent and n children

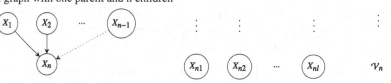

Fig. 4. A graph with n parents and one child   **Fig. 5.** A quasi-linear DAG

4.2   From Default Preference Rules to Conditional Preference Networks

While conditional preference graphs can be turned into default preference bases, we consider the reverse transformation, i.e., whether from any preference rule base, a network of conditional constraints can be generated. We show that this is generally not the case. Preference networks lead to very specific default preference statements. Contexts are always conjunctions of literals, which make it possible the construction of corresponding conditional data tables. But general preference statements admit more general forms of contexts. Moreover preferences in networks are local in the sense that they deal with values of single variables only. Finally, information in a preference base can be insufficient to build a conditional preference graph. Consider the following counter-example.

Example 2. *Considering the counterpart of the "penguin" example in non-monotonic reasoning [8,14]. Let c, r and s now stand for "Chicken (C)", "Red wine (R)" and "Spicy plate (S)". Preference rules are $\mathcal{R} = \{$ "With chicken, I prefer red wine", "If spicy, I prefer white wine" and "If spicy, I prefer chicken"\}, where "White wine" is the negation of "red wine". It corresponds to constraints $cr > c\bar{r}$, $s\bar{r} > sr$, $sc > s\bar{c}$ using the minimum specificity principle, it is well-known we get a well-ordered 3-partition with $E_0 = \bar{s} \wedge (\bar{c} \vee r)$, $E_1 = c \wedge \bar{r}$ and $E_2 = s \wedge (r \vee \bar{c})$. The rules indicate values of C and R depend on S and R depend on C, hence the graph of Fig. 6. However some information is missing to get a full preference graph. We miss the absolute preference between s and $\bar{s}$ on node S, the preference for chicken or not when the plate is not spicy is not given (represented by a question mark in Fig. 6). We also miss the preferences about wine when the chicken is spicy (from the rules this is a conflicting case, a double question mark in Fig. 6) and when the dish is not chicken nor spicy. In fact, S and C act as independent parents of R, which causes the conflict. It is forbidden in a preference* graph for a variable to have several parent groups. *The conflict between S and C is solved when applying minimum specificity ranking to the default rules (we conclude that $\bar{s} > s$, that $\bar{s}c > \bar{s}\bar{c}$, $sc\bar{r} > scr$ and no preference between $\bar{s}\bar{c}r$ and $\bar{s}\bar{c}\bar{r}$.*

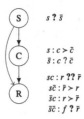

$$s\,?\,\bar{s}$$
$$s : c > \bar{c}$$
$$\bar{s} : c\,?\,\bar{c}$$
$$sc : r\,??\,\bar{r}$$
$$s\bar{c} : \bar{r} > r$$
$$\bar{s}c : r > \bar{r}$$
$$\bar{s}\bar{c} : f\,?\,\bar{r}$$

Fig. 6. Partial network from preference rules

Clearly there is a gap between general default preference rules and conditional preference networks. However from the well-ordered partition of solutions obtained by the minimum specificity principle, one can generate a conditional preference network. (This solves the question marks in the example).

4.3   Comparing π-pref Nets and Default Reasoning Approach

The preference between configurations in π-pref nets is based on the comparison between symbolic expressions obtained by the product chain rule. In this subsection, results are compared with those of the minimal specificity approach.

Example 3. *Consider a π-pref net expressing conditional preferences over 4 variables $\mathcal{V} = \{A, B, C, D\}$. Conditional distributions are derived from specifications of Fig. 7.*

The set S represents conditional preference specifications written under the form of defaults. $S=\{a, a \rightsquigarrow b, \bar{a} \rightsquigarrow \bar{b}, b \rightsquigarrow c, \bar{b} \rightsquigarrow \bar{c}, bc \rightsquigarrow d, \bar{b}\bar{c} \rightsquigarrow d, \bar{b}c \rightsquigarrow \bar{d}, \bar{b}\bar{c} \rightsquigarrow \bar{d}\}$. The well-ordered partition output by Algorithm 1 is given in Table 1 (left). To construct the ordering between configurations, let us now proceed by means the product-based chain rule on symbolic weights $\pi(ABCD) = \pi(A) \times \pi(B|A) \times \pi(C|B) \times \pi(D|BC)$. This leads to a 5-element well-ordered partition given in Table 1 (right).

$$\begin{array}{c} a > \bar{a} \\ a : b > \bar{b} \\ \bar{a} : \bar{b} > b \\ b : c > \bar{c} \\ \bar{b} : \bar{c} > c \\ bc : d > \bar{d} \\ b\bar{c} : d > \bar{d} \\ \bar{b}c : \bar{d} > d \\ \bar{b}\bar{c} : \bar{d} > d \end{array}$$

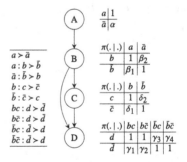

| | a | 1 |
|A| $\bar{a}$ | α |

$\pi(.\|.)$	a	$\bar{a}$
b	1	β_2
$\bar{b}$	β_1	1

$\pi(.\|.)$	b	$\bar{b}$
c	1	δ_2
$\bar{c}$	δ_1	1

$\pi(.\|.)$	bc	$b\bar{c}$	$\bar{b}c$	$\bar{b}\bar{c}$
d	1	1	γ_3	γ_4
$\bar{d}$	γ_1	γ_2	1	1

Fig. 7. Example of a π-pref net

Mind that in the minimum specificity column of Table 1, being in some subset means indifference while in the chain rule column it corresponds to incomparability. Taking a closer look at the obtained results of Table 1, we notice that, not only the chain rule results in more preference levels than minimal specificity (which is unsurprizing), but also the orderings are not in full agreement. The product chain rule agrees with the order of inclusion between subsets of nodes associated with violated preferences [4], and thus ranks solutions according to the number of violated nodes, whereas, the specificity algorithm just finds the most compact ordering where constraints are respected. Nevertheless, the two approaches lead to distinct results that are not fully compatible, since $\exists\, \omega\prime \in \Omega$ such that for the chain rule approach $\omega\prime > \omega$ whereas for the minimum specificity based approach $\omega > \omega\prime$. Indeed, the worst configuration $\bar{a}\bar{b}\bar{c}\bar{d}$ is ranked on the lowest level by the product chain rule, whereas it appears in the third level based on the minimum specificity approach. This is due to the fact that at some iteration the unsatisfied constraints do not prevent $\bar{a}\bar{b}\bar{c}\bar{d}$ from being higher than the remaining configurations $a\bar{b}cd, \bar{a}\bar{b}cd$.

Table 1. Well-ordered partitions based on three approaches

Minimum specificity	Levels	Maximum specificity	Chain rule
$\{abcd\}$	1	$\{a\bar{b}\bar{c}\bar{d}, \bar{a}\bar{b}\bar{c}\bar{d}\}$	$\{abcd\}$
$\{ab\bar{c}d, a\bar{b}\bar{c}d, a\bar{b}\bar{c}\bar{d}, \bar{a}\bar{b}\bar{c}\bar{d}\}$	2	$\{ab\bar{c}d, abc\bar{d}, ab\bar{c}\bar{d}, a\bar{b}\bar{c}d,$ $a\bar{b}cd, a\bar{b}c\bar{d}, \bar{a}b\bar{c}d, \bar{a}\bar{b}\bar{c}d, \bar{a}\bar{b}c\bar{d}\}$	$\{a\bar{b}\bar{c}\bar{d}, ab\bar{c}\bar{d}, a\bar{b}\bar{c}d, \bar{a}\bar{b}\bar{c}\bar{d}\}$
$\{a\bar{b}\bar{c}\bar{d}, a\bar{b}\bar{c}d, a\bar{b}c\bar{d}, a\bar{b}cd,$ $\bar{a}\bar{b}c\bar{d}, \bar{a}\bar{b}\bar{c}d, \bar{a}\bar{b}cd, \bar{a}b\bar{c}d, \bar{a}\bar{b}\bar{c}\bar{d}\}$	3	$\{a\bar{b}cd, \bar{a}\bar{b}cd, \bar{a}b\bar{c}d, \bar{a}\bar{b}cd\}$	$\{a\bar{b}\bar{c}\bar{d}, a\bar{b}\bar{c}d, a\bar{b}c\bar{d}, a\bar{b}cd,$ $\bar{a}\bar{b}c\bar{d}, \bar{a}\bar{b}\bar{c}d\}$
$\{a\bar{b}cd, \bar{a}\bar{b}cd\}$	4	$\{\bar{a}b\bar{c}d\}$	$\{a\bar{b}cd, \bar{a}\bar{b}cd, \bar{a}b\bar{c}d, \bar{a}\bar{b}cd\}$
$\{\emptyset\}$	5	$\{\emptyset\}$	$\{\bar{a}b\bar{c}d\}$

4.4   Maximal Specificity on Default Preference Statements

The minimum specificity algorithm outputs a well-ordered partition that clusters the worst configuration(s) with other less preferred ones all in the same set. This is due to the focus on the best models of formulas. It does not provide information on the least preferred models. In order to refine results of the optimistic preference interpretation approach, we can also exploit preference statements based on the maximum specificity principle using the *guaranteed possibility measure* $\Delta(.)$ (see end of Sect. 2.2). A procedure, symmetric to the one for minimal specificity can be devised, adapting step 2.1 of Algorithm 1, where we now assign to E_j solutions that do not appear on the *left-hand* side of remaining constraints. On the graph structure of Proposition 4, one can prove that again the maximum specificity well-ordered partition will have three elements. So the best solutions are left on a par using the maximum specificity method. It is tempting to consider the conjunction of the minimum and the maximum specificity rankings, say $>_{mM}$. Since the chain rule partial ordering is unquestionable (it corresponds to the Pareto dominance condition), it is interesting in the future to check whether the joint (partial) ordering $>_{mM}$ of configurations allows to refine the chain rule ordering without conflicting with it.

Example 4. *Let us consider again the π-pref net of Example 3. The result of applying both minimal and maximal specificity are given in Table 1. Note that the maximum specificity ranking partially contradicts the ordering of some elements put in levels 3 and 4 by the minimum specificity procedure. The reader can check that on this example, the joint (partial) ordering $>_{mM}$ is consistent with and does refine the chain rule ordering (adding the comparisons $\bar{a}\bar{b}\bar{c}\bar{d} >_{mM} a\bar{b}cd, a\bar{b}\bar{c}d >_{mM} a\bar{b}cd, \bar{a}\bar{b}\bar{c}d >_{mM} \bar{a}\bar{b}cd, a\bar{b}c\bar{d} >_{mM} \bar{a}\bar{b}cd, a\bar{b}c\bar{d} >_{mM} \bar{a}b\bar{c}d, \bar{a}\bar{b}c\bar{d} >_{mM} \bar{a}\bar{b}cd$ to it). On the other hand,* *some pairs are judged incomparable by* $>_{mM}$ $((abcd, a\bar{b}\bar{c}\bar{d}), (abcd, \bar{a}\bar{b}\bar{c}\bar{d}), (\bar{a}\bar{b}\bar{c}\bar{d}, \bar{a}bcd),$ $(\bar{a}\bar{b}\bar{c}\bar{d}, \bar{a}bcd))$ *while the chain rule ordering can order them (see Table 1). In this example it is thus possible to refine the chain rule ordering by* $>_{mM}$.

5 Conclusion

We investigate an approach to conditional preference graphs inspired by the possibilistic handling of default rules. First results indicate that it is not very discriminant on some graph structures. Ideally, two opposite information principles (minimal and maximal specificity) can be used jointly to refine the ranking of solutions obtained by the chain rule in π-pref nets. More work is needed to show that this is possible and to compare in a more detailed way the new approach with π-pref nets and CP nets.

References

1. Ben Amor, N., Dubois, D., Gouider, H., Prade, H.: Possibilistic preference networks. Inf. Sci. **460–461**, 401–415 (2018)
2. Ben Amor, N., Dubois, D., Prade, H., Saidi, S.: Representation of multiple agent preferences. In: Ciucci, D., Pasi, G., Vantaggi, B. (eds.) SUM 2018. LNCS (LNAI), vol. 11142, pp. 339–367. Springer, Cham (2018). https://doi.org/10.1007/978-3-030-00461-3_25
3. Ben Amor, N., Dubois, D., Gouider, H., Prade, H.: Expressivity of possibilistic preference networks with constraints. In: Moral, S., Pivert, O., Sánchez, D., Marín, N. (eds.) SUM 2017. LNCS (LNAI), vol. 10564, pp. 163–177. Springer, Cham (2017). https://doi.org/10.1007/978-3-319-67582-4_12
4. Ben Amor, N., Dubois, D., Gouider, H., Prade, H.: Preference modeling with possibilistic networks and symbolic weights: a theoretical study. In: ECAI 2016–22nd European Conference on Artificial Intelligence, 29 August-2 September 2016, The Hague, The Netherlands, pp. 1203–1211 (2016)
5. Ben Amor, N., Dubois, D., Gouider, H., Prade, H.: Graphical models for preference representation: an overview. In: Schockaert, S., Senellart, P. (eds.) SUM 2016. LNCS (LNAI), vol. 9858, pp. 96–111. Springer, Cham (2016). https://doi.org/10.1007/978-3-319-45856-4_7
6. Ben Amor, N., Dubois, D., Gouider, H., Prade, H.: Possibilistic conditional preference networks. In: Destercke, S., Denoeux, T. (eds.) ECSQARU 2015. LNCS (LNAI), vol. 9161, pp. 36–46. Springer, Cham (2015). https://doi.org/10.1007/978-3-319-20807-7_4
7. Benferhat, S., Dubois, D., Prade, H.: Towards a possibilistic logic handling of preferences. Appl. Intell. **14**, 303–317 (2001)
8. Benferhat, S., Dubois, D., Prade, H.: Representing default rules in possibilistic logic. In: Proceedings of the 3rd International Conference on Principles of Knowledge Representation and Reasoning (KR 1992), Cambridge, MA, USA, 25–29 October 1992, pp. 673–684 (1992)
9. Boutilier, C., Brafman, R.I., Hoos, H.H., Poole, D.: Reasoning with conditional Ceteris Paribus preference statements. In: UAI 1999: Proceedings of the Fifteenth Conference on Uncertainty in Artificial Intelligence, Stockholm, Sweden, 30 July – 1 August 1999, pp. 71–80 (1999)
10. Dubois, D., Kaci, S., Prade, H.: Expressing preferences from generic rules and examples – a possibilistic approach without aggregation function. In: Godo, L. (ed.) ECSQARU 2005. LNCS (LNAI), vol. 3571, pp. 293–304. Springer, Heidelberg (2005). https://doi.org/10.1007/11518655_26

11. Lehmann, D., Magidor, M.: What does a conditional knowledge base entail? Artif. Intell. **55**, 1–60 (1992)
12. Kaci, S.: Characterization of positive and negative information in comparative preference representation. In: ECAI 2012–20th European Conference on Artificial Intelligence, Montpellier, France, 27–31 August 2012, pp. 450–455 (2012)
13. Kaci, S., Van der Torre, L.W.N.: Reasoning with various kinds of preferences: logic, non-monotonicity, and algorithms. Ann. OR **163**, 89–114 (2008)
14. Kraus, S., Lehmann, D., Magidor, M., Preferential models and cumulative logics: Nonmonotonic reasoning. Artif. Intell. **44**, 167–207 (1990)

Conjunction of Conditional Events and t-Norms

Angelo Gilio[1]($\boxtimes$) and Giuseppe Sanfilippo[2]($\boxtimes$)

[1] Department SBAI, University of Rome "La Sapienza", Rome, Italy
angelo.gilio@sbai.uniroma1.it
[2] Department of Mathematics and Computer Science, University of Palermo,
Palermo, Italy
giuseppe.sanfilippo@unipa.it

Abstract. We study the relationship between a notion of conjunction among conditional events, introduced in recent papers, and the notion of Frank t-norm. By examining different cases, in the setting of coherence, we show each time that the conjunction coincides with a suitable Frank t-norm. In particular, the conjunction may coincide with the Product t-norm, the Minimum t-norm, and Lukasiewicz t-norm. We show by a counterexample, that the prevision assessments obtained by Lukasiewicz t-norm may be not coherent. Then, we give some conditions of coherence when using Lukasiewicz t-norm.

Keywords: Coherence · Conditional event · Conjunction · Frank t-norm

1 Introduction

In this paper we use the coherence-based approach to probability of de Finetti ([1, 2, 7, 9, 10, 13, 14, 16–18, 22, 34]). We use a notion of conjunction which, differently from other authors, is defined as a suitable conditional random quantity with values in the unit interval (see, e.g. [20, 21, 23, 24, 36]). We study the relationship between our notion of conjunction and the notion of Frank t-norm. For some aspects which relate probability and Frank t-norm see, e.g., [5, 6, 8, 11, 15, 33]. We show that, under the hypothesis of logical independence, if the prevision assessments involved with the conjunction $(A|H) \wedge (B|K)$ of two conditional events are coherent, then the prevision of the conjunction coincides, for a suitable $\lambda \in [0, +\infty]$, with the Frank t-norm $T_\lambda(x, y)$, where $x = P(A|H), y = P(B|K)$. Moreover, $(A|H) \wedge (B|K) = T_\lambda(A|H, B|K)$. Then, we consider the case $A = B$, by determining the set of all coherent assessment (x, y, z) on $\{A|H, A|K, (A|H) \wedge (A|K)\}$. We show that, under coherence, it holds

---

A. Gilio and G. Sanfilippo—Both authors contributed equally to the article and are listed alphabetically.
A. Gilio—Retired.

© Springer Nature Switzerland AG 2019
G. Kern-Isberner and Z. Ognjanović (Eds.): ECSQARU 2019, LNAI 11726, pp. 199–211, 2019.
https://doi.org/10.1007/978-3-030-29765-7_17

that $(A|H) \wedge (A|K) = T_\lambda(A|H, A|K)$, where $\lambda \in [1, +\infty]$. We also study the particular case where $A = B$ and $HK = \varnothing$. Then, we consider conjunctions of three conditional events and we show that to make prevision assignments by means of the Product t-norm, or the Minimum t-norm, is coherent. Finally, we examine the Lukasiewicz t-norm and we show by a counterexample that coherence is in general not assured. We give some conditions for coherence when the prevision assessments are made by using the Lukasiewicz t-norm.

2   Preliminary Notions and Results

In our approach, given two events A and H, with $H \neq \varnothing$, the conditional event $A|H$ is looked at as a three-valued logical entity which is true, or false, or void, according to whether AH is true, or $\bar{A}H$ is true, or $\bar{H}$ is true. We observe that the conditional probability and/or conditional prevision values are assessed in the setting of coherence-based probabilistic approach. In numerical terms $A|H$ assumes one of the values 1, or 0, or x, where $x = P(A|H)$ represents the assessed degree of belief on $A|H$. Then, $A|H = AH + x\bar{H}$. Given a family $\mathcal{F} = \{X_1|H_1, \ldots, X_n|H_n\}$, for each $i \in \{1, \ldots, n\}$ we denote by $\{x_{i1}, \ldots, x_{ir_i}\}$ the set of possible values of X_i when H_i is true; then, for each i and $j = 1, \ldots, r_i$, we set $A_{ij} = (X_i = x_{ij})$. We set $C_0 = \bar{H}_1 \cdots \bar{H}_n$ (it may be $C_0 = \varnothing$); moreover, we denote by $C_1, \ldots, C_m$ the constituents contained in $H_1 \vee \cdots \vee H_n$. Hence $\bigwedge_{i=1}^{n}(A_{i1} \vee \cdots \vee A_{ir_i} \vee \bar{H}_i) = \bigvee_{h=0}^{m} C_h$. With each C_h, $h \in \{1, \ldots, m\}$, we associate a vector $Q_h = (q_{h1}, \ldots, q_{hn})$, where $q_{hi} = x_{ij}$ if $C_h \subseteq A_{ij}$, $j = 1, \ldots, r_i$, while $q_{hi} = \mu_i$ if $C_h \subseteq \bar{H}_i$; with C_0 it is associated $Q_0 = \mathcal{M} = (\mu_1, \ldots, \mu_n)$. Denoting by $\mathcal{I}$ the convex hull of $Q_1, \ldots, Q_m$, the condition $\mathcal{M} \in \mathcal{I}$ amounts to the existence of a vector $(\lambda_1, \ldots, \lambda_m)$ such that: $\sum_{h=1}^{m} \lambda_h Q_h = \mathcal{M}$, $\sum_{h=1}^{m} \lambda_h = 1$, $\lambda_h \geq 0$, $\forall h$; in other words, $\mathcal{M} \in \mathcal{I}$ is equivalent to the solvability of the system (Σ), associated with $(\mathcal{F}, \mathcal{M})$,

$$(\Sigma) \quad \textstyle\sum_{h=1}^{m} \lambda_h q_{hi} = \mu_i, \, i \in \{1, \ldots, n\}, \sum_{h=1}^{m} \lambda_h = 1, \, \lambda_h \geq 0, \, h \in \{1, \ldots, m\}. \tag{1}$$

Given the assessment $\mathcal{M} = (\mu_1, \ldots, \mu_n)$ on $\mathcal{F} = \{X_1|H_1, \ldots, X_n|H_n\}$, let S be the set of solutions $\Lambda = (\lambda_1, \ldots, \lambda_m)$ of system (Σ). We point out that the solvability of system (Σ) is a necessary (but not sufficient) condition for coherence of $\mathcal{M}$ on $\mathcal{F}$. When (Σ) is solvable, that is $S \neq \varnothing$, we define:

$$I_0 = \{i : \max_{\Lambda \in S} \textstyle\sum_{h:C_h \subseteq H_i} \lambda_h = 0\}, \, \mathcal{F}_0 = \{X_i|H_i, i \in I_0\}, \, \mathcal{M}_0 = (\mu_i, i \in I_0). \tag{2}$$

For what concerns the probabilistic meaning of I_0, it holds that $i \in I_0$ if and only if the (unique) coherent extension of $\mathcal{M}$ to $H_i|(\bigvee_{j=1}^{n} H_j)$ is zero. Then, the following theorem can be proved ([3, Theorem 3]).

Theorem 1 [Operative characterization of coherence]. A conditional prevision assessment $\mathcal{M} = (\mu_1, \ldots, \mu_n)$ on the family $\mathcal{F} = \{X_1|H_1, \ldots, X_n|H_n\}$ is coherent if and only if the following conditions are satisfied: (i) the system (Σ) defined in (1) is solvable; (ii) if $I_0 \neq \varnothing$, then $\mathcal{M}_0$ is coherent.

Coherence can be related to proper scoring rules ([4, 19, 29–31]).

Definition 1. *Given any pair of conditional events $A|H$ and $B|K$, with $P(A|H) = x$ and $P(B|K) = y$, their conjunction is the conditional random quantity $(A|H) \wedge (B|K)$, with $\mathbb{P}[(A|H) \wedge (B|K)] = z$, defined as*

$$(A|H) \wedge (B|K) = \begin{cases} 1, & \text{if } AHBK \text{ is true,} \\ 0, & \text{if } \bar{A}H \vee \bar{B}K \text{ is true,} \\ x, & \text{if } \bar{H}BK \text{ is true,} \\ y, & \text{if } AH\bar{K} \text{ is true,} \\ z, & \text{if } \bar{H}\bar{K} \text{ is true.} \end{cases} \tag{3}$$

In betting terms, the prevision z represents the amount you agree to pay, with the proviso that you will receive the quantity $(A|H) \wedge (B|K)$. Different approaches to compounded conditionals, not based on coherence, have been developed by other authors (see, e.g., [26, 32]). We recall a result which shows that Fréchet-Hoeffding bounds still hold for the conjunction of conditional events ([23, Theorem 7]).

Theorem 2. *Given any coherent assessment (x, y) on $\{A|H, B|K\}$, with A, H, B, K logically independent, $H \neq \varnothing, K \neq \varnothing$, the extension $z = \mathbb{P}[(A|H) \wedge (B|K)]$ is coherent if and only if the following Fréchet Hoeffding bounds are satisfied:*

$$\max\{x + y - 1, 0\} = z' \leqslant z \leqslant z'' = \min\{x, y\}. \tag{4}$$

Remark 1. From Theorem 2, as the assessment (x, y) on $\{A|H, B|K\}$ is coherent for every $(x, y) \in [0, 1]^2$, the set Π of coherent assessments (x, y, z) on $\{A|H, B|K, (A|H) \wedge (B|K)\}$ is

$$\Pi = \{(x, y, z) : (x, y) \in [0, 1]^2, \max\{x + y - 1, 0\} \leqslant z \leqslant \min\{x, y\}\}. \tag{5}$$

The set Π is the tetrahedron with vertices the points $(1, 1, 1)$, $(1, 0, 0)$, $(0, 1, 0)$, $(0, 0, 0)$. For other definition of conjunctions, where the conjunction is a conditional event, some results on lower and upper bounds have been given in [35].

Definition 2. *Let be given n conditional events $E_1|H_1, \ldots, E_n|H_n$. For each subset S, with $\varnothing \neq S \subseteq \{1, \ldots, n\}$, let x_S be a prevision assessment on $\bigwedge_{i \in S}(E_i|H_i)$. The conjunction $\mathcal{C}_{1 \cdots n} = (E_1|H_1) \wedge \cdots \wedge (E_n|H_n)$ is defined as*

$$\mathcal{C}_{1 \cdots n} = \begin{cases} 1, & \text{if } \bigwedge_{i=1}^n E_i H_i, \text{ is true} \\ 0, & \text{if } \bigvee_{i=1}^n \bar{E}_i H_i, \text{ is true,} \\ x_S, & \text{if } \bigwedge_{i \in S} \bar{H}_i \bigwedge_{i \notin S} E_i H_i \text{ is true, } \varnothing \neq S \subseteq \{1, 2 \ldots, n\}. \end{cases} \tag{6}$$

In particular, $\mathcal{C}_1 = E_1|H_1$; moreover, for $\mathcal{S} = \{i_1, \ldots, i_k\} \subseteq \{1, \ldots, n\}$, the conjunction $\bigwedge_{i \in S}(E_i|H_i)$ is denoted by $\mathcal{C}_{i_1 \cdots i_k}$ and x_S is also denoted by $x_{i_1 \cdots i_k}$. In the betting framework, you agree to pay $x_{1 \cdots n} = \mathbb{P}(\mathcal{C}_{1 \cdots n})$ with the proviso that you will receive: 1, if all conditional events are true; 0, if at least one of the conditional events is false; the prevision of the conjunction of that conditional events which are void, otherwise. The operation of conjunction is associative and commutative. We observe that, based on Definition 2, when $n = 3$ we obtain

$$
\mathcal{C}_{123} = \begin{cases}
1, & \text{if } E_1 H_1 E_2 H_2 E_3 H_3 \text{ is true,} \\
0, & \text{if } \bar{E}_1 H_1 \vee \bar{E}_2 H_2 \vee \bar{E}_3 H_3 \text{ is true,} \\
x_1, & \text{if } \bar{H}_1 E_2 H_2 E_3 H_3 \text{ is true,} \\
x_2, & \text{if } \bar{H}_2 E_1 H_1 E_3 H_3 \text{ is true,} \\
x_3, & \text{if } \bar{H}_3 E_1 H_1 E_2 H_2 \text{ is true,} \\
x_{12}, & \text{if } \bar{H}_1 \bar{H}_2 E_3 H_3 \text{ is true,} \\
x_{13}, & \text{if } \bar{H}_1 \bar{H}_3 E_2 H_2 \text{ is true,} \\
x_{23}, & \text{if } \bar{H}_2 \bar{H}_3 E_1 H_1 \text{ is true,} \\
x_{123}, & \text{if } \bar{H}_1 \bar{H}_2 \bar{H}_3 \text{ is true.}
\end{cases} \tag{7}
$$

We recall the following result ([24, Theorem 15]).

Theorem 3. *Assume that the events $E_1, E_2, E_3, H_1, H_2, H_3$ are logically independent, with $H_1 \neq \varnothing, H_2 \neq \varnothing, H_3 \neq \varnothing$. Then, the set Π of all coherent assessments $\mathcal{M} = (x_1, x_2, x_3, x_{12}, x_{13}, x_{23}, x_{123})$ on $\mathcal{F} = \{\mathcal{C}_1, \mathcal{C}_2, \mathcal{C}_3, \mathcal{C}_{12}, \mathcal{C}_{13}, \mathcal{C}_{23}, \mathcal{C}_{123}\}$ is the set of points $(x_1, x_2, x_3, x_{12}, x_{13}, x_{23}, x_{123})$ which satisfy the following conditions*

$$
\begin{cases}
(x_1, x_2, x_3) \in [0, 1]^3, \\
\max\{x_1 + x_2 - 1, x_{13} + x_{23} - x_3, 0\} \leqslant x_{12} \leqslant \min\{x_1, x_2\}, \\
\max\{x_1 + x_3 - 1, x_{12} + x_{23} - x_2, 0\} \leqslant x_{13} \leqslant \min\{x_1, x_3\}, \\
\max\{x_2 + x_3 - 1, x_{12} + x_{13} - x_1, 0\} \leqslant x_{23} \leqslant \min\{x_2, x_3\}, \\
1 - x_1 - x_2 - x_3 + x_{12} + x_{13} + x_{23} \geqslant 0, \\
x_{123} \geqslant \max\{0, x_{12} + x_{13} - x_1, x_{12} + x_{23} - x_2, x_{13} + x_{23} - x_3\}, \\
x_{123} \leqslant \min\{x_{12}, x_{13}, x_{23}, 1 - x_1 - x_2 - x_3 + x_{12} + x_{13} + x_{23}\}.
\end{cases} \tag{8}
$$

Remark 2. As shown in (8), the coherence of $(x_1, x_2, x_3, x_{12}, x_{13}, x_{23}, x_{123})$ amounts to the condition

$$
\begin{aligned}
\max\{0, x_{12} + x_{13} - x_1, x_{12} + x_{23} - x_2, x_{13} + x_{23} - x_3\} &\leqslant x_{123} \\
\leqslant \min\{x_{12}, x_{13}, x_{23}, 1 - x_1 - x_2 - x_3 + x_{12} + x_{13} + x_{23}\}. &
\end{aligned} \tag{9}
$$

Then, in particular, the extension x_{123} on $\mathcal{C}_{123}$ is coherent if and only if $x_{123} \in [x'_{123}, x''_{123}]$, where $x'_{123} = \max\{0, x_{12} + x_{13} - x_1, x_{12} + x_{23} - x_2, x_{13} + x_{23} - x_3\}$, $x''_{123} = \min\{x_{12}, x_{13}, x_{23}, 1 - x_1 - x_2 - x_3 + x_{12} + x_{13} + x_{23}\}$.

Then, by Theorem 3 it follows [24, Corollary 1].

Corollary 1. *For any coherent assessment $(x_1, x_2, x_3, x_{12}, x_{13}, x_{23})$ on $\{\mathcal{C}_1, \mathcal{C}_2, \mathcal{C}_3, \mathcal{C}_{12}, \mathcal{C}_{13}, \mathcal{C}_{23}\}$ the extension x_{123} on $\mathcal{C}_{123}$ is coherent if and only if $x_{123} \in [x'_{123}, x''_{123}]$, where*

$$
\begin{aligned}
x'_{123} &= \max\{0, x_{12} + x_{13} - x_1, x_{12} + x_{23} - x_2, x_{13} + x_{23} - x_3\}, \\
x''_{123} &= \min\{x_{12}, x_{13}, x_{23}, 1 - x_1 - x_2 - x_3 + x_{12} + x_{13} + x_{23}\}.
\end{aligned} \tag{10}
$$

We recall that in case of logical dependencies, the set of all coherent assessments may be smaller than that one associated with the case of logical independence. However (see [24, Theorem 16]) the set of coherent assessments is the same when $H_1 = H_2 = H_3 = H$ (where possibly $H = \Omega$; see also [25, p. 232]) and a corollary similar to Corollary 1 also holds in this case. For a similar result based on copulas see [12].

3 Representation by Frank t-Norms for $(A|H) \wedge (B|K)$

We recall that for every $\lambda \in [0, +\infty]$ the Frank t-norm $T_\lambda : [0,1]^2 \to [0,1]$ with parameter λ is defined as

$$T_\lambda(u,v) = \begin{cases} T_M(u,v) = \min\{u,v\}, & \text{if } \lambda = 0, \\ T_P(u,v) = uv, & \text{if } \lambda = 1, \\ T_L(u,v) = \max\{u+v-1,0\}, & \text{if } \lambda = +\infty, \\ \log_\lambda(1 + \frac{(\lambda^u - 1)(\lambda^v - 1)}{\lambda - 1}), & \text{otherwise.} \end{cases} \quad (11)$$

We recall that T_λ is continuous with respect to λ; moreover, for every $\lambda \in [0, +\infty]$, it holds that $T_L(u,v) \leqslant T_\lambda(u,v) \leqslant T_M(u,v)$, for every $(u,v) \in [0,1]^2$ (see, e.g., [27,28]). In the next result we study the relation between our notion of conjunction and t-norms.

Theorem 4. *Let us consider the conjunction* $(A|H) \wedge (B|K)$*, with* A, B, H, K *logically independent and with* $P(A|H) = x$*,* $P(B|K) = y$*. Moreover, given any* $\lambda \in [0, +\infty]$*, let* T_λ *be the Frank t-norm with parameter* λ*. Then, the assessment* $z = T_\lambda(x,y)$ *on* $(A|H) \wedge (B|K)$ *is a coherent extension of* (x,y) *on* $\{A|H, B|K\}$*; moreover* $(A|H) \wedge (B|K) = T_\lambda(A|H, B|K)$*. Conversely, given any coherent extension* $z = \mathbb{P}[(A|H) \wedge (B|K)]$ *of* (x,y)*, there exists* $\lambda \in [0, +\infty]$ *such that* $z = T_\lambda(x,y)$*.*

Proof. We observe that from Theorem 2, for any given λ, the assessment $z = T_\lambda(x,y)$ is a coherent extension of (x,y) on $\{A|H, B|K\}$. Moreover, from (11) it holds that $T_\lambda(1,1) = 1$, $T_\lambda(u,0) = T_\lambda(0,v) = 0$, $T_\lambda(u,1) = u$, $T_\lambda(1,v) = v$. Hence,

$$T_\lambda(A|H, B|K) = \begin{cases} 1, & \text{if } AHBK \text{ is true,} \\ 0, & \text{if } \bar{A}H \text{ is true or } \bar{B}K \text{ is true,} \\ x, & \text{if } \bar{H}BK \text{ is true,} \\ y, & \text{if } \bar{K}AH \text{ is true,} \\ T_\lambda(x,y), & \text{if } \bar{H}\,\bar{K} \text{ is true,} \end{cases} \quad (12)$$

and, if we choose $z = T_\lambda(x,y)$, from (3) and (12) it follows that $(A|H) \wedge (B|K) = T_\lambda(A|H, B|K)$.

Conversely, given any coherent extension z of (x,y), there exists λ such that $z = T_\lambda(x,y)$. Indeed, if $z = \min\{x,y\}$, then $\lambda = 0$; if $z = \max\{x+y-1,0\}$, then $\lambda = +\infty$; if $\max\{x+y-1,0\} < z < \min\{x,y\}$, then by continuity of T_λ with respect to λ it holds that $z = T_\lambda(x,y)$ for some $\lambda \in]0, \infty[$ (for instance, if $z = xy$, then $z = T_1(x,y)$) and hence $(A|H) \wedge (B|K) = T_\lambda(A|H, B|K)$. □

Remark 3. As we can see from (3) and Theorem 4, in case of logically independent events, if the assessed values x, y, z are such that $z = T_\lambda(x, y)$ for a given λ, then the conjunction $(A|H) \wedge (B|K) = T_\lambda(A|H, B|K)$. For instance, if $z = T_1(x, y) = xy$, then $(A|H) \wedge (B|K) = T_1(A|H, B|K) = (A|H) \cdot (B|K)$. Conversely, if $(A|H) \wedge (B|K) = T_\lambda(A|H, B|K)$ for a given λ, then $z = T_\lambda(x, y)$. Then, the set Π given in (5) can be written as $\Pi = \{(x, y, z) : (x, y) \in [0, 1]^2, z = T_\lambda(x, y), \lambda \in [0, +\infty]\}$.

4   Conjunction of $(A|H)$ and $(A|K)$

In this section we examine the conjunction of two conditional events in the particular case when $A = B$, that is $(A|H) \wedge (A|K)$. By setting $P(A|H) = x$, $P(A|K) = y$ and $\mathbb{P}[(A|H) \wedge (A|K)] = z$, it holds that

$$(A|H) \wedge (A|K) = AHK + x\bar{H}AK + y\bar{K}AH + z\bar{H}\bar{K} \in \{1, 0, x, y, z\}.$$

Theorem 5. *Let A, H, K be three logically independent events, with $H \neq \varnothing$, $K \neq \varnothing$. The set Π of all coherent assessments (x, y, z) on the family $\mathcal{F} = \{A|H, A|K, (A|H) \wedge (A|K)\}$ is given by*

$$\Pi = \{(x, y, z) : (x, y) \in [0, 1]^2, T_P(x, y) = xy \leqslant z \leqslant \min\{x, y\} = T_M(x, y)\}. \tag{13}$$

Proof. Let $\mathcal{M} = (x, y, z)$ be a prevision assessment on $\mathcal{F}$. The constituents associated with the pair $(\mathcal{F}, \mathcal{M})$ and contained in $H \vee K$ are: $C_1 = AHK$, $C_2 = \bar{A}HK$, $C_3 = \bar{A}\bar{H}K$, $C_4 = \bar{A}H\bar{K}$, $C_5 = A\bar{H}K$, $C_6 = AH\bar{K}$. The associated points Q_h's are $Q_1 = (1, 1, 1), Q_2 = (0, 0, 0), Q_3 = (x, 0, 0), Q_4 = (0, y, 0), Q_5 = (x, 1, x), Q_6 = (1, y, y)$. With the further constituent $C_0 = \bar{H}\bar{K}$ it is associated the point $Q_0 = \mathcal{M} = (x, y, z)$. Considering the convex hull $\mathcal{I}$ (see Fig. 1) of $Q_1, \ldots, Q_6$, a necessary condition for the coherence of the prevision assessment $\mathcal{M} = (x, y, z)$ on $\mathcal{F}$ is that $\mathcal{M} \in \mathcal{I}$, that is the following system must be solvable

$$(\Sigma) \begin{cases} \lambda_1 + x\lambda_3 + x\lambda_5 + \lambda_6 = x, \quad \lambda_1 + y\lambda_4 + \lambda_5 + y\lambda_6 = y, \quad \lambda_1 + x\lambda_5 + y\lambda_6 = z, \\ \sum_{h=1}^{6} \lambda_h = 1, \quad \lambda_h \geqslant 0, \quad h = 1, \ldots, 6. \end{cases}$$

First of all, we observe that solvability of (Σ) requires that $z \leqslant x$ and $z \leqslant y$, that is $z \leqslant \min\{x, y\}$. We now verify that (x, y, z), with $(x, y) \in [0, 1]^2$ and $z = \min\{x, y\}$, is coherent. We distinguish two cases: (i) $x \leqslant y$ and (ii) $x > y$. Case (i). In this case $z = \min\{x, y\} = x$. If $y = 0$ the system (Σ) becomes

$$\lambda_1 + \lambda_6 = 0, \quad \lambda_1 + \lambda_5 = 0, \quad \lambda_1 = 0, \quad \lambda_2 + \lambda_3 + \lambda_4 = 1, \quad \lambda_h \geqslant 0, \quad h = 1, \ldots, 6.$$

which is clearly solvable. In particular there exist solutions with $\lambda_2 > 0, \lambda_3 > 0$, $\lambda_4 > 0$, by Theorem 1, as the set I_0 is empty the solvability of (Σ) is sufficient for coherence of the assessment $(0, 0, 0)$. If $y > 0$ the system (Σ) is solvable and a solution is $\Lambda = (\lambda_1, \ldots, \lambda_6) = (x, \frac{x(1-y)}{y}, 0, \frac{y-x}{y}, 0, 0)$. We observe that, if $x > 0$, then $\lambda_1 > 0$ and $I_0 = \varnothing$ because $\mathcal{C}_1 = HK \subseteq H \vee K$, so that $\mathcal{M} = (x, y, x)$ is coherent. If $x = 0$ (and hence $z = 0$), then $\lambda_4 = 1$ and $I_0 \subseteq \{2\}$. Then, as the sub-assessment $P(A|K) = y$ is coherent, it follows that the assessment $\mathcal{M} = (0, y, 0)$ is coherent too.

Case (ii). The system is solvable and a solution is $\Lambda = (\lambda_1, \ldots, \lambda_6) = (y, \frac{y(1-x)}{x}, \frac{x-y}{x}, 0, 0, 0)$. We observe that, if $y > 0$, then $\lambda_1 > 0$ and $I_0 = \varnothing$ because $\mathcal{C}_1 = HK \subseteq H \vee K$, so that $\mathcal{M} = (x, y, y)$ is coherent. If $y = 0$ (and hence $z = 0$), then $\lambda_3 = 1$ and $I_0 \subseteq \{1\}$. Then, as the sub-assessment $P(A|H) = x$ is coherent, it follows that the assessment $\mathcal{M} = (x, 0, 0)$ is coherent too. Thus, for every $(x, y) \in [0, 1]^2$, the assessment $(x, y, \min\{x, y\})$ is coherent and, as $z \leqslant \min\{x, y\}$, the upper bound on z is $\min\{x, y\} = T_M(x, y)$.

We now verify that (x, y, xy), with $(x, y) \in [0, 1]^2$ is coherent; moreover we will show that (x, y, z), with $z < xy$, is not coherent, in other words the lower bound for z is xy. First of all, we observe that $\mathcal{M} = (1-x)Q_4 + xQ_6$, so that a solution of (Σ) is $\Lambda_1 = (0, 0, 0, 1-x, 0, x)$. Moreover, $\mathcal{M} = (1-y)Q_3 + yQ_5$, so that another solution is $\Lambda_2 = (0, 0, 1-y, 0, y, 0)$. Then $\Lambda = \frac{\Lambda_1 + \Lambda_2}{2} = (0, 0, \frac{1-y}{2}, \frac{1-x}{2}, \frac{y}{2}, \frac{x}{2})$ is a solution of (Σ) such that $I_0 = \varnothing$. Thus the assessment (x, y, xy) is coherent for every $(x, y) \in [0, 1]^2$. In order to verify that xy is the lower bound on z we observe that the points Q_3, Q_4, Q_5, Q_6 belong to a plane π of equation $yX + xY - Z = xy$, where X, Y, Z are the axis' coordinates. Now, by considering the function $f(X, Y, Z) = yX + xY - Z$, we observe that for each constant k the equation $f(X, Y, Z) = k$ represents a plane which is parallel to π and coincides with π when $k = xy$. We also observe that $f(Q_1) = f(1, 1, 1) = x + y - 1 = T_L(x, y) \leqslant xy = T_P(x, y)$, $f(Q_2) = f(0, 0, 0) = 0 \leqslant xy = T_P(x, y)$, and $f(Q_3) = f(Q_4) = f(Q_5) = f(Q_6) = xy = T_P(x, y)$. Then, for every $\mathcal{P} = \sum_{h=1}^{6} \lambda_h Q_h$, with $\lambda_h \geqslant 0$ and $\sum_{h=1}^{6} \lambda_h = 1$, that is $\mathcal{P} \in \mathcal{I}$, it holds that $f(\mathcal{P}) = f\left(\sum_{h=1}^{6} \lambda_h Q_h\right) = \sum_{h=1}^{6} \lambda_h f(Q_h) \leqslant xy$. On the other hand, given any $a > 0$, by considering $\mathcal{P} = (x, y, xy - a)$ it holds that $f(\mathcal{P}) = f(x, y, xy - a) = xy + xy - xy + a = xy + a > xy$. Therefore, for any given $a > 0$ the assessment $(x, y, xy - a)$ is not coherent because $(x, y, xy - a) \notin \mathcal{I}$. Then, the lower bound on z is $xy = T_P(x, y)$. Finally, the set of all coherent assessments (x, y, z) on $\mathcal{F}$ is the set Π in (13).   □

Based on Theorem 5, we can give an analogous version for the Theorem 4 (when $A = B$).

Theorem 6. *Let us consider the conjunction $(A|H) \wedge (A|K)$, with A, H, K logically independent and with $P(A|H) = x$, $P(A|K) = y$. Moreover, given any $\lambda \in [1, +\infty]$, let T_λ be the Frank t-norm with parameter λ. Then, the assessment $z = T_\lambda(x, y)$ on $(A|H) \wedge (A|K)$ is a coherent extension of (x, y) on $\{A|H, A|K\}$; moreover $(A|H) \wedge (A|K) = T_\lambda(A|H, A|K)$. Conversely, given any coherent extension $z = \mathbb{P}[(A|H) \wedge (A|K)]$ of (x, y), there exists $\lambda \in [1, +\infty]$ such that $z = T_\lambda(x, y)$.*

The next result follows from Theorem 5 when H, K are incompatible.

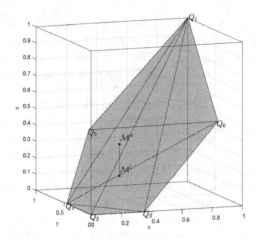

Fig. 1. Convex hull $\mathcal{I}$ of the points $Q_1, Q_2, Q_3, Q_4, Q_5, Q_6$. $\mathcal{M}' = (x, y, z')$, $\mathcal{M}'' = (x, y, z'')$, where $(x, y) \in [0, 1]^2$, $z' = xy$, $z'' = \min\{x, y\}$. In the figure the numerical values are: $x = 0.35$, $y = 0.45$, $z' = 0.1575$, and $z'' = 0.35$.

Theorem 7. *Let A, H, K be three events, with A logically independent from both H and K, with $H \neq \varnothing$, $K \neq \varnothing$, $HK = \varnothing$. The set Π of all coherent assessments (x, y, z) on the family $\mathcal{F} = \{A|H, A|K, (A|H) \wedge (A|K)\}$ is given by $\Pi = \{(x, y, z) : (x, y) \in [0, 1]^2, z = xy = T_P(x, y)\}$.*

Proof. We observe that

$$(A|H) \wedge (A|K) = \begin{cases} 0, & \text{if } \bar{A}\bar{H}K \vee \bar{A}H\bar{K} \text{ is true,} \\ x, & \text{if } \bar{H}AK \text{ is true,} \\ y, & \text{if } AH\bar{K} \text{ is true,} \\ z, & \text{if } \bar{H}\bar{K} \text{ is true.} \end{cases}$$

Moreover, as $HK = \varnothing$, the points Q_h's are $(x, 0, 0), (0, y, 0), (x, 1, x), (1, y, y)$, which coincide with the points $Q_3, \ldots, Q_6$ of the case $HK \neq \varnothing$. Then, as shown in the proof of Theorem 5, the condition $\mathcal{M} = (x, y, z)$ belongs to the convex hull of $(x, 0, 0), (0, y, 0), (x, 1, x), (1, y, y)$ amounts to the condition $z = xy$. □

Remark 4. From Theorem 7, when $HK = \varnothing$ it holds that $(A|H) \wedge (A|K) = (A|H) \cdot (A|K) = T_P(A|H, A|K)$, where $x = P(A|H)$ and $y = P(A|K)$.

5   Further Results on Frank t-Norms

In this section we give some results which concern Frank t-norms and the family $\mathcal{F} = \{\mathcal{C}_1, \mathcal{C}_2, \mathcal{C}_3, \mathcal{C}_{12}, \mathcal{C}_{13}, \mathcal{C}_{23}, \mathcal{C}_{123}\}$. We recall that, given any t-norm $T(x_1, x_2)$ it holds that $T(x_1, x_2, x_3) = T(T(x_1, x_2), x_3)$.

5.1   On the Product t-Norm

Theorem 8. *Assume that the events $E_1, E_2, E_3, H_1, H_2, H_3$ are logically independent, with $H_1 \neq \varnothing, H_2 \neq \varnothing, H_3 \neq \varnothing$. If the assessment $\mathcal{M} = (x_1, x_2, x_3, x_{12}, x_{13}, x_{23}, x_{123})$ on $\mathcal{F} = \{\mathcal{C}_1, \mathcal{C}_2, \mathcal{C}_3, \mathcal{C}_{12}, \mathcal{C}_{13}, \mathcal{C}_{23}, \mathcal{C}_{123}\}$ is such that $(x_1, x_2, x_3) \in [0,1]^3$, $x_{ij} = T_1(x_i, x_j) = x_i x_j$, $i \neq j$, and $x_{123} = T_1(x_1, x_2, x_3) = x_1 x_2 x_3$, then $\mathcal{M}$ is coherent. Moreover, $\mathcal{C}_{ij} = T_1(\mathcal{C}_i, \mathcal{C}_j) = \mathcal{C}_i \mathcal{C}_j$, $i \neq j$, and $\mathcal{C}_{123} = T_1(\mathcal{C}_1, \mathcal{C}_2, \mathcal{C}_3) = \mathcal{C}_1 \mathcal{C}_2 \mathcal{C}_3$.*

Proof. From Remark 2, the coherence of $\mathcal{M}$ amounts to the inequalities in (9). As $x_{ij} = T_1(x_i, x_j) = x_i x_j$, $i \neq j$, and $x_{123} = T_1(x_1, x_2, x_3) = x_1 x_2 x_3$, the inequalities (9) become

$$\max\{0, x_1(x_2 + x_3 - 1), x_2(x_1 + x_3 - 1), x_3(x_1 + x_2 - 1)\} \leqslant x_1 x_2 x_3 \tag{14}$$
$$\leqslant \min\{x_1 x_2, x_1 x_3, x_2 x_3, (1 - x_1)(1 - x_2)(1 - x_3) + x_1 x_2 x_3\}.$$

Thus, by recalling that $x_i + x_j - 1 \leqslant x_i x_j$, the inequalities are satisfied and hence $\mathcal{M}$ is coherent. Moreover, from (3) and (7) it follows that $\mathcal{C}_{ij} = T_1(\mathcal{C}_i, \mathcal{C}_j) = \mathcal{C}_i \mathcal{C}_j$, $i \neq j$, and $\mathcal{C}_{123} = T_1(\mathcal{C}_1, \mathcal{C}_2, \mathcal{C}_3) = \mathcal{C}_1 \mathcal{C}_2 \mathcal{C}_3$.   □

5.2   On the Minimum t-Norm

Theorem 9. *Assume that the events $E_1, E_2, E_3, H_1, H_2, H_3$ are logically independent, with $H_1 \neq \varnothing, H_2 \neq \varnothing, H_3 \neq \varnothing$. If the assessment $\mathcal{M} = (x_1, x_2, x_3, x_{12}, x_{13}, x_{23}, x_{123})$ on $\mathcal{F} = \{\mathcal{C}_1, \mathcal{C}_2, \mathcal{C}_3, \mathcal{C}_{12}, \mathcal{C}_{13}, \mathcal{C}_{23}, \mathcal{C}_{123}\}$ is such that $(x_1, x_2, x_3) \in [0,1]^3$, $x_{ij} = T_M(x_i, x_j) = \min\{x_i, x_j\}$, $i \neq j$, and $x_{123} = T_M(x_1, x_2, x_3) = \min\{x_1, x_2, x_3\}$, then $\mathcal{M}$ is coherent. Moreover, $\mathcal{C}_{ij} = T_M(\mathcal{C}_i, \mathcal{C}_j) = \min\{\mathcal{C}_i, \mathcal{C}_j\}$, $i \neq j$, and $\mathcal{C}_{123} = T_M(\mathcal{C}_1, \mathcal{C}_2, \mathcal{C}_3) = \min\{\mathcal{C}_1, \mathcal{C}_2, \mathcal{C}_3\}$.*

Proof. From Remark 2, the coherence of $\mathcal{M}$ amounts to the inequalities in (9). Without loss of generality, we assume that $x_1 \leqslant x_2 \leqslant x_3$. Then $x_{12} = T_M(x_1, x_2) = x_1$, $x_{13} = T_M(x_1, x_3) = x_1$, $x_{23} = T_M(x_2, x_3) = x_2$, and $x_{123} = T_M(x_1, x_2, x_3) = x_1$. The inequalities (9) become

$$\max\{0, x_1, x_1 + x_2 - x_3\} = x_1 \leqslant x_1 \leqslant x_1 = \min\{x_1, x_2, 1 - x_3 + x_1\}. \tag{15}$$

Thus, the inequalities are satisfied and hence $\mathcal{M}$ is coherent. Moreover, from (3) and (7) it follows that $\mathcal{C}_{ij} = T_M(\mathcal{C}_i, \mathcal{C}_j) = \min\{\mathcal{C}_i, \mathcal{C}_j\}$, $i \neq j$, and $\mathcal{C}_{123} = T_M(\mathcal{C}_1, \mathcal{C}_2, \mathcal{C}_3) = \min\{\mathcal{C}_1, \mathcal{C}_2, \mathcal{C}_3\}$.   □

Remark 5. As we can see from (15) and Corollary 1, the assessment $x_{123} = \min\{x_1, x_2, x_3\}$ is the unique coherent extension on $\mathcal{C}_{123}$ of the assessment $(x_1, x_2, x_3, \min\{x_1, x_2\}, \min\{x_1, x_3\}, \min\{x_2, x_3\})$ on $\{\mathcal{C}_1, \mathcal{C}_2, \mathcal{C}_3, \mathcal{C}_{12}, \mathcal{C}_{13}, \mathcal{C}_{23}\}$. We also notice that, if $\mathcal{C}_1 \leqslant \mathcal{C}_2 \leqslant \mathcal{C}_3$, then $\mathcal{C}_{12} = \mathcal{C}_1$, $\mathcal{C}_{13} = \mathcal{C}_1$, $\mathcal{C}_{23} = \mathcal{C}_2$, and $\mathcal{C}_{123} = \mathcal{C}_1$. Moreover, $x_{12} = x_1$, $x_{13} = x_1$, $x_{23} = x_2$, and $x_{123} = x_1$.

5.3   On Lukasiewicz t-Norm

We observe that in general the results of Theorems 8 and 9 do not hold for the Lukasiewicz t-norm (and hence for any given Frank t-norm), as shown in the example below. We recall that $T_L(x_1, x_2, x_3) = \max\{x_1 + x_2 + x_3 - 2, 0\}$.

Example 1. The assessment $(x_1, x_2, x_3, T_L(x_1, x_2), T_L(x_1, x_3), T_L(x_2, x_3),$ $T_L(x_1, x_2, x_3))$ on the family $\mathcal{F} = \{\mathcal{C}_1, \mathcal{C}_2, \mathcal{C}_3, \mathcal{C}_{12}, \mathcal{C}_{13}, \mathcal{C}_{23}, \mathcal{C}_{123}\}$, with $(x_1, x_2, x_3) = (0.5, 0.6, 0.7)$ is not coherent. Indeed, by observing that $T_L(x_1, x_2) = 0.1$ $T_L(x_1, x_3) = 0.2$, $T_L(x_2, x_3) = 0.3$, and $T_L(x_1, x_2, x_3) = 0$, formula (9) becomes $\max\{0, 0.1 + 0.2 - 0.5, 0.1 + 0.3 - 0.6, 0.2 + 0.3 - 0.7\} \leqslant$ $0 \leqslant \min\{0.1, 0.2, 0.3, 1 - 0.5 - 0.6 - 0.7 + 0.1 + 0.2 + 0.3\}$, that is: $\max\{0, -0.2\} \leqslant 0 \leqslant \min\{0.1, 0.2, 0.3, -0.2\}$; thus the inequalities are not satisfied and the assessment is not coherent.

More in general we have

Theorem 10. *The assessment* $(x_1, x_2, x_3, T_L(x_1, x_2), T_L(x_1, x_3), T_L(x_2, x_3))$ *on the family* $\mathcal{F} = \{\mathcal{C}_1, \mathcal{C}_2, \mathcal{C}_3, \mathcal{C}_{12}, \mathcal{C}_{13}, \mathcal{C}_{23}\}$, *with* $T_L(x_1, x_2) > 0$, $T_L(x_1, x_3) > 0$, $T_L(x_2, x_3) > 0$ *is coherent if and only if* $x_1 + x_2 + x_3 - 2 \geqslant 0$. *Moreover, when* $x_1 + x_2 + x_3 - 2 \geqslant 0$ *the unique coherent extension* x_{123} *on* $\mathcal{C}_{123}$ *is* $x_{123} = T_L(x_1, x_2, x_3)$.

Proof. We distinguish two cases: (i) $x_1 + x_2 + x_3 - 2 < 0$; (ii) $x_1 + x_2 + x_3 - 2 \geqslant 0$.

Case (i). From (8) the inequality $1 - x_1 - x_2 - x_3 + x_{12} + x_{13} + x_{23} \geqslant 0$ is not satisfied because $1 - x_1 - x_2 - x_3 + x_{12} + x_{13} + x_{23} = x_1 + x_2 + x_3 - 2 < 0$. Therefore the assessment is not coherent.
Case (ii). We set $x_{123} = T_L(x_1, x_2, x_3) = x_1 + x_2 + x_3 - 2$. Then, by observing that $0 < x_i + x_j - 1 \leqslant x_1 + x_2 + x_3 - 2$, $i \neq j$, formula (9) becomes $\max\{0, x_1 + x_2 + x_3 - 2\} \leqslant x_1 + x_2 + x_3 - 2 \leqslant \min\{x_1 + x_2 - 1,$ $x_1 + x_3 - 1, x_2 + x_3 - 1, x_1 + x_2 + x_3 - 2\}$, that is: $x_1 + x_2 + x_3 - 2 \leqslant$ $x_1 + x_2 + x_3 - 2 \leqslant x_1 + x_2 + x_3 - 2$. Thus, the inequalities are satisfied and the assessment $(x_1, x_2, x_3, T_L(x_1, x_2), T_L(x_1, x_3), T_L(x_2, x_3), T_L(x_1, x_2, x_3))$ on $\{\mathcal{C}_1, \mathcal{C}_2, \mathcal{C}_3, \mathcal{C}_{12}, \mathcal{C}_{13}, \mathcal{C}_{23}, \mathcal{C}_{123}\}$ is coherent and the sub-assessment $(x_1, x_2, x_3, T_L(x_1, x_2), T_L(x_1, x_3), T_L(x_2, x_3))$ on $\mathcal{F}$ is coherent too. □

A result related with Theorem 10 is given below.

Theorem 11. *If the assessment* $(x_1, x_2, x_3, T_L(x_1, x_2), T_L(x_1, x_3), T_L(x_2, x_3),$ $T_L(x_1, x_2, x_3))$ *on the family* $\mathcal{F} = \{\mathcal{C}_1, \mathcal{C}_2, \mathcal{C}_3, \mathcal{C}_{12}, \mathcal{C}_{13}, \mathcal{C}_{23}, \mathcal{C}_{123}\}$, *is such that* $T_L(x_1, x_2, x_3) > 0$, *then the assessment is coherent.*

Proof. We observe that $T_L(x_1, x_2, x_3) = x_1 + x_2 + x_3 - 2 > 0$; then $x_i > 0$, $i = 1, 2, 3$, and $0 < x_i + x_j - 1 \leqslant x_1 + x_2 + x_3 - 2$, $i \neq j$. Then formula (9) becomes: $\max\{0, x_1 + x_2 + x_3 - 2\} \leqslant x_1 + x_2 + x_3 - 2 \leqslant \min\{x_1 + x_2 - 1, x_1 + x_3 - 1, x_2 + x_3 - 1, x_1 + x_2 + x_3 - 2\}$, that is: $x_1 + x_2 + x_3 - 2 \leqslant x_1 + x_2 + x_3 - 2 \leqslant x_1 + x_2 + x_3 - 2$. Thus, the inequalities are satisfied and the assessment is coherent. □

6   Conclusions

We have studied the relationship between the notions of conjunction and of Frank t-norms. We have shown that, under logical independence of events and coherence of prevision assessments, for a suitable $\lambda \in [0, +\infty]$ it holds that $\mathbb{P}((A|H) \wedge (B|K)) = T_\lambda(x, y)$ and $(A|H) \wedge (B|K) = T_\lambda(A|H, B|K)$. Then, we have considered the case $A = B$, by determining the set of all coherent assessment (x, y, z) on $(A|H, B|K, (A|H) \wedge (A|K))$. We have shown that, under coherence, for a suitable $\lambda \in [1, +\infty]$ it holds that $(A|H) \wedge (A|K) = T_\lambda(A|H, A|K)$. We have also studied the particular case where $A = B$ and $HK = \varnothing$. Then, we have considered the conjunction of three conditional events and we have shown that the prevision assessments produced by the Product t-norm, or the Minimum t-norm, are coherent. Finally, we have examined the Lukasiewicz t-norm and we have shown, by a counterexample, that coherence in general is not assured. We have given some conditions for coherence when the prevision assessments are based on the Lukasiewicz t-norm. Future work should concern the deepening and generalization of the results of this paper.

Acknowledgments. We thank three anonymous referees for their useful comments.

References

1. Biazzo, V., Gilio, A.: A generalization of the fundamental theorem of de Finetti for imprecise conditional probability assessments. Int. J. Approximate Reasoning **24**(2–3), 251–272 (2000)
2. Biazzo, V., Gilio, A., Lukasiewicz, T., Sanfilippo, G.: Probabilistic logic under coherence: complexity and algorithms. Ann. Math. Artif. Intell. **45**(1–2), 35–81 (2005)
3. Biazzo, V., Gilio, A., Sanfilippo, G.: Generalized coherence and connection property of imprecise conditional previsions. In: Proceedings of IPMU 2008, Malaga, Spain, 22–27 June, pp. 907–914 (2008)
4. Biazzo, V., Gilio, A., Sanfilippo, G.: Coherent conditional previsions and proper scoring rules. In: Greco, S., Bouchon-Meunier, B., Coletti, G., Fedrizzi, M., Matarazzo, B., Yager, R.R. (eds.) IPMU 2012. CCIS, vol. 300, pp. 146–156. Springer, Heidelberg (2012). https://doi.org/10.1007/978-3-642-31724-8_16
5. Coletti, G., Gervasi, O., Tasso, S., Vantaggi, B.: Generalized bayesian inference in a fuzzy context: from theory to a virtual reality application. Comput. Stat. Data Anal. **56**(4), 967–980 (2012)
6. Coletti, G., Petturiti, D., Vantaggi, B.: Possibilistic and probabilistic likelihood functions and their extensions: common features and specific characteristics. Fuzzy Sets Syst. **250**, 25–51 (2014)
7. Coletti, G., Scozzafava, R.: Probabilistic Logic in a Coherent Setting. Kluwer, Dordrecht (2002)
8. Coletti, G., Scozzafava, R.: Conditional probability, fuzzy sets, and possibility: a unifying view. Fuzzy Sets Syst. **144**, 227–249 (2004)

9. Coletti, G., Scozzafava, R., Vantaggi, B.: Coherent conditional probability, fuzzy inclusion and default rules. In: Yager, R., Abbasov, A.M., Reformat, M.Z., Shahbazova, S.N. (eds.) Soft Computing: State of the Art Theory and Novel Applications. Studies in Fuzziness and Soft Computing, vol. 291, pp. 193–208. Springer, Heidelberg (2013). https://doi.org/10.1007/978-3-642-34922-5_14

10. Coletti, G., Scozzafava, R., Vantaggi, B.: Possibilistic and probabilistic logic under coherence: default reasoning and System P. Mathematica Slovaca **65**(4), 863–890 (2015)

11. Dubois, D.: Generalized probabilistic independence and its implications for utility. Oper. Res. Lett. **5**(5), 255–260 (1986)

12. Durante, F., Klement, E.P., Quesada-Molina, J.J.: Bounds for trivariate copulas with given bivariate marginals. J. Inequalities Appl. **2008**(1), 9 pages (2008). Article ID 161537

13. de Finetti, B.: La logique de la probabilité. In: Actes du Congrès International de Philosophie Scientifique, Paris (1935). pp. IV 1–IV 9 (1936)

14. de Finetti, B.: Theory of Probability, vol. 1, 2. Wiley, Chichester (1970/1974)

15. Flaminio, T., Godo, L., Ugolini, S.: Towards a probability theory for product logic: states, integral representation and reasoning. Int. J. Approximate Reasoning **93**, 199–218 (2018)

16. Gilio, A.: Probabilistic reasoning under coherence in system P. Ann. Math. Artif. Intell. **34**, 5–34 (2002)

17. Gilio, A.: Generalizing inference rules in a coherence-based probabilistic default reasoning. Int. J. Approximate Reasoning **53**(3), 413–434 (2012)

18. Gilio, A., Pfeifer, N., Sanfilippo, G.: Transitivity in coherence-based probability logic. J. Appl. Logic **14**, 46–64 (2016)

19. Gilio, A., Sanfilippo, G.: Coherent conditional probabilities and proper scoring rules. In: Proceedings of ISIPTA 2011, Innsbruck, pp. 189–198 (2011)

20. Gilio, A., Sanfilippo, G.: Conditional random quantities and iterated conditioning in the setting of coherence. In: van der Gaag, L.C. (ed.) ECSQARU 2013. LNCS (LNAI), vol. 7958, pp. 218–229. Springer, Heidelberg (2013). https://doi.org/10.1007/978-3-642-39091-3_19

21. Gilio, A., Sanfilippo, G.: Conjunction, disjunction and iterated conditioning of conditional events. In: Kruse, R., Berthold, M., Moewes, C., Gil, M., Grzegorzewski, P., Hryniewicz, O. (eds.) Synergies of Soft Computing and Statistics for Intelligent Data Analysis. AISC, vol. 190, pp. 399–407. Springer, Berlin (2013). https://doi.org/10.1007/978-3-642-33042-1_43

22. Gilio, A., Sanfilippo, G.: Quasi conjunction, quasi disjunction, t-norms and t-conorms: probabilistic aspects. Inf. Sci. **245**, 146–167 (2013)

23. Gilio, A., Sanfilippo, G.: Conditional random quantities and compounds of conditionals. Stud. Logica **102**(4), 709–729 (2014)

24. Gilio, A., Sanfilippo, G.: Generalized logical operations among conditional events. Appl. Intell. **49**(1), 79–102 (2019)

25. Joe, H.: Multivariate Models and Multivariate Dependence Concepts. Chapman and Hall/CRC, New York (1997)

26. Kaufmann, S.: Conditionals right and left: probabilities for the whole family. J. Philos. Logic **38**, 1–53 (2009)

27. Klement, E.P., Mesiar, R., Pap, E.: Triangular Norms. Springer, Dordrecht (2000). https://doi.org/10.1007/978-94-015-9540-7

28. Klement, E.P., Mesiar, R., Pap, E.: Triangular norms: basic notions and properties. In: Klement, E.P., Mesiar, R. (eds.) Logical, Algebraic, Analytic and Probabilistic Aspects of Triangular Norms, pp. 17–60. Elsevier Science B.V, Amsterdam (2005)

29. Lad, F., Sanfilippo, G., Agró, G.: Completing the logarithmic scoring rule for assessing probability distributions. In: AIP Conference Proceedings, vol. 1490, no. 1, pp. 13–30 (2012)
30. Lad, F., Sanfilippo, G., Agró, G.: Extropy: complementary dual of entropy. Stat. Sci. **30**(1), 40–58 (2015)
31. Lad, F., Sanfilippo, G., Agró, G.: The duality of entropy/extropy, and completion of the kullback information complex. Entropy **20**(8), 593 (2018)
32. McGee, V.: Conditional probabilities and compounds of conditionals. Philos. Rev. **98**, 485–541 (1989)
33. Navara, M.: Triangular norms and measures of fuzzy sets. In: Klement, E.P., Mesiar, R. (eds.) Logical, Algebraic, Analytic and Probabilistic Aspects of Triangular Norms., pp. 345–390. Elsevier, Amsterdam (2005)
34. Pfeifer, N., Sanfilippo, G.: Probability propagation in selected Aristotelian syllogisms. In: Kern-Isberner, G., Ognjanović, Z. (eds.) ECSQARU 2019. LNAI, vol. 11726, pp. 419–431. Springer, Cham (2019)
35. Sanfilippo, G.: Lower and upper probability bounds for some conjunctions of two conditional events. In: Ciucci, D., Pasi, G., Vantaggi, B. (eds.) SUM 2018. LNCS (LNAI), vol. 11142, pp. 260–275. Springer, Cham (2018). https://doi.org/10.1007/978-3-030-00461-3_18
36. Sanfilippo, G., Pfeifer, N., Over, D., Gilio, A.: Probabilistic inferences from conjoined to iterated conditionals. Int. J. Approximate Reasoning **93**(Supplement C), 103–118 (2018)

Reasoning About Exceptions in Ontologies: An Approximation of the Multipreference Semantics

Laura Giordano[1]($^{(\boxtimes)}$) and Valentina Gliozzi[2]

[1] DISIT - Università del Piemonte Orientale, Alessandria, Italy
laura.giordano@uniupo.it
[2] Dipartimento di Informatica, Università di Torino, Turin, Italy
valentina.gliozzi@unito.it

Abstract. Starting from the observation that rational closure has the undesirable property of being an "all or nothing" mechanism, we here consider a multipreferential semantics, which enriches the preferential semantics underlying rational closure in order to separately deal with the inheritance of different properties in an ontology with exceptions. We show that the MP-closure of an $\mathcal{ALC}$ knowledge base is a construction which is sound with respect to minimal entailment in the multipreference semantics for $\mathcal{ALC}$.

1 Introduction

Reasoning about exceptions in ontologies is one of the challenges the description logics community is facing, a challenge which is at the very roots of the development of non-monotonic reasoning in the 80's. Many non-monotonic extensions of Description Logics (DLs) have been developed incorporating non-monotonic features from most of the non-monotonic formalisms in the literature [2,6,7,12,15,16,20–23,29–31,34–36,41,43], and defining new constructions and semantics such as in [4,5,8,9].

In this paper we focus on the rational closure for DLs [14–16,19,31] and on its refinements. While the rational closure provides a simple and efficient approach for reasoning with exceptions, exploiting polynomial reductions to standard DLs [24,40], it is well known that it does not allow an independent handling of the inheritance of different defeasible properties of concepts: if a subclass of C is exceptional for a given aspect, it is exceptional tout court and does not inherit any of the typical properties of C. This problem was called by Pearl [42] "the blocking of property inheritance problem", and it is an instance of the "drowning problem" in [3].

To cope with this problem Lehmann [39] introduced the notion of the lexicographic closure, which was extended to Description Logics by Casini and Straccia [18], while in [19] the same authors develop an inheritance-based approach for defeasible DLs. In [13] Casini et al. also developed a closure construction weaker

© Springer Nature Switzerland AG 2019
G. Kern-Isberner and Z. Ognjanović (Eds.): ECSQARU 2019, LNAI 11726, pp. 212–225, 2019.
https://doi.org/10.1007/978-3-030-29765-7_18

than the lexicographic closure, called the Relevant Closure. In [33] Gliozzi developed a multi-preference semantics for defeasible inclusions in which models are equipped with several preference relations, providing a refinement of the rational closure semantics for $\mathcal{ALC}$. Some other proposal for non-monotonic reasoning in the literature, still based on a preferential semantics, also suffer from the problem of inheritance blocking such as, for instance, the typicality logic $\mathcal{ALC} + \mathbf{T}_{min}$ [30]. Its semantics, differently from the rational closure, is not based on ranked models. A multi-typicality version of this logic has been studied by Fernandez Gil [23] to address this problem. A logic which may build on the rational closure to determine specificity of defaults, but does not suffer from the problem of inheritance blocking, is the logic of overriding, $\mathcal{DL}^N$, proposed by Bonatti et al. [5,8].

In this paper, we reconsider the multi-preference semantics for $\mathcal{ALC}$ in [33] and we show that entailment in the multipreference semantics can be approximated by the MP-closure construction for $\mathcal{ALC}$. The idea of the multipreference semantics was to define a refinement of the rational closure for $\mathcal{ALC}$ in which preference with respect to specific aspects is considered. It is formulated in terms of enriched models, which also consider the preference relations $<_{A_i}$, associated with the different aspects (or concepts, like having feather or flying, being a sport lover or a swimmer), as for any two individuals, one may be more typical than the other one as a sport lover, but less typical as a swimmer.

Here, we refer to the definition of multipreference semantics in [27], which is slightly stronger than the one originally introduce by Gliozzi in [33] (although both of them lead to refinements of the rational closure), and we show that entailment in the multipreference semantics can be soundly approximated by the MP-closure, a notion of closure that is more cautious than the lexicographic closure. In the propositional case, the MP-closure has been studied as the natural alternative to the lexicographic closure when the Maximal Entropy approach is abandoned [28], and it has been proved to be weaker than the lexicographic closure, but stronger than the rational closure and the relevant closure. The MP-closure, as the lexicographic closure builds over the RC. Lehmann's lexicographic closure [39] strengthens the RC by allowing, roughly speaking, a class to inherit as many as possible of the defeasible properties of more general classes, giving preference to the more specific properties. It has been extended to the description logic $\mathcal{ALC}$ by Casini and Straccia in [18]. The idea underlying the construction of the MP-closure is similar to that of the lexicographic closure but, while the lexicographic ordering in [39] takes into consideration the size of the sets of defaults satisfied at each rank (and is modular), the MP-closure construction only compares sets of defaults based on subset inclusion.

A semantic characterization of the MP-closure for the description logic $\mathcal{ALC}$ was developed in [25] using bi-preferential (BP) interpretations, that is, preferential interpretations developed along the lines of the preferential semantics introduced by Kraus, Lehmann and Magidor [37,38], but containing two preference relations, the first $<_1$ playing the role of preference relations in the models of the RC, and the second $<_2$ representing a refinement of $<_1$. Instead, in the propositional case [28], we have considered a simpler semantic characterization of the MP-closure, more similar to the model-theoretic semantics of the lexicographic closure defined by Lehmann [39]. While we refer therein for details on the different semantics, here, we will only consider the multipreference semantics and the MP-closure construction used to prove the soundness result. The proofs can be found in [27].

2   The Rational Closure for $\mathcal{ALC}$

In this section we recall the extension of $\mathcal{ALC}$ with a typicality operator introduced in [29,31] under the preferential and ranked semantics. In particular, we recall the logic $\mathcal{ALC} + \mathbf{T_R}$ which is at the basis of a rational closure construction proposed in [31] for $\mathcal{ALC}$. The general idea is that of extending the description logic $\mathcal{ALC}$ with concepts of the form $\mathbf{T}(C)$, whose instances are the *typical* instances of concept C, thus distinguishing the properties that hold for all instances of concept C (given by *strict* inclusions $C \sqsubseteq D$), from the properties that hold for the typical instances of C (given by the *defeasible inclusions* $\mathbf{T}(C) \sqsubseteq D$). The extended language is defined as follows:

$$C_R := A \mid \top \mid \bot \mid \neg C_R \mid C_R \sqcap C_R \mid C_R \sqcup C_R \mid \forall R.C_R \mid \exists R.C_R$$
$$C_L := C_R \mid \mathbf{T}(C_R),$$

where A is a concept name and R a role name. A knowledge base K is a pair $(\mathcal{T}, \mathcal{A})$, where the TBox $\mathcal{T}$ contains a finite set of concept inclusions $C_L \sqsubseteq C_R$, and the ABox $\mathcal{A}$ contains a finite set of assertions of the form $C_R(a)$ and $R(a, b)$, for a, b individual names, and R role name. The TBox contains two kinds of inclusions: strict inclusions $C \sqsubseteq D$ (the C's are D's), where C and D are $\mathcal{ALC}$ concepts, and typicality inclusions $\mathbf{T}(C) \sqsubseteq D$ (the typical C's are D's), corresponding to KLM conditionals $C \vdash D$.

The semantics of $\mathcal{ALC}$ with typicality is defined in terms of preferential models, extending to $\mathcal{ALC}$ the preferential semantics by Kraus, Lehmann and Magidor in [37,38]: ordinary models of $\mathcal{ALC}$ are equipped with a *preference relation* $<$ on the domain, whose intuitive meaning is to compare the "typicality" of domain elements: $x < y$ means that x is more typical than y. The instances of $\mathbf{T}(C)$ are the instances of concept C that are minimal with respect to $<$. The preference relation $<$ is assumed to be *well-founded* (i.e., there is no infinite $<$-descending chain, so that, if $S \neq \emptyset$, also $min_<(S) \neq \emptyset$). In ranked models, which characterize $\mathcal{ALC} + \mathbf{T_R}$, $<$ is further assumed to be *modular* (i.e., for all $x, y, z \in \Delta$, if $x < y$ then either $x < z$ or $z < y$). Ranked models characterize $\mathcal{ALC} + \mathbf{T_R}$. Let us shortly recap their definition.

Definition 1 (Preferential and ranked interpretations of $\mathcal{ALC} + \mathbf{T}$). *A preferential interpretation $\mathcal{M}$ is any structure $\mathcal{M} = \langle \Delta, <, I \rangle$ where: Δ is the domain; $<$ is an irreflexive, transitive and well-founded relation over Δ. I is an interpretation function that maps each concept name $C \in N_C$ to $C^I \subseteq \Delta$, each role name $R \in N_R$ to $R^I \subseteq \Delta^I \times \Delta^I$ and each individual name $a \in N_I$ to $a^I \in \Delta$. For concepts of $\mathcal{ALC}$, C^I is defined in the usual way in $\mathcal{ALC}$ interpretations*

[1]. In particular: $\top^I = \Delta$, $\bot^I = \emptyset$, $(\neg C)^I = \Delta\backslash C^I$, $(C \sqcap D)^I = C^I \cap D^I$, $(C \sqcup D)^I = C^I \cup D^I$ *and*

$$(\forall R.C)^I = \{x \in \Delta \mid \text{ for all } y, (x,y) \in R^I \text{ implies } y \in C^I\}$$
$$(\exists R.C)^I = \{x \in \Delta \mid \text{ for some } y \ (x,y) \in R^I \text{ and } y \in C^I\}$$

For the **T** *operator, we have* $(\mathbf{T}(C))^I = min_<(C^I)$.
When the interpretation I is also modular, I is called a ranked *interpretation.*

The notion of satisfiability of a KB in an interpretation is defined as usual. Given an $\mathcal{ALC}$ interpretation $\mathcal{M} = \langle \Delta, <, I \rangle$:

- I satisfies an inclusion $C \sqsubseteq D$ if $C^I \subseteq D^I$;
- I satisfies an assertion $C(a)$ if $a^I \in C^I$;
- I satisfies an assertion $R(a,b)$ if $(a^I, b^I) \in R^I$.

Definition 2 (Model of a KB [29]). *A preferential (ranked) model of a knowledge base $K = (\mathcal{T}, \mathcal{A})$ is a preferential (ranked) interpretation $\mathcal{M}$ that satisfies all inclusions in $\mathcal{T}$ and all assertions in $\mathcal{A}$.*

A *query* F (either an assertion $C_L(a)$ or an inclusion relation $C_L \sqsubseteq C_R$) is preferentially (rationally) entailed by a knowledge base K, written $K \models_{\mathcal{ALC}+\mathbf{T}} F$ (resp., $K \models_{\mathcal{ALC}+\mathbf{T_R}} F$) if F is satisfied in all the models (resp., ranked models) of K.

In particular, the definition of the rational closure for $\mathcal{ALC}$ and its semantics in [31] exploit the extension of $\mathcal{ALC}$ with typicality under a ranked semantics, which was called $\mathcal{ALC} + \mathbf{T_R}$. As shown therein, the logic $\mathcal{ALC} + \mathbf{T_R}$ enjoys the finite model property and finite $\mathcal{ALC} + \mathbf{T_R}$ models can be equivalently defined by postulating the existence of a function $k_{\mathcal{M}} : \Delta \longmapsto \mathbb{N}$, where $k_{\mathcal{M}}$ assigns a finite rank to each world: the rank $k_{\mathcal{M}}$ of a domain element $x \in \Delta$ is the length of the longest chain $x_0 < \cdots < x$ from x to a minimal x_0 (s. t. there is no x' with $x' < x_0$). The rank $k_{\mathcal{M}}(C_R)$ of a concept C_R in $\mathcal{M}$ is $i = min\{k_{\mathcal{M}}(x) : x \in C_R^I\}$.

In [31,32] a non monotonic construction of rational closure has been defined for $\mathcal{ALC} + \mathbf{T_R}$, extending the construction of rational closure introduced by Lehmann and Magidor [38] to the description logic $\mathcal{ALC}$ (alternative constructions have been studied in [15,16]). Its definition is based on the notion of exceptionality. Roughly speaking $\mathbf{T}(C) \sqsubseteq D$ holds in the rational closure of K if C is less exceptional than $C \sqcap \neg D$. We shortly recall the construction of rational closure of TBox and refer to [31] for details.

Definition 3 (Exceptionality of concepts and inclusions). *Let E be a TBox and C a concept. C is exceptional for E if and only if $E \models_{\mathcal{ALC}+\mathbf{T_R}} \mathbf{T}(\top) \sqsubseteq \neg C$. An inclusion $\mathbf{T}(C) \sqsubseteq D$ is exceptional for E if C is exceptional for E. The set of inclusions which are exceptional for E will be denoted by $\mathcal{E}(E)$.*

Given a TBox $\mathcal{T}$, it is possible to define a sequence of non increasing subsets of the TBox $\mathcal{T}$ ordered according to the exceptionality of the elements $E_0 \supseteq E_1 \supseteq E_2 \ldots$ by letting $E_0 = \mathcal{T}$ and, for $i > 0$, $E_i = \mathcal{E}(E_{i-1}) \cup \{C \sqsubseteq D \in \mathcal{T}$ s.t.

T does not occur in C}. Observe that, being knowledge base finite, there is an $n \geq 0$ such that, for all $m > n, E_m = E_n$ or $E_m = \emptyset$. A concept C has *rank i* in the rational closure (denoted $rank(C) = i$) for TBox, iff i is the least natural number for which C is not exceptional for E_i. If C is exceptional for all E_i then $rank(C) = \infty$ (C has no rank in the rational closure). The rank of a typicality inclusion $\mathbf{T}(C) \sqsubseteq D$ is $rank(C)$. Observe that, for $i < j$, E_i contains less specific defeasible properties then E_j.

Example 1. Consider a knowledge base $K = (\mathcal{T}, \mathcal{A})$, where $\mathcal{A} = \emptyset$ and $\mathcal{T}$ contains the following inclusions:

$\mathbf{T}(Bird) \sqsubseteq Fly$	$Penguin \sqsubseteq Bird$
$\mathbf{T}(Penguin) \sqsubseteq \neg Fly$	$\mathbf{T}(Penguin) \sqsubseteq BlackFeather$
$BabyPenguin \sqsubseteq Penguin$	$\mathbf{T}(BabyPenguin) \sqsubseteq \neg BlackFeather$

stating that normally birds fly, normally penguins (which are birds) do not fly and have black feather, while normally baby penguins do not have black feather. The rational closure construction, assigns rank 0 to *Bird*, rank 1 to *Penguin* and rank 2 to *BabyPenguin*. In particular, *Penguin* has rank 1, as it is exceptional w.r.t. the property that birds typically fly.

Rational closure builds on this notion of exceptionality:

Definition 4 (Rational closure of TBox). *Let $K = (\mathcal{T}, \mathcal{A})$ be a DL knowledge base. The rational closure of TBox is defined as:*

$$RC(\mathcal{T}) = \{\mathbf{T}(C) \sqsubseteq D \in \mathcal{T} \mid \text{ either } rank(C) < rank(C \sqcap \neg D) \text{ or}$$
$$rank(C) = \infty\} \cup \{C \sqsubseteq D \in \mathcal{T} \mid KB \models_{\mathcal{ALC}+\mathbf{T}_R} C \sqsubseteq D\}$$

where C and D are $\mathcal{ALC}$ concepts.

For instance, in Example 1, $\mathbf{T}(Penguin \sqcap Antarticus) \sqsubseteq \neg Fly$ is in $RC(\mathcal{T})$, as $rank\ (Penguin \sqcap Antarticus) = 1 < rank(Penguin \sqcap Antarticus \sqcap Fly) = 2$.

In [31] it is shown that deciding if an inclusion $\mathbf{T}(C) \sqsubseteq D$ belongs to the rational closure of TBox is a problem in ExpTime and that the semantics corresponding to rational closure can be given in terms of *minimal canonical $\mathcal{ALC}+\mathbf{T}_R$* models. In such models the rank of domain elements is minimized to make each domain element as typical as possible. This is expressed by the following definitions.

Definition 5 (Minimal models of K). *Given $\mathcal{M} = \langle \Delta, <, I \rangle$ and $\mathcal{M}' = \langle \Delta', <', I' \rangle$, we say that $\mathcal{M}$ is preferred to $\mathcal{M}'$ ($\mathcal{M} \prec \mathcal{M}'$) if: $\Delta = \Delta'$, $C^I = C^{I'}$ for all (non-extended) concepts C, for all $x \in \Delta$, it holds that $k_{\mathcal{M}}(x) \leq k_{\mathcal{M}'}(x)$ whereas there exists $y \in \Delta$ such that $k_{\mathcal{M}}(y) < k_{\mathcal{M}'}(y)$.*

Given a knowledge base $K = (\mathcal{T}, \mathcal{A})$, we say that $\mathcal{M}$ is a minimal model of K (with respect to TBox) if it is a model satisfying K and there is no $\mathcal{M}'$ model satisfying K such that $\mathcal{M}' \prec \mathcal{M}$.

The models corresponding to rational closure are required to be canonical. This property, expressed by the next definition, is needed when reasoning about the (relative) rank of the concepts: it is important to have them all represented by some instance in Δ.

Definition 6 (Canonical model). *Given $K = (\mathcal{T}, \mathcal{A})$, a model $\mathcal{M} = \langle \Delta, <, I \rangle$ satisfying K is canonical if for each set of concepts $\{C_1, C_2, \ldots, C_n\}$ consistent with K, there exists (at least) a domain element $x \in \Delta$ such that $x \in (C_1 \sqcap C_2 \sqcap \cdots \sqcap C_n)^I$.*

Definition 7 (Minimal canonical models (with respect to TBox)). $\mathcal{M}$ *is a minimal canonical model of K, if it is a canonical model of K and it is minimal with respect $\prec$ (see Definition 5) among the canonical models of K.*

The correspondence between minimal canonical models and rational closure is established by the following theorem.

Theorem 1 ([31]). *Let $K = (\mathcal{T}, \mathcal{A})$ be a knowledge base and $C \sqsubseteq D$ a query. Let $RC(\mathcal{T})$ be the rational closure of K w.r.t. TBox. We have that $C \sqsubseteq D \in RC(\mathcal{T})$ if and only if $C \sqsubseteq D$ holds in all minimal canonical models of K with respect to TBox.*

Furthermore: the rank of a concept C in any minimal canonical model of K is exactly the rank $rank(C)$ assigned by the rational closure construction, when $rank(C)$ is finite. Otherwise, $rank(C) = \infty$ and concept C is not satisfiable in any model of the TBox.

It can be seen that, in Example 1, the defeasible inclusion $\mathbf{T}(BabyPenguin) \sqsubseteq \neg Fly$ is not minimally entailed from K and, consistently, this inclusion does not belong to the rational closure of $\mathcal{T}$. Indeed, baby penguins are exceptional penguins, as they violates the defeasible property of penguins that, normally, they have black feather. For this reason, *BabyPenguin* does not inherit "any" of the defeasible properties of *Penguin*, the well-known "blocking of property inheritance problem" [42].

To overcome this weakness of the rational closure, Lehmann introduced the notion of lexicographic closure [39], which strengthens the rational closure by allowing a class to inherit as many as possible of the defeasible properties of more general classes, giving preference to the more specific properties. The lexicographic closure has been extended to the description logic $\mathcal{ALC}$ by Casini and Straccia in [18]. In the example above, the property that penguins do not fly would be inherited by baby penguins, as it is consistent with all (strict and defeasible) properties of baby penguins.

3   The Multipreference Semantics

The aim of the multipreference semantics in [33] is to define a refinement of the ranked models of the rational closure of a knowledge base K in which the modular preference relation $<$ satisfies the following additional condition on the preference relations $<_{A_i}$:

(a) **If** $x <_{A_i} y$, for some A_i, and there is no A_j such that $y <_{A_j} x$, **then** $x < y$.

The intended meaning of $x <_{A_i} y$ is that x satisfies some default for A_i which is instead violated by y. More precisely, $<_{A_i}$ is the preference relation in a ranked model of a knowledge base K_i containing only the defaults of the form $\mathbf{T}(C) \sqsubseteq A_i \in K$. In the minimal canonical ranked models $\mathcal{M}_i = \langle \Delta, <_{A_i}, I \rangle$ of K_i (according to Definition 7), $x <_{A_i} y$ has precisely the meaning that x satisfies some default for A_i which is violated by y[1]. Condition (a) alone, however, is to weak to capture refinements of the models of rational closure, and a *specificity condition* was added to define enriched models. Here, we refer to the definition of *S-enriched rational models* in [27], which is slightly stronger than the one in [33].

Definition 8 (S-Enriched rational models of K). $\mathcal{M} = \langle \Delta, <_{A_1}, \ldots, <_{A_n}, <, I \rangle$ *is a strongly enriched model of K if the following conditions hold:*

- $\langle \Delta, <, I \rangle$ *is a ranked model of K (as in Definition 2, Sect. 2);*
- *for all* $\mathbf{T}(C) \sqsubseteq A_i \in K$, *for all* $w \in \Delta$, *if* $w \in Min^{\mathcal{M}}_{<_{A_i}}(C)$ *then* $\mathcal{M}, w \models A_i$;
- *the preference relation* $<$ *satisfies the conditions* (a) *above, and the following* specificity condition: *for all* $x, y \in \Delta$

$$x < y \text{ if } \quad (i) \text{ } y \text{ violates some defeasible inclusion satisfied by } x \text{ and}$$
$$(ii) \text{ for all } \mathbf{T}(C_j) \sqsubseteq D_j \in K, \text{ which is violated by } x \text{ and not by } y,$$
$$\text{there is a } \mathbf{T}(C_k) \sqsubseteq D_k \in K, \text{ which is violated by } y \text{ and not by } x,$$
$$\text{such that } k_{\mathcal{M}}(C_j) < k_{\mathcal{M}}(C_k).$$

In (i) and (ii) the ranking function $k_{\mathcal{M}}$ of model $\mathcal{M}$ is the one associated with $<$ itself, and the intended meaning of the specificity condition is that preference should be given to the worlds that falsifies less specific defaults (defaults with lower ranks). Namely, the defaults violated by x are less serious than the defaults violated by y, as formula C_k is more specific than C_j.

A simplification to the notion of S-enriched models comes from the fact that the semantics in [27,33] considers the *minimal S-enriched models*, among all the S-enriched models if K, which are obtained by first minimizing the $<_{A_i}$ and then minimizing $<$ (as done for the ranked models of the rational closure), in this order, thus giving preference to models with lower ranks. It was proved in [27] (Proposition 1 therein) that, in minimal S-enriched models, the specificity condition is strong enough to enforce condition (a). As a consequence, one can simplify the definition of S-enriched rational models from the beginning, by removing condition (a) as well as the preference relations $<_{A_1}, \ldots, <_{A_n}$, thus starting from the following simplified notion of enriched model. That is, the multipreference semantics in [27] collapses into a semantics without multiple preferences.

---

[1] Indeed, it is easy to see that, for a satisfiable K_i, in the minimal ranked models $\mathcal{M}_i$ of K_i, which are the models of the rational closure of K_i, two elements $x, y \in \Delta$ either have rank 0, and satisfy all the conditionals $\mathbf{T}(C) \sqsubseteq A_i$ in K_i, or have rank 1, and falsify at least some conditional $\mathbf{T}(C) \sqsubseteq A_i$ in K_i.

Definition 9 (Simplified-enriched models of K). *A simplified-enriched model of K is a ranked model* $\mathcal{M} = \langle \Delta, <, I \rangle$ *of K (according to Definition 2 in Sect. 2) such that the preference relation* $<$ *satisfies the specificity condition.*

With this simplification, minimal canonical S-enriched models of K in [27] correspond to minimal canonical simplified-enriched rational models of K (according to the notion of minimal canonical model in Definition 7, Sect. 2). In the following we will write: $K \models^{min}_{\mathcal{ALC}^{R}\mathbf{T}_{SE}} C \sqsubseteq D$ to mean that $C \sqsubseteq D$ holds in all *minimal canonical* S-enriched models of K.

Example 2. To see that, in the KB in Example 1, the typical baby penguins inherit the defeasible property of not flying, although the property of having *BlackFeather* is overridden, let us consider two domain elements z and w which are both baby penguins and have a non black feather. Suppose that z flies and w doesn't. Then z violates the defeasible property that penguins typically do not fly, while w violates the defeasible property that birds typically fly. As $z <_{Fly} w$ and $w <_{\neg Fly} z$, condition (a) neither allows to conclude $w < z$, nor $z < w$. However, z violates a more specific defeasible property than w and, hence, by the specificity condition (4) of S-enriched models in Definition 8, we can conclude that $w < z$ holds. Indeed, the S-enriched minimal model semantics allows us to conclude that $\mathbf{T}(BabyPenguin) \sqsubseteq \neg Fly$, as wanted.

Although in this example the lexicographic closure comes to the same conclusions as the multipreference semantics, it can be seen that it does not define a weaker notion of entailment wrt. minimal entailment in S-enriched models.

4   A Sound Closure Construction for the Multipreference Semantics

The lexicographic closure strengthens the rational closure by allowing, roughly speaking, a class to inherit as many as possible of the defeasible properties of more general classes, giving preference to the more specific properties. The next example, adapted from Example 1 in [8], shows that the lexicographic closure is not weaker than the multipreference semantics.

Example 3. Suppose that project coordinators are both administrative staff and research staff. Typical administrative staff are allowed to sign payments, while typical research staff are not. Also, normally, administrative staff have no publications, while researchers have publications, and normally the project coordinator defines milestones.

1. $\mathbf{T}(Admin) \sqsubseteq \exists hasRight.Sign$
2. $\mathbf{T}(Admin) \sqsubseteq \neg HasPublication$
3. $\mathbf{T}(Research) \sqsubseteq \neg \exists hasRight.Sign \sqcap HasPublication$
4. $\mathbf{T}(PrjCrd) \sqsubseteq DefineMilestones$
5. $PrjCrd \sqsubseteq Admin \sqcap Research$

Defaults 1, 2 and 3 have rank 0 in RC, while default 4 has rank 1. There is a single basis $\{1,2,4\}$ in the lexicographic closure. Indeed, the set of defaults $\{1,2,4\}$ is preferred to $\{3,4\}$ as both of them contain one default with rank 1, but the first set contains two defaults with rank 0, while the second set just one. The lexicographic closure would then conclude that a typical project coordinator has no publications, i.e., $\mathbf{T}(PrjCrd) \sqsubseteq \neg HasPublication$. The multipreference semantics, instead, is more cautious and would not conclude this.

As an alternative refinement of the rational closure, the MP-closure, a variant of the lexicographic closure which has been considered in [25,27], would not come to the conclusion that a typical project coordinator has no publications. In this example, the MP-closure would consider the alternative sets of defaults $\{1,2\}$ and $\{3\}$, as being incomparable (using subset inclusion), and both $\{1,2,4\}$ and $\{3,4\}$ are bases for $PrjCrd$. The notion of MP-closure is therefore a good candidate as a sound construction for minimal entailment in the multipreference semantics.

Observe that, although in this example the MP-closure, as the multipreference semantics, appears to be more cautious and less syntax dependent than the lexicographic closure, all such refinements of the rational closure are syntax dependent to some extent (and the specificity condition used in Definitions 8 and 9 is syntax dependent).

Let n be the number of finite ranks in the rational closure construction, for a given TBox $\mathcal{T}$. Let D_i to be the set of all typicality inclusions with rank i and D_∞ the set of defeasible inclusions with rank ∞. Given a set S of typicality inclusions in $\mathcal{T}$, we let: $S_i = S \cap D_i$, for all ranks $i = 0, \ldots, n$ in the rational closure, and $S_\infty = S \cap D_\infty$ thus defining a partition of S, according to the ranks of the defaults. We represent such a partition of S with a tuple $\langle S_\infty, S_n, \ldots, S_1, S_0 \rangle$, where each set contains defaults with decreasing rank. Considering the (strict) subset inclusion relation $\subset$ among sets, we consider a natural lexicoghaphic ordering $\prec$ on the tuples $\langle S_\infty, S_n, \ldots, S_1, S_0 \rangle$, which is a strict partial ordering and we call $\prec$ MP-ordering, to distinguish it from the ordering used in the lexicographic closure. The MP-ordering $\prec$ is not necessarily modular. Instead, the lexicographic ordering used in the definition of the lexicographic closure is a strict modular partial ordering [39].

For instance, in the previous example, $n = 1$ and a set of defaults S can be represented as a tuple $\langle S_\infty, S_1, S_0 \rangle$. In particular, $\langle \emptyset, \emptyset, \emptyset \rangle \prec \langle \emptyset, \{4\}, \{1\} \rangle \prec \langle \emptyset, \{4\}, \{1,2\} \rangle$. The two bases $\{1,2,4\}$ and $\{3,4\}$ are not comparable, as neither $\langle \emptyset, \{4\}, \{1,2\} \rangle \prec \langle \emptyset, \{4\}, \{3\} \rangle$, nor $\langle \emptyset, \{4\}, \{3\} \rangle \prec \langle \emptyset, \{4\}, \{1,2\} \rangle$. Instead, in the lexicographic closure, the base $\{1,2,4\}$ is preferred to $\{3,4\}$. In fact, the base $\{1,2,4\}$ is associated with the tuple $\langle 0,1,2 \rangle$, the base $\{3,4\}$ is associated with the tuple $\langle 0,1,1 \rangle$, and $\langle 0,1,1 \rangle \prec \langle 0,2,1 \rangle$ in the lexicographic order.

Let us introduce the definition of an MP-basis for the typicality extension of $\mathcal{ALC}$.

Definition 10. *Let B be a concept such that $rank(B) = k$ and let S be a set of typicality inclusions in $\mathcal{T}$.*

- *S is consistent with B iff $E_k \not\models_{\mathcal{ALC}+\mathbf{T}_R} \mathbf{T}(\top) \sqcap \tilde{S} \sqsubseteq \neg B$.*
- *S is an MP-basis for B w.r.t. $\mathcal{T}$, if S is consistent with B and S is maximal with respect to the MP-ordering $\prec$ for this property, i.e. there is no $S' \subseteq \delta(TBox)$ such that S' is consistent with B and $S \prec S'$ (S' is preferred to S).*

where $\tilde{S}$ is the materialization of S, i.e., $\tilde{S} = \sqcap\{(\neg C \sqcup D) \mid \mathbf{T}(C) \sqsubseteq D \in S\}$.

Informally, S is and MP-basis for B if it is a maximal set of defeasible inclusions consistent with B: there is no set S' which is consistent with B and is preferred to S as it contains more specific defeasible inclusions. Remember that E_k also contains the strict inclusions in $\mathcal{T}$. The construction is similar to that of a basis in the lexicographic closure [18,39], although, the lexicographic ordering is different.

A *subsumption* $\mathbf{T}(B) \sqsubseteq D$ *follows from the MP-closure of* $\mathcal{T}$ if D holds in all the MP-bases for B, i.e., for all MP-bases S for B w.r.t. $\mathcal{T}$:

$$E_k \models_{\mathcal{ALC}+\mathbf{T}_R} \mathbf{T}(\top) \sqcap \tilde{S} \sqsubseteq (\neg B \sqcup D)$$

It can be proved (see [27]) that the MP-closure is a sound construction for capturing entailment in minimal canonical S-enriched models.

Proposition 1. *If $\mathbf{T}(B) \sqsubseteq D$ is in the MP-closure of K, then $K \models^{min}_{\mathcal{ALC}^R\mathbf{T}_{SE}} \mathbf{T}(B) \sqsubseteq D$.*

The MP-closure is not complete for minimal entailment in the S-enriched semantics. In fact, in an S-enriched model $\mathcal{M}$, $<$ depends on $k_{\mathcal{M}}$, and it may occur that there are two concepts C_j and C_k, such that $k_{\mathcal{M}}(C_j) < k_{\mathcal{M}}(C_k)$ although in the rational closure (whose ranking is used by the MP-closure) $rank(C_j) = rank(C_k)$. This may cause additional conclusions in the S-enriched semantics (see [27]) with an iterative flavor.

5   Conclusions and Related Work

In this paper, we reconsider the multi-preference semantics proposed for $\mathcal{ALC}$ [33] and we show that entailment in the multipreference semantics can be approximated by the MP-closure construction for $\mathcal{ALC}$. The idea of the multipreference semantics was to define a refinement of the rational closure for $\mathcal{ALC}$, considering preference with respect to specific aspects and reconciling it with a global notion of preference and a single typicality operator. We consider the MP-closure for $\mathcal{ALC}$, a closure construction which is a valiant of the lexicographic closure defined by Lehmann [39] and extended to description logics by Casini and Straccia [18], and we show that it provides a weaker entailment wrt. the multipreference semantics.

The multi-preference closure (MP-closure for short) was first introduced in [27] as a sound approximation of Gliozzi's multi-preference semantics [33]. As the lexicographic closure, it builds over the rational closure but it defines a preferential, not necessarily ranked, semantics, using a different lexicographic order to compare sets of defaults. A semantic characterization of the MP-closure for the description logic $\mathcal{ALC}$ was developed in [25] using bi-preferential (BP) interpretations, preferential interpretations developed along the lines of the preferential semantics introduced by Kraus, Lehmann and Magidor [37,38], but containing two preference relations. The skeptical closure [26] was shown to be a weaker variant of the MP-closure in [25].

The relevant closure [13] is based on the idea of relevance of subsumptions to a query, to overcome the limitation of the weakness of rational closure. Relevance is determined based on justifications and, in minimal relevant closure, the idea is that subsumptions with lower ranks are removed first. Another refinement of the rational closure, which also deal with the problem of inheritance blocking, is the inheritance-based rational closure in [17,19], a closure construction which is defined by combining the rational closure with defeasible inheritance networks.

In the propositional case, the relations of the MP-closure with the Relevant Closure and with the Lexicographic closure have been explored in [28], where it has been shown that the MP-closure is weaker than the Lexicographic closure but stronger than the Relevant Closure. Similar relations may be expected to hold in the description logic setting, in which it is already known from [13] that the relevant closure is a weaker construction than the lexicographic closure. Further investigation is needed.

The idea of having different preference relations, was first exploited by Gil [23] to define a multi-typicality formulation of the preferential logic $\mathcal{ALC} + \mathbf{T}_{min}$ [30], a logic with a preferential but not a ranked minimal model semantics. As a further difference, here we consider a single typicality operator. An extension of DLs with defeasible roles and defeasible role subsumptions has been studied by Britz and Varzinczak in [10,11].

The logic $\mathcal{DL}^N$, proposed by Bonatti et al. in [5,8], captures a form of "inheritance with overriding": a defeasible inclusion is inherited by a more specific class if it is not overridden by more specific (conflicting) properties. The logic $\mathcal{DL}^N$ is not necessarily applied starting from the ranking given by the rational closure but, when it does, it provides another approach to deal with the problem of inheritance blocking in the rational closure. The approach in $\mathcal{DL}^N$ is a skeptical and polynomial one (it builds a single base). An unsolved conflict among different defeasible inclusions gives rise to an inconsistent prototype. In such cases, the MP-closure, the Relevant Closure and the Lexicographic closure all silently ignore the conflicting defaults. In $\mathcal{DL}^N$, instead, unresolved conflicts have to be detected and fixed by modifying the knowledge base.

Bozzato et al. in [9] present an extension of the CKR framework in which defeasible axioms are allowed in the global context and can be overridden by knowledge in a local context. Exceptions have to be justified in terms of semantic consequence. A translation of extended CHRs (with knowledge bases in $\mathcal{SROIQ}$-RL) into Datalog programs under the answer set semantics is also defined.

Acknowledgement. This research is partially supported by INDAM-GNCS Project 2018 "Metodi di prova orientati al ragionamento automatico per logiche non-classiche".

References

1. Baader, F., Calvanese, D., McGuinness, D.L., Nardi, D., Patel-Schneider, P.F.: The Description Logic Handbook - Theory, Implementation, and Applications, 2nd edn. Cambridge University Press, New York (2007)
2. Baader, F., Hollunder, B.: Priorities on defaults with prerequisites, and their application in treating specificity in terminological default logic. J. Autom. Reason. (JAR) **15**(1), 41–68 (1995)
3. Benferhat, S., Dubois, D., Prade, H.: Possibilistic logic: from nonmonotonicity to logic programming. In: Clarke, M., Kruse, R., Moral, S. (eds.) ECSQARU 1993. LNCS, vol. 747, pp. 17–24. Springer, Heidelberg (1993). https://doi.org/10.1007/BFb0028177
4. Bonatti, P.A.: Rational closure for all description logics. Artif. Intell. **274**, 197–223 (2019)
5. Bonatti, P.A., Faella, M., Petrova, I., Sauro, L.: A new semantics for overriding in description logics. Artif. Intell. **222**, 1–48 (2015)
6. Bonatti, P.A., Faella, M., Sauro, L.: Defeasible inclusions in low-complexity DLs. J. Artif. Intell. Res. (JAIR) **42**, 719–764 (2011)
7. Bonatti, P.A., Lutz, C., Wolter, F.: The complexity of circumscription in DLs. J. Artif. Intell. Res. (JAIR) **35**, 717–773 (2009)
8. Bonatti, P.A., Sauro, L.: On the logical properties of the nonmonotonic description logic DL^N. Artif. Intell. **248**, 85–111 (2017)
9. Bozzato, L., Eiter, T., Serafini, L.: Enhancing context knowledge repositories with justifiable exceptions. Artif. Intell. **257**, 72–126 (2018)
10. Britz, K., Varzinczak, I.: Rationality and context in defeasible subsumption. In: Ferrarotti, F., Woltran, S. (eds.) FoIKS 2018. LNCS, vol. 10833, pp. 114–132. Springer, Cham (2018). https://doi.org/10.1007/978-3-319-90050-6_7
11. Britz, A., Varzinczak, I.: Contextual rational closure for defeasible ALC (extended abstract). In: Proceedings of the 32nd International Workshop on Description Logics, Oslo, Norway, 18–21 June 2019 (2019)
12. Britz, K., Heidema, J., Meyer, T.: Semantic preferential subsumption. In: Brewka, G., Lang, J. (eds.) Principles of Knowledge Representation and Reasoning: Proceedings of the 11th International Conference (KR 2008), Sidney, Australia, September 2008, pp. 476–484. AAAI Press (2008)
13. Casini, G., Meyer, T., Moodley, K., Nortjé, R.: Relevant closure: a new form of defeasible reasoning for description logics. In: Fermé, E., Leite, J. (eds.) JELIA 2014. LNCS (LNAI), vol. 8761, pp. 92–106. Springer, Cham (2014). https://doi.org/10.1007/978-3-319-11558-0_7
14. Casini, G., Meyer, T., Moodley, K., Sattler, U., Varzinczak, I.: Introducing defeasibility into OWL ontologies. In: Arenas, M., et al. (eds.) ISWC 2015. LNCS, vol. 9367, pp. 409–426. Springer, Cham (2015). https://doi.org/10.1007/978-3-319-25010-6_27
15. Casini, G., Meyer, T., Varzinczak, I.J., Moodley, K.: Nonmonotonic reasoning in description logics: rational closure for the ABox. In: DL 2013, 26th International Workshop on Description Logics, volume 1014 of CEUR Workshop Proceedings, pp. 600–615. CEUR-WS.org (2013)

16. Casini, G., Straccia, U.: Rational closure for defeasible description logics. In: Janhunen, T., Niemelä, I. (eds.) JELIA 2010. LNCS (LNAI), vol. 6341, pp. 77–90. Springer, Heidelberg (2010). https://doi.org/10.1007/978-3-642-15675-5_9

17. Casini, G., Straccia, U.: Defeasible inheritance-based description logics. In: Walsh, T. (ed.) Proceedings of the 22nd International Joint Conference on Artificial Intelligence (IJCAI 2011), Barcelona, Spain, July 2011, pp. 813–818. Morgan Kaufmann (2011)

18. Casini, G., Straccia, U.: Lexicographic closure for defeasible description logics. In: Proceedings of Australasian Ontology Workshop, vol. 969, pp. 28–39 (2012)

19. Casini, G., Straccia, U.: Defeasible inheritance-based description logics. J. Artif. Intell. Res. (JAIR) **48**, 415–473 (2013)

20. Donini, F.M., Nardi, D., Rosati, R.: Description logics of minimal knowledge and negation as failure. ACM Trans. Comput. Logic (ToCL) **3**(2), 177–225 (2002)

21. Eiter, T., Ianni, G., Lukasiewicz, T., Schindlauer, R.: Well-founded semantics for description logic programs in the semantic web. ACM Trans. Comput. Log. **12**(2), 11 (2011)

22. Eiter, T., Ianni, G., Lukasiewicz, T., Schindlauer, R., Tompits, H.: Combining answer set programming with description logics for the semantic web. Artif. Intell. **172**(12–13), 1495–1539 (2008)

23. Fernandez Gil, O.: On the Non-Monotonic Description Logic ALC+T_{min}. CoRR, abs/1404.6566 (2014)

24. Giordano, L., Gliozzi, V.: Encoding a preferential extension of the description logic $\mathcal{SROIQ}$ into $\mathcal{SROIQ}$. In: Esposito, F., Pivert, O., Hacid, M.-S., Raś, Z.W., Ferilli, S. (eds.) ISMIS 2015. LNCS (LNAI), vol. 9384, pp. 248–258. Springer, Cham (2015). https://doi.org/10.1007/978-3-319-25252-0_27

25. Giordano, L., Gliozzi, V.: Reasoning about exceptions in ontologies: from the lexicographic closure to the skeptical closure. CoRR, abs/1807.02879 (2018)

26. Giordano, L., Gliozzi, V.: Reasoning about exceptions in ontologies: from the lexicographic closure to the skeptical closure. In: Proceedings of the Second Workshop on Logics for Reasoning about Preferences, Uncertainty, and Vagueness, PRUV@IJCAR 2018, Oxford, UK, 19 July 2018 (2008)

27. Giordano, L., Gliozzi, V.: Reasoning about multiple aspects in DLs: semantics and closure construction. CoRR, abs/1801.07161 (2018)

28. Giordano, L., Gliozzi, V.: A reconstruction of the multipreference closure. CoRR, abs/1905.03855 (2019)

29. Giordano, L., Gliozzi, V., Olivetti, N., Pozzato, G.L.: ALC+T: a preferential extension of description logics. Fundamenta Informaticae **96**, 1–32 (2009)

30. Giordano, L., Gliozzi, V., Olivetti, N., Pozzato, G.L.: A NonMonotonic description logic for reasoning about typicality. Artif. Intell. **195**, 165–202 (2013)

31. Giordano, L., Gliozzi, V., Olivetti, N., Pozzato, G.L.: Semantic characterization of rational closure: from propositional logic to description logics. Artif. Intell. **226**, 1–33 (2015)

32. Giordano, L., Gliozzi, V., Olivetti, N., Pozzato, G.L.: Minimal model semantics and rational closure in description logics. In: 26th International Workshop on Description Logics (DL 2013), August 2013, vol. 1014, pp. 168–180 (2013)

33. Gliozzi, V.: Reasoning about multiple aspects in rational closure for DLs. In: Proceedings of AI*IA 2016 - XVth International Conference of the Italian Association for Artificial Intelligence, Genova, Italy, 29 November–1 December 2016, pp. 392–405 (2016)

34. Gottlob, G., Hernich, A., Kupke, C., Lukasiewicz, T.: Stable model semantics for guarded existential rules and description logics. In: Proceedings of KR 2014 (2014)

35. Ke, P., Sattler, U.: Next steps for description logics of minimal knowledge and negation as failure. In: Baader, F., Lutz, C., Motik, B. (eds.) Proceedings of Description Logics, volume 353 of CEUR Workshop Proceedings, Dresden, Germany, May 2008 CEUR-WS.org (2008)
36. Knorr, M., Hitzler, P., Maier, F.: Reconciling owl and non-monotonic rules for the semantic web. In: ECAI 2012, pp. 474–479 (2012)
37. Kraus, S., Lehmann, D., Magidor, M.: Nonmonotonic reasoning, preferential models and cumulative logics. Artif. Intell. **44**(1–2), 167–207 (1990)
38. Lehmann, D., Magidor, M.: What does a conditional knowledge base entail? Artif. Intell. **55**(1), 1–60 (1992)
39. Lehmann, D.J.: Another perspective on default reasoning. Ann. Math. Artif. Intell. **15**(1), 61–82 (1995)
40. Moodley, K.: Practical reasoning for defeasible description logics. Ph.D. Thesis, University of Kwazulu-Natal (2016)
41. Motik, B., Rosati, R.: Reconciling description logics and rules. J. ACM **57**(5), 1–62 (2010)
42. Pearl, J.: System Z: a natural ordering of defaults with tractable applications to nonmonotonic reasoning. In: Parikh, R. (ed.) TARK 3rd Conference on Theoretical Aspects of Reasoning about Knowledge, pp. 121–135. Morgan Kaufmann, Pacific Grove, CA, USA (1990)
43. Straccia, U.: Default inheritance reasoning in hybrid KL-one-style logics. In: Bajcsy, R. (ed.) Proceedings of the 13th International Joint Conference on Artificial Intelligence (IJCAI 1993), Chambéry, France, August 1993. pp. 676–681, Morgan Kaufmann (1993)

Computation of Closures
of Nonmonotonic Inference Relations
Induced by Conditional Knowledge Bases

Steven Kutsch$^{(\boxtimes)}$ and Christoph Beierle

Department of Computer Science, FernUniversität in Hagen, Hagen, Germany
{steven.kutsch,christoph.beierle}@fernuni-hagen.de

Abstract. Conditionals of the form "If A, then usually B" are often used to define nonmonotonic inference relations. Many ways have been proposed to inductively complete a knowledge base consisting of a finite set of conditionals to a complete inference relation. Implementations of these semantics are usually used to answer specific queries on demand. However, for some applications it is necessary or advantageous to compute the closure of the inference relation induced by a knowledge base. In this paper, we propose an approach to computing complete inference relations using implementations of inference systems for single queries. Our approach exploits special characteristics of conditionals and inference properties like Right Weakening in order to reduce the amount of costly query answering to a minimum.

1 Introduction

A common way to model nonmonotonic inference relations is using conditional statements of the form "If A, then usually B". Several semantics have been proposed for knowledge bases $\mathcal{R}$ containing conditionals, for example Lewis' system of spheres [15], conditional objects evaluated using boolean intervals [9], possibility distributions [8], ranking functions [17,18], or special classes of ranking functions like c-representations [13]. Under each of these semantics, $\mathcal{R}$ induces a nonmonotonic inference relation $\vdash$. Implementations like InfOCF [3] realize these inference relations by taking a knowledge base $\mathcal{R}$ as input, and allowing the user to ask questions of the form "In the context of $\mathcal{R}$, does $A \vdash B$ hold?" Equivalently, one may ask whether under the given semantics $\mathcal{R}$ entails the conditional $(B|A)$. If this is the case, we formally denote this by $A \vdash_{\mathcal{R}} B$. The system then answers the question in the context of the knowledge base $\mathcal{R}$.

For some applications it is however necessary or advantageous to compute the closure of an inference relation. One such application is the comparison of two inference relations for empirical purposes (e.g. [5]). In order to automatically compare two inference relations induced by different knowledge bases or defined using different semantics (e.g. different sets of models), the answer to every possible query needs to be available.

© Springer Nature Switzerland AG 2019
G. Kern-Isberner and Z. Ognjanović (Eds.): ECSQARU 2019, LNAI 11726, pp. 226–237, 2019.
https://doi.org/10.1007/978-3-030-29765-7_19

Another example is the empirical discovery of counterexamples with respect to inference properties. For instance, for the inference property Rational Monotony

$$\frac{A \mathrel{\vmid\sim} B \qquad A \mathrel{\not\vmid\sim} \overline{C}}{AC \mathrel{\vmid\sim} B} \qquad \text{(RM)}$$

formulas A, B and C need to be determined such that $A \mathrel{\vmid\sim} B$ and $A \mathrel{\not\vmid\sim} \overline{C}$, but not $AC \mathrel{\vmid\sim} B$. This can be achieved by iterating systematically for A, B and C over all formulas and checking the inclusion of the required conditionals in $\mathrel{\vmid\sim}$, which requires access to the full closure of $\mathrel{\vmid\sim}$ (assuming that no particularities of the inference property can be exploited).

In this paper we present an approach to calculate the closure of a nonmonotonic inference relation defined by conditional statements. The approach uses a decision procedure to answer arbitrary queries of the form "Does $A \mathrel{\vmid\sim}_{\mathcal{R}} B$ hold?". Since in general these decision procedures may be computationally expensive, our approach minimizes the number of calls to this procedure and exploits particular properties of conditionals and nonmonotonic inference relations to determine the inclusion of most conditionals without answering a query.

Note that, while we use similar terminology, our approach can not only be used for calculating the rational closure of a conditional knowledge base [14]. Lehmann and Magidor's rational closure is just one inference relation defined for sets of conditionals, that is implemented via a decision procedure that answers the question "Is the conditional $(B|A)$ contained in the rational closure of the knowledge base $\mathcal{R}$?". Our approach uses any such decision procedure to calculate the closure of the inference relation implemented by that decision procedure. Thus, besides abstracting from the knowledge base $\mathcal{R}$ inducing the inference relation $\mathrel{\vmid\sim}_{\mathcal{R}}$, our approach also abstracts from the different ways how an inference relation could be induced by $\mathcal{R}$ (see e.g. [1,2,14,16]).

The idea of calculating complete closures of inference relations is related to saturation [11], a popular technique in the area of automated theorem proving. Although the aim of our approach is different, techniques for increasing efficiency originating in saturation-based theorem provers will also be employed here.

2 Background

Let $\Sigma = \{v_1, ..., v_m\}$ be a propositional alphabet. A *literal* is the positive (v_i) or negated ($\overline{v_i}$) form of a propositional variable. From these we obtain the propositional language $\mathcal{L}$ as the set of formulas of Σ closed under negation $\neg$, conjunction $\wedge$, and disjunction $\vee$. For shorter formulas, we abbreviate conjunction by juxtaposition (i.e., AB stands for $A \wedge B$), and negation by overlining (i.e., $\overline{A}$ is equivalent to $\neg A$). Let Ω_Σ denote the set of possible worlds over $\mathcal{L}$; Ω_Σ will be taken here simply as the set of all propositional interpretations over $\mathcal{L}$ and can be identified with the set of all complete conjunctions over Σ; we will often just write Ω instead of Ω_Σ. For $\omega \in \Omega$, $\omega \models A$ means that the propositional formula $A \in \mathcal{L}$ holds in the possible world ω. For any propositional formula A let $\Omega_A = \{\omega \in \Omega \mid \omega \models A\}$ be the set of all possible worlds satisfying A.

A *conditional* $(B|A)$ with $A, B \in \mathcal{L}$ encodes the defeasible rule "if A then usually B" and is a trivalent logical entity with the evaluation [10,13]

$$[\![(B|A)]\!]_\omega = \begin{cases} true & \text{iff} \quad \omega \models AB & \text{(verification)} \\ false & \text{iff} \quad \omega \models A\overline{B} & \text{(falsification)} \\ undefined & \text{iff} \quad \omega \models \overline{A} & \text{(not applicable)} \end{cases} \quad (1)$$

We say $(B|A)$ is conditionally equivalent to $(B'|A')$ if $A \equiv A'$ and $AB \equiv A'B'$. With $(\mathcal{L}|\mathcal{L})$ we denote the set of all conditionals over the language $\mathcal{L}$.

For inference relations $\vdash\!\!\!\!\sim$ defined over sets of conditionals, we will assume that $\vdash\!\!\!\!\sim$ satisfies the KLM postulates [14]. While several semantics for conditionals satisfy these postulates, we will use Spohn's ranking functions as an example. A *Ranking Function* (Ordinal Conditional Function, OCF) [17] is a function $\kappa : \Omega \to \mathbb{N}_0 \cup \{\infty\}$ that assigns to each world $\omega \in \Omega$ an implausibility rank $\kappa(\omega)$: the higher $\kappa(\omega)$, the more surprising ω is. OCFs have to satisfy the normalization condition that there has to be a world that is maximally plausible, i.e., $\kappa^{-1}(0) \neq \emptyset$. The rank of a formula A is defined by $\kappa(A) = \min\{\kappa(\omega) \mid \omega \models A\}$ where $\min \emptyset = \infty$. An OCF κ *accepts* a conditional $(B|A)$, denoted by $\kappa \models (B|A)$, iff the verification of the conditional is less surprising than its falsification, i.e., iff $\kappa(AB) < \kappa(A\overline{B})$. This can also be understood as a nonmonotonic inference relation between the premise A and the conclusion B: We say that A κ-*entails* B, written $A \vdash\!\!\!\!\sim^\kappa B$, iff κ accepts the conditional $(B|A)$ or if $A \equiv \bot$:

$$A \vdash\!\!\!\!\sim^\kappa B \quad \text{iff} \quad A \equiv \bot \text{ or } \kappa(AB) < \kappa(A\overline{B}). \quad (2)$$

Note that κ-entailment is based on the total preorder on possible worlds induced by a ranking function κ as $A \vdash\!\!\!\!\sim^\kappa B$ iff for all $\omega' \in \Omega_{A\overline{B}}$, there is a $\omega \in \Omega_{AB}$ such that $\kappa(\omega) < \kappa(\omega')$. Note also that, in order to ensure supraclassicality, explicit handling of the case $A \equiv \bot$ is necessary when using the formulation via the ranks of verification and falsification directly, as in (2).

The acceptance relation is extended as usual to a set $\mathcal{R}$ of conditionals, called a *knowledge base*, by defining $\kappa \models \mathcal{R}$ iff $\kappa \models (B|A)$ for all $(B|A) \in \mathcal{R}$. This is synonymous to saying that κ is *admissible* with respect to $\mathcal{R}$ [12], or that κ is a *ranking model* of $\mathcal{R}$. Then A entails B in the context of $\mathcal{R}$ if for every model κ of $\mathcal{R}$, A κ-entails B:

$$A \vdash\!\!\!\!\sim_\mathcal{R} B \quad \text{iff} \quad A \vdash\!\!\!\!\sim^\kappa B \text{ for all } \kappa \models \mathcal{R}.$$

3   Notations and Running Example

In the rest of this paper, we will consider the signature $\Sigma_{ab} = \{a, b\}$ for illustrating our approach. There are four possible worlds over Σ_{ab}, namely $\Omega_{\Sigma_{ab}} = \{ab, a\overline{b}, \overline{a}b, \overline{a}\overline{b}\}$. There are therefore 16 formulas in the form of disjunctions over complete conjunctions (i.e. disjunctions over elements of Ω_Σ) given in

Table 1. All formulas over $\Sigma_{ab} = \{a, b\}$ given as disjunctions over complete conjunctions and clauses of possible worlds. The order is given as in [6].

n	Formula	Sets of interpretations	n	Formula	Sets of interpretations
1	$\bot$	$\{\}$	9	$a\bar{b} \vee \bar{a}b$	$\{a\bar{b}, \bar{a}b\}$
2	ab	$\{ab\}$	10	$a\bar{b} \vee \bar{a}\bar{b}$	$\{a\bar{b}, \bar{a}\bar{b}\}$
3	$a\bar{b}$	$\{a\bar{b}\}$	11	$\bar{a}b \vee \bar{a}\bar{b}$	$\{\bar{a}b, \bar{a}\bar{b}\}$
4	$\bar{a}b$	$\{\bar{a}b\}$	12	$ab \vee a\bar{b} \vee \bar{a}b$	$\{ab, a\bar{b}, \bar{a}b\}$
5	$\bar{a}\bar{b}$	$\{\bar{a}\bar{b}\}$	13	$ab \vee a\bar{b} \vee \bar{a}\bar{b}$	$\{ab, a\bar{b}, \bar{a}\bar{b}\}$
6	$ab \vee a\bar{b}$	$\{ab, a\bar{b}\}$	14	$ab \vee \bar{a}b \vee \bar{a}\bar{b}$	$\{ab, \bar{a}b, \bar{a}\bar{b}\}$
7	$ab \vee \bar{a}b$	$\{ab, \bar{a}b\}$	15	$a\bar{b} \vee \bar{a}b \vee \bar{a}\bar{b}$	$\{a\bar{b}, \bar{a}b, \bar{a}\bar{b}\}$
8	$ab \vee \bar{a}\bar{b}$	$\{ab, \bar{a}\bar{b}\}$	16	$ab \vee a\bar{b} \vee \bar{a}b \vee \bar{a}\bar{b}$	$\{ab, a\bar{b}, \bar{a}b, \bar{a}\bar{b}\}$

Table 1. With $[\mathcal{L}_\Sigma]$ we denote the list of formulas over Σ_{ab} in the order given in Table 1.

For the signature Σ_{ab} we have therefore 16^2 possible conditionals using this representation of propositional formulas. In the following, we will illustrate inference relations over $\mathcal{L}_{ab}$ as 16×16 grids called closure matrices (cf. Fig. 1a). Each conditional $(B|A)$ is represented by one cell in the grid. The column indicates the antecedent A and the row indicates the consequent B. The numbers indexing the rows/columns correspond to the numbers of formulas given in Table 1.

With CM we denote the data structure holding a grid as given in Fig. 1. Then CM is a representation of the closure of the inference relation $\vdash$ if $CM(A, B) = 1$ iff $(B|A) \in \vdash$ and $CM(A, B) = 0$ iff $(B|A) \notin \vdash$. CM will be computed incrementally, and it will be initiated with a template described in Sect. 5.

For a signature Σ, the set $Cond(\Sigma)$ denotes the finite sets of all conditionals over $\mathcal{L}_\Sigma$ with antecedent and consequent represented as disjunctions of possible worlds over Σ. In reference to the set representation of formulas, we write $|A|$ to denote the number of possible worlds present in the disjunction A.

4 Trivial Conditionals

We first discuss conditionals that are contained in all inference relations $\vdash$ or that are not contained in any inference relation.

Due to supra classicality (which in turn follows from (Ref) and (RW)), all conditionals $(B|A)$ with $A \models B$ are included in every nonmonotonic inference relation that satisfy the KLM postulates. We call these conditionals *self-fulfilling*.

Definition 1 (self-fulfilling). *A conditional $(B|A)$ is called* self-fulfilling *if $A \models B$ holds. The set of all self-fulfilling condtionals in $Cond(\Sigma)$ is denoted by $Cond^+(\Sigma)$.*

Proposition 1. *For every self-fulfilling conditional $(B|A)$ and every nonmonotonic inference relation $\vdash$ satisfying KLM it holds that $A \vdash B$.*

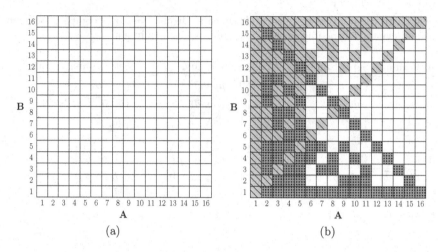

Fig. 1. (a) An empty grid representation of a nonmonotonic inference relation over Σ_{ab}. Every cell represents a conditional $(B|A)$ in $Cond(\Sigma_{ab})$ with the column indicating the antecedent A and the row indicating the consequent B. The inclusion of all conditionals in the inference relation to be represented is undetermined. **(b)** Every trivial conditional marked in an inference relation. Self-fulfilling conditionals are marked green (lined) as being contained in every inference relation and contradictory conditionals are marked red (dotted) as not being contained in any inference relation. (Color figure online)

Another large portion of conditionals over Σ is *contradictory*.

Definition 2 (contradictory). *A conditional $(B|A)$ is called* contradictory *if $A \not\equiv \bot$ and $AB \equiv \bot$ holds. The set of all contradictory condtionals in $Cond(\Sigma)$ is denoted by $Cond_-(\Sigma)$.*

Proposition 2. *For every contradictory conditional $(B|A)$ and every nonmonotonic inference relation $\vdash\!\!\sim$ it holds that $A \not\vdash\!\!\sim B$.*

Proof. Let $A \not\equiv \bot$ and $AB \equiv \bot$. Then for every ranking function κ it holds that $\kappa(AB) = \infty$. It also holds that $A\overline{B} \equiv A$ and therefore $\kappa(A\overline{B}) = \kappa(A) < \infty$, which implies $\kappa(AB) > \kappa(A\overline{B})$.   □

Again, we proved Proposition 2 for ranked inference relations, but the result can be translated to similar semantics.

We denote the set of all trivial conditionals in $Cond(\Sigma)$ with $Cond^{\pm}_-(\Sigma) = Cond^+(\Sigma) \cup Cond_-(\Sigma)$.

Because of Propositions 1 and 2, we can fill in a portion of CM for every inference relation with self-fulfilling conditionals being contained in every inference relation and contradictory conditionals being contained in no inference relation. Figure 1b shows the grid representation of CM containing entries for all trivial conditionals in $Cond(\Sigma)$.

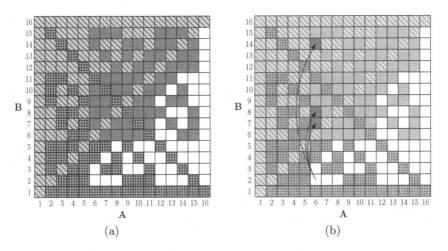

(a)   (b)

Fig. 2. (a) CM-template for Σ_{ab}. Self-fulfilling conditionals are marked green (lined), contradictory conditionals are marked red (dotted). Non-trivial Conditionals not in conditional normal form are marked in gray (unpatterned). Normal form conditionals are marked white. All gray conditionals are equivalent to a normal form conditional in the same column. (b) The arrows indicate the links in CM-template from the conditional $(ab|ab \vee a\overline{b})$ (cf. (5)) represented in cell $(6, 2)$ to its three syntactic variants $(ab \vee \overline{a}b|ab \vee a\overline{b})$, $(ab \vee \overline{a}\overline{b}|ab \vee a\overline{b})$ and $(ab \vee \overline{a}b \vee \overline{a}\overline{b}|ab \vee a\overline{b})$, represented by the cells $(6, 7)$, $(6, 8)$ and $(6, 14)$, respectively. (Color figure online)

5   Normal Form Conditionals and CM-Templates

In [6] the concept of normal form conditionals was introduced. Normal form conditionals are both verifiable and falsifiable. The set $NFC(\Sigma)$ also does not contain any conditional that is pairwise equivalent to any other conditional in $NFC(\Sigma)$.

Definition 3 (Normal Form Conditionals $NFC(\Sigma)$). *Let Σ be a propositional signature. The set of normal form conditionals over Σ, denoted by $NFC(\Sigma)$, is defined as*

$$NFC(\Sigma) = \{(B|A) \in Cond(\Sigma) \mid A \subseteq \Omega_{\Sigma},\, B \subsetneqq A,\, B \neq \emptyset\}. \tag{3}$$

Recall that A and B in Definition 3 are sets of interpretations. Figure 2 shows the CM grid representation of all 50 normal form conditionals over Σ_{ab} marked in white. Every conditional $(B|A)$ marked in gray can be assigned to a normal form conditional $(AB|A)$ in the same column. Note that the normal form $(AB|A)$ of a conditional is conditionally equivalent to the original syntactic variant $(B|A)$, because both conditionals partition the set of possible worlds according to (1) in the same way, i.e.

For example, the conditional

$$(ab \vee \overline{a}b \vee \overline{a}\overline{b}|ab \vee a\overline{b}) \tag{4}$$

represented by cell $(6, 14)$ has the conditional normal from

$$(ab|ab \vee a\bar{b}) \tag{5}$$

represented by the cell $(6, 2)$ in Fig. 2.

Because of their conditional equivalence, for a conditional $(B|A)$ and its conditional normal form $(AB|A)$ and every nonmonotonic inference relation $\vdash$, it holds that

$$A \vdash B \quad \text{iff} \quad A \vdash AB. \tag{6}$$

Proposition 3 $(NFC(\Sigma)$ [6]$)$. *For the set $NFC(\Sigma)$ the following holds:*

(nontrivial) $NFC(\Sigma)$ *does not contain any trivial conditional.*
(complete) *For every nontrivial conditional over Σ there is an equivalent conditional in $NFC(\Sigma)$.*
(minimal) *All conditional in $NFC(\Sigma)$ are pairwise non-equivalent.*

Using the notation $nNFC(\Sigma) = Cond(\Sigma) \setminus (Cond_-^+(\Sigma) \cup NFC(\Sigma))$ we thus obtain a partition of $Cond(\Sigma)$ in the form

$$Cond(\Sigma) = Cond^+(\Sigma) \cup Cond_-(\Sigma) \cup NFC(\Sigma) \cup nNFC(\Sigma). \tag{7}$$

Using this partition, we can build a template for CM, in which every self-fulfilling conditional is set to 1, every contradictory conditional is set to 0 and every conditional that is not in normal form is linked to its normal form. This directly corresponds to Fig. 2, where every cell marked in gray represents a link to the corresponding normal form conditional marked in white. The procedure producing this CM-template is called CMTEMPLATE(Σ).

CMTEMPLATE(Σ) effectively partitions the grid, and therefore the set $Cond(\Sigma)$, into three components:

$Cond^+(\Sigma)$: all self-fulfilling conditionals over Σ (green),
$Cond_-(\Sigma)$: all contradictory conditionals over Σ (red),
$NFC(\Sigma)$: all normal form conditionals over Σ (white), and
$nNFC(\Sigma)$: all non-trivial conditionals over Σ not in normal form (gray).

This partition only depends on Σ and is identical for all inference relations over $\mathcal{L}_\Sigma$. CMTEMPLATE(Σ) can therefore be computed in advance and can be reused for all inference relations over $\mathcal{L}_\Sigma$.

The majority of the grid is in the components $Cond_-^+(\Sigma)$ and $nNFC(\Sigma)$. For Σ_{ab} there are only 50 normal form conditionals but 206 conditionals not in normal form (including trivial conditionals). For $\Sigma_{abc} = \{a, b, c\}$ there are 6,050 normal form conditionals and 59,486 conditionals from $Cond(\Sigma_{abc})$ not in normal form (including trivial conditionals).

Thus, for computing the closure of an inference relation $\vdash$, we need to determine whether $A \vdash B$ holds only for $(B|A) \in NFC(\Sigma)$. We may however reduce the amount of explicit query answering further by taking general inference properties into account.

6   Exploiting Right Weakening to Calculate CM

A well known property of nonmonotonic inference relations defined over conditional statements is Right Weakening,

$$\frac{A \mathrel{|\!\sim} B \qquad B \models B'}{A \mathrel{|\!\sim} B'} \tag{RW}$$

stating that, if the conditional $(B|A)$ is contained in $\mathrel{|\!\sim}$, then all conditionals $(B'|A)$ with $B \models B'$ are also contained in $\mathrel{|\!\sim}$. We will call a conditional $(B'|A)$ a right weakening of a conditional $(B|A)$ if $B \models B'$.

Some normal form conditionals are right weakenings of other normal form conditionals. In particular, every normal form conditional with $|B| > 1$ is a right weakening of a normal form conditional with $|B| = 1$.

Definition 4 (Single   Consequent   Normal   Form   Conditionals, $NFC^*(\Sigma)$**).** *Let Σ be a propositional signature. The set of* single consequent *normal form conditionals* over Σ, denoted by $NFC^*(\Sigma)$, *is defined as*

$$NFC^*(\Sigma) = \{(B|A) \in NFC(\Sigma) \mid |B| = 1\} \tag{8}$$

If we know that the inference relation $\mathrel{|\!\sim}$ satisfies (RW), it is sufficient to explicitly determine the inclusion of all single consequent normal form conditionals, and then determine the inclusion of right weakenings of single consequent normal form conditionals that are not contained in $\mathrel{|\!\sim}$.

This allows us to determine the inclusion of normal form conditionals with larger consequents, if their stronger version is already determined to be included in $\mathrel{|\!\sim}$.

We can ensure that the strongest versions of normal form conditionals are checked first by iterating over the set $NFC(\Sigma)$ in the order defined in [6]. Intuitively this order is given by first sorting $NFC(\Sigma)$ by antecedents from smallest to largest (as given in Table 1) and then sorting by consequent among the conditionals with equal antecedents in the same manner. The list of normal form conditionals over a signature $\mathcal{L}_\Sigma$ in this order is denoted by $[NFC(\Sigma)]$. In our grid representation this corresponds to iterating over all normal form conditionals from left to right and from bottom to top.

The   algorithm   CALCCLOSURE   (Algorithm 1)   uses   a   procedure ANSWERQUERY(A, B) to determine if the conditional $(B|A)$ is contained in $\mathrel{|\!\sim}$. Usually, this procedure uses a knowledge base $\mathcal{R}$ (e.g. [4,7,14]), but since we abstract from a concrete inference relation and knowledge base, we omit the parameter $\mathcal{R}$. It starts by initializing CM with a template as described in Sect. 5. It then iterates over $NFC(\Sigma)$ in the order described above. If $CM(A, B)$ is already set to 1 (Line 4), then $(B|A)$ is a right weakening of a normal form conditional that has already been determined to be included in $\mathrel{|\!\sim}$. In Line 6 the inclusion of the current normal form conditional $(B|A)$ is determined by calling ANSWERQUERY(A, B). If $A \mathrel{|\!\sim} B$ holds, $CM(A, B)$ is set to 1. Note that due to the preprocessing in CMTEMPLATE(Σ), this also effects all syntactic variants $(B'|A)$ of $(B|A)$, such

---

Algorithm 1. CALCCLOSURE(Σ, ANSWERQUERY)

Input : Signature Σ, Boolean valued procedure ANSWERQUERY for
           inference relation $\mathrel|\!\sim$
Output: Closure CM for $\mathrel|\!\sim$

```
 1 begin
 2 │   CM ← CMTEMPLATE(Σ)
 3 │   foreach (B|A) ∈ [NFC(Σ)] do                    // NF conditionals
 4 │   │   if CM(A, B) = 1 then                       // already included
 5 │   │   │   break
 6 │   │   if ANSWERQUERY(A, B) then                  // determine A |∼ B?
 7 │   │   │   CM(A, B) ← 1
 8 │   │   │   foreach (B'|A) ∈ NFC(Σ) with B ⊂ B' do  // (RW)
 9 │   │   │   │   CM(A, B') ← 1
10 │   │   else
11 │   │   │   CM(A, B) ← 0                           // cond. not in |∼
12 │   return CM
```

---

that $CM(B', A)$ is also set to 1. In Lines 8 and 9 the inclusion of every right weakening of the current normal form conditional is then set in CM. These conditionals will be skipped in subsequent iterations (Line 4). If $A \not\mathrel|\!\sim B$ holds, $CM(A, B)$ is set to 0 in Line 11. As above, this effects all syntactic variants via the links introduced in CMTEMPLATE(Σ).

Proposition 4. *Given a signature Σ and a decision procedure* ANSWERQUERY *for a nonmonotonic inference relation $\mathrel|\!\sim$ over $\mathcal{L}_\Sigma$ satisfying (RW) and defined over conditional statements, the Algorithm* CALCCLOSURE *terminates and returns CM, for which the following holds:*

(full representation) *For all A and B it holds that $CM(A, B) = 1$ or $CM(A, B) = 0$.*
(soundness) *If $CM(A, B) = 1$ then $A \mathrel|\!\sim B$.*
(completeness) *If $A \mathrel|\!\sim B$ then $CM(A, B) = 1$.*

     Further, Algorithm CALCCLOSURE *has the following properties:*

(no trivial queries) *For all $(B|A) \in Cond_-^+(\Sigma)$ there is no call* ANSWERQUERY(A, B).
(no non normal form queries) *For all $(B|A) \in nNFC(\Sigma)$ there is no call* ANSWERQUERY(A, B).
(RW exploitation) *If $A \mathrel|\!\sim B$, then for all $B \subsetneq B'$ there is no call* ANSWERQUERY(A, B').
(no repeated queries) *For all $(B|A) \in NFC(\Sigma)$, there is at most one call* ANSWERQUERY(A, B).

Using Propositions 1–3 the proof of Proposition 4 is obtained by formalizing the observations given above in the description of the algorithm CALCCLOSURE.

In the case of Σ_{ab}, compared to the naive approach of calling ANSWERQUERY for all 256 conditionals, the algorithm CALCCLOSURE only calls ANSWERQUERY at most for the 50 normal form conditionals. Due to the order in which normal form conditionals are checked, in the best case only single consequent normal form conditionals need to be checked. For Σ_{ab} there are only 28 single consequent normal form conditionals with the remaining 22 normal form conditionals being right weakenings.

More generally, for a signature Σ with $|\Sigma| = n$, the naive approach of calling ANSWERQUERY(A, B) for every possible conditional $(B|A)$ over Σ requires $f(n) = 2^{2^n \times 2}$ calls of ANSWERQUERY(A, B). The algorithm CALCCLOSURE only calls ANSWERQUERY for the

$$g(n) = \sum_{i=2}^{2^n} \binom{2^n}{i} (2^i - 2)$$

normal form conditionals. In the best case only

$$\sum_{i=2}^{2^n} \binom{2^n}{i} i$$

single consequent normal form conditionals are checked. As n grows, the difference between the number of conditionals and the number of normal form conditionals becomes arbitrarily large, i.e.

$$\lim_{n \to \infty} \frac{g(n)}{f(n)} = 0.$$

7 Conclusions and Future Work

Some applications in the domain of nonmonotonic inference research require access to the complete inference closures. We presented an approach for calculating the closure of nonmonotonic inference relations defined over conditional statements. Our approach uses a simple grid representation, employs a strong preprocessing step that only depends on the signature and can be reused for every inference relation over that signature, and it reduces the number of explicit query answering by exploiting properties of conditionals and inference relations defined over them.

In our current work, we are using our approach for empirically checking and evaluating inference relations implemented within the InfOCF system [3] with respect to various postulates proposed for nonmonotonic inference relation.

References

1. Beierle, C., Eichhorn, C., Kern-Isberner, G., Kutsch, S.: Skeptical, weakly skeptical, and credulous inference based on preferred ranking functions. In: Kaminka, G.A. (eds.) Proceedings 22nd European Conference on Artificial Intelligence, ECAI-2016. Frontiers in Artificial Intelligence and Applications, vol. 285, pp. 1149–1157. IOS Press (2016)

2. Beierle, C., Eichhorn, C., Kern-Isberner, G., Kutsch, S.: Properties of skeptical C-inference for conditional knowledge bases and its realization as a constraint satisfaction problem. Annal. Math. Artif. Intell. **83**(3–4), 247–275 (2018)

3. Beierle, C., Eichhorn, C., Kutsch, S.: A practical comparison of qualitative inferences with preferred ranking models. KI - Künstliche Intelligenz **31**(1), 41–52 (2017)

4. Beierle, C., Kutsch, S., Sauerwald, K.: Compilation of static and evolving conditional knowledge bases for computing induced nonmonotonic inference relations. Annal. Math. Artif. Intell. (2019, to appear)

5. Beierle, C., Kutsch, S.: Computation and comparison of nonmonotonic skeptical inference relations induced by sets of ranking models for the realization of intelligent agents. Appl. Intell. **49**(1), 28–43 (2019). https://doi.org/10.1007/s10489-018-1203-5

6. Beierle, C., Kutsch, S.: Systematic generation of conditional knowledge bases up to renaming and equivalence. In: Calimeri, F., Leone, N., Manna, M. (eds.) JELIA 2019. LNCS (LNAI), vol. 11468, pp. 279–286. Springer, Cham (2019). https://doi.org/10.1007/978-3-030-19570-0_18

7. Beierle, C., Kutsch, S., Sauerwald, K.: Compilation of conditional knowledge bases for computing C-inference relations. In: Ferrarotti, F., Woltran, S. (eds.) FoIKS 2018. LNCS, vol. 10833, pp. 34–54. Springer, Cham (2018). https://doi.org/10.1007/978-3-319-90050-6_3

8. Benferhat, S., Dubois, D., Prade, H.: Possibilistic and standard probabilistic semantics of conditional knowledge bases. J. Logic Comput. **9**(6), 873–895 (1999)

9. Dubois, D., Prade, H.: Conditional objects as nonmonotonic consequence relationships. IEEE Trans. Syst. Man Cybern. **24**(12), 1724–1740 (1994). Special Issue on Conditional Event Algebra

10. de Finetti, B.: La prévision, ses lois logiques et ses sources subjectives. Ann. Inst. H. Poincaré 7(1), 1–68 (1937). English translation. Kyburg, H., Smokler, H.E. (eds.) Studies in Subjective Probability, pp. 93–158. Wiley, New York (1974)

11. Ganzinger, H.: Saturation-based theorem proving (abstract). In: Meyer, F., Monien, B. (eds.) ICALP 1996. LNCS, vol. 1099, pp. 1–3. Springer, Heidelberg (1996). https://doi.org/10.1007/3-540-61440-0_113

12. Goldszmidt, M., Pearl, J.: Qualitative probabilities for default reasoning, belief revision, and causal modeling. Artif. Intell. **84**(1–2), 57–112 (1996)

13. Kern-Isberner, G.: Conditionals in Nonmonotonic Reasoning and Belief Revision. LNCS (LNAI), vol. 2087. Springer, Heidelberg (2001). https://doi.org/10.1007/3-540-44600-1

14. Lehmann, D., Magidor, M.: What does a conditional knowledge base entail? Artif. Intell. **55**, 1–60 (1992)

15. Lewis, D.: Counterfactuals. Harvard University Press, Cambridge (1973)

16. Pearl, J.: System Z: a natural ordering of defaults with tractable applications to nonmonotonic reasoning. In: Proceedings of the 3rd Conference on Theoretical Aspects of Reasoning About Knowledge (TARK 1990), pp. 121–135. Morgan Kaufmann Publ. Inc., San Francisco, CA, USA (1990)

17. Spohn, W.: Ordinal conditional functions: a dynamic theory of epistemic states. In: Harper, W., Skyrms, B. (eds.) Causation in Decision, Belief Change, and Statistics, II, pp. 105–134. Kluwer Academic Publishers, Dordrecht (1988)
18. Spohn, W.: The Laws of Belief: Ranking Theory and Its Philosophical Applications. Oxford University Press, Oxford (2012)

Solving Word Analogies: A Machine Learning Perspective

Suryani Lim[1], Henri Prade[2], and Gilles Richard[2(✉)]

[1] Federation University, Churchill, Australia
`suryani.lim@federation.edu.au`
[2] IRIT, Toulouse, France
{`henri.prade,gilles.richard`}`@irit.fr`

Abstract. Analogical proportions are statements of the form 'a is to b as c is to d', formally denoted $a : b :: c : d$. This means that the way a and b (resp. b and a) differ is the same as c and d (resp. d and c) differ, as revealed by their logical modeling. The postulates supposed to govern such proportions entail that when $a : b :: c : d$ holds, then seven permutations of a, b, c, d still constitute valid analogies. It can also be derived that $a : a :: a : b$ does not hold except if $a = b$. From a machine learning perspective, this provides guidelines to build training sets of positive and negative examples. We then suggest improved methods to classify word-analogies and also to solve analogical equations. Viewing words as vectors in a multi-dimensional space, we depart from the traditional parallelogram view of analogy to adopt a purely machine-learning approach. In some sense, we learn a functional definition of analogical proportions without assuming any pre-existing formulas. We mainly use the logical properties of proportions to define our training sets and to design proper neural networks, approximating the hidden relations. Using a GloVe embedding, the results we get show high accuracy and improve state of the art on words analogy-solving problems.

1 Introduction

Analogical proportions are statements of the form a is to b as c is to d, usually denoted $a : b :: c : d$, such as "the calf is to the cow as the foal is to the mare". Such statements have been considered for a long time as a linguistic counterpart to numerical proportions. Although they naturally emerge when stating parallels between two situations regarded as analogous (e.g., "electrons are to the atom nucleus as planets are to the sun"), they were rarely studied in the mainstream literature. However, in the last two decades we have witnessed an increased interest in the development of original research directions.

At least two branches can be roughly distinguished in this recent series of work. First, a numerically oriented trend based on an arithmetic proportion-based modeling having its roots in a seminal paper by Rumelhart and Abrahamson [20] and illustrations in recent works such as [10,15]. In a visual

© Springer Nature Switzerland AG 2019
G. Kern-Isberner and Z. Ognjanović (Eds.): ECSQARU 2019, LNAI 11726, pp. 238–250, 2019.
https://doi.org/10.1007/978-3-030-29765-7_20

context, analogical proportions have also been used for expressing constraints in multi-class categorization tasks and attribute transfer [4,5].

The second branch covers propositional logic modelings of analogical proportions [13,18] which originates in various formal models [7,21,26], and computational linguistics [1,6,8].

The logical modeling of analogical proportions has revealed that analogy is as much a matter of dissimilarity as a matter of similarity. Indeed the logical expression of a is to b as c is to d exactly says that a differs from b as c differs from d and that b differs from a as d differs from c. This is also made clear in the numerical modeling which corresponds to the constraint $a - b = c - d$ when a, b, c, d refers to numbers rather than to Boolean variables. Importantly enough, dissimilarity and similarity are put on a par by the logical modeling which is also equivalent to "what a and d have in common (positively or negatively), b and c have it also" [13,18]. The logical modeling straightforwardly extends from Boolean variables to vectors thereof in a component-wise way. Then given three vectors $\overrightarrow{a}, \overrightarrow{b}, \overrightarrow{c}$, one may look for a fourth one $\overrightarrow{d}$, if it exists, s.t. $\overrightarrow{a} : \overrightarrow{b} :: \overrightarrow{c} : \overrightarrow{d}$ holds. It is the basis of analogical proportion-based inference [3,21].

In case of a vector-based encoding, the inference is based on the apparent consequence of the arithmetic proportion-based view, which leads to computing $\overrightarrow{d}$ as $\overrightarrow{c} + \overrightarrow{b} - \overrightarrow{a}$ Items such as calf, cow, foal, mare can be represented in terms of Boolean features (such as bovid, equid, male, female, adult, young, ...), or in a numerical manner using word embedding techniques. Then it is natural to bridge the gap between the logical and numerical viewpoints, or at least to take advantage of one for improving the inferential result of the other.

The paper is structured as follows. Section 2 recalls the postulates characterizing analogical proportions and identifies a method of enlarging a set of examples and counterexamples. Section 3 provides a new approach to the recognition of analogical proportions, and the solving of analogical proportion equations, between natural language words. Sections 4 and 5 present the experimental setting and report promising results showing the interest of the approach. Related works are discussed in Sect. 6, before concluding.

2   Analogical Proportions: What They Are

Basic postulates. Taking inspiration from the properties of numerical proportions, such as geometric proportions (i.e., $\frac{a}{b} = \frac{c}{d}$), or arithmetic proportion (i.e., $a - b = c - d$), analogical proportions are quaternary relations, supposed to obey the three following postulates (e.g., [7]): $\forall a, b, c, d$,

1. $a : b :: a : b$ (*reflexivity*);
2. $a : b :: c : d \rightarrow c : d :: a : b$ (*symmetry*);
3. $a : b :: c : d \rightarrow a : c :: b : d$ (*central permutation*).

Other properties are direct consequences of these postulates like $a : a :: b : b$ (*identity*); $a : b :: c : d \rightarrow b : a :: d : c$ (*inside pair reversing*); $a : b :: c : d \rightarrow d : b :: c : a$ (*extreme permutation*).

Repeated applications of postulates (2) and (3) show that an analogical proportion has exactly *eight* equivalent forms:
$a : b :: c : d = c : d :: a : b = c : a :: d : b = d : b :: c : a =$
$d : c :: b : a = b : a :: d : c = b : d :: a : c = a : c :: b : d.$

Moreover a postulate stronger than (1) may be expected:

4. $\forall a, b, x, \; a : b :: a : x \rightarrow (x = b)$ (*unicity*).

The failure of such postulate would lead to proportions such as $b : a :: c : a$, which might seem acceptable (e.g., one can say "Bob is to Peter as Ric is to Peter" as soon as, e.g., Bob and Ric are Peter's sons), but $a : a :: b : c$ looks troublesome since a and a are identical, while b and c are clearly not.

Consequences of postulates (2)–(3)–(4) that can be derived.

Fact 1. $a : a :: b : x \implies x = b$

Proof: $a : a :: b : x \implies a : b :: a : x$ (3) $\implies x = b$ (4) □

Fact 2. $a : b :: c : c \implies a = b$

Proof: $a : b :: c : c \implies c : c :: a : b$ (2) $\implies a = b$ (Fact 1) □

Fact 3. $a : b :: c : b \implies a = c$

Proof: $a : b :: c : b \implies a : c :: b : b$ (2) $\implies a = c$ (Fact 2) □

Fact 4. $a : b :: c : d \wedge (a \neq b) \implies c \neq d$

Proof: Suppose $c = d$. Then $a : b :: c : d$ is $a : b :: c : c$, and by symmetry, $c : c :: a : b$ which implies $a = b$ (from (Fact 1)). Which contradicts the hypothesis $a \neq b$. □

However, from $a : b :: b : c$, we cannot infer $a = c$: it is called a *continuous* analogical proportion, and for arithmetic proportions, it just means that b is the middle of the segment $[a, c]$. Clearly, all the above properties are satisfied by arithmetic proportions.

Given 4 distinct items a, b, c, d, they can be ordered in $4! = 24$ different ways. This indicates that among these 24 permutations there are 3 classes of 8 permutations each that are stable under the postulates of analogical proportions, since $a : b :: c : d$ can be written in 8 equivalent forms.

As a consequence $b : a :: c : d$ and $a : d :: c : b$ do not belong to the same class as $a : b :: c : d$ and are in fact elements of two different classes. If an element of a class is a valid (resp. not valid) proportion, then the 7 remaining ones are also valid (resp. not valid). Although it does not follow from postulates (2)–(3)–(4), one can consider that if $a : b :: c : d$ holds then neither $b : a :: c : d$ nor $a : d :: c : b$ hold as valid analogical proportions. This indeed can be observed on the example $a = \texttt{calf}$, $b = \texttt{cow}$, $c = \texttt{foal}$, $d = \texttt{mare}$, and is also incompatible with a function-based view of analogical proportion where $a : b :: c : d$ holds iff $\exists f, b = f(a), d = f(c)$.

Despite their obvious semantics, these fundamental properties of analogical proportions are rarely used in practice. However, they have implications when it comes to machine learning, as seen in Sect. 3.

3   Machine Learning for Analogical Proportions

We take advantage of the previous theoretical analysis for revisiting the problems of identifying and solving analogical proportions expressed in natural language and proposing a new approach to this issue which has been widely investigated by the NLP community [7,9,10,14,24]. It is now common to convert words into numerical vectors for computational purposes, a process known as word embedding. If V is the target vector space and W the corpus of words, we denote $embed(W)$ the subset of V representing the words of W.

There are two well-known embedding models: word2vec [14,15] from Google and GloVe [17] from Stanford University. Both models learn geometrical encodings (vectors) of words from their co-occurrence information. They differ in how they learn this information, and a detailed overview of the available options with their advantages and drawbacks can be found in [11]. In both techniques, $V = \mathbb{R}^n$ with $n \in \{50, 100, 200, 300\}$. They also found that the qualities of the embeddings are almost on-par if tuned using the correct parameters, but the qualities of the analogical testings differ depending on the formulas used. Based on the observations in Sect. 2, we suggest improved methods to classify word-analogies and also to solve analogical equations. Viewing words as vectors in a multi-dimensional space, we depart from the traditional parallelogram view of analogy to adopt a purely machine-learning approach. For all our experiments, we use GloVe.

3.1   Analogy Testing as a Classification Problem

As far as we know, the problem of deciding if a quadruple of words is a valid analogical proportion or not is a binary classification task, and it has not been widely investigated (see [2] for instance). One method that has been successful in classification tasks is convolutional neural network (CNN) - it is believed that this network can capture high-level features not easily extracted otherwise, especially within pictures. So, we believe CNN can also capture hidden semantical links underlying a valid analogy.

Embedding 4 words via GloVe in the space of dimension n and stacking together the 4 vectors, we get a matrix $n \times 4$ that we can consider as an *image*. From this point of view, our problem is to build a binary classifier for *images* using CNN. Using the permutation properties described in Sect. 2, we provide a proper dataset with both positive and negative examples to avoid semantic loss.

3.2   Analogy Solving as a Regression Problem

In natural language processing, analogy-solving is the problem of having a set of triple words a, b, c and looking for d such that $a : b :: c : d$ is a valid analogy [1]. It has been shown in [10] that cosine similarity multiplication ($3CosMul$) is the

---

[1] From now on, we use lower case letter a to denote the word or the vector. Moreover we alleviate the notation by writing a instead of $\vec{a}$ and so on for b, c, d.

state-of-the-art approach of finding d. Still, there are plenty of unsatisfactory results (that the solution fails to form an analogy). We feel that the problem comes from the fact that the authors consider analogy as a matter of *similarity* only, introducing in the various suggestions, similarities between the candidate solution d and the input variables a, b and c. As explained in the beginning, analogical proportions, as expressed by our indicators, cannot be reduced to equalities of similarities: dissimilarities matter!

That is why our approach is entirely different. We agree that the solution d should be a function of a, b, and c but we do not suggest any formula. We want to learn the hidden function f such that $f(a, b, c) = d$ where d is the solution, and we do not assume a predefined analytic form for d. In that case, we can consider it is a multi-variable regression problem where the training set is just built from the initial set of analogies $a : b :: c : d$ as pair $(a, b, c), d$. As multi-layer neural networks are universal function approximators, it is appropriate to use such a network to estimate our target function.

4   Experimental Settings

The driving force behind our ideas is to improve classification accuracy and analogy solving success.

4.1   Initial Datasets

To get a dataset of analogies, we started from Google dataset (questions-words) containing exactly 19,544 analogies, each one involving 4 distinct words. These analogies are classified in diverse categories such as capital-common-countries like *Athens : Greece : Ottawa : Canada*, country-currency, or opposite like *acceptable : unacceptable :: aware : unaware*, etc. Considering a quadruple of words, like *(engine, car, heart, human)* in that order, we want to build a classifier able to tell us if the proportion *engine:car::heart:human* is a *valid* analogy. *Valid* means that a human considers it as an analogy, without being able to give us a concise definition of an analogy.

GloVe provides $\mathbb{R}^n$ words embedding where $n \in \{50, 100, 200, 300\}$. Ultimately, our dataset was loaded as a CSV file, and each row was compiled into a real-valued matrix of dimension $n \times 4$. The GloVe embedding originally contained 400,000 words, but we removed non-alphabetical words, so we were left with 317,544 words.

4.2   Extended Datasets

Due to the properties of analogical proportions viewed in Sect. 2, it is easy to rigorously extend our dataset by applying the permutation properties. Doing so, we end up with $19{,}544 \times 8 = 156{,}352$ valid analogies. We get invalid analogies just by using the previous dataset and permuting only the 2 first elements of a valid analogy. Starting from 19,544 valid analogies, we then get $19{,}544 \times 16$

Table 1. Positive/negative examples with $m = 19,544$

	$+$	$-$	$-$	$-$
	$a : b :: c : d$	$b : a :: c : d$	$c : b :: a : d$	$a : a :: c : d$
Theory	$8 \times m$	$8 \times m$	$8 \times m$	$8 \times m$
Practice	156, 352	156, 352	156, 352	156, 352

invalid analogies. Permuting the 1st and the 3rd elements still provides an invalid analogy, and we end up with 19,544 × 16 invalid analogies. It has also been seen in Sect. 2 that when we have 3 *distinct* elements a, b, c, then $a : a :: c : d$ is an invalid analogy. It is then appropriate to add a set of negative examples such as $a : a :: c : d$, getting 19,544 × 8 more negative examples. All in all, we get a final dataset of 19,544 × 32 = 625,408 examples, among which only 156,352 are valid analogies, and the remaining 469,056 examples are invalid analogies.

We summarize in Table 1 the process of building a dataset having positive and negative examples.

4.3 Neural Networks Approach

Classification Task. Stacking together the 4 vectors corresponding to a quadruple a, b, c, d, we get a matrix $n \times 4$ considered as an *image*. We design and train a CNN to classify between valid analogies/invalid analogies. With filters respecting the boundaries of the 2 pairs, this is the structure of the CNN:

- 1st layer (convolutional): 128 filters of size $h \times w = 1 \times 2$ with strides $(1, 2)$ and relu activation.
- 2nd layer (convolutional): 64 filters of size $(2, 2)$ with strides $(2, 2)$ and relu activation.
- 3rd layer (dense): one output and sigmoid activation as we want a score between 0 and 1.

The structure of this network can be seen in (Fig. 1), and the results with this approach are given in Subsect. 5.1.

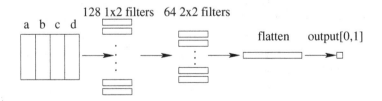

Fig. 1. Structure of the CNN as a classifier.

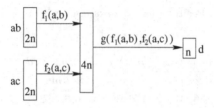

Fig. 2. Structure of the neural network for regression, that is given $a : b :: c$, the network is to find d. All a, b and c used $\mathbb{R}^n$ words embedding where $n \in \{50, 100, 200, 300\}$, where n is the number of dimensions for embedding the words.

Regression Problem. Regarding the equation solving problem, we first consider a, b, c as the input of our network, and d as the output. A neural net with $3 \times n$ input neurons and n output neurons was first designed to approximate the function f such that $f(a, b, c,) = d$. We obtained inferior results, probably because this way of processing does not take into account the similarity/dissimilarity:

1. between a and b on one side and
2. between a and c on the other side

A high-level of description would then be as follows:

1. The hidden link between a and b is described via a (unknown) function f_1
2. The hidden link between a and c is described via a (unknown) function f_2
3. The final solution is then a function g of $f_1(a, b)$ and $f_2(a, c)$ i.e.: $d = g(f_1(a, b), f_2(a, c))$.

Starting from this point of view, we approximate this function $g(f_1, f_2)$ via two networks approximating f_1 (resp. f_2). A final neural network approximating g received as input two values: the output of f_1 and the output of f_2. The output of $g(f_1, f_2)$ is d, is then an approximation of $d = g(f_1(a, b), f_2(a, c))$. Figure 2 describes the structure of the regression neural network. The output of the network, d, is unlikely to be a GloVe embedding, so we have to find the word closest to d i.e. the nearest neighbor of d in $embed(W)$. This word is assumed to be the word represented by d, and could the correct answer.

We compared the results of the neural network regression with $3CosMul$, the state of the art analogy recovery method [10,11], and it is defined as:

$$argmax_{d \in embed(W)} \frac{d \cdot b \times d \cdot c}{d \cdot a + \epsilon} \qquad (1)$$

5   Experimental Results

The data and networks described above were designed to:

1. classify whether a quadruple of words is a valid analogical proportion or not (binary classification task): in that case, the output is either true or false (see Sect. 5.1).
2. solve an equation to find the valid word to complete an analogy (regression task): in that case, the output is a word (see Sect. 5.2).

We then ran a batch of experiments. The results are provided and commented in the following sections. Obviously, the baseline here is the parallelogram approach where $a : b :: c : d$ is considered as a valid analogy iff a, b, c, d define a parallelogram in the word embedding space.

5.1   Analogy Testing as a Classification Task

With 10 fold cross-validation, the network was trained to minimize binary cross-entropy, and its performance was measured via its accuracy. As we used 10-fold cross-validation, we ended up with training sets of size $562, 848 = 32 \times 17, 589$ and testing sets of size $62, 560 = 32 \times 1, 955$. The network's performance on several dimensions is given in Table 2, which shows the performance of the CNN classifier with $\mathbb{R}^n$ words embedding where $n \in \{50, 100, 200, 300\}$.

As shown in Table 2, the accuracy increases as the number of dimensions and epochs increase, with the highest accuracy of 99.34% (300 dimension vectors trained for 10 epochs). The accuracy increases significantly when increasing the number of dimensions from 50 to 100. For 50 dimensions, the accuracy is 91.02%, but at 100 dimensions, the accuracy increases by over 6% to 96.79%. However, the gain appears to be less significant for the higher dimensions, as the accuracy for 200 dimensions is only 2.4% higher than 100 dimensions, and the accuracy for 300 dimensions is just 0.1% higher than 200 dimensions. It appears that 100 dimensions are a good compromise for efficient and effective analogy classification task.

Table 2. Accuracy of the CNN for analogy classification: impact of word embedding dimensions and the number of epochs.

#epochs	Average accuracy (std dev.)			
	50	100	200	300
1	83.9% (± 6.4)	81.1% (± 8.5)	75.5 (± 1.2)	80.26 (± 11.13)
3	81.19% (± 6.87)	81.40% (±12.01)	84.27% (± 8.19)	83.84% (± 7.77%)
5	90.68% (± 5.77)	93.07% (± 6.83)	93.19% (± 6.21)	95.22% (± 4.90%)
10	91.02% (± 7.67)	96.79% (± 5.05)	99.24% (± 1.17)	99.34% (± 0.79%)

5.2   Equation Solving as a Regression Task

In this experiment, we used $\mathbb{R}^n$ word embedding where $n \in \{50, 100, 200, 300\}$. The network was trained for 50 epochs 10 fold cross-validation (training and testing split of 90/10). Recall that the total number of the dataset was 156,352 (19,544 × 8 permutations). To ensure that there was enough independent data for testing and training for each fold, the permutation was applied after splitting. The results of the experiment are given in Table 3, which shows the accuracy of neural network regression and $3CosMul$ when embedding the words in 50, 100, 200, and 300 dimensions. For the neural network, the best overall performance is given by 100 dimensions at 79.0% of accuracy, and the performance drops as the number of dimensions increases. It is also interesting to note that the MSE for regression decreases as the number of dimensions increases, yet the accuracy drops after the 100 dimensions. For $3CosMul$, the 300 dimensions give the best performance (68.1% accuracy), but it is nearly 11% lower than the best results from the neural network. Table 3 also shows the accuracy for each category within the dataset. It can be seen that the regression for 100 dimensions outperforms $3CosMul$ for 12 of the 14 categories, on par for one category (nationality) and below par for one category (gender).

The formula $3CosMul$, although it is so far the most accurate equation, suffers from some drawbacks. A literal implementation often provides b or c as a candidate solution for $a : b :: c : x$. The implementation in [11] removes b, c and also a from the candidate solution - our implementation of $3CosMul$ also follows this practice. Also, $3CosMul$ is not robust to reverse analogies (i.e. even if we

Table 3. Comparing the effectiveness of the neural network and the formula $3CosMul$ for all categories and each category. The total number of analogies is 19,544, and the "Common Capital" category has 506 analogies.

	Neural network regression				3CosMul			
	50	100	200	300	50	100	200	300
Overall	63.9%	79.0%	75.4%	71.4%	36.2%	56.7%	65.0%	68.1%
Common capitals (506)	96.3%	98.8%	97.4%	96.9%	63.8%	80.0%	86.9%	88.9%
All capitals (4524)	86.7%	97.1%	97.1%	90.9%	50.0%	77.0%	87.5%	90.1%
Currencies (866)	50.2%	63.4%	61.2%	56.3%	5.0%	15.0%	22.5%	24.4%
US cities (2467)	32.4%	53.5%	62.0%	59.5%	6.9%	17.0%	29.8%	36.5%
Gender (506)	53.2%	44.3%	40.7%	34.4%	61.5%	79.4%	86.6%	88.6%
Adj to adverb (992)	34.5%	55.1%	31.9%	30.1%	10.8%	22.0%	22.9%	23.4%
Opposite (812)	24.9%	43.1%	26.3%	23.3%	6.2%	17.1%	21.5%	25.4%
Comparative (1332)	74.6%	89.2%	85.6%	83.2%	41.4%	71.9%	79.8%	83.3%
Superlative (1122)	73.1%	86.4%	78.9%	76.2%	18.6%	50.1%	67.5%	73.7%
Base to gerund (1056)	43.9%	78.3%	67.9%	70.3%	35.2%	65.6%	68.1%	71.0%
Nationalities (1599)	93.5%	94.4%	96.3%	94.3%	84.7%	89.1%	94.1%	94.6%
Gerund to past (1560)	52.2%	80.8%	69.5%	66.7%	27.3%	53.1%	59.6%	62.5%
Plurals (1332)	78.3%	87.1%	88.1%	74.6%	48.2%	68.5%	74.1%	76.5%
Base to 3rd person (870)	46.2%	74.8%	59.2%	56.4%	28.8%	59.1%	64.9%	68.4%
MSE (train)	0.1	0.07	0.06	0.05				
MSE (test)	0.1	0.07	0.06	0.05				

are not far from finding the solution d of $a : b :: c : x$, one can by far missing the solution c of $b : a :: d : x$). The neural network regression does not seem to suffer from these two drawbacks: it does not return b or c as the candidate solution, and it appears to be more robust with respect to permutation.

6 Related Work

Our work makes use of recent word embedding systems to develop machine learning solutions to deal with word analogies. Nevertheless, the problem of recognizing word analogies has been attempted well before the emergence of effective words embedding systems: we can cite [19, 22–25] for instance. The dataset was generally limited with regard to the current standard. For instance [24] uses a corpus of 374 questions coming from the SAT analogy questions: the higher accuracy on this set is 56.1% knowing that a random guess yields an accuracy of 20% and senior high school students get 57%. The results are quite interesting as the dataset mainly consists of semantic analogies like *mason:stone::carpenter:wood*. More recently, researchers were more interested in working on the equation solving problem, as with the word embedding techniques, numerical tools became widely available. Instead of dealing with discrete space of words, they deal with continuous space language models.

One of the landmark paper is [16] where the representation is learned via a recurrent neural network. These vectors representations attempt to capture syntactic and semantic regularities in the English language, and the relationships between words can be characterized by vector offsets. Still solution of the equation $a : b :: c : x$ is supposed to be (close to) $b + c - a$ (offset method), where "close to" relation is cosine similarity of normalized vectors. The best equation approximating this relation is the $3CosMul$ formula described in Eq. 1. In Sect. 5.2, we showed a neural network regression based solution outperforms $3CosMul$. From another perspective, the BART (for Bayesian Analogy with Relational Transformations) model [12], is also neural network-based but more dedicated to finding a pair (c, d) when a pair (a, b) is given.

7 Conclusion

We have investigated the axioms of analogical proportions to derive new properties rarely used in practice and showing that analogy is as much a matter of dissimilarity as a matter of similarity. When we refer to numbers or vectors, an analogy is often viewed as $a - b = c - d$ (arithmetic analogy) and may be unable to capture the complexity of similarity/dissimilarity. To try to bridge the gap between the Boolean view and the numerical one, we focus on natural language analogies using word embedding techniques. This leads us to suggest radically new approaches not only for classifying whether a quadruple of words is an analogy but also for solving analogical equations.

Previous attempts rely on predefined formulas: this can be restrictive as it assumes that we have a clear understanding of what an analogy is in natural language. In fact, we have no such understanding. Our methodology is quite different: it takes advantages of the theoretical analysis and recent word embedding systems to learn the unknown relationship amongst the words in an analogy. Embedding the words using GloVe into a multi-dimensional vector space $\mathbb{R}^n$, the matrix constituted of 4 vectors can be viewed as an image of dimension $n \times 4$. From a machine learning perspective, the permutation properties of analogical proportion lead to an increase of the dataset size by 8 for positive examples and by 24 for negative examples. Using analogies from Google as the dataset, we have implemented a CNN classifier with an accuracy higher than 94%.

Regarding the equation solving process, we also depart from the classical views derived from the arithmetic analogy using a formula. Instead, we learned such a definition via neural networks. Considering that the solution of $a : b :: c : x$ is related to hidden high-level relationships between the pairs (a, b) and (a, c), we try to capture these relationships in the network. Experiment results show that the neural network outperforms the formula-based methods and might be useful for solving more semantic analogies. Unlike previous methods, the network does not return b or c as candidate solutions, and it appears to be more robust with respect to permutation.

More in-depth investigations are needed to confirm our view: especially using other datasets and other word embedding systems. For instance, we could generate a more effective word embedding dictionary, and take advantage of the work in [11] which provides tips on how to properly tune the hyper-parameters of such a process. Also, the neural networks we used were not entirely optimized in terms of structure and parameters. Doing so could lead to substantial improvements. In addition, for the classification task, we could compare our approach, which uses the convolutional network, to fully connected layers as we are unsure if the translation invariance of convolution brings any value to our work.

Finally, the target dataset does not include truly semantic analogies, and it is worth investigating such analogies as they are part of the day to day language. These are tracks for future work.

Acknowledgements. This work was partially supported by ANR-11-LABX-0040-CIMI (Centre International de Mathématiques et d'Informatique) within the program ANR-11-IDEX-0002-02, project ISIPA.

References

1. Ando, S., Lepage, Y.: Linguistic structure analysis by analogy: its efficiency. In: NLPRS, Phuket (1997)
2. Bayoudh, M., Prade, H., Richard, G.: Evaluation of analogical proportions through Kolmogorov complexity. Knowl.-Based Syst. **29**, 20–30 (2012)
3. Bounhas, M., Prade, H., Richard, G.: Analogy-based classifiers for nominal or numerical data. Int. J. Approximate Reasoning **91**, 36–55 (2017)
4. Hwang, S.J., Grauman, K., Sha, F.: Analogy-preserving semantic embedding for visual object categorization. In: Proceedings of 30th ICML, vol. 28, III-639-647 (2013)

5. Jing, L., Lu, Y., Gang, H., Sing Bing, K.: Visual attribute transfer through deep image analogy. ACM Trans. Graph. (Proceedings of Siggraph) **36**(4), 120:1–120:15 (2017)
6. Langlais, P., Patry, A.: Translating unknown words by analogical learning. In: Joint Conference on Empirical Method in NLP (EMNLP) and Conference on Computational Natural Langauge Learning (CONLL), Prague, pp. 877–886 (2007)
7. Lepage, Y.: Analogy and formal languages. Electr. Notes Theor. Comput. Sci. **53**, 189–191 (2001)
8. Lepage, Y., Denoual, E.: Purest ever example-based machine translation: detailed presentation and assessment. Mach. Transl. **19**(3–4), 251–282 (2005)
9. Lepage, Y., Migeot, J., Guillerm, E.: A measure of the number of true analogies between chunks in Japanese. In: Vetulani, Z., Uszkoreit, H. (eds.) LTC 2007. LNCS (LNAI), vol. 5603, pp. 154–164. Springer, Heidelberg (2009). https://doi.org/10.1007/978-3-642-04235-5_14
10. Levy, O., Goldberg, Y.: Dependency-based word embeddings. In: Proceedings of 52nd Annual Meeting of Association Computational Linguistics (Vol 2: Short Papers), pp. 302–308 (2014)
11. Levy, O., Goldberg, Y., Dagan, I.: Improving distributional similarity with lessons learned from word embeddings. Trans. Ass. Comput. Ling. **3**, 211–225 (2015)
12. Lu, H., Wu, Y., Holyoak, K.H.: Emergence of analogy from relation learning. Proc. Nat. Acad. Sci. **116**, 4176–4181 (2019)
13. Miclet, L., Prade, H.: Handling analogical proportions in classical logic and funny logics settings. In: Sossai, C., Chemello, G. (eds.) ECSQARU 2009. LNCS (LNAI), vol. 5590, pp. 638–650. Springer, Heidelberg (2009). https://doi.org/10.1007/978-3-642-02906-6_55
14. Mikolov, T., Chen, K., Corrado, G.S., Dean, J.: Efficient estimation of word representations in vector space. CoRR, abs/1301.3781 (2013)
15. Mikolov, T., Sutskever, I., Chen, K., Corrado, G.S., Dean, J.: Distributed representations of words and phrases and their compositionality. In: Burges, C.J.C. et al., (eds.) Advances in Neural Information Processing Systems, vol. 26, pp. 3111–3119. Curran Associates Inc. (2013)
16. Mikolov, T., Yih, W., Zweig, G.: Linguistic regularities in continuous space word representations. In: Proceedings of the 2013 Conference of the North American Chapter of the Association for Computational Linguistics: Human Language Technologies, pp. 746–751 (2013)
17. Pennington, J., Socher, R., Manning, C.: Glove: global vectors for word representation. In: Proceedings of Conference on Empirical Methods in Natural Language Processing (EMNLP), vol. 14, pp. 1532–1543. Association for Computational Linguistics, January 2014
18. Prade, H., Richard, G.: From analogical proportion to logical proportions. Logica Univers. **7**, 441–505 (2013)
19. Reitman, W.R.: Cognition and thought. an information processing approach. Psych. in the Schools, vol. 3, no. 2 (1965)
20. Rumelhart, D.E., Abrahamson, A.A.: A model for analogical reasoning. Cogn. Psychol. **5**, 1–28 (2005)
21. Stroppa, N., Yvon, F.: Analogical learning and formal proportions: Definitions and methodological issues. Technical Report D004, ENST-Paris (2005)
22. Turney, P.D.: Similarity of semantic relations. Comput. Linguist. **32**(3), 379–416 (2006)
23. Turney, P.D.: The latent relation mapping engine: algorithm and experiments. JAIR **33**, 615–655 (2008)

24. Turney, P.D.: A uniform approach to analogies, synonyms, antonyms, and associations. In: Proceedings of 22nd International Conference on Computational Linguistics - vol. 1 (COLING 2008), pp. 905–912. Association for Computational Linguistics (2008)
25. Veale, T.: Re-representation and creative analogy: a lexico-semantic perspective. New Gener. Comput. **24**(3), 223–240 (2006)
26. Yvon, F., Stroppa, N., Delhay, A., Miclet, L.: Solving analogical equations on words. Technical report, Ecole Nationale Supérieure des Télécommunications (2004)

Decrement Operators in Belief Change

Kai Sauerwald$^{(\boxtimes)}$ ⓘ and Christoph Beierle$^{(\boxtimes)}$

FernUniversität in Hagen, 58084 Hagen, Germany
{kai.sauerwald,christoph.beierle}@fernuni-hagen.de

Abstract. While research on iterated revision is predominant in the field of iterated belief change, the class of iterated contraction operators received more attention in recent years. In this article, we examine a non-prioritized generalisation of iterated contraction. In particular, the class of weak decrement operators is introduced, which are operators that by multiple steps achieve the same as a contraction. Inspired by Darwiche and Pearl's work on iterated revision the subclass of decrement operators is defined. For both, decrement and weak decrement operators, postulates are presented and for each of them a representation theorem in the framework of total preorders is given. Furthermore, we present two sub-types of decrement operators.

Keywords: Belief revision · Belief contraction ·
Non-prioritized change · Gradual change · Forgetting ·
Decrement operator

1 Introduction

Changing beliefs in a rational way in the light of new information is one of the core abilities of an agent - and thus one of the main concerns of artificial intelligence. The established AGM theory [1] deals with desirable properties of rational belief change. The AGM approach provides properties for different types of belief changes. If new beliefs are incorporated into an agent's beliefs while maintaining consistency, this is called a revision. Expansion adds a belief unquestioned to an agent's beliefs, and contraction removes a belief from an agent's beliefs. Building upon the characterisations of these kinds of changes and the underlying principle of minimal change, the theory fanned out in different directions and sub-fields.

The field of iterated belief revision examines the properties of belief revision operators which, due to their nature, can be applied iteratively. In this sub-field, one of the most influential articles is the seminal paper [6] by Darwiche and Pearl (DP), establishing the insight that belief sets are not a sufficient representation for iterated belief revision. An agent has to encode more information about her belief change strategy into her *epistemic state* - where the revision strategy deeply corresponds with conditional beliefs. This requires additional postulates that guarantee intended behaviour in forthcoming changes. The common way of encoding, also established by Darwiche and Pearl [6], is an extension

© Springer Nature Switzerland AG 2019
G. Kern-Isberner and Z. Ognjanović (Eds.): ECSQARU 2019, LNAI 11726, pp. 251–262, 2019.
https://doi.org/10.1007/978-3-030-29765-7_21

of Katsuno and Mendelzon's characterisation of AGM revision in terms of plausibility orderings [11], where it is assumed that the epistemic states contain an order over worlds (or interpretations).

Similar work has been done in recent years for iterated contraction. Chopra, Ghose, Meyer and Wong [5] contributed postulates for contraction on epistemic states. Caridroit, Konieczny and Marquis [4] provided postulates for contraction in propositional logic and a characterisation with plausibility orders in the style of Katsuno and Mendelzon. By this characterisation, the main characteristic of a contraction with α is that the worlds of the previous state remain plausible and that the most plausible counter-models of α become plausible.

However, in the sub-field of non-prioritised belief change, or more specifically, in the field of gradual belief change much work remains to be done on contraction. An important generalisation of iterated revision operators are the class of improvement operators by Konieczny and Pino Pérez [13], which achieve the state of an revision by multiple steps in a gradual way. These kind of changes where intensively studied by Konieczny, Pino Pérez, Booth, Fermé and Grespan [3,12]. A counterpart of improvement operators for the case of contraction is missing. This article fills this gap. We investigate the contraction analogon to improvement operators, which we call decrement operators. The leading idea is to examine a class of operators which lead, after enough consecutive applications, to the same states as an (iterative) contraction would do.

The research presented in this paper is also motivated by the quest for a formalisation of forgetting operators within the field of knowledge representation and reasoning (KRR). In a recent survey article by Eiter and Kern-Isberner [7] the connection between contraction and forgetting of a belief is dealt with from a KRR point of view. Steps towards a general framework for kinds of forgetting in common-sense based belief management, revealing links to well-known KRR methods, are taken in [2]. However, for the fading out of rarely used beliefs that takes places in humans gradually over time, or for the change of routines, e.g. in established workflows, often requiring many iterations and the intentional forgetting of the previous routines, counterparts in the formal methods of KRR are missing. With our work on decrement operators, we provide some basic building blocks that may prove useful for developing a formalisation of these psychologically inspired forgetting operations.

In summary, the main contributions of this paper are[1]:

- Postulates for operators which allow one to perform contractions gradually.
- Representation theorems for these classes in the framework or epistemic states and total preorders.
- Define two special types of decrement operators.

---

[1] The complete version of this paper contains full proofs for all theorems given here. Due to the lack of space, the proofs are not included in this version.

The rest of the paper is organised as follows. Section 2 briefly presents the required background on belief change. Section 3 introduces the main idea and the postulates along with a representation theorem for weak decrement operators. In Sect. 4 the weak decrement operators are restricted by DP-like iteration postulates, leading to the class of decrement operators; we give also a representation theorem for the class of decrement operators. In Sect. 5 two special types of decrement operators are specified. We close the paper with a discussion and point out future work in Sect. 6.

2 Background

Let Σ be a propositional signature. The propositional language $\mathcal{L}_\Sigma$ is the smallest set, such that $a \in \mathcal{L}_\Sigma$ for every $a \in \mathcal{L}_\Sigma$ and $\neg\alpha \in \mathcal{L}_\Sigma$, $\alpha \wedge \beta, \alpha \vee \beta \in \mathcal{L}_\Sigma$ if $\alpha, \beta \in \mathcal{L}_\Sigma$. We omit often Σ and write $\mathcal{L}$ instead of $\mathcal{L}_\Sigma$. We write formulas in $\mathcal{L}$ with lower Greek letters $\alpha, \beta, \gamma, \ldots$, and propositional variables with lower case letters $a, b, c, \ldots \in \Sigma$. The set of propositional interpretations Ω, also called set of worlds, is identified with the set of corresponding complete conjunctions over Σ. Propositional entailment is denoted by $\models$, with $[\![\alpha]\!]$ we denote the set of models of α, and $Cn(\alpha) = \{\beta \mid \alpha \models \beta\}$ is the deductive closure of α. This is lifted to a set X by defining $Cn(X) = \{\beta \mid X \models \beta\}$. For two sets of formulas X, Y we say X is equivalent to Y with respect to the formula α, written $X =_\alpha Y$, if $Cn(X \cup \{\alpha\}) = Cn(Y \cup \{\alpha\})^2$. For two sets of interpretations $\Omega_1, \Omega_2 \subseteq \Omega$ we say Ω_1 is equivalent to Ω_2 with respect to the formula α, written $\Omega_1 =_\alpha \Omega_2$, if Ω_1 and Ω_2 contain the same set of models of α, i.e. $\{\omega_1 \in \Omega_1 \mid \omega_1 \models \alpha\} = \{\omega_2 \in \Omega_2 \mid \omega_2 \models \alpha\}$. For a set of worlds $\Omega' \subseteq \Omega$ and a total preorder $\leq$ (reflexive and transitive relation) over Ω, we denote with $\min(\Omega', \leq) = \{\omega \mid \omega \in \Omega' \text{ and } \forall\omega' \in \Omega' \; \omega \leq \omega'\}$ the set of all worlds in the lowest layer of $\leq$ that are elements in Ω'. For a total preorder $\leq$, we denote with $<$ its strict variant, i.e. $x < y$ iff $x \leq y$ and $y \not\leq x$; with $\ll$ the direct successor variant, i.e. $x \ll y$ iff $x < y$ and there is no z such that $x < z < y$; and we write $x \simeq y$ iff $x \leq y$ and $y \leq x$.

2.1 Epistemic States and Belief Changes

Every agent is equipped with an *epistemic state*, sometimes also called belief state, that maintains all necessary information for her belief apparatus. With $\mathcal{E}$ we denote the set of all epistemic states. Without defining what a epistemic state is, we assume that for every epistemic state $\Psi \in \mathcal{E}$ we can obtain the set of plausible sentences $\mathrm{Bel}\,(\Psi) \subseteq \mathcal{L}$ of Ψ, which is deductively closed. We write $\Psi \models \alpha$ iff $\alpha \in \mathrm{Bel}\,(\Psi)$ and we define $[\![\Psi]\!] = \{\omega \mid \omega \models \alpha \text{ for each } \alpha \in \mathrm{Bel}\,(\Psi)\}$. A belief change operator over $\mathcal{L}$ is a (left-associative) function $\circ : \mathcal{E} \times \mathcal{L} \to \mathcal{E}$. We denote with $\Psi \circ^n \alpha$ the n-times application of α by $\circ$ to Ψ [13].

---

[2] $Cn(X \cup \{\alpha\})$ matches belief expansion with α on belief sets. However, in the context here, the context of iterative changes, we understand this purely technically. The problem of expansion in this context is more complex [8].

Darwiche and Pearl [6] propose that an epistemic state ψ should be equipped with an ordering $\leq_\Psi$ of the worlds (interpretations), where the compatibility with $\mathrm{Bel}\,(\Psi)$ is ensured by the so-called faithfulness. Based on the work of Katsuno and Medelezon [11], a mapping $\Psi \mapsto \leq_\Psi$ is called faithful assignment if the following is satisfied [6]:

$$\text{if } \omega_1 \in [\![\Psi]\!] \text{ and } \omega_2 \in [\![\Psi]\!], \text{ then } \omega_1 \simeq_\Psi \omega_2$$
$$\text{if } \omega_1 \in [\![\Psi]\!] \text{ and } \omega_2 \notin [\![\Psi]\!], \text{ then } \omega_1 <_\Psi \omega_2$$

Konieczny and Pino Pérez give a stronger variant of faithful assignments for iterated belief change [13], which ensures that the mapping $\Psi \mapsto \leq_\Psi$ is compatible with the belief change operator with respect to syntax independence.

Definition 1 (Strong Faithful Assignment [13]). *Let $\circ$ be a belief change operator. A function $\Psi \mapsto \leq_\Psi$ that maps each epistemic state to a total preorder on interpretations is said to be a strong faithful assignment with respect to $\circ$ if:*

(SFA1)   *if $\omega_1 \in [\![\Psi]\!]$ and $\omega_2 \in [\![\Psi]\!]$, then $\omega_1 \simeq_\Psi \omega_2$*

(SFA2)   *if $\omega_1 \in [\![\Psi]\!]$ and $\omega_2 \notin [\![\Psi]\!]$, then $\omega_1 <_\Psi \omega_2$*

(SFA3)   *if $\alpha_1 \equiv \beta_1, \ldots, \alpha_n \equiv \beta_n$, then $\leq_{\Psi \circ \alpha_1 \circ \ldots \circ \alpha_n} = \leq_{\Psi \circ \beta_1 \circ \ldots \circ \beta_n}$*

We will make use of strong faithful assignments for the characterisation theorems.

2.2   Iterated Contraction

Postulates for AGM contraction in the framework of epistemic states were given by Chopra, Ghose, Meyer and Wong [5] and by Konieczny and Pino Pérez [14]. We give here the formulation by Chopra et al. [5]:

(C1) $\mathrm{Bel}\,(\Psi - \alpha) \subseteq \mathrm{Bel}\,(\Psi)$

(C2)   if $\alpha \notin \mathrm{Bel}\,(\Psi)$, then $\mathrm{Bel}\,(\Psi) \subseteq \mathrm{Bel}\,(\Psi - \alpha)$

(C3)   if $\alpha \not\equiv \top$, then $\alpha \notin \mathrm{Bel}\,(\Psi - \alpha)$

(C4) $\mathrm{Bel}\,(\Psi) \subseteq Cn(\mathrm{Bel}\,(\Psi - \alpha) \cup \alpha)$

(C5)   if $\alpha \equiv \beta$, then $\mathrm{Bel}\,(\Psi - \alpha) = \mathrm{Bel}\,(\Psi - \beta)$

(C6) $\mathrm{Bel}\,(\Psi - \alpha) \cap \mathrm{Bel}\,(\Psi - \beta) \subseteq \mathrm{Bel}\,(\Psi - (\alpha \wedge \beta))$

(C7)   if $\beta \notin \mathrm{Bel}\,(\Psi - (\alpha \wedge \beta))$, then $\mathrm{Bel}\,(\Psi - (\alpha \wedge \beta)) \subseteq \mathrm{Bel}\,(\Psi - \beta)$

For an explanation of these postulates we refer to the article of Caridroit et al. [4]. A characterisation in terms of total preorders on epistemic states is given by the following proposition.

Proposition 1 (AGM Contraction for Epistemic State [14]). *A belief change operator* $-$ *fulfils the postulates* (C1) *to* (C7) *if and only if there is a faithful assignment* $\Psi \mapsto \leq_\Psi$ *such that:*

$$(1) \qquad [\![\Psi - \alpha]\!] = [\![\Psi]\!] \cup \min([\![\neg\alpha]\!], \leq_\Psi)$$

In addition to the postulates (C1) to (C7), Konieczny and Pino Pérez give DP-like postulates for intended iteration behaviour of contraction [14]. In the following, we call these class of operators iterated contraction operators, which are characterized by the following proposition.

Proposition 2 (Iterated Contraction [14]). *Let* $-$ *be a belief change operator* $-$ *which satisfies* (C1) *to* (C7). *Then* $-$ *is an* iterated contraction operator *if and only if there exists a faithful assignment* $\Psi \mapsto \leq_\Psi$ *such that* (1) *holds and the following is satisfied:*

if $\omega_1, \omega_2 \in [\![\alpha]\!]$, *then* $\omega_1 \leq_\Psi \omega_2 \Leftrightarrow \omega_1 \leq_{\Psi-\alpha} \omega_2$

if $\omega_1, \omega_2 \in [\![\neg\alpha]\!]$, *then* $\omega_1 \leq_\Psi \omega_2 \Leftrightarrow \omega_1 \leq_{\Psi-\alpha} \omega_2$

if $\omega_1 \in [\![\neg\alpha]\!]$ *and* $\omega_2 \in [\![\alpha]\!]$, *then* $\omega_1 <_\Psi \omega_2 \Rightarrow \omega_1 <_\Psi$ $_\alpha$ ω_2

if $\omega_1 \in [\![\neg\alpha]\!]$ *and* $\omega_2 \in [\![\alpha]\!]$, *then* $\omega_1 \leq_\Psi \omega_2 \Rightarrow \omega_1 \leq_{\Psi-\alpha} \omega_2$

2.3 Improvement Operators

The idea of (weak) improvements is to split the process of an AGM revision for epistemic states [6, p. 7ff] into multiple steps of an operator $\circ$. For such a gradual operator $\circ$ define $\Psi \bullet \alpha = \Psi \circ^n \alpha$, where $n \in \mathbb{N}$ is smallest integer such that $\alpha \notin \mathrm{Bel}(\Psi \circ^n \alpha)$. In the initial paper about improvement operators [13], Konieczny and Pino Pérez gave postulates for $\circ$, such that $\bullet$ is an AGM revision for epistemic states. Due to space reasons, we refer the interested reader to the original paper for the postulates [13]. The following representation theorem gives an impression on weak improvement operators.

Proposition 3 (Weak Improvement Operator [13, Thm. 1]). *A belief change operator* $\circ$ *is a weak improvement operator if and only if there exists a strong faithful assignment* $\Psi \mapsto \leq_\Psi$ *such that:*

$$[\![\Psi \bullet \alpha]\!] = \min([\![\alpha]\!], \leq_\Psi)$$

Furthermore, the class of weak improvement operators is restricted by DP-like iteration postulates to the so-called improvement operators [13], which are unique[3]. Again, we refer to the work of Konieczny and Pino Pérez [13] for these postulates, and only present the characterisation in the framework of total preorders.

---

[3] Note that the notion of improvement operators is not used consistently in the literature. For instance, the improvement operators as defined in [12] are not unique.

Proposition 4 (Improvement Operator [13, **Thm. 2**]**).** *A weak improvement operator* $\square$ *is an improvement operator if and only if there exists a strong faithful* $\Psi \mapsto \leq_\Psi$ *assignment such that*

(S1)   *if* $\omega_1, \omega_2 \in [\![\alpha]\!]$, *then* $\omega_1 \leq_\Psi \omega_2 \Leftrightarrow \omega_1 \leq_{\Psi \square \alpha} \omega_2$

(S2)   *if* $\omega_1, \omega_2 \in [\![\neg\alpha]\!]$, *then* $\omega_1 \leq_\Psi \omega_2 \Leftrightarrow \omega_1 \leq_{\Psi \square \alpha} \omega_2$

(S3)   *if* $\omega_1 \in [\![\alpha]\!]$ *and* $\omega_2 \in [\![\neg\alpha]\!]$, *then* $\omega_1 \leq_\Psi \omega_2 \Rightarrow \omega_1 <_{\Psi \square \alpha} \omega_2$

(S4)   *if* $\omega_1 \in [\![\alpha]\!]$ *and* $\omega_2 \in [\![\neg\alpha]\!]$, *then* $\omega_1 <_\Psi \omega_2 \Rightarrow \omega_1 \leq_{\Psi \square \alpha} \omega_2$

(S5)   *if* $\omega_1 \in [\![\alpha]\!]$ *and* $\omega_2 \in [\![\neg\alpha]\!]$, *then* $\omega_2 \ll_\Psi \omega_1 \Rightarrow \omega_1 \leq_{\Psi \square \alpha} \omega_2$

holds and the following is satisfied:

$$[\![\Psi \blacksquare \alpha]\!] = \min([\![\alpha]\!], \leq_\Psi)$$

In the following section we use the basic ideas of (weak) improvement operators as a starting point for developing the weak decrement operators.

3   Weak Decrement Operators

A property of a contraction operator $-$ is that the success condition of contraction is instantaneously achieved, i.e., if α is believed in a state ($\alpha \in \text{Bel}(\Psi)$) then after the contraction with α, it is not believed any more ($\alpha \notin \text{Bel}(\Psi - \alpha)$). As a generalisation, we define hesitant contractions as operators who achieve the success condition of contraction after multiple consecutive applications.

Definition 2. *A belief change operator* $\circ$ *is called a* hesitant contraction operator *if the following postulate is fulfilled:*

(hesitance)   *if* $\alpha \not\equiv \top$, *then there exists* $n \in \mathbb{N}_0$ *such that* $\alpha \notin \text{Bel}(\Psi \circ^n \alpha)$

If $\circ$ is an hesitant contraction operator, then we define a corresponding operator $\bullet$ by $\Psi \bullet \alpha = \Psi \circ^n \alpha$, where $n = 0$ if $\alpha \equiv \top$, otherwise n is the smallest integer such that $\alpha \notin \text{Bel}(\Psi \circ^n \alpha)$.

The following Example 1 shows a modelling application for hesitant belief change operators.

Example 1. Addison bought a new mobile with much easier handling. She does no longer have to press a sequence of buttons to access her favourite application. However, it takes multiple changes of her epistemic state before she contracts the belief of having to press the sequence of buttons for her favourite application.

We now introduce weak decrement operators, which fulfil AGM-like contraction postulates, adapted for the decrement of beliefs.

Definition 3 (Weak Decrement Operator). *A belief change operator $\circ$ is called a* weak decrement operator *if the following postulates are fulfilled:*

(D1) $Bel(\Psi \bullet \alpha) \subseteq Bel(\Psi)$

(D2) *if* $\alpha \notin Bel(\Psi)$*, then* $Bel(\Psi) \subseteq Bel(\Psi \bullet \alpha)$

(D3) $\circ$ *is a hesitant contraction operator*

(D4) $Bel(\Psi) \subseteq Cn(Bel(\Psi \bullet \alpha) \cup \{\alpha\})$

(D5) *if* $\alpha_1 \equiv \beta_1, ..., \alpha_n \equiv \beta_n$*, then* $Bel(\Psi \circ \alpha_1 \circ ... \circ \alpha_n) = Bel(\Psi \circ \beta_1 \circ ... \circ \beta_n)$

(D6) $Bel(\Psi \bullet \alpha) \cap Bel(\Psi \bullet \beta) \subseteq Bel(\Psi \bullet (\alpha \wedge \beta))$

(D7) *if* $\beta \notin Bel(\Psi \bullet (\alpha \wedge \beta))$*, then* $Bel(\Psi \bullet (\alpha \wedge \beta)) \subseteq Bel(\Psi \bullet \beta)$

The postulates (D1) to (D7) correspond to the postulates (C1) to (C7). By (D1) a weak decrement does not add new beliefs, and together with (D2) the beliefs of an agent are not changed if α is not believed priorly. (D3) ensures that after enough consecutive application a belief α is removed. (D4) is the recovery postulate, stating that removing α and then adding α again recovers all initial beliefs. The postulate (D5) ensures syntax independence in the case of iteration. (D6) and (D7) state that a contraction of a conjunctive belief is constrained by the results of the contractions with each of the conjuncts alone.

For the class of weak decrement operators the following representation theorem holds:

Theorem 1 (Representation Theorem: Weak Decrement Operators).
Let $\circ$ be a belief change operator. Then the following items are equivalent:

(a) $\circ$ is a weak decrement operator
(b) there exists a strong faithful assignment $\Psi \mapsto \leq_\Psi$ with respect to $\circ$ such that:

   (decrement sucess)
   there exists $n \in \mathbb{N}_0$ such that $[\![\Psi \circ^n \alpha]\!] = [\![\Psi]\!] \cup \min([\![\neg \alpha]\!], \leq_\Psi)$
   and n is the smallest integer such that $[\![\Psi \circ^n \alpha]\!] \not\subseteq [\![\alpha]\!]$

From Theorem 1 we easily get the following corollary:

Corollary 1. *If $\circ$ is a weak decrement operator, then $\bullet$ fulfils (C1) to (C7). Furthermore, every belief change operator that fulfils (C1) to (C7) and (D5) is a weak decrement operator.*

This shows that weak decrement operators are (up to (D5)) a generalisation of AGM contraction for epistemic states in the sense of Proposition 1.

4 Decrement Operators

We now introduce an ordering on the formulas in order to shorten our notion in the following postulates.

Definition 4. *Let* $\circ$ *be a hesitant contraction operator, then we define for every epistemic state* Ψ *and every two formula* α, β:

$$\alpha \preceq_\Psi^\circ \beta \text{ iff } Bel(\Psi \bullet \alpha\beta) \subseteq Bel(\Psi \bullet \alpha)$$

With $\prec_\Psi^\circ$ *we denote the strict variant of* $\preceq_\Psi^\circ$ *and define* $\alpha \lll_\Psi^\circ \beta$ *if* $\alpha \prec_\Psi^\circ \beta$ *and there is no* γ *such that* $\alpha \prec_\Psi^\circ \gamma \prec_\Psi^\circ \beta$.

Intuitively $\alpha \prec_\Psi^\circ \beta$ means that in the state Ψ the agent is more willing to give up the belief α than the belief β.

For the iteration of decrement operators we give the following postulates:

(D8)   if $\neg\alpha \models \beta$, then $Bel(\Psi \circ \alpha \bullet \beta) =_\alpha Bel(\Psi \bullet \beta)$

(D9)   if $\alpha \models \beta$, then $Bel(\Psi \circ \alpha \bullet \beta) =_{\neg\beta} Bel(\Psi \bullet \beta)$

(D10)   if $\alpha \models \gamma$, then $\Psi \circ \alpha \bullet \beta \models \gamma \Rightarrow \Psi \bullet \beta \models \gamma$

(D11)   if $\neg\alpha \models \gamma$, then $\Psi \bullet \beta \models \gamma \Rightarrow \Psi \circ \alpha \bullet \beta \models \gamma$

(D12)   if $\alpha \models \beta$ and $\neg\alpha \models \gamma$, then $\gamma \lll_\Psi^\circ \beta \Rightarrow \beta \preceq_{\Psi \circ \alpha}^\circ \gamma$

(D13) $Bel(\Psi \circ \alpha) \subseteq Bel(\Psi)$

(D8) states that a prior decrement with α does not influence the beliefs of an decrement with β if $\neg\alpha \models \beta$. states that a prior decrement with α does not influence the beliefs of an decrement with β if $\alpha \models \beta$. The postulate (D10) states that if a belief in γ is believed after a decrement of α and the removal of β, then only a removal of β does not influence the belief in γ if $\alpha\neg\gamma$ implies β. By (D11), if γ and α do not share anything, then a decrease of α does not influence this belief. By (D12), if in the state Ψ the agent prefers removing a consequence of $\neg\alpha$ minimally more than removing a consequence of α, then after a decrement of α, she is more willing to remove the consequence of α. The postulate (D13) axiomatically enforces that a single step does not add beliefs.

We call operators that fulfil these postulates decrement operators.

Definition 5 (Decrement Operator). *A* $\circ$ *weak* decrement *operator is called a* decrement operator *if* $\circ$ *satisfies* (D8) – (D13).

On the semantic side, we define a specific form of strong faithful assignment which implements decrementing on total preorders.

Definition 6 (Decreasing Assignment). *Let* $\circ$ *be a hesitant belief change operator. A strong faithful assignment* $\Psi \mapsto \leq_\Psi$ *with respect to* $\circ$ *is said to be a* decreasing assignment *(with respect to* $\circ$) *if the following postulates are satisfied:*

(DR8)   if $\omega_1, \omega_2 \in [\![\alpha]\!]$, then $\omega_1 \leq_\Psi \omega_2 \Leftrightarrow \omega_1 \leq_{\Psi \circ \alpha} \omega_2$

(DR9)   if $\omega_1, \omega_2 \in [\![\neg\alpha]\!]$, then $\omega_1 \leq_\Psi \omega_2 \Leftrightarrow \omega_1 \leq_{\Psi \circ \alpha} \omega_2$

(DR10)   if $\omega_1 \in [\![\neg\alpha]\!]$ and $\omega_2 \in [\![\alpha]\!]$, then $\omega_1 \leq_\Psi \omega_2 \Rightarrow \omega_1 \leq_{\Psi \circ \alpha} \omega_2$

(DR11)   if $\omega_1 \in [\![\neg\alpha]\!]$ and $\omega_2 \in [\![\alpha]\!]$, then $\omega_1 <_\Psi \omega_2 \Rightarrow \omega_1 <_{\Psi \circ \alpha} \omega_2$

(DR12)   if $\omega_1 \in [\![\neg\alpha]\!]$ and $\omega_2 \in [\![\alpha]\!]$, then $\omega_2 \ll_\Psi \omega_1 \Rightarrow \omega_1 \leq_{\Psi \circ \alpha} \omega_2$

(DR13)   if $\omega_1 \in [\![\neg\alpha]\!]$, $\omega_2 \in [\![\alpha]\!]$ and $\omega_2 \leq_\Psi \omega_3$ for all ω_3, then $\omega_2 \leq_{\Psi \circ \alpha} \omega_1$

The postulates (DR8) to (DR11) are the same as given by Konieczny and Pino Pérez [14] for iterated contraction (cf. Proposition 2). The postulate (DR12) states that a world of $\neg\alpha$ which is minimally less plausible than a world of α should be made at least as plausible as this world of α. (DR13) ensures that (together with the other postulates) that world in $[\![\Psi]\!]$ stays plausible after a decrement.

The main result is that decrement operators are exactly those which are compatible with a decreasing assignment.

Theorem 2 (Representation Theorem: Decrement Operators). *Let $\circ$ be a belief change operator. Then the following items are equivalent:*

(a) $\circ$ is a decrement operator

(b) there exists a decreasing assignment $\Psi \mapsto \leq_\Psi$ with respect to $\circ$ that satisfies (decrement sucess), i.e.:

$$\text{there exists } n \in \mathbb{N}_0 \text{ such that } [\![\Psi \circ^n \alpha]\!] = [\![\Psi]\!] \cup \min([\![\neg\alpha]\!], \leq_\Psi)$$
$$\text{and } n \text{ is the smallest integer such that } [\![\Psi \circ^n \alpha]\!] \not\subseteq [\![\alpha]\!]$$

Table 1. Example changes by two decrement operators $\circ_1$ and $\circ_2$.

	Ψ_1		$\Psi_1 \circ_1 a$		$\Psi_1 \circ_2 a$	
Layer 2	$a\neg b$	$\neg a \neg b$			$a\neg b$	
Layer 1		$\neg ab$	$a\neg b$	$\neg a \neg b$		$\neg a \neg b$
Layer 0 $[\![\Psi]\!]$	ab		ab	$\neg ab$	ab	$\neg ab$

The following proposition presents a nice property of decrement operators: Like AGM contraction for epistemic sates (cf. Proposition 1) a decrement operators keeps plausible worlds; and only the least unplausible counter-worlds may become plausible.

Proposition 5. *Let $\circ$ be a hesitant belief change operator. If there exists a decreasing assignment $\Psi \mapsto \leq_\Psi$ with respect to $\circ$, then we have:*

(partial success) $[\![\Psi]\!] \subseteq [\![\Psi \circ \alpha]\!] \subseteq [\![\Psi]\!] \cup \min([\![\neg\alpha]\!], \leq_\Psi)$

5 Specific Decrement Operators

Unlike improvement operators [13], there is no unique decrement operator. The reason for this is, that if $w_2 \simeq_\Psi w_1$ for $w_1 \in [\![\neg\alpha]\!]$ and $w_2 \in [\![\alpha]\!]$, and it is not required otherwise by (DR12), then the relative plausibility of w_1 and w_2 might not be changed by a decrement operator $\circ$, i.e. $w_2 \simeq_{\Psi \circ \alpha} w_1$. Example 2 demonstrates this.

Example 2. Let $\Sigma = \{a, b\}$ and Ψ_1 be an epistemic state as given in Table 1. Then the change from Ψ_1 to $\Psi_1 \circ_2 a$ in Table 1 is a valid change by a decrement operator. Likewise, the change from Ψ_1 to $\Psi_1 \circ_2 a$ from Table 1 is also a valid change for a decrement operator.

We capture this observation by two types of decrement operators. In the first case, the decrement operator improves the plausibility of a counter-model whenever it is possible.

Definition 7 (Type-1 Decrement Operator). *A decrement operator $\circ$ is a type-1 decrement operator if there exists a decreasing assignment $\Psi \mapsto \leq_\Psi$ with:*

(DR14)   *if $\omega_1 \in [\![\neg\alpha]\!]$ and $\omega_2 \in [\![\alpha]\!]$, then $\omega_2 \simeq_\Psi \omega_1 \Rightarrow \omega_1 \ll_{\Psi\circ\alpha} \omega_2$*

The second type of decrement operators keeps the order $\omega_1 \simeq_\Psi \omega_2$ whenever possible. We capture the cases when this is possible by the following notion. If $\leq\ \subseteq \Omega \times \Omega$ is a total preorder on worlds, we say ω_1 is *frontal* with respect to α, if (1) there is no $\omega_3 \in [\![\alpha]\!]$ such that $\omega_3 \ll \omega_1$, and (2) there is no $\omega_3 \in [\![\neg\alpha]\!]$ such that $\omega_1 \ll \omega_3$. We define the second type of decrement operators as follows.

Definition 8 (Type-2 Decrement Operator). *A decrement operator $\circ$ is a type-2 decrement operator if there exists a decreasing assignment $\Psi \mapsto \leq_\Psi$ with:*

(DR15)
if $\omega_1 \in [\![\neg\alpha]\!]$, $\omega_2 \in [\![\alpha]\!]$ and ω_1 is frontal w.r.t α, then $\omega_2 \simeq_\Psi \omega_1 \Rightarrow \omega_2 \simeq_{\Psi\circ\alpha} \omega_1$

Example (continuation of Example 2). The change from Ψ_1 to $\Psi_1 \circ_1 a$ in Table 1 can be made by a type-1 decrement operator, but not by a type-2 decrement operator. Conversely, the change from Ψ_1 to $\Psi_1 \circ_2 a$ from Table 1 can be made by a type-2 decrement operator, but not by a type-1 decrement operator.

6   Discussion and Future Work

We provide postulates and representation theorems for gradual variants of AGM contractions in the Darwich-Pearl framework of epistemic states. These so-called weak decrement operators are a generalisation of AGM contraction for epistemic states. Additionally, we give postulates for intended iterative behaviour of these operators, forming the class of decrement operators. For both classes of operators we presented a representation theorem in the framework of total preorders. For the definition of the postulates, the new relation $\preceq_\Psi^\circ$ (see Definition 4) is introduced. While $\preceq_\Psi^\circ$ is related to epistemic entrenchment [9], it can be shown that $\preceq_\Psi^\circ$ is not an epistemic entrenchment. The exploration of the exact nature of $\preceq_\Psi^\circ$ remains an open task.

The next natural step will be to investigate the interrelation between (weak) decrement operators and (weak) improvement operators. One approach is to generalize the Levi identity [15] and Haper identity [10] to these operators. Another approach could be the direct definition of a contraction operator from improvement operators, as suggested by Konieczny and Pino Pérez [13]. For such operators, after achieving success, a next improvement may make certain models unplausible, while a decrement operator keeps the plausibility. While this already indicated a difference between the operators, the study of their specific interrelationship is part of future work. Another goal for future work is to generalize (weak) decrement operators to a more general class of gradual change operators [16]. Such operators are candidates for a formalisation of psychologically inspired forgetting operations. An immediate target towards this goal is to take a closer look at subclasses and interrelate them with the taxonomy of improvement operators [12].

Acknowledgements. We thank the reviewers for their valuable hints and comments that helped us to improve the paper and we thank Gabriele Kern-Isberner for fruitful discussions and her encouragement to follow the line of research leading to this paper. This work was supported by DFG Grant BE 1700/9-1 given to Christoph Beierle as part of the priority program "Intentional Forgetting in Organizations" (SPP 1921). Kai Sauerwald is supported by this Grant.

References

1. Alchourrón, C.E., Gärdenfors, P., Makinson, D.: On the logic of theory change: partial meet contraction and revision functions. J. Symb. Log. **50**(2), 510–530 (1985)
2. Beierle, C., Kern-Isberner, G., Sauerwald, K., Bock, T., Ragni, M.: Towards a general framework for kinds of forgetting in common-sense belief management. KI - Künstliche Intelligenz **33**(1), 57–68 (2019)
3. Booth, R., Fermé, E.L., Konieczny, S., Pino Pérez, R.: Credibility-limited improvement operators. In: Schaub, T., Friedrich, G., O'Sullivan, B. (eds.) ECAI 2014–21st European Conference on Artificial Intelligence, Frontiers in Artificial Intelligence and Applications, 18–22 August 2014, Prague, Czech Republic, vol. 263, pp. 123–128. IOS Press (2014)
4. Caridroit, T., Konieczny, S., Marquis, P.: Contraction in propositional logic. In: Destercke, S., Denoeux, T. (eds.) ECSQARU 2015. LNCS (LNAI), vol. 9161, pp. 186–196. Springer, Cham (2015). https://doi.org/10.1007/978-3-319-20807-7_17
5. Chopra, S., Ghose, A., Meyer, T.A., Wong, K.-S.: Iterated belief change and the recovery axiom. J. Philosophical Logic **37**(5), 501–520 (2008)
6. Darwiche, A., Pearl, J.: On the logic of iterated belief revision. Artif. Intell. **89**, 1–29 (1997)
7. Eiter, T., Kern-Isberner, G.: A brief survey on forgetting from a knowledge representation and reasoning perspective. KI - Künstliche Intelligenz **33**(1), 9–33 (2019)
8. Fermé, E., Wassermann, R.: On the logic of theory change: iteration of expansion. J. Braz. Comp. Soc. **24**(1), 8:1–8:9 (2018)
9. Gärdenfors, P.: Knowledge in Flux: Modeling the Dynamics of Epistemic States. MIT Press, Cambridge (1988)
10. Harper, W.L.: Rational conceptual change. In: PSA: Proceedings of the Biennial Meeting of the Philosophy of Science Association, vol. 1976, pp. 462–494 (1976)
11. Katsuno, H., Mendelzon, A.O.: Propositional knowledge base revision and minimal change. Artif. Intell. **52**(3), 263–294 (1992)

12. Konieczny, S., Grespan, M.M., Pérez, R.P.: Taxonomy of improvement operators and the problem of minimal change. In: Lin, F., Sattler, U., Truszczynski, M., (eds.) Principles of Knowledge Representation and Reasoning: Proceedings of the Twelfth International Conference, KR 2010, Toronto, Ontario, Canada, 9–13 May, 2010. AAAI Press (2010)
13. Konieczny, S., Pérez, R.P.: Improvement operators. In: Brewka, G., Lang, J., (eds.) Principles of Knowledge Representation and Reasoning: Proceedings of the Eleventh International Conference, KR 2008, Sydney, Australia, 16–19 September, 2008, pp. 177–187. AAAI Press (2008)
14. Konieczny, S., Pino Pérez, R.: On iterated contraction: syntactic characterization, representation theorem and limitations of the levi identity. In: Moral, S., Pivert, O., Sánchez, D., Marín, N. (eds.) SUM 2017. LNCS (LNAI), vol. 10564, pp. 348–362. Springer, Cham (2017). https://doi.org/10.1007/978-3-319-67582-4_25
15. Levi, I.: Subjunctives, dispositions and chances. Synthese **34**(4), 423–455 (1977)
16. Sauerwald, K.: Student research abstract: modelling the dynamics of forgetting and remembering by a system of belief changes. In: The 34th ACM/SIGAPP Symposium on Applied Computing (SAC 2019), 8–12 April , 2019, Limassol, Cyprus, New York. ACM (2019)

Learning and Decision Making

A Novel Document Generation Process for Topic Detection Based on Hierarchical Latent Tree Models

Peixian Chen[1,2], Zhourong Chen[1], and Nevin L. Zhang[1(✉)]

[1] The Hong Kong University of Science and Technology, Kowloon, Hong Kong
zchenbb@connect.ust.hk, lzhang@cse.ust.hk
[2] Lenovo Machine Intelligence Center, Shenzhen, Hong Kong
pchen15@lenovo.com

Abstract. We propose a novel document generation process based on hierarchical latent tree models (HLTMs) learned from data. An HLTM has a layer of observed word variables at the bottom and multiple layers of latent variables on top. For each document, the generative process first samples values for the latent variables layer by layer via logic sampling, then draws relative frequencies for the words conditioned on the values of the latent variables, and finally generates words for the document using the relative word frequencies. The motivation for this work is to take word counts into consideration with HLTMs. In comparison with LDA-based hierarchical document generation processes, the new process achieves drastically better model fit with much fewer parameters. It also yields more meaningful topics and topic hierarchies. It is the new state-of-the-art for the hierarchical topic detection.

Keywords: Latent Tree Models · Topic detection

1 Introduction

The objective of *hierarchical topic detection* is, given a corpus of documents, to obtain a tree of topics with more general topics at high levels of the tree and more specific topics at low levels of the tree. Several hierarchical topic detection methods have been proposed based on latent Dirichlet allocation (LDA) [7], including the hierarchical latent Dirichlet allocation (hLDA) model [5], the hierarchical Pachinko allocation model (hPAM) [15,20], and the nested hierarchical Dirichlet process (nHDP) [23].

A very different method named *hierarchical latent tree analysis (HLTA)* is recently proposed by [10,11,16]. HLTA essentially learns a Bayesian network such as the one shown in Fig. 1, where there is a layer of observed variables at the bottom, and one or more layers of latent variables on top. The variables are connected up to form a tree. The model is hence called a *hierarchical latent tree model (HLTM)*. The observed variables are binary and represent the absence or presence of words in documents. The latent variables are also binary and

© Springer Nature Switzerland AG 2019
G. Kern-Isberner and Z. Ognjanović (Eds.): ECSQARU 2019, LNAI 11726, pp. 265–276, 2019.
https://doi.org/10.1007/978-3-030-29765-7_22

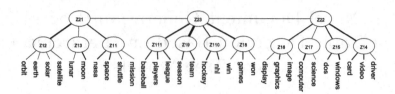

Fig. 1. An example hierarchical latent tree model from [11], learned by HLTA on a toy text dataset.

introduced during data analysis to explain co-occurrence patterns. For example, $z14$ explains the probabilistic co-occurrences of the words `card`, `video` and `driver`; $z16$ explains the co-occurrences of `display`, `graphics` and `image`; and $z22$ explains the probabilistic co-occurrence of the patterns represented by $z14$, $z15$, $z16$ and $z17$.

HLTMs is a generalization of latent class models (LCMs) [3], which is a type of finite mixture models for discrete data. In a finite mixture model, there is one latent variable and it is used to partition objects into soft clusters. Similarly, in an HLTM, each latent variable z partitions all the documents into two clusters. One of the clusters (corresponds to latent state $z = 1$) consists of the documents where the words in subtree rooted at the latent variable occur with much higher probabilities. It is interpreted as a topic. The other cluster ($z = 0$) is viewed as the background. Hence each latent variable gives one topic.

Topics given by some of the latent variables in Fig. 1 are listed below. For example, $z14$ gives a topic that consists of 12% of the documents, and the words `card`, `video` and `driver` occur with relatively high probabilities inside the topic and relatively low probabilities outside. Note that, for $z22$, only a subset of words in its subtree are used when characterizing the topic. The reader is referred to [10] for how the words for characterizing a topic are picked and ordered.

> $z22$ [0.24] `windows card graphics video dos`
> $z14$ [0.12] `card video driver`
> $z15$ [0.15] `windows dos`
> $z16$ [0.10] `graphics display image`
> $Z17$ [0.09] `computer science`

In general, latent variables at high levels of an HLTM capture "long-range" word co-occurrence patterns and hence give thematically more general topics, while those at low levels capture "short-range" word co-occurrence patterns with thematically more specific topics. For example, the topic given by $z22$ concerns several aspects of computers, while its subtopics are each concerned with only one aspect of computers. Hence HLTA is a tool for hierarchical topic detection.

HLTA differs fundamentally from the LDA-based methods. However, comparisons between them are still possible. The reason is that they both define distributions over documents and characterize topics using lists of words. Empirical results reported by [10,11] show that HLTA significantly outperforms the

LDA-based methods in terms of model quality as measured by held-out likelihood, and it finds more meaningful topics and topic hierarchies.

It should be noted, however, that the aforementioned comparisons were conducted only on binary data. The reason is that HLTA is unable to take word counts into consideration. In the experiments, documents were represented as binary vectors over the vocabulary for HLTA. For the LDA-based methods, they were represented as bags of words, where duplicates were removed such that no word appears more than once. The two representations are equivalent.

To amend the serious drawback, this paper extends HLTA so as to take word counts into consideration. Specifically, we propose a document generation model based on the model structure learned by HLTA from binary data, design a parameter learning algorithm for the new model, and give an importance sampling method for model evaluation. The new method is named HLTA-c, where the letter "c" stands for count data. We present empirical results to show that, on count data, HLTA-c also significantly outperforms LDA-based methods in terms of both model quality and meaningfulness of topics and topic hierarchies.

2 Related Work

Detecting topics and topic hierarchies from large archives of documents has been one of the most active research areas in last decade. The most commonly used method is latent Dirichlet allocation (LDA) [7]. LDA has been extended in various ways for additional modeling capabilities. Topic correlations are considered in [14,15]; topic evolution is modeled in [1,6,27]; topic hierarchies are built in [4,15,20]; side information is exploited in [2,13]; and so on.

A fundamental difference between HLTA/HLTA-c and the LDA-based methods for hierarchical topic detection is that observed variables in HLTA/HLTA-c correspond to words in the vocabulary, while those in the LDA-based methods correspond to tokens in the documents. The use of word variables allows the detection and representation of patterns of word co-occurrences qualitatively using model structures as illustrated in Fig. 1.

Another important difference is in the definition and characterization of topics. Topics in the LDA-based methods are probabilistic distributions over a vocabulary. When presented to users, a topic is characterized using a few words with the highest probabilities. In contrast, topics in HLTA/HLTA-c are clusters of documents. For presentation to users, a topic is characterized using the words that not only occur with high probabilities in the topic but also occur with low probabilities outside the topic.

The document generation process of HLTA/HLTA-c differs fundamentally with LDA and its variants. Generally, HLTA/HLTA-c assumes a document is generated in two steps: (1) Decide to which of the clusters it belongs to; and (2) generate individual words according to the characteristics of the clusters.

3   Document Generation Process

Our HLTM-c model for document generation is illustrated in Fig. 2(b). It is based on an HLTM M_b learned from binary data.

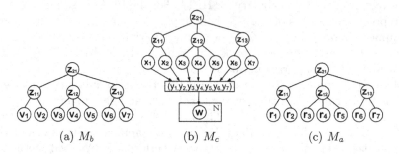

(a) M_b (b) M_c (c) M_a

Fig. 2. The model M_b is learned by HLTA from binary data. Each v_i is binary variable and corresponds to the i-th word in the vocabulary. Its distribution given its parent is a Bernoulli distribution. The model M_c is for modeling count data and defines a document generation process. Each x_i takes its value from the interval $[0, 1]$ and its distribution given its parent is a truncated Normal distribution. The y'_i are obtained by normalizing the x_i's. The model M_a is a auxiliary model used in parameter estimation and model evaluation. It shares the same parameters as M_c, except that the variables r_i are observed and are not restricted to the interval $[0, 1]$.

We regard M_b as a model that generates binary vectors over words as follows: First, pick values for the latent variables using logic sampling [24]. Specifically, sample a value for the root z_{21} from its marginal distribution $P(z_{21})$ and sample values for other latent variables from their conditional distributions given their respective parents. For example, the value for z_{11} is sampled from the distribution $P(z_{11}|z_{21})$. Then, sample a binary value for each of the word variables $v_1, \ldots, v_7$ from the its conditional distribution given its parent such as $P(v_1|z_{11})$. Note that all distributions mentioned here are Bernoulli distributions.

In the HLTM-c model M_c, we generate values for the latent variables in the same way as in M_b. However, we do not sample binary values for word variables. Instead, we draw a real value x_i for each word v_i. The value x_i is restricted to lie in the interval $[0, 1]$ and it is meant to be, approximately, the relative frequency of the word v_i in a document. Let z be the parent of x_i. We draw x_i from a truncated normal distribution $P(x_i|z) = \mathcal{TN}(\mu_{iz}, \sigma_{iz}^2, 0, 1)$ with mean μ_{iz} and standard deviation σ_{iz}. Note that, for a fixed x_i, both the notation μ_{iz} and σ_{iz} refer to two numerical values, one for $z = 0$ and another for $z = 1$.

Different x_i's are drawn independently and hence there is no guarantee that they sum to 1. We normalize them to get a multinomial distribution $(y_1, \ldots, y_7)$. Finally, we draw words W for a document from this multinomial distribution.

In general, suppose there is a collection D of documents and there are V words in the vocabulary. Assume an HLTM M_b with binary word variables has been learned from the binary version of the data using the HLTA algorithm described in [10]. To take word counts into consideration, we turn M_b into a document generation model M_c by replacing the binary word variables v_i with real-valued variables x_i, and we assume a document of length N is generated using M_c as follows:1, Draw values of the binary latent variables via logic sampling; 2, For each $i \in \{1, \ldots, V\}$, draw x_i from the conditional distribution $p(x_i|pa(x_i))$, a truncated normal distribution, of x_i given its parent $pa(x_i)$; 3, For each $i \in \{1, \ldots, V\}$, set $y_i = x_i / \sum_{i=1}^{n} x_i$; 4, For each $n \in \{1, \ldots, N\}$, draw the n-th word of the document from $Multi(y_1, \ldots, y_n)$.

In the generation process, Step 2 generates the *unnormalized relative frequencies (URF)* $\mathbf{x} = (x_1, \ldots, \mathbf{x}_V)$, while Step 3 obtains the (normalized) *relative frequencies* $\mathbf{y} = (y_1, \ldots, y_n)$. The first three steps define a distribution over all possible the relative frequency vectors, i.e., over the probability simplex $S = \{(y_1, \ldots, y_n)|y_i \geq 0, \sum_{i=1}^{n} y_i = 1\}$. We denote the distribution as $p(\mathbf{y}|M_c, \theta)$. The parameter vector θ includes the parameters for the distributions of all binary latent variables, and the means and standard deviations for the truncated normal distributions for the URF variables.

Let N_i be the number of times the i-th word from the vocabulary occurs in a document d. The count vector $\mathbf{c} = (N_1, N_2, \ldots, N_V)$ can be used as a representation of d. The sum of those counts is the document length, i.e., $N = \sum_{i=1}^{V} N_i$. The conditional probability of d given the relative frequencies $\mathbf{y}$ is:

$$P(d|\mathbf{y}) = \frac{N!}{N_1! \ldots N_V!} \prod_{i=1}^{V} y_i^{N_i}. \tag{1}$$

The entire generation process defines a distribution over documents. The probability of a document d is:

$$P(d|M_c, \theta_c) = \int_{\mathbf{y}} P(d|\mathbf{y}) p(\mathbf{y}|M_c, \theta) d\mathbf{y}. \tag{2}$$

In HLTA-c, a cluster is characterized using the relative frequencies x_i that word v_i occurs. For each latent variable z, $z = 1$ corresponds to the document cluster on topic, where the words in subtree appear more often, and $z = 0$ corresponds to the documents off the topic. The relative frequencies y_i of words v_i in a document is obtained by normalizing the relevant $x_i's$. Note that the $x_i's$ are w.r.t document clusters, and the $y_i's$ are w.r.t a document to be generated.

The document generation process given here is very different in flavor from the generation processes one typically sees in the LDA literature. Nonetheless, it is a well-defined generation process by defining a distribution over count-vector representations of documents. An LDA-based model, on the other hand, defines a distribution over bag-of-words representations of documents. Because the count-vector representation is equivalent to the bag-of-words representation, the two methods define distributions over the same collection of objects and hence can be compared with each other.

4   Parameter Estimation

We now consider how to estimate the parameters θ of the mode M_c. The log likelihood function of θ given a collection of documents D is:

$$\log P(D|M_c, \theta) = \sum_{d \in D} \log P(d|M_c, \theta). \tag{3}$$

The objective of parameter estimation is to find the value of θ that maximizes the likelihood function. This task is difficult because of the use of truncated normal distributions and the normalization step in the document generation process.

We propose an approximate method based on two ideas. First, notice that the model parameters θ influence the relative frequencies $\mathbf{y}$ and the word counts $\mathbf{c}$ indirectly through the URF variables $\mathbf{x}$. Given $\mathbf{x}$, $\mathbf{c}$ and $\mathbf{y}$ are independent of θ. Our first idea is to obtain a point estimate of $\mathbf{x}$ from $\mathbf{c}$ and regard $\mathbf{x}$ as observed variables afterwards.

It is well known that, given the word counts $\mathbf{c}$, the maximum likelihood estimation (MLE) of $\mathbf{y}$ is $y_i = N_i/N$ for all i, i.e., the empirical relative word frequencies. Ignoring the normalization step, we also use the empirical relative word frequencies as a point estimation for URF variables $\mathbf{x}$, i.e., we assume $\mathbf{x} = (N_1/N, \ldots, N_V/N)$.

The second idea is to relax the restriction that x_i must be from the interval $[0, 1]$, and to assume that x_i is sampled from a normal distribution $\mathcal{N}(\mu_{iz}, \sigma_{iz}^2)$ instead of a truncated normal distribution $\mathcal{TN}(\mu_{iz}, \sigma_{iz}^2, 0, 1)$.

Those two considerations turn the problem of estimating θ in M_c with data represented as count vectors into the problem of estimating θ in a related model, denoted as M_a, with data represented as vectors of relative frequencies. As shown in Fig. 2(c), the auxiliary model M_a is the same as the top part of M_c, except that the URF variables x_i's are replaced with real-value variables r_i's. The conditional distribution of each r_i given its parent z is a normal distribution, i.e., $p(r_i|z) = \mathcal{N}(\mu_{iz}, \sigma_{iz}^2)$.

Use d_f to denote the vector of relative word frequencies in a document d, i.e., $d_f = (N_1/N, \ldots, N_V/N)$. Moreover, use D_f to denote the entire data set when represented as vectors of relative frequencies. In the auxiliary model M_a, the log likelihood of θ given D_f is

$$\log P(D_f|M_a, \theta) = \sum_{d_f \in D_f} \log P(d_f|M_a, \theta). \tag{4}$$

Maximizing this likelihood function is relatively easy because M_a is a tree model. It can be done using the EM algorithm.

There are strong reasons to believe that maximizing (4) would result in high quality parameter estimation for the generative model M_c due to the way the approximation is derived. Empirical results to be presented later show that the method does produce good enough parameter estimations for M_c to achieve substantially higher held-out likelihood than the LDA-based methods.

Although M_a is a tree model, EM can still be very time consuming when the sample size is large. Here we use stepwise EM [8,25], which scales much better by applying the idea of stochastic gradient descent to EM.

5   Model Evaluation

After obtaining an estimation θ^* of the parameters in the document generation model M_c, we need to evaluate it by calculating its log likelihood on a test set D_t:

$$\log P(D_t | M_g, \theta^*)) = \sum\nolimits_{d \in D_t} \log P(d | M_c, \theta^*).$$

To calculate the probability $\log P(d | M_c, \theta^*)$ of a test document d, we need to approximately compute the integral in (2). Since the distribution of $P(\mathbf{y} | M_c, \theta^*)$ is defined through a generative process, it is straightforward to obtain samples of $\mathbf{y}$ by running the process multiple times. Suppose K samples $\mathbf{y}^{(1)}, \ldots, \mathbf{y}^{(K)}$ of $\mathbf{y}$ are obtained. We can estimate $P(d | M_c, \theta^*)$ as follows:

$$P(d | M_c, \theta^*) \approx \frac{1}{K} \sum\nolimits_{k=1}^{K} P(d | \mathbf{y}^{(k)}). \tag{5}$$

Unfortunately, there is a well-known problem with this naive method [26]. When the document d is long, the integrand $p(d | \mathbf{y})$ as a function of $\mathbf{y}$ is highly peaked around the MLE of $\mathbf{y}$ and is very small elsewhere in the probability simplex. Because the document d is not taken into consideration when drawing samples of $\mathbf{y}$, it is unlikely for the samples to hit the high value area. This can easily lead to underestimation and high variance. Unless K is extremely large, there could be large differences in the estimates one obtains at different runs.

A standard way to solve this problem is to use *importance sampling* [22], and to utilize a proposal distribution that is related to d and has its density concentrated in the region where the integrand function $P(d | \mathbf{y})$ is not close to zero. We derive such a distribution using the auxiliary model M_a.

Let $\mathbf{z}$ be the set of all latent variables in the auxiliary model M_a on the level right above the r_i's. For each latent variable z in $\mathbf{z}$, it is easy to compute the posterior distribution $p(z | d_f, M_a, \theta^*)$ of z given a document d (represented as a vector of relative word frequencies d_f) because M_a is a tree model. In fact, it can be done in linear time using message propagation. We define

$$q(\mathbf{z} | d) = \prod\nolimits_{z \in \mathbf{z}} p(z | d_f, M_a, \theta^*). \tag{6}$$

Note that $\mathbf{z}$ (defined in M_a) is the same as the set of all latent variables in the generative model M_c on the level right above the x_i's. We rewrite (2) as follows for the test document d: $P(d | M_c, \theta^*) = \int \sum_{\mathbf{z}} P(d | \mathbf{y}) p(\mathbf{y} | \mathbf{z}) p(\mathbf{z} | M_c, \theta^*) d\mathbf{y}$. Note that $p(\mathbf{z} | M_c, \theta^*) = p(\mathbf{z} | M_a, \theta^*)$. Inserting $\frac{q(\mathbf{z} | d)}{q(\mathbf{z} | d)}$ into the right hand side and rearranging terms, we get $P(d | M_c, \theta^*) = \int \sum_{\mathbf{z}} P(d | \mathbf{y}) \frac{p(\mathbf{z} | M_a, \theta^*)}{q(\mathbf{z} | d)} p(\mathbf{y} | \mathbf{z}) q(\mathbf{z} | d) d\mathbf{y}$.

This expression implies that we can sample a sequence of pairs $(\mathbf{y}^{(1)}, \mathbf{z}^{(1)}), \ldots, (\mathbf{y}^{(K)}, \mathbf{z}^{(K)})$ from $p(\mathbf{y} | \mathbf{z}) q(\mathbf{z} | d)$, and estimate $P(d | M_c, \theta^*)$ as follows:

$$P(d | M, \theta^*) \approx \frac{1}{K} \sum\nolimits_{k=1}^{K} P(d | \mathbf{y}^{(k)}) \frac{p(\mathbf{z}^{(k)} | M_a, \theta^*)}{q(\mathbf{z}^{(k)})}. \tag{7}$$

Note that $P(d|\mathbf{y}^{(k)})$ can be calculated using (1) and the term $p(\mathbf{z}^{(k)}|M_a,\theta^*)$ is obtained using message propagation in M_a. As mentioned earlier, the term $q(\mathbf{z}^{(k)})$ can also be easily computed in M_a.

In comparison with (5), the use of (7) improves estimation accuracy and reduces the variance because the sample points $(\mathbf{y}^{(k)},\mathbf{z}^{(k)})$ are generated by taking the test document d into consideration. Hence, the samples are more likely to hit the area where the integrand function $P(d|\mathbf{y})$ has high values. In the experiments, we set $K = 300$.

6   Empirical Results

In this section, we present empirical results to compare HLTA-c with common LDA-based methods for hierarchical topic detection, including hPAM [20], hLDA [5] and nHDP [23]. The comparisons are in terms of both model quality and the quality of topics and topic hierarchies. Model quality can be measured using held-out likelihood on test data as pointed out in [10,17]. The quality of topics is assessed using topic coherence [21] and topic compactness [12]. Example branches of the topic hierarchies obtained by nHDP and HLTA-c are also included for qualitative comparisons.

Table 1. Per-document held-out log likelihood scores. The sign "-" indicates non-termination after 96 h.

	NIPS-1k	NIPS-5k	NIPS-10k	News-1k	News-5k	NYT	AI
HLTA-c	−1,182±2	−2,658±1	−3,249±2	−183±1	−383±2	−1,255±3	−3,216±3
hLDA	−2,951±35	−5,626±117	—	—	—	—	—
nHDP	−3,273±6	−7,169±11	−8,318±18	−262±1	−565±3	−2,070±6	−7,606± 12
hPAM	−3,196±3	−6,759±15	−7,922 ± 12	−255±2	−556±4	—	—

Table 2. Average coherence scores.

	NIPS-1k	NIPS-5k	NIPS-10k	News-1k	News-5k	NYT	AI
HLTA-c	−6.46±0.01	−8.20±0.02	−8.93±0.04	−12.50±0.08	−13.43±0.15	−12.70 ± 0.18	−16.18 ± 0.15
hLDA	−7.46±0.31	−9.03±0.16	—	—	—	—	—
nHDP	−7.66±0.23	−9.70±0.19	−10.89±0.38	−13.51±0.08	−13.93±0.21	−12.90±0.16	−18.66±0.21
hPAM	−6.86±0.08	−8.89±0.04	−9.74±0.04	−11.74±0.14	−14.06 ±0.09	—	—

6.1   Datasets and Settings

We used four datasets in our experiments: (1) NIPS dataset, which consists of 1,955 articles published at the NIPS conference between 1988 and 1999; (2) 20 Newsgroup dataset consisting of 19,940 newsgroup posts; (3) New York Times (NYT) dataset[1] with 300,000 articles published on New York Times between

---

[1]   NIPS: http://www.cs.nyu.edu/~roweis/data.html,   News: http://qwone.com/~jason/20Newsgroups/,   NYT: http://archive.ics.uci.edu/ml/datasets/Bag+of+Words.

1987 and 2007; and (4) AI dataset which includes all 24,307 papers published at seven AI conferences and three AI journals between 2000 to 2017.

To have some variabilities on the vocabulary size, we created different versions of the NIPS and the Newsgroup datasets by choosing vocabularies with different sizes using average TF-IDF. The NIPS dataset has three versions with vocabulary sizes of 1,000, 5,000 and 10,000 respectively, and the Newsgroup dataset has two versions with vocabulary sizes of 5,000 and 10,000. The NYT and AI datasets each have only one version with vocabulary size 10,000.

Implementations of HLTA and the LDA baselines were obtained from their authors[2]. HLTA-c was implemented on top of HLTA. The implementation will be released along with the publication of this paper. HLTA-c determined the height of topic hierarchy and the number of nodes at each level by running the HLTA at its default parameter settings on the binary version of a dataset. We tuned the parameters of the LDA-based baselines in such a way that they would yield roughly the same total number of topics as HLTA-c. The other parameters of the baselines were left at their default values.

HLTA-c needs to call stepwise EM in the parameter estimation step. Stepwise EM has a parameter called stepwise η_t, which we set as $\eta_t = (t + 2)^{-0.75}$ as is usually done in the literature.

Table 3. Average compactness scores.

	NIPS-1k	NIPS-5k	NIPS-10k	News-1k	News-5k	NYT	AI
HLTA-c	**0.228±0.001**	**0.255±0.001**	**0.243±0.001**	**0.219±0.001**	**0.226±0.001**	**0.288±0.009**	**0.229±0.001**
hLDA	0.163±0.003	0.153±0.001	—	—	—	—	—
nHDP	0.164±0.005	0.147±0.006	0.138±0.002	0.150±0.003	0.148±0.004	0.250±0.003	0.144±0.001
hPAM	0.211±0.003	0.167±0.001	0.141±0.002	0.210±0.006	0.178±0.002	—	—

6.2 Model Quality

We randomly divided each dataset into a training set with 80% of the data, and a test set with 20% of the data. The per-document log likelihood scores on test data are reported in Table 1.

The held-out likelihood scores for HLTA-c are drastically higher than those for all the baseline methods. On the NYT and AI datasets, the models produced by HLTA-c have scores of −1,255 and −3,216 respectively, while the models by nHDP have scores of −2,070 and −7,606. This implies that the models obtained by HLTA-c can predict unseen data much better. HLTA-c not only achieved much higher held-out likelihood scores than the baselines, but also did so with much fewer parameters.

---

[2] github.com/kmpoon/hlta; github.com/blei-lab/hlda; www.columbia.edu/~jwp21 28/code/nHDP.zip; www.arbylon.net/projects/knowceans-lda-cgen/Hpam2pGibbs Sampler.java.

6.3   Topic Quality

After an HLTM-c has been learned, we extract topics from it as described in [9]. Both HLTA-c and the LDA-based methods characterize topics using lists of words when presenting them to users. Direct comparisons are therefore possible. We measure the quality of a topic using two metrics. The first one is the *topic coherence score* [21]. The intuition behind this metric is that words in a good topic should tend to co-occur in the documents. Suppose a topic t is characterized by a list $\{w_1, w_2, \ldots, w_M\}$ of M words. The coherence score of t is given by: $\texttt{coherence}(t) = \sum_{i=2}^{M} \sum_{j=1}^{i-1} \log \frac{D(w_i, w_j) + 1}{D(w_j)}$, where $D(w_i)$ is the number of documents containing the word w_i, and $D(w_i, w_j)$ is the number of documents containing both w_i and w_j. Higher coherence score means better topic quality.

The second metric is the *topic compactness score* [12]. It is calculated on the basis of the word2vec model that was trained on a part of the Google News dataset [18,19]. The word2vec model maps each word into a vector that captures the semantic meaning of the word. The intuition behind the compactness score is that words in a good topic should be closely related semantically. The compactness score of a topic t is given by: $\texttt{compactness}(t) = \frac{2}{M(M-1)} \sum_{i=2}^{M} \sum_{j=1}^{i-1} S(w_i, w_j)$ where $S(w_i, w_j)$ is the cosine similarity between the vector representations of the words w_i and w_j. Words that do not occur in the *word2vec* model were simply skipped. Higher compactness score means better topic quality.

Both of the scores decrease with the length M of the word list. Some of the topics produced by HLTA-c consist of only 4 words. Hence, we set $M = 4$. Using a higher value for M would put the LDA-based methods at a disadvantage.

The average coherence and compactness scores are shown in Tables 2 and 3. HLTA-c achieved the highest compactness score in all cases. It also achieved the highest coherence score in most datasets. The differences between the scores for HLTA-c and other methods are often large.

6.4   Selected Branches of Topic Hierarchies

Figure 3 shows branches of the topic trees produced by HLTA-c and nHDP on the AI dataset that are related to neural networks and deep learning. In the HLTA-c topic tree, there is a topic on *neural network* and *deep learning*. The subtopic *neural network* in turn has subtopics on network *architecture* and training algorithms (*contrastive divergence* and *stochastic gradient descent*). The topic on *deep learning* has subtopics on *convolutional neural network, restricted Boltzmann machine, deep neural network*, and *autoencoder*. The names of several prominent deep learning authors appear in the topic descriptions. The topics *recurrent neural network* and *lstm* are placed in the first group instead of the second, which indicates that they co-occur more often with the *neural network* topics than the *deep learning* topics. A similar statement can be made about *word embedding*. The last three topics of nHDP clearly do not fit well with the other topics in the group.

neural-network layer deep architecture hinton deep-learning stochastic-gradient
  neural-network layer architecture stochastic-gradient hidden-layer
    contrastive-divergence hamiltonian welling boltzmann-machine visible
    stochastic-gradient-descent stochastic-optimization minibatch mini_batch
  ---
    layer architecture activation activate hidden-node two_layer
    neural-network hidden-layer hidden-unit recurrent-neural-network
    validation lstm schmidhuber encoder decoder recurrent-neural long_term
    learning-rate perceptron multilayer learn-rate perceptron-algorithm decay

deep hinton deep-learning pool convolution softmax convolutional deep-network
  softmax sutskever krizhevsky fine_tuning pre_trained representation-
  pool convolution convolutional imagenet convolutional-neural-network
  ---
    hinton salakhutdinov initialization restricted-boltzmann-machine deep-belief
    deep deep-learning deep-network deep-neural-network
    larochelle autoencoder vincent tetris cross_entropy deep-architecture

document text query category content precision wikipedia information-retrieval
  word-representation word-embedding mikolov vector-space

Layer neural-network deep architecture hinton deep-learning hidden-layer
  hidden-unit restricted-boltzmann-machine visible-unit contrastive –divergence
  neuron layer convolutional-layer compression activation imagenet residual
  filter layer pool convolution convolutional architecture ransformation
  stochastic variational generative-model estimator gradient recognition
  autoencoder reconstruction encoder code generative auto_encoder
  domain-adaptation transfer deep target-domain adaptation unsupervised
  epoch hyperparameter decay learning-rate schedule step-size momentum
  hash hash-code quantization pairwise hash-method deep hash-function

attention sentence generation token lstm recurrent-neural-network softmax
  embedding word-embedding recursive phrase word-vector sentence
  recurrent lstm gate recurrent-neural-network memory schmidhuber
  answer question-answering question-answer sentence convolution
  neural-network neuron connectionist binding cognitive-science
  embedding entity rank segment triple loss ensemble query
  curvature natural-gradient quasi_newton hessian adagrad
  neuron brain block response layer code activity spike activation
  memory store associative active register unitary item extended
  multi_task multitask-learning shared multi_task-learning multiple-task

Fig. 3. Selected branches of the topic trees produced by HLTA-c (left) and nHDP (right) on the AI dataset.

7 Concluding Remarks

HLTA is a recently proposed method for hierarchical topic detection. It can only deal with binary data. In this paper we extend HLTA to HLTA-c so as to take word counts into consideration. It is achieved by proposing a document generation process based on the model structure learned by HLTA. In comparison with LDA-based methods, HLTA-c achieves far better held-out likelihood with much fewer parameters, with significantly better topics and topic hierarchies. The work on HLTA and this paper serve to illustrate a strategy where one uses a simpler form of data for model structure learning and the full data for parameter learning. Such a strategy can lead to superior performances when compared with the practice where ones relies on manually constructed model structures.

Acknowledgements. Research on this article was supported by Hong Kong Research Grants Council under grants 16202515.

References

1. Ahmed, A., Xing, E.: Dynamic non-parametric mixture models and the recurrent Chinese restaurant process: with applications to evolutionary clustering. In: ICDM (2008)
2. Andrzejewski, D., Zhu, X., Craven, M.: Incorporating domain knowledge into topic modeling via Dirichlet forest priors. In: ICML (2009)
3. Bartholomew, D.J., Knott, M.: Latent Variable Models and Factor Analysis, 2nd edn. Arnold, New York (1999)
4. Blei, D.M., Griffiths, T., Jordan, M., Tenenbaum, J.: Hierarchical topic models and the nested Chinese restaurant process. In: NIPS (2004)
5. Blei, D.M., Griffiths, T., Jordan, M.: The nested Chinese restaurant process and Bayesian nonparametric inference of topic hierarchies. J. ACM **57**(2), 7:1–7:30 (2010)
6. Blei, D.M., Lafferty, J.D.: Dynamic topic models. In: ICML (2006)

7. Blei, D.M., Ng, A.Y., Jordan, M.I.: Latent Dirichlet allocation. J. Mach. Learn. Res. **3**, 993–1022 (2003)

8. Cappé, O., Moulines, E.: On-line expectation-maximization algorithm for latent data models. J. Roy. Stat. Soc. Seri. B (Stat. Method.) **71**(3), 593–613 (2009)

9. Chen, P., Chen, Z., Zhang, N.L.: A novel document generation process for topic detection based on hierarchical latent tree models. arXiv preprint arXiv:1712.04116 (2018)

10. Chen, P., Zhang, N.L., Liu, T., Poon, L.K., Chen, Z., Khawar, F.: Latent tree models for hierarchical topic detection. Artif. Intell. **250**, 105–124 (2017)

11. Chen, P., Zhang, N.L., Poon, L.K., Chen, Z.: Progressive EM for latent tree models and hierarchical topic detection. In: AAAI (2016)

12. Chen, Z., Zhang, N.L., Yeung, D., Chen, P.: Sparse Boltzmann machines with structure learning as applied to text analysis. In: AAAI (2017)

13. Jagarlamudi, J., Daumé III, H., Udupa, R.: Incorporating lexical priors into topic models. In: EACL (2012)

14. Lafferty, J., Blei, D.M.: Correlated topic models. In: NIPS (2006)

15. Li, W., McCallum, A.: Pachinko allocation: DAG-structured mixture models of topic correlations. In: ICML (2006)

16. Liu, T., Zhang, N.L., Chen, P.: Hierarchical latent tree analysis for topic detection. In: Calders, T., Esposito, F., Hüllermeier, E., Meo, R. (eds.) ECML PKDD 2014. LNCS (LNAI), vol. 8725, pp. 256–272. Springer, Heidelberg (2014). https://doi.org/10.1007/978-3-662-44851-9_17

17. Lubke, G., Neale, M.C.: Distinguishing between latent classes and continuous factors: resolution by maximum likelihood? Multivariate Behav. Res. **41**(4), 499–532 (2006). https://doi.org/10.1207/s15327906mbr4104_4. pMID: 26794916

18. Mikolov, T., Chen, K., Corrado, G., Dean, J.: Efficient estimation of word representations in vector space. In: International Conference on Learning Representations Workshops (2013)

19. Mikolov, T., Sutskever, I., Chen, K., Corrado, G.S., Dean, J.: Distributed representations of words and phrases and their compositionality. In: NIPS (2013)

20. Mimno, D., Li, W., McCallum, A.: Mixtures of hierarchical topics with pachinko allocation. In: ICML (2007)

21. Mimno, D., Wallach, H.M., Talley, E., Leenders, M., McCallum, A.: Optimizing semantic coherence in topic models. In: EMNLP (2011)

22. Owen, A.B.: Monte carlo theory, methods and examples. Monte Carlo Theory, Methods and Examples. Art Owen (2013)

23. Paisley, J., Wang, C., Blei, D.M., Jordan, M., et al.: Nested hierarchical Dirichlet processes. IEEE Trans. Pattern Anal. Mach. Intell. **37**(2), 256–270 (2015)

24. Pearl, J.: Probabilistic Reasoning in Intelligent Systems: Networks of Plausible Inference. Morgan Kaufmann, Saint Paul (1988)

25. Sato, M.A., Ishii, S.: On-line EM algorithm for the normalized Gaussian network. Neural Comput. **12**(2), 407–432 (2000)

26. Wallach, H.M., Murray, I., Salakhutdinov, R., Mimno, D.: Evaluation methods for topic models. In: ICML (2009)

27. Wang, X., McCallum, A.: Topics over time: a non-Markov continuous-time model of topical trends. In: KDD (2006)

Fast Structure Learning for Deep Feedforward Networks via Tree Skeleton Expansion

Zhourong Chen[⊠], Xiaopeng Li, Zhiliang Tian, and Nevin L. Zhang[⊠]

The Hong Kong University of Science and Technology, Hong Kong, China
{zchenbb,xlibo,ztianac,lzhang}@cse.ust.hk

Abstract. Despite the popularity of deep learning, structure learning for deep models remains a relatively under-explored area. In contrast, structure learning has been studied extensively for probabilistic graphical models (PGMs). In particular, an efficient algorithm has been developed for learning a class of tree-structured PGMs called hierarchical latent tree models (HLTMs), where there is a layer of observed variables at the bottom and multiple layers of latent variables on top. In this paper, we propose a simple unsupervised method for learning the structures of feedforward neural networks (FNNs) based on HLTMs. The idea is to expand the connections in the tree skeletons from HLTMs and to use the resulting structures for FNNs. Our method is very fast and it yields deep structures of virtually the same quality as those produced by the very time-consuming grid search method.

Keywords: Fast structure learning · Feedforward neural networks

1 Introduction

Deep learning has achieved great successes in the past few years [10,15,17,22]. More and more researchers are now starting to investigate the possibility of learning structures for deep models instead of constructing them manually [3,6,24,31]. There are three main objectives in structure learning: improving model performance, reducing model size, and saving manual labor and/or computation time. Most previous methods focus on the first and second objectives. For example, the goal of constructive algorithms [16] and neural architecture search [31] is to find network structures which can achieve good performance for specific tasks. Network pruning [9,18], on the other hand, aims to learn models which contain fewer parameters but still achieve comparable performance compared with dense models.

In this paper, we focus on the third objective. In practice, people usually determine model structure by manual tuning or grid-search. This is time-consuming as there can be a large number of hyper-parameter combinations to consider. We propose a fast unsupervised structure learning method for neural

© Springer Nature Switzerland AG 2019
G. Kern-Isberner and Z. Ognjanović (Eds.): ECSQARU 2019, LNAI 11726, pp. 277–289, 2019.
https://doi.org/10.1007/978-3-030-29765-7_23

networks. Our method determines the bulk of a network automatically, while allowing minor adjustments. It also learns the sparse connectivity between adjacent layers.

Our work is carried out in the context of standard feedforward neural networks (FNNs). While convolutional neural networks (CNNs) and recurrent neural networks (RNNs) are designed for spatial and sequential data respectively, standard FNNs are used for data that are neither spatial nor sequential. The structures of CNNs and RNNs are relatively more sophisticated than those of FNNs. For example, a neuron at a convolutional layer is connected only to neurons in a small receptive field at the level below. The underlying assumption is that neurons in a small spatial region tend to be strongly correlated in their activations. In contrast, a neuron in an FNN is connected to all neurons at the level below. We aim to learn sparse FNN structures where a neuron is connected to only a small number of strongly correlated neurons at the level below.

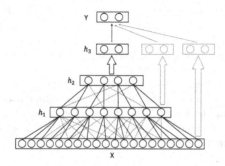

Fig. 1. Model structure of our Tree Skeleton Expansion Networks (TSE-Nets). The PGM core includes the bottom three layers $x - h_2$. The solid connections make up the skeleton and the dashed connections are added during the expansion phase. The black part of the model is called the Backbone, while the red part provides narrow skip-paths from the PGM core to the output layer.

Our work is built upon *hierarchical latent tree analysis* (HLTA) [5,20], an algorithm for learning tree-structured PGMs where there is a layer of observed variables at the bottom and multiple layers of latent variables on top. HLTA first partitions all the observed variables into groups such that the variables in each group are strongly correlated and the correlations can be better modelled using a single latent variable than using two. It then introduces a latent variable to explain the correlations among the variables in each group. After that it converts the latent variables into observed variables via data completion and repeats the process to produce a hierarchy.

To learn a sparse FNN structure, we assume data are generated from a PGM with multiple layers of latent variables and we try to approximately recover the structure of the generative model. To do so, we first run HLTA to obtain a tree model and use it as a *skeleton*. Then we expand it with additional edges to model salient probabilistic dependencies not captured by the skeleton. The result is a PGM structure and we call it a *PGM core*. To use the PGM core for classification, we further introduce a small number of neurons for each layer, and we connect them to all the units at the layers and all output units. This is to allow features from all layers to contribute to classification directly.

Figure 1 illustrates the result of our method. The PGM core includes the bottom three layers $x - h_2$. The solid connections make up the skeleton and the dashed connections are added during the expansion phase. The neurons at layer h_3 and the output units are added at the last step. The neurons at layer h_3 can be conceptually divided into two groups: those connected to the top layer of the PGM core and those connected to other layers. The PGM core, the first group at layer h_3 and the output units together form the *Backbone* of the model, while the second group at layer h_3 provide narrow *skip-paths* from low layers of the PGM core to the output layer. As the structure is obtained by expanding the connections of a tree skeleton, our model is called *Tree Skeleton Expansion Network* (TSE-Net).

2   Related Works

The primary goal in structure learning is to find a model with optimal or close-to-optimal generalization performance. Brute-force search is not feasible because the search space is large and evaluating each model is costly as it necessitates model training. Early works in the 1980's and 1990's have focused on what we call the *micro expansion* approach where one starts with a small network and gradually adds new neurons to the network until a stopping criterion is met [2, 4, 16]. The word "micro" is used here because at each step only one or a few neurons are added. This makes learning large model computationally difficult as reaching a large model would require many steps and model evaluation is needed at each step. In addition, those early methods typically do not produce layered structures that are commonly used nowadays. Recently, a *macro expansion* method [19] has been proposed where one starts from scratch and repeatedly add layers of hidden units until a threshold is met.

Other recent efforts have concentrated on what we call the *contraction* approach where one starts with a larger-than-necessary structure and reduces it to the desired size. Contraction can be done either by repeatedly pruning neurons and/or connections [9, 18, 26], or by using regularization to force some of the weights to zero [28]. From the perspective of structure learning, the contraction approach is not ideal because it requires a complex model as input. After all, a key motivation for a user to consider structure learning is to avoid building models manually.

A third approach is to explore the model space stochastically. One way is to place a prior over the space of all possible structures and carry out MCMC sampling to obtain a collection of models with high posterior probabilities [1]. Another way is to encode a model structure as a sequence of numbers, use a reinforcement learning meta model to explore the space of such sequences, learn a good meta policy from the sequences explored, and use the policy to generate model structures [31]. An obvious drawback of such *stochastic exploration* method is that they are computationally very expensive.

All the aforementioned methods learn model structures from supervised feedback. While useful, class labels contain far less information than model structures. As pointed out by [8], "The process of classification discards most of the information in the input and produces a single output (or a probability distribution over values of that single output)." In other words, there are rich information in data beyond class labels that we can make use of. As such, there are severe limitations if one relies only on supervised information to determine model structures. In this paper, we propose a novel structure learning method that makes use of unsupervised information in data. The method is called *skeleton expansion*. We first learn a tree-structured model based on correlations among variables, and then add a certain number of new units and connections to it in one shot. The method has two advantages: First, learning tree models is easier than learning non-tree models; Second, we need to train only one non-tree model, i.e., the final model.

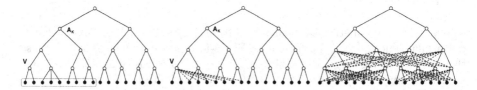

Fig. 2. Tree skeleton expansion. Left: A multi-layer tree skeleton is first learned. A_K is the ancestor of V which is $K = 2$ layers above V. Nodes in the blue circle are the descendants of A_K at the layer below V. Middle: New connections are then added to connect V to all the descendants of A_K at the layer below V. Right: Expansion is conducted on all the layers exception the top K layers. Read Sect. 4 for more details.

The skeleton expansion idea has been used in [6] to learn structures for restricted Boltzmann machines, which have only one hidden layer. This is the first time that the idea is applied to and tested on multi-layer feedforward networks.

3 Learning Tree Skeleton via HLTA

The first step of our method is to learn a tree-structured probabilistic graphical model $\mathcal{T}$ (an example $\mathcal{T}$ is shown in the left panel in Fig. 2). Let $\mathbf{X}$ be the set of observed variables at the bottom and $\mathbf{H}$ be the latent variables. Then $\mathcal{T}$ defines a joint distribution over all the variables:

$$P(\mathbf{X}, \mathbf{H}) = \prod_{v \in \{\mathbf{X}, \mathbf{H}\}} p(v|pa(v)),$$

where $pa(v)$ denotes the parent variable of v in $\mathcal{T}$. The distribution of $\mathbf{X}$ can be computed as:

$$P(\mathbf{X}) = \sum_{\mathbf{H}} P(\mathbf{X}, \mathbf{H}).$$

When learning the structure of $\mathcal{T}$, the objective is to maximize the BIC score [25] of $\mathcal{T}$ over data:

$$BIC(\mathcal{T}|D) = \log P(D|\theta^*) - \frac{d}{2}\log(N),$$

where θ^* is the maximum likelihood estimate of the parameters, d denotes the number of free parameters and D denotes the training data with N training samples. Guided by the BIC score, HLTA builds the structure in a layer-wise manner. It first partitions the observed variables into groups and learns a latent class model (LCM) [14] for each group. Let S denotes the set of observed variables which haven't been included into any variable groups. HLTA computes the mutual information for each pair of variables in S. Then it picks the pair with the highest mutual information and uses them as the seeds of a new variable group G. Other variables from S are then added to G one by one in descending order of their mutual information with variables already in G. Each time when a new variable is added into G, HLTA builds two models ($\mathcal{M}_1$ and $\mathcal{M}_2$). The two models are the best models with one single latent variable and two latent variables respectively, as shown in Fig. 3. HLTA computes the BIC scores of the two models and tests whether the following condition is met:

$$BIC(\mathcal{M}_2|D) - BIC(\mathcal{M}_1|D) \le \delta, \tag{1}$$

where δ is a threshold which is always set at 3 in our experiments. When the condition is met, the two latent variable model $\mathcal{M}_2$ is not significantly better than the one latent variable model $\mathcal{M}_1$. Correlations among variables in G are still well modelled using a single latent variable. Then HLTA keeps on adding new variables to G. If the test fails, HLTA takes the subtree in $\mathcal{M}_2$ which doesn't contain the newly added variable and identifies the observed variables in it as a finalized variable group. The group are then removed from S. And the above process is repeated on S until all the variables are partitioned into disjoint groups (Fig. 4(b)).

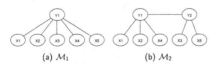

(a) $\mathcal{M}_1$ (b) $\mathcal{M}_2$

Fig. 3. Test whether five observed variables should be grouped together: (a) The best model with one latent variable. (b) The best model with two latent variables.

Next, HLTA introduces a latent variable for each variable group to explain the correlations among variables in the group. This results in a collection of latent class models, which are sometimes referred to as islands (Fig. 4(c), Left). To build the next layer, HLTA links up the islands and obtains what is called a flat model (Fig. 4(c), Right), and it turns the latent variables into observed variables by carrying out data completion within the flat model. Linking up the islands and carrying out data completion in the connected model is time-consuming. In this paper, we propose not to link up the islands. Instead, we carry out data completion in each individual island, which is crucial to speeding up the structure learning process. After converting the latent variables into observed variables $\mathbf{X}'$, the above process is repeated over $\mathbf{X}'$ to obtain another layer of latent variables (Fig. 4(d)). And this is repeated until there is only one new island (Fig. 4(e)). All the variables are then linked up as a multi-layer tree skeleton $\mathcal{T}$ with the last latent variable as the root (Fig. 4(f)).

4   Expanding Tree Skeleton to PGM Core

We have restricted the structure of $\mathcal{T}$ to be a tree, as parameter estimation in tree-structured PGMs is relatively efficient. However, this restriction in return also hurts the model's expressiveness. For example, in text analysis, the word *Apple* is highly correlated with both fruit words and technology words concep-tually. But *Apple* is directly connected to only one latent variable in $\mathcal{T}$ and it is difficult for the single latent variable to express both the two concepts, which may cause severe underfitting. On the other hand, in standard FNNs, units at a layer are always fully connected to those at the previous layer, resulting in high connection redundancies.

In this paper, we aim to learn sparse connections between adjacent layers, such that they are neither as sparse as those in a tree, nor as dense as those in an FNN. To this end, the sparse connections should capture only the most important correlations among the observed variables. Thus we propose to use $\mathcal{T}$ as a structure skeleton and expand it to a denser structure $\mathcal{G}$ which we call the *PGM core*.

Our tree skeleton expansion method works as follows. For a node V at layer L, it finds the ancestor A_K of V that is K layers above V, and connects V to all descendants of A_K at layer $L-1$. We call K the up-looking parameter and set it to 2 in all our experiments. This means that each node is connected to all nodes in the "next generation" who descend from the same grandparent of the node (See Figure 2). We call it the "Grandparent expansion rule". This expansion phase is carried out over all the adjacent layers except the top K layers, as shown in Fig. 2 (Right). After the expansion phase, we take the bottom M layers and use the resulting sparse deep structure as the *PGM core* $\mathcal{G}$. M is a hyper-parameter that we need to determine in experiments.

5   Constructing Sparse FNNs from PGM Core

Our tree expansion method learns a multi-layer sparse structure $\mathcal{G}$ in an unsu-pervised manner. To utilize the resulting structure in a discriminative model, we convert each latent variable h in $\mathcal{G}$ to a hidden unit by defining:

$$h = o(\mathbf{W}'\mathbf{x} + b),$$

where $\mathbf{x}$ denotes a vector of the units directly connected to h at the layer below, $\mathbf{W}$ and b are connection weights and bias respectively, and o denotes a non-linear activation function, e.g. ReLU [7,23]. In this way, we convert $\mathcal{G}$ into a sparse multi-layer neural network. Next we discuss how we use it as a feature extractor in supervised learning tasks. Our model contains two parts, the *Backbone* and the *skip-paths*.

The Backbone. For a specific classification or regression task, we introduce a fully-connected layer on the top of $\mathcal{G}$, which we call the *feature layer*, followed by a output layer. As shown in Fig. 1, the feature layer serves as a feature "aggregator", aggregating the features extracted by $\mathcal{G}$ and feeding them to the output layer. We call the whole resulting module ($\mathcal{G}$, feature layer and output layer together) the Backbone, as it is supposed to be the major module of our model. The user needs to determine the number of units U at the feature layer. We set it to 100 in all our experiments.

The Skip-paths. As the structure of $\mathcal{G}$ is sparse and is learned to capture the strongest correlations in data, some weak but useful correlations may easily be missed. More importantly, different tasks may rely on different weak correlations and this cannot be taken into consideration during the unsupervised structure learning. To remedy this, we consider allowing the model to contain some narrow fully-connected paths to the feature layer such that they can capture those missed features. More specifically, as there are M layers of units in $\mathcal{G}$, we introduce $M-1$ more groups of units into the feature layer, with each group fully connected from

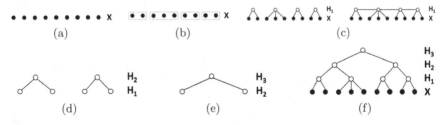

Fig. 4. The structure learning procedure for multi-layer tree skeleton. Black nodes represent observed variables while white nodes represent latent variables. (a) A set of observed variables X. (b) Partition the observed variables into groups. (c) Two options: (Left) Introduce a latent variable for each group. (Right) Introduce a latent variable for each group and link up the latent variables. The first option is much faster and is used in this paper. (d) Convert the layer-1 latent variables H_1 into observed variables and repeat the previous process on them. (e) Convert the layer-2 latent variables H_2 into observed variables and repeat (a)-(c) on them. (f) Stack the LCMs up to form a multi-layer tree skeleton.

a layer in $\mathcal{G}$ except the top one. In this way, each layer except the top one in $\mathcal{G}$ has both a sparse path (the Backbone) and a fully-connected path to the feature layer. The fully-connected paths are supposed to capture those minor features during parameter learning. These new paths are called *skip-paths*. Each group of units in the feature layer contains U units.

As shown in Fig. 1, the Backbone and the skip-paths together form our final model, named *Tree Skeleton Expansion Network* (TSE-Net). The model can then be trained like a normal neural network using back-propagation.

6   Discussion on Hyper-Parameters

There are four hyper-parameters, δ, K, M and U, in our method. The threshold δ is set at 3, which is determined from a threshold [5] suggested for Bayes factor. As for the up-looking parameter K and feature layer group size U, we suggest fixing $K = 2$ and $U = 100$ which work well on a range of datasets. The only hyper-parameter that the user needs to tune during training is the depth M of PGM core. The value depends heavily on the specific dataset. Tuning M allows the user to introduce minor adjustments to the model structure depending on the task and data.

7   Experiments

7.1   Datasets

We evaluate our method in 17 classification tasks. Table 1 gives a summary of the datasets. We choose 12 tasks of chemical compounds classification and 5 tasks of text classification. All the datasets are published by previous researchers.

Tox21 Challenge Dataset[1]. There are about 12,000 environmental chemical compounds in the dataset, each represented as its chemical structure. The tasks are to predict 12 different toxic effects for the chemical compounds [13,21]. We treat them as 12 binary classification tasks. We filter out sparse features which are present in fewer than 5% compounds, and rescale the remaining 1,644 features to zero mean and unit variance. The validation set is randomly sampled and removed from the original training set.

Text Classification Datasets[2]. We use 5 text classification datasets from [30]. After removing stop words, the top 10,000 frequent words in each dataset are selected as the vocabulary respectively and each document is represented as bag-of-words over the vocabulary. The validation set is randomly sampled and removed from the training samples.

---

[1] https://github.com/bioinf-jku/SNNs.
[2] https://github.com/zhangxiangxiao/Crepe.

Table 1. Statistics of all the datasets.

Dataset	Classes	Training samples	Validation samples	Test samples
Tox21	2	~ 9,000	500	~600
Yelp Review Full	5	640,000	10,000	50,000
DBPedia	14	549,990	10,010	70,000
Sogou News	5	440,000	10,000	60,000
Yahoo!Answer	10	1,390,000	10,000	60,000
AG's News	4	110,000	10,000	7,600

7.2 Experiment Setup

We compare our model TSE-Net with standard FNN. For fair comparison, we treat the number of units and number of layers as hyper-parameters of an FNN and optimize them via grid-search over all the defined combinations using validation data. Table 2 shows the space of network configurations considered, following the setup in [13]. In TSE-Net, the number of layers and number of units at each layer are determined by the algorithm.

We also compare our model with pruned FNN whose connections are sparse. We take the best FNN as the initial model and perform pruning as in [9]. As micro expansion and stochastic exploration methods are not learning layered FNNs and are computationally expensive, they are not included in comparison.

We use ReLUs [23] as the non-linear activation functions in all the networks. Dropout [11,27] with rate 0.5 is applied after each non-linear projection. We use Adam [12] as the network optimizer. Codes will be released after the paper is accepted.

7.3 Results

Training Time and Effective FLOP. The training time and effective FLOP (floating point operations) of TSE-Net, Backbone and FNN are reported in Table 3. The Total Time column shows the total time of structure learning/validation and network training in seconds. Effective FLOP is derived for the final model by calculating the number of non-zero weights, which can show us the computation saving when using the sparse model.

Table 2. Hyper-parameters for the structure of FNNs.

Hyper-parameter	Values considered
Number of units per layer	{512, 1024, 2048}
Number of hidden layers	{1,2,3,4}
Network shape	{Rectangle, Conic}

From the table we can see that, the training of TSE-Net and Backbone with unsupervised structure learning is significantly faster than that of FNN with grid-search, especially on large datasets. On the largest dataset, the training time ratio of TSE-Net w.r.t FNN is only 3.5%. And these differences can even be larger if we slightly increase the grid-search space. Moreover, our method is learning sparse models. The FLOP ratios of TSE-Net w.r.t FNN range from 7.01% to 37%, which means that our model can also save a significant part of computations in test time, given appropriate hardware support. In addition, our model is also containing much fewer parameters than the best FNN, which can save memory use in practice.

Classification. Classification results are reported in Table 4. All the experiments are run for three times and we report the average classification AUC scores/accuracies with standard deviations.

TSE-Nets vs FNNs. From the table we can see that, TSE-Net performs very close to the best FNN in 4 out of the 6 datasets, and achieves a 2.01% relative improvement on the Tox21 dataset. In our experiments, TSE-Net achieves better AUC scores than FNN in 9 out of the 12 tasks in the Tox21 dataset. It should be emphasized that, the structure of TSE-Net is trained in an unsupervised manner and it contains much fewer parameters than FNN, while the structure of FNN is manually optimized over the validation data. The results show that, the structure of TSE-Net successfully captures the crucial correlations in data and greatly reduces parameter number without significant performance loss.

It is worth noting that pure FNNs are not the state-of-the-art models for the tasks here. For example, [21] proposes an ensemble of FNNs, random forests and SVMs with expert knowledge for the Tox21 dataset. [13] tests different normalization techniques for FNNs on the Tox21 dataset. They both achieve an average AUC score around 0.846. Complicated RNNs [29] with attention also achieve better results than FNNs for the 5 text datasets. However, the goal of our paper is to learn sparse structure for FNNs, instead of proposing state-of-the-art methods for any specific tasks. Their methods are all much more complex and even task-specific, and hence it is not fair to include their results as comparison. Moreover, their methods can also be combined with ours to give better results.

Contribution of the Backbone. To validate our assumption that the Backbone in TSE-Net captures most of the crucial correlations in data and acts as a main part of the model, we remove the narrow skip-paths in TSE-Net and train the model to test its performance.

As we can see from the results, the Backbone path alone already achieves AUC scores or accuracies which are only slightly worse than those of TSE-Net. Note that the number of parameters in the Backbone is even much smaller than that of TSE-Net. The Backbone contains only 2%~17% of the parameters in FNN. The results not only show the importance of the Backbone in TSE-Net, but also show that our structure learning for the Backbone path is effective.

Table 3. Time and sparsity. Total time contains time for structure learning/validation and network training in seconds. FLOP% column shows the FLOP ratio w.r.t FNN.

Task	TSE-Net (Ours)		Backbone		FNN	
	Total time	FLOP%	Total time	FLOP%	Total time	FLOP
Tox21 Average	128	17.25%	144	2.90%	154	1.64M
AG's News	1,107	7.01%	1,071	3.13%	11,099	28.88M
DBPedia	1,161	19.80%	1,327	8.98%	24,570	10.36M
Yelp Review Full	1,253	37.00%	1,332	16.06%	18,949	5.38M
Yahoo!Answer	1,315	36.70%	1,480	16.97%	37,475	5.39M
Sogou News	1,791	14.90%	1,806	6.38%	27,654	13.39M

Table 4. Classification results. All the experiments are run for three times and we report the average classification AUC scores/accuracies with standard deviations.

Task	TSE-Net (Ours)	Backbone	FNN	Pruned FNN
Tox21 Average	**0.8168**±0.0037	0.7856±0.0066	0.8010±0.0017	0.7998±0.0034
AG's News	91.49%±0.05%	91.54%±0.05%	**91.61%**±0.01%	91.49%±0.09%
DBPedia	**98.04%**±0.01%	97.74%±0.02%	97.00%±0.04%	97.95%±0.02%
Yelp Review Full	58.98%±0.09%	58.38%±0.07%	**59.13%**±0.14%	58.83%±0.01%
Yahoo!Answer	71.48%±0.12%	70.72%±0.02%	**71.84%**±0.07%	71.74%±0.05%
Sogou News	95.91%±0.01%	95.44%±0.06%	96.11%±0.06%	**96.20%**±0.06%

TSE-Nets vs Pruned FNNs. We also compare our method with a baseline method [9] for obtaining sparse FNNs. The pruning method provides regularization over the weights of a network. The regularization is even stronger than $l1/l2$ norm as it is producing many weights being exactly zeros. We start from the fully pretrained FNNs reported in Table 4, and prune the weak connections with the smallest absolute weight values. The pruned networks are then retrained again to compensate for the removed connections. After pruning, the number of remaining parameters in each FNN is the same as that in the corresponding TSE-Net for the same task. As shown in Table 4, TSE-Net and pruned FNN achieve pretty similar results. Note again that pruned FNN took much longer time than TSE-Net. Without any supervision or pre-training over connection weights, our unsupervised structure learning successfully identifies important connections and learns sparse structures. TSE-Net also achieves better interpretability as shown in https://arxiv.org/pdf/1803.06120.pdf.

8 Conclusions

It is important to the applications of deep learning to quickly learn a model structure appropriate for the problem at hand. A fast unsupervised structure learning method is proposed and investigated in this paper. In comparison with standard FNN, our model contains much fewer parameters and it takes much shorter time to learn. It also achieves comparable classification results in a range of tasks. Our method is also shown to learn models with better interpretability. In the future, we will generalize our method to networks like RNNs and CNNs.

Acknowledgments. Research on this article was supported by Hong Kong Research Grants Council under grants 16212516.

References

1. Adams, R.P., Wallach, H.M., Ghahramani, Z.: Learning the structure of deep sparse graphical models. In: AISTATS (2010)
2. Ash, T.: Dynamic node creation in backpropagation networks. Connection Sci. **1**(4), 365–375 (1989)
3. Baker, B., Gupta, O., Naik, N., Raskar, R.: Designing neural network architectures using reinforcement learning. In: ICLR (2017)
4. Bello, M.G.: Enhanced training algorithms, and integrated training/architecture selection for multilayer perceptron networks. IEEE Trans. Neural Networks **3**, 864–875 (1992)
5. Chen, P., Zhang, N.L., Liu, T., Poon, L.K., Chen, Z., Khawar, F.: Latent tree models for hierarchical topic detection. Artif. Intell. **250**, 105–124 (2017)
6. Chen, Z., Zhang, N.L., Yeung, D.Y., Chen, P.: Sparse Boltzmann machines with structure learning as applied to text analysis. In: AAAI (2017)
7. Glorot, X., Bordes, A., Bengio, Y.: Deep sparse rectifier neural networks. In: AISTATS (2011)
8. Goodfellow, I., Bengio, Y., Courville, A.: Deep Learning. MIT Press, Cambridge (2016). http://www.deeplearningbook.org
9. Han, S., Pool, J., Tran, J., Dally, W.: Learning both weights and connections for efficient neural network. In: NIPS (2015)
10. Hinton, G.E., et al.: Deep neural networks for acoustic modeling in speech recognition: the shared views of four research groups. IEEE Sig. Process. Mag. **29**(6), 82–97 (2012)
11. Hinton, G.E., Srivastava, N., Krizhevsky, A., Sutskever, I., Salakhutdinov, R.R.: Improving neural networks by preventing co-adaptation of feature detectors. arXiv preprint arXiv:1207.0580 (2012)
12. Kingma, D., Ba, J.: Adam: a method for stochastic optimization. arXiv preprint arXiv:1412.6980 (2014)
13. Klambauer, G., Unterthiner, T., Mayr, A., Hochreiter, S.: Self-normalizing neural networks. In: NIPS (2017)
14. Knott, M., Bartholomew, D.J.: Latent Variable Models and Factor Analysis (1999)
15. Krizhevsky, A., Sutskever, I., Hinton, G.E.: Imagenet classification with deep convolutional neural networks. In: NIPS (2012)
16. Kwok, T.Y., Yeung, D.Y.: Constructive algorithms for structure learning in feedforward neural networks for regression problems. IEEE Trans. Neural Networks **8**(3), 630–645 (1997)
17. LeCun, Y., Bengio, Y., Hinton, G.: Deep learning. Nature **521**(7553), 436–444 (2015)

18. Li, H., Kadav, A., Durdanovic, I., Samet, H., Graf, H.P.: Pruning filters for efficient convnets. In: ICLR (2017)
19. Liu, J., Gong, M., Miao, Q., Wang, X., Li, H.: Structure learning for deep neural networks based on multiobjective optimization. IEEE Trans. Neural Networks Learn. Syst. **29**, 2450–2463 (2017)
20. Liu, T., Zhang, N.L., Chen, P.: Hierarchical latent tree analysis for topic detection. In: Calders, T., Esposito, F., Hüllermeier, E., Meo, R. (eds.) ECML PKDD 2014. LNCS (LNAI), vol. 8725, pp. 256–272. Springer, Heidelberg (2014). https://doi.org/10.1007/978-3-662-44851-9_17
21. Mayr, A., Klambauer, G., Unterthiner, T., Hochreiter, S.: Deeptox: toxicity prediction using deep learning. Front. Environ. Sci. **3**, 80 (2016)
22. Mikolov, T., Deoras, A., Povey, D., Burget, L., Černocký, J.: Strategies for training large scale neural network language models. In: IEEE Workshop on Automatic Speech Recognition and Understanding, pp. 196–201 (2011)
23. Nair, V., Hinton, G.E.: Rectified linear units improve restricted Boltzmann machines. In: ICML (2010)
24. Real, E., Moore, S., Selle, A., Saxena, S., Suematsu, Y.L., Le, Q., Kurakin, A.: Large-scale evolution of image classifiers. In: ICML (2017)
25. Schwarz, G., et al.: Estimating the dimension of a model. Ann. Stat. **6**(2), 461–464 (1978)
26. Srinivas, S., Babu, R.V.: Data-free parameter pruning for deep neural networks. In: Proccedings of the British Machine Vision Conference (2015)
27. Srivastava, N., Hinton, G.E., Krizhevsky, A., Sutskever, I., Salakhutdinov, R.: Dropout: a simple way to prevent neural networks from overfitting. JMLR **15**(1), 1929–1958 (2014)
28. Wen, W., Wu, C., Wang, Y., Chen, Y., Li, H.: Learning structured sparsity in deep neural networks. In: NIPS (2016)
29. Yang, Z., Yang, D., Dyer, C., He, X., Smola, A.J., Hovy, E.H.: Hierarchical attention networks for document classification. In: HLT NAACL (2016)
30. Zhang, X., Zhao, J., LeCun, Y.: Character-level convolutional networks for text classification. In: NIPS (2015)
31. Zoph, B., Le, Q.V.: Neural architecture search with reinforcement learning. In: ICLR (2017)

An Effective Maintenance Policy for a Multi-component Dynamic System Using Factored POMDPs

İpek Kıvanç[1]([✉]) and Demet Özgür-Ünlüakın[2]

[1] Boğaziçi University, 34342 Bebek, Istanbul, Turkey
ipek.kivanc@boun.edu.tr
[2] Işık University, 34980 Şile, Istanbul, Turkey
demet.unluakin@isikun.edu.tr

Abstract. With the latest advances in technology, almost all systems are getting substantially more uncertain and complex. Since increased complexity costs more, it is challenging to cope with this situation. Maintenance optimization plays a critical role in ensuring effective decision-making on the correct maintenance actions in multi-component systems. A Partially Observable Markov Decision Process (POMDP) is an appropriate framework for such problems. Nevertheless, POMDPs are rarely used for tackling maintenance problems. This study aims to formulate and solve a factored POMDP model to tackle the problems that arise with maintenance planning of multi-component systems. An empirical model consisting of four partially observable components deteriorating in time is constructed. We resort to Symbolic Perseus solver, which includes an adapted variant of the point-based value iteration algorithm, to solve the empirical model. The obtained maintenance policy is simulated on the empirical model in a finite horizon for many replications and the results are compared to the other predefined maintenance policies. Drawing upon the policy results of the factored representation, we present how factored POMDPs offer an effective maintenance policy for the multi-component systems.

Keywords: POMDP · Factored representations · Maintenance

1 Introduction

In recent years, technology and state-of-the-art methods in the area of the industry have led an undeniable wave of change in the industrial world. Due to the latest advances in technology, the complexity of systems has increased. Every system used in the manufacturing sector has a certain life span and the systems need maintenance during their life cycle. Therefore, to ensure the continuity

Supported by the Scientific and Technological Research Council of Turkey (TUBITAK) under grant no 117M587.

© Springer Nature Switzerland AG 2019

G. Kern-Isberner and Z. Ognjanović (Eds.): ECSQARU 2019, LNAI 11726, pp. 290–300, 2019.
https://doi.org/10.1007/978-3-030-29765-7_24

of the system, the role, and importance of the maintenance has increased. It is important to keep up with innovative technologies in the selection of maintenance policies and to cope with increased cost as well. In other words, it is essential to know how to handle associated challenges for implementing efficient planning and management of maintenance activities. There is uncertainty in the majority of maintenance problems. Hence, the maintenance problem can be said to be a real-life example problem that can be observed with partial observability and modeled with the POMDP framework. In this study, we investigate the applicability of the maintenance problem in the context of factored POMDPs. POMDPs are used to make sequential decisions under uncertainty in decision problems where system states cannot be fully observed throughout the planning horizon. In typical POMDPs, flat representation is used via a single node which has multiple states. However, multi-component POMDP models in real life may require a compact representation because belief space size grows exponentially with the number of states while using flat representations. In literature, factored representations [2] have been proposed to simplify the complexity of the states by taking advantage of the factored structure already available in the nature of the problem. The main idea of factored POMDPs is that they can be compactly modeled through dynamic Bayesian networks (DBNs), which are graphical representations for stochastic processes, by exploiting the structure of this representation. In the flat representation, it is required that all states and all transitions are explicitly enumerated in the model. However, factored POMDPs can be modeled and solved by taking advantage of compact data structure representations for problems with large state space spaces. Decision trees [2] and algebraic decision diagrams (ADDs) [9] can be used to represent data compactly. Symbolic Perseus [17] and Heuristic Search Value Iteration (HSVI) [19] are proposed to solve the factored POMDP models by using ADDs.

Recently, POMDPs are getting more popular in the domain of maintenance. A machine may have several interacting components having different dependencies on each other, and degrading or deteriorating over time. POMDPs are generally adopted for maintenance planning of civil structures. In [11], POMDPs are applied to decision-making for highway pavement. In [7] and [6], POMDP models were formulated to generate a policy for bridge inspection. In [3], seasonal dependent situation-based maintenance policies are obtained for wind turbines with finite horizon. Papakonstantinou et al. conducted a comprehensive literature review in the area of inspection scheduling and maintenance planning using POMDPs [15]. Also, they obtained inspection and maintenance policies for corroded structures and erected concrete structures respectively using POMDPs [13]. Although many maintenance optimization approaches are present in literature that are used to model sequential decision problems, factored-POMDPs have not been well studied yet. However, they can be well-suited frameworks for obtaining optimal or near-optimal maintenance policies.

The rest of this paper is structured as follows: Sect. 2 introduces the preliminaries of POMDPs including definitions and the methodology. Section 3 presents the empirical POMDP model and the proposed maintenance policies. Section 4 covers the experimental evaluations. Finally, conclusion and future studies are provided in Sect. 5.

2   Preliminaries

Markov Decision Processes (MDPs) are used to model decision-making problems in uncertain environments as a sequential decision-making model [18]. However, MDPs are limited in models with further state uncertainty. For this reason, MDPs have been extended to POMDPs which take into account the partial observability of the states. POMDPs are stochastic models with better results in planning optimal policies under uncertainty in real life domain [14]. A POMDP is defined by the 6-tuple: $< S, A, T, R, \theta, O >$. S is a finite set of system state. A is a finite set of all alternative actions that can be chosen. The agent aims to maximize his/her reward by the correct actions. $T : S \times A \times S \to \Delta(S)$ is the transition function that gives the probability of a transition from current state s to the state s' after executing action a. $R: S \times A \times S \times \theta \to \Re$ is the reward after executing an action, making a state transition and receiving an observation. θ is a finite set of observations that the agent can observe. $O : S \times A \times \theta \to \Delta(O)$ is the observation probabilities describing the relationships among observations, states, and actions. A typical POMDP representation is illustrated in Fig. 1.

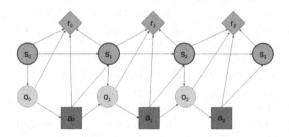

Fig. 1. A typical POMDP representation.

Point-based approaches are approximate POMDP methods that have come popular recently due to their ability to solve much larger problems than exact methods. Modern point based solvers can solve thousands of state problems. An important breakthrough for this type of solvers is the approximation of the belief space by selective belief point sampling. Thus, the exponential growth of the value function is prevented. For this reason, much less computing power is required for the solution of the long-term horizon and large state space POMDPs.

Point-based value iteration (PBVI) [16] is based solely on the idea of creating a limited set of belief space. The main point of point-based value iteration is that a representative belief points can be selected to cover the reachable areas of belief with high approximation accuracy. In each iteration, PBVI expands its belief subsets by choosing the new accessible beliefs that are as far away from the current belief points.

Perseus: Randomized point-based value iteration for POMDPs [5] is a randomized version of point-based value iteration. By reducing the number of belief updates required for each iteration, Perseus increases the computational efficiency in the value iteration process, without any loss in the value function approximation. Perseus initially creates a fixed set of accessible belief points. This set of belief points is formed in a less complex way compared to other point-based algorithms that use heuristic approaches. In each iteration, it backs up minimum number of belief points through working on a subset of the set of created belief points only at the beginning. This process continues until it ensures that the value function approach is improved for all points in the initial set of beliefs [14].

Symbolic Perseus is an adapted variant of the point-based value iteration algorithm to solve the factored POMDP models [17]. The distinctive feature of Symbolic Perseus is a limitation of the number of the vector representing the value function without a loss of quality. This reduces the cost of calculation of backup operations. ADDs used for belief states and vectors provide significant savings in required memory and calculations. Real-life problems require scalable POMDP algorithms that will be powerful against two major challenges, such as the curse of dimensionality and the curse of history. Symbolic Perseus can handle large state-space complexity with a limited number of backups of reachable beliefs. It also solves the curse of dimensionality by using ADDs to represent the alpha vectors and beliefs. Alpha-vectors are a set of hyperplanes that define belief functions. The curse of history can be handled by limiting the number of vectors for the value function. The classic value iteration application with ADDs was implemented by [9, 10].

3   Maintenance Policies with POMDPs

In this study, an empirical model consisting of four partially observable components deteriorating in time, three processes and one observation node is constructed to symbolize the maintenance problem of a multi-component system. It is not possible to observe the system, i.e., the components and the processes, directly. However, the state of processes and components can be estimated by the observation node. The relationship between components and processes is shown in Fig. 2. There are three states of all components ("working", "deteriorate" and "fail"), two states of all processes ("working" and "fail"), and three states of the observation node ("green", "yellow" and "red").

The assumptions about the empirical maintenance model are as follows:

– Processes 1 and 2 are defined as the result of the interaction between their predecessor components. P1 with the interaction of C1 and C2, P2 with the interaction of C3 and C4. The main process node, P3, which points to the performance of the system and is directly influenced by the process nodes P1 and P2.

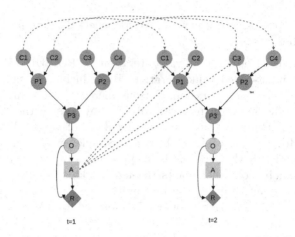

Fig. 2. Empirical POMDP model for two time slots.

- P3 is directly linked to the observable node O which is used to gather information from the process node. Conditional probabilities of the observation node are given in Table 1.
- All components (C1, C2, C3, C4) are replaceable, and at most one of them can be replaced at a day. Hence, there are 5 action states ("Do nothing", "Replace C1", "Replace C2", "Replace C3", "Replace C4") at a time.
- All components are in their "working" state initially.
- Rewards -costs for the maintenance models- are collected from actions and observations. Maintenance costs depend on the observations received and the maintenance activity executed. The total maintenance cost consists of the cost of the production loss and the repair of the relevant component.
- Replacement costs of components are {100, 200, 300, 400} when a green or yellow signal is observed, and {200, 400, 600, 800} when a red signal is observed respectively.
- A downtime cost of 2,500 incurs, when a green or yellow signal is observed, also 7,500 incurs as a downtime cost when a red signal is observed.

Table 1. Conditional probabilities of the observation.

P3	W	NW
Green	0.90	0
Yellow	0.09	0.01
Red	0.01	0.99

Maintenance costs of each component including both the cost of production losses depending on the observation received and also the cost of the action taken are given in Table 2.

Table 2. Maintenance costs.

Action	Observation		
	G	Y	R
Do nothing	0	0	7,500
Replace C1	2,600	2,600	7,700
Replace C2	2,700	2,700	7,900
Replace C3	2,800	2,800	8,100
Replace C4	2,900	2,900	8,300

There are two uncertainties in the model. The current system state is not known to the agent; however, it can be estimated by the observations received. Furthermore, the system state after the action is taken is not known precisely as in MDPs. For this reason, POMDPs fit well for solving this kind of maintenance problem.

4 Experimental Evaluations

We use Symbolic Perseus which is an adapted variant of the PBVI algorithm to solve the factored POMDP models to solve the empirical maintenance model [17]. Classical POMDPs are limited to solve problems because the size of the state spaces grows exponentially as the number of components increases. The empirical model works well with the factored representation due to the inherent factored structure of the maintenance problem. Thus, an approximate policy is obtained via Symbolic Perseus (SP) for the empirical maintenance problem. Then, this policy is simulated using DBNs to evaluate its performance and sensitivity analysis is conducted under different cost parameters. The simulation algorithm used is given in Fig. 3.

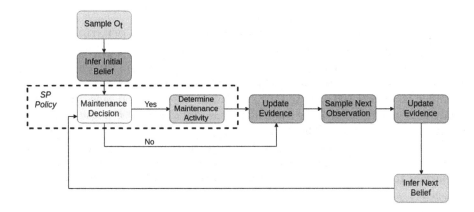

Fig. 3. Maintenance policy simulation.

4.1   Comparison of Various Maintenance Policies

The maintenance problem is first formulated as a factored POMDP and solved by Symbolic Perseus to obtain an approximate maintenance policy. Model is discounted by a factor of 0.99. The policy is then simulated using DBNs for a planning horizon of 100 days and the simulation is repeated 50 times. Sensitivity analysis is performed with different downtime cost values. The results of the sensitivity analysis are given in Table 3; where TCost is the total cost of the given horizon, TRed is the number of total red signals observed and TRep is the total number of replacement on the same horizon. The averages and the standard deviations of these measures are reported in the table. Also, the average total replacements for each component is given in the table. As can be seen from the table, the average total cost increases naturally as the downtime cost increases. In addition, as the downtime cost increases, the average total number of red signals decreases while the average total number of replacements increases. It can be said that as the downtime cost increases, the policy behaves more proactively without waiting for a red signal.

Table 3. Sensitivity analysis under different downtime costs - SP.

Downtime Cost	TCost		TRed		TRep		Avg. Comp. Rep.			
	Avg	Std	Avg	Std	Avg	Std	C1	C2	C3	C4
2,500	72,190	12,972	24.98	4.44	24.98	4.44	12.02	5.52	4.12	3.32
5,000	136,698	22,423	23.96	4.16	26,68	4.18	11.84	5.68	4.62	4.54
7,500	190,874	33,829	20.72	3.94	30.94	5.14	12.44	6.72	6.18	5.60
10,000	233,002	42,878	17.90	3.58	35.44	6.03	13.66	7.96	6.94	6.88
12,500	281,028	48,529	17.00	3.59	39.78	4.90	13.06	8.42	9.08	9.22
15,000	310,562	65,675	15.94	3.84	40.06	6.75	13.00	10.16	8.32	8.58
17,500	364,738	67,988	15.80	3.66	45.96	5.54	13.54	11.56	10.60	10.26

The maintenance problem is also formulated as a classical POMDP with a flat representation and run by SARSOP [1] which is one of the approximate point-based solvers. The model is discounted by a factor of 0.99 and run for two hours to obtain an approximate policy. The policy is then simulated using DBNs for a planning horizon of 100 days and the simulation is repeated 50 times. Sensitivity analysis is performed with the same downtime cost values used in the sensitivity analysis of the SP policy. The results are given in Table 4. According to Tables 3 and 4, it can be concluded SARSOP and SP policies behave similarly and there is no significant difference between the TCost values of SP and SARSOP. It is important to note here that no exact solution is found with "POMDP-solve" [4] which is a well known exact POMDP solver.

Table 4. Sensitivity analysis under different downtime costs - SARSOP.

Downtime cost	TCost		TRed		TRep		Avg. Comp. Rep.			
	Avg	Std	Avg	Std	Avg	Std	C1	C2	C3	C4
2,500	77,404	12,032	27.24	4.24	20.94	4.27	9.88	4.92	3.36	2.78
5,000	135,230	24,605	20.04	3.77	30.28	5.49	12.36	6.70	6.16	5.06
7,500	193,246	31,912	18.62	3.51	35.92	5.43	14.06	7.28	7.86	6.72
10,000	238,452	35,806	17.44	3.20	36.68	4.62	13.68	8.64	8.78	7.58
12,500	292,710	45,874	17.16	3.40	43.56	4.70	13.98	10.34	10.24	9.00
15,000	324,478	58,937	15.84	3.61	45.42	5.62	13.88	10.70	10.76	10.08
17,500	366,034	66,325	13.98	3.83	56.78	3.55	22.02	11.24	12.46	11.06

In addition, the performance of the policies for the empirical model is compared with some predefined practical corrective and proactive maintenance strategies. The predefined maintenance strategies and their explanations are given in Table 5 where R, Y, G denote red, yellow and green signal respectively. In the table, the observations where maintenance can be done according to the respective output policy for SP and SARSOP are marked with cross sign. Furthermore, the observations where maintenance is done are also marked for the other predefined strategies. For the downtime cost of 7,500, the results are given in Table 6.

In the Random and Order algorithms, the components are selected randomly and in order for the maintenance respectively. In the Fault Effect Myopic (FEM) algorithm (similar to FEM_{fp} in [12]), the component with the maximum efficiency measure based on the worst state probability and maintenance cost is selected for the maintenance. The Fault Effect Look-ahead (FEL) algorithm (similar to FEL_{fp} in [12]) also takes into account the worst state probability of components as in the FEM method, however, the component with the maximum efficiency measure based on the worst state probability and maintenance cost in the next period is selected for maintenance.

For a comparison of the cost means of the methods in Table 6, the one-way analysis of variance is used. The assumption of normality and the assumption of homoscedasticity have been checked. Although residuals fit the normal distribution, the assumption of homoscedasticity is violated. Thus, the comparison is further performed by the Games Howell post hoc test [8] which does not assume equal variances. The results of the analysis are given in Table 7. The strategies using the random method give the highest cost as expected. Within the others, SARSOP and SP policy costs are the least. However, SP requires a shorter solution time (approximately 0.5 min) than SARSOP (120 min) in generating policy.

Table 5. Overview of the predefined maintenance policies.

Method	Method explanation	Obs			Algorithms
		G	Y	R	
SP	Symbolic Perseus	X	X	X	SP
SARSOP	SARSOP	X	X	X	SARSOP
CorRND	Corrective Random Selection			X	Random
ProCorRND	Proactive Corrective Random Selection		X	X	Random
CorFEM	Corrective Fault Effect Myopic			X	FEM
ProCorFEM	Proactive Corrective Fault Effect Myopic		X	X	FEM
CorFEL	Corrective Fault Effect Look-ahead			X	FEL
ProCorFEL	Proactive Corrective Fault Effect Look-ahead		X	X	FEL
CorORD	Corrective Order Method			X	Order
CorProORD	Proactive Corrective Order		X	X	Order

We also evaluate the approximation quality of the policies obtained from SP for different values of parameters such as the number of belief points and the number of α-vectors since these parameters directly affect the quality of the policy. Belief space is sampled for 100 and 500 belief points and for each iteration, the maximum number of alpha vectors used in the algorithm is taken as 100 and 500 respectively. The results are given in Table 8. Average total cost doesn't differ in the two parameter sets for this size of the problem. However, the computational time required for the solver increases considerably by the number of selected belief points.

Table 6. Comparison of the maintenance policies when the downtime cost is 7,500.

Method	TCost		TRed		TRep		Avg. Comp. Rep.			
	Avg	Std	Avg	Std	Avg	Std	C1	C2	C3	C4
SARSOP	193,246	31,912	18.62	3.51	35.92	5.43	14.06	7.28	7.86	6.72
SP	190,874	33,830	20.72	3.94	30.94	5.14	12.44	6.72	6.18	5.60
CorRND	255,322	47,131	31.90	5.86	31.90	5.86	7.82	7.92	7.94	8.22
ProCorRND	266,432	50,793	31.04	6.39	37.68	6.51	9.58	9.68	9.18	9.24
CorFEM	210,322	35,433	27.02	4.51	27.02	4.51	19.70	3.90	2.82	0.60
ProCorFEM	227,060	32,995	26.7	4.46	33.92	4.30	24.26	5.00	3.74	0.92
CorFEL	218,588	34,687	28.20	4.45	28.20	4.45	21.40	6.36	0.44	0.00
ProCorFEL	226,158	31,298	26.64	4.22	34.04	4.14	26.12	5.88	1.98	0.06
CorORD	211,106	37,682	26.42	4.71	26.42	4.71	7.00	6.74	6.42	6.26
ProCorORD	211,598	37,713	24.10	5.05	31.02	4.36	8.18	7.88	7.66	7.30

Table 7. Games-Howell Post-Hoc Test.

Method	Avg. Cost	Std. Dev.	95% CI	Games-Howell group
ProCorRND	266,432	50,793	(241,927; 268,717)	A
CorRND	255,322	47,131	(208,730; 228,446)	A
ProCorFEM	227,060	32,995	(200,397; 221,815)	B
ProCorFEL	226,158	31,298	(251,997; 280,867)	B
CorFEL	218,588	34,687	(205,450, 229,350)	B
ProCorORD	211,598	37,713	(217,683; 236,437)	B,C
CorORD	211,106	37,682	(217,263; 235,053)	B,C
CorFEM	210,322	35,433	(200,880; 222,316)	B,C
SARSOP	193,246	31,912	(181,260; 200,488)	C
SP	190,874	33,830	(184,177; 202,315)	C

Table 8. Comparison of different parameters for the solution of SP.

Parameters	TCost		TRed		TRep		Time
	Avg	Std	Avg	Std	Avg	Std	Sec
100 Belief Points-100 Alpha Vectors	190,874	33,829	20.72	3.94	30.94	5.14	26.80
500 Belief Points-500 Alpha Vectors	190,228	35,035	19.86	4.10	32.46	5.28	75.05

5   Conclusion

We formulate a factored POMDP offers a near-optimal maintenance policy for an empirical maintenance model of the four-component partially observable dynamic system. The sensitivity of the methodology is also analyzed under different downtime cost values. We compare ten different policies. The first two are the computed POMDP policies generated by SP and SARSOP, while the other eight are predefined policies based on corrective or proactive maintenance strategies. The results show that the computed policies from the POMDP model are superior to the others. Statistical results clearly show that the performances of the two POMDP solvers do not differ. However, SP provides a compact representation due to the factored nature of the problem and uses this property to result in a shorter solution time than SARSOP in generating policy. Future studies involve the implementation of a real-life maintenance problem.

References

1. http://bigbird.comp.nus.edu.sg/pmwiki/farm/appl/
2. Boutilier, C., Poole, D.: Computing optimal policies for partially observable decision processes using compact representations. In: Proceedings of the National Conference on Artificial Intelligence, pp. 1168–1175. Citeseer (1996)

3. Byon, E., Ntaimo, L., Ding, Y.: Optimal maintenance strategies for wind turbine systems under stochastic weather conditions. IEEE Trans. Reliab. **59**(2), 393–404 (2010)
4. Cassandra, A.R.: Home. http://www.pomdp.org/pomdp/index.shtml
5. Cassandra, A.R., Littman, M.L., Zhang, N.L.: Incremental pruning: a simple, fast, exact method for partially observable Markov decision processes. CoRR abs/1302.1525 (2013)
6. Corotis, R.B., Hugh Ellis, J., Jiang, M.: Modeling of risk-based inspection, maintenance and life-cycle cost with partially observable decision processes. Struct. Infrastruct. Eng. **1**(1), 75–84 (2005)
7. Ellis, H., Jiang, M., Corotis, R.B.: Inspection, maintenance, and repair with partial observability. J. Infrastruct. Syst. **1**(2), 92–99 (1995)
8. Games, P.A., Howell, J.F.: Pairwise multiple comparison procedures with unequal N's and/or variances: a Monte Carlo study. J. Educ. Stat. **1**(2), 113–125 (1976). https://doi.org/10.2307/1164979
9. Hansen, E.A., Feng, Z.: Dynamic programming for POMDPs using a factored state representation. In: AIPS (2000)
10. Hoey, J., St-Aubin, R., Hu, A., Boutilier, C.: SPUDD: stochastic planning using decision diagrams. In: Proceedings of the Fifteenth Conference on Uncertainty in Artificial Intelligence, pp. 279–288. Morgan Kaufmann Publishers Inc. (1999)
11. Madanat, S., Ben-Akiva, M.: Optimal inspection and repair policies for infrastructure facilities. Transp. Sci. **28**(1), 55–62 (1994)
12. Özgür-Ünlüakın, D., Türkali, B., Karacaörenli, A., Aksezer, S.Ç.: A DBN based reactive maintenance model for a complex system in thermal power plants. Reliab. Eng. Saf. **190** (2019). https://doi.org/10.1016/j.ress.2019.106505
13. Papakonstantinou, K., Shinozuka, M.: Optimum inspection and maintenance policies for corroded structures using partially observable decision processes and stochastic, physically based models. Probab. Eng. Mech. **37**, 93–108 (2014)
14. Papakonstantinou, K., Shinozuka, M.: Planning structural inspection and maintenance policies via dynamic programming and Markov processes. Part I: Theory. Reliab. Eng. Syst. Saf. **130**(C), 202–213 (2014)
15. Papakonstantinou, K.G., Shinozuka, M.: Planning structural inspection and maintenance policies via dynamic programming and Markov processes. Part II: POMDP implementation. Reliab. Eng. Syst. Saf. **130**, 214–224 (2014)
16. Pineau, J., Gordon, G., Thrun, S., et al.: Point-based value iteration: an anytime algorithm for POMDPs. In: IJCAI, vol. 3, pp. 1025–1032 (2003)
17. Poupart, P.: Exploiting structure to efficiently solve large scale partially observable decision processes. Ph.D. thesis, Toronto, Ont., Canada, Canada (2005). aAINR02727
18. Puterman, M.L.: Markov Decision Processes: Discrete Stochastic Dynamic Programming. Wiley, New York (2014)
19. Smith, T., Simmons, R.: Heuristic search value iteration for POMDPs. In: Proceedings of the 20th Conference on Uncertainty in Artificial Intelligence, pp. 520–527. AUAI Press (2004)

Epistemic Sets Applied to Best-of-n Problems

Jonathan Lawry[1(✉)], Michael Crosscombe[1], and David Harvey[2]

[1] University of Bristol, Bristol BS8 1UB, UK
{j.lawry,m.crosscombe}@bristol.ac.uk
[2] Research Technology and Innovation, Thales, Reading, UK
david.harvey@uk.thalesgroup.com

Abstract. Epistemic sets are a simple and efficient way of representing uncertain beliefs in AI, in which an agent identifies those states or worlds that she deems to be possible. We investigate their application to multi-agent distributed learning and decision making, in particular to best-of-n problems in which a population of agents must reach a consensus by identifying the best out of n possible alternatives or choices, each of different quality. We show that, despite their limited representational power, epistemic sets can be effectively deployed by agents engaged in a learning process in which they receive evidence directly from the environment and also pool or fuse their beliefs with those of other agents, in order to solve a best-of-n problem. We describe an analytical model of such a system based on ordinary differential equations and conduct a fixed point analysis so as to obtain insights into macro-level convergence properties. We then conduct a series of agent-based simulation experiments to investigate the robustness of the epistemic set approach. The results suggest that when applied to best-of-n problems epistemic sets are robust to noise and scalable to large state spaces, even when the population size is relatively small. This in turn supports the claim that they have potential applications in decentralised AI and swarm robotics at a range of different scales.

Keywords: Epistemic sets · Multi-agent systems · Consensus

1 Introduction

Epistemic sets are one of the simplest formalisms for representing uncertainty in AI, in which an agent's belief is represented by the set of states or worlds that she deems possible [5,6]. The concept dates back at least to Hintikka's possible worlds semantics [8] with early applications in computer science and AI proposed by Vardi [21] and Ruspini [17]. Epistemic sets are equivalent to Boolean possibility distributions [4] in that they can be characterised by a function from states into $\{0,1\}$, but they lack the quantitative aspect of general possibility theory [22] or indeed many of the established theories of uncertainty such as probability theory, imprecise probabilities or Dempster-Shafer theory. Despite

© Springer Nature Switzerland AG 2019
G. Kern-Isberner and Z. Ognjanović (Eds.): ECSQARU 2019, LNAI 11726, pp. 301–312, 2019.
https://doi.org/10.1007/978-3-030-29765-7_25

their limited expressiveness, in this paper we aim to show that when used in a dynamic multi-agent setting, epistemic sets can enable effective learning and problem solving at the population level, which is surprisingly robust to noise and scalable to large state spaces. In particular, we will consider a type of distributed learning problem referred to as best-of-n and show that it can be solved by a population of agents who iteratively update epistemic sets upon receiving evidence and who also regularly pool or fuse their beliefs with those of other agents. In this respect we are not concerned with proposing new fusion operators or with studying the axiomatic properties of particular fusion operators at a local level (see [6] for an overview of the latter). Instead our main contribution is to demonstrate the useful macro or population level properties of a system of agents applying well-known operators and updating rules in the framework of epistemic sets.

Best-of-n is a general class of problem in collective decision-making and learning which is particularly important in areas such as swarm robotics [20]. The aim in such problems is for a population of agents to collectively identify the best out of a set of n possible alternatives based only on local feedback and interactions. These alternatives might correspond to physical locations such as in decentralised search and rescue in which a robot swarm must identify the region of a search area in which the most casualties are located [14]. Another application of this kind is pollution treatment swarms which could be deployed after an oil spill and which would need to identify the region where there is highest concentration of pollutants [9]. On the other hand the n alternatives could also refer to different possible control strategies in swarm flocking, or different routes in a routing problem. A variety of different methods have been applied to the best-of-n problem many of which are inspired by the behaviour of social insects such as honeybees or ants [16]. For example, the weighted voter model [19] is inspired by the behaviour of honeybees and Tembothorax ants when searching for nest sites. However, in most cases the assumption is that a large population of agents are deployed to solve a problem in which n is relatively low, e.g. $n \leq 10$. In contrast, we will show that epistemic sets provide a computationally efficient means of solving best-of-n problems with much larger values of n and varying population sizes.

Emergent behaviour where groups of agents reach consensus on the basis of only local interactions is also extensively studied in the context of social network analysis and opinion dynamics [2]. Opinion diffusion logic is of particular relevance to this paper in that it sometimes employs a semantic model of belief which is similar or equivalent to epistemic sets e.g. [7] and [18]. For example, an agent's belief may be represented by a logical formula F or equivalently by the epistemic set consisting of those states (or interpretations) in which F is true [3]. In such studies, however, the focus is often on a network of agents each of whose beliefs is affected by a restricted set of 'influencers' to whom they are connected. In contrast, the type of swarm robotics or decentralised AI applications on which we are focusing will involve individuals moving independently through an environment and encountering a variety of different agents at different times.

This may be better modelled by a system where there is free mixing between agents, i.e., a totally connected graph of agents, but where there are only relatively few interactions at any given time. This is referred to as a 'well-stirred' system in [13], corresponding to the assumption that each agent is equally likely to interact with any other agent in the population and that such interactions are independent events.

An outline of the rest of the paper is as follows: Sect. 2 introduces a pooling operator and an evidential updating method for epistemic sets relevant to a decentralised agent-based setting. In Sect. 3 we write down a set of ordinary differential equations to describe the rate of change of beliefs in a population of agents applying these operators to tackle best-of-n, and then carry out a fixed point stability analysis to provide insight into macro-level convergence properties of the system. Section 4 then presents an extensive study of robustness and scalability of the proposed model using agent-based simulations. Finally, in Sect. 5 we give some discussions and conclusions.

2   Decentralised Pooling and Updating of Epistemic Sets

In this section we formulate the best-of-n problem within the framework of epistemic sets. Let $\mathbb{S} = \{s_1, \ldots, s_n\}$ be a finite set of all possible states of the world and for each state s_i we assume that there is an associated quality value q_i which we take to be in the interval $[0, 1]$. Without loss of generality we will assume that the states are enumerated so that $q_1 < q_2 < \ldots < q_n$. Let $\mathcal{A} = \{a_1, \ldots, a_k\}$ denote a population of k agents, where each agent's belief is represented by an epistemic set $B \subseteq \mathbb{S}$ such that $B \neq \emptyset$, indicating which she believes to be the possible states of the world. In the best-of-n problem we conceive of agents as exploring their environment, interacting and pooling their beliefs with other agents, as well as receiving evidence directly. In particular, we will focus on the following well-known intersection & union operator as a mechanism for combining agent beliefs.

Definition 1. *Intersection & Union Pooling Operator*
For $\emptyset \neq B_1, B_2 \subseteq \mathbb{S}$;

$$B_1 \odot B_2 = \begin{cases} B_1 \cap B_2 : B_1 \cap B_2 \neq \emptyset \\ B_1 \cup B_2 : B_1 \cap B_2 = \emptyset \end{cases}$$

This is a merging operator in the sense of [11] and in an extensive overview of information fusion under uncertainty, Dubois et al. [6] show that Definition 1 is the only operator for combining epistemic sets which satisfies the following properties: *optimism* which requires that the combination of two intersecting epistemic sets B_1 and B_2 to be a subset of both of them, *unanimity* according to which the combination of B_1 and B_2 is a superset of $B_1 \cap B_2$ and a subset of $B_1 \cup B_2$, and *minimal commitment* by which the combination of B_1 and B_2 is the largest epistemic set satisfying both *optimism* and *unanimity*. Notice that this operator is not associative and consequently that the requirement of associativity is not consistent with optimism, unanimity and minimal commitment. Furthermore, the operator is non-monotonic in the sense that as $|B_1 \cap B_2|$ decreases, the combined epistemic set represents an increasingly precise belief until the two sets are inconsistent, at which point the precision of the combined belief decreases. Nonetheless, the experimental results presented in this paper will suggest that the failure of associativity has little impact on population level convergence in the best-of-n problem, at least in the case where there is free mixing between agents, as is often assumed in swarm robotics applications[1]. Furthermore, the decrease in precision when combining inconsistent beliefs seems to provide a mechanism by which errors can be corrected and an overall consensus can be reached.

In addition to pooling their beliefs with others, agents also receive evidence directly from the environment. Here we will assume that evidence is received in the form of a direct comparison between the quality values of two states. In this case, a piece of evidence E could be a comparison between the quality values for states s_i and s_j and if $q_i > q_j$ then this can be represented by the epistemic set $E_j = \mathbb{S} - \{s_j\}$ expressing the information that s_j is not the best state. On receiving evidence we propose that agents update their beliefs according to the following general belief updating rule, which is an evidential updating rule in the sense of [1][2].

Definition 2. *Evidential Updating*
For $\emptyset \neq B, E \subseteq \mathbb{S}$;

$$B|E = \begin{cases} B \cap E : B \cap E \neq \emptyset \\ B : B \cap E = \emptyset \end{cases}$$

This form of negative updating in which certain states are ruled out as part of the learning process has already been applied effectively in swarm robotics. For example, in [12] a swarm of robots must identify the best location from a number of options. Applying probabilistic pooling and updating they visit two locations at a time, receive quality feedback from both, and then use negative updating, as above, to rule one of them out. This approach was shown to have much lower convergence times than other updating approaches even when the cost of visiting two sites during an exploration is taken into account.

3   Fixed Point Analysis

We adopt a discrete time model in order to study the macro-level convergence properties of a population of agents each attempting to individually solve a

---

[1] In some cases swarms alternate between periods of exploration and periods of pooling where for the latter they move to a common location to ensure mixing of different beliefs.

[2] We recognize that there are other possible approaches to updating in this context, especially when the agent's belief is inconsistent with the evidence [10]. Here we simply adopt this updating method as one viable possibility.

best-of-n problem formulated using epistemic sets as outlined in Sect. 2. At each time step t two agents are selected at random to combine their beliefs by applying the operator given in Definition 1. Also, within a time-step each agent selects a pair of distinct states to investigate by picking them randomly from their current epistemic set. There is then a probability ρ, referred to as the evidence rate, that the agent will succeed in sampling both relevant quality values and update their beliefs accordingly. Hence, we have two stochastic functions applied consecutively to each agent in $\mathcal{A}$. There is a pooling function $\mathcal{P}$ such that for epistemic state B;

$$\mathcal{P}(B) = \begin{cases} B : \text{with probability } 1 - \pi \\ B \odot B' : \text{ for } \emptyset \neq B' \subseteq \mathbb{S} \text{ with probability } \pi x_{B'} \end{cases}$$

where π is the probability that an individual agent will be selected for pooling corresponding to $\pi = \frac{2}{k} - \frac{1}{k^2}$ in the proposed model, and for $\emptyset \neq B' \subseteq \mathbb{S}$, $x_{B'}$ denotes the proportion of agents with epistemic set B' as their belief. There is then an evidential updating function $\mathcal{U}$ such that for epistemic state B;

$$\mathcal{U}(B) = \begin{cases} B : \text{with probability } 1 - \rho \\ B|E_j : \text{with probability } \rho p_j^B \end{cases}$$

where p_j^B is the probability that two states are selected at random from B such that one is s_j and the other is a state with quality greater than q_j. Hence, at each time step we can model the combined process of evidential updating and pooling by the composition $\mathcal{U} \circ \mathcal{P}$ applied to the agents in $\mathcal{A}$. This compositional mapping can then be used to determine a set of ordinary differential equations for the rate of change of proportions for each epistemic set, and stability analysis of the fixed points of these equations can then give an insight into the macro level convergence properties of the system. In the remainder of this section we present this analysis for the case when $n = 3$. This is an extension of earlier work by Perron and Vasudevan [15] which studied the convergence properties of the $n = 2$ case in which there is pooling but no evidence. Interestingly in [15] they present their model as one in which agents are attempting to determine the truth value of a single proposition, but where there was a third truth value corresponding to 'unknown'. This can be translated into our state based model by taking s_1 to be the state in which the proposition is true, s_2 to be the state in which the proposition is false, and the epistemic set $\{s_1, s_2\}$ to representing unknown.

For notational simplicity we enumerate the proportions x_B in the $n = 3$ case such that $x_1 = x_{\{s_1,s_2,s_3\}}$, $x_2 = x_{\{s_1,s_2\}}$, $x_3 = x_{\{s_1,s_3\}}$, $x_4 = x_{\{s_2,s_3\}}$, $x_5 = x_{\{s_1\}}$, $x_6 = x_{\{s_2\}}$ and $x_7 = x_{\{s_3\}}$, and where since these represent proportions we have the constraint that $\sum_{i=1}^{7} x_i = 1$. The composite stochastic function $\mathcal{U} \circ \mathcal{P}$ then generates a set of ODEs defining the rate of change of the proportion of each epistemic set, as follows;

$$\dot{x}_1 = 2\pi(1-\rho)x_7x_2 + 2\pi(1-\rho)x_6x_3 + 2\pi(1-\rho)x_5x_3$$
$$+ x_1\left(-(1-\pi)\rho - \pi + \pi(1-\rho)x_1\right)$$
$$\dot{x}_2 = 2\pi(1-\rho)x_5x_6 + 2\pi(1-\rho)x_1x_2 + x_2\left(-(1-\pi)\rho - \pi + \pi(1-\rho)x_2\right)$$
$$\dot{x}_3 = x_1\tfrac{1}{3}\left((1-\pi)\rho + \pi\rho x_1\right) + 2\pi(1-\rho)x_1x_3 + \tfrac{2}{3}\pi\rho x_2x_7 + \tfrac{2}{3}\pi\rho x_3x_6 + \pi\rho x_4x_5$$
$$+ 2\pi(1-\rho)x_5x_7 + x_3\left(-(1-\pi)\rho - \pi + \pi(1-\rho)x_3\right)$$
$$\dot{x}_4 = x_1\tfrac{2}{3}\left((1-\pi)\rho + \pi\rho x_1\right) + 2\pi(1-\rho)x_1x_4 + \tfrac{4}{3}\pi\rho x_2x_7 + \tfrac{4}{3}\pi\rho x_3x_6 + \tfrac{4}{3}\pi\rho x_4x_5$$
$$+ 2\pi(1-\rho)x_6x_7 + x_4\left(-(1-\pi)\rho - \pi + \pi(1-\rho)x_4\right)$$
$$\dot{x}_5 = \pi x_1x_5 + x_2 2\pi\left(x_3 + x_5\right) + x_3\pi\left(x_2 + x_5\right) + x_5\pi\left(-x_4 - x_6 - x_7\right)$$
$$\dot{x}_6 = x_1\pi\left(\rho x_2 + x_6\right) + x_2\left((1-\pi)\rho + \pi\rho x_1 + \pi\rho x_2 + \pi x_4 + \pi x_6\right) + x_4\pi\left(x_2 + x_6\right)$$
$$+ \pi\rho x_5x_6 + x_6\pi\left(x_3 - (1-\rho)x_5 - x_7\right)$$
$$\dot{x}_7 = x_1\pi\left(\rho x_3 + \rho x_4 + x_7\right) + x_3\left((1-\pi)\rho + \pi\rho x_1 + \pi\rho x_3 + \pi x_4 + \pi x_7\right)$$
$$+ x_4\left((1-\pi)\rho + \pi\rho x_1 + \pi x_3 + \pi\rho x_4 + \pi x_7\right) + \pi\rho x_5x_7 + \pi\rho x_6x_7$$
$$+ x_7\pi\left(-x_2 - (1-\rho)x_5 - (1-\rho)x_6\right)$$

The fixed points for this system of equations are those values of $x_1, \ldots, x_7$ satisfying $\dot{x}_1 = \ldots = \dot{x}_7 = 0$. Stability is then determined from the Jacobian $J = \left(\frac{\partial \dot{x}_i}{\partial x_j}\right)$ by evaluating J at each fixed point and finding the eigenvalues. A fixed point is stable if and only if the real parts of all the eigenvalues are negative. For the above system of ODEs the only stable fixed points are those for which the entire population converges on epistemic sets with only a single element[3], but which of these are stable depends on the values of ρ and π (and therefore indirectly on the population size k). This is summarised in Fig. 1 in which the black region corresponds to those values of ρ and k for which all of $x_{\{s_1\}} = 1$, $x_{\{s_2\}} = 1$ and $x_{\{s_3\}} = 1$ are all stable fixed points, while the white region consists of those values of ρ and k for which only $x_{\{s_3\}} = 1$ is a stable fixed point. Hence, for parameter values in the white region of Fig. 1 we can expect stable convergence to the best state across a range of initial conditions.

In fact the epistemic set model is rather more effective for very low evidence rates than is suggested by the above fixed point analysis. This seems to be particularly true if agents are initialised as being completely ignorant, so that their beliefs' at $t = 0$ correspond to the most general epistemic set i.e. $B = \mathbb{S}$. This is reasonable in many multi-agent and swarm robotics applications where, in the absence of any prior evidence, agents would begin by considering all states as being equally possible. We can generate a discrete time simulation of the above set of ODEs under these initial conditions by applying the following mapping at each time step;

$$x_B \to x_B + \Delta t\, \dot{x}_B$$

---

[3] In other words, the only possible stable fixed points are $(x_1, \ldots, x_7) = (0,0,0,0,1,0,0)$, $(0,0,0,0,0,1,0)$ or $(0,0,0,0,0,0,1)$ corresponding to $x_{\{s_1\}} = 1$, $x_{\{s_2\}} = 1$ and $x_{\{s_3\}} = 1$ respectively.

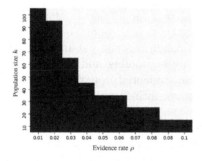

Fig. 1. A map showing the stability of fixed points. In the white region the only stable fixed point is $x_{\{s_3\}} = 1$. In the black region all three fixed points, $x_{\{s_1\}} = 1$, $x_{\{s_2\}} = 1$ and $x_{\{s_3\}} = 1$, are stable.

Figure 2 shows the resulting time series trajectories of the proportions of various epistemic sets for $k = 100$ and $\rho = 0.01$. Even though this combination of k and ρ values is in the black region of Fig. 1 there is nonetheless full convergence to the highest quality state $\{s_3\}$ with agents adopting intermediate beliefs corresponding to the epistemic sets $\{s_1, s_3\}$ and $\{s_2, s_3\}$.

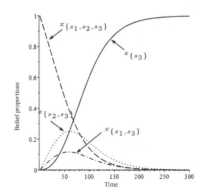

Fig. 2. Trajectories of the proportions holding different beliefs as function of time generated from the ODEs given in Sect. 3 with $k = 100$, $\rho = 0.01$ and initialised so that $x_{\{s_1,s_2,s_3\}} = 1$ at time $t = 0$.

In the next section we will describe the results from a more systematic study of this model using agent-based simulations under different parameter values and in the presence of noise.

4 Agent-Based Simulations

We now present the results from a series of agent-based simulation experiments of the discrete time model described in Sect. 3 with a completely ignorant initial population, i.e. so that $B = \mathbb{S}$ for all agents at time $t = 0$. The aim here is to demonstrate the effectiveness and robustness of the epistemic set approach under a range of different conditions and parameter values. We begin by noting that the combination of agent pooling and evidential updating tends to lead to faster convergence than evidential updating alone. In other words, pooling plays a positive role in propagating evidence across the population of agents. For instance, Fig. 3 shows the proportion of the population with belief $\{x_5\}$, i.e. the best state, plotted against time for $n = 5$, $k = 100$ and for an evidence rate of $\rho = 0.01$. Results are averaged over 100 independent runs of the simulation and error bars show 90% confidence intervals. In this case we see that the combination of pooling and updating (black line) converges after only 400 iterations, while evidence only (grey line) requires over 1000 iterations.

To study the effect of noise on this system we introduce a model of noise in which the probability of confusing the order of quality values q_i and q_j is a decreasing function of their difference. That is we define a decreasing error function $f : [0, 1] \rightarrow [0, 1]$ such that if an agent with belief B receives evidence E corresponding to a comparison between states s_i and s_j where $q_i > q_j$ then their updated belief will be given by;

$$B|E = \begin{cases} B|E_j : \text{with probability } 1 - f(q_i - q_j) \\ B|E_i : \text{with probability } f(q_i - q_j) \end{cases}$$

where, as in Sect. 2, $E_i = \mathbb{S} - \{s_i\}$ and $E_j = \mathbb{S} - \{s_j\}$. We now propose a parametrised error function f, allowing for error regimes of greater or lesser severity.

Definition 3. *Error Function*
For $d \in [0, 1]$ and $\lambda \in \mathbb{R}$,

$$f(d) = \begin{cases} \dfrac{0.5(e^{-\lambda d} - e^{-\lambda})}{1 - e^{-\lambda}} : \lambda \neq 0 \\ 0.5(1 - d) : \lambda = 0 \end{cases}$$

In the following experiments we will assume that quality values are uniformly distributed over the interval $[0, 1]$ so that $q_i = \frac{i}{n+1}$ for $i = 1, \ldots, n$. Figure 4a shows plots of the error function for different values of the parameter λ from which we can see that increasing the value of λ decreases the severity of the error model. In the following we will consider error models including those of up to a severity level given by $\lambda = 0$ i.e. as corresponding to linear error. In this context Fig. 4b shows the proportion of agents with belief $\{s_5\}$ after 1500 iterations plotted against the error parameter λ when $k = 100$, $n = 5$ and $\rho = 0.01$. With the exception of very low values of λ, i.e. $\lambda < 2$, we see that the presence of pooling significantly improves the robustness to error with close to total convergence to the best state for $\lambda \geq 5$. In fact for only a slight increase in the evidence rate ρ, Fig. 4c suggests that with pooling, the epistemic set model can be robust even for a $\lambda = 0$ error regime. Here we see that for $\lambda = 0$

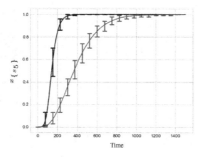

Fig. 3. Time series plot of the proportion of agents with belief $\{s_5\}$ averaged over 100 runs of the simulation. Parameter values are $k = 100$, $n = 5$ and $\rho = 0.01$. The grey line is for evidential updating only and the black line is for both updating and pooling.

the combination of pooling and updating results in convergence to an average proportion $x_{\{s_5\}}$ of over 0.95 for $\rho \geq 0.05$.

For this model we also investigate the scalability of epistemic sets to varying population sizes k and number of states n. In the swarm literature, studies of the best-of-n problem have predominantly focussed on scenarios in which k is high and n is relatively low [20]. However, in many application domains it may be more realistic to consider smaller swarms deployed to solve very high-dimensional problems. For example, for search and rescue we might expect swarms sizes of order tens of robots to be deployed to search a large and complex environment. Figure 5 shows heat plots of epistemic set proportions after 1500 iterations under different λ and ρ, as k and n vary. Figure 5a is a heat map of the proportion of agents with epistemic set $\{s_n\}$ after 1500 iterations with an evidence rate of $\rho - 0.01$ and assuming no errors in evidential updating. In this case the approach is very robust to varying population size and number of states with performance only declining for a population size of 5 and where there are more than 75 states. However, Figs. 5b and c suggest that robustness declines significantly under increasingly severe error regimes

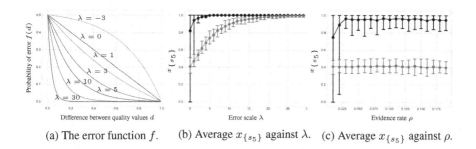

(a) The error function f.   (b) Average $x_{\{s_5\}}$ against λ.   (c) Average $x_{\{s_5\}}$ against ρ.

Fig. 4. Performance under different error regimes. In Figs. 4b and c the grey line shows the results for evidence only updating while the black line is for pooling and updating combined. Results are averaged over 100 simulation runs with $k = 100$ and $n = 5$. For Fig. 4b $\rho = 0.01$ and for Fig. 4c $\lambda = 0$. All results are after 1500 iterations.

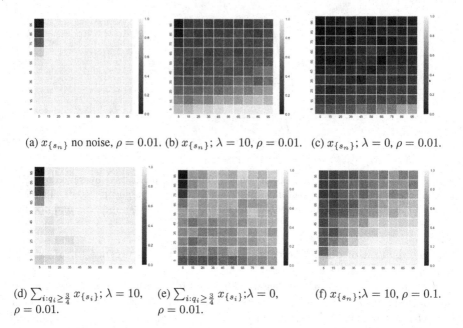

(a) $x_{\{s_n\}}$ no noise, $\rho = 0.01$. (b) $x_{\{s_n\}}$; $\lambda = 10$, $\rho = 0.01$. (c) $x_{\{s_n\}}$; $\lambda = 0$, $\rho = 0.01$.

(d) $\sum_{i:q_i \geq \frac{3}{4}} x_{\{s_i\}}$; $\lambda = 10$, $\rho = 0.01$.

(e) $\sum_{i:q_i \geq \frac{3}{4}} x_{\{s_i\}}$; $\lambda = 0$, $\rho = 0.01$.

(f) $x_{\{s_n\}}$; $\lambda = 10$, $\rho = 0.1$.

Fig. 5. Heat maps showing robustness to varying population size and number of states under different error regimes.

with good performance only achieved for higher numbers of agents and lower numbers of states. Part of this effect is likely to be an artefact of the way in which quality values have been defined for different values of n. Recall that we define $q_i = \frac{i}{n+1}$ and hence, for example, the difference between consecutive quality values, given by $d = (q_{i+1} - q_i) = \frac{i}{n+1}$, decreases with n. Furthermore, since the error function is a decreasing function of d for all λ, then this implies that the probability of confusing the order of consecutive quality values increases with n. One way to adjust for this would be to assess performance in terms of a quality threshold. For example, we could evaluate the proportion of the population who have converged to an epistemic state $\{s_i\}$ with quality greater than or equal to some specified threshold. Figures 5d and e are heat maps showing the value of $\sum_{i:q_i \geq \frac{3}{4}} x_{\{s_i\}}$ for $\lambda = 10$ and $\lambda = 0$ respectively, after 1500 iterations and with an evidence rate of $\rho = 0.01$. In comparison to Figs. 5b and c we note that the degradation of performance under increasingly severe error regimes is much less stark[4]. Alternatively, Fig. 4c suggests that another route to increasing robustness would be to increase the evidence rate ρ. Figure 5f shows $x_{\{s_n\}}$ after 1500 iterations for $\lambda = 10$ and an increased evidence rate of $\rho = 0.1$. In comparison to Fig. 5b which shows $x_{\{s_n\}}$ for the same level of error, i.e. $\lambda = 10$, but at the lower evidence rate of $\rho = 0.01$, we see that there are much better results for higher values of n and lower values of k.

---

[4] Arguably in this case the agents are not strictly solving best-of-n since they may not be identifying the best choice but rather just choices of high quality.

5   Discussion and Conclusions

Epistemic sets are a very simple way of representing uncertain beliefs in AI. They provide a computationally efficient approach to belief revision, and despite their representational limitations, we have shown that when deployed in a multi-agent system setting, they can provide a framework in which a whole population of agents can efficiently solve best-of-n problems. By combining updating based on direct evidence with belief pooling between agents, the agent population is able to compensate for sparsity of evidence, and also correct errors resulting from noise in the evidence collection process.

In future work we will investigate the application of epistemic sets to a broader class of decentralised learning problems, operating under different local interaction rules. In particular, we will weaken the well-stirred system assumption by imposing restrictions on the communications between agents. We will also consider other forms of evidential updating in which agents aim to discover the true state of the world rather than the highest quality state.

Acknowledgments. This work was funded and delivered in partnership between the Thales Group, University of Bristol and with the support of the UK Engineering and Physical Sciences Research Council, ref. EP/R004757/1 entitled "Thales-Bristol Part nership In Hybrid Autonomous Systems Engineering (T-B PHASE)."

References

1. Alchourrón, C., Gardenfors, P., Makinson, D.: On the logic of theory change: partial meet contraction and revision functions. J. Symbolic Logic **50**(2), 510–520 (1985)
2. Baroncholli, A.: The emergence of consensus: a primer. Roy. Soc. Open Sci. **5** (2018). https://doi.org/10.1098/rsos.172189, https://royalsocietypublishing.org/doi/full/10.1098/rsos.172189
3. Cholvy, L.: Opinion diffusion and influence: a logical approach. Int. J. Approximate Reasoning **93**, 24–39 (2018)
4. Ciucci, D., Dubois, D., Lawry, J.: Borderline vs. unknown: comparing three-valued representations of imperfect information. Int. J. Approximate Reasoning **44**, 1866–1889 (2014). https://doi.org/10.1016/j.ijar.2014.07.004. https://www.sciencedirect.com/science/article/pii/S0888613X14001157
5. Couso, I., Dubois, D.: Statistical reasoning with set-valued information: ontic vs epistemic views. Int. J. Approximate Reasoning **55**, 1502–1518 (2014)
6. Dubois, D., Liu, W., Ma, J., Prade, H.: The basic principles of uncertain information fusion. An organised review of merging rules in different representation frameworks. Inf. Fusion **32**, 12–39 (2016)
7. Grandi, U., Lorini, E., Perrussel, L.: Propositional opinion diffusion. In: Proceedings of the 2015 International Conference on Autonomous Agents and Multiagent Systems, AAMAS 2015, pp. 989–997. International Foundation for Autonomous Agents and Multiagent Systems (2015)
8. Hintikka, J.: Knowledge and Belief. Cornell University Press, Ithaca (1962)
9. Kakalis, N., Ventikos, Y.: Robotic swarm concept for efficient oil spill confrontation. J. Hazard. Mater. **32**, 12–39 (2008)

10. Katsuno, H., Mendelzon, A.: On the difference between updating a knowledge base and revising it. In: KR 1991 Proceedings of the Second International Conference on Principles of Knowledge Representation and Reasoning, pp. 387–394. Morgan Kaufmann (1991)
11. Konieczny, S., Pino Pérez, R.: Logic based merging. J. Philos. Logic **40**(2), 239–270 (2011)
12. Lee, C., Lawry, J., Winfield, A.: Negative updating combined with opinion pooling in the best-of-n problem in swarm robotics. In: Dorigo, M., Birattari, M., Blum, C., Christensen, A.L., Reina, A., Trianni, V. (eds.) ANTS 2018. LNCS, vol. 11172, pp. 97–108. Springer, Cham (2018). https://doi.org/10.1007/978-3-030-00533-7_8
13. Parker, C., Zhang, H.: Cooperative decision-making in multiple-robot systems: the best-of-n problem. IEEE Trans. Mechatron. **14**, 240–251 (2009)
14. Peleg, D.: Distributed coordination algorithms for mobile robot swarms: new directions and challenges. In: Pal, A., Kshemkalyani, A.D., Kumar, R., Gupta, A. (eds.) IWDC 2005. LNCS, vol. 3741, pp. 1–12. Springer, Heidelberg (2005). https://doi.org/10.1007/11603771_1
15. Perron, E., Vasudevan, D., Vojnovic, M.: Using three states for binary consensus on complete graphs. In: IEEE INFOCOM 2009, pp. 2527–2535 (2009)
16. Reina, A., Marshall, J., Trianni, V., Bose, T.: Model of the best-of-n nest-site selection process in honeybees. Phys. Rev. E **95** (2017)
17. Ruspini, E.H.: Epistemic logics, probability, and the calculus of evidence. In: Proceedings of the International Joint Conference on Artificial Intelligence, pp. 924–931 (1987)
18. Schwind, N., Inoue, K., Bourgne, G., Konieczny, S., Marquis, P.: Belief revision games. In: AAAI 2015, pp. 1590–1596 (2015)
19. Valentini, G., Ferrante, E., Dorigo, M.: Self-organized collective decision making: the weighted voter model. In: Proceedings of AAMAS 14, pp. 45–52 (2014)
20. Valentini, G., Ferrante, E., Dorigo, M.: The best-of-n problem in robot swarms: formalization, state of the art, and novel perspectives. Front. Rob. AI **4**, 9 (2017)
21. Vardi, M.Y.: On the complexity of epistemic reasoning. In: [1989] Proceedings. Fourth Annual Symposium on Logic in Computer Science, pp. 243–252, June 1989
22. Zadeh, L.: Fuzzy sets as a basis for a theory of possibility. Fuzzy Sets Syst. **1**, 3–28 (1978)

Multi-task Transfer Learning
for Timescale Graphical Event Models

Mathilde Monvoisin and Philippe Leray[(✉)]

Université de Nantes, LS2N UMR CNRS 6004, Nantes, France
mathilde.monvoisin@etu.univ-nantes.fr, philippe.leray@univ-nantes.fr

Abstract. Graphical Event Models (GEMs) can approximate any smooth multivariate temporal point processes and can be used for capturing the dynamics of events occurring in continuous time for applications with event logs like web logs or gene expression data. In this paper, we propose a multi-task transfer learning algorithm for Timescale GEMs (TGEMs): the aim is to learn the set of k models given k corresponding datasets from k distinct but related tasks. The goal of our algorithm is to find the set of models with the maximal posterior probability. The procedure encourages the learned structures to become similar and simultaneously modifies the structures in order to avoid local minima. Our algorithm is inspired from an universal consistent algorithm for TGEM learning that retrieves both qualitative and quantitative dependencies from event logs. We show on a toy example that our algorithm could help to learn related tasks even with limited data.

Keywords: Graphical Event Model (GEM) · Transfer learning ·
Multi-task learning (MTL) · Multivariate temporal point process ·
Process mining

1 Introduction

While probabilistic graphical models such as Dynamic Bayesian Networks [5,7] allow modeling of temporal dependencies in discrete time, some recent works are dedicated to modeling continuous time processes, with for instance, Continuous Time Bayesian Networks [10], Poisson Networks [12], Conjoint Piecewise-Constant Conditional Intensity Models [11].

In [6], Gunawardana and Meek have introduced Graphical Event Models (GEMs) that generalize such models, and Timescale GEM (TGEMs) which are GEMs where the temporal range and granularity of each temporal dependency is made explicit. TGEMs provide a way to understand temporal relationships between some variables, through a graph whose nodes are those variables and whose edges are the dependencies between them. In the case that the observed phenomena is a sequence of events, we can call it a process, so nodes are events and an edge between $node_i$ and $node_j$ means that the appearance of $event_i$ has some influence on the occurrence frequency of $event_j$.

© Springer Nature Switzerland AG 2019
G. Kern-Isberner and Z. Ognjanović (Eds.): ECSQARU 2019, LNAI 11726, pp. 313–323, 2019.
https://doi.org/10.1007/978-3-030-29765-7_26

In the same work, they have proposed an asymptotically consistent greedy algorithm to learn the structure and parameters of one single TGEM from an event log file. However, one may want to learn multiple processes that might be close. In order to complete this goal, multi-task transfer learning [3] is useful since it allows to learn k related models from k corresponding data sets.

In this paper, we propose an algorithm for transfer learning with TGEM, to allow simultaneous Multi Task Learning (MTL), inspired from Niculescu's method for MTL [8,9] with Bayesian Networks. Section 2 is a recall of the background elements useful afterwards, which include Timescale Graphical Event Models definition and current learning methods. Section 3 explains the global strategy used for learning multiple TGEMs, and proposes a method for likelihood and prior calculation in order to find the k structures that maximize the posterior probability of the structures given the data. Finally, a toy example in Sect. 4 illustrates the interest of MTL on TGEMs and Sect. 5 concludes on the contribution of this paper and the perspectives of research afterwards.

2   Background

This section is a reminder about formal definition of TGEMs and about the greedy search algorithm used for TGEM learning. More details about TGEM definition and learning can be found in Ref. [6].

The data D we use for learning consists in a timed sequence of events until time t^*:

$$D = \{(t_1, l_1), ..., (t_i, l_i), ..., (t_n, l_n)\}, \qquad . \qquad (1)$$

where $t_0 = 0 < t_i < t_{i+1} < t^*$ and $1 \leq i \leq n - 1$. l_i are labels from a finite vocabulary. The history $h(t)$ at any time t is the subset of events that occurred before t.

2.1   Timescale Graphical Event Models

A Timescale Graphical Event Model $\mathcal{M} = (\mathcal{G}, \mathcal{T})$ is a probabilistic graphical model that can represent data D as given above, using conditional intensity functions. The directed graph $\mathcal{G} = (\mathcal{L}, E)$ represents the dependencies between events, with $\mathcal{L}$ the labels of the events, E the edges of the graph. $\mathcal{T} = \{T_e\}_{e \in E}$ associates each edge e to a list of consecutive timescales T_e where $|T_e| \geq 1$. A timescale has the form $(a, b]$, with $a \geq 0$ and $b > a$.

We call *temporal range* the moment during which the timescales of some parent has an impact on the child node. On Fig. 1, the *temporal range* of A on C takes place between t and $t - 2$ with a certain intensity and between $t - 2$ and $t - 4$ with another one.

In all the models generalizing in the GEM family, the conditional intensity function is used to specify how the present depends on the past in an evolutionary process. This conditional intensity λ_l of a given event is usually a piecewise-constant function and varies according to the history of the parents in the model.

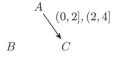

Fig. 1. One example of TGEM. $\mathcal{L} = \{A, B, C\}$, $E = \{AC\}$ and $\mathcal{T} = \{\mathcal{T}_{AC} = (0, 2], (2, 4]\}$. The occurrence of event C at time t will depend on possible occurrence of A in time windows $[t - 4, t - 2)$ and $[t - 2, t)$. The occurrences of A and B are independent from other events.

$\lambda_l(t|h) = \lambda_{l, C_l(h,t)}$ where the index $C_l(h, t)$ is the *parent count vector* of l: the number of occurrences of the parents in the timescales. For the entire paper, we consider that every element of $C_l(h, t)$ is either 0 or 1, thus only the fact that a parent has occurred or not within the corresponding timescale is important.

The marginal likelihood of a TGEM $\mathcal{M}$ according to data D can be computed at any time t, as defined in [6]:

$$p(D|\mathcal{M}, \lambda) = \prod_{l \in \mathcal{L}} \prod_{j \in pcv} \lambda_{l,j}^{n_{t,l,j}(D)} e^{-\lambda_{l,j} d_{t,l,j}(D)}, \tag{2}$$

with $n_{t,l,j}(D)$ and $d_{t,l,j}(D)$ respectively the total, at time t, number of occurrences of the event l within its parents configuration j, and duration of this configuration.

2.2   Learning TGEM for a Single Task

Single Task Learning (STL) consists in finding the optimal TGEM (its graph $\mathcal{G}$ and its timescales $\mathcal{T}$) from a dataset D as defined in Sect. 2.1.

A greedy BIC procedure for TGEM structure learning has been proven as asymptotically consistent in [6]. This strategy is to maximize the BIC score by performing the search on two stages, a *Forward search* by adding edges and refining the suitability of the timescales, and a *Backward search* which simplifies the model and deletes unnecessary edges.

The *Forward search* starts from the empty model $\mathcal{M}_0$ and computes the neighborhood until convergence to finally reach the model $\mathcal{M}_{\mathcal{FS}}$. The neighborhood $\mathcal{N}_{FS}(\mathcal{M})$ of $\mathcal{M}$ is computed with the three operators (*add, split* and *extend*) defined below. $\mathcal{M}' \in \mathcal{N}_{FS}(\mathcal{M}) \Leftrightarrow \exists O \in \mathcal{O} = \{O_{add}(e), O_{split}(\mathcal{T}_e), O_{extend}(e)\}$ such as $O(\mathcal{M}) = \mathcal{M}'$.

The *Backward search* starts with $\mathcal{M}_{\mathcal{FS}}$ and generates all neighbors $\mathcal{M}' \in \mathcal{N}_{BS}(\mathcal{M})$ such as $O(\mathcal{M}') = \mathcal{M}$ until convergence.

The BIC score used for the structure learning procedure is, at time t^*:

$$\text{BIC}_{t^*}(\mathcal{M}) = \log p(D|\mathcal{M}, \lambda_{t^*}(D)) - \sum_{l \in \mathcal{L}} |C_l| \log t^*, \tag{3}$$

with $\lambda_{t^*}(D)$ the optimal parameters obtained by likelihood estimation and $|C_l|$ the number of distinct parents configurations of node l.

The subfamily of TGEMs used by the structure learning procedure is called Recursive Timescale Graphical Event Models. A RTGEM refers to any TGEM that can be reached by performing *recursively* the following operators, starting from an empty model.

The *add edge* operator $O_{add}(e)$ takes as input an edge to be added to the graph with the default timescale $\mathcal{T} = (0, h_{def}]$ with h_{def} the default horizon. The *split timescale* operator $O_{split}(\mathcal{T}_e)$ takes as input a timescale $(a, b]$ of a specific existing edge and substitutes it by $(a, \frac{a+b}{2}], (\frac{a+b}{2}, b]$. The *extend horizon* operator $O_{extend}(e)$ takes as input an existing edge with horizon h and adds a timescale to this edge $(h, 2h]$ to double its horizon.

Gunawardana and Meek [6] have also proven than RTGEMs can approximate any non-explosive non-deterministic smooth marked point process with finite horizon.

2.3   Distance Between Two RTGEMs

In order to estimate the distance between two RTGEMs, Antakly et al. [1] have proposed an extension of the usual Structural Hamming Distance. The distance between two RTGEMs $\mathcal{M}_1 = ((\mathcal{L}, E_1), \mathcal{T}_1)$ and $\mathcal{M}_2 = ((\mathcal{L}, E_2), \mathcal{T}_2)$ with the same set of labels, is defined by:

$$d(\mathcal{M}_1, \mathcal{M}_2) = \sum_{e \in E_{sd}} 1 + \sum_{e \in E_{inter}} d(\mathcal{T}_{1,e}, \mathcal{T}_{2,e}), \tag{4}$$

where E_{sd} are the edges that are present in just one of the two models, and E_{inter} are present in both models. $\mathcal{T}_{i,e}$ are the timescales for edge e in model $\mathcal{M}_i$ and v_i is the list of endpoints[1] of model $\mathcal{M}_i$. The distance between the timescales is defined by:

$$d(\mathcal{T}_{1,e}, \mathcal{T}_{2,e}) = \frac{v_{nid}}{v_{nid} + v_{id}}, \tag{5}$$

where $v_{nid} = |v_1 \backslash v_2| + |v_2 \backslash v_1|$ and $v_{id} = |v_1 \cap v_2|$ is the number of endpoints that exist respectively in one and two of timescales $\mathcal{T}_{1,e}$ and $\mathcal{T}_{2,e}$.

2.4   Example of Single Task Learning

The aim of this example is to illustrate the previously introduced notions. We consider the two models of user behavior on e-banking sites described by the underlying RTGEMs $\mathcal{M}_1^R$ (Fig. 2) and $\mathcal{M}_2^R$ (Fig. 3). We also consider that we have (relatively small) web logs D_1 and D_2 for both sites.

The procedure when considering 2 models separately is to apply the *Forward-Backward search* introduced in Sect. 2.2, with a default horizon of edge $h = 10$. Figure 4 (resp. Fig. 5) describes the RTGEM $\mathcal{M}_1^{STL}$ (resp. $\mathcal{M}_2^{STL}$) obtained at the end of this STL algorithm applied to the event log D_1 (resp. D_2).

---

[1] It is another way of representing timescales. $\mathcal{T} = (0, a], (a, b], (b, c]$ is equivalent to $v = [0, a, b, c]$.

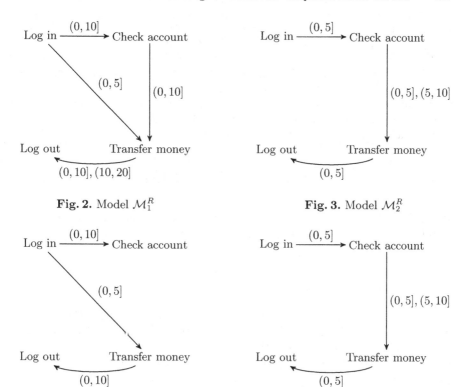

Fig. 2. Model $\mathcal{M}_1^R$

Fig. 3. Model $\mathcal{M}_2^R$

Fig. 4. Model $\mathcal{M}_1^{STL}$ learned from D_1

Fig. 5. Model $\mathcal{M}_2^{STL}$ learned from D_2

This single task learning doesn't take advantage of similarities between both tasks, and can lead to inaccurate results when there is a lack of data. For instance, in this toy example, the event log D_1 is not sufficient to identify the dependence between *Check Account* and *Transfer Money* in $\mathcal{M}_1^{STL}$.

3   Learning Multiple RTGEMs for Related Tasks

3.1   Problem Statement

In the previous section, we were interested in learning one single RTGEM from one single dataset. We now want to learn a *set* $\mathcal{S}_{best} = \{\mathcal{M}_1^*, \ldots, \mathcal{M}_k^*\}$ of k RTGEMs from k datasets $\mathcal{D} = \{D_1, \ldots, D_k\}$. The datasets contains event logs as defined in Sect. 2.1, with overlapping labels $\mathcal{L} = \bigcap_{q=1}^k \mathcal{L}_q \neq \emptyset$.

We are then interested in maximizing the posterior probability of the set of models given the data:

$$\mathcal{S}_{best} = argmax_{\mathcal{M}_1, \ldots, \mathcal{M}_k}(p(\mathcal{M}_1, \ldots, \mathcal{M}_k | D_1, \ldots, D_k)). \qquad (6)$$

According to Bayes rules, this posterior probability is proportional to the prior of the models and the marginal likelihood of the set:

$$p(\mathcal{M}_1, \ldots, \mathcal{M}_k | D_1, \ldots, D_k) \propto p(\mathcal{M}_1, \ldots, \mathcal{M}_k) p(D_1, \ldots, D_k | \mathcal{M}_1, \ldots, \mathcal{M}_k). \quad (7)$$

When considering *a priori* parameters independence, the marginal likelihood over the set of models can be factorized into the product of the marginal likelihood of each data set, and our problem statement can now be expressed as:

$$\mathcal{S}_{best} = argmax_{\mathcal{M}_1, \ldots, \mathcal{M}_k} (p(\mathcal{M}_1, \ldots, \mathcal{M}_k) \prod_{q=1}^{k} p(D_q | \mathcal{M}_q)). \quad (8)$$

In order to solve this task, we have to compute the marginal likelihood of each model $\mathcal{M}_q$, as well as the prior of the joint distribution over the models $\mathcal{M}_1 \ldots \mathcal{M}_k$ and finally we need a strategy to find the best set.

3.2 Marginal Likelihood

It was demonstrated by Chickering and Heckerman in [4] that the marginal log-likelihood of a Bayesian Network can be approximated by its BIC score. We will conjecture in this paper that the same approximation can be made for Timescale Graphical Event Models, which is the approximation made by [6] and [2]. For a model $\mathcal{M}_q$ at time t^*, the marginal log-likelihood $\log p(D_q | \mathcal{M}_q)$ can be approximated by the BIC score defined in Eq. (3) (Sect. 2.2).

3.3 Prior

The probability $p(\mathcal{M}_1, \ldots, \mathcal{M}_k)$ is called the prior because it represents the *a priori* knowledge of how similar the models might be. The two extreme cases are therefore, if the models have to be:

- independent: $p(\mathcal{M}_1, \ldots, \mathcal{M}_k) = \prod_{q=1}^{k} p(\mathcal{M}_q)$,
- equal: $p(\mathcal{M}_1, \ldots, \mathcal{M}_k)$ should be 1 if there is no difference between models, and 0 otherwise.

The solution offered in [8] for Bayesian Networks is to use a constant $\delta \in [0, 1]$ that penalizes every difference between the models structure when calculating the prior. Niculescu-Mizil and Caruana propose two different priors: one of them considers the minimum number of modifications necessary to make each edge the same in every structure (Edit Prior), and the other one considers the differences per pair of structures (Paired Prior). However, finding the minimum of edits to make all the edges the same is more difficult in TGEMs than in Bayesian Networks. Indeed, there are only two possibilities (present, not present) when considering an arc of a Bayesian Network, while the search space for a single arc of a TGEM is infinite because of the timescales that can always be split or extended. For this reason, the prior we suggest for TGEM learning is an adaptation of the Paired Prior and is defined as follows:

$$p(\mathcal{M}_1, ..., \mathcal{M}_k) = Z_{\delta,k} \prod_{1 \le q \le k} p(\mathcal{M}_q)^{\frac{1}{1+(k-1)\delta}} \prod_{1 \le q < q' \le k} (1-\delta)^{\frac{d(\mathcal{M}_q, \mathcal{M}_{q'})}{k-1}}, \quad (9)$$

where $Z_{\delta,k}$ is a normalization constant and $d(\mathcal{M}_q, \mathcal{M}_{q'})$ is the distance between two RTGEMs introduced in Sect. 2.3. In transfer learning context, all the models may not have identical labels. However, for the distance computing, we will only consider shared labels from both models $\mathcal{L}_{inter} = \mathcal{L}_1 \cap \mathcal{L}_2$ where $\mathcal{M}_1 = ((\mathcal{L}_1, E_1), \mathcal{T}_1)$ and $\mathcal{M}_2 = ((\mathcal{L}_2, E_2), \mathcal{T}_2)$.

The choice of the penalty δ affects the prior such as the higher δ, the closer the models have to be. When $\delta = 0$, the differences $d(\mathcal{M}_q, \mathcal{M}_{q'})$ will not affect $p(\mathcal{M}_1, \ldots, \mathcal{M}_k)$, so the models are considered as independent. When $\delta = 1$, any distance other than zero between the models makes $p(\mathcal{M}_1, \ldots, \mathcal{M}_k) = 0$ so the models have to be equal if we want a non-zero prior.

3.4 Finding the Best Set

The strategy named *MTL Forward-Backward search* that we propose to learn multiple TGEMs is inspired from the one proposed for Single Task Learning in Sect. 2.2. The strategy uses two steps, one *MTL Forward search* (Algorithm 1) that starts from an empty set $\mathcal{S}_0$ (i.e. a set of empty graphs) and one *MTL Backward search* (Algorithm 2) that starts with the set $\mathcal{S}_{FS}$ resulting from the *MTL Forward search*.

The scoring function $p(\mathcal{S}|\mathcal{D})$ optimized here is obtained from Eq. (8) with the posterior distribution defined in Eq. 9 and a marginal log-likelihood approximated by the BIC score defined in Eq. (3).

Algorithm 1. MTL Forward search	**Algorithm 2.** MTL Backward search
Input: $\mathcal{D} = \{D_1, \cdots D_k\}, \mathcal{S}_0$	**Input:** $\mathcal{D} = \{D_1, \cdots D_k\}, \mathcal{S}_{FS}$
Output: $\mathcal{S}_{FS}$	**Output:** $\mathcal{S}_{BS}$
1: $\mathcal{S} \leftarrow \mathcal{S}_0$	1: $\mathcal{S} \leftarrow \mathcal{S}_{BS}$
2: **repeat**	2: **repeat**
3: $refined \leftarrow$ **false**	3: $coarsened \leftarrow$ **false**
4: **for** $\mathcal{S}' \in \mathcal{N}_{FS}(\mathcal{S})$ **do**	4: **for** $\mathcal{S}' \in \mathcal{N}_{BS}(\mathcal{S})$ **do**
5: **if** $p(\mathcal{S}'|\mathcal{D}) > p(\mathcal{S}|\mathcal{D})$ **then**	5: **if** $p(\mathcal{S}'|\mathcal{D}) > p(\mathcal{S}|\mathcal{D})$ **then**
6: $\mathcal{S} \leftarrow \mathcal{S}'$	6: $\mathcal{S} \leftarrow \mathcal{S}'$
7: $refined \leftarrow$ **true**	7: $refined \leftarrow$ **true**
8: **end if**	8: **break**
9: **end for**	9: **end if**
10: **until** not $refined$	10: **end for**
11: $\mathcal{S}_{FS} \leftarrow \mathcal{S}$	11: **until** not $coarsened$
12: **return** $\mathcal{S}_{FS}$	12: $\mathcal{S}_{BS} \leftarrow \mathcal{S}$
	13: **return** $\mathcal{S}_{BS}$

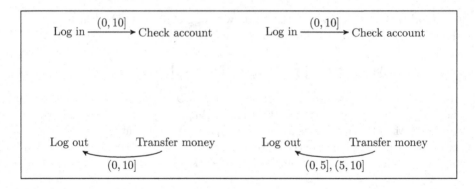

Fig. 6. $S = \{M_1, M_2\}$, set of models obtained during *MTL Forward search*

As highlighted in [9], changing only one model in the set at each iteration will usually weakly increase the score function or will lead to local optima, so our greedy algorithm has to test modifications in several models at the same time. For this reason, the neighborhoods $\mathcal{N}_{FS}(\mathcal{S})$ or $\mathcal{N}_{BS}(\mathcal{S})$ are generated thanks to the three operators (*add, split* and *extend*) introduced in Sect. 2.2, but applied to all the possible subsets of models in $\mathcal{S}$.

As also observed by Niculescu-Mizil and Caruana for Bayesian Networks, the size of a set of models neighborhood grows much faster than the size of a single model neighborhood. However, the search of the best solution in each neighborhood can be optimized by using a Branch and Bound algorithm, in a similar way to [9], that we can not describe here due to a lack of space.

4 Toy Example of Multi Task Learning

Let us take the simple example (labels are identical) used in Sect. 2.4, and consider now a Multi Task Learning by applying the *MTL Forward-Backward search* proposed in Sect. 3. It is not necessary that the labels are the same, just that they overlap. Figure 6 describes the set of RTGEMs $\{\mathcal{M}_1, \mathcal{M}_2\}$) jointly obtained at the end of the third iteration of the *MTL Forward phase*. The optimal sequence of operators was:

1. $\mathcal{M}_{1,2}$:O_{add}(Log in, Check account) (adding the edge in both models),
2. $\mathcal{M}_{1,2}$:O_{add}(Transfer money, Log out) (adding the edge in both models),
3. $\mathcal{M}_2$:O_{split}(Transfer money, Log out, $(0, 10]$) (splitting the edge in $\mathcal{M}_2$ only).

Let us develop now the next step of this phase. As usual in greedy algorithms, the neighborhood of $\mathcal{S}$, $\mathcal{N}_{FS}(\mathcal{S})$, will be explored in order to find the next considered set of models. This neighborhood consists in all the pairs of models $\{\mathcal{M}_1, \mathcal{M}_2\}$ generated from $\mathcal{S}$ by applying one single operator to $\mathcal{M}_1$ only, $\mathcal{M}_2$ only, and both $\mathcal{M}_1$ and $\mathcal{M}_2$.

Each box in Fig. 7 contains one neighbor of $\mathcal{S}$ corresponding to the operator O_{add}(Check account, Transfer money) applied to $\mathcal{M}_1$ only, $\mathcal{M}_2$ only, or both $\mathcal{M}_1$ and $\mathcal{M}_2$ (and respectively leading to $\mathcal{S}_1, \mathcal{S}_2$ and $\mathcal{S}_{12}$).

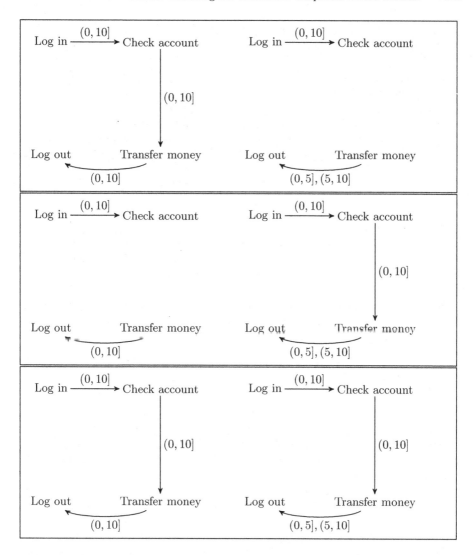

Fig. 7. Some neighbors of S during *MTL Forward search*: considering to add an edge between Check account and Transfer money (O_{add}(Check account, Transfer money)) to $\mathcal{M}_1$ only, $\mathcal{M}_2$ only, or both $\mathcal{M}_1$ and $\mathcal{M}_2$ and respectively leading to (from top to bottom) S_1, S_2 and S_{12}.

We consider now our objective function (Eq. (8)) with the Paired prior of Eq. (9) for our set of two models to determine which of the neighbors will be the one selected for the next step of the phase.

In a need for simplicity, we assume that $p(\mathcal{M}_1) = p(\mathcal{M}_2)$. To select the most likely set, we look for

$$argmax_{\mathcal{M}_1,\mathcal{M}_2}(p(D_1|\mathcal{M}_1) \cdot p(D_2|\mathcal{M}_2) \cdot (1 - \delta)^{d(\mathcal{M}_1,\mathcal{M}_2)}) \ . \tag{10}$$

The distance between the models $\mathcal{M}_1$ and $\mathcal{M}_2$ on the sets $\mathcal{S}, \mathcal{S}_1, \mathcal{S}_2$ and $\mathcal{S}_{12}$. From Figs. 6 and 7 are $d_{\mathcal{S}_1}(\mathcal{M}_1, \mathcal{M}_2) = d_{\mathcal{S}_2}(\mathcal{M}_1, \mathcal{M}_2) = \frac{4}{3}$ and $d_{\mathcal{S}}(\mathcal{M}_1, \mathcal{M}_2) = d_{\mathcal{S}_{12}}(\mathcal{M}_1, \mathcal{M}_2) = \frac{1}{3}$.

$p_{\mathcal{S}_1}(D_1|\mathcal{M}_1) < p_{\mathcal{S}}(D_1|\mathcal{M}_1)$, and $\mathcal{M}_2$ is the same in both $\mathcal{S}_1$ and $\mathcal{S}$. From previous calculations of distances, we know that the penalty term $(1 - \delta)^{d_{\mathcal{S}}(\mathcal{M}_1, \mathcal{M}_2)}$ is higher than $(1 - \delta)^{d_{\mathcal{S}_1}(\mathcal{M}_1, \mathcal{M}_2)}$, so $\mathcal{S}_1$ has a lower *posterior* than $\mathcal{S}$.

We assume that there is a strong dependency between Check account and Transfer money in D_2, that makes the presence of the arc from Check account to Transfer Money in $\mathcal{M}_2$ more likely than its absence in $\mathcal{M}_1$. We can express it with:

$$\frac{p_{\mathcal{S}_{12}}(D_1|\mathcal{M}_1)}{p_{\mathcal{S}_2}(D_1|\mathcal{M}_1)} > \frac{p_{\mathcal{S}_{12}}(D_2|\mathcal{M}_2)}{p_{\mathcal{S}_2}(D_2|\mathcal{M}_2)}. \tag{11}$$

Therefore, from $(1 - \delta)^{\frac{1}{3}} > (1 - \delta)^{\frac{4}{3}}$ and Eq. (11), $\mathcal{S}_{12}$ happens to be more likely than $\mathcal{S}_2$, and both are better sets than $\mathcal{S}$. Finally, $\mathcal{S}_{12}$ is selected for the next step of the *MTL Forward search* and the edge between Check Account and Transfer Money is now present in $\mathcal{M}_1$ when it was not considered in $\mathcal{M}_1^{STL}$ because of the lack of data.

We can see in this example that using our Multi-task learning algorithm can help to learn several related tasks even with limited data by using information from their related tasks.

5   Conclusion

Multi Task Learning is one kind of Transfer Learning, well studied in Machine Learning, but no so developed for probabilistic graphical models such as Bayesian Networks. Graphical Event Models are probabilistic graphical models dedicated to modeling continuous time processes. Single Task Learning such models from event logs have been very recently studied in a few works.

In this paper we proposed an algorithm for Multi Task Learning with Timescale Graphical Event Models. This algorithm, *MTL Forward-Backward search*, is an adaptation of the one proposed for Bayesian networks by [9] that also combines the efficient TGEM structure learning method proposed by [6] and the TGEM distance recently proposed in [1]. In this preliminary work, we also illustrated this algorithm with a simple toy example in order to give the intuition of its interest.

In the future, we plan to finalize the implementation of our algorithm, and to apply it on real world case studies in computer security. We also look forward to generalize this approach to another very recent GEM approach (Proximal GEM) [2].

We are also interested in studying beyond Multi-task learning and looking at other Transfer Learning tasks for Graphical Event Models, and dealing with both Incremental and Transfer Learning.

References

1. Antakly, D., Delahaye, B., Leray, P.: Graphical event model learning and verification for security assessment. In: Wotawa, F., Friedrich, G., Pill, I., Koitz-Hristov, R., Ali, M. (eds.) IEA/AIE 2019. LNCS, vol. 11606, pp. 245–252. Springer, Cham (2019). https://doi.org/10.1007/978-3-030-22999-3_22
2. Bhattacharjya, D., Subramanian, D., Gao, T.: Proximal graphical event models. In: Bengio, S., Wallach, H., Larochelle, H., Grauman, K., Cesa-Bianchi, N., Garnett, R. (eds.) Advances in Neural Information Processing Systems, vol. 31, pp. 8136–8145. Curran Associates, Inc. (2018)
3. Caruana, R.: Multitask learning: a knowledge-based source of inductive bias. In: Proceedings of the Tenth International Conference on Machine Learning, pp. 41–48. Morgan Kaufmann (1993)
4. Chickering, D., Heckerman, D.: Efficient approximation for the marginal likelihood of incomplete data given a Bayesian network. In: UAI 1996, pp. 158–168. Morgan Kaufmann (1996)
5. Dean, T., Kanazawa, K.: A model for reasoning about persistence and causation. Comput. Intell. **5**(2), 142–150 (1998)
6. Gunawardana, A., Meek, C.: Universal models of multivariate temporal point processes. In: Gretton, A., Robert, C.C. (eds.) Proceedings of the 19th International Conference on Artificial Intelligence and Statistics. Proceedings of Machine Learning Research, vol. 51, pp. 556–563. PMLR, Cadiz, 09–11 May 2016
7. Murphy, K.: Dynamic Bayesian networks: representation, inference and learning. Ph.D. thesis, University of California, Berkeley (2002)
8. Niculescu-Mizil, A., Caruana, R.: Inductive transfer for Bayesian network structure learning. In: Meila, M., Shen, X. (eds.) Proceedings of the Eleventh International Conference on Artificial Intelligence and Statistics. Proceedings of Machine Learning Research, vol. 2, pp. 339–346. PMLR, San Juan, 21–24 March 2007
9. Niculescu-Mizil, A., Caruana, R.: Inductive transfer for Bayesian network structure learning. In: Guyon, I., Dror, G., Lemaire, V., Taylor, G., Silver, D. (eds.) Proceedings of ICML Workshop on Unsupervised and Transfer Learning. Proceedings of Machine Learning Research, vol. 27, pp. 167–180. PMLR, Bellevue, 02 July 2012
10. Nodelman, U., Shelton, C., Koller, D.: Continuous time Bayesian networks. In: Proceedings of the Eighteenth Conference on Uncertainty in Artificial Intelligence (UAI), pp. 378–387 (2002)
11. Parikh, A.P., Meek, C.: Conjoint modeling of temporal dependencies in event streams. In: UAI Bayesian Modelling Applications Workshop, August 2012
12. Rajaram, S., Graepel, T., Herbrich, R.: Poisson-networks: a model for structured point processes. In: Proceedings of the Tenth International Workshop on Artificial Intelligence and Statistics, January 2005

Explaining Completions Produced
by Embeddings of Knowledge Graphs

Andrey Ruschel[⊠], Arthur Colombini Gusmão, Gustavo Padilha Polleti,
and Fabio Gagliardi Cozman

Universidade de São Paulo, São Paulo, SP, Brazil
{andrey.ruschel,fgcozman}@usp.br

Abstract. Advanced question answering typically employs large-scale
knowledge bases such as DBpedia or Freebase, and are often based on
mappings from entities to real-valued vectors. These mappings, called
embeddings, are accurate but very hard to explain to a human subject. Although interpretability has become a central concern in machine
learning, the literature so far has focused on non-relational classifiers
(such as deep neural networks); embeddings, however, require a whole
range of different approaches. In this paper, we describe a combination
of symbolic and quantitative processes that explain, using sequences of
predicates, completions generated by embeddings.

Keywords: Knowledge graph · Knowledge base · Explainable AI ·
Embedding · Interpretability

1 Introduction

Query answering systems and chatbots have benefited from symbolic facts stored
in knowledge graphs (KGs) such as NELL [16], YAGO [22], Freebase [2]. Even
though KGs contain many facts, typically stored as triples "subject, relationship,
object", KGs are far from complete, and a broad range of *completion* techniques
have emerged recently. These techniques often resort to *embeddings* that turn the
symbolic data into quantitative vectors, modeling relations between entities by
numeric operations over vectors [19]. Completion of a KG then relies on deciding
whether a particular triple is predicted through these numeric operations [25].

While embeddings usually offer the most accurate way to predict relationships between entities, they are rather hard to be interpreted by human users.
Consider an example that provides background on what it means to "interpret
an embedding". Take, for instance, a chatbot answering the question "Is Paris

The work has been supported by Itaú Unibanco S.A. through the Itaú Scholarship
Program (first and third authors are recipients). The last author is partially supported
by CNPq grant 312180/2018-7. The work has also been supported by FAPESP grant
2016/18841-0, and in part by the Coordenação de Aperfeiçoamento de Nivel Superior
(CAPES) - finance code 001.

© Springer Nature Switzerland AG 2019
G. Kern-Isberner and Z. Ognjanović (Eds.): ECSQARU 2019, LNAI 11726, pp. 324–335, 2019.
https://doi.org/10.1007/978-3-030-29765-7_27

the capital of France?". Suppose the chatbot uses a KG containing countries and cities and the relation *is capital of*, but the triple "Paris, is capital of, France" is not in the KG. And suppose chatbot returns YES: why is it? An acceptable reason might be that another triple shows that Paris is a capital, and Paris as located in France. Another possible explanation to the YES-answer would be one focusing on properties of the embedding itself: we might learn that every time we have a city and a country that map into vectors aligned in some particular direction, the former is the capital of the latter—and that this is happening with Paris and France. Note that the purpose of the latter explanation is to understand the behavior of the embedding when answering a particular question. Both explanations require insights that are more sophisticated than existing techniques to explain classifiers based on detecting which features are most relevant [13,20], as indicating that a particular dimension of an embedding strongly affects the decision does not, in itself, provides any clue as to what the embedding is doing.

In previous work [10], we proposed a framework that can produce explanations for KG completion tasks performed by any embedding model (model-agnostic). In essence, it works by mining significant paths in the KG to build a feature matrix from which explanations are extracted through logistic regression. We present here novel insights and refinements that significantly increase the fidelity of explanations. We start by providing in the next section some background knowledge about KBs and KGs, as well as a brief description of KG completion, graph features and latent features. We then summarize a few notions about interpretability and explanations and present our approach. Later we describe experiments and results and then close the paper with a discussion of possible future work.

2   Background

In this section, we provide a short review of the required concepts about knowledge graphs and knowledge graph completion.

2.1   Knowledge Graphs and Their Completion Tasks

Several large knowledge bases have been created to store information in triples of the form ⟨entity, predicate, entity⟩ (loosely following the RDF framework [23]). One may use also *head* for the subject, *relation* for the predicate, and *tail* for the object. For instance, information about the religion of the king Francis II of the Two Sicilies would be represented as a triple ⟨Francis II of the Two Sicilies, Religion, Catholic⟩. A set of triples can naturally be depicted as a *directed acyclic graph*, with an edge from the head to the tail of a triple.

Large KBs, generally built by extracting facts from unstructured text, suffer from incorrect/incomplete information. A fundamental task that involves KGs is *link prediction*. This is the task of, given a specific entity and a relation, finding a matching entity. For instance, for a given head and relation, to predict the tail $\langle e_h, r, ? \rangle$, or, given a tail and a relation, to predict the matching head

$\langle ?, r, e_t \rangle$. One may be interested instead in finding a relation between two entities $\langle e_h, ?, e_t \rangle$, a challenge sometimes called *relation prediction*. Another task is, given a triple $\langle e_h, r, e_t \rangle$ not previously seen on the KG, to evaluate whether this triple is true or false. This is referred to as *triple classification*. Finally, in *entity resolution* one must detect the same entity in different bits of information. For instance, one can find Barak Obama represented as *Barak Obama* and *B. Obama*.

Completion problems have been investigated within statistical relational learning [7,19]. Some useful notation from that literature will be employed in this paper. Let $\mathcal{E} = \{e_1, ..., e_N\}$ represent the set of all possible entities and $\mathcal{R} = \{r_1, ..., r_M\}$ represent the set of all possible relations in a KG. A possible triple is represented by $x_{h,r,t} = \langle e_h, r, e_t \rangle$, where e_h, r and e_t stand for *head, relation* and *tail*, respectively. Note that we use the *open world assumption* in this paper, meaning that facts not present in the KG are only considered unknown. We denote the set of all possible triples (or facts) in $\mathcal{G}$ by $\mathcal{T} = \mathcal{E} \times \mathcal{R} \times \mathcal{E}$.

We now contrast two approaches to KG completion: graph feature models and embedding models.

2.2   Graph Feature Models

Graph feature models aim to perform KG completion by observing characteristics of the graph to infer new facts, often by resorting to rules or similar symbolic manipulation [19].

With the PRA algorithm [11], Lao and Cohen suggested that triples can be predicted by a feature matrix constructed with random walks of bounded length in a KG. Based on PRA, Gardner et al. [6] proposed the *Subgraph Feature Extraction (SFE)* algorithm that we now summarize.

The idea is to focus on a relation r at a time. Denote by $\mathcal{D}^+$ the triples that have r as relation, to emphasize that these are "positive" triples. Note that $\mathcal{D}^+$ depends on the particular r, but we simplify notation by not explicitly referring to r. To train the model, a set of "negative" triples $\mathcal{D}^-$ is built by corrupting positive triples in the KG by randomly replacing one of its entities (head or tail) [25]. For a generic triple $\langle e_h, r, e_t \rangle$ we take a path (a sequence of edges) π from e_h to e_t with at most L edges; each edge in a path corresponds to a relation or the inverse of a relation. A triple is then associated with the set of all path patterns connecting its head to tail. In fact, not all possible paths from head to tail are generated; the SFE algorithm runs random walks to sample such paths.

Denote by $\pi_L(h, r, t)$ a path type of maximum length L connecting entity e_h to e_t. The set of all encountered paths $\pi_L(h, r, t)$ between those entities is represented by $\Pi_L(h, r, t)$. Denote by z_π a binary variable that indicates existence or not of a given path π. The feature vector extracted for a given triple is represented by $\phi_{hrt}^{SFE} = [z_\pi : \pi \in \Pi_L(h, r, t)]$. For a given relation r, using the latter expression, the SFE algorithm constructs a feature matrix combining the feature vectors ϕ_{hrt}^{SFE} extracted for each training example $\langle e_h, r, e_t \rangle \in \{\mathcal{D}^+ \cup \mathcal{D}^-\}$. This feature matrix can be used as input to any classifier; if one chooses a logistic regressor, a parameter matrix w_r is then obtained for the relation r.

We then calculate the probability of existence of the triple $\langle e_a, r, e_b \rangle \notin \mathcal{D}^+$ with $f_{abc}^{SFE} := w_r^T \phi_{abc}^{SFE}$.

To extract paths, the SFE algorithm builds subgraphs departing from each entity e_h and e_t with k steps. If two subgraphs $\mathcal{G}_h$ and $\mathcal{G}_t$ contain paths $\pi_{h,i}$ and $\pi_{t,i}$ departing from each entity and arriving at some intermediate node i, then a path type $\pi_{h,i} \cup \pi_{i,t}$ is stored in the feature vector. Gardner et al. [6] adopted random walks for the SFE algorithm but also proposed to construct the subgraphs via a *breadth-first search (BFS)*, to increase the number of extracted features. To keep the search computationally tractable during BFS in large KGs, they proposed to skip the expansion of nodes with a high out-degree (the number of incoming/outcoming edges). So if a path departing from e_h reaches a node with degree higher than a given number (a parameter of the model), that node will not be expanded in further steps, but it will still be considered as an intermediate node i that can later be merged to the subgraph departing from e_t. This strategy significantly increases the number of extracted features.

2.3 Embedding Models

Latent feature models map semantically rich entities and relations into real-valued vectors. The mappings are referred to as *embeddings*. The state-of-the-art in KG tasks uses this idea because operations and gradients can be easily run in numerical spaces. Usually, an embedding model represents entities as vectors of an arbitrary dimension and relations as operations within the same vector space. The symbolic data is then manipulated through numeric operations over those vectors. The "plausibility" of a triple $x_{h,r,t}$ is represented by a scoring function $f(x_{h,r,t} \mid \Theta)$ [25], where Θ represents the set of parameters of the model. Basically, there are two major families of embeddings: the first one, called *translational distance models*, focus in distance-base scoring functions like *TransE* and its extensions [17,25]. The second is formed of *semantic matching models*, that rely on scoring functions that measure semantic similarity, being *ANALOGY* and *RESCAL* examples of this family [17,25].

Although the interpretability techniques we propose in this paper are agnostic to any particular embedding, we will focus on TransE, a very popular embedding model proposed by Bordes et al. [3], inspired by *Word2Vec* [14]. In TransE, entities and relations are represented by vectors of an arbitrary dimension, and relations are translations within the vector space. A triple $\langle e_h, r, e_t \rangle$ is deemed true when $e_h + r = e_t$ (or rather, when this equality is approximately true within some threshold). The various vectors are obtained by optimization, taking into account all the information in the KG of interest. In short, a vector representation of entities and relations is learned so as to minimize the loss function $\|e_h + r - e_t\|$ for all facts in the knowledge base.

3 Explaining Embeddings

Currently, embeddings offer the most accurate way to complete KGs, but they are difficult to understand as they strip the underlying KG of its semantic content. On the other hand, graph feature models capture some of the structure of the KG, thus offering decisions that can be related to semantic properties but that are less accurate than the ones produced by embeddings. Thus one naturally asks whether it is possible to automatically explain completions produced by embeddings, perhaps using the symbolic features of the KG as a source of semantic guidance. This is our strategy in this paper.

First, a word on "Explainable AI", a topic that has received significant attention. Even though there has been work on explaining neural networks since at least the nineties [1,4], the recent emergence of very complex classifiers, for instance, ones based on very deep neural networks or very wide random forests, has led to many insights concerning the interpretation of automated decisions by classifiers [8]. It should be noted that the notion of "interpretability" is not a simple one [12]; it is certainly not an absolute notion as it depends on the target customer. A classifier can be interpreted through mathematical equations if the customer is a data scientist, but it should be explained in a textual manner if the customer is a lawyer in the auditing department. In any practical scenario, one may have "degrees" of interpretability depending on the understanding of causes for the prediction at hand [15]. Also, the interpretation of a model is related to the trust assigned by the user to the model; it is hard to trust a decision that cannot be adequately explained. Another perspective is this: when explaining a prediction, do we want to explain it "absolutely" in the sense that we want to justify why it makes sense, or do we want to explain is "relatively" to the model, focusing on the reasons why the model made the decision even if they are not logically perfect? The former seems useful in all circumstances, but the latter can be even more critical when the intended user is a data scientist trying to figure out the behavior of a classifier, or an auditing specialist trying to determine whether a classifier is biased or not. Simple metrics like accuracy are of no help when interpretability is needed [5]; in fact, there is a natural tension between accuracy and interpretability: more accuracy in the presence of large datasets tends to require more complex models that lead to less interpretable decisions.

There are two distinct basic approaches to interpreting classifiers. First, *decompositional* approaches extract rules and explanations by taking into account the specific structure of the classifier of interest [1]. Second, *pedagogical* or *agnostic* approaches consider the classifier as a black-box and implement a simpler classifier to mimic the outputs from the complex one and also provide explanations about the outputs. In this paper, we focus on agnostic techniques as we wish to provide tools that can be useful for a variety of embedding frameworks.

Interpretations are often generated by detecting which features are most important, through various sensitivity analyses, or perhaps by detecting which data points are most influential [24].

Alas, such approaches cannot work in interpreting embeddings. Indeed, embeddings turn a semantically rich input into numeric vectors, and one cannot operate in the vector space that is actually used in classification. The transfor-

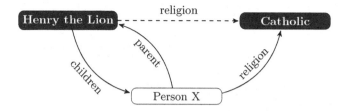

Fig. 1. Example produced from FB13. Entities of interest are in blue; dashed edge is assigned by TransE (other edges belong to FB13). (Color figure online)

mation from semantic entities to vectors is indeed a crucial part of the whole model, and should also be explained.

Our idea is to consider the embedding model we are trying to explain as a black-box and to implement an interpretable classifier around it, by extracting features (path patterns) from the original graph, using the labels predicted by the embedding. This interpretable classifier is then used to produced symbolic explanations, in the form of weighted *Horn clauses*, that are regarded as easily interpretable [19]. The result is a set of symbolic explanations obtained by graph features for each completion produced by the embedding. We pursued this basic idea in a previous publication [10], but our previous proposals had rather low fidelity. Here *fidelity* refers to the fraction of completions where the extracted graph feature model agrees with the original embedding. Of course one should aim at 100% fidelity when interpreting embeddings: there is no point in hiring an "interpreter" that may provide reasons for decisions that were *not* made.

The contribution of this paper is to present a framework able to produce explanations that correctly mimic the embedding classifier for every prediction. Before we plunge into a description of our contributions, it is worth considering a pair of examples generated by our implementation and depicted in Figs. 1 and 2. These examples were generated with data from two popular KGs, namely FB13 [2,21] and NELL186 [9,16]. Each explanation consists of a subgraph containing entities in the KGs; in some cases the specific entity is irrelevant ("Person X", and so on). However, a symbolic explanation can be easily produced: for instance, we see that Henry the Lion is considered catholic by TransE in the FB13 dataset for a simple reason: his child is catholic (a fact that is in FB13). Similarly, Fig. 2 describes why TransE determines UIC Flames to play in the Ice Hockey league using data in NELL186.

We now describe our method in detail; to do so, in this paragraph, we review the XKE-TRUE algorithm by Gusmão et al. [10]. Consider a KG $\mathcal{G}$ with $\mathcal{T} = \mathcal{E} \times \mathcal{R} \times \mathcal{E}$ representing the set of all possible triples of this KG. Denote by $g : \mathcal{T} \to \{0, 1\}$ the function of the embedding black-box classifier. Define $\Pi_{\mathcal{G}}$ as the set of all possible paths connecting two entities, and $P(\Pi_{\mathcal{G}})$ its power set. The feature extraction function performed by the SFE algorithm for a given triple $x_{h,r,t} \in \mathcal{T}$, and for the given graph $\mathcal{G}$, is represented by $SFE : \mathcal{T} \to P(\Pi_{\mathcal{G}})$. The result of applying the SFE algorithm to a triple $x_{h,r,t}$ is denoted by $\Pi_{h,r,t|\mathcal{G}} \in P(\Pi_{\mathcal{G}})$.

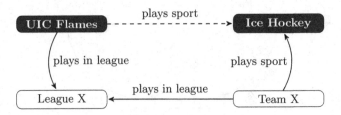

Fig. 2. Example extracted from NELL186. Entities of interest are in blue; the dashed edge is assigned by TransE (other edges belong to NELL186). The explanation for the fact that the team UIC Flames plays Ice Hockey is the fact that they play in a league, and another team that also plays in the same league also plays Ice Hockey. (Color figure online)

Then XKE-TRUE builds an auxiliary training set

$$\mathbb{D} = \{(SFE(x_{h,r,t} \mid \mathcal{G}), g(x_{h,r,t})) \mid x_{h,r,t} \in \mathcal{D}\} \tag{1}$$

and trains a interpretable classifier, in our case a logistic regressor, $g'' : P(\Pi_{\mathcal{G}}) \to \{0,1\}$ using $\mathbb{D}$, from which explanations are drawn in the form of weighted Horn clauses, where each rule (feature) is a path type extracted from $\mathcal{G}$, preceded by a weight assigned by the logistic regressor.

Even though the XKE-TRUE algorithm just outlined works reasonably, it has a severe drawback: its fidelity is far from 100%, thus making it fail to provide explanations in many cases. We now present two enhancements to the original XKE-TRUE to improve its fidelity. The first one allows SFE to extract more features from the KG, leading to a substantial increase in fidelity, while the second deals with the case where the interpretable classifier contradicts an embedding prediction.

3.1   Modified SFE

We have implemented a new version of the SFE algorithm, following the same principles proposed by Gardner et al. [6]. Gardner's SFE implementation with *BFS* makes use of a parameter of the model that specifies the maximum node out-degree, and nodes with out-degree above that value will not be expanded. In our BFS implementation, starting nodes with out-degree higher than the maximum value will be expanded only one step away. This single step turned out to be very useful to build many more paths sequences to be used as features.

3.2   XKE-e

Despite the modified SFE algorithm described above, for a reasonable amount of training examples, BFS was not able to find any feature in the KG simply because there was no path of length L connecting the triples entities. This hinders the fidelity of the resulting logistic regressor. One way to overcome this issue would be to increase L and find paths of greater length, but in large KGs, this would be computationally very expensive. Instead, we propose to leverage the knowledge obtained by applying SFE and logistic regression: for the triples with inconsistent prediction between the logistic regression and the embedding, we build new paths (using the embedding for KG completion) connecting the entities using the most critical paths according to the weights assigned by the logistic regression.

Let us denote by $\mathbb{D}_0$ the subset of $\mathbb{D}$ in which the SFE function applied to the triple resulted in an empty set $\Pi_{h,r,t|\mathcal{G}} = \varnothing$. For each triple $x_{h,r,t} \in \mathbb{D}_0$ we get the active rules w_c of the trained logistic regression and construct this path using $g(\cdot)$, following the algorithm below:

Algorithm 1.1 XKE-e

1: **procedure** BUILD-EXTENDED-SET(g, $\mathcal{G}$, $x_{h,r,t}$)
2:     $\hat{\Pi} \leftarrow \{\}$ ▷ Set of path types for the triple
3:     $\hat{g} \leftarrow \{\}$ ▷ Set of new found facts
4:     **for all** $\pi_{h,r,t} \in \Pi_{h,r,t|\mathcal{G}}^{(w \neq 0)}$ **do**
5:         **for each edge** $\in \pi_{h,r,t}$ **do**
6:             **if** $g(\pi_{i,j,k}) = 1$ **then**   ▷ If edge holds
7:                 $\hat{g} \leftarrow \pi_{i,j,k} \cup \hat{g}$
8:                 **if** $\pi_{h,r,t} = TRUE$ **then**   ▷ If path holds
9:                     $\hat{\Pi} \leftarrow \pi_{h,r,t} \bigcup \hat{\Pi}$
        return $\hat{\Pi}$, $\hat{g}$
10: **procedure** XKE-E
11:     $\Pi_{h,r,t|\mathcal{G}} \leftarrow SFE(x_{h,r,t} \mid \mathcal{G})$
12:     **for all** $x_{h,r,t} \in \mathbb{D}_0$ **do**
13:         $\hat{\Pi}, \hat{g} \leftarrow$ BUILD-EXTENDED-SET(g, $\mathcal{G}$, $x_{h,r,t}$)
14:         $\Pi_{h,r,t|\mathcal{G}} \leftarrow \hat{\Pi}$
15:         $\mathcal{G} \leftarrow \mathcal{G} \cup \hat{g}$
16:     Using $\mathbb{D}$, train an interpretable classifier $g' : P(\Pi_{\mathcal{G}}) \leftarrow \{0, 1\}$
17:     Draw explanations from g' in the form of Horn clauses

---

4   Experiments

Here we present the results obtained with our novel approach, comparing them with results obtained by Gusmão et al. [10]. We will evaluate them according to the following metrics:

- **Accuracy:** ratio of correct predictions by the logistic regressor (logit);
- **Fidelity:** ratio of prediction matches between logit and embedding;
- **F1:** the classical definition, obtained for accuracy and fidelity;
- **Average # of features per example:** is the average number of features extracted by SFE per example in the test set;
- **% of Examples with # of features > 0:** represents the proportion of cases in the test set with at least one feature extracted by SFE;

Table 1. Results (micro-average) for XKE-TRUE/TransE. Results marked with * were extracted from Gusmão et al. [10]. Results highlighted in bold are the best for each dataset.

Dataset	FB13			NELL186	
Embedding accuracy (%)	82.55			86.40	
Maximum path length (SFE)	4*	4	6	4*	4
Expand initial node (SFE)	No	Yes	Yes	No	Yes
Maximum node out-degree (SFE)	100	100	100	100	unl
Avg # of features per example	2.91	4.15	55.89	70.66	78.39
% Examples with # features > 0	54.73	87.09	91.41	50.01	55.47
Explanation mean rules	2.29	3.60	18.86	105.30	102.06
Explanation mean body rule length	3.09	3.01	4.62	3.86	3.86
Fidelity (%)	73.26	75.54	**83.04**	86.55	**88.59**
Accuracy (%)	73.43	75.49	**79.98**	89.10	**92.33**
F1 Fidelity (%)	76.66	76.24	**80.47**	83.19	**86.53**
F1 Accuracy (%)	77.35	77.25	**77.69**	86.89	**91.40**

For interpretability, we consider the following metrics:

- **Explanation Mean # of Rules:** the average number of rules per example that the logistic regressor assigned a weight different than zero;
- **Explanation Mean Body Rule Length:** the average number of relations forming active rules with weight different than zero assigned by the logistic regressor.

Below we describe all model parameters used in our experiments. To fairly compare each SFE strategy, we used the same embeddings trained by Gusmão et al. [10] as proposed by [18]; negative examples were generated via Bernoulli distribution at a 1–1 rate. Model training was limited to 1,000 epochs, splitted into 100 mini-batches and using SGD with Adagrad optimizer. The best model accuracy was obtained with a learning rate $\eta = 1$, ℓ_2 norm, margin $\gamma = 1$ and embedding dimension $k = 100$ for *FB13* and $k = 50$ for *NELL*. For the feature extraction with our implementation of the SFE algorithm, we now describe the parameters. The logistic regressor for each relation was trained using *SGD* to minimize the log loss with elastic net regularization and a grid search was run to find the best fidelity using the following parameters: $\eta = 1$ regularization ratio $\gamma = \{0.1, 0.7, 0.7, 0.9, 0.95, 0.99, 1.0\}$, regularization weight $\alpha = \{0.1, 0.001, 0.0001\}$ and stopping criteria $\epsilon = 0.001$; class weights were inversely proportional to their frequency to properly balance classes.

We can see from Table 1 that relaxing the parameters of the feature extraction deployed by the SFE algorithm helped to increase the number of features; more importantly, this increase led to an improvement in the fidelity results for both datasets. We were able to generate a scenery using paths of length 6 with *FB13*, which turned out to provide the highest amount of examples with at least one feature, and also the best fidelity in mimicking the embeddings predictions.

For the NELL186 dataset, we found the same positive effect, mainly because we were able to run tests with *unlimited* out-degree. We can see here that the *accuracy* of the interpretable model was better than the accuracy of the embedding model itself, indicating that one could use SFE+logistic regression as a primary tool for KG completion, without the use of the embedding.

As for the interpretability metrics, we saw no significant changes, except for the FB13 scenery with paths of maximum size 6, where the number of active rules per example exploded. In contrast, the explanation mean body rule length only grew from 3 to 4.63, which still can be regarded as easily interpretable.

Table 2. Example of explanation generated by XKE-TRUE, extracted from. [10].

Head	Henry the Lion
Relation	Religion
Tail	Catholic
Reason #1	(0.649) parent^{-1}, religion
Reason #2	(0.500) children, religion
Bias	(0.681)
XKE	0.862
Embedding	1

Table 2 brings the same example from Fig. 1, now showing the active rules obtained by the logistic regressor with its weights, explaining the prediction of the embedding. The intuition behind this explanation is that, by applying the SFE algorithm followed by a logistic regression, XKE-e generalized this fact from the KG attributing a weight for that reason. Indeed it is a much more convincing explanation than a statement saying that de dimension *43* of the entity vectors in $\mathbb{R}^{100}$ was the one that contributed to the correct answer.

5   Conclusion and Future Work

We have presented a novel method to produce explanations for completions generated by embeddings. We started with the XKE-TRUE algorithm [10]; our purpose was to increase the fidelity of that algorithm. We did this by introducing changes to the SFE algorithm and by adding various steps to XKE-TRUE (Algorithm 1.1). The resulting XKE-e algorithm is a novel scheme that has high fidelity, and that produces intuitive and plausible explanations, as we have shown through experiments and through examples.

Despite the advances described here, a significant concern when dealing with operations in knowledge graphs is the exponential growth of possible paths with the number of entities. In future work, we would like to investigate local explanations through the extraction of more expressive paths, as we believe that this would even further improve the fidelity of our explanations.

References

1. Andrews, R., Diederich, J., Tickle, A.B.: Survey and critique of techniques for extracting rules from trained artificial neural networks. Knowl.-Based Syst. **8**, 373–389 (1995)
2. Bollacker, K., Evans, C., Paritosh, P., Sturge, T., Taylor, J.: Freebase: a collaboratively created graph database for structuring human knowledge. In: ACM SIGMOD International Conference on Management of Data, pp. 1247–1250 (2008)
3. Bordes, A., Usunier, N., García-Durán, A., Weston, J.: Translating embeddings for modeling multi-relational data. In: Advances in Neural Information Processing Systems, pp. 2787–2795 (2013)
4. Craven, M.W., Shavlik, J.W.: Extracting thee-structured representations of thained networks. In: Advances in Neural Information Processing Systems, pp. 24–30 (1996)
5. Doshi-Velez, F., Kim, B.: Towards a rigorous science of interpretable machine learning. Harvard J. Law Technol. **31**, 841–887 (2017)
6. Gardner, M., Mitchell, T.: Efficient and expressive knowledge base completion using subgraph feature extraction. In: Proceedings of the 2015 Conference on Empirical Methods in Natural Language Processing, pp. 1488–1498. Association for Computational Linguistics (2015)
7. Getoor, L., Taskar, B.: Introduction to Statistical Relational Learning (Adaptive Computation and Machine Learning). The MIT Press, Cambridge (2007)
8. Gunning, D.: Broad Agency Announcement Explainable Artificial Intelligence (XAI). Technical report (2016)
9. Guo, S., Wang, Q., Wang, B., Wang, L., Guo, L.: Semantically Smooth Knowledge Graph Embedding. In: Proceedings of the 53rd Annual Meeting of the Association for Computational Linguistics and the 7th International Joint Conference on Natural Language Processing, pp. 84–94 (2015)
10. Gusmão, A.C., Correia, A.C., De Bona, G., Cozman, F.G.: Interpreting embedding models of knowledge bases : a pedagogical approach. In: 2018 ICML Workshop on Human Interpretability in Machine Learning (WHI 2018), pp. 79–86, no. Whi (2018)
11. Lao, N., Cohen, W.W.: Relational retrieval using a combination of path-constrained random walks. Mach. Learn. **81**(n.1), 53–67 (2010)
12. Lipton, Z.C.: The mythos of model interpretability. In: ICML Workshop on Human Interpretability in Machine Learning, pp. 96–100 (2016)
13. Lundberg, S.M., Allen, P.G., Lee, S.I.: A unified approach to interpreting model predictions. In: Advances in Neural Information Processing Systems, pp. 4765–4774 (2017)
14. Mikolov, T., Chen, K., Corrado, G., Dean, J.: Efficient estimation of word representations in vector space. CoRR abs/1301.3, 1–12 (2013)
15. Miller, T.: Explanation in artificial intelligence: insights from the social sciences. Artif. Intell. **267** (2017)
16. Mitchell, T., et al.: Never-ending learning. Commun. ACM **61**(1), 2302–2310 (2015)

17. Nguyen, D.Q.: An overview of embedding models of entities and relationships for knowledge base completion. CoRR abs/1703.0 (2017)
18. Nguyen, D.Q., Sirts, K., Qu, L., Johnson, M.: Neighborhood mixture model for knowledge base completion. In: Proceedings of the 20th SIGNLL Conference on Computational Natural Language Learning, Berlin, Germany, pp. 40–50 (2016)
19. Nickel, M., Murphy, K., Tresp, V., Gabrilovich, E.: A review of relational machine learning for knowledge graphs. IEEE **104**, 11–33 (2015)
20. Ribeiro, M.T., Singh, S., Guestrin, C.: "Why should i trust you?": explaining the predictions of any classifier. In: 22nd ACM SIGKDD International Conference on Knowledge Discovery and Data Mining, pp. 1135–1144 (2016)
21. Socher, R., Chen, D., Manning, C.D., Ng, A.Y.: Reasoning with neural tensor networks for knowledge base completion. In: Neural Information Processing Systems (2003), pp. 926–934 (2013)
22. Suchanek, F.M., Kasneci, G., Weikum, G.: YAGO: a core of semantic knowledge unifying WordNet and Wikipedia. In: WWW 2007 (2007)
23. W3: RDF 1.1 Concepts and Abstract Syntax. https://www.w3.org/TR/2014/REC-rdf11-concepts-20140225/
24. Wachter, S., Mittelstadt, B., Russell, C.: Couterfactual explanations without opening the black-box: automated decisions and the GDPR. Harvard J. Law Technol. **31**, 841 (2017)
25. Wang, Q., Mao, Z., Wang, B., Guo, L.: Knowledge graph embedding : a survey of approaches and applications. IEEE Trans. Knowl. Data Eng. **29**(12), 2724–2743 (2017)

Efficient Algorithms for Minimax Decisions Under Tree-Structured Incompleteness

Thijs van Ommen[1(✉)], Wouter M. Koolen[2], and Peter D. Grünwald[2,3]

[1] Utrecht University, Utrecht, The Netherlands
t.vanommen@uu.nl
[2] Centrum Wiskunde & Informatica, Amsterdam, The Netherlands
{wmkoolen,pdg}@cwi.nl
[3] Leiden University, Leiden, The Netherlands

Abstract. When decisions must be based on incomplete (coarsened) observations and the coarsening mechanism is unknown, a minimax approach offers the best guarantees on the decision maker's expected loss. Recent work has derived mathematical conditions characterizing minimax optimal decisions, but also found that computing such decisions is a difficult problem in general. This problem is equivalent to that of maximizing a certain conditional entropy expression. In this work, we present a highly efficient algorithm for the case where the coarsening mechanism can be represented by a tree, whose vertices are outcomes and whose edges are coarse observations.

Keywords: Coarse data · Incomplete observations ·
Minimax decision making · Maximum entropy

1 Introduction

Suppose a decision maker needs to choose an action a, and will suffer an amount of loss determined by a and an unobserved random variable X. The decision maker knows the distribution of X, and receives some information on the realized value $X = x$ in the form of a coarsened observation: a set $Y = y$ that includes x but also other, unrealized outcomes. Here, x lies in a finite set $\mathcal{X}$, and y is a member of some family $\mathcal{Y} \subset 2^{\mathcal{X}}$; both $\mathcal{X}$ and $\mathcal{Y}$ are also known to the decision maker, but importantly, the distribution $P(Y \mid X)$ of the coarsening mechanism is not.

One of the most well-known examples illustrating this setting is the Monty Hall puzzle [14]: In a game show, the contestant is faced with three doors $\mathcal{X} = \{1, 2, 3\}$. X indicates which of these hides a prize. The contestant initially picks a door; we will assume w.l.o.g. this is door 2. Then the quizmaster opens either door 1 or 3, revealing a goat. When both doors could be opened (i.e. if $X = 2$), one is chosen by the quizmaster's unknown coarsening mechanism. (In our

© Springer Nature Switzerland AG 2019
G. Kern-Isberner and Z. Ognjanović (Eds.): ECSQARU 2019, LNAI 11726, pp. 336–347, 2019.
https://doi.org/10.1007/978-3-030-29765-7_28

notation, Y is the set of the two doors that are still closed at this point; its possible values are the two members of $\mathcal{Y} = \{\{1, 2\}, \{2, 3\}\}$.) The contestant is now offered the option to switch to another door. The surprising insight is that the strategy of switching doors results in a larger probability of winning the prize.

We adopt a minimax (or worst-case) approach: we want to find a strategy (a function that maps each $y \in \mathcal{Y}$ to an action) for which the maximum expected loss over all possible coarsening mechanisms is as small as possible. Such a strategy does not require us to make any assumptions on the coarsening mechanism, so is a robust choice when the mechanism is unknown. (In the case of the Monty Hall puzzle, this means we do not need to assume anything about the distribution of Y given $X = 2$.)

In this paper, we propose efficient algorithms for a decision problem that generalizes the Monty Hall puzzle in the following way: to any number of outcomes, any distribution $P(X)$ over them, a very general class of loss functions, and any family $\mathcal{Y}$ that is the set of edges of an undirected tree over vertices $\mathcal{X}$. In other words, each set $y \in \mathcal{Y}$ consists of two elements of $\mathcal{X}$, and for each pair $x_a, x_b \in \mathcal{X}$, there exists a unique sequence $(x_i)_{i=1}^k$ of distinct elements from $\mathcal{X}$ with $x_1 = x_a$, $x_k = x_b$, and $\{x_1, x_2\}, \{x_2, x_3\}, \ldots, \{x_{k-1}, x_k\} \in \mathcal{Y}$. We will call $\mathcal{Y}$ the message structure, and its elements $y \in \mathcal{Y}$ messages.

To illustrate this generalization, consider a version of the Monty Hall game show with a row of doors $\mathcal{X} = \{1, 2, \ldots, n\}$, where the quizmaster will pick two adjacent doors that he leaves shut, revealing a goat behind each other door (so $\mathcal{Y} = \{\{i, i+1\} \mid i = 1, \ldots, n-1\}$). If the number of doors n is odd and the distribution on X is uniform, we find (see Sect. 2) that, upon observing $Y = y$, a cautious decision maker should assign probability $(n + 1)/(2n) > 1/2$ to the door in y with the odd index, and $(n - 1)/(2n) < 1/2$ to the one with the even index. The case $n = 3$ is the original Monty Hall puzzle, where a contestant who always switches to door 1 or 3 will have a probability of $2/3$ of winning the prize. The generalization we consider also extends to distributions over X other than uniform, for which the problem becomes computationally trickier.

The above message structure $\mathcal{Y}$ (which we call a *path graph*) may occur in practice as the message structure of a decision problem when, for example, a real-valued quantity of interest is reported to us as an integer, but we do not know if the value was rounded up or down. Then outcomes $x \in \mathcal{X}$ correspond to the intervals $(a_i, a_i + 1)$ between consecutive integers (we assume that the true value is a.s. not an integer), and messages to unions of two adjacent intervals.

This type of decision problem with incomplete information was introduced in [16], where minimax optimal strategies are characterized for arbitrary $\mathcal{Y}$, but where the question of how to compute them is not addressed. This computational problem was considered previously in [17], where it was demonstrated that the problem is hard for general $\mathcal{Y}$, but a direct formula could be given for finding a minimax optimal strategy in the special case that $\mathcal{Y}$ forms a partition matroid.[1]

---

[1] We refer to that paper for a definition, but remark that if each message consists of two outcomes, the class of partition matroids coincides with complete bipartite graphs. The message structure of the Monty Hall puzzle is an example.

Related work dealing with coarse data sometimes proceeds by making assumptions about the coarsening mechanism; for example, the CAR (coarsened at random) assumption [9] or the superset assumption [10] (see Sect. 2 for an explanation). Neither of these assumptions is compatible with the Monty Hall setting; see e.g. [5,7].

The approach we consider is more closely related to the maximin strategy described in [3] and studied specifically in [8]. However, it differs in some respects, e.g.: our objective does not feature a marginal or joint, but a conditional distribution of outcomes given messages (see (1)); and our objective is not interpreted as a (generalized) likelihood function, as we do not have a data set. This reflects that we are interested in making a decision pertaining to an unknown outcome *given* a *single* message (which is a coarsened observation of the outcome).

The general problem of finding a minimax optimal strategy can be solved using convex optimization. Reasonably efficient algorithms exist for this task [2], but they converge to the solution rather than computing it exactly. The algorithms in this paper are *strongly polynomial* [13]. A strongly polynomial algorithm finds the exact solution in a number of steps polynomial in the number of elements in the input, regardless of the precision of any numeric elements.

The rest of this paper is structured as follows. In Sect. 2, relevant results from [16] are summarized. We consider in Sect. 3 the special case where $(\mathcal{X}, \mathcal{Y})$ is a path graph, where the solution is given by a surprising and intuitive algorithm. This algorithm is extended to the case of arbitrary trees in Sect. 4. All proofs are in the appendix. The results on path graphs were previously described in [15]; the results on trees have not appeared elsewhere.

2   Preliminaries

A *decision problem with tree-structured incompleteness* is given by a finite set $\mathcal{X}$, a family $\mathcal{Y}$ of two-element subsets of $\mathcal{X}$ such that the undirected graph $(\mathcal{X}, \mathcal{Y})$ forms a tree, a distribution p over $\mathcal{X}$ having $p_x > 0$ everywhere, and a loss function $L : \mathcal{X} \times \mathcal{A} \to [0, \infty]$, where $\mathcal{A}$ is the set of actions available to the decision maker. We assume that the loss function L satisfies the conditions in Theorem 18 of [16].[2] A *coarsening mechanism* for this problem is an (unknown) joint distribution P on $\mathcal{X} \times \mathcal{Y}$ that satisfies $P(x, y) = 0$ whenever $x \notin y$, and $P(x) = \sum_{y \ni x} P(x, y) = p_x$ for each x.

The minimax approach may be viewed as a game: first the decision maker chooses a strategy $A : \mathcal{Y} \to \mathcal{A}$, then the opponent chooses a coarsening mechanism P, and finally the decision maker's expected loss $\sum_{x,y} P(x, y)L(x, A(y))$ is evaluated. The opponent's goal is to make the expected loss as large as possible (i.e. it is a zero-sum game). If the opponent were to move first, their best strategy would be the *maximin optimal* coarsening mechanism. In the case that L is

---

[2] Namely, that the generalized entropy H_L [6] is finite and continuous; further, H_L or an affine transformation of it is invariant under permutation of x_1 and x_3 whenever $\{x_1, x_2\}, \{x_2, x_3\} \in \mathcal{Y}$. See [16] for definitions.

logarithmic loss, this is the P that maximizes the expected conditional entropy

$$\sum_{y \in \mathcal{Y}} P(y) H(P(\cdot \mid y)) = \sum_{y \in \mathcal{Y}} P(y) \sum_{x \in y} -P(x \mid y) \log P(x \mid y). \tag{1}$$

It was found in [16] that if the action space is rich enough, this game has a Nash equilibrium, so that neither player benefits from knowing the other's strategy before picking their own. We concentrate on finding a maximin P, because once it is known, minimax optimal strategies A are typically easily determined.

For the setting considered in this paper, it was shown in [16, Theorem 18] that a coarsening mechanism P is maximin optimal if for some vector $q \in [0,1]^{\mathcal{X}}$ it satisfies the *RCAR condition*:

$$q_x = P(x \mid y) \text{ for all } y \in \mathcal{Y}, x \in y \text{ with } P(y) > 0, \text{ and}$$

$$\sum_{x \in y} q_x \leq 1 \text{ for all } y \in \mathcal{Y}. \tag{2}$$

Note that the second equation holds with equality for y with $P(y) > 0$. The vector q is called the RCAR vector; it exists and is unique by [16, Lemma 11]. We remark that the loss function L does not feature in this condition; this implies that a coarsening mechanism P that satisfies the RCAR condition is maximin optimal *regardless* of what loss function we are interested in,[3] and that the entropy maximization (1) is relevant for any L, not just logarithmic loss.

RCAR stands for 'reverse CAR', because the first line of (2) mimics the form of the CAR assumption [9], but with x and y switched. It is also similar in form to (part of) the superset assumption [10], but both CAR and the superset assumption look at $P(y \mid x)$, while RCAR considers $P(x \mid y)$. But the most crucial difference with these is that RCAR is not an assumption, but rather a condition we can check to verify if a coarsening mechanism is maximin optimal.

The question is: how do we find maximin optimal/RCAR coarsening mechanism? One computational challenge in finding the maximin optimal coarsening mechanism is that for some coarsening mechanisms P (including the maximin one), some y may have $P(y) = 0$. At such a point P, that y's contribution to the expected conditional entropy is nondifferentiable. This means for instance that standard convex optimization algorithms will converge slowly to such a point. The algorithms in this paper overcome this challenge.

3   Path Graphs and the Taut String Algorithm

Even though the results in this section will be superseded later when we give an algorithm for general trees, we devote this section to the subclass of decision problems for which the graph $(\mathcal{X}, \mathcal{Y})$ consists of just a single path. The reason is that for such problems, the minimax optimal coarsening mechanisms turn out to be described by an intuitive physical problem, for which an efficient algorithm is already known.

---

[3] This property is called *loss invariance*: see [16, Section 5.5].

3.1   Correspondence

Consider a decision problem where the messages form a path: for the $n \geq 2$ outcomes $\mathcal{X} := \{1, 2, \ldots, n\}$, the messages $\mathcal{Y}$ are $y_1 = \{1, 2\}, \ldots, y_{n-1} = \{n-1, n\}$. (A graph of the form $(\mathcal{X}, \mathcal{Y})$ is called a *path graph*.) Then the solution corresponds to that of a *taut string problem*. Imagine a string is constrained to pass above certain points (say, pins on a board), and below others. Then the string is pulled taut. The taut string will follow the shortest allowed path between its endpoints, going in straight line segments between the points it is pushed against.

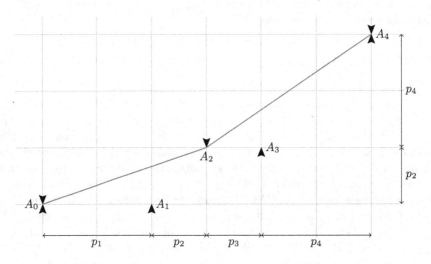

Fig. 1. The taut string problem corresponding to the decision problem with $\mathcal{X} = \{1, 2, 3, 4\}$, $\mathcal{Y} = \{y_1, y_2, y_3\}$ with $y_i = \{i, i+1\}$, and marginal on the outcomes $p = (1/3, 1/6, 1/6, 1/3)$. The arrowheads at the points $A_0, \ldots, A_4$ show on what side the string must pass. We see that the string, when pulled taut, touches the point A_2; its slope is $1/3$ to the left of A_2 and $2/3$ to the right.

Taut strings have been considered in the statistics literature before; see for example [1, 4, 11], where taut strings appear as a way of defining simple functions approximating noisy regression data. In these applications, all pins come in pairs, but we do not restrict the placement of pins in this way.

The taut string problem we are interested in uses the constraining points $A_0, A_1, \ldots, A_n$, with $A_0 = (0, 0)$ and

$$A_k = \left(\sum_{i \leq k} p_i, \sum_{\substack{i \leq k, \\ i \text{ even}}} p_i \right) \tag{3}$$

for $k \in \{1, \ldots, n\}$. The string must pass through the points A_0 and A_n; above points A_k with k odd; and below A_k for k even. See Fig. 1 for an example.

The following theorem relates the solution of this taut string problem instance to the maximin optimal P (the proof is in the appendix).

Theorem 1. *Given a decision problem on a path graph, find the solution of the taut string problem described in (3). Then a maximin optimal coarsening mechanism P is given by:*

- *For $0 < k < n$ such that the string touches the point A_k, we have $P(y_k) = 0$;*
- *For $0 < k < n$ such that the string does not touch A_k, $P(y_k) = |\delta_k|/(\alpha_k(1 - \alpha_k))$, where δ_k is the vertical distance between A_k and the string, and α_k is the slope of the string at that point;*
- *Also for k such that the string does not touch the point A_k (so $P(y_k) > 0$), the conditional distribution $P(\cdot \mid y_k)$ puts mass on the even outcome in y_k equal to the slope of the string as it passes above or below A_k.*

For the decision problem and corresponding taut string problem displayed in Fig. 1, we conclude that:

- $P(y_2) = 0$;
- For y_1, $|\delta_1| = 1/9$ (it is two-thirds of $p_2 = 1/6$) and $\alpha_1 = 1/3$, so $P(y_1) = (1/9)/(2/9) = 1/2$. In the same way, we find $P(y_3) = 1/2$.
- Using that the slope of the string above A_1 equals $1/3$, we find $P(2 \mid y_1) = 1/3$ (so $P(1 \mid y_1 = 2/3)$. Above A_3, the slope equals $2/3$, so $P(4 \mid y_3) = 2/3$ and $P(3 \mid y_3) = 1/3$.

3.2   Algorithm

We can now find maximin optimal coarsening mechanisms for decision problems on path graphs efficiently using the taut string algorithm, in $O(n)$ time [4]. This is clearly much more efficient than using a general purpose convex optimization algorithm. We list the taut string algorithm in Algorithm 1; for a more detailed explanation, we refer to [4] and [1].

The algorithm keeps track of three sequences of points. These represent piecewise linear functions: K is the solution, specified as the sequence of points the taut string pushes against; G is the *greatest convex minorant* (the pointwise maximum convex function respecting the upper bounds) of the part of the input that has been read but not added to the solution yet: and S the *smallest concave majorant* of that part of the input. Each of these sequences is a subsequence of the input points $A_0, \ldots, A_n$, and at each step of the algorithm, the first points of G and S are equal to the last point of K. The algorithm operates only on the beginning and end of each of these sequences, so these operations can be implemented efficiently without the aid of complex data structures.

We denote the number of elements in a sequence by $|K|$, use zero-based indices (so K[0] is the first element of K), and use negative indices to refer to the end of a sequence: K[−1] is the last element, K[−2] the second-to-last, etc. We write $\alpha(A_i, A_j)$ for the slope of the line segment from A_i to A_j, with $i < j$. For the points (3) used in the taut string problem corresponding to a decision problem, these slopes are given by $\alpha(A_i, A_j) = \sum_{i<k\leq j, k \text{ even}} p_k / \sum_{i<k\leq j} p_k$.

Algorithm 1: The taut string algorithm [4]

Input: Points $A_0, \ldots, A_n$ constraining the string, sorted from left to right
Output: Sequence of points K where the taut string pushes against the constraints

Let the sequences G, S, and K each consist of the single point A_0;
for i from 1 to n **do**
    if i is even or i = n **then**
        // A_i is an upper bound
        Append A_i to G;
        while $|G| \geq 3$ and $\alpha(G[-3], G[-2]) \geq \alpha(G[-2], G[-1])$ **do**
            | Delete second-to-last point from G;
        end
        // G is now convex
    end
    if i is odd or i = n **then**
        // A_i is a lower bound
        Append A_i to S;
        while $|S| \geq 3$ and $\alpha(S[-3], S[-2]) \leq \alpha(S[-2], S[-1])$ **do**
            | Delete second-to-last point from S;
        end
        // S is now concave
    end
    while $|G| \geq 2$, $|S| \geq 2$, and $\alpha(G[0], G[1]) < \alpha(S[0], S[1])$ **do**
        // No straight path remains between G and S
        Remove first point from G and from S;
        if (new) first point in G is to the left of first point in S **then**
            | Append first point in G to K, and prepend it to S;
        else
            | Append first point in S to K, and prepend it to G;
        end
    end
end
Append A_n to K.

---

4 Generalization to Trees

Our main contribution is the generalization of the taut string algorithm to tree-shaped coarsening mechanisms. The result is displayed as Algorithm 2.

4.1 Mathematical Description of the Algorithm

Algorithm 2 takes an arbitrary node $r \in \mathcal{X}$ to be the root of the tree. In reference to this root, we write $\mathrm{Ch}(x)$ for the children of node x, $\mathrm{pa}(x)$ for the (unique) parent of $x \neq r$, and $\mathrm{De}(x)$ for the descendants of x, which include x itself.

The f_x and $\bar{f}_x$ in the algorithm are piecewise linear functions from $[0, 1]$ to $[0, 1]$, which allows them to be stored efficiently; see Sect. 4.2 for details. If the graph is a path and one of its endpoint is chosen as the root, the computations can be simplified to those of Algorithm 1.

Theorem 2. *The P computed by Algorithm 2 is a coarsening mechanism that satisfies the RCAR condition (2).*

Algorithm 2: The taut tree algorithm

Treating the tree as a bipartite graph, call one of its parts S;
Pick an arbitrary node $r \in \mathcal{X}$ to be the root of the tree;
Recursively determine:

$$p_x^S = \sum_{x' \in S \cap De(x)} p_{x'};$$

$$f_x(\alpha) = p_x \alpha + \sum_{c \in Ch(x)} \bar{f}_c(\alpha);$$

$$\bar{f}_x(\alpha) = \begin{cases} \min(f_x(\alpha), p_x^S) & \text{if } x \in S; \\ \max(f_x(\alpha), p_x^S) & \text{if } x \notin S; \end{cases}$$

Find α_r such that $f_r(\alpha_r) = p_r^S$, and α_x such that $f_x(\alpha_x) = \bar{f}_x(\alpha_{pa(x)})$;

For $x \neq r$, let $P(\{x, pa(x)\}) = \begin{cases} \frac{1}{\alpha_x(1-\alpha_x)}|\bar{f}_x(\alpha_x) - p_x^S| & \text{if } \alpha_x = \alpha_{pa(x)}; \\ 0 & \text{otherwise}; \end{cases}$

Let $P(x \mid \cdot) = \begin{cases} \alpha_x & \text{for } x \in S; \\ 1 - \alpha_x & \text{for } x \notin S. \end{cases}$

4.2 Efficient Implementation of the Algorithm

An efficient algorithm would require a data structure that allows us to construct the piecewise linear function f_r (or enough of it to determine α_r) quickly. For computing the α_x's afterwards, it suffices to store in each node the value of α where $f_x(\alpha) = p_x^S$.

A function f can be represented using a double-ended priority queue whose elements represent the bends in f, keyed by α and with values equal to the change in slope at that point. Alongside this priority queue, $f(0)$, $f(1)$, $f'(0)$ and $f'(1)$ are stored. Because the function is increasing (see the proof of Theorem 2), taking a min (max) with a constant of an increasing function represented this way can be done by testing and discarding the smallest (largest) element repeatedly. Summing two functions requires merging their priority queues, so ideally we would use a priority queue that supports an efficient merge operation. By using for example a pointer-based min-max-pair heap [12], a worst-case time complexity of $O(n \log n)$ can be achieved.

5   Conclusion

In this paper, we showed how to efficiently find a coarsening mechanism that maximizes the conditional entropy (1), for two special cases of the message structure $\mathcal{Y}$. In the case where $(\mathcal{X}, \mathcal{Y})$ is a path graph, the problem can be reduced to a taut string problem, and solved in $O(n)$. We then generalized this algorithm to the case that $(\mathcal{X}, \mathcal{Y})$ is a tree, and showed that this allows the solution to be found in $O(n \log n)$.

Acknowledgments. This research was supported by Vici grant 639.073.04 and Veni grant 639.021.439 from the Netherlands Organization for Scientific Research (NWO).

Appendix

Proof (Theorem 1). Define $q \in \mathbf{R}^n$ as follows: for x even, let q_x equal the slope of the string between A_{x-1} and A_x; for x odd, let q_x equal one minus this slope. For any message $y_k \in \mathcal{Y}$ and outcome $x \in y_k$ such that the string does not push against A_k, we see that $P(x \mid y) = q_x$ since both are determined by the slope of the string as it passes above or below A_k. Thus P is RCAR with vector q.

For any message y_k with $P(y_k) = 0$, we need to verify that P satisfies $q_k + q_{k+1} \leq 1$. Note that the string does not touch A_1, because all other points are above the line through A_0 and A_1. By the same argument (replacing 'above' by 'below' if n is odd) the string does not touch A_{n-1}. If k is even, the string may be pushed down at A_k, so the slope to the left of that point, which equals q_k, must be smaller than or equal to the slope to the right, which equals $1 - q_{k+1}$. If k is odd, we similarly find $1 - q_k \geq q_{k+1}$. In both cases, we conclude $q_k + q_{k+1} \leq 1$.

What remains is to show that the marginal of P on the outcomes given in the theorem agrees with p. We do this by first deriving from p a formula for the marginal of P on the messages.

Consider two points A_a, A_b with $a < b$ such that the string touches these points but no points in between (thus the string follows a straight line between points A_a and A_b). Using the notation p_S for $\sum_{x \in S} p_x$, the slope of this segment of the string equals

$$\frac{p_{(a,b],\text{even}}}{p_{(a,b]}}.$$

This quantity equals q_x for any even $a < x \leq b$, so we call it q_{even}, and define $q_{\text{odd}} := p_{(a,b],\text{odd}} / p_{(a,b]} = 1 - q_{\text{even}}$.

For $a < x \leq b$, the marginal constraints $\sum_{y \ni x} P(y)P(x \mid y) = p_x$ are equivalent to $\sum_{y \ni x} P(y) = p_x / q_x$. By defining $P(y_0)$ and $P(y_n)$ as 0 (note that there are no such elements in $\mathcal{Y}$), we can write $\sum_{y \ni x} P(y) = P(y_{x-1}) + P(y_x)$. For $a < k \leq b$, we must have $P(y_k) = p_k / q_k - P(y_{k-1})$ by the marginal constraint on $x = k$. Using $P(y_a) = 0$ and applying this recursion repeatedly, we find that the following choice of marginal on messages satisfies all marginal constraints for $a < x \leq b$:

$$P(y_k) = (-1)^k \left(\frac{p_{(a,k],\text{even}}}{q_{\text{even}}} - \frac{p_{(a,k],\text{odd}}}{q_{\text{odd}}} \right) \qquad \text{for } a < k \leq b.$$

(Note that we get $P(y_b) = 0$ as required.) Meanwhile in string land, the point A_k is at height $p_{(0,k],\text{even}}$, and the string intersects the vertical line through A_k at height $p_{(0,a],\text{even}} + p_{(a,k]}q_{\text{even}}$; the (signed) difference is

$$\delta_k := p_{(a,k],\text{even}} - p_{(a,k]}q_{\text{even}} = p_{(a,k],\text{even}} - (p_{(a,k],\text{even}} + p_{(a,k],\text{odd}})q_{\text{even}}$$

$$= p_{(a,k],\text{even}}q_{\text{odd}} - p_{(a,k],\text{odd}}q_{\text{even}}.$$

This is positive at even k where the string passes below A_k, and negative at odd k. Thus the choice of marginal we found above equals the choice given in the theorem:

$$P(y_k) = (-1)^k \frac{\delta_k}{q_{\text{even}}q_{\text{odd}}} = \frac{|\delta_k|}{q_{\text{even}}q_{\text{odd}}}.$$

which is positive for all $a < k < b$. Because P also satisfies all marginal constraints, it follows that P is a probability distribution.

Proof (Theorem 2). Write $q_x = P(x \mid \cdot)$; we will show that q is the RCAR vector to P. We first prove the following claims by induction:

1. for all x, f_x is a strictly increasing function of α;
2. for all $x \neq r$, $\bar{f}_x$ is a nondecreasing function of α;
3. for all x such that either $x = r$ or $\alpha_x \neq \alpha_{\text{pa}(x)}$, $f_x(\alpha_x) = p_x^S = \sum_{x' \in \text{De}(x)} p_{x'}\alpha_{x'}$.

For the first two claims, both the base case and induction step are straightforward.

For the third claim, first observe that if $x = r$, α_x is chosen to satisfy $f_x(\alpha_x) = p_x^S$. The other case is $\alpha_x \neq \alpha_{\text{pa}(x)}$; this happens only if $f_x(\alpha_{\text{pa}(r)}) \neq \bar{f}_x(\alpha_{\text{pa}(x)})$, hence if $\bar{f}_x(\alpha_{\text{pa}(x)}) = p_x^S$. Then we also see that α_x is chosen to satisfy $f_x(\alpha_x) = p_x^S$. A base case for the induction occurs when x is such that all $x' \in \text{De}(x)$ have the same $\alpha_{x'}$. In such a base case, we have for all $x' \in \text{De}(x)$ that $\bar{f}_{x'}(\alpha_{x'}) = f_{x'}(\alpha_{x'})$, so that $f_x(\alpha_x) = \sum_{x' \in \text{De}(x)} p_{x'}\alpha_{x'}$. For the induction step, we can use that for $x' \in \text{De}(x)$ with $\alpha_{x'} \neq \alpha_{\text{pa}(x')}$, the claim holds by induction. Let T_x be the descendants of x that are not descendants of such an x'. Then for all $t \in T_x$, we again have $\bar{f}_t(\alpha_t) = f_t(\alpha_t)$, so that the total contribution to $f_x(\alpha_x)$ from terms $p_t\alpha$ with $t \in T_x$ equals $\sum_{t \in T_x} p_t\alpha_t$; combined with the induction hypothesis, we find $f_x(\alpha_x) = \sum_{x' \in \text{De}(x)} p_{x'}\alpha_{x'}$. This completes the proof of claim 3.

For given x, we get $\sum_{y \ni x} P(y)q_x = p_x$ (and $P(\{x, x'\}) = 0$ whenever $q_x + q_{x'} \neq 1$) if the marginal distribution $P(Y)$ satisfies

$$P(\{x, \text{pa}(x)\}) = \begin{cases} \frac{p_x}{q_x} - \sum_{x' \in \text{Ch}(x)} P(\{x', x\}) & \text{if } q_{\text{pa}(x)} = q_x; \\ 0 & \text{otherwise.} \end{cases}$$

For $x \in S$, this equals, and for $x \notin S$, this equals the negative of,

$$\sum_{x' \in T_x \cap S} \frac{p_{x'}}{q_{x'}} - \sum_{x' \in T_x \backslash S} \frac{p_{x'}}{q_{x'}} = \frac{1 - \alpha_x}{\alpha_x(1 - \alpha_x)} \sum_{x' \in T_x \cap S} p_{x'} - \frac{\alpha_x}{\alpha_x(1 - \alpha_x)} \sum_{x' \in T_x \cap S} p_{x'}$$

$$= \frac{1}{\alpha_x(1 - \alpha_x)} \left[\sum_{x' \in T_x \cap S} p_{x'} - \alpha_x \sum_{x' \in T_x} p_{x'} \right] = \frac{1}{\alpha_x(1 - \alpha_x)} \left[p_x^S - f_x(\alpha_x) \right],$$

where the final equality follows by applying claim 3 to $x' \notin T_x$. We see that this is equal (in both magnitude and sign) to the value assigned to $P(\{x, \mathrm{pa}(x)\})$ by the algorithm. Because $\sum_x p_x = 1$, it follows in particular that the algorithm's output is a probability distribution. The remaining aspects of the RCAR condition are now easy to verify.

References

1. Barlow, R.E., Bartholomew, D.J., Bremmer, J.M., Brunk, H.D.: Statistical Inference Under Order Restrictions: The Theory and Application of Isotonic Regression. Wiley, New York (1972)
2. Boyd, S., Vandenberghe, L.: Convex Optimization. Cambridge University Press, Cambridge (2004)
3. Couso, I., Dubois, D.: A general framework for maximizing likelihood under incomplete data. Int. J. Approximate Reasoning **93**, 238–260 (2018)
4. Davies, P.L., Kovac, A.: Local extremes, runs, strings and multiresolution. Ann. Stat. **29**, 1–65 (2001)
5. Gill, R.D., Grünwald, P.D.: An algorithmic and a geometric characterization of coarsening at random. Ann. Stat. **36**, 2409–2422 (2008)
6. Grünwald, P.D., Dawid, A.P.: Game theory, maximum entropy, minimum discrepancy and robust Bayesian decision theory. Ann. Stat. **32**, 1367–1433 (2004)
7. Grünwald, P.D., Halpern, J.Y.: Updating probabilities. J. Artif. Intell. Res. **19**, 243–278 (2003)
8. Guillaume, R., Couso, I., Dubois, D.: Maximum likelihood with coarse data based on robust optimisation. In: Antonucci, A., Corani, G., Couso, I., Destercke, S. (eds.) Proceedings of the Tenth International Symposium on Imprecise Probability: Theories and Applications (ISIPTA), pp. 169–180 (2017)
9. Heitjan, D.F., Rubin, D.B.: Ignorability and coarse data. Ann. Stat. **19**, 2244–2253 (1991)
10. Hüllermeier, E., Cheng, W.: Superset learning based on generalized loss minimization. In: Appice, A., Rodrigues, P.P., Santos Costa, V., Gama, J., Jorge, A., Soares, C. (eds.) ECML PKDD 2015. LNCS (LNAI), vol. 9285, pp. 260–275. Springer, Cham (2015). https://doi.org/10.1007/978-3-319-23525-7_16
11. Mammen, E., van de Geer, S.: Locally adaptive regression splines. Ann. Stat. **25**, 387–413 (1997)
12. Olariu, S., Overstreet, C.M., Wen, Z.: A mergeable double-ended priority queue. Comput. J. **31**(5), 423–427 (1991)
13. Schrijver, A.: Combinatorial Optimization: Polyhedra and Efficiency, vol. 24. Springer, Heidelberg (2003)
14. Selvin, S.: A problem in probability. Am. Stat. **29**, 67 (1975). Letter to the editor

15. Van Ommen, T.: Better predictions when models are wrong or underspecified. Ph.D. thesis, Mathematical Institute, Faculty of Science, Leiden (2015)
16. Van Ommen, T., Koolen, W.M., Feenstra, T.E., Grünwald, P.D.: Robust probability updating. Int. J. Approximate Reasoning **74**, 30–57 (2016)
17. Van Ommen, T.: Computing minimax decisions with incomplete observations. In: Antonucci, A., Corani, G., Couso, I., Destercke, S. (eds.) Proceedings of the Tenth International Symposium on Imprecise Probability: Theories and Applications (ISIPTA), pp. 358–369 (2017)

Precise and Imprecise Probabilities

A Probabilistic Graphical Model-Based Approach for the Label Ranking Problem

Juan Carlos Alfaro$^{(\boxtimes)}$, Enrique González Rodrigo, Juan Ángel Aledo,
and José Antonio Gámez

School of Computer Science and Engineering, University of Castilla-la Mancha,
02071 Albacete, Spain
{JuanCarlos.Alfaro,Enrique.GRodrigo,JuanAngel.Aledo,Jose.Gamez}@uclm.es

Abstract. The goal of the *Label Ranking (LR) Problem* is to learn *preference models* that predict the preferred ranking of class labels for a given unlabelled instance. Different well-known machine learning algorithms have been adapted to deal with the LR problem. In particular, fine-tuned instance-based algorithms have exhibited a remarkable performance, specially when the model is trained with *complete* rankings, while model-based algorithms (e.g. decision trees) have been proved to be more robust when some data is missing, that is, the model is trained with *incomplete* rankings.

Probabilistic Graphical Models (*PGMs*, e.g. *Bayesian networks*) have not been considered to deal with this problem because of the difficulty to model permutations in that framework. In this paper, we propose a *Hidden Naive Bayes classifier* (*HNB*) to cope with the LR problem. By introducing the hidden variable we can design a hybrid Bayesian network in which several types of distributions can be combined, in particular, the *Mallows* distribution, which is a well-known distribution to deal with permutations. The experimental evaluation shows that the HNB classifier is competitive in accuracy when compared with *Label Ranking* (*decision*) *Trees*, being, moreover, considerably faster.

Keywords: Naive Bayes · Label Ranking · Machine learning

1 Introduction

Preferences are comparative judgments about a set of alternatives or choices. The *Label Ranking (LR) Problem* [9] is a well-known non standard supervised classification problem [7,17], whose goal is to learn *preference models* that predict the preferred ranking over a set of class labels for a given unlabelled instance.

Formally, we consider a problem domain defined over n *predictive variables* or *attributes*, $X_1, \ldots, X_n$, and a *class variable* C with k labels, $dom(C) = \{c^1, \ldots, c^k\}$. We are interested in predicting the ranking π of the labels for a given unlabelled instance $x = (x_1, \ldots, x_n) \in dom(X_1) \times \cdots \times dom(X_n)$ from a dataset $\mathbf{D} = \{(x_1^j, \ldots, x_n^j, \pi^j)\}_{j=1}^N$ with N labelled instances. Hence, the LR

© Springer Nature Switzerland AG 2019
G. Kern-Isberner and Z. Ognjanović (Eds.): ECSQARU 2019, LNAI 11726, pp. 351–362, 2019.
https://doi.org/10.1007/978-3-030-29765-7_29

problem consists in learning a LR-Classifier $\mathcal{C}$ from $\mathbf{D}$ which generalized well on unseen data. In other words, the goal of the LR problem is to induce a model able to predict permutations by taking advantage of all the available information in the learning process. In the literature we can find different approaches to tackle this problem:

- *Transformation methods.* They transform the whole problem into a set of single-class classifiers: *labelwise* [29] and *pairwise* approaches [15,19], *chain classifiers* [16], etc.
- *Adaptation methods.* They adapt well-known machine learning algorithms to cope with the new class structure. Cheng et al. in [9] introduce a *model-based algorithm* that induces a decision tree and a *model-free algorithm* which uses k-nearest neighbors. Other techniques, like association rules [26] or neural networks [25], have been also adapted.
- *Ensemble methods.* Recently, different tree-based aggregation approaches like *Random Forests* [5] and *Bagging predictors* [4] have been successfully applied to the LR problem [1,28,30].

In this paper we propose a new model-based LR-classifier focusing on adaptation methods. Our motivation is twofold:

- Although fine-tuned instance-based algorithms have exhibited a remarkable performance, specially when the model is trained with *complete rankings* (i.e., *permutations*), model-based algorithms have been proved to be more robust when some data is missing, that is, when the model is trained with *incomplete rankings*.
- *Probabilistic Graphical Models* (*PGMs*), e.g. *Bayesian networks* [22], have not been used in this problem because of the difficulty to model permutations in this framework [8,9].

The proposed LR-classifier is modelled by using a *hybrid Bayesian network* [12] where different probability distributions are used to conveniently model variables of distinct nature: multinomial for discrete variables, Gaussian for numerical variables and *Mallows model* for permutations [24]. The Mallows probability distribution is usually considered for permutations, and is, in fact, the core of the *decision tree algorithm* (*Label Ranking Trees*, *LRT*) proposed in [9].

To overcome the constraints regarding the topology of the network when dealing with different types of variables, we propose a *Naive Bayes* structure where the root is a *hidden discrete variable*. In that way, only univariate probability distributions have to be estimated for each state of the hidden variable. We design a learning algorithm based on the well-known *Expectation-Maximization* (EM) estimation principle and we provide several inference schemes which combine methods to tackle the *Kemeny Ranking Problem* (*KRP*) [21] with probabilistic inference.

The rest of the paper is structured as follows. In Sect. 2 we review some basic notions needed to deal with rank data. In Sect. 3 we formally describe the proposed *Hidden Naive Bayes* (*HNB*) as well as the algorithms to induce it from data and carry out inference. In Sect. 4 we set forth the empirical study carried out to evaluate the method designed in this paper. Finally, in Sect. 5, we provide the conclusions and future research lines.

2 Background

In this section, we review some notions regarding the *Kemeny Ranking Problem* [21], the *Mallows probability distribution* [24] and the *Naive Bayes model* [22].

2.1 Kemeny Ranking Problem

The *Kemeny Ranking Problem (KRP)* [21] consists in obtaining the *consensus permutation (mode)* $\pi_0 \in \mathbb{S}_k$ that best represents a sample with N permutations $\Pi = \{\pi_1, \ldots, \pi_N\}$, $\pi_i \in \mathbb{S}_k$. Here, $\mathbb{S}_k$ stands for the set of permutations of k elements. Formally, the KRP looks for the consensus permutation $\pi_0 \in \mathbb{S}_k$ that minimizes

$$\pi_0 = \underset{\pi \in \mathbb{S}_k}{argmin} \sum_{i=1}^{N} D(\pi_0, \pi_i)$$

where $D(\pi, \tau)$, $\pi, \tau \in \mathbb{S}_k$, is a distance measure between two permutations π and τ. Normally, the *Kendall distance* is taken, and the (greedy) *Borda count algorithm* [3] is employed to solve the KRP, due to its trade-off between efficiency and accuracy. Borda count algorithm basically assign n points to the item ranked first, $n - 1$ to the second one and so on. Once all the input rankings have been computed, the items are sorted according to the number of accumulated points.

2.2 Mallows Probability Distribution

The *Mallows probability distribution (Mallows model)* [24] is an exponential probability distribution over permutations based on distances. The Mallows model is parametrized by two parameters, the *central permutation (mode)* $\pi_0 \in \mathbb{S}_k$ and the *spread parameter (dispersion)* $\theta \in [0, +\infty)$. Given a distance D in $\mathbb{S}_k$, the probability assigned to a permutation $\pi \in \mathbb{S}_k$ by a Mallows distribution with $\pi_0 \in \mathbb{S}_k$ and $\theta \in [0, +\infty)$ is

$$P(\pi; \pi_0, \theta) = \frac{e^{-\theta \cdot D(\pi, \pi_0)}}{\Psi(\theta)}$$

where $\Psi(\theta)$ is a normalization constant. The spread parameter θ quantifies the concentration of the distribution around π_0. For $\theta = 0$, a uniform distribution is obtained, while for $\theta = +\infty$, the model assigns a probability equal to 1 to π_0 and equal to 0 to the rest of the permutations. In our work, we take D as the Kendall distance.

Parameter estimation can be done by using Borda count method for π_0 and, although there is no closed form to estimate θ, numerical algorithms, e.g. Newton-Raphson, can be used to accurately estimate it. Therefore, both parameters can be efficiently estimated (polinomial time) [20].

2.3 Naïve Bayes

Naive Bayes models [22] are well-known *probabilistic classifiers* based on the conditional independence hypothesis, that is, every pair of features are considered conditionally independent given the class variable. As most of the probabilistic classifiers, Naive Bayes models follow the *maximum a posteriori (MAP)* principle, that is, they return the most probable class for any input instance. Formally, given an input instance $x = (x_1, \ldots x_n) \in dom(X_1) \times \cdots \times dom(X_n)$ and being C the class variable with $dom(C) = \{c^1, \ldots, c^n\}$, a *Naive Bayes Classifier* $\mathcal{C}$ returns

$$\mathcal{C}(x) = \underset{c \in dom(C)}{argmax} P(c \mid x) = \underset{c \in dom(C)}{argmax} P(x, c) = \underset{c \in dom(C)}{argmax} \prod_{i=1}^{n} P(x_i \mid c) \cdot P(c)$$

according to the *Bayes' theorem* and the conditional independence hypothesis, respectively. The conditional distributions above may be multinomial for discrete attributes and Gaussian for continuous attributes.

3 Hidden Naïve Bayes LR-Classifier

This section presents the proposed model, defining the structure as well as the parameter estimation process.

3.1 Model Definition

To overcome the constraints regarding the topology of the network when dealing with different types of variables, the model proposed here is a mixture model with the Naive Bayes assumption. The root element of the model is the discrete hidden variable, which we will denote as $z \in 1\ldots Z$, where Z is the total number of mixture models. The rest of the variables are observed variables. We consider two types of observed variables:

- The *feature variables*, observed both in the training and in the test phase. We consider two kinds: discrete variables, denoted as d_j, and continuous variables, denoted as x_k.
- The *ranking variable*, denoted as π, which is only present at training time. This is the one to infer.

Figure 1 offers a representation of the model with the different types of variables described above. The model assumes that each of these variable types follow a different conditional distribution given the root variable:

- Continuous variables follow a Gaussian distribution, $P(x_k|z) = \mathcal{N}(x_k; \mu_k^z, \sigma_k^z)$
- Discrete variables follow a Multinomial distribution, $P(d_j|z) = Mult(d_j; \mathbf{a_j^z})$
- The ranking variable follows a Mallows distribution, $P(x_k|z) = \mathcal{M}(\pi; \pi_0^z, \theta^z)$
- The hidden variable follows a Multinomial distribution, $P(z) = Mult(z; \mathbf{w})$

The parameters for each of the conditional distributions need to be estimated to perform inference using the model.

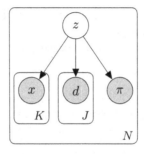

Fig. 1. The proposed Hidden Naïve Bayes model

3.2   Parameter Estimation

Due to the fact that the model has one hidden variable, we use the EM algorithm to estimate jointly the parameters of both the observed and the hidden variables. The EM algorithm consists of two steps: the Expectation step (E step), where the value for the hidden variable is estimated; and the Maximization step (M step), where the parameters for the conditional distributions are obtained.

E step. Under the assumption that the parameters of the model (μ_k^z, σ_k^z, $\mathbf{a_j^z}$, π_0^z, θ^z, $\mathbf{w}$) are known, the probability of an example to be in a mixture is

$$P(z_i|\mathbf{d_i}, \mathbf{x_i}, \pi_i) \propto P(z_i)P(\mathbf{d_i}, \mathbf{x_i}, \pi_i|z_i) = P(z)P(\pi_i|z)\prod_j P(d_{ij}|z_i)\prod_k P(x_{ik}|z_i)$$

Normalizing the above expression for all values of the hidden variable we obtain the probability of an example to be in the mixture.

M step. Under the assumption that the probabilities of belonging to each mixture for all examples are known, the parameters of the model can be estimated as follows:

- Multinomial parameters for the discrete variables, $P(d_j|z)$. MLE estimation is done, where the count for each instance is weighted by the probability of that instance given a mixture $z = z_i$.
- Gaussian distribution parameters for the continuous variables, $P(x_k|z)$. Analogous to the previous case but using a Gaussian distribution $\mathcal{N}(x_k; \mu_k^z, \sigma_k^z)$ for each $z = z_i$.
- Mallows distribution parameters for the ranking variable. For each $z = z_i$ a Mallows distribution $\mathcal{M}(\pi; \pi_0^z, \theta^z)$ must be estimated. In particular $\pi_0^{z_i}$ is computed by applying a weighted version of Borda count algorithm (points assigned to items are weighted by the probability of that instance given the mixture z_i), and θ^{z_i} is calculated by using a numerical optimization process (e.g. Newton-Raphson).
- The mixture model probabilities $P(z)$ are computed according to the weights $P(z|\mathbf{d_i}, \mathbf{x_i}, \pi_i)$ for each mixture $z = z_i$.

Stop Condition. Although the model can easily be extended to use several types of stop conditions, we propose to check the convergence on the probabilities with which the samples belong to each mixture.

3.3 Learning Process

The learning procedure includes several executions of the EM algorithm with an increasing number of mixtures. This process is based on the one proposed in [23]. The algorithm starts with a predefined number of mixtures (a hyperparameter), and at each iteration the number of mixtures is increased according to a given parameter (in our case, one by one).

For each of these iterations, the mixtures must be initialized as a previous step to the EM algorithm. For each new mixture added, a sample with replacement of the dataset is used for parameter estimation. The parameters for the conditional probabilities given this mixture are calculated as if all the data points had a probability of 1 to belong to the sample. After that, the data points which were not used for the initialization of the new mixtures are used for the parameter optimization procedure. If the solution obtained does not improve upon the previous solution (using the *Kendall coefficient* τ_K as score over a validation set), the algorithm returns the best solution. If the solution improves, the algorithm continues adding new mixtures.

3.4 Inference

In the inference process, the method needs to predict the best consensus ranking for a new data point. In our proposal we do that by marginalizing variables until obtaining an expression for the probability of a ranking

$$P(\pi_s|\mathbf{d_r}, \mathbf{x_r}) \propto \sum_{z_i} P(z_i)P(\pi_s|z_i) \prod_{j}^{J} P(d_{rj}|z_i) \prod_{k}^{K} P(x_{rk}|z_i)$$

To estimate the best permutation $\widetilde{\pi}$, we take the one that maximizes the score

$$\pi^* = \underset{\pi_s \in \mathbb{S}_k}{argmax}\ P(\pi_s|\mathbf{d_r}, \mathbf{x_r})$$

However, due to the number of values of π, an approximation may be obtained by aggregating the rankings weighted by the factor given by the marginalization

$$P(z_i|\mathbf{d_r}, \mathbf{x_r}) \propto P(z_i) \prod_{j}^{J} P(d_{rj}|z_i) \prod_{k}^{K} P(x_{rk}|z_i)$$

Then, we apply weighted Borda count by using the consensus permutation identified for each component z_i of the mixture and using probabilities $P(z_i|\mathbf{d_r}, \mathbf{x_r})$ as weights.

4   Experimental Evaluation

In this section we carry out an experimental evaluation to assess the performance of the proposed algorithm. Next, we describe the employed datasets, the tested algorithms, the methodology and the results.

4.1   Datasets

We used the 21 datasets proposed in [9,19]. The first 16 may be considered semi-synthetic since they were obtained by transforming 8 multi-class (type A) and 8 regression datasets (type B) to the LR problem, while the last 5 correspond to real-world biological problems. Table 1 provides the main characteristics of each dataset. The columns #rankings and max #rankings correspond to the number of different rankings in the dataset and the maximum number of rankings that can be generated for such dataset, respectively.

Table 1. Datasets description.

Dataset	type	#instances	#features	#labels	#rankings	max #rankings
authorship	A	841	70	4	17	4!
bodyfat	B	252	7	7	236	7!
calhousing	B	20640	4	4	24	4!
cpu-small	B	8192	6	5	119	5!
elevators	B	16599	9	9	131	9!
fried	B	40769	9	5	120	5!
glass	A	214	9	6	30	6!
housing	B	506	6	6	112	6!
iris	A	150	4	3	5	3!
pendigits	A	10992	16	10	2081	10!
segment	A	2310	18	7	135	7!
stock	B	950	5	5	51	5!
vehicle	A	846	18	4	18	4!
vowel	A	528	10	11	294	11!
wine	A	178	13	3	5	3!
wisconsin	B	194	16	16	194	16!
spo	-	2465	24	11	2361	11!
heat	-	2465	24	6	622	6!
dtt	-	2465	24	4	24	4!
cold	-	2465	24	4	24	4!
diau	-	2465	24	7	967	7!

4.2   Algorithms

The algorithms involved in the experimental evaluation were the following ones:

- The model-based *Label Ranking Trees* (*LRT*) algorithm [9], based on decision tree induction [6]. To avoid overfitting, we fixed the minimum number of instances for splitting an inner node to twice the number of class labels.
- The model-free *Instance-Based Label Ranking* (*IBLR*) algorithm [9], which follows the nearest neighbors paradigm [10]. The nearest neighbors were identified by using the Euclidean distance. The number of nearest neighbors were adjusted by applying an inner five-fold cross validation method (5-cv) over the training fold.
- Our model-based proposal of *Hidden Naive Bayes LR-classifier* (*HNB*).

4.3   Methodology

We adopted the following design decisions:

- The algorithms were assessed by using a five repetitions of a ten-fold cross validation method (5×10-cv).
- The *Kendall coefficient* τ_K was used as goodness score (see [1] for details).
- The algorithms were implemented in Python 3.6.5 and the experiments executed in computers running CentOS Linux 7 with CPU Intel(R) Xeon(R) E5-2630 running at 2.40 GHz and 16 GB of RAM memory.

4.4   Results

Next, we provide the accuracy and time results, as well as their corresponding statistical analysis.

The accuracy results are shown in Table 2. Each cell contains the mean and the standard deviation of the Kendall coefficient τ_K for the test folds over the 5×10-cv. The boldfaced cells correspond to the algorithm(s) that obtain(s) the best result for each dataset.

To properly analyze the results, we applied the standard statistical analysis procedure for machine learning algorithms described in [11,14] by using the exreport package [2]. This procedure can be divided in two steps:

- First, a *Friedman test* [13] was applied using a significance level of $\alpha = 0.05$. The obtained $p-value$ was $3.253e^{-5}$, and so we rejected the null hypothesis (H_0) that all the algorithms were equivalent in terms of accuracy in favour of the alternative (H_1), that is, at least one of them was different.
- Second, taking as control the algorithm ranked first by the Friedman test (IBLR), we performed a post-hoc test with the *Holm's procedure* [18], also using a significance level of $\alpha = 0.05$. This test compares all the algorithms with the one taking as control to discover the outstanding methods. The results for the post-hoc test are shown in Table 3. The *win*, *tie* and *loss* columns stand for the number of datasets in which the control algorithm wins, ties and losses with respect to the one on the column *Method*.

According to these results, we can conclude that:

- The Friedman test ranked first the IBLR algorithm, which was taken as control for the post-hoc test. LRT was ranked second, and HNB third.
- The post-hoc test revealed that the IBLR algorithm was statistically different in terms of accuracy with respect to HNB and LRT.
- Regarding LRT and HNB, the pairwise Shaffer's post-hoc test [27] ($\alpha = 0.05$) obtained a $p-value$ of $8.774e^{-1}$. Therefore, we can not reject the null hypothesis (H_0) that these algorithms were equivalent in terms of accuracy.

Table 2. Accuracy results for each algorithm.

Dataset	IBLR	LRT	HNB
aut	**0.932 ($\pm$ 0.013)**	0.862 ($\pm$ 0.033)	0.918 ($\pm$ 0.018)
bod	**0.224 ($\pm$ 0.067)**	0.159 ($\pm$ 0.070)	0.116 ($\pm$ 0.073)
cal	0.337 ($\pm$ 0.010)	**0.340 ($\pm$ 0.011)**	0.183 ($\pm$ 0.026)
cpu	**0.501 ($\pm$ 0.013)**	0.445 ($\pm$ 0.015)	0.429 ($\pm$ 0.014)
ele	0.728 ($\pm$ 0.007)	**0.753 ($\pm$ 0.008)**	0.683 ($\pm$ 0.021)
fii	**0.975 ($\pm$ 0.001)**	0.893 ($\pm$ 0.003)	0.747 ($\pm$ 0.120)
gla	**0.838 ($\pm$ 0.072)**	0.829 ($\pm$ 0.064)	0.837 ($\pm$ 0.077)
hou	0.721 ($\pm$ 0.0339)	**0.757 ($\pm$ 0.033)**	0.418 ($\pm$ 0.255)
iri	0.955 ($\pm$ 0.042)	0.924 ($\pm$ 0.056)	**0.958 ($\pm$ 0.044)**
pen	**0.941 ($\pm$ 0.002)**	0.924 ($\pm$ 0.003)	0.863 ($\pm$ 0.006)
seg	**0.951 ($\pm$ 0.006)**	0.943 ($\pm$ 0.007)	0.773 ($\pm$ 0.055)
sto	**0.921 ($\pm$ 0.011)**	0.894 ($\pm$ 0.018)	0.888 ($\pm$ 0.018)
veh	**0.854 ($\pm$ 0.027)**	0.811 ($\pm$ 0.044)	0.805 ($\pm$ 0.039)
vow	**0.870 ($\pm$ 0.016)**	0.718 ($\pm$ 0.037)	0.748 ($\pm$ 0.039)
win	**0.945 ($\pm$ 0.039)**	0.885 ($\pm$ 0.071)	0.934 ($\pm$ 0.049)
wis	**0.491 ($\pm$ 0.047)**	0.373 ($\pm$ 0.046)	0.386 ($\pm$ 0.051)
spo	**0.148 ($\pm$ 0.017)**	0.105 ($\pm$ 0.016)	0.144 ($\pm$ 0.018)
hea	**0.061 ($\pm$ 0.024)**	0.035 ($\pm$ 0.020)	0.052 ($\pm$ 0.022)
dtt	**0.127 ($\pm$ 0.032)**	0.075 ($\pm$ 0.038)	0.117 ($\pm$ 0.034)
col	**0.076 ($\pm$ 0.028)**	0.051 ($\pm$ 0.027)	0.065 ($\pm$ 0.034)
dia	**0.225 ($\pm$ 0.027)**	0.151 ($\pm$ 0.025)	0.217 ($\pm$ 0.028)

Table 3. Post-hoc test for the accuracy results.

Method	Rank	$p-value$	Win	Tie	Loss
IBLR	1.19	-	-	-	-
LRT	2.38	$1.205e^{-4}$	18	0	3
HNB	2.43	$1.205e^{-4}$	20	0	1

In this work, we deal with model-free and model-based methods for the LR problem, whose CPU requirements are clearly different. Therefore, to make a fair comparison, the time for the whole process (learning with the training dataset and validating with the test one) was gathered. The improvement ratios (time) of HNB with respect to IBLR and LRT are shown in Table 4.

Table 4. Time results for each algorithm.

Method	aut	bod	cal	cpu	ele	fri	gla	hou	iri	pen
IBLR	2.007	1.664	74.086	35.242	60.933	444.556	0.535	2.485	0.428	24.819
LRT	11.477	0.868	1.222	1.213	2.237	4.169	2.210	1.042	0.086	1.729

Method	seg	sto	veh	vow	win	wis	spo	hea	dtt	col	dia
IBLR	8.704	1.834	1.820	1.216	0.864	0.874	23.612	11.223	17.486	10.410	24.706
LRT	15.667	1.024	1.024	2.061	6.081	2.515	14.479	8.724	9.012	8.334	10.997

In light of these results, we can highlight that the HNB classifier is two times faster than the IBLR algorithm for smaller datasets, while for larger ones this value is multiplied by a factor of ten. Regarding the LRT method, we observe that the HNB classifier is two and ten times faster for smaller and larger datasets, respectively. Therefore, we may sacrifice a bit of time to improve the parameters of the HNB model.

Finally, it should be remarked that there are some datasets (e.g., segment or pendigits) where the HNB model fails in the prediction task when compared with the IBLR and LRT algorithms. To find an explanation, we decided to apply the corresponding algorithms for some of these datasets but in the classification setup. When examining these results, we realized that the Naive Bayes algorithm also failed, while decision trees and instance-based methods succeeded. Thus, we think that the assumption that all the features follow a Gaussian distribution restrict the predictive power of PGMs. Our suspicions were confirmed when we observed that the results of the Naive Bayes model strongly improved when the features were discretized. Therefore, we expect that the HNB model also improves when we properly apply multinomial probability distributions instead of Gaussian ones.

5   Conclusions

In this paper, we cope with the LR problem. Based on the EM estimation principle, we have defined a Naive Bayes structure where the root is a hidden discrete variable, used to model the different probability distributions that must be managed in such problem (multinomial and Gaussian for the features and Mallows for the permutations).

From the experimental evaluation, we can conclude that our proposal of Hidden Naive Bayes is clearly faster than the LRT and IBLR methods while being competitive in accuracy with the first one.

As future research we plan to introduce a discretization method to treat as discrete variables those features that does not follow a Gaussian probability distribution. Also, we will deal with a more general approach where incomplete rankings are allowed on the training dataset.

Acknowledgements. This work has been partially funded by FEDER funds, the Spanish Government (AEI/MINECO) through the project TIN2016-77902-C3-1-P and the Regional Government (JCCM) by SBPLY/17/180501/000493.

References

1. Aledo, J., Gámez, J., Molina, D.: Tackling the supervised label ranking problem by bagging weak learners. Inf. Fusion **35**, 38–50 (2017)
2. Arias, J., Cózar, J.: ExReport: fast, reliable and elegant reproducible research (2015). http://exreport.jarias.es/
3. Borda, J.: Memoire sur les elections au scrutin. Histoire de l'Academie Royal des Sciences (1770)
4. Breiman, L.: Bagging predictors. Mach. Learn. **24**(2), 123–140 (1996)
5. Breiman, L.: Random forests. Mach. Learn. **45**, 5–32 (2001)
6. Breiman, L., Friedman, J., Stone, C., Olshen, R.: Classification and Regression Trees. Wadsworth Inc., Wadsworth (1984)
7. Charte, D., Charte, F., García, S., Herrera, F.: A snapshot on nonstandard supervised learning problems: taxonomy, relationships, problem transformations and algorithm adaptations. Prog. Artif. Intell. **8**(1), 1–14 (2019)
8. Cheng, W., Dembczynski, K., Hüllermeier, E.: Label ranking methods based on the Plackett-Luce model. In: Proceedings of the 27th International Conference on Machine Learning, pp. 215–222 (2010)
9. Cheng, W., Hühn, J., Hüllermeier, E.: Decision tree and instance-based learning for label ranking. In: Proceedings of the 26th Annual International Conference on Machine Learning, pp. 161–168. ACM (2009)
10. Cover, T., Hart, P.: Nearest neighbor pattern classification. IEEE Trans. Inf. Theory **13**, 21–27 (1967)
11. Demšar, J.: Statistical comparisons of classifiers over multiple data sets. J. Mach. Learn. Res. **7**, 1–30 (2006)
12. Fernández, A., Gámez, J.A., Rumí, R., Salmerón, A.: Data clustering using hidden variables in hybrid Bayesian networks. Prog. Artif. Intell. **2**, 141–152 (2014)
13. Friedman, M.: A comparison of alternative tests of significance for the problem of m rankings. Ann. Math. Stat. **11**, 86–92 (1940)
14. Garcša, S., Herrera, F.: An extension on "statistical comparisons of classifiers over multiple data sets" for all pairwise comparisons. J. Mach. Learn. Res. **9**, 2677–2694 (2008)
15. Gurrieri, M., Fortemps, P., Siebert, X.: Alternative decomposition techniques for label ranking. In: Laurent, A., Strauss, O., Bouchon-Meunier, B., Yager, R.R. (eds.) IPMU 2014. CCIS, vol. 443, pp. 464–474. Springer, Cham (2014). https://doi.org/10.1007/978-3-319-08855-6_47

16. Har-Peled, S., Roth, D., Zimak, D.: Constraint classification for multiclass classification and ranking. In: Proceedings of the 15th International Conference on Neural Information Processing Systems, pp. 785–792 (2002)
17. Hernández, J., Inza, I., Lozano, J.A.: Weak supervision and other non-standard classification problems: a taxonomy. Pattern Recogn. Lett. **69**, 49–55 (2016)
18. Holm, S.: A simple sequentially rejective multiple test procedure. Scand. J. Stat. **6**, 65–70 (1979)
19. Hüllermeier, E., Fürnkranz, J., Cheng, W., Brinker, K.: Label ranking by learning pairwise preferences. Artif. Intell. **172**, 1897–1916 (2008)
20. Irurozk, E., Calvo, B., Lozano, J.A.: PerMallows: an R package for mallows and generalized mallows models. J. Stat. Softw. **71**(12), 1–30 (2016)
21. Kemeny, J., Snell, J.: Mathematical Models in the Social Sciences. MIT Press, Cambridge (1972)
22. Koller, D., Friedman, N.: Probabilistic Graphical Models: Principles and Techniques - Adaptive Computation and Machine Learning. MIT Press, Cambridge (2009)
23. Lowd, D., Domingos, P.: Naive Bayes models for probability estimation. In: Proceedings of the 22nd International Conference on Machine Learning, pp. 529–536. ACM (2005)
24. Mallows, C.L.: Non-null ranking models. Biometrika **44**, 114–130 (1957)
25. Ribeiro, G., Duivesteijn, W., Soares, C., Knobbe, A.: Multilayer perceptron for label ranking. In: Villa, A.E.P., Duch, W., Érdi, P., Masulli, F., Palm, G. (eds.) ICANN 2012. LNCS, vol. 7553, pp. 25–32. Springer, Heidelberg (2012). https://doi.org/10.1007/978-3-642-33266-1_4
26. de Sá, C.R., Soares, C., Jorge, A.M., Azevedo, P., Costa, J.: Mining association rules for label ranking. In: Huang, J.Z., Cao, L., Srivastava, J. (eds.) PAKDD 2011. LNCS (LNAI), vol. 6635, pp. 432–443. Springer, Heidelberg (2011). https://doi.org/10.1007/978-3-642-20847-8_36
27. Shaffer, J.P.: Multiple hypothesis testing. Annu. Rev. Psychol. **46**(1), 561–584 (1995)
28. de Sá, C.R., Soares, C., Knobbe, A.J., Cortez, P.: Label ranking forests. Expert Systems **34**(1), e12166 (2017)
29. Cheng, W., Henzgen, S., Hüllermeier, E.: Labelwise versus pairwise decomposition in label ranking. In: Proceedings of the Workshop on Lernen, Wissen & Adaptivität, pp. 129–136 (2013)
30. Zhou, Y., Qiu, G.: Random forest for label ranking. Expert Syst. Appl. **112**, 99–109 (2018)

Sure-Wins Under Coherence:
A Geometrical Perspective

Stefano Bonzio[1]($\boxtimes$), Tommaso Flaminio[2], and Paolo Galeazzi[3]

[1] Polytechnic University of the Marche,
Via Tronto 10/a, 60126 Torrette di Ancona, Italy
s.bonzio@univpm.it
[2] Artificial Intelligence Research Institute (IIIA - CSIC),
Campus UAB, 08193 Bellaterra, Spain
tommaso@iiia.csic.es
[3] Center for Information and Bubble Studies, University of Copenhagen,
Karen Blixens Plads 8, 2300 København S, Denmark
pagale87@gmail.com

Abstract. In this contribution we will present a generalization of de Finetti's betting game in which a gambler is allowed to buy and sell unknown events' betting odds from more than one bookmaker. In such a framework, the sole coherence of the books the gambler can play with is not sufficient, as in the original de Finetti's frame, to bar the gambler from a sure-win opportunity. The notion of *joint coherence* which we will introduce in this paper characterizes those coherent books on which sure-win is impossible. Our main results provide geometric characterizations of the space of all books which are jointly coherent with a fixed one. As a consequence we will also show that joint coherence is decidable.

Keywords: Coherence · Sure-win · De Finetti's betting game · Geometry of coherence · Decidability

1 Introduction

The logical foundations of subjective probability theory find in the work of de Finetti, started with [1] and culminated with [2], a solid ground which, especially in the last years has been the object of a deep study and several generalizations (see for instance [6,8,9]).

To set the scene of de Finetti's approach to probability, let us consider a bookmaker who fixes a finite number of events $e_1, \ldots, e_k$ which are represented by sentences of classical propositional logic and a book β on them, i.e., a complete assignment $\beta \colon \{e_1, \ldots, e_k\} \to [0,1]$ of betting odds $\beta(e_i) = \beta_i$. In order to bet on the events, a gambler chooses *stakes* $\sigma_1, \ldots, \sigma_k \in \mathbb{R}$, one for each event, and pays the bookmaker the amount $\sigma_i \cdot \beta_i$ for each e_i (with $i \in \{1, \ldots, k\}$). Note that σ_i may be negative, in which case, paying $\sigma_i \cdot \beta_i$ means receiving $-\sigma_i \cdot \beta_i$, as money transfer is orientated from the gambler to the bookmaker. When a (classical propositional) valuation h determines the truth-value of each e_i, the gambler

© Springer Nature Switzerland AG 2019
G. Kern-Isberner and Z. Ognjanović (Eds.): ECSQARU 2019, LNAI 11726, pp. 363–373, 2019.
https://doi.org/10.1007/978-3-030-29765-7_30

gains σ_i if $h(e_i) = 1$, i.e., the event e_i has actually occurred, and 0 otherwise[1]. The book β is said to be *coherent* if there is no choice of stakes $\sigma_1, \ldots, \sigma_k \in \mathbb{R}$ which forces gambler's balance not to be strictly positive under every valuation h. In other words, coherent books are those which bar the gambler from a *sure-win* opportunity, i.e., a strictly positive gain, independently of the truth-value of the events involved.

A slightly more general, yet completely realistic, situation is the one in which a gambler is allowed to place her stakes on two or more coherent books[2]. As we are going to point out in the present contribution, in such a case the sole coherence of the books the gambler decides to play with is not sufficient to bar her from a sure-win. Consider, for instance, the following very elementary example: Two bookmakers B_1 and B_2 fix betting odds to the events "Heads" and "Tails" of the typical coin-tossing game according to the following scheme: B_1 assigns $1/2$ to both "Heads" and "Tails", while B_2 assigns $1/3$ to "Heads" and $2/3$ to "Tails". Notwithstanding the coherence of the two assignments, buying "Tails" from B_1 for $\sigma_1 = 1$ euro and "Heads" from B_2 for $\sigma_2 = 1$ euro leads the gambler to a sure-win.

Situations of this kind have been studied by Nau and his collaborators (see [10, 11]) in the context of noncooperative games. There, given a set of players, a (conjoined) strategy is said to be *jointly coherent*, if it does not expose the group to arbitrage. In other words, "players who subscribe to the standard of joint coherence, are those who do not let themselves be used *collectively* as a money pump" (see [10, p. 426]).

In this paper we deepen this research line sticking within de Finetti's original betting framework and we move the first steps towards a logico-mathematical formalization of those coherent books which avoid sure-win (i.e., arbitrage) opportunities. They will be called *jointly coherent* books. In particular, we will give an answer to the following question: given a coherent book, which other coherent books, if any, bar a malicious gambler from a sure-win opportunity? More precisely, for every coherent book β, we will provide a geometric characterization of the set of all (coherent) books which are jointly coherent with it.

This paper is organized as follows: in the next section we will recall, in a more precise way, de Finetti's coherence criterion, de Finetti's theorem and, in particular, we will focus on its geometric version. In Sect. 3, we will formally introduce the concepts of sure-win and joint coherence of a book. In Sect. 4 we will study the geometry of joint coherence and provide the main result of the paper.

2   Preliminaries

Along this paper we fix a finite set of events that we denote by $\Phi = \{e_1, \ldots, e_k\}$. As we recalled in Sect. 1, a book β on Φ is *coherent* if for each choice of stakes

---

[1] For details, see for instance [5].

[2] The maybe unrealistic assumption which sees the bookmakers to consider exactly the same set of events can indeed be relaxed with an inessential modification.

$\sigma_1, \ldots, \sigma_k \in \mathbb{R}$ there exists at least a possible world h such that gambler's total balance $\sum_{i=1}^{k} \sigma_i(h(e_i) - \beta(e_i)) \leq 0$.

A book β is said to be *incoherent* if it is not coherent. Obviously, incoherent books are those which allow the gambler for a sure-win opportunity.

Recall that a finitely additive *probability measure* over a Boolean algebra **A** is a map $P \colon \mathbf{A} \to [0,1]$ such that $P(1) = 1$ and $P(a \vee b) = P(a) + P(b)$, provided that $a \wedge b = 0$. De Finetti's theorem then states that a book $\beta : \Phi \to [0,1]$ is coherent iff it extends to a finitely additive probability measure over the Boolean algebra generated by the events in Φ, denoted by $\mathbf{B}_\Phi$, see [2].

This result admits an equivalent geometrical formulation (see [12]), which we are going to recall here. Any finite set of events $\Phi = \{e_1, \ldots, e_k\}$ determines a polytope in $[0,1]^k$ by the following construction. Let $h_1, \ldots, h_t$ be the homomorphisms from $\mathbf{B}_\Phi$ to the two element Boolean algebra $\mathbf{2} = \langle \{0,1\}, \wedge, \vee, \neg, 0 \rangle$. For every $j = 1, \ldots, t$ let q_j be the point of $\{0,1\}^k$

$$q_j = (h_j(e_1), \ldots, h_j(e_k)). \tag{1}$$

Finally, let $\mathscr{C}_\Phi$ be the polytope of $[0,1]^k$ generated by $q_1, \ldots, q_t$:

$$\mathscr{C}_\Phi = \mathrm{co}(\{q_j \mid j = 1, \ldots, t\}),$$

where co stands for *convex hull*.

A book $\beta : \Phi \to [0,1]$ determines a point $\beta = (\beta(e_1), \ldots, \beta(e_k)) \in [0,1]^k$. The following result, whose proof can be found in [6, Lemma 6.3] and [12, Theorem 2], provides a geometric characterization of coherent books.

Theorem 1. *For a book $\beta \colon \Phi \to [0,1]$ the following conditions are equivalent:*

1. β is coherent;
2. $\beta \in \mathscr{C}_\Phi$.

The construction illustrated above is better visualized towards an example.

Example 1. Consider $\Phi = \{e_1, e_2\}$, where $e_1 = p$ and $e_2 = p \vee q$ in a language with two propositional variables p, q. Following de Finetti [1], the above mentioned events may be thought as of referring to a horse race: the atomic event p can be interpreted as "Horse number 1 is the winner", while $p \vee q$ could stand for "An Italian horse is winning", under the assumption that only two horses are Italian (and one of them is actually the number 1).

The algebra $\mathbf{B}_\Phi$ counts of four homomorphisms to $\mathbf{2}$, namely those maps $h_1, h_2, h_3, h_4 \colon \{p, q\} \to \{0,1\}$ which assign, respectively, to p and q the values $(0,0)$, $(0,1)$, $(1,0)$ and $(1,1)$. Therefore, we obtain the following points $q_1, \ldots, q_4 \in \mathbb{R}^2$:

$$q_1 = (h_1(e_1), h_1(e_2)) = (h_1(p), h_1(p \vee q)) = (0,0);$$
$$q_2 = (h_2(e_1), h_2(e_2)) = (h_2(p), h_2(p \vee q)) = (0,1);$$
$$q_3 = (h_3(e_1), h_3(e_2)) = (h_3(p), h_3(p \vee q)) = (1,1);$$
$$q_4 = (h_4(e_1), h_4(e_2)) = (h_4(p), h_4(p \vee q)) = (1,1).$$

Since $q_3 = q_4$, we have:

$$\mathscr{C}_\Phi = \mathrm{co}(\{q_1, q_2, q_3\}) = \mathrm{co}(\{(0,0), (0,1), (1,1)\}),$$

depicted as in Fig. 1 below. Theorem 1 tells us that a book β is coherent if and only if it is a convex combination of q_1, q_2 and q_3.

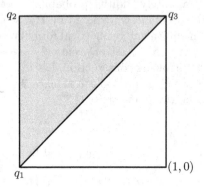

Fig. 1. The convex set $\mathscr{C}_\Phi$ (in gray) of all coherent books on events $e_1 = p$ and $e_2 = p \vee q$.

3   Sure-Wins and Jointly Coherent Books

As mentioned in the Introduction, we are interested in situation where a gambler has the opportunity of betting on two (or ideally more) books over the same set of events. The gambler's concept of *sure-win* opportunity becomes wider and it is made precise in the following.

Definition 1. *Let β_1, β_2 be coherent books on the set of events $\Phi = \{e_1, \ldots, e_k\}$. We say that a gambler has a sure-win opportunity on β_1, β_2 if there exists a total map $\xi : \{1, \ldots, k\} \to \{1, 2\}$ such that, the book*

$$\beta : e_i \mapsto \beta_{\xi(i)}(e_i)$$

is incoherent. If such function ξ does not exist, then β_1 and β_2 are said to be jointly coherent.

Therefore, a gambler has a sure-win opportunity on β_1, β_2 if there exists a map $\xi : \{1, \ldots, k\} \to \{1, 2\}$ and stakes $\sigma_1, \ldots, \sigma_k \in \mathbb{R}$ such that in very every possible world h, gambler's balance

$$\sum_{i=1}^{k} \sigma_i(h(e_i) - \beta_{\xi(i)})) > 0,$$

where $\beta : e_i \mapsto \beta_{\xi(i)}(e_i)$ is as in Definition 1.

Remark 1. Notice that two coherent books β_1 and β_2 are jointly coherent if any book in the set

$$\Xi(\beta_1, \beta_2) = \{\beta_1(e_1), \beta_2(e_1)\} \times \{\beta_1(e_2), \beta_2(e_2)\} \times \ldots \times \{\beta_1(e_k), \beta_2(e_k)\}$$

is coherent as well. For any pair of coherent books β_1, β_2 we will call $\Xi(\beta_1, \beta_2)$ the set of *crossed-books* of β_1 and β_2.

Also notice that, by Definition 1, a gambler is not allowed to buy (or sell) a bet on the same event from both β_1 and β_2. This restriction is imposed in order to not trivialize her opportunities of sure-win. Indeed, since β_1 and β_2 are distinct, there always exists at least an event e such that $\beta_1(e) \neq \beta(e)$. Thus, assuming that $\beta_1(e) < \beta_2(e)$ without loss of generality, buying $\beta_1(e)$ for 1 euro and $\beta_2(e)$ for -1 euro (i.e., selling $\beta_2(e)$ for 1 euro) would immediately ensure the gambler a sure-win.

Example 2. Let $\Phi = \{e_1, e_2\}$ as in Example 1 and consider the books:

1. $\beta_1(e_1) = 1/2$ and $\beta_1(e_2) = 2/3$;
2. $\beta_2(e_1) = 1/4$ and $\beta_2(e_2) = 2/3$;
3. $\beta_3(e_1) = \beta_3(e_2) = 1/3$.

Then, β_1 is jointly coherent with β_2, which is joint coherent with β_3. On the other hand β_1 is not jointly coherent with β_3. Indeed, the book $\alpha \in \Xi(\beta_1, \beta_3)$ defined by $e_1 \mapsto \beta_1(e_1) = 1/2, e_2 \mapsto \beta_3(e_2) = 1/3$ is incoherent (see Fig. 2). Therefore, a gambler who is allowed to choose, for each event, which book to bet with has a sure-win opportunity if the books into play are β_1 and β_3.

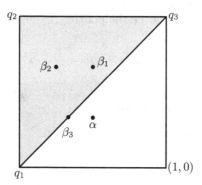

Fig. 2. The convex hull $\mathscr{C}_\Phi$ (gray); the coherent books β_1, β_2, β_3 and the incoherent book $\alpha \in \Xi(\beta_1, \beta_3)$.

We now present a first easy result. Recall that a subset $B = \{b_1, \ldots, b_r\}$ of a Boolean algebra $\mathbf{A}$ is a *partition* if $\bigvee_{i=1}^{r} b_i = \top$ and, for all $i \neq j$, $b_i \wedge b_j = \bot$.

Proposition 1. *If $\Phi = \{e_1, \ldots, e_k\}$ is a partition of $\mathbf{B}_\Phi$, then for any two coherent books β_1, β_2 on Φ the following conditions are equivalent:*

1. $\beta_1 \neq \beta_2$;
2. β_1 and β_2 are not jointly coherent, i.e. the gambler has a sure-win opportunity on β_1, β_2.

Proof. The direction $(2) \Rightarrow (1)$ is trivial. In order to prove $(1) \Rightarrow (2)$, observe that, since by hypothesis Φ is a partition of $\mathbf{B}_\Phi$, any book β on Φ is coherent if and only if

$$\sum_{i=1}^{k} \beta(e_i) = 1. \tag{2}$$

Now, since $\beta_1 \neq \beta_2$, there exists $e_i \in \Phi$ such that $\beta_1(e_i) \neq \beta_2(e_i)$. Let us assume, without loss of generality, that $\beta_1(e_i) < \beta_2(e_i)$ and let us consider the book $\beta \colon \Phi \to [0,1]$ defined as follows: for every $e \in \Phi$,

$$\beta(e) = \begin{cases} \beta_1(e) & \text{if } e \neq e_i, \\ \beta_2(e) & \text{otherwise.} \end{cases}$$

Notice that $\beta \in \Xi(\beta_1, \beta_2)$ and it is not coherent since $\sum_{j=1}^{k} \beta(e_j) < 1$. Therefore, β_1 and β_2 are not jointly coherent. This settles the claim. $\qquad\square$

4 The Geometry of Joint Coherence

We are interested in providing a full characterization of all those coherent books which are jointly coherent with a fixed one. In this section, we will give geometric characterizations of the space of these books.

We set the background for proving this result. For every book $\beta \colon \Phi \to [0,1]$ and for every $i = 1, \ldots, k$, let δ_i be the pair $(d_i^+, d_i^-) \in \mathbb{R}^2$ be such that:

1. $d_i^\pm \geq 0$;
2. the books $\beta_{d_i^+} = (\beta_1, \ldots, \beta_{i-1}, \beta_i + d_i^+, \beta_{i+1}, \ldots, \beta_k)$ and $\beta_{d_i^-} = (\beta_1, \ldots, \beta_{i-1}, \beta_i - d_i^-, \beta_{i+1}, \ldots, \beta_k)$ are coherent;
3. for all $\varepsilon > 0$, $(\beta_1, \ldots, \beta_{i-1}, \beta_i + d_i^+ + \varepsilon, \beta_{i+1}, \ldots, \beta_k)$ and $(\beta_1, \ldots, \beta_{i-1}, \beta_i - d_i^- - \varepsilon, \beta_{i+1}, \ldots, \beta_k)$ are incoherent.

Let us hence define the rectangle

$$\mathscr{R}_\beta = \{ \gamma \in \mathbb{R}^k \mid (\forall i = 1, \ldots, n)\ d_i^- \leq |\gamma_i - \beta_i| \leq d_i^+ \},$$

and the convex set

$$\mathscr{C}_\beta = \mathscr{C}_\Phi \cap \mathscr{R}_\beta. \tag{3}$$

Obviously $\mathscr{C}_\beta$ is nonempty iff β is coherent.

Example 3. Let Φ and $\beta_1 \colon \Phi \to [0,1]$ be as in Example 2. Thus, $\beta_1(p) = 1/2$ and $\beta_1(p \vee q) = 2/3$. The vertices (extreme points) of the rectangle $\mathscr{R}_{\beta_1}$ are easy to compute: $v_1 = (2/3, 1/2)$; $v_2 = (0, 1/2)$; $v_3 = (0, 1)$; $v_4 = (2/3, 1)$. Thus, $\mathscr{C}_{\beta_1} = \mathscr{C}_\Phi \cap \mathscr{R}_{\beta_1}$ coincides with the gray region as in Fig. 3.

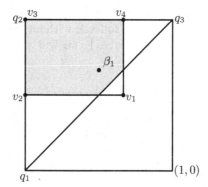

Fig. 3. The coherent book β_1 from Example 2 and the convex set $\mathscr{C}_{\beta_1}$ (the gray region) obtained by intersecting $\mathscr{C}_\Phi$ (the triangle of vertices q_1, q_2 and q_3) and the rectangle $\mathscr{R}_{\beta_1}$ (whose vertices are v_1, v_2, v_3 and v_4).

The following result shows that, for every fixed coherent book β, the convex set $\mathscr{C}_\beta$ characterizes all the coherent books which are jointly coherent with β.

Proposition 2. *Let* $\beta, \beta' : \Phi \to [0, 1]$ *be two coherent books. Then the following conditions are equivalent:*

1. β' *is jointly coherent with* β;
2. $\beta' \in \mathscr{C}_\beta$.

Proof. $(1) \Rightarrow (2)$. Assume that β and β' are jointly coherent. Since β' is coherent then, by Theorem 1, $\beta' \in \mathscr{C}_\Phi$. We only have to show that $\beta' \in \mathscr{R}_\beta$. Suppose, by contradiction, that $\beta' \notin \mathscr{R}_\beta$, i.e. there exists $1 \le i \le k$ such that $| \beta_i - \beta'_i | > d_i^+$ (or $| \beta_i - \beta'_i | < d_i^-$, but the reasoning is analogous). The definition of $\mathscr{R}_\beta$ immediately implies that β' is not coherent, a contradiction.

$(2) \Rightarrow (1)$. Let $\beta' \in \mathscr{C}_\beta = \mathscr{C}_\Phi \cap \mathscr{R}_\beta$, i.e. β' is a coherent book which satisfies the above conditions 1.-3. Let α be any book in $\Xi(\beta, \beta')$. Since β is coherent, we have that $\beta \in \mathscr{C}_\beta$ and, by assumption, $\beta' \in \mathscr{C}_\beta$, which together imply that $\alpha \in \mathscr{C}_\beta$. Thus $\alpha \in \mathscr{C}_\Phi$, which, by Theorem 1, implies that is coherent, therefore β and β' are jointly coherent books. $\qquad\square$

It is immediate to see that the relation of being jointly coherent is symmetric. Therefore, from Proposition 2 above β and β' are jointly coherent iff $\beta' \in \mathscr{C}_\beta$ iff $\beta \in \mathscr{C}_{\beta'}$. Therefore the following is immediate.

Corollary 1. *Let* $\beta, \beta' : \Phi \to [0, 1]$ *be two coherent books. Then* β *and* β' *are jointly coherent iff* $\beta, \beta' \in \mathscr{C}_\beta \cap \mathscr{C}_{\beta'}$.

In the next we will show, for every coherent book β, another geometric characterization of $\mathscr{C}_\beta$. Recalling that every closed convex subsets of $\mathbb{R}^k$ is an intersection of halfspaces (see [4, Theorem 3.8]), for every polytope $\mathscr{P} \subseteq \mathbb{R}^k$, there are linear polynomials $f_1, \ldots, f_n$ such that $\mathscr{P} = \{(x_1, \ldots, x_k) \in \mathbb{R}^k \mid \forall\, i = 1, \ldots, n, f_i(x_1, \ldots, x_k) \geq 0\}$. In what follows, without danger of confusion, for every finite set of events $\Phi = \{e_1, \ldots, e_k\}$, we will write $f_1, \ldots, f_n$ for those polynomials such that

$$\mathscr{C}_\Phi = \{(x_1, \ldots, x_k) \in \mathbb{R}^k \mid \forall\, i = 1, \ldots, n, f_i(x_1, \ldots, x_k) \geq 0\}. \tag{4}$$

For the sake of a lighter notation, we will denote by K the index set $\{1, \ldots, k\}$.

Let us fix two points $a = (a_1, \ldots, a_k)$ and $b = (b_1, \ldots, b_k)$ of $\mathbb{R}^k$, and a subset J of K. Then, we denote by $(a_J, b_{K \setminus J})$ the tuple obtained by substituting b_j by a_j, in b, for each $j \in J$.

Theorem 2. *Let $\mathscr{C}_\Phi$, $f_1, \ldots, f_n$ be as in (4) and let β, β' be two coherent books. Then, $\beta' \in \mathscr{C}_\beta$ iff β' is solution of the following system,*

$$\mathscr{S}(\beta) = \{f_i(\beta_J, x_{K \setminus J}) \geq 0 \mid i = 1, \ldots, n, \ J \subseteq K\}.$$

In other words, $\mathscr{C}_\beta$ coincides with the set of solutions of $\mathscr{S}(\beta)$.

Proof. ($\Leftarrow$). Assume, by contraposition, that $\beta' \notin \mathscr{C}_\beta$. Thus, by Proposition 2, β and β' are not jointly coherent and hence, by Remark 1, there exists a $\hat\beta \in \Xi(\beta, \beta')$ which is not coherent. This means, by Theorem 1, that $\hat\beta \notin \mathscr{C}_\Phi$. Therefore, by (4), there exists an index $i \in \{1, \ldots, n\}$ such that $f_i(\hat\beta_1, \ldots, \hat\beta_k) < 0$ and in particular $\hat\beta$ is not a solution of $\mathscr{S}(\beta)$.

($\Rightarrow$). Assume, again by contraposition, that β' is not solution of $\mathscr{S}(\beta)$. Thus, there exists a $J \subseteq K$ and an index $i \in \{1, \ldots, n\}$ such that $f_i(\beta_J, \beta'_{K \setminus J}) < 0$. Therefore, the claim immediately follows by observing that $(\beta_J, \beta'_{K \setminus J}) \in \Xi(\beta, \beta')$ and $(\beta_J, \beta'_{K \setminus J}) \notin \mathscr{C}_\Phi$ proving that β and β' are not jointly coherent. $\square$

An immediate consequence of the above theorem is the decidability of the problem determining if two rational-books are jointly coherent. For the following result to make sense, we will hence assume that the books involved take value into the rational unit interval $[0, 1] \cap \mathbb{Q}$.

Corollary 2. *Given two rational-valued books $\beta_1, \beta_2 \in \mathscr{C}_\Phi$, the problem of determining if β_1 and β_2 are jointly coherent is decidable.*

Proof. The following (sketched) procedure, which takes in input the events $e_1, \ldots, e_k$ and the rational numbers $\beta_1(e_i)$'s and $\beta_2(e_i)$'s, decides if β_1 and β_2 are jointly coherent.

Step 1: Determine the extremal points of $\mathscr{C}_\Phi$ by computing, for each truth-assignment h_j, $q_j = (h_j(e_1), \ldots, h_j(e_k))$ as in (1).

Step 2: Let $q_{t_1}, \ldots, q_{t_r}$ be, among the q_j's, the extremal points of a face $\mathscr{F}_t$ of $\mathscr{C}_\Phi$ and let $f_t(x_1, \ldots, x_k)$ the be equation of the hyperplane through $q_{t_1}, \ldots, q_{t_r}$.

Step 3: Iterate **Step 2** for all faces $\mathscr{F}_1, \ldots, \mathscr{F}_n$ of $\mathscr{C}_\Phi$ and hence determine $f_1, \ldots, f_n$ such that $\mathscr{C}_\Phi = \{(x_1, \ldots, x_k) \in \mathbb{R}^k \mid \forall\, i = 1, \ldots, n, f_i(x_1, \ldots, x_k) \geq 0\}$ as in (4).

Step 4: Introduce the system of inequalities $\mathscr{S}(\beta_1)$ as in the statement of Theorem 2.

Therefore, in the end, check if $\beta_2 = (\beta_2(e_1), \ldots, \beta_2(e_k))$ is a (rational) solution of $\mathscr{S}(\beta_1)$. □

Our last example applies the result of Theorem 2 and exemplifies also the procedure sketched in the proof of Corollary 2.

Example 4. Let $\Phi = \{p, p \vee q\}$ and $\beta_1(p) = \frac{1}{2}$, $\beta_1(p \vee q) = \frac{2}{3}$. Then

$$\mathscr{C}_\Phi = \{(x_1, x_2) \in \mathbb{R}^2 : 0 \leq x_1, x_2 \leq 1, x_1 \leq y_2\}.$$

A book β is jointly coherent with β_1 if and only if it satisfies the system given by the following inequalities

- for $J = \emptyset$, we have $x_1 \geq 0$, $1 - x_1 \geq 0$, $x_2 \geq 0$, $1 - x_2 \geq 0$ and $x_2 - x_1 \geq 0$;
- for $J = \{1\}$, we have $1/2 \geq 0$, $1 - 1/2 \geq 0$, $x_2 \geq 0$, $1 - x_2 \geq 0$ and $x_2 - 1/2 \geq 0$;
- for $J = \{2\}$, we have $x_1 \geq 0$, $1 - x_1 \geq 0$, $2/3 \geq 0$, $1 - 2/3 \geq 0$ and $2/3 - x_1 \geq 0$;
- for $J = \{1, 2\}$, we have $1/2 \geq 0$, $1 - 1/2 \geq 0$, $2/3 \geq 0$, $1 - 2/3 \geq 0$ and $2/3 - 1/2 \geq 0$.

Notice that the inequalities obtained for $J = \emptyset$ just describes $\mathscr{C}_\Phi$ and it is also immediate to see that for $J = \{1, 2\}$, we get the inequalities which assure that β_1 is a coherent book.

5 Conclusion and Future Work

The present work is motivated by the observation that when two different book-makers assign betting quotes over the same set of events, the notion of coherence is not enough to prevent a gambler who is allowed to bet on both assignments from a sure-win opportunity. We thereby proposed the notion of *joint coherence* of two books.

Our main results consist of geometrical characterizations of the space of books which are jointly coherent with a given one. Such a space is a closed convex subset of the set of all coherent books and it identifies which books can be considered "safe" once β has been fixed.

We believe that the mathematics of joint coherence as well as its computational aspects deserve further attention. In particular, since (two) jointly coherent books are necessarily coherent, we wonder what is the effect of the property of being jointly coherent on the sets of probability measures which extend them, by de Finetti's theorem, on the Boolean algebra generated by the events. Further, still on this line, it would be interesting to extend the notion of joint coherence to more general theories of uncertainty and, in particular, to Walley's definition of coherence for imprecise probabilities where negative betting rates are forbidden [13].

We are interested also in providing different characterizations of the notion of joint coherence. In particular, following [7], where it is shown that (strict) coherence admits three characterizations (algebraic, geometrical and logical), one of the aims for the future is to extend such characterizations to joint coherence as well.

A very natural question that computer scientists may rise is whether it is possible to establish a computational bound to the problem of determining whether two books are jointly coherent. Although we proved that checking joint coherence of two books is decidable, providing a NP-bound for the same seems challenging and it will be object of further investigation.

Joint coherence arises from allowing a multiplicity of bookmakers publishing coherent books, who can be viewed as individually *rational agents*. Grounding on this consideration, it is reasonable to think that this notion may suggest an alternative way to approach *collective judgments* (see for instance [3]) and *collective rationality*. This will also be addressed in our future work.

Acknowledgments. The first author acknowledges the support of the European Research Council, ERC Starting Grant GA:639276: "Philosophy of Pharmacology: Safety, Statistical Standards, and Evidence Amalgamation". Flaminio acknowledges partial support by the Spanish Ramon y Cajal research program RYC-2016-19799; the Spanish FEDER/MINECO project TIN2015- 71799-C2-1-P and the SYSMICS project (EU H2020-MSCA-RISE-2015, Project 689176).

References

1. de Finetti, B.: Sul significato soggettivo della probabilità. Fundamenta Mathematicae **17**, 289–329 (1931)
2. de Finetti, B.: Theory of Probability, vol. 1. John Wiley and Sons, New York (1974)
3. Dietrich, F., List, C.: Probabilistic opinion pooling. The Oxford Handbook of Probability and Philosophy (Hájek, A., Hitchcock, C. (eds.)) (2016)
4. Ewald, G.: Combinatorial Convexity and Algebraic Geometry. Springer, New York (1996). https://doi.org/10.1007/978-1-4612-4044-0
5. Flaminio, T., Godo, L., Hosni, H.: On the logical structure of de Finetti's notion of event. J. Appl. Logic **12**(3), 279–301 (2014)
6. Flaminio, T., Hosni, H., Montagna, F.: Strict coherence on many-valued events. J. Symbolic Logic **83**(1), 55–69 (2018)
7. Flaminio, T.: Three characterizations of strict coherence on infinite-valued events (Submitted)
8. Mundici, D.: Bookmaking over infinite-valued events. Int. J. Approximate Reasoning **43**(3), 223–240 (2006)
9. Mundici, D.: De Finetti coherence and the product law for independent events. Synthese **196**(1), 265–271 (2019)
10. Nau, R., MaCardle, K.F.: Coherent Behavior in Noncooperative Games. J. Econ. Theory **50**, 424–444 (1990)
11. Nau, R.: Joint coherence in games of incomplete information. Manage. Sci. **38**(3), 374–387 (1992)

12. Paris, J.: A note on the Dutch Book method. In: De Cooman, G., Fine, T., Seidenfeld, T. (eds.) Proceedings of the Second International Symposium on Imprecise Probabilities and their Applications, ISIPTA 2001, pp. 301–306. Shaker Publishing Company, Ithaca, NY, USA (2001)
13. Walley, P.: Statistical Reasoning with Imprecise Probabilities. Chapman and Hall, London (1991)

Conditioning of Imprecise Probabilities Based on Generalized Credal Sets

Andrey G. Bronevich[1(✉)] and Igor N. Rozenberg[2]

[1] National Research University Higher School of Economics,
Myasnitskaya 20, 101000 Moscow, Russia
`brone@mail.ru`
[2] JSC "Research and Design Institute for Information Technology,
Signalling and Telecommunications on Railway Transport",
Orlikov per.5, building 1, 107996 Moscow, Russia
`I.Rozenberg@gismps.ru`

Abstract. Recently, generalized credal sets have been introduced for modeling contradiction (incoherence) in the information. In previous papers, we did not discuss how such information could be updated if some events occur. In this paper, we show that it can be done by the conjunctive rule based on generalized credal sets. We show that the application of generalized credal sets results in several types of conditioning for imprecise probabilities.

Keywords: Generalized credal sets ·
Contradictory (incoherent) lower previsions · Incoherence correction ·
Updating imprecise probabilities

1 Introduction

The processing of incoherent (inconsistent) information was considered in several papers [4,14], and the main approach was to eliminate incoherence and to return to consistent information in order to use well-developed methods from imprecise probabilities after that. However, there exists another way, in which we try to model inconsistency in the information. In our opinion, this way can be more suitable and it can allow us not to lose some important information like the conflict between our prior information and real data [17]. In this paper, we will demonstrate this observation showing possible conditioning for imprecise probabilities based on generalized credal sets. We show that this type of conditioning is very close to conditioning proposed by Cattaneo [5], where he tried to describe the set of possible probability distributions using the likelihood function. Analogous ideas were used in earlier papers [10,11], where the evidential information was described by possibility distributions defined on sets of probability measures.

In this paper, we give a new interpretation of generalized credal sets. They can be viewed as sets of probability distributions with a degree how they contradict to given data. This degree is closely related to the corresponding likelihood function. Based on the possible ways of transforming contradictory information to the

© Springer Nature Switzerland AG 2019
G. Kern-Isberner and Z. Ognjanović (Eds.): ECSQARU 2019, LNAI 11726, pp. 374–384, 2019.
https://doi.org/10.1007/978-3-030-29765-7_31

consistent one, we can introduce several approaches to updating information represented by imprecise probabilities and among of them, there are known ones, in particular, the updating based on Dempster's rule [7,15] or the maximal likelihood updating [9].

2   Monotone Measures: Some Useful Constructions

Let X be a finite universal set and 2^X be the power set of X. The set function $\mu : 2^X \to [0,1]$ is called

- normalized if $\mu(\emptyset) = 0$ and $\mu(X) = 1$;
- monotone if $A \subseteq B$, $A, B \in 2^X$ implies $\mu(A) \leqslant \mu(B)$;
- a monotone measure [6] if μ is normalized and monotone;
- a belief function [15] if there is a set function $m : 2^X \to [0,1]$ with $\sum_{B \in 2^X} m(B) = 1$ called the basic belief assignment (bba) such that $\mu(A) = \sum_{B \in 2^X | B \subseteq A} m(B)$.

We write $\mu_1 \leqslant \mu_2$ for set functions on 2^X if $\mu_1(A) \leqslant \mu_2(A)$ for all $A \in 2^X$. The set function ν is called the dual of μ if $\nu(A) = 1 - \mu(A^c)$ for all $A \in 2^X$, where A^c denotes the complement of A. To indicate the dual of μ, we use the upper index d, i.e. μ^d denotes the dual of μ. A set function μ is a convex sum of set functions μ_1 and μ_2 if there is an $a \in [0,1]$ such that $\mu(A) = a\mu_1(A) + (1-a)\mu_2(A)$ for all $A \in 2^X$. This fact is denoted by $\mu = a\mu_1 + (1-a)\mu_2$.

Let Bel be a belief function with the bba m, then $A \in 2^X$ is called a focal element for Bel if $m(A) > 0$. A belief function is called categorical if it has the only one focal element B. This function is denoted by $\eta_{\langle B \rangle}$ and clearly

$$\eta_{\langle B \rangle}(A) = \begin{cases} 1, & B \subseteq A, \\ 0, & \text{otherwise}, \end{cases} \quad A \in 2^X.$$

Every belief function Bel with the bba m can be represented as a convex sum of categorical belief functions as

$$Bel = \sum_{B \in 2^X} m(B)\eta_{\langle B \rangle}.$$

3   Modeling Uncertainty with Imprecise Probabilities

One among the traditional models of uncertainty is based on probability measures. A belief function P is a probability measure if its body of evidence consists of singletons, i.e. every probability measure P on 2^X can be represented as

$$P = \sum_{i=1}^{n} a_i \eta_{\langle \{x_i\} \rangle}, \tag{1}$$

where $\sum_{i=1}^{n} a_i = 1$, $a_i \geqslant 0$, $i = 1, ..., n$. Thus, the bba m of P is

$$m(A) = \begin{cases} a_i, & A = \{x_i\}, \\ 0, & \text{otherwise}. \end{cases}$$

We denote the set of all possible probability measures on 2^X by M_{pr}. Assume that an experiment is described by a probability measure P described by (1) and a function $f : X \to \mathbb{R}$ shows us the award that we get after conducting the experiment, i.e. $f(x_i)$ is the award if the outcome results in x_i. Then, in a frequentist's view, the value

$$E_P(f) = \sum_{i=1}^{n} f(x_i) P(\{x_i\}) = \sum_{i=1}^{n} a_i f(x_i)$$

gives us the expected award. This award can be considered as the mean value of awards obtained during the series of the same independent experiments as follows from fundamentals of the classical probability theory. If the measure P is not exactly known, then it is possible to describe an experiment by a set of probability measures $\mathbf{P}$, and we know the lower bound $\underline{E}_{\mathbf{P}}(f)$ and upper bound $\overline{E}_{\mathbf{P}}(f)$ of the expected award defined by

$$\underline{E}_{\mathbf{P}}(f) = \inf_{P \in \mathbf{P}} E_P(f), \ \overline{E}_{\mathbf{P}}(f) = \sup_{P \in \mathbf{P}} E_P(f).$$

Let K be the linear space of all possible real valued functions on X. Then functionals $\underline{E}_{\mathbf{P}}$ and $\overline{E}_{\mathbf{P}}$ on K have the following properties [1,16]:

(1) $\underline{E}_{\mathbf{P}}(af + c) = a\underline{E}_{\mathbf{P}}(f) + c$, $\overline{E}_{\mathbf{P}}(af + c) = a\overline{E}_{\mathbf{P}}(f) + c$ for every $a \geqslant 0$, $c \in \mathbb{R}$ and $f \in K$;
(2) $\underline{E}_{\mathbf{P}}(f_1 + f_2) \geqslant \underline{E}_{\mathbf{P}}(f_1) + \underline{E}_{\mathbf{P}}(f_2)$, $\overline{E}_{\mathbf{P}}(f_1 + f_2) \leqslant \overline{E}_{\mathbf{P}}(f_1) + \overline{E}_{\mathbf{P}}(f_2)$, for every $f_1, f_2 \in K$;
(3) $\underline{E}_{\mathbf{P}}(f_1) \leqslant \underline{E}_{\mathbf{P}}(f_2)$ for every $f_1, f_2 \in K$ with $f_1(x) \leqslant f_2(x)$ for all $x \in X$;
(4) $\overline{E}_{\mathbf{P}}(f) = -\underline{E}_{\mathbf{P}}(-f)$ for every $f \in K$.

By some reasons [1,16], in the theory of imprecise probabilities such non-empty sets of probability measures are assumed to be closed and convex sets known in the literature as credal sets. Assume that the functional $\Phi : K \to \mathbb{R}$ obeys properties (1), (2) and (3) as the functional $\underline{E}_{\mathbf{P}}$, then there is a credal set $\mathbf{P}$ such that $\Phi = \underline{E}_{\mathbf{P}}$ on K. Thus, there is a bijection between credal sets and functionals $\underline{E}_{\mathbf{P}}$.

In the theory of imprecise probabilities, there are several ways of presenting uncertain information. Let $K' \subseteq K$, then every functional $\Phi : K' \to \mathbb{R}$ is called a lower prevision [1,16] if its values $\Phi(f)$ can be viewed as lower bounds of $E_P(f)$. A lower prevision $\Phi : K' \to \mathbb{R}$ is called non-contradictory or consistent, if it defines the credal set

$$\mathbf{P}(\Phi) = \{P \in M_{pr} | \forall f \in K' : E_P(f) \geqslant \Phi(f)\}.$$

Otherwise, if the set $\mathbf{P}(\Phi)$ is empty, then Φ is called contradictory (incoherent or inconsistent) lower prevision. In the sequel, we assume that every lower prevision $\Phi : K' \to \mathbb{R}$ obeys the property

$$\Phi(f) \leqslant \max_{x \in X} f(x).$$

We can analogously introduce upper previsions. Every functional $\Phi' : K' \to \mathbb{R}$ is called an upper prevision if its values $\Phi'(f)$ can be viewed as upper bounds of $E_P(f)$ and Φ is called non-contradictory (or consistent), if it defines the credal set

$$\mathbf{P}(\Phi') = \{P \in M_{pr} | \forall f \in K' : E_P(f) \leqslant \Phi'(f)\}.$$

In the sequel, we assume that every upper prevision $\Phi : K' \to \mathbb{R}$ obeys the property $\Phi'(f) \geqslant \min\limits_{x \in X} f(x)$.

Models based on lower previsions and based on upper previsions are equivalent, because if we have a lower prevision $\Phi : K' \to \mathbb{R}$, then we can get the upper prevision Φ' on $\{-f | f \in K'\}$ using the formula $\Phi'(f) = -\Phi(-f)$, where $-f \in K'$, and, clearly, the corresponding credal sets coincide for both models.

We can also describe uncertain information using set functions (or monotone measures). In this case, the set K' consists of characteristic functions $1_A(x) = 1$ if $x \in A$ and $1_A(x) = 0$ otherwise. Thus, every Φ on $K' = \{1_A | A \in 2^X\}$ can be considered as a set function $\mu(A) = \Phi(1_A)$, $A \in 2^X$. A monotone measure μ is called a lower probability if its values are viewed as lower bounds of probabilities. A lower probability μ is called non-contradictory, if it defines the credal set $\mathbf{P}(\Phi) = \{P \in M_{pr} | P \geqslant \mu\}$. Otherwise, if $\mathbf{P}(\Phi)$ is empty, then μ is called contradictory. Analogously, a monotone measure μ' is called an upper probability if its values are viewed as upper bounds of probabilities. An upper probability μ' is called non-contradictory if μ' defines the credal set $\mathbf{P}(\mu') = \{P \in M_{pr} | P \leqslant \mu'\}$. Otherwise, if $\mathbf{P}(\mu') = \emptyset$, then μ' is a contradictory upper probability.

4   Contradiction Correction Based on Generalized Credal Sets

Assume $\Phi : K' \to \mathbb{R}$ is an upper prevision. Following [2], Φ can be represented as the convex sum of two functionals:

$$\Phi = (1 - a)\Phi^{(1)} + a\Phi^{(2)}, \tag{2}$$

where $a \in [0, 1]$, $\Phi^{(1)}$ is a non-contradictory upper prevision and $\Phi^{(2)}$ is a contradictory upper prevision. This representation always exists if K' is such that the upper prevision

$$V_{\min}(f) = \min\limits_{x \in X} f(x), \quad f \in K',$$

is contradictory. The exact lower bound of a in representation (2) is called the amount of contradiction in Φ, and this value is denoted by $Con(\Phi)$. Because, by definition, $V_{\min}(f) \leqslant \Phi^{(2)}(f)$, $f \in K'$, for every upper prevision $\Phi^{(2)}$, and $\Phi^{(1)}$ is a non-contradictory lower prevision if there is a $P \in M_{pr}$ such that $E_P(f) \leqslant \Phi^{(1)}(f)$ for all $f \in K'$ the amount of contradiction can be computed by

$$Con(\Phi) = 1 - \sup\{a \in [0, 1]|$$
$$\exists P \in M_{pr}, \forall f \in K' : aE_P(f) + (1 - a)V_{\min}(f) \leqslant \Phi(f)\}.$$

We can simplify the above formula, introducing so-called normalized functions. A function $f \in K$ is called normalized if $\min_{x \in X} f(x) = 0$. We can normalize every function $f \in K$ assuming that the corresponding normalized function is $\underline{f}(x) = f(x) - \min_{x \in X} f(x)$, $x \in X$. Let us define $\underline{\Phi}(f) = \Phi(f) - \min_{x \in X} f(x)$, $x \in X$. Then the inequality $aE_P(f) + (1-a)V_{\min}(f) \leqslant \Phi(f)$ is equivalent to $aE_P(\underline{f}) \leqslant \underline{\Phi}(f)$ and we rewrite the expression for $Con(\Phi)$ as

$$Con(\Phi) = 1 - \sup \left\{ a \in [0,1] | \exists P \in M_{pr}, \forall f \in K' : aE_P(\underline{f}) \leqslant \underline{\Phi}(f) \right\}.$$

Let $X = \{x_1, ..., x_n\}$ and denote $aE_P(\underline{f}) = \sum_{i=1}^n a_i \underline{f}(x_i)$, where $a_i = aP(\{x_i\})$, $i = 1, ..., n$, and assume that the set K' is finite, then the value $Con(\Phi)$ can be computed by solving the following linear programming problem:

$$Con(\Phi) = 1 - \sum_{i=1}^n a_i \to \min$$

$$\begin{cases} \sum_{i=1}^n \underline{f}(x_i)a_i \leqslant \underline{\Phi}(f), \ f \in K', \\ \sum_{i=1}^n a_i \leqslant 1, \ a_i \geqslant 0, \ i = 1, ..., n. \end{cases}$$

In particular, if μ is an upper probability, then the amount of contradiction $Con(\mu)$ can be computed as

$$Con(\Phi) = 1 - \sum_{i=1}^n a_i \to \min$$

$$\begin{cases} \sum_{x_i \in A} a_i \leqslant \mu(A), \ A \in 2^X \backslash \{\emptyset\}, \\ a_i \geqslant 0, \ i = 1, ..., n. \end{cases}$$

An upper prevision Φ is called fully contradictory if $Con(\Phi) = 1$. In decision process, we identify the fully contradictory information with the full ignorance that can be modeled by the upper prevision $V_{\max}(f) = \max_{x \in X} f(x)$ or by the credal set $\mathbf{P}(V_{\max}) = M_{pr}$ that contains all possible probability measures. If the information described by the upper prevision Φ is not fully contradictory, then $Con(\Phi) = a \in [0,1)$ and Φ can represented as

$$\Phi = (1-a)\Phi^{(1)} + aV_{\min},$$

where $\Phi^{(1)}$ is a non-contradictory upper prevision, and for making decisions we need to correct the contradictory information. Clearly, the decision process should be based on the consistent information in $\Phi^{(1)}$ and depend on the amount of contradiction $a = Con(\Phi)$. We propose in [2,3] to take the natural extension $n.ext(\Phi^{(1)})^1$ of $\Phi^{(1)}$ and transform the fully contradictory information, represented by $V_{\min}$, to the full ignorance $V_{\max}$. This transformation gives us the

---

[1] The value $n.ext(\Phi^{(1)})(f)$ for $f \in K$ is computed by $n.ext(\Phi^{(1)})(f) = \overline{E}_{\mathbf{P}(\Phi^{(1)})}(f)$, where $\mathbf{P}(\Phi^{(1)})$ is the usual credal set that corresponds to $\Phi^{(1)}$.

coherent upper prevision

$$\Phi' = (1 - a)n.ext(\Phi^{(1)}) + aV_{\max}. \tag{3}$$

We can describe the above transformation using so-called generalized credal sets. At first, consider monotone measures of the type

$$P = a_0\eta_{\langle X \rangle} + \sum_{i=1}^{n} a_i\eta_{\langle \{x_i\} \rangle}, \tag{4}$$

where $\sum_{i=0}^{n} a_i = 1$, and $a_i \geqslant 0$, $i = 1, ..., n$, viewed as upper probabilities. We see that the set function $\eta_{\langle X \rangle}$ is the counterpart of $V_{\min}$, since $\eta_{\langle X \rangle}(A) = V_{\min}(1_A)$, $A \in 2^X$. If $a_0 < 1$, then $P^{(1)} = \frac{1}{1-a_0}\sum_{i=1}^{n} a_i\eta_{\langle \{x_i\} \rangle}$ is in M_{pr} and $P = a_0\eta_{\langle X \rangle} + (1-a_0)P^{(1)}$. Thus, $Con(P) = a_0$. We denote the set of all monotone measures of the type (4) by M_{cpr}^d. We can extend any P on 2^X to K by

$$\overline{E}_P(f) = a_0 V_{\min}(f) + (1 - a_0)E_{P^{(1)}}(f) = a_0\min_{x \in X} f(x) + \sum_{i=1}^{n} a_i f(x_i).$$

Definition 1. *A non-empty subset* $\mathbf{P} \subseteq M_{cpr}^d$ *is called a lower generalized credal set (LG-credal set) iff the following conditions hold*
(1) $P_1 \in \mathbf{P}$ *implies* $P_2 \in \mathbf{P}$ *for every* $P_2 \leqslant P_1$ *in* M_{cpr}^d;
(2) $\mathbf{P}$ *is a convex subset of* M_{cpr}^d, *i.e.* $P_1, P_2 \in \mathbf{P}$ *implies* $aP_1 + (1-a)P_2 \in \mathbf{P}$ *for any* $a \in [0, 1]$;
(3) $\mathbf{P}$ *is a closed subset of* $\mathbb{R}^n$ *(every* P *defined by (4) is considered as a point* $(a_1, ..., a_n)$ *of* $\mathbb{R}^n$*).*

LG-credal sets are generalizations of usual credal sets. It can be shown using profiles of LG-credal sets. Let $\mathbf{P}$ be a LG-credal set, then the set of all maximal elements in $\mathbf{P}$ is called the profile of $\mathbf{P}$ and denoted by $profile(\mathbf{P})$. If we consider the usual credal set $\mathbf{P}'$, then we identify it with the LG-credal set

$$\mathbf{P} = \{P \in M_{cpr}^d | \exists P' \in \mathbf{P}' : P \leqslant P'\}, \tag{5}$$

whose profile is $\mathbf{P}'$. The amount of contradiction in every LG-credal set $\mathbf{P}$ can be computed by $Con(\mathbf{P}) = \inf\{Con(P) | P \in \mathbf{P}\}$. Every upper prevision $\Phi : K' \to \mathbb{R}$ can be described by the LG-credal set $\mathbf{P}(\Phi)$ defined by

$$\mathbf{P}(\Phi) = \left\{P \in M_{cpr}^d | \forall f \in K' : \overline{E}_P(f) \leqslant \Phi\right\}. \tag{6}$$

Proposition 1. *Let* $\Phi : K' \to \mathbb{R}$ *be an upper prevision with* $Con(\Phi) = a_0 < 1$, *and let* $\mathbf{P}(\Phi)$ *be the LG-credal set defined by (6), then the coherent upper prevision* Φ' *from (6) coincides with*

$$\Phi''(f) = \sup\{\overline{E}_{P^d}(f) | P \in \mathbf{P}(\Phi), Con(P) = a_0\},$$

where $\overline{E}_{P^d}(f)$ *for* $f \in K$ *and* $P^d = a_0\eta_{\langle X \rangle}^d + \sum_{i=1}^{n} a_i\eta_{\langle \{x_i\} \rangle}$ *is defined by* $\overline{E}_{P^d}(f) = a_0\max_{x \in X} f(x) + \sum_{i=1}^{n} a_i f(x_i)$.

5   Basic Aggregation Rules in the Theory of Imprecise Probabilities

In the literature, a reader can find several approaches of information aggregation based on imprecise probabilities [8,12,13], but most of them can be considered as a combination of conjunctive, disjunctive and mixture rules. Assume that sources of information are described by upper previsions $\Phi_1, ..., \Phi_m$ on $K' \subseteq K$. If these sources of information are assumed to be reliable, then we can use the conjunctive rule:

$$\Phi(f) = \min_{i=1,...,m} \Phi_i(f), f \in K'.$$

If we describe these sources of information by LG-credal sets $\mathbf{P}(\Phi_i)$, $i = 1, ..., m$, then

$$\mathbf{P}(\Phi) = \bigcap_{i=1}^{m} \mathbf{P}(\Phi_i).$$

If we assume that at least one source of information is reliable, then we can use the disjunctive rule defined by $\Phi(f) = \max_{i=1,...,m} \Phi_i(f)$, $f \in K'$. The counterpart of it based on LG-credal sets is

$$\mathbf{P} = \left\{ \sum_{i=1}^{m} a_i P_i | P_i \in \mathbf{P}(\Phi_i), \sum_{i=1}^{m} a_i = 1, \ a_i \geqslant 0, \ i = 1, ..., m \right\},$$

where obviously $\mathbf{P} \subseteq \mathbf{P}(\Phi)$. The mixture rule is used if we can evaluate the reliability of information sources by real positive numbers r_i, $i = 1, ..., m$. Then the result of the mixture rule is the upper prevision $\Phi(f) = \sum_{i=1}^{m} r_i \Phi_i(f)$, $f \in K'$. The counterpart of it based on LG-credal sets is

$$\mathbf{P} = \left\{ \sum_{i=1}^{m} r_i P_i | P_i \in \mathbf{P}(\Phi_i) \right\},$$

where, obviously, $\mathbf{P} \subseteq \mathbf{P}(\Phi)$.

We can also define analogous aggregation rules based on usual credal sets, but in this case each source of information should be consistent and the conjunctive rule can be applied, if the intersection of corresponding credal sets is not empty, i.e. when there is no contradiction among information sources.

6   Updating Information Based on the Conjunctive Rule

In probability theory, the updating information has a high importance. Let P be a probability measure on 2^X and we know that an event $B \subseteq X$ occurs, then we can update probabilities using the conditional probability measure

$$P_B(A) = P(A \cap B)/P(B), A \in 2^X.$$

The conditional probability measure P_B is not defined if the event B fully contradicts to the probability measure P, i.e. $P(B) = 0$. Let us find what would be the result if we aggregate these sources of information using the conjunctive rule for LG-credal sets. In this case, the information that outcome is in the set B can be described by the upper probability $\eta^d_{\langle B \rangle}$. Assume that $P = \sum_{i=1}^{n} a_i \eta_{\langle \{x_i\} \rangle}$, then the result of the conjunctive rule is $\mu = \min\{P, \eta^d_{\langle B \rangle}\}$.

Proposition 2. *Let $P = \sum_{i=1}^{n} a_i \eta_{\langle \{x_i\} \rangle}$ be in M_{pr} and $\mu = \min\{P, \eta^d_{\langle B \rangle}\}$ for a $B \in 2^X \backslash \{\emptyset\}$. Then the LG-credal set $\mathbf{P}(\mu) = \{P \in M^d_{cpr} | P \leqslant \mu\}$ has the profile $\{P^*_B\}$, where $P^*_B = a_0 \eta_{\langle X \rangle} + \sum_{x_i \in B} a_i \eta_{\langle \{x_i\} \rangle}$, where $a_0 = 1 - \sum_{x_i \in B} a_i$.*

We can generalize Proposition 2 as follows.

Proposition 3. *Let $P = a_0 \eta_X + \sum_{i=1}^{n} a_i \eta_{\langle \{x_i\} \rangle}$ be in M^d_{cpr} and $\mu = \min\{P, \eta^d_{\langle B \rangle}\}$ for a $B \in 2^X \backslash \{\emptyset\}$. Then the LG-credal set $\mathbf{P}(\mu) = \{P \in M^d_{cpr} | P \leqslant \mu\}$ has the profile $\{P^*_B\}$, where $P^*_B = b_0 \eta_{\langle X \rangle} + \sum_{x_i \in B} a_i \eta_{\langle \{x_i\} \rangle}$ and $b_0 = 1 - \sum_{x_i \in B} a_i$.*

We will keep the notation from Propositions 2 and 3, assuming that for every $P = a_0 \eta_X + \sum_{i=1}^{n} a_i \eta_{\langle \{x_i\} \rangle}$ in M^d_{cpr} and $B \in 2^X$

$$P^*_B = b_0 \eta_{\langle X \rangle} + \sum_{x_i \in B} a_i \eta_{\langle \{x_i\} \rangle},$$

where $b_0 = 1 - \sum_{x_i \in B} a_i$, and the value $P^*_B(A)$ for $A \in 2^X$ can be computed by $P^*_B(A) = \begin{cases} P(A \cap B), & A \in 2^X \backslash \{X\}, \\ 1, & A = X. \end{cases}$ The measure P^*_B can be viewed as the result of the conditioning P given B based on generalized credal sets. We see that the result of conditioning is always defined, even in the case, when $P(B) = 0$ (or, in particular, $B = \emptyset$). In this case, $P^*_B = \eta_{\langle X \rangle}$, and the event B fully contradicts to the measure P.

Consider an arbitrary LG-credal set $\mathbf{P} \subseteq M^d_{cpr}$. Then the conditioning of $\mathbf{P}$ given $B \in 2^X$ is the subset of M^d_{cpr} defined by $\mathbf{P}^*_B = \{P^*_B | P \in \mathbf{P}\}$.

Lemma 1. *Let $\mathbf{P} \subseteq M^d_{cpr}$ be a LG-credal set and $B \in 2^X$. Then $\mathbf{P}^*_B = \mathbf{P} \cap (M^d_{cpr})^*_B$, where $(M^d_{cpr})^*_B = \{P^*_B | P \in M^d_{cpr}\}$.*

Proposition 4. *The subset $\mathbf{P}^*_B$ of M^d_{cpr} defined above is the LG-credal set.*

Thus, by Proposition 4, we define for every LG-credal set $\mathbf{P}$ and every event $B \in 2^X$ the conditional LG-credal set $\mathbf{P}^*_B$. Then we need to make the correction and to have the consistent information after that. Notice that the way, considered in Sect. 4, is not suitable for us, because it does give us the usual result for probability measures. This correction for measures in M^d_{cpr} should be

$$\varphi\left(a_0 \eta_{\langle X \rangle} + \sum_{i=1}^{n} a_i \eta_{\langle \{x_i\} \rangle}\right) = \frac{1}{1 - a_0} \sum_{i=1}^{n} a_i \eta_{\langle \{x_i\} \rangle}$$

for every $P = a_0 \eta_{\langle X \rangle} + \sum_{i=1}^{n} a_i \eta_{\langle \{x_i\} \rangle}$ in M^d_{cpr} with $a_0 < 1$.

For LG-credal sets correction that are not fully contradictory, we have the following major possibilities:

1) $\varphi^{(1)}(\mathbf{P}) = \{\varphi(P) | P \in \mathbf{P}, Con(P) = Con(\mathbf{P})\}$;
2) $\varphi^{(2)}(\mathbf{P}) = \{\varphi(P) | P \in profile(\mathbf{P})\}$.

Remark 1. It is easy to see that $\varphi^{(1)}(\mathbf{P})$ is the usual credal set for every LG-credal set $\mathbf{P}$ with $Con(\mathbf{P}) < 1$. Assume that $\mathbf{P}$ is a LG-credal set with the profile, which is the usual credal set $\mathbf{P}'$. Then the updating $\varphi^{(1)}(\mathbf{P}^*_B)$ for $B \in 2^X$ exists if $\overline{E}_{\mathbf{P}'}(1_B) > 0$ and the conditioning $\varphi^{(1)}(\mathbf{P}^*_B)$ is called the maximal likelihood conditioning because

$$\varphi^{(1)}(\mathbf{P}^*_B) = \left\{ P_B | P \in \mathbf{P}', P(B) = \overline{E}_{\mathbf{P}'}(1_B) \right\}.$$

Remark 2. Because the profile of a LG-credal set is not necessarily a convex set, the set $\varphi^{(2)}(\mathbf{P})$ is not a usual credal set in general. For decision-making, we don't need to describe $profile(\mathbf{P})$ explicitly, we need to know the extreme points of $\mathbf{P}$ that belong to $profile(\mathbf{P})$. This problem is investigated in [3] (see Proposition 2). Assume that $\mathbf{P}$ is a LG-credal set with the profile, which is the usual credal set $\mathbf{P}'$. Then the updating $\varphi^{(2)}(\mathbf{P}^*_B)$ for $B \in 2^X$ exists if $\overline{E}_{\mathbf{P}'}(1_B) > 0$. Assume that $\mathbf{P}'$ has a finite set of extreme points $\{P_1, ..., P_k\}$. Then we can find all extreme points in $profile\{\mathbf{P}^*_B\}$ in the set $\{(P_1)^*_B, ..., (P_k)^*_B\}$ as shown in the next example.

Example 1. Assume that $X = \{x_1, ..., x_5\}$ and $\mathbf{P}$ is the LG-credal set whose profile is the usual credal set described by extreme points $P_1 = (0.5, 0.3, 0.1, 0.1, 0)$, $P_2 = (0.2, 0.3, 0.1, 0.1, 0.3)$, $P_3 = (0.4, 0.2, 0.2, 0.1, 0.1)$. Consider the event $B = \{x_1, x_2\}$. Then $(P_1)^*_B = (0.5, 0.3, 0, 0, 0)$, $(P_2)^*_B = (0.2, 0.3, 0, 0, 0)$, $(P_3)^*_B = (0.4, 0.2, 0, 0, 0)$. Because $(P_1)^*_B \geqslant (P_2)^*_B$ and $(P_1)^*_B \geqslant (P_3)^*_B$, the LG-credal set $\mathbf{P}^*_B$ has the profile $\{(P_1)^*_B\}$. Thus, in this case $\varphi^{(1)}(\mathbf{P}^*_B) = \varphi^{(2)}(\mathbf{P}^*_B) = \{(P_1)_B\}$.

Analogously, let $C = \{x_1, x_2, x_3\}$ then $(P_1)^*_C = (0.5, 0.3, 0.1, 0, 0)$, $(P_2)^*_C = (0.2, 0.3, 0.1, 0, 0)$, $(P_3)^*_C = (0.4, 0.2, 0.2, 0, 0)$. Because $(P_1)^*_C \geqslant (P_2)^*_C$, the LG-credal set $\mathbf{P}^*_C$ has the profile $\{a(P_1)^*_C + (1 - a)(P_3)^*_C | a \in [0, 1]\}$. In this case, $\varphi^{(1)}(\mathbf{P}^*_C) = \{(P_1)_C\}$ and $\varphi^{(2)}(\mathbf{P}^*_C) = \{a(P_1)_C + (1 - a)(P_3)_C | a \in [0, 1]\}$.

Remark 3. Let $\mathbf{P}'$ be the usual credal set. Then the following updating rule is often used in the theory of imprecise probabilities [1,16]: $(\mathbf{P}')_B = \{P_B | P \in \mathbf{P}'\}$ for $B \in 2^X$. This rule is defined iff $\underline{E}_{\mathbf{P}'}(1_B) > 0$. It is well-known that if $\mathbf{P}'$ has a finite set of extreme points $\{P_1, ..., P_k\}$, then extreme points of $(\mathbf{P}')_B$ are in the set $\{(P_1)_B, ..., (P_k)_B\}$. Thus, $(\mathbf{P}')_B = \left\{ \sum_{i=1}^{k} a_i (P_i)_B \middle| \sum_{i=1}^{k} a_i = 1, \ a_i \geqslant 0, \ i = 1, ..., k \right\}.$

Let us consider the conditioning when the information is presented by an upper prevision $\Phi : K' \to \mathbb{R}$. Because there are many upper previsions that define the same LG-credal set, we can consider many rules of updating information. The simplest way seems to be the following: let $K'' = K' \cup \{1_{B^c}\}$ and

$$\Phi^*_B(f) = \begin{cases} 0, & f = 1_{B^c}, \\ \Phi(f), & f \in K'' \setminus \{1_{B^c}\}. \end{cases}$$

Then the upper prevision $\Phi_B^* : K'' \to \mathbb{R}$ defines the LG-credal set $\mathbf{P}(\Phi)_B^*$, since $\mathbf{P}(\Phi)_B^* = \mathbf{P}(\Phi_B^*)$.

The following proposition shows how updating information is produced for monotone measures viewed as upper probabilities.

Proposition 5. *Let μ be an upper probability on 2^X, $B \in 2^X$, and let μ_B^* be an upper probability on 2^X defined by*

$$\mu_B^*(A) = \begin{cases} \mu(A \cap B), & A \neq X, \\ 1, & A = X. \end{cases}$$

Then $(\mathbf{P}(\mu))_B^* = \mathbf{P}(\mu_B^*)$.

Proposition 6. *Let μ be an upper probability on 2^X and $\mu(B) > 0$ for some $B \in 2^X$. Then $(\mathbf{P}(\mu))_B^* = \{(1 - \mu(B))\eta_X + \mu(B)P | P \in \mathbf{P}(\mu_B)\}$, where $\mu_B(A) = \mu(A \cap B)/\mu(B)$, $A \in 2^X$.*

Corollary 1. *Let μ be an 2-alternating upper probability[2] on 2^X and $\mu(B) > 0$ for some $B \in 2^X$, and we use the notation from Proposition 5. Then*

$$\varphi^{(1)}((\mathbf{P}(\mu))_B^*) = \varphi^{(2)}((\mathbf{P}(\mu))_B^*) - \{P \subset M_{pr} | P \leqslant \mu_B\}.$$

Remark 4. Let us remind that plausibility functions are the dual of belief functions. We see that the conditioning based on generalized credal sets coincides with the conditioning based on Dempster's rule that for a plausibility function Pl on 2^X gives the result $Pl_B(A) = Pl(A \cap B)/Pl(B)$ for $A \in 2^X$ and $B \in 2^X$ such that $Pl(B) > 0$.

7 Conclusion

As one can see from results in Sect. 6, the conditioning based on generalized credal sets looks simpler than in the traditional theory of imprecise probabilities. The updating rule is applicable in more cases and comparing with the traditional rule the conditioning is based only on probability measures that are more plausible, this allows us to implement some learning process based on statistical data.

Acknowledgment. This work was partially supported by the grant 18-01-00877 of RFBR (Russian Foundation for Basic Research).

---

[2] A monotone measure μ is called 2-alternating if $\mu(A) + \mu(B) \geqslant \mu(A \cap B) + \mu(A \cup B)$ for all $A, B \in 2^X$.

References

1. Augustin, T., Coolen, F.P.A., de Cooman, G., Troffaes, M.C.M. (eds.): Introduction to Imprecise Probabilities. Wiley, New York (2014)
2. Bronevich, A.G., Rozenberg, I.N.: Incoherence correction and decision making based on generalized credal sets. In: Antonucci, A., Cholvy, L., Papini, O. (eds.) ECSQARU 2017. LNCS (LNAI), vol. 10369, pp. 271–281, Springer, Cham (2017). https://doi.org/10.1007/978-3-319-61581-3_25
3. Bronevich, A.G., Rozenberg, I.N.: Modelling uncertainty with generalized credal sets: application to conjunction and decision. Int. J. Gen. Syst. **27**(1), 67–96 (2018)
4. Brozzi, A., Capotorti, A., Vantaggi, B.: Incoherence correction strategies in statistical matching. Int. J. Approx. Reason. **53**, 1124–1136 (2012)
5. Cattaneo, M.: Likelihood decision functions. Electron. J. Stat. **7**, 2924–2946 (2013)
6. Denneberg, D.: Non-Additive Measure and Integral. Kluwer, Dordrecht (1997)
7. Dempster, A.: Upper and lower probabilities induced by a multivalued mapping. Ann. Math. Stat. **38**, 325–339 (1967)
8. Destercke, S., Antoine, V.: Combining imprecise probability masses with maximal coherent subsets: application to ensemble classification. In: Kruse, R., Berthold, M.R., Moewes, C., Gil, M.A., Grzegorzewski, P., Hryniewicz, O. (eds.) Synergies of Soft Computing and Statistics for Intelligent Data Analysis Advances in Intelligent Systems and Computing, vol. 190, pp. 27–35. Springer, Heidelberg (2013). https://doi.org/10.1007/978-3-642-33042-1_4
9. Good, I.J.: Good Thinking: The Foundations of Probability and Its Applications. University of Minnesota Press Minneapolis, Minn (1983)
10. Moral, S., De Campos, L.M.: Updating uncertain information. In: Bouchon-Meunier, B., Yager, R.R., Zadeh, L.A. (eds.) IPMU 1990. LNCS, vol. 521, pp. 58–67. Springer, Heidelberg (1991). https://doi.org/10.1007/BFb0028149
11. Moral, S.: Calculating uncertainty intervals from conditional convex sets of probabilities. In: Dubois, D., Wellman, M.P., D'Ambrosio, B. Smets, Ph. (eds.) Proceedings of 8th Conference on Uncertainty in Artificial Intelligence, pp. 199–206, Stanford University (1992)
12. Moral, S., Sagrado, J.: Aggregation of imprecise probabilities. In: Bouchon Meunier, B. (ed.) Aggregation and Fusion of Imperfect Information, pp. 162–188. Physica-Verlag, Heidelberg (1997)
13. Nau, R.: The aggregation of imprecise probabilities. J. Stat. Plan. Inference **105**(1), 265–282 (2002)
14. Quaeghebeur, E.: Characterizing coherence, correcting incoherence. Int. J. Approx. Reason. **56**, 208–223 (2015). (Part B)
15. Shafer, G.: A Mathematical Theory of Evidence. Princeton University Press, Princeton (1976)
16. Walley, P.: Statistical Reasoning with Imprecise Probabilities. Chapman and Hall, London (1991)
17. Walter, G., Augustin, T.: Imprecision and prior-data conflict in generalized Bayesian inference. J. Stat. Theory Pract. **3**, 255–271 (2009)

Probabilistic Logic for Reasoning About Actions in Time

Šejla Dautović[1] and Dragan Doder[2](✉)

[1] Matematički Institut SANU, Belgrade, Serbia
shdautovic@mi.sanu.ac.rs
[2] IRIT – Paul Sabatier University, Toulouse, France
dragan.doder@gmail.com

Abstract. In this paper, we develop a probabilistic logic for reasoning about preconditions, postconditions and execution of actions in time. The language of our logic allows statements like "precondition of the action A will hold in the next moment" and uncertain information like "probability that the precondition of A will hold in the next moment is at least one half." We axiomatize this logic, provide corresponding semantics built on branching-time temporal models, and prove that the axiomatization is sound and strongly complete.

Keywords: Probabilistic logic · Temporal logic · Action

1 Introduction

In the last few decades, uncertain reasoning has become a popular topic of investigation for the researchers in the fields of computer science and artificial intelligence. A particular line of research concerns the formalization in terms of logics for reasoning about probability [6,11,14].

In this paper, we focus on the problem how to apply probability logic to model uncertainty about action and time, whose interplay is fundamental for the design and development of intelligent systems such as autonomous systems, robotic applications, and service agents [5,19,22]. Reasoning about actions and time has received a lot of attention in the last couple of decades [1,8,10,16]. Several temporal logical systems have been developed for reasoning about the pre and postconditions of actions with explicit time points, such as the Event Calculus [10], Temporal Action Logics [8], extensions to the Fluent Calculus [18], and extensions to the Situation Calculus [15], and a simple parametrized-time action logic PAL [20,22] which is shown useful for revision of beliefs and intentions [21].

The starting point for this work was PAL logic from [22]. We first extend the language of PAL by employing the full power of CTL* [17], allowing expressions like "precondition of action a will always hold". While actions are assumed to be deterministic, different selection of available actions might lead to different future moments, naturally leading to a branching-time semantic, in which transitions between time moments are labeled by actions. In the second step, we apply probability operators to those formulas, in order to represent uncertainty about future. Semantically, that leads o probability spaces over CTL*-like models.

© Springer Nature Switzerland AG 2019
G. Kern-Isberner and Z. Ognjanović (Eds.): ECSQARU 2019, LNAI 11726, pp. 385–396, 2019.
https://doi.org/10.1007/978-3-030-29765-7_32

Our central technical result is strong completeness of the logic, which we obtain using Henkin's method, modifying extensively techniques presented in [2–4,12,13]. Since neither temporal nor probability logics are compact, it is known that there is no finitary strongly complete axiomatization [7]. We obtain completeness using infinitary rules of inference, thus keeping formulas finite.

2   Syntax and Semantics

In this section, we present the syntax and semantics of our logic, which we denote by $pCTL_A^*$. The logic contains two types of formulas: temporal formulas without probabilities, which represents statements about the current moment and future moment and actions that may be performed, and formulas which deal with uncertainty in time, obtained by applying probability operators to the first type of formulas.

Syntax. Let $\mathcal{P}$ be a nonempty set of propositional letters, and let $Act = \{a, b, c, \ldots\}$ be a finite set of deterministic primitive actions. We will denote the elements of $\mathcal{P}$ with p and q, possibly with subscripts. We use negation and conjunction as a complete list of Boolean connectives. We use the usual abbreviations for the other classical connectives and the symbols $\top$ and $\bot$.

Definition 1 (CTL_A^*−formula). *The set of formulas $For_{CTL_A^*}$ with the set of primitive propositions $Prop = \mathcal{P} \cup \{pre(a), post(a) \mid a \in Act\}$ is inductively defined in the following way:*

$$\alpha ::= \chi \mid do(a) \mid \bigcirc\alpha \mid A\alpha \mid \alpha U\alpha \mid \alpha \wedge \alpha \mid \neg\alpha$$

with $\chi \in Prop$ and $a \in Act$. We will denote the formulas from $For_{CTL_A^*}$ with α, β and γ, possibly with subscripts.

The temporal operators $\bigcirc$ (next), U (until) and A (universal path operator) are standard operators of CTL^* [17]. Other temporal operators F (sometimes), G (always) and E (existential path quantifier) are defined as abbreviations: $F\alpha \equiv \top U\alpha$, $G\alpha \equiv \neg F\neg\alpha$ and $E\alpha \equiv \neg A\neg\alpha$. If T is a set of formulas, then $\bigcirc T$ denotes $\{\bigcirc\alpha \mid \alpha \in T\}$, while AT denotes $\{A\alpha \mid \alpha \in T\}$. Furthermore, for $k \in w$, $\bigcirc^{k+1}\alpha$ is an abbreviation for $\bigcirc(\bigcirc^k\alpha)$.

Example 1. The expression

$$(pre(a) \wedge G(pre(a) \rightarrow \neg pre(b)) \rightarrow \neg E \bigcirc post(b)$$

is a CTL_A^* formula. Its meaning is that if the precondition of action a holds and if preconditions of a and b are incompatible at all future moments, then it is not possible that the postcondition of b will hold in the next moment.

Now we introduce our probabilistic formulas. We use the list of probability operators of the form $P_{\geq r}$, for every $r \in \mathbb{Q} \cap [0, 1]$, which can be applied to CTL_A^*−formulas.

Definition 2 (Probabilistic formula). *A* probabilistic formula *is any Boolean combination of the formulas of the form* $P_{\geq r}\alpha$, *where* $\alpha \in For_{\text{CTL}^*_A}$. *We denote by* For_P *the set of all probabilistic formulas, and we denote arbitrary probabilistic formulas by* ϕ *and* ψ, *indexed if necessary.*

We use the following abbreviations to introduce other types of operators: $P_{<s}\alpha$ is $\neg P_{\geq s}\alpha$, $P_{\leq s}\alpha$ is $P_{\geq 1-s}\neg\alpha$, $P_{>s}\alpha$ is $\neg P_{\leq s}\alpha$, and $P_{=s}\alpha$ is $P_{\geq s}\alpha \wedge P_{\leq s}\alpha$.

Example 2. The expression

$$P_{\geq \frac{1}{2}}(\bigcirc pre(a)) \rightarrow P_{\leq \frac{1}{2}}\bigcirc do(b)$$

is a probabilistic formula. Its meaning is that if the precondition of a will hold in the next moment is at least a half, then the probability that b will be executed in the next moment is at most one half.

Definition 3 (Formula). $For_{p\text{CTL}^*_A} = For_{\text{CTL}^*_A} \cup For_P$. *We denote arbitrary formulas by* Φ *and* Ψ *(possibly with subscripts).*

Obviously, mixing of pure propositional formulas and probability formulas is not allowed. In the paper, we denote both $\alpha \wedge \neg\alpha$ and $\phi \wedge \neg\phi$ by $\bot$ (and similarly for $\top$), and we let the context determine the meaning.

Semantics. Now we define models of our logic. First we need to introduce the structures in which CTL^*_A−formulas are evaluated.

Definition 4 (Transition System). *A transition system* TS *is tuple* (S, R, v) *where:*

- *S is a non- empty set of states;*
- $v : S \to 2^{Prop}$ *is a valuation function from states to sets of propositions;*
- $R = \bigcup R_a$, $a \in Act$, R_a *relation on* $S \times S$, *such that the following conditions hold:*
  - *(a) R is serial;*
  - *(b) For all action* $a \in Act$: *If* sR_as' *and* sR_as'' *then* $s' = s''$;
  - *(c) If* $pre(a) \in v(s), a \in Act, s \in S$ *then there existis* $s' \in S, sR_as'$;
  - *(d) If* sR_as' *then* $post(a) \in v(s')$.

Definition 5 (Path). *A path,* π *in* TS *is an infinite sequence of states of* TS *and actions,* $\pi = (s_0, a_0, s_1, a_1, \ldots)$, *such that for each i,* $s_iR_{a_i}s_{i+1}$. *For a path* $\pi = (s_0, a_0, s_1, a_1, \ldots)$, *we write* π_k *for state* s_k, $\pi_{\geq i}$ *for the path* $(s_i, a_i, s_{i+1}, a_{i+1}, \ldots)$ *and* $act(\pi, k) = a_k$ *(action on the path* π *in the time k is* a_k).

We define what it means for a formula α to be satisfied at a path π in a transition system TS, denoted by TS, $\pi \models_c \alpha$, as follows:

- TS, $\pi \models_c \chi$ iff $\chi \in v(\pi_0)$, $\chi \in Prop$;
- TS, $\pi \models_c do(\alpha)$ iff $act(\pi, 0) = a$;
- TS, $\pi \models_c \neg\alpha$ iff $M, \pi \not\models_c \alpha$;

- TS, $\pi \models_c \alpha \wedge \beta$ iff $M, \pi \models_c \alpha$ and $M, \pi \models_c \beta$;
- TS, $\pi \models_c \bigcirc \alpha$ iff $M, \pi_{\geq 1} \models_c \alpha$;
- TS, $\pi \models_c \alpha U \beta$ iff for some $i \geq 0$, $M, \pi_{\geq i} \models_c \beta$ and for all j, $0 \leq j < i$ $M, \pi_{\geq j} \models_c \alpha$;
- TS, $\pi \models_c A\alpha$ iff for all π', $\pi_0 = \pi'_0$, $M, \pi' \models_c \alpha$.

The logic notions of satisfiability, validity and semantics consequences are defined as usual. We denote the corresponding value of α in a path π in a transition system TS by $v(\alpha, \text{TS}, \pi)$:

$$v(\alpha, \text{TS}, \pi) = 1, \text{ if TS}, \pi \models_c \alpha,$$
$$v(\alpha, \text{TS}, \pi) = 0 \text{ otherwise.}$$

The transition systems are just an intermediate step towards our class of models. It is based on the possible-world approach. It captures our approach in which CTL_A^*−formulas represent certain information, while uncertainty is represented by probabilistic formulas.

Definition 6 ($p\text{CTL}_A^*$−model). *A $p\text{CTL}_A^*$−model is a tuple $M = \langle W, H, \mu, \sigma \rangle$ where:*

- *W is a nonempty set of worlds,*
- *$\langle W, H, \mu \rangle$ is a probability space, i.e.,*
  - *H is an algebra of subsets of W, i.e., a set of subsets of W with the property:*
    - *$W \in H$,*
    - *If $A, B \in H$, then $W \setminus A \in H$ and $A \cup B \in H$.*
    The elements of H are called measurable sets.
  - *$\mu : H \longrightarrow [0, 1]$ is a finitely additive measure, i.e.,*
    - *$\mu(W) = 1$,*
    - *If $A, B \in H$ and $A \cap B = \emptyset$, then $\mu(A \cup B) = \mu(A) + \mu(B)$.*
- *σ provides for each world $w \in W$ a transition system and a path, i.e., $\sigma(w) = (\text{TS}_w, \pi_w)$.*

For a $p\text{CTL}_A^*$−model is a tuple $M = \langle W, H, \mu, \sigma \rangle$, we define

$$[\alpha]_M = \{w \in W \mid v(\alpha, \text{TS}_w, \pi_w) = 1\}.$$

We say that M is *measurable*, if $[\alpha]_M \in H$ for every $\alpha \in For_{\text{CTL}_A^*}$. We denote the class of all measurable $p\text{CTL}_A^*$−models with $p\text{CTL}_A^*{}^{Meas}$.

Now we define the satisfiability of a formula in a model from $p\text{CTL}_A^*{}^{Meas}$.

Definition 7 (Satisfiability). *Let $M = \langle W, H, \mu, \pi \rangle$ be a PL_{LTL} structure. We define the satisfiability relation $\models \subseteq PL_{LTL}^{Meas} \times For$ recursively as follows:*

- *$M \models \alpha$ iff $v(\alpha, \text{TS}_w, \pi_w) = 1$ for every $w \in W$,*
- *$M \models P_{\geq r}\alpha$ if $\mu([\alpha]) \geq r$,*
- *$M \models \neg\phi$ iff $M \not\models \phi$,*
- *$M \models \phi \wedge \psi$ iff $M \models \phi$ and $M \models \psi$.*

Definition 8 (Model of a formula, semantical consequence). *For a measurable structure M and a set of formulas T, we say that M is a* model *of T and write $M \models T$ iff $M \models \Phi$ for every $\Phi \in T$. We also say that T is* satisfiable, *if there is M such that $M \models T$.*

We say that a set of formulas T entails *a formula Φ and write $T \models \Phi$, if all models of T are models of Φ. A formula Φ is* valid *if $\emptyset \models \Phi$.*

3 Axiomatization

In this section we present an axiomatization of our logic, which we denote $Ax(p\text{CTL}^*_\text{A})$. First we need to introduce two useful notions.

Definition 9 (k-nested implication). *A k-nested implication $\Phi_k(\tau, \overline{\gamma})$ for the formula τ, based on the sequence $\overline{\gamma} = (\gamma_0, \ldots, \gamma_k)$ of formulas, is defined recursively as follows:*

$$\Phi_0(\tau, \overline{\gamma}) = \gamma_0 \rightarrow \tau$$

$$\Phi_k(\tau, \gamma) = \gamma_k \rightarrow A(\Phi_{k-1}(\tau, (\gamma_0, \ldots, \gamma_{k-1}))).$$

For example, $\Phi_2(\alpha, (\gamma_0, \gamma_1, \gamma_2))$ is the formula

$$\gamma_2 \rightarrow A(\gamma_1 \rightarrow A(\gamma_0 \rightarrow \alpha)).$$

Definition 10 (State formula). *A formula is a* state formula *if it is a Boolean combination of elements of Prop and formulas od the form $A\alpha$. We denote the set of all state formulas by St.*

Also, we introduce the operator U_n in the following way:

$$\alpha U_n \beta := (\bigwedge_{k=0}^{n-1} \bigcirc^k \alpha) \wedge \bigcirc^v \beta.$$

Our axiomatization contains the following axiom schemas and inference rules.

Axiom schemas:

(A1) All instances of classical propositional tautologies for both CTL^*_A–formulas and probabilistic formulas.

(A2) $\bigcirc(\alpha \rightarrow \beta) \rightarrow (\bigcirc\alpha \rightarrow \bigcirc\beta)$

(A3) $\alpha U \beta \rightarrow \beta \vee (\alpha \wedge \bigcirc(\alpha U \beta))$

(A4) $\alpha \rightarrow A\alpha$ where $\alpha \in Prop$

(A5) $E\alpha \rightarrow \alpha$ where $\alpha \in Prop$

(A6) $A\alpha \rightarrow \alpha$

(A7) $A(\alpha \rightarrow \beta) \rightarrow (A\alpha \rightarrow A\beta)$

(A8) $A\alpha \rightarrow AA\alpha$

(A9) $E\alpha \rightarrow AE\alpha$

(A10) $\bigvee_{a \in Act} do(a)$

(A11) $do(a) \rightarrow \bigwedge_{b \neq a} \neg do(b)$
(A12) $pre(a) \rightarrow E do(a)$
(A13) $do(a) \rightarrow \bigcirc post(a)$
(A14) $E(do(a) \wedge \bigcirc \alpha) \rightarrow A(do(a) \rightarrow \bigcirc \alpha)$, where α is a state formula.
(A15) $P_{\geq 0} \alpha$
(A16) $P_{\leqslant r} \alpha \rightarrow P_{<s} \alpha$ whenever $r < s$
(A17) $P_{<r} \alpha \rightarrow P_{\leqslant r} \alpha$
(A18) $(P_{\geq r} \alpha \wedge P_{\geq s} \beta \wedge P_{\geq 1}(\neg \alpha \vee \neg \beta)) \rightarrow P_{\geq \min\{1, r+s\}}(\alpha \vee \beta)$
(A19) $(P_{\leqslant r} \alpha \wedge P_{<s} \beta) \rightarrow P_{<r+s}(\alpha \vee \beta)$, whenever $r + s \leqslant 1$

Inference rules:

(R1) From $\{\Phi, \Phi \rightarrow \Psi\}$ infer Ψ
(R2) From α infer $\bigcirc \alpha$
(R3) From α infer $A\alpha$
(R4) From the set of premises $\{\Phi_k(\neg(\alpha U_n \beta), \overline{\gamma}) \mid n \in \mathbb{N}\}$ infer $\Phi_k(\neg(\alpha U \beta), \overline{\gamma})$
(R5) From α infer $P_{\geq 1} \alpha$.
(R6) From the set of premises $\{\phi \rightarrow P_{\geq r - \frac{1}{k}} \alpha \mid k \in \mathbb{N}, k \geq \frac{1}{r}\}$ infer $\phi \rightarrow P_{\geq r} \alpha$.

The axioms A2–A9 are temporal axioms; A10–A14 are actions for pre and postconditions, modified from [22]. A14 describes deterministic behavior of actions. The axioms A16–A19 are standard probabilistic axioms [14]. R2, R3 and R5 are three forms of Necessitation. The rule R4 is a novel rule that generalize a rule form [9]; R6 is so called Archimedean rule.

A formula α is a *theorem*, denoted by $\vdash \alpha$, if there is an at most countable sequence of formulas $\alpha_0, \alpha_1, \ldots, \alpha$, such that every α_i is an axiom, or it is derived from the preceding formulas by an inference rule. A formula α is *deducible* from a set T ($T \vdash \alpha$) if there is at most countable sequence of formulas $\alpha_0, \alpha_1, \ldots, \alpha$, such that every α_i is an axiom or a formula from T, or it is derived from the preceding formulas by an inference rules, with exception that the inference (R2) and (R3) can be applied to the theorems only. The sequence $\alpha_0, \alpha_1, \ldots, \alpha$ is the *proof* of $T \vdash \alpha$. A set T of formulas is *consistent* if there is at least one formula which is not deducible from T, otherwise T is *inconsistent*.

A set T of formula is *maximal consistent set* (mcs) if it is consistent and any proper subset of T is inconsistent.

Theorem 1 (Soundness). *The axiomatization is sound with respect to the class of models* $p\mathrm{CTL}_A^{* \, Meas}$.

Now we show that Deduction theorem holds in $p\mathrm{CTL}_A^*$.

Theorem 2 (Deduction theorem). *Let T be a set of* CTL_A^**–formulas. and let α and β be two* CTL_A^**–formulas. Then $T, \alpha \vdash \beta$ iff $T \vdash \alpha \rightarrow \beta$.*

Proof. The direction from right to left it trivial. For left to right we use the transfinite induction on the length of the inference. Here we will only consider the case when β is obtained by the rule R4, i.e $\beta = \Phi_k(\neg(\alpha' U_n \beta'), \overline{\gamma})$, where $\overline{\gamma} = (\gamma_0, \ldots, \gamma_k)$. Then we have

$T, \alpha \vdash \gamma_k \rightarrow A(\Phi_{k-1}(\neg(\alpha' U_n \beta'), (\gamma_0, ..., \gamma_{k-1})))$, for all n (definition of Φ_k)

$T \vdash (\alpha \wedge \gamma_k) \rightarrow A(\Phi_{k-1}(\neg(\alpha' U_n \beta'), (\gamma_0, ..., \gamma_{k-1})))$, for all n (by (A1))

$T \vdash \Phi_k(\neg(\alpha' U_n \beta'), (\gamma_0, ..., \gamma_{k-1}, \alpha \wedge \gamma_k))$, for all n

$T \vdash \Phi_k(\neg(\alpha' U \beta'), (\gamma_0, ..., \gamma_{k-1}, \alpha \wedge \gamma_k))$ (by R4)

$T \vdash (\alpha \wedge \gamma_k) \rightarrow A(\Phi_{k-1}(\neg(\alpha' U \beta'), (\gamma_0, ..., \gamma_{k[1]})))$

$T \vdash \alpha \rightarrow (\gamma_k \rightarrow A(\Phi_{k-1}(\neg(\alpha' U \beta'), (\gamma_0, ..., \gamma_{k-1})))$

$T \vdash \alpha \rightarrow \Phi_k(\neg(\alpha' U_n \beta'), \overline{\gamma})$

$T \vdash \alpha \rightarrow \beta$.

Theorem 3 (Lindenbaum's Theorem). *Every consistent set of formulas can be extended to a maximal consistent set.*

Proof. Let $\Psi_0, \Psi_1, ...$ be an enumeration of all formulas. For a given consistent set T, we define the sequence of sets T_i, $i = 0, 1, 2, ...$ and the set T^* in the following way:

1. $T_0 = T$,
2. for every $i \geq 0$,
   (a) if $T_i \cup \{\Psi_i\}$ is consistent, then $T_{i+1} = T_i \cup \{\Psi_i\}$, otherwise
   (b) if Ψ_i is of the form $\Phi_k(\neg(\alpha U \beta), \overline{\gamma})$, then $T_{i+1} = T_i \cup \{ \Phi_k(\neg(\alpha U_n \beta), \overline{\gamma})\}$, where n is the smallest nonnegative integer such that T_{i+1} is consistent, otherwise
   (c) if Ψ_i is of the form $\phi \rightarrow P_{\geq r}\alpha$, then $T_{i+1} = T_i \cup \{\phi \rightarrow P_{<r-\frac{1}{k}}\alpha\}$, where k is the smallest positive integer such that $r - \frac{1}{k} \geq 0$ and T_{i+1} is consistent, otherwise
   (d) $T_{i+1} = T_i$.
3. $T^* = \bigcup_{i=0}^{\infty} T_i$.

Note that using Deduction Theorem we can prove that T^* is correctly defined: there exist n from the parts 2(b) and 2(c) of the construction. Each set T_i is consistent by construction. The steps (1) and (2) guarantee that T^* is maximal. Also, T^* obviously doesn't contain all formulas. Finally, one can show that T^* is deductively closed set. Consequently, we have that T^* is consistent, since otherwise $\perp \in T^*$.

4   Completeness

In this section we will prove our main result: the axiomatization $Ax(pCTL_A^*)$ is strongly complete for the class of models $pCTL_A^{*\ Meas}$. As a part of the proof, we first need to prove completeness of the sublogic CTL_A^*.

Completeness of CTL_A^*. We consider the non-probabilistic part of our logic, with CTL_A^*–formulas and with transition systems as semantics. Let us denote by $Ax(pCTL_A^*)$ the subsystem of $Ax(pCTL_A^*)$ which consists of the axioms (A1)–(A14) and (R1)–(R4). Of course, (A1) is now restricted to CTL_A^*–formulas only. We start with some auxiliary statements.

Lemma 1. *For any maximal consistent set T, the following are true:*

1. *If T is a mcs set of CTL_A^*–formulas then $T' = \{\alpha \mid \bigcirc\alpha \in T\}$ is also a mcs.*
2. *$T \vdash \alpha$ then $\bigcirc T \vdash \bigcirc\alpha$.*
3. *$T \vdash \alpha$ then $AT \vdash A\alpha$.*
4. *If $A\alpha \notin T$ then there exists $T' \in [T]$, such that $\alpha \notin T'$.*
5. *If $E\alpha \in T$ then there exists $T' \in [T]$, such that $\alpha \in T'$.*

Proof. Let us prove 3. and 4.

3. We will use the induction on the depth of the derivation of α from T. Suppose that $T \vdash \Phi_k(\neg(\alpha U\beta), \gamma)$ iz obtained from $T \vdash \Phi_k(\neg(\alpha U_n\beta), \gamma)$, for all $n \in w$, by the inference rule R4. Then we have

$AT \vdash A(\Phi_k(\neg(\alpha U_n\beta), \gamma))$ for all $n \in w$ (by the induction hypothesis),
$AT \vdash \top \rightarrow A(\Phi_k(\neg(\alpha U_n\beta), \gamma))$ for all $n \in w$
$AT \vdash \Phi_{k+1}(\neg(\alpha U_n\beta), \overline{\gamma})$ for all $n \in w$, where $\overline{\gamma} = (\gamma, \top)$
$AT \vdash \Phi_{k+1}(\neg(\alpha U\beta), \overline{\gamma})$ (by R4)
$AT \vdash \top \rightarrow A(\Phi_k(\neg(\alpha U\beta), \gamma))$
$AT \vdash A(\Phi_k(\neg(\alpha U\beta), \gamma))$

The other cases can be solved in a similar way.

4. Let $T' = T \cap St$. If $T \cup \{\neg\alpha\}$ is consistent then, by Lindenbaum's lemma, it can be extended to a mcs T^*. Since $T^* \cap St = T'$, then $T^* \in [T]$. If $T' \cup \{\neg\alpha\}$ is inconsistent then $T' \vdash \alpha$. By lemma 4, $AT' \vdash A\alpha$. Since $T' \subseteq St$, by the axioms A4, A8 and A9, $T' \vdash A\alpha$. Thus, $A\alpha \in T$ which contradicts the assumption.

Now we define an equivalence relation on maximal consistent sets.

Definition 11. *For two maximal consistent sets T and T' of CTL_A^*–formulas, we define the equivalence relation between them, denoted by $T \equiv T'$, as follows: $T \equiv T'$ iff $T \cap St = T' \cap St$. For the mcs T, $[T]$ is the set of all mcs's that are equivalent with T, i.e, $[T] = \{T' \mid T' \equiv T\}$.*

Using the relation $\equiv$, we define the canonical transition system.

Definition 12 (Canonical transition system). *We define a tuple $TS_{can} = (S, R, v)$ such that:*

1. $S = \{[T] \mid T \text{ is a mcs}\}$
2. $R = \bigcup R_a$, $a \in Act$, *such that* sR_as' *iff exists mcs* $T' \in s$, $do(a) \in T'$ *and* $T'' \in s'$, $T'' = \{\alpha \mid \bigcirc\alpha \in T'\}$

Theorem 4. TS_{can} *is a transition system.*

Proof. First, note that TS_{Can} is well defined, since T'' is a mcs, by Lemma 1(1). We need to show that R from TS_{can} satisfies the conditions $(a) - (d)$ from Definition 4.

(a) For $s \in S$, there is a mcs T_1, $T_1 \in s$. By A10, for some $a \in Act$, $do(a) \in T_1$, by 2. from definition of TS_{can} it follows that R is serial.

(b) Let $a \in Act$, and $sR_a s'$ and $sR_a S''$. By definition of TS_{can}, there exists $T \in s$, $do(a) \in T$ and $T' \in s'$, $T' = \{\alpha \mid \bigcirc \alpha \in T\}$, i.e, there exists $T_1 \in s$, $do(a) \in T_1$ and $T_1' \in s''$, $T_1' = \{\alpha \mid \bigcirc \alpha \in T_1\}$. From A14 it follows that $T' \equiv T_1'$, i.e, $s' = s''$.

(c) If $pre(a) \in v(s)$, s from TS_{can}, then there is mcs $T \in s$ such that $pre(a) \in T$. By A12, $Edo(a) \in T$, and by Lemma 1.5, there exists $T' \in [T]$, such that $do(a) \in T'$. From definition of TS_{can}, $sR_a s'$.

(d) If $sR_a s'$ then by definition of M_{can}, exists $T \in s$, $do(a) \in T$ and $T' \in s'$, $T' = \{\alpha \mid \bigcirc \alpha \in T\}$. By A13, $\bigcirc post(a) \in T$ then $post(a) \in T'$. Therefore, $post(a) \in v(s')$.

Now we introduce the operator $\bigcirc^{-n}$ on sets of formulas: $\bigcirc^{-n}T := \{\alpha \mid \bigcirc^n \alpha \in T\}$. Note that if T is a mcs, then $\bigcirc^{-n}T$ is also a mcs, by Lemma 1(1).

Definition 13. *For a given mcs T, we define the path $\pi_T = (s_0, a_0, s_1, a_1, \ldots)$ as follows:*

- $s_i = [\bigcirc^{-i}T]$,
- $a_i = a$ *such that* $\bigcirc^i do(a) \in T$.

It is easy to check that π_T is well defined. Moreover, it can be shown that the mapping from maximal consistent sets to paths of TS_{Can} is a bijection.

Lemma 2. *Let's Σ be a set of all paths in TS_{can} and MCS set of all maximal consistent sets of CTL_A^*–formulas. Then, the function $f : MCS \to \Sigma$, such that $f(T) = \pi_T$, is bijection.*

Lemma 3. *(a) For two mcs, T and T', $T \equiv T'$ iff $(\pi_T)_0 = (\pi_{T'})_0$;*
(b) For a mcs T, $\pi_{(\bigcirc^{-k}T)} = (\pi_T)_{\geq k}$.

Now we can prove the completeness result.

Theorem 5 (Strong completeness of CTL_A^*). *A set of CTL_A^*–formulas is consistent if and only if it is satisfiable in a path of a transition system.*

Proof. The ($\Leftarrow$)-direction can be easily checked. In order to prove ($\Rightarrow$)-direction we extend given consistent set to a maximal consistent set T, using Lindenbaum's Theorem, and we construct the canonical transition system $TS = TS_{can}$. We need to show that for every formula α, $\alpha \in T$ iff $TS, \pi_T \models_c \alpha$. The proof is by induction on the complexity of α.

The cases when $\alpha \in Prop$ and $\alpha = do(a)$ follow directly from the definition. The case when formulas are negations and conjunctions can be proved as usual.

If $\alpha = \bigcirc \beta$, $\alpha \in T$ iff $\beta \in \bigcirc^{-1}T$ iff (by the induction hypothesis) $TS, \pi_{(\bigcirc^{-1}T)} \models_c \beta$ iff $TS, \pi_T \models_c \bigcirc \beta$ (by Lemma 3(b)) iff $TS, \pi_T \models_c \alpha$.

If $\alpha = \beta U \gamma$, $TS, \pi_T \models_c \beta U \gamma$ iff (for some $i \geq 0$, $TS, (\pi_T)_{\geq i} \models_c \gamma$ and for all j, $0 \leq j < i$, $TS, (\pi_T)_{\geq j} \models_c \beta$) iff (for some $i \geq 0$, $TS, \pi_{\bigcirc^{-i}T} \models_c \gamma$ and for all j, $0 \leq j < i$, $TS, \bigcirc^{-i}T \models_c \beta$) (by Lemma 3(b)) iff (for some $i \geq 0$, $\gamma \in \bigcirc^{-i}T$ and for all j, $0 \leq j < i$, $\beta \in \bigcirc^{-i}T$ (by the induction hypothesis) iff $\beta U \gamma \in T$ (by (A2), (A3) and (R4)).

Let $\alpha = A\beta$. Suppose that TS, $\pi_T \models_c A\beta$. Then for all π' such that $\pi_0' = (\pi_T)_0$, TS, $\pi' \models_c \beta$. By Lemma 3, for all mcs T' such that $T' \equiv T$, TS, $\pi_{T'} \models_c \beta$. By the induction hypothesis, for all T' such that $T' \equiv T$, $\beta \in T'$. Then, by Lemma 5.4, $A\beta \in T$. For the other direction, suppose that TS, $\pi_T \not\models_c A\beta$. Then, there exists π', such that $\pi_0' = (\pi_T)_0$ and TS, $\pi' \not\models_c \beta$. By Lemma 3, there is a mcs T', such that $T' \equiv T$ and $\pi' = \pi_{T'}$. By induction hypothesis, $\beta \notin T'$, so by A6, $A\beta \notin T'$, and finally $A\beta \notin T$, since $T' \equiv T$.

Completeness of $p\text{CTL}_A^*$ Now we consider full logical language. Similarly as above, starting from a mcs, we define a canonical model.

Definition 14 (Canonical model). *For a maximal consistent set T^*, we define $M_{T^*} = \langle W, H, \mu, \sigma \rangle$, such that:*

1. $W = \{(\text{TS}, \pi) \mid v(\alpha, \text{TS}, \pi) = 1$ *for all* $\alpha \in T^* \cap For_{\text{CTL}_A^*}\}$,
2. $H = \{[\alpha] \mid \alpha \in For_{\text{CTL}_A^*}\}$, *where* $[\alpha] = \{(\text{TS}, \pi) \in W \mid v(\alpha, \text{TS}, \pi) = 1\}$,
3. $\mu([\alpha]) = \sup\{r \in \mathcal{Q} \mid T^* \vdash P_{\geq r}\alpha\}$, *for every* $\alpha \in For_{\text{CTL}_A^*}$,
4. $\sigma(w) = w$ *for every* $w \in W$.

Lemma 4. *M_{T^*} is a measurable model.*

Theorem 6 (Strong completeness). *A set of formulas $T \subseteq For$ is consistent iff it is satisfiable.*

Proof. The direction from right to left follows from Theorem 1. For the other direction, it is sufficient to show that a consistent set of formulas T has a model. First we use Lindenbaum's theorem to extend T to a maximal consistent set T^*, and then we construct the canonical model M_{T^*}. We show that M_{T^*} is a model of T^*, and, consequently, a model of T. It is sufficient to prove that for all $\Phi \in For$, $T^* \vdash \Phi$ iff $M_{T^*} \models \Phi$.

Let $\Phi = \alpha \in For_{\text{CTL}_A^*}$. If $\alpha \in T^*$, then by the definition of W from M_{T^*}, $M_{T^*} \models \alpha$. Conversely, if $M_{T^*} \models \alpha$, by Theorem 5, $\alpha \in T^*$.

If $\Phi \in For_P$, we proceed by induction on the complexity of Φ. If $\Phi = P_{\geq r}\alpha$, the proof is similar to the one presented in [14]. Conjunction and negation are treated in a standard way.

5   Conclusion

In this work introduced a logic for reasoning about action and time in presence of uncertainty, where uncertainty is modeled by probability. The logic can represent agent's quantitative belief about the pre and postconditions of actions, and execution of actions in current moment and future moments. We proposed an axiomatic system for the logic and proved strong completeness for the considered class of Kripke-like models, using Henkin's construction.

Acknowledgments. This work was supported by the Serbian Ministry of Education and Science through project ON174026, and by ANR-11-LABX-0040-CIMI.

References

1. Broersen, J.: A complete STIT logic for knowledge and action, and some of its applications. In: Baldoni, M., Son, T.C., van Riemsdijk, M.B., Winikoff, M. (eds.) DALT 2008. LNCS (LNAI), vol. 5397, pp. 47–59. Springer, Heidelberg (2009). https://doi.org/10.1007/978-3-540-93920-7_4

2. Doder, D., Marković, Z., Ognjanović, Z., Perović, A., Rašković, M.: A probabilistic temporal logic that can model reasoning about evidence. In: Link, S., Prade, H. (eds.) FoIKS 2010. LNCS, vol. 5956, pp. 9–24. Springer, Heidelberg (2010). https://doi.org/10.1007/978-3-642-11829-6_4

3. Doder, D., Ognjanovic, Z.: A probabilistic logic for reasoning about uncertain temporal information. In: Proceedings of UAI 2015, Amsterdam, The Netherlands, pp. 248–257 (2015)

4. Doder, D., Ognjanovic, Z., Markovic, Z.: An axiomatization of a first-order branching time temporal logic. J. UCS **16**(11), 1439–1451 (2010)

5. Doherty, P., Kvarnström, J., Heintz, F.: A temporal logic-based planning and execution monitoring framework for unmanned aircraft systems. Auton. Agent. Multi-Agent Syst. **19**(3), 332–377 (2009)

6. Fagin, R., Halpern, J.Y., Megiddo, N.: A logic for reasoning about probabilities. Inf. Comput. **87**(1/2), 78–128 (1990)

7. van der Hoek, W.: Some considerations on the logic PDF. J. Appl Non-Classical Logics **7**(3) (1997)

8. Kvarnström, J.: TALplanner and other extensions to Temporal Action Logic. Ph.D. thesis, Linköpings universitet (2005)

9. Marinkovic, B., Ognjanovic, Z., Doder, D., Perovic, A.: A propositional linear time logic with time flow isomorphic to ω^2. J. Appl. Logic **12**(2), 208–229 (2014)

10. Mueller, E.T.: Commonsense Reasoning. Morgan Kaufmann, Burlington (2010)

11. Nilsson, N.J.: Probabilistic logic. Artif. Intell. **28**(1), 71–87 (1986)

12. Ognjanović, Z., Doder, D., Marković, Z.: A branching time logic with two types of probability operators. In: Benferhat, S., Grant, J. (eds.) SUM 2011. LNCS (LNAI), vol. 6929, pp. 219–232. Springer, Heidelberg (2011). https://doi.org/10.1007/978-3-642-23963-2_18

13. Ognjanovic, Z., Markovic, Z., Raskovic, M., Doder, D., Perovic, A.: A propositional probabilistic logic with discrete linear time for reasoning about evidence. Ann. Math. Artif. Intell. **65**(2–3), 217–243 (2012)

14. Ognjanović, Z., Rašković, M., Marković, Z.: Probability Logics. Springer, Cham (2016). https://doi.org/10.1007/978-3-319-47012-2

15. Papadakis, N., Plexousakis, D.: Actions with duration and constraints: The ramification problem in temporal databases. Int. J. Artif. Intell. Tools **12**(3), 315–353 (2003)

16. Reiter, R.: Knowledge in Action: Logical Foundations for Specifying and Implementing Dynamical Systems. MIT Press, Cambridge (2001)

17. Reynolds, M.: An axiomatization of full computation tree logic. J. Symb. Log. **66**(3), 1011–1057 (2001)

18. Thielscher, M.: The concurrent, continuous fluent calculus. Studia Logica **67**(3), 315–331 (2001)

19. Wooldridge, M.: Reasoning about Rational Agents. MIT Press, Cambrdge (2000)

20. van Zee, M., Dastani, M., Doder, D., van der Torre, L.: Consistency conditions for beliefs and intentions. In: Twelfth International Symposium on Logical Formalizations of Commonsense Reasoning (2015)

21. van Zee, M., Doder, D.: AGM-style revision of beliefs and intentions. In: ECAI 2016, The Hague, pp. 1511–1519 (2016)
22. van Zee, M., Doder, D., Dastani, M., van der Torre, L.: AGM Revision of Beliefs about Action and Time. In: Proceedings of the International Joint Conference on Artificial Intelligence (2015)

Towards a Standard Completeness for a Probabilistic Logic on Infinite-Valued Events

Tommaso Flaminio[✉]

Artificial Intelligence Research Institute (IIIA - CSIC),
Campus de la Univ. Autònoma de Barcelona s/n, 08193 Bellaterra, Spain
tommaso@iiia.csic.es

Abstract. MV-algebras with internal states, or SMV-algebras for short, are the equivalent algebraic semantics of the logic SFP(Ł, Ł) which allows to represent and reason about the probability of infinite-valued events. In this paper we will make the first steps towards establishing completeness for SFP(Ł, Ł) with respect to the class of standard SMV-algebras, a problem which has been left open since the first paper on SMV-algebras was published. More precisely we will prove that, if we restrict our attention to a particular, yet expressive, subclass of formulas, then theorems of SFP(Ł, Ł) are the same as tautologies of a class of SMV-algebras that can be reasonably called "standard".

Keywords: MV-algebras · States · Internal states ·
Standard completeness · Probabilistic logic

1 Introduction

MV-algebras with an internal state, or SMV-algebras, have been introduced in [6] and they consist of an MV-algebra **A** (cf. [2]) and a unary map σ on **A** axiomatized so as to preserve some basic properties of a *state* of **A** (cf. [5,12]), i.e., a $[0, 1]$-valued, normalized and finitely additive function. These structures form a variety, denoted by SMV, which is the equivalent algebraic semantics for the probabilistic logic SFP(Ł, Ł) which permits to represent and reason about the probability of infinite-valued events. In that logic, in particular, events are described by formulas of the infinite-valued Łukasiewicz calculus (cf. [2]).

From the perspective of reasoning about uncertainty, the interest of Łukasiewicz events is twofold: whilst capturing properties of the world which are better described as *gradual* rather than *yes-or-no*, they also mimic bounded random variables. Indeed, any Łukasiewicz event e may be regarded as a $[0, 1]$-valued continuous function e^* on a compact Hausdorff space (see [2, Theorem 9.1.5] and Sect. 2) and any state of e coincides with the expected value of e^* (see [4, Remark 2.8] and Sect. 3).

SMV-algebras provide a framework to treat a generalization of classical probability theory in a universal-algebraic setting. However, notwithstanding their meaningful expressive power, several problem concerning their universal-algebraic

© Springer Nature Switzerland AG 2019
G. Kern-Isberner and Z. Ognjanović (Eds.): ECSQARU 2019, LNAI 11726, pp. 397–407, 2019.
https://doi.org/10.1007/978-3-030-29765-7_33

properties are still open. Here, we will concentrate on the following issue: determining a proper subclass of SMV-algebras which generates 𝕊𝕄𝕍 and which could be regarded as a class of "measure-theoretical" models. In other words, the above problem is to establishing a "standard completeness" for the logic SFP(Ł, Ł).

In particular we will show that, upon restricting our language to a special, yet sufficiently expressive subset of well-formed formulas, the logic SFP(Ł, Ł) turns out to be sound and complete with respect to a prominent subclass of SMV-algebras which can be reasonably called "standard". Indeed, these standard SMV-algebras are grounded on the same MV-algebra of continuous real-valued functions and their internal states are obtained by integrating the continuous functions which correspond to Łukasiewicz events by a regular, Borel (and hence σ-additive) probability measure. Although our main result only offers a partial solution to the problem of establishing standard completeness for SFP(Ł, Ł), the idea behind its proof is promising and it suggests ways to tackle the general problem. Indeed, the class of formulas to which we will restrict our attention, called *purely probabilistic formulas*, form the ground probabilistic language upon which the all formulas of SFP(Ł, Ł) are inductively defined.

This paper is organizes as follows: in the next section we will recall some basic definitions and needed results about Łukasiewicz logic and MV-algebras. Section 3 is dedicated to a brief introduction to generalized probability theory on Łukasiewicz logic, i.e., state theory, while in Sect. 4 we will focus on SMV-algebras and the probabilistic logic SFP(Ł, Ł). In Sect. 5 we will prove the main results of this paper and in particular that a purely probabilistic formula ϕ is a theorem of SFP(Ł, Ł) iff ϕ holds in all standard SMV-algebras.

2   Łukasiewicz Logic and MV-algebras

The language of the infinite-valued Łukasiewicz calculus Ł is made of a countable set of propositional variables $\{q_1, q_2, \ldots\}$, the binary connective $\oplus$, the unary connective $\neg$ and the constant $\bot$. Formulas are defined by the usual inductive rules. Further useful connectives are definable as follows:

$$\top = \neg\bot; \; \varphi \odot \psi = \neg(\neg\varphi \oplus \neg\psi); \; \varphi \to \psi = \neg\varphi \oplus \psi; \; \varphi \wedge \psi = \varphi \odot (\varphi \to \psi);$$
$$\varphi \vee \psi = \neg(\neg\varphi \wedge \neg\psi); \; \varphi \ominus \psi = \neg(\varphi \to \psi); \; d(\varphi, \psi) = (\varphi \to \psi) \oplus (\psi \to \varphi).$$

We invite the reader to consult [2,8] for an axiomatization of Łukasiewicz logic. For the sake of the present paper it is important to recall that this calculus has an equivalent algebraic semantics: the variety 𝕄𝕍 of *MV-algebras*. These are structures of the form $\mathbf{A} = (A, \oplus, \neg, \bot)$ (the same language of Łukasiewicz logic), where $(A, \oplus, \bot)$ is a commutative monoid and, defining further operations as above, the following equations hold:

$$x \oplus \top = \top$$
$$(x \to y) \to y = (y \to x) \to x.$$

The algebraizability of Łukasiewicz logic with respect to 𝕄𝕍 implies that Ł is sound and complete w.r.t. the class of MV-algebras. This means the following:

for every MV-algebra **A**, define an **A**-*valuation* as a map v from the propositional variables q_i's to A which extends to all formulas by truth functionality; then, a formula φ is a theorem of Łukasiewicz logic iff for every MV-algebra **A** and every **A**-valuation v, $v(\varphi) = \top$.

Furthermore, every formula φ on propositional variables $q_1, \ldots, q_k$ can be regarded, on the algebraic side, as a term $t(x_1, \ldots, x_k)$ on the same number of variables. For every MV-algebra **A** and for every term t, we denote by $t^{\mathbf{A}}$ the *interpretation* (or *semantics*) of t in **A**.

In every MV-algebra the relation $x \leq y$ iff $x \to y = \top$ determines a lattice-order which coincides with the one given by the operations $\wedge$ and $\vee$. If in **A** the order $\leq$ is linear, we will say that **A** is an *MV-chain*.

The following collects two typical examples of MV-algebras which will play a central role in this paper.

Example 1. (1) The *standard* MV-algebra $[0,1]_{MV}$ has support on the real unit interval $[0,1]$ and for all $x, y \in [0,1]$, $x \oplus y = \min\{1, x+y\}$, $\neg x = 1 - x$ and $\perp = 0$. The semantics, in $[0,1]_{MV}$, of the Łukasiewicz connectives we defined above is as follows:

$$\top = 1;\ x \odot y = \max\{0, x+y-1\};\ x \to y = \min\{1, 1-x+y\};$$
$$x \wedge y = \min\{x, y\};\ x \vee y = \max\{x, y\};\ x \ominus y = \max\{0, x-y\};$$
$$d(x,y) = |x-y| \text{ (the usual Euclidean distance).}$$

Chang's theorem [1] shows that $\mathbb{MV}$ is generated by $[0,1]_{MV}$, whence Łukasiewicz logic is complete w.r.t. to $[0,1]_{MV}$.

(2) For every $k \in \omega$, let $\mathcal{F}(k)$ be the set of functions $f : [0,1]^k \to [0,1]$ which are continuous, piecewise linear and such that each piece has integer coefficient, i.e., the set of k-*variable McNaughton functions* [2]. The point-by-point application of the operations of $[0,1]_{MV}$ makes $\mathcal{F}(k)$ an MV-algebra which coincides, up to isomorphism, with the free MV-algebra over k-variables. For every formula ψ of Łukasiewicz language, we will denote by f_ψ its corresponding McNaughton function. The free MV-algebra over infinitely-many generators will be henceforth denoted by $\mathcal{F}(\omega)$.

Let **A** and **B** be two MV-algebras. An *MV-homomorphism* is a function $h : A \to B$ such that, adopting without danger of confusion the same symbols for the operations of both algebras: (1) $h(\perp) = \perp$; (2) $h(x \oplus y) = h(x) \oplus h(y)$; (3) $h(\neg x) = \neg h(x)$. If S is a subset of A, a map $h : S \to B$ is a *partial homomorphism* provided that the above conditions (1–3) holds for the elements of S. Injective (partial) homomorphisms are called *(partial) embeddings*. (Partial) embeddings of **A** to **B** will be denoted by $\mathbf{A} \hookrightarrow \mathbf{B}$ ($\mathbf{A} \hookrightarrow^p \mathbf{B}$, respectively).

Let **A** be an MV-algebra and let S be a subset of A. We denote $\langle S \rangle_{\mathbf{A}}$ the MV-subalgebra of **A** generated by S. In particular, if S is finite, for every $b \in \langle S \rangle_{\mathbf{A}}$, there are $a_1, \ldots, a_n \in S$ and an n-ary term t such that $b = t^{\mathbf{A}}(a_1, \ldots, a_n)$.

Lemma 1. *(1) For every MV-chain* **A** *and for every finite subset* S *of* A*, there exists a partial embedding* $h_S : S \hookrightarrow^p [0,1]_{MV}$*.*
(2) Let **A**, **B** *be MV-algebras, let* S *be a finite subset of* A*. Every partial homomorphism* $h_S : S \subseteq A \to B$ *extends to a homomorphisms* $\hat{h}_S : \langle S \rangle_{\mathbf{A}} \to \mathbf{B}$*.*

Proof. (1) Immediately follows from Gurevich-Kokorin theorem (see for instance [8, Theorem 1.6.17]).

(2) For every element $b \in \langle S \rangle_{\mathbf{A}}$, there is an term t and $a_1, \ldots, a_n \in S$ such that $b = t^{\mathbf{A}}(a_1, \ldots, a_n)$. Define $\hat{h}_S : \langle S \rangle_{\mathbf{A}} \to \mathbf{B}$ as follows:

$$\hat{h}(b) = \hat{h}(t^{\mathbf{A}}(a_1, \ldots, a_n)) = t^{\mathbf{B}}(h(a_1), \ldots, h(a_n)).$$

It is easy to see that $\hat{h}$ commutes with the operations of $\mathbf{A}$ and $\mathbf{B}$, whence it is an MV-homomorphism.   □

An *ideal* of an MV-algebra $\mathbf{A}$ is a subset I of A such that: (1) $\bot \in I$; (2) if $a \in I$ and $b \in I$, then $a \oplus b \in I$; (3) if $a \in I$ and $b \leq a$, then $b \in I$. Every ideal I of $\mathbf{A}$ determines a congruence Θ_I of $\mathbf{A}$:

$$(a, b) \in \Theta_I \text{ iff } d(a, b) \in I.$$

For every MV-algebra $\mathbf{A}$ and every ideal I of $\mathbf{A}$, we denote by $\mathbf{A}/I$ the quotient of $\mathbf{A}$ by the congruence Θ_I. Further, for every $a \in A$, we denote by a/I the equivalence class of a modulo Θ_I: $a/I = \{b \in A \mid d(a, b) \in I\}$. The set of ideals of any MV-algebra, ordered by the set-theoretical inclusion, forms a lattice whose maximal elements are called *maximal ideals*. We write $Max(\mathbf{A})$ for the set of maximal ideals of $\mathbf{A}$. The intersection of all maximal ideals of an MV-algebra $\mathbf{A}$ is an ideal, called the *radical* of $\mathbf{A}$ and denoted by $Rad(\mathbf{A})$.

In the rest of this paper two kinds of structures will be of particular interest: *simple* and *semisimple* MV-algebras. The latter are characterized as those MV-chains which, up to isomorphism, are MV-subalgebra of $[0, 1]_{MV}$ (and therefore can be regarded as algebras of real numbers) [2, Theorem 3.5.1]; the former are algebras of continuous $[0, 1]$-valued functions defined on a compact Hausdorff space [2, Corollary 3.6.8]. It is worth to recall that semisimple MV-algebras are those structures whose radical coincides with $\{\bot\}$. Moreover, for every MV-algebra $\mathbf{A}$, the quotient $\mathbf{A}/Rad(\mathbf{A})$ is semisimple (see [2, Lemma 3.6.6]) and it will be called the *most general semisimple quotient* of $\mathbf{A}$.

3   Probability Theory on Łukasiewicz Events

States of MV-algebras were introduced by Mundici in [12] as averaging values of Łukasiewicz truth-valuations.

Definition 1. *A state of an MV-algebra $\mathbf{A}$ is a map $s : A \to [0, 1]$ satisfying the following conditions:*

(1) $s(\top) = 1$,
(2) forall $x, y \in A$ such that $x \odot y = \bot$, $s(x \oplus y) = s(x) + s(y)$.

While condition (1) says that every state is normalized, (2) is usually called *additivity* with respect to Łukasiewicz sum $\oplus$. Indeed, the requirement $x \odot y = 0$ is analogous to disjointness of a pair of elements in a Boolean algebra: if $\mathbf{A}$ is a

Boolean algebra, then $x \odot y = \perp$ iff $x \wedge y = \perp$. Thus states generalize finitely additive probabilities to the realm of MV-algebras: finitely additive probabilities on a Boolean algebra are states as a special case of the above definition.

The following results collect properties of states which will be used in the proof of the main result of this paper.

Proposition 1 ([5, **Proposition 3.1.7(ii)]**). *For every MV-algebra* **A**, *for every state* s *of* **A** *and for every* $a, b \in A$, *if* $a/Rad(\mathbf{A}) = b/Rad(\mathbf{A})$, *then* $s(a) = s(b)$.

Proposition 1 above implies that every MV-algebra **A** and its most general semisimple quotient $\mathbf{A}/Rad(\mathbf{A})$ have the same class of states. Thus, every state s of any MV-algebra **A** assigns, for every $a, b \in A$ with $b \in a/Rad(\mathbf{A})$, the same value $s(a) \in [0, 1]$.

The following has been proved in [11, Theorem 6] and provides an MV-analogous of the well-known Horn-Tarski extension theorem [9].

Proposition 2. *Let* **A** *and* **B** *be MV-algebras and let* **B** *be an MV-subalgebra of* **A**. *Every state* $s_B : B \to [0, 1]$ *extends (not uniquely) to a state* $s_A : A \to [0, 1]$.

Remark 1. Every homomorphism h of an MV-algebra **B** to $[0, 1]_{MV}$ is a state. Thus, if **B** is an MV-subalgebra of **A**, Proposition 2 shows that h extends (not uniquely) to a state of **A**. In other words, for every pairs of MV-algebras **B** and **A** such that **B** is an MV-subalgebra of **A** and for every homomorphism $h : \mathbf{B} \to [0, 1]_{MV}$, there exists a state s of **A** such that for every $b \in B$, $s(b) = h(b)$.

The following theorem, independently proved by Kroupa [10] and Panti [14] (see also [5, §4]), represents every state s of an MV-algebra **A** as the Lebesgue integral given by a unique regular, Borel probability measure. More precisely, every state of an MV-algebra **A** is the Lebesgue integral on the continuous functions a^*'s of its most general semisimple quotient, defined on the compact Hausdorff space $Max(\mathbf{A})$ of maximal ideals of **A**.

Theorem 1. *For every MV-algebra* **A** *and for every state* s *of* **A** *there exists a unique regular, Borel probability measure* μ *on the Borel subsets of* $Max(\mathbf{A})$ *such that, for every* $a \in A$,

$$s(a) = \int_{Max(\mathbf{A})} a^* \, d\mu$$

4   A Probabilistic Logic on Łukasiewicz Events

In this section we will recall basic definitions and properties for the logics SFP(Ł, Ł) and its algebraic semantics[1]. The language of SFP(Ł, Ł) is obtained by expanding that of Łukasiewicz infinite-valued calculus with a unary modality P. The set **SPFm** of well-formed formulas in this language is defined by

---

[1] We invite the reader to consult [6,7] and [5, §6] for a more exhaustive introduction to SMV-algebras, fuzzy probabilistic logics and their relation with uncertain reasoning.

induction as usual. Axioms and rules of SFP(Ł, Ł) are as follows: all instances of axioms and rules of Łukasiewicz logic, plus the following axioms and rules for the modality P

- $P(\bot) \rightarrow \bot$ (Normalization);
- $P(\varphi \rightarrow \psi) \rightarrow (P(\varphi) \rightarrow P(\psi))$ (Monotonicity);
- $P(P(\varphi) \oplus P(\psi)) \leftrightarrow P(\varphi) \oplus P(\psi)$ (Idempotency);
- $P(\varphi \oplus \psi) \leftrightarrow [P(\varphi) \oplus P(\psi \ominus (\varphi \odot \psi))]$ (Additivity);
- from ϕ derive $P(\phi)$ (Necessitation).

Proofs are defined with no modification from the classical definition and $\vdash_{SFP} \phi$ will be used to denote that ϕ is a *theorem* of SFP(Ł, Ł). If $\Gamma \cup \{\phi\}$ is a subset of **SPFm**, we will write $\Gamma \vdash_{SFP} \phi$ if there is a proof of ϕ from Γ.

The logic SFP(Ł, Ł) is algebraizable and its equivalent algebraic semantics is the variety $\mathbb{SMV}$ of *MV-algebras with internal state* (or *SMV-algebras* for short).

Definition 2 ([6]). *An SMV-algebra is a pair $(\mathbf{A}, \sigma)$ where $\mathbf{A}$ is an MV-algebra and the internal state $\sigma : A \rightarrow A$ satisfies the following equations:*

$(\sigma 1)$ $\sigma(\bot) = \bot$;
$(\sigma 2)$ $\sigma(x \rightarrow y) \leq \sigma(x) \rightarrow \sigma(y)$;
$(\sigma 3)$ $\sigma(\sigma(x \oplus y))) = \sigma(x \oplus y)$;
$(\sigma 5)$ $\sigma(x \oplus y) = \sigma(x) \oplus \sigma(y \ominus (x \odot y))$.

Example 2. (1) Every idempotent endomorphism σ of an MV-algebra $\mathbf{A}$ makes $(\mathbf{A}, \sigma)$ an SMV-algebra.

(2) Let $\mathcal{F}_{\mathbb{R}}(\omega)$ be the set of all continuous and piecewise linear functions with *real* coefficients. The pointwise application of $\oplus$ and $\neg$ of $[0,1]_{MV}$ makes $\mathcal{F}_{\mathbb{R}}(\omega)$ into an MV-algebra which contains all constant functions $\overline{\alpha}$ for every $\alpha \in [0,1]$. These functions are continuous and hence Riemann-integrable. Let $\sigma : \mathcal{F}_{\mathbb{R}}(\omega) \rightarrow \mathcal{F}_{\mathbb{R}}(\omega)$ be defined as follows: for every $f \in \mathcal{F}_{\mathbb{R}}(\omega)$ with k variables,

$$\sigma(f) = \overline{\int_{[0,1]^k} f \, dx}.$$

That is, $\sigma(f)$ is the function which is constantly equal to the Riemann integral of f. This map σ is an internal state of $\mathcal{F}_{\mathbb{R}}(\omega)$.

Proposition 3 ([6]). *For every SMV-algebra $(\mathbf{A}, \sigma)$ the following properties hold:*

(1) the image $\sigma(A)$ of A under σ, endowed with the operations inherited from $\mathbf{A}$ forms an MV-subalgebra $\sigma(\mathbf{A})$ of $\mathbf{A}$;

(2) if $(\mathbf{A}, \sigma)$ is a subdirectly irreducible, then $\sigma(\mathbf{A})$ is totally ordered. As a consequence the variety $\mathbb{SMV}$ is generated by its elements $(\mathbf{A}, \sigma)$ such that $\sigma(\mathbf{A})$ is an MV-chain.

Valuations to an SMV-algebra $(\mathbf{A}, \sigma)$ are defined in the usual way. A valuation v is said to be a *model* of a formula ϕ, if $v(\phi) = \top$. A formula $\phi \in \mathbf{SPFm}$ is said to be an *SMV-tautology* ($\models_{SMV} \phi$ in symbols), if $v(\phi) = \top$ for every SMV-algebra and every valuation to it. Further, for every (finite or infinite) subset $\Gamma \cup \{\phi\}$ of $\mathbf{SPFm}$, we will write $\Gamma \models_{SMV} \phi$ if for every SMV-algebra $(\mathbf{A}, \sigma)$ and every valuation v to $(\mathbf{A}, \sigma)$ which is a model of each formulas in Γ, then v is a model of ϕ as well.

Proposition 4 ([6, **Theorem 4.5**]). *For every (finite or infinite) subset* $\Gamma \cup \phi$ *of* $\mathbf{SPFm}$, $\Gamma \vdash_{SFP} \phi$ *iff* $\Gamma \models_{SMV} \phi$.

We end this section with the following observation.

Remark 2. One of the most important properties for a t-norm based fuzzy logic, is completeness with respect to the class of its *standard* algebras. By "standard" we usually mean linearly ordered structures, order-embeddable into the real unit interval. Although SFP(Ł, Ł) is grounded on Łukasiewicz calculus which enjoys a standard completeness theorem, SFP(Ł, Ł) is not *standard complete* in the usual sense. In fact, for instance, the equation $\sigma(x \wedge y) = \sigma(x) \wedge \sigma(y)$ holds in every SMV-chain, but it does not hold in every SMV-algebra and in particular it does not hold in the algebra $(\mathcal{T}_{\mathbb{R}}(\omega), \upsilon)$ of Example 2(2).

5   Toward a Standard Completeness for SFP(Ł, Ł)

In this section we will present a class of SMV-algebras that, in the light of Remark 2, it is reasonable to call *standard*. Our main result will show that, for a restricted, yet quite expressive, subset of $\mathbf{SPFm}$, a formula is provable in SFP(Ł, Ł) iff it is true in these standard algebras.

Definition 3. *The set* $\mathbf{PFm}$ *of purely probabilistic formulas is the smallest subset of* $\mathbf{SPFm}$ *which contains all the modal formulas* $P(\psi)$ *(for* ψ *a Łukasiewicz formula) and which is closed under the connectives of Łukasiewicz logic.*

Notice that each formula of $\mathbf{PFm}$ is in the form $t[P(\psi_1), \ldots, P(\psi_k)]$ for t being a(n MV-)term. Thus, for instance $P(\varphi) \to (P(\psi) \oplus P(\gamma)) \in \mathbf{PFm}$, but neither $P(\psi \to P(\varphi))$, nor $P(\psi) \to \gamma$ belong to $\mathbf{PFm}$. It is also worth to point out that $\mathbf{PFm}$ forms the language for the weaker logic, with respect to SFP(Ł, Ł), introduced in [3] and denoted by FP(Ł, Ł).

First of all, we will recall the definition of a class of SMV-algebras which was introduced in [7].

Definition 4. *An SMV-algebra* $(\mathbf{A}, \sigma)$ *is said to be* σ-simple *if* $\mathbf{A}$ *is semisimple and* $\sigma(\mathbf{A})$ *is simple.*

In the light of the last comment of Subsection 2 a σ-simple SMV-algebra $(\mathbf{A}, \sigma)$ can be regarded as an algebra of continuous functions on a compact Hausdorff space endowed with a normalized, idempotent, finitely additive and real-valued internal state σ.

The notions of *valuation* in a σ-simple SMV-algebra and *tautology* are as in Sect. 4. We will write $\models_\sigma \phi$ to denote that the formula ϕ is a tautology in this setting.

In the proof of the following result we will make use of the construction of *MV-tensor product* which was introduced by Mundici in [13] and heavily used in [6] to *internalize* a state of an MV-algebra. In what follows we will restrict the attention to those MV-algebras $[0,1]_{MV} \otimes \mathbf{A}$ defined by taking the tensor product of the standard MV-algebra $[0,1]_{MV}$ with an MV-algebra $\mathbf{A}$. For the sake of the present paper, it is important to recall the following properties: (1) the generators of $[0,1]_{MV} \otimes \mathbf{A}$ are of the form $\alpha \otimes a$ for $\alpha \in [0,1]$ and $a \in A$; (2) the maps $a \in A \mapsto 1 \otimes a$ and $\alpha \mapsto \alpha \otimes \top$ respectively are embeddings of $\mathbf{A}$ and $[0,1]_{MV}$ into $[0,1]_{MV} \otimes \mathbf{A}$; (3) the top element of $[0,1]_{MV} \otimes \mathbf{A}$ is $1 \otimes \top$. By (2), $[0,1]_{MV} \otimes \mathbf{A}$ contains isomorphic copies of both $[0,1]_{MV}$ and $\mathbf{A}$.[2]

Theorem 2. *For every formula $\phi \in \mathbf{PFm}$, $\vdash_{\mathrm{SFP}} \phi$ iff $\models_\sigma \phi$.*

Proof. The left-to-right direction follows from Proposition 4.

In order to prove the right-to-left direction, assume $\nvdash_{\mathrm{SFP}} \phi$. Thus, from Proposition 4 there exists an SMV-algebra $(\mathbf{A}, \sigma)$ and a valuation v in A such that $v(\phi) < \top$. Without loss of generality, by Proposition 3, we can assume $(\mathbf{A}, \sigma)$ to be subdirectly irreducible and hence $\sigma(\mathbf{A})$ linearly ordered. Since $\phi \in \mathbf{PFm}$, it will be in the form

$$t[P(\psi_1), \ldots, P(\psi_k)]$$

where t is a term in the language of MV-algebras. Let us denote by S_ϕ the following subset of A:

$$S_\phi = \{v(\phi), v(P(\psi_1)), \ldots, v(P(\psi_k))\} = \{v(\phi), \sigma(v(\psi_1)), \ldots, \sigma(v(\psi_k))\}.$$

Therefore S_ϕ is a finite subset of $\sigma(A)$ and since $\sigma(\mathbf{A})$ is an MV-chain, there is a partial embedding h_ϕ of S_ϕ into the standard MV-algebra $[0,1]_{MV}$ (see Lemma 1 (1)):

$$h_\phi : S_\phi \hookrightarrow^p [0,1]_{MV}.$$

By Lemma 1 (2), h_ϕ extends to a homomorphism, that we still denote by h_ϕ, of the MV-subalgebra $\mathbf{S} = \langle S_\phi \rangle_{\sigma(\mathbf{A})}$ of $\sigma(\mathbf{A})$, generated by S_ϕ, to $[0,1]_{MV}$:

$$h_\phi : \mathbf{S} \hookrightarrow [0,1]_{MV}.$$

Notice that $h_\phi(v(\phi)) < 1$. By Remark 1, h_ϕ extends to a state $s_\phi : \sigma(\mathbf{A}) \to [0,1]$. Thus, let $s : A \to [0,1]$ be the map

$$s : a \in A \mapsto s_\phi(\sigma(a)).$$

---

[2] We invite the reader to consult [13] for an exhaustive description of the MV-tensor product construction and [6,7] for its application to the theory of SMV-algebras.

Notice that s and s_ϕ agree on $\sigma(A)$. Indeed by the idempotency of σ (equation (σ3) of Definition 2), if $a = \sigma(b) \in \sigma(A)$, $s(a) = s(\sigma(b)) = s_\phi(\sigma\sigma(b)) = s_\phi(\sigma(b)) = s_\phi(a)$. Therefore, for every subformula $P(\psi_i)$ of ϕ, since s_ϕ extends h_ϕ,

$$s(v(\psi_i)) = s_\phi(\sigma(v(\psi_i))) = h_\phi(v(P(\psi_i))), \tag{1}$$

which leads to the following fact.

Fact 1. $t^{[0,1]}[s(v(\psi_1)), \ldots, s(v(\psi_k))] < 1$.

As a matter of fact, by (1)

$$
\begin{aligned}
t^{[0,1]}[s(v(\psi_1)), \ldots, s(v(\psi_k))] &= t^{[0,1]}[h_\phi(v(P(\psi_1))), \ldots, h_\phi(v(P(\psi_k)))] \\
&= h_\phi(t^{\mathbf{A}}[v(P(\psi_1))), \ldots, v(P(\psi_k))]) \\
&= h_\phi(v(t[P(\psi_1), \ldots, P(\psi_k)])) \\
&= h_\phi(v(\phi)) < 1.
\end{aligned}
$$

In order to get the claim, we have now to define a σ-simple SMV-algebra $(\mathbf{M}, \sigma_M)$ and a valuation w to M such that $w(\phi) < 1$. To this aim, let $\mathbf{N} = [0,1]_{MV} \otimes \mathbf{A}$ and $s_N : N \to [0,1]$ be such that $s_N(\alpha \otimes a) = \alpha \cdot s(a)$. Than s_N is a state of $\mathbf{N}$ which extends o (indeed, $\mathbf{A}$ is a subalgebra of $\mathbf{N}$). Regardless of the semisimplicity of $\mathbf{A}$, $\mathbf{N}$ needs not to be semisimple (see [13, Theorem 3.3]). Thus, let $\mathbf{M}$ its most general semisimple quotient:

$$\mathbf{M} = \mathbf{N}/Rad(\mathbf{N}) \quad \text{and} \quad \sigma_M : m/Rad(\mathbf{N}) \in M \mapsto s_N(m).$$

Notice that σ_M is well-defined because of Proposition 1. Moreover, $\mathbf{M}$ is semisimple, σ_M is an internal state of $\mathbf{M}$ (because the range $[0,1]_{MV}$ of σ_M is a subalgebra of $\mathbf{M}$) and $\sigma_M(\mathbf{M})$ is clearly simple. Thus, $(\mathbf{M}, \sigma_M)$ is σ-simple.

Let w be the following valuation to M: for each propositional variable q, $w(q) = 1 \otimes v(q)$ and $w(P(\gamma)) = \sigma_M(v(\gamma)) \otimes \top$. Notice that, for every subformula $P(\psi_i)$ of ϕ, one has

$$w(P(\psi_i)) = \sigma_M(v(\psi_i)) \otimes \top = s_N(v(\psi_i)) \otimes \top = s(v(\psi_i)) \otimes \top,$$

and hence, by Fact 1, $w(\phi) = t^{[0,1]}[s(v(\psi_1)), \ldots, s(v(\psi_k))] \otimes \top < 1 \otimes \top$. □

We now introduce a further refined semantics for SFP(L, Ł). To this end let $\mathcal{F}_\mathbb{R}(\omega)$ be the MV-algebra of Example 2(2). It is worth to notice that $\mathcal{F}_\mathbb{R}(\omega)$ is semisimple [7, Proposition 2.4]. Further, every state $s : \mathcal{F}_\mathbb{R}(\omega) \to [0,1]$ can be easily internalized by considering the map $\sigma_s : \mathcal{F}_\mathbb{R}(\omega) \to \mathcal{F}_\mathbb{R}(\omega)$ defined as follows: for every $a \in \mathcal{F}_\mathbb{R}(\omega)$

$$\sigma_s(a) = \overline{s(a)}. \tag{2}$$

Then $(\mathcal{F}_\mathbb{R}(\omega), \sigma_s)$ is a σ-simple SMV-algebra. Notice that, thanks to Theorem 1 the internal state σ_s assigns to every $f \in \mathcal{F}_\mathbb{R}(\omega)$, the real number $\int f \, d\mu$: the Lebesgue integral of f by a regular, Borel probability measure.

Definition 5. A standard SMV-algebra is a pair $(\mathcal{F}_\mathbb{R}(\omega), \sigma_s)$ where $s : \mathcal{F}_\mathbb{R}(\omega) \to [0,1]$ is a state and σ_s is defined as in (2).

Taking into account that the free ω-generated MV-algebra of Example 1(2) is a subalgebra of $\mathcal{F}_{\mathbb{R}}(\omega)$ (see [7, Proposition 2.4]), a *valuation* v of SFP(Ł, L)-formulas in $(\mathcal{F}_{\mathbb{R}}(\omega), \sigma_s)$ can be inductively defined as follows:

- $v(\psi) = f_\psi$ for ψ a Łukasiewicz formula;
- v commutes with Łukasiewicz connectives;
- $v(P(\psi)) = \sigma_s(v(\psi))$.

We will write $\models_{St} \phi$ to denote that ϕ is a tautology for this class of algebras.

Theorem 3. *For every formula $\phi \in \mathbf{PFm}$, $\models_\sigma \phi$ iff $\models_{St} \phi$. In other words, σ-simple SMV-algebras and standard SMV-algebras share the same* $\mathbf{PFm}$-*tautologies.*

Proof. Since every standard SMV-algebra is σ-simple, $\models_\sigma \phi$ implies $\models_{St} \phi$. Let hence assume that $\not\models_\sigma \phi$ and let v be a valuation in a σ-simple SMV-algebra $(\mathbf{A}, \sigma)$ such that $v(\phi) < 1$. Define $s : \mathcal{F}(\omega) \to [0,1]$ by the following stipulation: for every Łukasiewicz formula ψ, $s(f_\psi) = \sigma(v(\psi))$.

Clearly s is a state of $\mathcal{F}(\omega)$ which, since $\mathcal{F}(\omega)$ is an MV-subalgebra of $\mathcal{F}_{\mathbb{R}}(\omega)$, extends to a state $\hat{s}$ of $\mathcal{F}_{\mathbb{R}}(\omega)$ by Proposition 2. Define $\sigma_{\hat{s}}$ by (2) and the valuation w on $(\mathcal{F}_{\mathbb{R}}(\omega), \sigma_{\hat{s}})$ by $w(q) = f_q$. Let us prove that w is not a model of ϕ. Since $\phi \in \mathbf{PFm}$, it is in the form $t[P(\psi_1), \ldots, P(\psi_k)]$ and because s and $\hat{s}$ agree on $\mathcal{F}(\omega)$, so do σ_s and $\sigma_{\hat{s}}$ by definition. Thus, by (2) one has:

$$
\begin{aligned}
w(\phi) &= t^{[0,1]}[\sigma_{\hat{s}}(w(\psi_1)), \ldots, \sigma_{\hat{s}}(w(\psi_k))] \\
&= t^{[0,1]}[\sigma_{\hat{s}}(f_{\psi_1}), \ldots, \sigma_{\hat{s}}(f_{\psi_k})] \\
&= t^{[0,1]}[\sigma_s(f_{\psi_1}), \ldots, \sigma_s(f_{\psi_k})] \\
&= t^{[0,1]}[s(f_{\psi_1}), \ldots, s(f_{\psi_k})] \\
&= v(\phi) < 1
\end{aligned}
$$

Therefore, $\not\models_{St} \phi$ which settles the claim. $\square$

The following result is a direct consequence of Theorems 2 and 3.

Corollary 1. *For every formula $\phi \in \mathbf{PFm}$, $\vdash_{SFP} \phi$ iff $\models_{St} \phi$.*

6   Conclusion and Future Work

In this paper we presented a partial solution to the problem of establishing a a standard completeness theorem for the probabilistic logic SFP(Ł, L) introduced in [6]. In particular we proved that, for a restricted class of formulas, theorems of SFP(Ł, L) are tautologies for a class of SMV-algebras which are defined from real-valued metric spaces. In our future work on this argument we plan to extend the ideas and constructions which led to the proof of our main theorems to provide a standard completeness theorem for the whole language of SFP(Ł, L).

Acknowledgments. The author acknowledges partial support by the Spanish Ramon y Cajal research program RYC-2016-19799; the Spanish FEDER/MINECO project TIN2015-71799-C2-1-P and the SYSMICS project (EU H2020-MSCA-RISE-2015, Project 689176).

References

1. Chang, C.C.: Algebraic analysis of many-valued logics. Trans. Am. Math. Soc. **88**, 467–490 (1958)
2. Cignoli, R., D'Ottaviano, I.M.L., Mundici, D.: Algebraic Foundationsof Many-valued Reasoning. Kluwer, Alphen aan den Rijn (2000)
3. Flaminio, T., Godo, L.: A logic for reasoning about the probability of fuzzy events. Fuzzy Sets Syst. **158**(6), 625–638 (2007)
4. Flaminio, T., Hosni, H., Lapenta, S.: Convex MV-algebras: Many-valued logics meet decision theory. Stud. Logica **106**(5), 913–945 (2018)
5. Flaminio, T., Kroupa, T.: States of MV-algebras. Chapter XVII of Handbook of Mathematical Fuzzy Logic - volume 3. In: Fermüller, C., Cintula, P., Noguera, C. (Eds.) Studies in Logic, Mathematical Logic and Foundations, vol. 58. College Publications, London (2015)
6. Flaminio, T., Montagna, F.: MV-algebras with internal states and probabilistic fuzzy logics. Int. J. Approximate Reasoning **50**(1), 138–152 (2009)
7. Flaminio, T., Montagna, F.: Models for many-valued probabilistic reasoning. J. Logic Comput. **21**(3), 447–464 (2011)
8. Hájek, P.: Metamathematics of Fuzzy Logics. Kluwer Academic Publishers, Dordrecht (1998)
9. Horn, A., Tarski, A.: Measures on Boolean algebras. Trans. Am. Math. Soc. **64**, 467–497 (1948)
10. Kroupa, T.: Every state on semisimple MV-algebra is integral. Fuzzy Sets Syst. **157**(20), 2771–2787 (2006)
11. Kroupa, T.: Representation and extension of states on MV-algebras. Arch. Math. Logic **45**, 381–392 (2006)
12. Mundici, D.: Averaging the truth-value in Łukasiewicz logic. Stud. Logica **55**, 113–127 (1995)
13. Mundici, D.: Tensor products and the Loomis-Sikorski theorem for MV-algebras. Adv. Appl. Math. **22**, 227–248 (1999)
14. Panti, G.: Invariant measures on free MV-algebras. Commun. Algebra **36**(8), 2849–2861 (2009)

Bayesian Confirmation and Justifications

Hamzeh Mohammadi[1] and Thomas Studer[2(✉)]

[1] Department of Mathematical Science,
Isfahan University of Technology, Isfahan, Iran
hamzeh.mohammadi@math.iut.ac.ir
[2] Institute of Computer Science, University of Bern, Bern, Switzerland
tstuder@inf.unibe.ch

Abstract. We introduce a family of probabilistic justification logics that feature Bayesian confirmations. Our logics include new justification terms representing evidence that make a proposition firm in the sense of making it more probable. We present syntax and semantics of our logic and establish soundness and strong completeness. Moreover, we show how to formalize in our logic the screening-off condition for transitivity of Bayesian confirmations.

Keywords: Epistemic logic · Justification logic · Bayesian confirmation

1 Introduction

Justification logic is a type of logic that explicitly includes justifications why something is known or believed [6,18]. The first justification logic, the Logic of Proofs [2], has been developed to provide a classical provability semantics for intuitionistic logic. In that approach, justifications represent proofs in a formal system like Peano arithmetic [17]. Later justification logic was introduced into formal epistemology where justifications can represent not only proofs but evidence in general [4]. For instance, an agent's knowledge may be justified by direct observation or by communication with another agent. In this context, notions like common knowledge [3,8] and public announcements [7,9] have been studied in detail.

Milnikel [19] was the first to investigate uncertain justifications. This lead to several further frameworks that model uncertain reasoning in justification logic: fuzzy justification logics [12,21], possibilistic justification logics [11,28], probabilistic justification logics [13,14,20], and logics for combining evidence and uncertainty [1,25].

Having logics that contain justifications for belief as well as operators for conditional probabilities, it is natural to extend them to a framework in which justifications can represent Bayesian confirmations [29]. The main principle of Bayesian confirmation theory says that (for simplicity we do not consider a background theory here) evidence E confirms hypothesis H if the prior probability of H conditional on E is greater than the prior unconditional probability

© Springer Nature Switzerland AG 2019
G. Kern-Isberner and Z. Ognjanović (Eds.): ECSQARU 2019, LNAI 11726, pp. 408–418, 2019.
https://doi.org/10.1007/978-3-030-29765-7_34

of H, that is if $\mathsf{P}(H|E) > \mathsf{P}(H)$. Carnap [10] calls this condition *confirmation as increase in firmness*.

We aim at a probabilistic justification logic that implements the above idea, that is in which something like

$$\mathsf{P}(H|E) > \mathsf{P}(H) \quad \text{entails} \quad \mathsf{j}(E) : H \tag{1}$$

holds, where $\mathsf{j}(E)$ is a term that represents the evidence E. Hence in this logic we read the formula $e : F$ as *evidence e confirms F*.

In order to model this relationship between conditional probability and evidence, we need a way to consider formulas as evidence terms. In (1) this is the role of the operator j. It takes a formula E and produces an evidence term $\mathsf{j}(E)$ representing the evidence E.

A similar kind of justification operator has been considered in the treatment of public announcements [16] where the operator up transforms formulas to evidence terms. We will use a similar strategy for the j-operator of (1). Further we will employ operators for conditional probabilities $\mathsf{CP}_{\geq s}$ as in [20,22] and operators for the degree of confirmation $D_{\geq s}$ as in [26]. A formula $\mathsf{CP}_{\geq s}(A, B)$ means that *the conditional probability of A given B is at least s* and a formula $D_{\geq s}(A, B)$ means that *the difference between the conditional probability of A given B and the probability of A is at least r*.

The paper is organized as follows. In the next section we introduce syntax and semantics of Bayesian justification logic, i.e. we present the deductive system $\mathsf{BJ_{CS}}$ and we introduce the class of measurable Bayesian models. Then in Sect. 3 we establish soundness and completeness of $\mathsf{BJ_{CS}}$ with respect to those models. Section 4 discusses transitivity of Bayesian confirmations in the framework of justification logic. Finally, Sect. 5 concludes the paper.

2   Bayesian Justification Logic BJ

2.1   Syntax

We start with countably many constants c_i, countably many variables x_i, and countably many atomic propositions p_i. Further, we define $S := \mathbb{Q} \cap [0, 1]$ and $S^* := \mathbb{Q} \cap [-1, 1]$, where $\mathbb{Q}$ is the set of all rational numbers. The (evidence) terms and formulas of the language of BJ are defined by simultaneous induction as follows:

- Evidence terms.
    - Every constant c_i and every variable x_i is an atomic term. If A is a formula, then j_A is an atomic term. Every atomic term is a term.
    - If t and s are terms, then $t \cdot s$ is a term and $!t$ is a term.
- Formulas.
    - Every atomic proposition p_i is a formula.
    - $\bot$ is a formula.
    - If A and B are formulas, t is a term, $s \in S$, and $r \in S^*$, then $A \to B$, $t : A$, $\mathsf{CP}_{\geq s}(A, B)$, and $D_{\geq r}(A, B)$ are formulas.

The set of all constants is denoted by Con, the set of all terms is Tm, and we use $t, s, u, v, \ldots$ to denote terms. The set of atomic propositions and the set of justification formulas are denoted by Prop and Fml, respectively. We use $A, B, \ldots$ to denote formulas. The classical Boolean connectives $\neg, \vee, \wedge, \leftrightarrow$ are defined as usual and we set

$$\mathsf{CP}_{\leq s}(B,C) := \mathsf{CP}_{\geq 1-s}(\neg B, C) \qquad \text{and} \qquad \mathsf{D}_{\leq r}(B,C) := \mathsf{D}_{\geq -r}(\neg B, C)$$

for $s \in S$ and $r \in S^*$. Moreover, we use the standard abbreviations, see [22], for the following formulas:

$$\mathsf{CP}_{<s}(A,B) \quad \mathsf{CP}_{>s}(A,B) \quad \mathsf{CP}_{=s}(A,B) \quad \mathsf{P}_{\rho s}(A) \quad \text{for } \rho \in \{\geq, \leq, >, <, =\}$$

and similarly for $\mathsf{D}_{<s}(A,B)$, $\mathsf{D}_{>s}(A,B)$ and $\mathsf{D}_{=s}(A,B)$.

The axiom schemes of BJ are the following where $\bigcirc \in \{\mathsf{CP}, \mathsf{D}\}$:

1. all classical tautologies
2. $t : (A \to B) \to (s : A \to t \cdot s : B)$
3. $\mathsf{CP}_{\geq 0}(A, B)$
4. $\bigcirc_{\leq s}(A, B) \to \bigcirc_{<t}(A, B)$, for $t > s$
5. $\bigcirc_{<s}(A, B) \to \bigcirc_{\leq s}(A, B)$
6. $\mathsf{P}_{\geq 1}(A \leftrightarrow B) \to (\mathsf{P}_{=s}A \to \mathsf{P}_{=s}B)$
7. $\mathsf{P}_{=s}A \wedge \mathsf{P}_{=t}B \wedge \mathsf{P}_{\geq 1}\neg(A \wedge B)) \to \mathsf{P}_{=\min\{1,s+t\}}(A \vee B)$
8. $\mathsf{P}_{=0}B \to \mathsf{CP}_{=1}(A, B)$
9. $\mathsf{P}_{>s}(A \wedge B) \wedge \mathsf{P}_{\leq t}B \to \mathsf{CP}_{\geq \frac{s}{t}}(A, B)$, for $t \neq 0$
10. $\mathsf{P}_{\leq s}(A \wedge B) \wedge \mathsf{P}_{\geq t}B \to \mathsf{CP}_{\leq \frac{s}{t}}(A, B)$, for $t \neq 0$
11. $\mathsf{CP}_{\geq s}(A, B) \wedge \mathsf{P}_{\leq t}A \to \mathsf{D}_{\geq s-t}(A, B)$
12. $\mathsf{CP}_{\leq s}(A, B) \wedge \mathsf{P}_{\geq t}A \to \mathsf{D}_{\leq s-t}(A, B)$
13. $j_B : A \leftrightarrow \mathsf{D}_{>0}(A, B)$

Axioms 1 to 10 come from justification logic with conditional probabilities, see [20]. The main difference is that we replaced the axiom

$$\mathsf{P}_{=s}(A \wedge B) \wedge \mathsf{P}_{=t}B \to \mathsf{CP}_{=\frac{s}{t}}(A, B) \quad \text{for } t \neq 0$$

from [20] with our axioms 9 and 10, which yields a slightly stronger system. This additional power is needed to prove Lemma 4. Axioms 11 and 12 formalize the relationship between conditional probabilities and degrees of confirmation as in [26]. Axiom 13 finally states that terms j_B represent Bayesian confirmations.

A *constant specification* is any set CS that satisfies

$$\mathsf{CS} \subseteq \{(c, A) \mid c \text{ is a constant and}$$

$$A \text{ is an instance of some axiom of BJ}\}.$$

Let CS be any constant specification. The deductive system $\mathsf{BJ_{CS}}$ is the Hilbert system obtained by adding to the axioms of BJ the rules (MP), (CE), (ST.1), (ST.2) and (AN!) as given in Fig. 1.

Note that (ST.1) and (ST.2) are infinitary rules, which we need to obtain strong completeness. Observe also the difference in the definitions of rules (MP), (ST.1), (ST.2), and (CE) in Fig. 1. Rule (CE) can only be applied to theorems of BJ (i.e. formulas that are deducible from the empty set), whereas (MP), (ST.1), and (ST.2) can always be applied.

$$\boxed{\begin{array}{l} \quad \text{axioms of BJ} \\ \quad\quad + \\ \textbf{(AN!)} \ \vdash\ !^n c : !^{n-1}c : \cdots : !c : c : A, \text{ where } (c,A) \in \mathsf{CS} \text{ and } n \in \mathbb{N} \\ \textbf{(MP)} \ \text{if } T \vdash A \text{ and } T \vdash A \to B \text{ then } T \vdash B \\ \textbf{(CE)} \ \text{if } \vdash A \text{ then } \vdash \mathsf{P}_{\geq 1} A \\ \textbf{(ST.1)} \ \text{if } T \vdash A \to \mathsf{CP}_{\geq s - \frac{1}{k}}(B,C) \text{ for every integer } k \geq \frac{1}{s} \text{ and } s > 0 \\ \quad\quad \text{then } T \vdash A \to \mathsf{CP}_{\geq s}(B,C) \\ \textbf{(ST.2)} \ \text{if } T \vdash A \to \mathsf{D}_{\geq r - \frac{1}{k}}(B,C) \text{ for every integer } k \geq \frac{1}{1+r} \text{ and } r > -1 \\ \quad\quad \text{then } T \vdash A \to \mathsf{D}_{\geq r}(B,C) \end{array}}$$

Fig. 1. System $\mathsf{BJ_{CS}}$

2.2 Semantics

To introduce semantics for $\mathsf{BJ_{CS}}$, we begin with the notion of a basic evaluation, which is the cornerstone for many interpretations of justification logic [5,15]. In the following we use $\mathcal{P}(X)$ to denote the power set of a set X.

Definition 1 (Basic Evaluation). *Let* CS *be a constant specification. A basic evaluation for* CS, *or a basic* CS-*evaluation, is a function* $*$ *that maps atomic propositions to truth values and maps justification terms to subsets of* Fml, *i.e.*

$$* : \mathsf{Prop} \to \{\mathsf{T},\mathsf{F}\} \quad and \quad * : \mathsf{Tm} \to \mathcal{P}(\mathsf{Fml}),$$

such that for $u,v \in \mathsf{Tm}$, *for* $c \in \mathsf{Con}$ *and* $A,B \in \mathsf{Fml}$ *we have:*

1. $(A \to B \in u^* \text{ and } A \in v^*) \implies B \in (u \cdot v)^*$
2. *if* $(c,A) \in \mathsf{CS}$ *then for all* $n \in \mathbb{N}$ *we have[1]:*

$$!^{n-1}c : !^{n-2}c : \cdots :!c : c : A \in (!^n c)^*$$

We usually write t^* *and* p^* *instead of* $*(t)$ *and* $*(p)$, *respectively.*

Definition 2 (Algebra over a Set). *Let* W *be a non-empty set and let* H *be a non-empty subset of* $\mathcal{P}(W)$. *We call* H *an* algebra over W *iff the following hold:*

- $W \in H$
- $U,V \in H \implies U \cup V \in H$
- $U \in H \implies W \setminus U \in H$

Definition 3 (Finitely Additive Measure). *Let* H *be an algebra over* W *and* $\mu : H \to [0,1]$. *We call* μ *a* finitely additive measure *iff the following hold:*

1. $\mu(W) = 1$

---

[1] We agree to the convention that the formula $!^{n-1}c : !^{n-2}c : \cdots : !c : c : A$ represents the formula A for $n = 0$.

2. *for all* $U, V \in H$:

$$U \cap V = \emptyset \implies \mu(U \cup V) = \mu(U) + \mu(V)$$

Definition 4 (Probability Space). *A* probability space *is a triple*

$$P = \langle W, H, \mu \rangle,$$

where:

- W *is a non-empty set*
- H *is an algebra over* W
- $\mu : H \to [0,1]$ *is a finitely additive measure*

Definition 5 (Model). *Let* CS *be a constant specification. A* $\mathsf{BJ_{CS}}$-*model is a quintuple* $M = \langle U, W, H, \mu, * \rangle$ *where:*

1. U *is a non-empty set of objects called worlds*
2. W, H, μ *and* $*$ *are functions, which have* U *as their domain, such that for every* $w \in U$:
   - $\langle W(w), H(w), \mu(w) \rangle$ *is a probability space with* $W(w) \subseteq U$
   - $*_w$ *is a basic* CS-*evaluation*[2]

The ternary satisfaction relation $\models$ is defined between models, worlds, and formulas. We will use μ_w for $\mu(w)$, p_w^* for p^{*w}, and t_w^* for t^{*w}.

Definition 6 (Truth in a $\mathsf{BJ_{CS}}$-model). *Let* CS *be a constant specification and let* $M = \langle U, W, H, \mu, * \rangle$ *be a* $\mathsf{BJ_{CS}}$-*model. We define by simultaneous induction*

1. *what it means for a formula to hold in* M *at a world* $w \in U$ *and*
2. *what it means for a formula to be measurable in* M *at a world* $w \in U$

as follows:

- $M, w \models p$ *iff* $p_w^* = \mathsf{T}$ *for* $p \in$ Prop*;*
- $M, w \not\models \bot$*;*
- $M, w \models A \to B$ *iff* $M, w \not\models A$ *or* $M, w \models B$*;*
- $M, w \models t : A$ *iff* $A \in t_w^*$*;*
- $M, w \models \mathsf{CP}_{\geq s}(A, B)$ *iff* $A \wedge B$ *and* B *are measurable at* w *and either* $\mu_w([B]) = 0$*, or* $\mu_w([B]) > 0$ *and* $\frac{\mu_w([A \wedge B])}{\mu_w([B])} \geq s$*;*
- $M, w \models \mathsf{D}_{\geq s}(A, B)$ *iff* A *and* B *are measurable at* w *and either* $\mu_w([B]) = 0$ *and* $1 - \mu_w([A]) \geq s$*, or* $\mu_w([B]) > 0$ *and* $\frac{\mu_w([A \wedge B])}{\mu_w([B])} - \mu_w([A]) \geq s$.

We say a formula B *is measurable in* M *at a world* $w \in U$ *if the set*

$$[B]_{M,w} := \{ x \in W(w) \mid M, x \models B \}$$

is an element of $H(w)$.

---

[2] We will usually write $*_w$ instead of $*(w)$.

Definition 7 (Measurable Model). *Let* CS *be a constant specification and let* $M = \langle U, W, H, \mu, * \rangle$ *be a* $\mathsf{BJ_{CS}}$*-model. M is called measurable iff every formula A is measurable at each $w \in U$.* $\mathsf{BJ_{CS,Meas}}$ *denotes the class of measurable* $\mathsf{BJ_{CS}}$ *models.*

Finally, we call a model Bayesian if terms of the form j_A represent Bayesian evidence.

Definition 8. *A* $\mathsf{BJ_{CS}}$*-model $M = \langle U, W, H, \mu, * \rangle$ is called Bayesian model if at each $w \in U$,*

$$M, w \models \mathrm{D}_{>0}(A, B) \quad \textit{iff} \quad M, w \models j_B : A.$$

The class of Bayesian $\mathsf{BJ_{CS}}$-models is denoted by $\mathsf{BJ_{CS,Bayes}}$. The class of Bayesian measurable $\mathsf{BJ_{CS}}$-models is denoted by $\mathsf{BJ_{CS,Meas,Bayes}}$.

For a model $M = \langle U, W, H, \mu, * \rangle$, $M \models A$ means that $M, w \models A$ for all $w \in U$. Let $T \subseteq \mathsf{Fml}$. Then $M \models T$ means that $M \models A$ for all $A \in T$. Further $T \models A$ means that for all $M \in \mathsf{BJ_{CS,Meas,Bayes}}$, $M \models T$ implies $M \models A$.

To be precise we should write $T \vdash_{\mathsf{CS}} A$ and $T \models_{\mathsf{CS}} A$ instead of $T \vdash A$ and $T \models A$, respectively, since these two notions depend on a given constant specification CS. However, CS will always be clear from the context and we thus omit it.

3   Soundness and Completeness for Bayesian Justification Logic

Soundness of $\mathsf{BJ_{CS}}$ can be proved by induction on the depth of derivations. To establish completeness, we make use of a canonical model construction. For lack of space, however, we cannot give a detailed completeness proof here. We will only present a series of definitions and lemmas (without proofs) that leads to the completeness result.

Theorem 1 (Soundness). *Let* CS *be a constant specification. The axiomatic system* $\mathsf{BJ_{CS}}$ *is sound with respect to the class of* $\mathsf{BJ_{CS,Meas,Bayes}}$*-models, i.e., for any formula A and any set $T \subseteq \mathsf{Fml}$ we have*

$$T \vdash A \quad \Longrightarrow \quad T \models A.$$

Now we define the notion of a $\mathsf{BJ_{CS}}$-consistent sets.

Definition 9 ($\mathsf{BJ_{CS}}$-Consistent Sets). *Let* CS *be any constant specification and let T be a set of formulas.*

- *T is said to be $\mathsf{BJ_{CS}}$-consistent if and only if $T \nvdash_{\mathsf{BJ_{CS}}} \bot$. Otherwise T is said to be $\mathsf{BJ_{CS}}$-inconsistent.*
- *T is said to be maximal if and only if for every $A \in \mathsf{Fml}$ either $A \in T$ or $\neg A \in T$.*
- *T is said to be maximal $\mathsf{BJ_{CS}}$-consistent if and only if it is maximal and $\mathsf{BJ_{CS}}$-consistent.*

We have the following deduction theorem for $\mathsf{BJ_{CS}}$. The proof is similar to the one given in [13, 22].

Theorem 2 (Deduction Theorem for $\mathsf{BJ_{CS}}$). *Let T be a set of formulas and A and B be formulas. We have*

$$T, A \vdash B \quad \textit{iff} \quad T \vdash A \rightarrow B.$$

The deduction theorem makes it possible to establish the following property of consistent sets of formulas, see [13, Lemma 27].

Lemma 1. *Let CS be a constant specification and let T be a $\mathsf{BJ_{CS}}$-consistent set of formulas.*

1. *If $\neg(B \rightarrow \mathsf{CP}_{\geq s}(A, C)) \in T$ for $s > 0$, then there is some integer $n \geq \frac{1}{s}$ such that $T, \neg(B \rightarrow \mathsf{CP}_{\geq s - \frac{1}{n}}(A, C))$ is $\mathsf{BJ_{CS}}$-consistent.*
2. *If $\neg(B \rightarrow \mathsf{D}_{\geq r}(A, C)) \in T$ for $r > -1$, then there is some integer $n \geq \frac{1}{r+1}$ such that $T, \neg(B \rightarrow \mathsf{D}_{\geq r - \frac{1}{n}}(A, C))$ is $\mathsf{BJ_{CS}}$-consistent.*

The Lindenbaum lemma for probabilistic justification logics has been established in [13]. The proof for $\mathsf{BJ_{CS}}$ is similar.

Lemma 2 (Lindenbaum). *Let CS be a constant specification. Every $\mathsf{BJ_{CS}}$-consistent set of formulas can be extended to a maximal $\mathsf{BJ_{CS}}$-consistent set.*

Definition 10 (Canonical Model). *Let CS be a constant specification. The canonical model for $\mathsf{BJ_{CS}}$ is given by the quintuple $M = \langle U, W, H, \mu, * \rangle$, defined as follows:*

- $U = \{w \mid w$ *is a maximal $\mathsf{BJ_{CS}}$-consistent set of formulas*$\}$
- *for every $w \in U$ the probability space $\langle W(w), H(w), \mu(w) \rangle$ is defined as follows:*
  1. $W(w) = U$
  2. $H(w) = \{(A)_M \mid A \in \mathsf{Fml}\}$ *where* $(A)_M = \{x \mid x \in U, A \in x\}$
  3. *for all $A \in \mathsf{Fml}$, $\mu(w)((A)_M) = \sup_s \{\mathsf{P}_{\geq s} A \in w\}$*
- *for every $w \in W$ the basic CS-evaluation $*_w$ is defined as follows:*
  1. *for all $p \in \mathsf{Prop}$:*

$$p_w^* = \begin{cases} \mathsf{T} & \textit{if } p \in w \\ \mathsf{F} & \textit{if } \neg p \in w \end{cases}$$

  2. *for all $t \in \mathsf{Tm}$:*

$$t_w^* = \{A \mid t : A \in w\}$$

Lemma 3. *Let CS be a constant specification. The canonical model for $\mathsf{BJ_{CS}}$ is a $\mathsf{BJ_{CS}}$-model.*

The following lemma is proved by induction on the complexity of the formula A where we make use of a complexity measure such that the complexity of $\mathsf{CP}_{\geq s}(B, C)$ and $\mathsf{D}_{\geq s}(B, C)$ is greater than the complexity of $B \wedge C$.

Lemma 4. *Let $M = \langle U, W, H, \mu, * \rangle$ be the canonical model for $\mathsf{BJ_{CS}}$. Then we have*

$$(\forall A \in \mathsf{Fml})(\forall w \in U)\big[[A]_{M,w} = (A)_M\big].$$

From Lemma 4 we get the following corollary.

Corollary 1. *Let CS be any constant specification. The canonical model for $\mathsf{BJ_{CS}}$ is a $\mathsf{BJ_{CS,Meas}}$-model.*

Making use of the properties of maximal consistent sets, we can establish the truth lemma.

Lemma 5 (Truth Lemma). *Let CS be a constant specification and let $M = \langle U, W, H, \mu, * \rangle$ be the canonical model for $\mathsf{BJ_{CS}}$. For every $A \in \mathsf{Fml}$ and any $w \in U$ we have:*

$$A \in w \quad \Longleftrightarrow \quad M, w \models A.$$

Using the truth lemma we find that the canonical model satisfies the condition for Bayesian models, i.e. we have the following corollary.

Corollary 2. *Let CS be any constant specification. The canonical model for $\mathsf{BJ_{CS}}$ is a $\mathsf{BJ_{CS,Meas,Bayes}}$-model.*

Finally, we get the completeness theorem as usual.

Theorem 3 (Strong Completeness for BJ). *Let CS be a constant specification, let $T \subseteq \mathsf{Fml}$ and let $A \in \mathsf{Fml}$. Then we have:*

$$T \models A \quad \Longrightarrow \quad T \vdash A.$$

4   Transitivity

It is well known that Bayesian confirmation is not transitive, i.e., the following principle is not valid

$$\mathsf{P}(B|A) > \mathsf{P}(B) \text{ and } \mathsf{P}(C|B) > \mathsf{P}(C) \quad \Longrightarrow \quad \mathsf{P}(C|A) > \mathsf{P}(C) \ . \tag{2}$$

We refer to, e.g., [24, 27] for examples where transitivity fails.

It turns out, however, that there are conditions under which (2) holds. Shogenji [27] introduces the following condition, called *screening-off condition*,

$$\mathsf{P}(C|A \wedge B) = \mathsf{P}(C|B) \quad \text{and} \quad \mathsf{P}(C|A \wedge \neg B) = \mathsf{P}(C|\neg B) \tag{3}$$

and shows that transitivity holds under it. Intuitively, (3) means that once truth or falsity of B is known, A is irrelevant to the probability of C. In other words, A affects the probability of C only indirectly through its impact on B [24].

Roche [23] presents the following weakening of (3)

$$\mathsf{P}(C|A \wedge B) \geq \mathsf{P}(C|B) \quad \text{and} \quad \mathsf{P}(C|A \wedge \neg B) \geq \mathsf{P}(C|\neg B) \ . \tag{4}$$

and shows that transitivity also holds under this weaker condition.

We are now going to formalize this result in Bayesian justification logic. We show that we can represent (4) in BJ and that this condition entails transitivity of Bayesian justifications.

Theorem 4. *Let A, B, and C be formulas of* Fml. *Let T be the set of formulas that consists of:*

1. $\mathsf{CP}_{=r}(C, B) \to \mathsf{CP}_{\geq r}(C, A \wedge B)$ *for all $r \in S$,*
2. $\mathsf{CP}_{=r}(C, \neg B) \to \mathsf{CP}_{\geq r}(C, A \wedge \neg B)$ *for all $r \in S$,*
3. $\mathsf{P}_{\neq 0} A$, $\mathsf{P}_{\neq 0}(A \wedge B)$, $\mathsf{P}_{\neq 0}(A \wedge \neg B)$, $\mathsf{P}_{\neq 0} B$, *and* $\mathsf{P}_{\neq 0} \neg B$.

Then we have that
$$T \vdash j_A : B \wedge j_B : C \to j_A : C \ .$$

Let M be any $\mathsf{BJ}_{\mathsf{CS},\mathsf{Meas},\mathsf{Bayes}}$-model such that $M \models T$. We observe that since M satisfies all formulas in T, the model M also satisfies condition (4). Thus we can show that $M \models j_A : B \wedge j_B : C \to j_A : C$ by essentially following the original proof that transitivity holds under (4) given in [23]. The theorem follows by strong completeness of $\mathsf{BJ}_{\mathsf{CS}}$.

5   Conclusion

In this paper we have introduced $\mathsf{BJ}_{\mathsf{CS}}$, a family of justification logics that feature Bayesian confirmations. Because the language of Bayesian justification logics includes both probability operators and explicit justifications, we were able to define a class of models that satisfies condition (1). Hence $\mathsf{BJ}_{\mathsf{CS}}$ not only includes justification terms built up from variables and constants, i.e. terms that represent assumptions and logical axioms, but also terms that represent Bayesian confirmations. In particular, a formula $j_A : B$, i.e. j_A justifies B, can be read as evidence A confirms B in the sense of *increase in firmness*.

We have established soundness and completeness of $\mathsf{BJ}_{\mathsf{CS}}$ with respect to Bayesian models. Further we have shown that we can formalize the screening-off condition and that this condition entails transitivity of confirmation in Bayesian models.

Future work includes studying the computational properties of Bayesian justification logic, i.e., establishing decidability and complexity results, as well as developing a corresponding proof theory.

Acknowledgements. Hamzeh Mohammadi has been supported by the Ministry of Science, Research and Technology of Iran and part of the research was carried out during a visit at University of Bern.

Thomas Studer has been supported by the Swiss National Science Foundation grant 200021_165549.

References

1. Artemov, S.: On aggregating probabilistic evidence. In: Artemov, S., Nerode, A. (eds.) LFCS 2016. LNCS, vol. 9537, pp. 27–42. Springer, Cham (2016). https://doi.org/10.1007/978-3-319-27683-0_3
2. Artemov, S.N.: Explicit provability and constructive semantics. Bull. Symbolic Logic **7**(1), 1–36 (2001)

3. Artemov, S.N.: Justified common knowledge. Theoret. Comput. Sci. **357**(1–3), 4–22 (2006). https://doi.org/10.1016/j.tcs.2006.03.009
4. Artemov, S.N.: The logic of justification. Rev. Symbolic Logic **1**(4), 477–513 (2008). https://doi.org/10.1017/S1755020308090060
5. Artemov, S.N.: The ontology of justifications in the logical setting. Studia Logica **100**(1–2), 17–30 (2012). https://doi.org/10.1007/s11225-012-9387-x. Published online February 2012
6. Artemov, S.N., Fitting, M.: Justification logic. In: Zalta, E.N. (ed.) The Stanford Encyclopedia of Philosophy. Fall 2012 edn. (2012). http://plato.stanford.edu/archives/fall2012/entries/logic-justification/
7. Bucheli, S., Kuznets, R., Renne, B., Sack, J., Studer, T.: Justified belief change. In: Arrazola, X., Ponte, M. (eds.) Proceedings LogKCA10, pp. 135–155. University of the Basque Country Press, Vitoria-Gasteiz (2010)
8. Bucheli, S., Kuznets, R., Studer, T.: Justifications for common knowledge. J. Appl. Non-classical Logic **21**(1), 35–60 (2011). https://doi.org/10.3166/JANCL.21.35-60
9. Bucheli, S., Kuznets, R., Studer, T.: Realizing public announcements by justifications. J. Comput. Syst. Sci. **80**(6), 1046–1066 (2014). https://doi.org/10.1016/j.jcss.2014.04.001
10. Carnap, R.: Logical Foundations of Probability, 2nd edn. University of Chicago Press, Chicago (1962)
11. Fan, T., Liau, C.: A logic for reasoning about justified uncertain beliefs. In: Yang, Q., Wooldridge, M. (eds.) Proceedings IJCAI 2015, pp. 2948–2954. AAAI Press, Menlo Park (2015)
12. Ghari, M.: Pavelka-style fuzzy justification logics. Logic J. IGPL **24**(5), 743–773 (2016)
13. Kokkinis, I., Maksimović, P., Ognjanović, Z., Studer, T.: First steps towards probabilistic justification logic. Logic J. IGPL **23**(4), 662–687 (2015). https://doi.org/10.1093/jigpal/jzv025
14. Kokkinis, I., Ognjanović, Z., Studer, T.: Probabilistic justification logic. In: Artemov, S., Nerode, A. (eds.) LFCS 2016. LNCS, vol. 9537, pp. 174–186. Springer, Cham (2016). https://doi.org/10.1007/978-3-319-27683-0_13
15. Kuznets, R., Studer, T.: Justifications, ontology, and conservativity. In: Bolander, T., Braüner, T., Ghilardi, S., Moss, L. (eds.) Advances in Modal Logic, vol. 9, pp. 437–458. College Publications, Cambridge (2012)
16. Kuznets, R., Studer, T.: Update as evidence: belief expansion. In: Artemov, S., Nerode, A. (eds.) LFCS 2013. LNCS, vol. 7734, pp. 266–279. Springer, Heidelberg (2013). https://doi.org/10.1007/978-3-642-35722-0_19
17. Kuznets, R., Studer, T.: Weak arithmetical interpretations for the logic of proofs. Logic J. IGPL **24**(3), 424–440 (2016)
18. Kuznets, R., Studer, T.: Logics of Proofs and Justifications. College Publications, Cambridge (2019)
19. Milnikel, R.S.: The logic of uncertain justifications. APAL **165**(1), 305–315 (2014). https://doi.org/10.1016/j.apal.2013.07.015
20. Ognjanović, Z., Savić, N., Studer, T.: Justification logic with approximate conditional probabilities. In: Baltag, A., Seligman, J., Yamada, T. (eds.) LORI 2017. LNCS, vol. 10455, pp. 681–686. Springer, Heidelberg (2017). https://doi.org/10.1007/978-3-662-55665-8_52
21. Pischke, N.: A note on strong axiomatization of Gödel-justification logic. E-print 1809.09608, arXiv.org (2018)

22. Rašković, M., Marković, Z., Ognjanović, Z.: A logic with approximate conditional probabilities that can model default reasoning. Int. J. Approximate Reasoning **49**(1), 52–66 (2008). https://doi.org/10.1016/j.ijar.2007.08.006
23. Roche, W.: A weaker condition for transitivity in probabilistic support. Eur. J. Philos. Sci. **2**(1), 111–118 (2012). https://doi.org/10.1007/s13194-011-0033-7
24. Roche, W., Shogenji, T.: Confirmation, transitivity, and moore: the screening-off approach. Philos. Stud. **168**(3), 797–817 (2014). https://doi.org/10.1007/s11098-013-0161-3
25. Schechter, L.M.: A logic of plausible justifications. Theoret. Comput. Sci. **603**, 132–145 (2015). https://doi.org/10.1016/j.tcs.2015.07.018
26. Schlesinger, G.N.: Measuring degrees of confirmation. Analysis **55**(3), 208–212 (1995). https://doi.org/10.1093/analys/55.3.208
27. Shogenji, T.: A condition for transitivity in probabilistic support. Br. J. Philos. Sci. **54**(4), 613–616 (2003)
28. Su, C.P., Fan, T.F., Liau, C.J.: Possibilistic justification logic: reasoning about justified uncertain beliefs. ACM Trans. Comput. Logic **18**(2), 15:1–15:21 (2017). https://doi.org/10.1145/3091118
29. Talbott, W.: Bayesian epistemology. In: Zalta, E.N. (ed.) The Stanford Encyclopedia of Philosophy. Summer 2015 edn. (2015)

Probability Propagation in Selected Aristotelian Syllogisms

Niki Pfeifer[1]([✉]) and Giuseppe Sanfilippo[2]([✉])

[1] Department of Philosophy, University of Regensburg, Regensburg, Germany
niki.pfeifer@ur.de
[2] Department of Mathematics and Computer Science, University of Palermo,
Palermo, Italy
giuseppe.sanfilippo@unipa.it

Abstract. This paper continues our work on a coherence-based probability semantics for Aristotelian syllogisms (Gilio, Pfeifer, and Sanfilippo, 2016; Pfeifer and Sanfilippo, 2018) by studying Figure III under coherence. We interpret the syllogistic sentence types by suitable conditional probability assessments. Since the probabilistic inference of $P|S$ from the premise set $\{P|M, S|M\}$ is not informative, we add $p(M|(S \vee M)) > 0$ as a probabilistic constraint (i.e., an "existential import assumption") to obtain probabilistic informativeness. We show how to propagate the assigned premise probabilities to the conclusion. Thereby, we give a probabilistic meaning to all syllogisms of Figure III. We discuss applications like generalised quantifiers (like Most S are P) and (negated) defaults.

Keywords: Aristotelian syllogisms · Coherence · Conditional events · Figure III · Imprecise probability · Default reasoning

1 Motivation and Outline

Aristotelian syllogisms constitute one of the oldest logical reasoning systems. Given the over two millennia long history, not many authors proposed *probabilistic* semantics for Aristotelian syllogisms (see, e.g., [7,8,11,16,30]) to overcome formal restrictions inherited by deductive logic, like its *monotonicity* (i.e., the inability to retract conclusions in the light of new evidence) or its *bivalence* (i.e., the inability to express *degrees of belief*). This paper continues our work on categorical Aristotelian syllogisms within coherence-based probability logic (see, e.g., [5,10,12,16,39]; for other approach to probability logic see, e.g., [1,2,24,32]). We aim to manage nonmonotonicity and degrees of belief, which are necessary for the formalisation of commonsense reasoning. We have studied Figure I, which have transitive structures [16] and Figure II, where the middle term constitutes the consequents of both premises [41]. We extend this work by studying Figure III under coherence. The middle term constitutes the antecedents of the premises of Figure III syllogisms (see Table 1). After recalling some preliminary notions

---

N. Pfeifer and G. Sanfilippo—contributed equally to the article and are listed alphabetically.

© Springer Nature Switzerland AG 2019
G. Kern-Isberner and Z. Ognjanović (Eds.): ECSQARU 2019, LNAI 11726, pp. 419–431, 2019.
https://doi.org/10.1007/978-3-030-29765-7_35

Table 1. Traditional (logically valid) Aristotelian syllogisms of Figure III (term order: M—P, M—S, *therefore* S—P). [*] denotes syllogisms which require implicit existential import assumptions for logical validity (since universally quantifiers statements could be vacuously true, M must not be "empty", i.e., $\exists x M x$).

AII	Datisi	*Every M is P, some M is S, therefore some S is P*
AAI[*]	Darapti	*Every M is P, every M is S, therefore some S is P*
EIO	Ferison	*No M is P, some M is S, therefore some S is not P*
EAO[*]	Felapton	*No M is P, every M is S, therefore some S is not P*
IAI	Disamis	*Some M is P, every M is S, therefore some S is P*
OAO	Bocardo	*Some M is not P, every M is S, therefore some S is not P*

and results in Sect. 2, we show how to propagate the assigned probabilities to the sequence of conditional events $(P|M, S|M, M|(S \vee M))$ to the conclusion $P|S$ in Sect. 3. This result is applied in Sect. 4, where we firstly give a probabilistic meaning to the traditionally valid syllogisms of Figure III (see Table 1). Secondly, we connect Aristotelian syllogistics with nonmonotonic reasoning by constructing syllogisms in terms of defaults and negated defaults. Section 5 concludes by remarks on further applications and future work.

2   Preliminary Notions and Results

In this section we recall selected key features of coherence (for more details see, e.g., [4,9,10,19,20,34,45]). Given two events E and H, with $H \neq \bot$, the *conditional event* $E|H$ is defined as a three-valued logical entity which is *true* if EH (i.e., $E \wedge H$) is true, *false* if $\bar{E}H$ is true, and *void* if H is false. In betting terms, assessing $p(E|H) = x$ means that, for every real number s, you are willing to pay an amount $s \cdot x$ and to receive s, or 0, or $s \cdot x$, according to whether EH is true, or $\bar{E}H$ is true, or $\bar{H}$ is true (i.e., the bet is called off), respectively. In these cases the random gain (that is, the difference between the (random) amount that you receive and the amount that you pay) is $\mathcal{G} = (sEH + 0\bar{E}H + sx\bar{H}) - sx = sEH + sx(1 - H) - sx = sH(E - x)$. More generally speaking, consider a real-valued function $p : \mathcal{K} \to \mathbb{R}$, where $\mathcal{K}$ is an arbitrary (possibly not finite) family of conditional events. Let $\mathcal{F} = (E_1|H_1, \ldots, E_n|H_n)$ be a sequence of conditional events, where $E_i|H_i \in \mathcal{K}$, $i = 1, \ldots, n$, and let $\mathcal{P} = (p_1, \ldots, p_n)$ be the vector of values $p_i = p(E_i|H_i)$, where $i = 1, \ldots, n$. We denote by $\mathcal{H}_0$ the disjunction $H_1 \vee \cdots \vee H_n$. With the pair $(\mathcal{F}, \mathcal{P})$ we associate the random gain $\mathcal{G} = \sum_{i=1}^{n} s_i H_i (E_i - p_i)$, where $s_1, \ldots, s_n$ are n arbitrary real numbers. $\mathcal{G}$ represents the net gain of n transactions. Let $\mathcal{G}_{\mathcal{H}_0}$ denote the set of possible values of $\mathcal{G}$ restricted to $\mathcal{H}_0$, that is, the values of $\mathcal{G}$ when at least one conditioning event is true.

Definition 1. *Function p defined on $\mathcal{K}$ is coherent if and only if, for every integer n, for every sequence $\mathcal{F}$ of n conditional events in $\mathcal{K}$ and for every $s_1, \ldots, s_n$, it holds that:* $\min \mathcal{G}_{\mathcal{H}_0} \leqslant 0 \leqslant \max \mathcal{G}_{\mathcal{H}_0}$.

Intuitively, Definition 1, means in betting terms that a probability assessment is coherent if and only if, in any finite combination of n bets, it cannot happen that the values in $G_{\mathcal{H}_0}$ are all positive, or all negative (*no Dutch Book*). Coherence can be also characterized in terms of proper scoring rules [6], which can be related to the notion of entropy and extropy in information theory [28, 29]. Coherence can be checked, for example, by applying [13, Algorithm 1] or by the CkC-package [3]. We recall the fundamental theorem of de Finetti for conditional events, which states that a coherent assessment of premises can always be coherently extended to a conclusion [4, 9, 25, 31, 43, 46]:

Theorem 1. *Let a coherent probability assessment* $\mathcal{P} = (p_1, \ldots, p_n)$ *on a sequence* $\mathcal{F} = (E_1|H_1, \ldots, E_n|H_n)$ *be given. Then, for a given further conditional event* $E_{n+1}|H_{n+1}$, *there exists a suitable closed interval* $[z', z''] \subseteq [0,1]$ *such that the extension* $(\mathcal{P}, z)$ *of* $\mathcal{P}$ *to* $(\mathcal{F}, E_{n+1}|H_{n+1})$ *is coherent if and only if* $z \in [z', z'']$.

Definition 2. *An* imprecise, *or set-valued, assessment* $\mathcal{I}$ *on a finite sequence of* n *conditional events* $\mathcal{F}$ *is a (possibly empty) set of precise assessments* $\mathcal{P}$ *on* $\mathcal{F}$.

Definition 2, introduced in [13], states that an *imprecise (probability) assessment* $\mathcal{I}$ on a finite sequence $\mathcal{F}$ of n conditional events is just a (possibly empty) subset of $[0,1]^n$. We recall the notions of g-coherence and total-coherence for imprecise (in the sense of set-valued) probability assessments [16].

Definition 3. *Let a sequence of* n *conditional events* $\mathcal{F}$ *be given. An imprecise assessment* $\mathcal{I} \subseteq [0,1]^n$ *on* $\mathcal{F}$ *is g-coherent if and only if there exists a coherent precise assessment* $\mathcal{P}$ *on* $\mathcal{F}$ *such that* $\mathcal{P} \in \mathcal{I}$.

Definition 4. *An imprecise assessment* $\mathcal{I}$ *on* $\mathcal{F}$ *is* totally coherent *(t-coherent) if and only if the following two conditions are satisfied: (i)* $\mathcal{I}$ *is non-empty; (ii) if* $\mathcal{P} \in \mathcal{I}$, *then* $\mathcal{P}$ *is a coherent precise assessment on* $\mathcal{F}$.

We denote by Π the set of *all coherent precise* assessments on $\mathcal{F}$. We recall that if there are no logical relations among the events $E_1, H_1, \ldots, E_n, H_n$ involved in $\mathcal{F}$, that is $E_1, H_1, \ldots, E_n, H_n$ are logically independent, then the set Π associated with $\mathcal{F}$ is the whole unit hypercube $[0,1]^n$. If there are logical relations, then the set Π *could be* a strict subset of $[0,1]^n$. As it is well known $\Pi \neq \varnothing$; therefore, $\varnothing \neq \Pi \subseteq [0,1]^n$.

Remark 1. Note that: $\mathcal{I}$ is g-coherent $\iff \Pi \cap \mathcal{I} \neq \varnothing$; $\mathcal{I}$ is t-coherent $\iff \varnothing \neq \Pi \cap \mathcal{I} = \mathcal{I}$. Then: $\mathcal{I}$ is t-coherent $\Rightarrow \mathcal{I}$ is g-coherent. Thus, g-coherence is weaker than t-coherence. For further details and relations to coherence see [16].

Given a g-coherent assessment $\mathcal{I}$ on a sequence of n conditional events $\mathcal{F}$, for each coherent precise assessment $\mathcal{P}$ on $\mathcal{F}$, with $\mathcal{P} \in \mathcal{I}$, we denote by $[z'_{\mathcal{P}}, z''_{\mathcal{P}}]$ the interval of coherent extensions of $\mathcal{P}$ to $E_{n+1}|H_{n+1}$; that is, the assessment $(\mathcal{P}, z)$ on $(\mathcal{F}, E_{n+1}|H_{n+1})$ is coherent if and only if $z \in [z'_{\mathcal{P}}, z''_{\mathcal{P}}]$. Then, defining the set $\Sigma = {}_{\mathcal{P} \in \Pi \cap \mathcal{I}}[z'_{\mathcal{P}}, z''_{\mathcal{P}}]$, for every $z \in \Sigma$, the assessment $\mathcal{I} \times \{z\}$ is a g-coherent extension of $\mathcal{I}$ to $(\mathcal{F}, E_{n+1}|H_{n+1})$; moreover, for every $z \in [0,1] \setminus \Sigma$, the extension $\mathcal{I} \times \{z\}$ of $\mathcal{I}$ to $(\mathcal{F}, E_{n+1}|H_{n+1})$ is not g-coherent. We say that Σ is the *set of coherent extensions* of the imprecise assessment $\mathcal{I}$ on $\mathcal{F}$ to the conditional event $E_{n+1}|H_{n+1}$.

3   Figure III: Propagation of Probability Bounds

We observe that the probabilistic inference of $C|A$ from the premise set $\{C|B, A|B\}$, which corresponds to the general form of syllogisms of Figure III, is probabilistically non-informative. Therefore, we add the probabilistic constraint $p(B|A \vee B) > 0$ to obtain probabilistic informativeness. This constraint serves as an "existential import assumption" (see also [16,41]). Contrary to first order monadic predicate logic, which requires existential import assumptions for Darapti and Felapton only (see Table 1), our probabilistic existential import assumption is required for all valid syllogisms of Figure III.

Remark 2. Let A, B, C be logically independent events. It can be proved that the assessment (x, y, z) on $(C|B, A|B, C|A)$ is coherent for every $(x, y, z) \in [0,1]^3$, that is, the imprecise assessment $\mathcal{I} = [0,1]^3$ on $(C|B, A|B, C|A)$ is totally coherent. Moreover, it can also be proved that the assessment (x, y, t) on $(C|B, A|B, B|A \vee B)$ is coherent for every $(x, y, t) \in [0,1]^3$, that is, the imprecise assessment $\mathcal{I} = [0,1]^3$ on $(C|B, A|B, B|A \vee B)$ is totally coherent. It is sufficient to check the coherence of each vertex of the unit cube [13].

Consider a coherent probability assessment (x, y, t) on the sequence of conditional events $(C|B, A|B, B|A \vee B)$. The next result allows for computing the lower and upper bounds, z' and z'' respectively, for the coherent extension $z = p(C|A)$.

Theorem 2. *Let A, B, C be three logically independent events and $(x, y, t) \in [0,1]^3$ be a (coherent) assessment on the family $(C|B, A|B, B|A \vee B)$. Then, the extension $z = p(C|A)$ is coherent if and only if $z \in [z', z'']$, where*

$$z' = \begin{cases} 0, & \text{if } t(x+y-1) \leqslant 0, \\ \dfrac{t(x+y-1)}{1-t(1-y)}, & \text{if } t(x+y-1) > 0, \end{cases} \quad z'' = \begin{cases} 1, & \text{if } t(y-x) \leqslant 0, \\ 1 - \dfrac{t(y-x)}{1-t(1-y)}, & \text{if } t(y-x) > 0. \end{cases}$$

Proof. In order to compute the lower and upper probability bounds on the further event $C|A$ (i.e., the conclusion), we exploit Theorem 1 by applying [16, Algorithm 1] (which is originally based on [4, Algorithm 2]) in a symbolic way.

Computation of the lower probability bound z' on $C|A$.

Input. The assessment (x, y, t) on $\mathcal{F} = (C|B, A|B, B|A \vee B)$ and the event $C|A$.

Step 0. The constituents associated with $(C|B, A|B, B|A \vee B, C|A)$ are $C_0 = \bar{A}\bar{B}$, $C_1 = ABC$, $C_2 = A\bar{B}C$, $C_3 = AB\bar{C}$, $C_4 = A\bar{B}\bar{C}$, $C_5 = \bar{A}BC$, $C_6 = \bar{A}B\bar{C}$. We observe that $\mathcal{H}_0 = A \vee B$; then, the constituents contained in $\mathcal{H}_0$ are

$C_1, \ldots, C_6$. We construct the starting system with the unknowns $\lambda_1, \ldots, \lambda_6, z$:

$$
\begin{cases}
\lambda_1 + \lambda_2 = z(\lambda_1 + \lambda_2 + \lambda_3 + \lambda_4), \\
\lambda_1 + \lambda_5 = x(\lambda_1 + \lambda_3 + \lambda_5 + \lambda_6), \\
\lambda_1 + \lambda_3 = y(\lambda_1 + \lambda_3 + \lambda_5 + \lambda_6), \\
\lambda_1 + \lambda_3 + \lambda_5 + \lambda_6 = t(\sum_{i=1}^{6} \lambda_i), \\
\sum_{i=1}^{6} \lambda_i = 1, \; \lambda_i \geqslant 0, \; i = 1, \ldots, 6,
\end{cases}
\Longleftrightarrow
\begin{cases}
\lambda_1 + \lambda_2 = z(\lambda_1 + \lambda_2 + \lambda_3 + \lambda_4), \\
\lambda_1 + \lambda_5 = xt, \\
\lambda_1 + \lambda_3 = yt, \\
\lambda_1 + \lambda_3 + \lambda_5 + \lambda_6 = t, \\
\sum_{i=1}^{6} \lambda_i = 1, \; \lambda_i \geqslant 0, \; i = 1, \ldots, 6.
\end{cases}
\tag{1}
$$

Step 1. By setting $z = 0$ in System (1), we obtain

$$
\begin{cases}
\lambda_1 + \lambda_2 = 0, \; \lambda_3 = yt, \; \lambda_5 = xt, \\
\lambda_3 + \lambda_5 + \lambda_6 = t, \\
\lambda_3 + \lambda_4 + \lambda_5 + \lambda_6 = 1, \\
\lambda_i \geqslant 0, \; i = 1, \ldots, 6.
\end{cases}
\Longleftrightarrow
\begin{cases}
\lambda_1 = \lambda_2 = 0, \\
\lambda_3 = yt, \; \lambda_4 = 1 - t, \; \lambda_5 = xt, \\
\lambda_6 = t(1 - x - y), \\
\lambda_i \geqslant 0, \; i = 1, \ldots, 6.
\end{cases}
\tag{2}
$$

As $(x, y, t) \in [0, 1]^3$, the conditions $\lambda_h \geqslant 0$, $h = 1, \ldots, 5$, in System (2) are all satisfied. Then, System (2), i.e. System (1) with $z = 0$, is solvable if and only if $\lambda_6 = t(1 - x - y) \geqslant 0$. We distinguish two cases: (i) $t(1 - x - y) < 0$ (i.e. $t > 0$ and $x + y > 1$); (ii) $t(1 - x - y) \geqslant 0$, (i.e. $t = 0$ or $(t > 0) \wedge (x + y \leqslant 1)$). In Case (i), System (2) is not solvable and we go to Step 2 of the algorithm. In Case (ii), System (2) is solvable and we go to Step 3.

Case (i). By Step 2 we have the following linear programming problem: *Compute* $\gamma' = \min(\sum_{i : C_i \subseteq AC} \lambda_r) = \min(\lambda_1 + \lambda_2)$ *subject to:*

$$
\begin{cases}
\lambda_1 + \lambda_5 = x(\lambda_1 + \lambda_3 + \lambda_5 + \lambda_6), \; \lambda_1 + \lambda_3 = y(\lambda_1 + \lambda_3 + \lambda_5 + \lambda_6), \\
\lambda_1 + \lambda_3 + \lambda_5 + \lambda_6 = t(\sum_{i=1}^{6} \lambda_i), \; \lambda_1 + \lambda_2 + \lambda_3 + \lambda_4 = 1, \\
\lambda_i \geqslant 0, \; i = 1, \ldots, 6.
\end{cases}
\tag{3}
$$

We notice that y is positive since $x + y > 1$ (and $(x, y, t) \in [0, 1]^3$). Then, also $1 - t(1 - y)$ is positive and the constraints in (3) can be rewritten as

$$
\begin{cases}
\lambda_1 + \lambda_5 = xt(1 + \lambda_5 + \lambda_6), \\
\lambda_1 + \lambda_3 = yt(1 + \lambda_5 + \lambda_6), \\
\lambda_5 + \lambda_6 = (t - yt)(1 + \lambda_5 + \lambda_6) \\
\lambda_1 + \lambda_2 + \lambda_3 + \lambda_4 = 1, \\
\lambda_i \geqslant 0, \; i = 1, \ldots, 6,
\end{cases}
\Longleftrightarrow
\begin{cases}
\lambda_5 + \lambda_6 = \frac{t(1-y)}{1 - t(1-y)}, \\
\lambda_1 + \lambda_5 = xt(1 + \frac{t(1-y)}{1-t(1-y)}) = \frac{xt}{1-t(1-y)}, \\
\lambda_1 + \lambda_3 = yt(1 + \frac{t(1-y)}{1-t(1-y)}) = \frac{yt}{1-t(1-y)}, \\
\lambda_1 + \lambda_2 + \lambda_3 + \lambda_4 = 1, \\
\lambda_i \geqslant 0, \; i = 1, \ldots, 6,
\end{cases}
$$

$$
\Longleftrightarrow
\begin{cases}
\max\{0, \frac{t(x+y-1)}{1-t(1-y)}\} \leqslant \lambda_1 \leqslant \min\{x, y\} \frac{t}{1-t(1-y)}, \\
0 \leqslant \lambda_2 \leqslant \frac{1-t}{1-t(1-y)}, \quad \lambda_3 = \frac{yt}{1-t(1-y)} - \lambda_1, \quad \lambda_4 = \frac{1-t}{1-t(1-y)} - \lambda_2, \\
\lambda_5 = \frac{xt}{1-t(1-y)} - \lambda_1, \quad \lambda_6 = \frac{t(1-x-y)}{1-t(1-y)} + \lambda_1.
\end{cases}
\tag{4}
$$

Thus, by recalling that $x + y - 1 > 0$, the minimum γ' of $\lambda_1 + \lambda_2$ subject to (3), or equivalently subject to (4), is obtained at $(\lambda_1', \lambda_2') = (\frac{t(x+y-1)}{1-t(1-y)}, 0)$. The *procedure stops* yielding as *output* $z' = \gamma' = \lambda_1' + \lambda_2' = \frac{t(x+y-1)}{1-t(1-y)}$.

Case (ii). We take Step 3 of the algorithm. We denote by Λ and $\mathcal{S}$ the vector of unknowns $(\lambda_1, \ldots, \lambda_6)$ and the set of solutions of System (2), respectively. We consider the following linear functions (associated with the conditioning events $H_1 = H_2 = B, H_3 = A \vee B, H_4 = A$) and their maxima in $\mathcal{S}$:

$$\Phi_1(\Lambda) = \Phi_2(\Lambda) = \sum_{r:C_r \subseteq B} \lambda_r = \lambda_1 + \lambda_3 + \lambda_5 + \lambda_6,$$
$$\Phi_3(\Lambda) = \sum_{r:C_r \subseteq A \vee B} \lambda_r = \lambda_1 + \lambda_2 + \lambda_3 + \lambda_4 + \lambda_5 + \lambda_6,$$
$$\Phi_4(\Lambda) = \sum_{r:C_r \subseteq A} \lambda_r = \lambda_1 + \lambda_2 + \lambda_3 + \lambda_4, \quad M_i = \max_{\Lambda \in \mathcal{S}} \Phi_i(\Lambda), \quad i = 1, 2, 3, 4 .$$
(5)

By (2) we obtain: $\Phi_1(\Lambda) = \Phi_2(\Lambda) = 0 + yt + xt + t - xt - yt = t$, $\Phi_3(\Lambda) = 1$, $\Phi_4(\Lambda) = yt + 1 - t = 1 - t(1 - y)$, $\forall \Lambda \in \mathcal{S}$. Then, $M_1 = M_2 = t$, $M_3 = 1$, and $M_4 = 1 - (1 - y)t$. We consider two subcases: $t < 1$; $t = 1$. If $t < 1$, then $M_4 = yt + 1 - t > yt \geqslant 0$; so that $M_4 > 0$ and we are in the first case of Step 3 (i.e., $M_{n+1} > 0$). Thus, the *procedure stops* and yields $z' = 0$ as *output*. If $t = 1$, then $M_1 = M_2 = M_3 = 1 > 0$ and $M_4 = y$. Hence, we are in the first case of Step 3 (when $y > 0$) or in the second case of Step 3 (when $y = 0$). Thus, the *procedure stops* and yields $z' = 0$ as *output*.

Computation of the upper probability bound z'' on $C|A$. Input and *Step 0* are the same as in the proof of z'. *Step 1.* By setting $z = 1$ in System (1), we obtain

$$\begin{cases} \lambda_1 + \lambda_2 = \lambda_1 + \lambda_2 + \lambda_3 + \lambda_4, \ \lambda_1 + \lambda_5 = xt, \ \lambda_1 + \lambda_3 = yt, \\ \lambda_1 + \lambda_3 + \lambda_5 + \lambda_6 = t, \ \lambda_1 + \lambda_2 + \lambda_3 + \lambda_4 + \lambda_5 + \lambda_6 = 1, \ \lambda_i \geqslant 0, \ i = 1, \ldots, 6, \end{cases}$$

or equivalently

$$\begin{cases} \lambda_3 = \lambda_4 = 0, \ \lambda_1 + \lambda_5 = xt, \\ \lambda_1 = yt, \ \lambda_1 + \lambda_5 + \lambda_6 = t, \\ \lambda_1 + \lambda_2 + \lambda_5 + \lambda_6 = 1, \\ \lambda_i \geqslant 0, \ i = 1, \ldots, 6; \end{cases} \iff \begin{cases} \lambda_1 = yt, \ \lambda_2 = 1 - t, \ \lambda_3 = \lambda_4 = 0, \\ \lambda_5 = (x - y)t, \ \lambda_6 = t(1 - x), \\ \lambda_i \geqslant 0, \ i = 1, \ldots, 6. \end{cases} \quad (6)$$

As $(x, y, t) \in [0, 1]^3$, the inequalities $\lambda_h \geqslant 0$, $h = 1, 2, 3, 4, 6$ are satisfied. Then, System (6), i.e. System (1) with $z = 1$, is solvable if and only if $\lambda_5 = (x - y)t \geqslant 0$. We distinguish two cases: *(i)* $(x - y)t < 0$, i.e. $x < y$ and $t > 0$; *(ii)* $(x - y)t \geqslant 0$, i.e. $x \geqslant y$ or $t = 0$. In Case *(i)*, System (6) is not solvable and we go to Step 2 of the algorithm. In Case *(ii)*, System (6) is solvable and we go to Step 3.

Case (i). By Step 2 we have the following linear programming problem: *Compute* $\gamma'' = \max(\lambda_1 + \lambda_2)$ *subject to the constraints in (3).* As $(x, y, t) \in [0, 1]^3$ and $x < y$, it follows that $\min\{x, y\} = x$ and $y > 0$. Then, in this case the quantity $1 - t(1 - y)$ is positive and the constraints in (3) can be rewritten as in (4). Thus, the maximum γ'' of $\lambda_1 + \lambda_2$ subject to (4), is obtained at $(\lambda_1'', \lambda_2'') = (\frac{xt}{1 - t(1 - y)}, \frac{1 - t}{1 - t(1 - y)})$. The *procedure stops* yielding as *output* $z'' = \gamma'' = \lambda_1'' + \lambda_2'' = \frac{xt}{1 - t(1 - y)} + \frac{1 - t}{1 - t(1 - y)} = \frac{1 - t + xt}{1 - t + yt} = 1 - \frac{t(y - x)}{1 - t + yt}$.

Case (ii). We take Step 3 of the algorithm. We denote by Λ and $\mathcal{S}$ the vector of unknowns $(\lambda_1, \ldots, \lambda_6)$ and the set of solutions of System (6), respectively. We consider the functions $\Phi_i(\Lambda)$ and the maxima M_i, $i = 1, 2, 3, 4$, given in

(5). From System (6), we observe that the functions $\Phi_1, \ldots, \Phi_4$ are constant for every $\Lambda \in \mathcal{S}$, in particular it holds that $\Phi_1(\Lambda) = \Phi_2(\Lambda) = t$, $\Phi_3(\Lambda) = 1$ and $\Phi_4(\Lambda) = yt + 1 - t + 0 + 0 = 1 - t(1 - y)$ for every $\Lambda \in \mathcal{S}$. So that $M_1 = M_2 = t$, $M_3 = 1$, and $M_4 = 1 - t(1 - y)$. We consider two subcases: $t < 1$; $t = 1$.

If $t < 1$, then $M_4 = yt + 1 - t > yt \geqslant 0$; so that $M_4 > 0$ and we are in the first case of Step 3 (i.e., $M_{n+1} > 0$). Thus, the *procedure stops* and yields $z'' = 1$ as *output*.

If $t = 1$, then $M_1 = M_2 = M_3 = 1 > 0$ and $M_4 = y$. Hence, we are in the first case of Step 3 (when $y > 0$) or in the second case of Step 3 (when $y = 0$). Thus, the *procedure stops* and yields $z'' = 1$ as *output*. □

Remark 3. From Theorem 2, we obtain $z' > 0$ if and only if $t(x + y - 1) > 0$. Moreover, we obtain $z'' < 1$ if and only if $t(y - x) > 0$.

Based on Theorem 2, the next result presents the set of coherent extensions of a given interval-valued probability assessment $\mathcal{I} = ([x_1, x_2] \times [y_1, y_2] \times [t_1, t_2]) \subseteq [0, 1]^3$ on $(C|B, A|B, B|A \vee B)$ to the further conditional event $C|A$.

Theorem 3. *Let* A, B, C *be three logically independent events and* $\mathcal{I} = ([x_1, x_2] \times [y_1, y_2] \times [t_1, t_2]) \subseteq [0, 1]^3$ *be an imprecise assessment on* $(C|B, A|B, B|A \vee B)$. *Then, the set* Σ *of the coherent extensions of* $\mathcal{I}$ *on* $C|A$ *is the interval* $[z^*, z^{**}]$, *where*

$$z^* = \begin{cases} 0, & \text{if } t_1(x_1 + y_1 - 1) \leqslant 0, \\ \dfrac{t_1(x_1 + y_1 - 1)}{1 - t_1(1 - y_1)}, & \text{if } t_1(x_1 + y_1 - 1) > 0, \end{cases} \quad \text{and}$$

$$z^{**} = \begin{cases} 1, & \text{if } t_1(y_1 - x_2) \leqslant 0, \\ 1 - \dfrac{t_1(y_1 - x_2)}{1 - t_1(1 - y_1)}, & \text{if } t_1(y_1 - x_2) > 0. \end{cases}$$

Proof. Since the set $[0, 1]^3$ on $(C|B, A|B, B|A \vee B)$ is totally coherent (Remark 2), it follows that $\mathcal{I}$ is also totally coherent. For every precise assessment $\mathcal{P} = (x, y, t) \in \mathcal{I}$, we denote by $[z'_\mathcal{P}, z''_\mathcal{P}]$ the interval of the coherent extension of $\mathcal{P}$ on $C|A$, where $z'_\mathcal{P}$ and $z''_\mathcal{P}$ coincide with z' and z'', respectively, as defined in Theorem 2. Then, $\Sigma = {}_{\mathcal{P} \in \mathcal{I}}[z'_\mathcal{P}, z''_\mathcal{P}] = [z^*, z^{**}]$, where $z^* = \inf_{\mathcal{P} \in \mathcal{I}} z'_\mathcal{P}$ and $z^{**} = \sup_{\mathcal{P} \in \mathcal{I}} z''_\mathcal{P}$.

Concerning the computation of z^* we distinguish the following alternative cases: (i) $t_1(x_1 + y_1 - 1) \leqslant 0$; (ii) $t_1(x_1 + y_1 > 1) > 0$. Case (i). By Theorem 2 it holds that $z'_\mathcal{P} = 0$ for $\mathcal{P} = (x_1, y_1, t_1)$. Thus, $\{z'_\mathcal{P} : \mathcal{P} \in \mathcal{I}\} \supseteq \{0\}$ and hence $z^* = 0$.

Case (ii). We note that the function $t(x + y - 1) : [0, 1]^3$ is nondecreasing in the arguments x, y, t. Then, $t(x + y - 1) \geqslant t_1(x_1 + y_1 - 1) > 0$ for every $(x, y, t) \in \mathcal{I}$. Hence by Theorem 2, $z'_\mathcal{P} = \frac{t(x+y-1)}{1-t(1-y)}$ for every $\mathcal{P} \in \mathcal{I}$. Moreover, the function $\frac{t(x+y-1)}{1-t(1-y)}$ is nondecreasing in the arguments x, y, t over the restricted domain $\mathcal{I}$; then, $\frac{t(x+y-1)}{1-t(1-y)} \geqslant \frac{t_1(x_1+y_1-1)}{1-t_1(1-y_1)}$. Thus, $z^* = \inf\{z'_\mathcal{P} : \mathcal{P} \in \mathcal{I}\} = \inf\left\{\frac{t(x+y-1)}{1-t(1-y)} : (x, y, z) \in \mathcal{I}\right\} = \frac{t_1(x_1+y_1-1)}{1-t_1(1-y_1)}$.

Concerning the computation of z^{**} we distinguish the following alternative cases: (i) $t_1(y_1 - x_2) \leqslant 0$; (ii) $t_1(y_1 - x_2) > 0$. Case (i). By Theorem 2 it holds that $z_{\mathcal{P}}'' = 1$ for $\mathcal{P} = (x_2, y_1, t_1) \in \mathcal{I}$. Thus, $\{z_{\mathcal{P}}'' : \mathcal{P} \in \mathcal{I}\} \supseteq \{1\}$ and hence $z^{**} = 1$.

Case (ii). We observe that $t(y - x) \geqslant t_1(y - x) \geqslant t_1(y_1 - x) \geqslant t_1(y_1 - x_2) > 0$ for every $(x, y, t) \in \mathcal{I}$. Then, the condition $t(y - x) > 0$ is satisfied for every $\mathcal{P} = (x, y, t) \in \mathcal{I}$ and hence by Theorem 2, $z_{\mathcal{P}}'' = 1 - \frac{t(y-x)}{1-t(1-y)}$ for every $\mathcal{P} \in \mathcal{I}$. The function $1 - \frac{t(y-x)}{1-t(1-y)}$ is nondecreasing in the argument x and it is nonincreasing in the arguments y, t over the restricted domain $\mathcal{I}$. Thus, $1 - \frac{t(y-x)}{1-t(1-y)} \leqslant 1 - \frac{t(y-x_2)}{1-t(1-y)} \leqslant 1 - \frac{t_1(y_1-x_2)}{1-t_1(1-y_1)}$ for every $(x, y, t) \in \mathcal{I}$. Then $z^{**} = \sup\{z_{\mathcal{P}}'' : \mathcal{P} \in \mathcal{I}\} = \sup\left\{1 - \frac{t(y-x)}{1-t(1-y)} : (x, y, z) \in \mathcal{I}\right\} = 1 - \frac{t_1(y_1-x_2)}{1-t_1(1-y_1)}$. □

4   Selected Syllogisms of Figure III

In this section we consider examples of probabilistic categorical syllogisms of Figure III (Datisi, Darapti, Ferison, Felapton, Disamis, Bocardo) by suitable instantiations in Theorem 2. We consider three events P, M, S corresponding to the predicate, middle, and the subject term, respectively.

Datisi. The direct probabilistic interpretation of the categorical syllogism *"Every M is P, Some M is S, therefore Some S is P"* would correspond to infer $p(P|S) > 0$ from the premises $p(P|M) = 1$ and $p(S|M) > 0$; however, this inference is not justified. Indeed, by Remark 2, a probability assessment $(1, y, z)$ on $(P|M, S|M, P|S)$ is coherent for every $(y, z) \in [0, 1]^2$. In order to construct a probabilistically informative version of Datisi, a further constraint of the premise set is needed. Based on [16,41] we use $p(M|S \vee M) > 0$ as a further constraint (i.e., our existential import assumption). Then, by instantiating S, M, P in Theorem 2 for A, B, C with $x = 1, y > 0$ and $t > 0$, as $t(x + y - 1) = ty > 0$, it follows that $z' = \frac{t(x+y-1)}{1-t(1-y)} = \frac{ty}{1-t(1-y)} > 0$. Then,

$$p(P|M) = 1, \ p(S|M) > 0, \ \text{and} \ p(M|S \vee M) > 0 \Longrightarrow p(P|S) > 0. \qquad (7)$$

Therefore, inference (7) is a probabilistically informative version of Datisi.

Darapti. From (7) it follows that

$$p(P|M) = 1, \ p(S|M) = 1, \ \text{and} \ p(M|S \vee M) > 0 \Longrightarrow p(P|S) > 0. \qquad (8)$$

which is a probabilistically informative interpretation of Darapti (*Every M is P, Every M is S, therefore Some S is P*) under the existential import assumption ($p(M|S \vee M) > 0$).

Ferison. By instantiating S, M, P in Theorem 2 for A, B, C with $x = 0, y > 0$ and $t > 0$, as $t(y - x) = ty > 0$, it follows by Remark 3 that $z'' < 1$. Then, $p(P|M) = 0$, $p(S|M) > 0$, and $p(M|S \vee M) > 0 \Longrightarrow p(P|S) < 1$, which can be rewritten as

$$p(\bar{P}|M) = 1, \ p(S|M) > 0, \ \text{and} \ p(M|S \vee M) > 0 \Longrightarrow p(\bar{P}|S) > 0. \qquad (9)$$

Inference (9) is a probabilistically informative version of Ferison (*No M is P, Some M is S, therefore Some S is not P*) under the existential import.

Felapton. From (9) it follows that

$$p(\bar{P}|M) = 1, \ p(S|M) = 1, \ \text{and} \ p(M|S \vee M) > 0 \Longrightarrow p(\bar{P}|S) > 0, \qquad (10)$$

which is a probabilistically informative interpretation of Felapton (*No M is P, Every M is S, therefore Some S is not P*) under the existential import.

Disamis. The direct probabilistic interpretation of the categorical syllogism "*Some M is P, Every M is S, therefore Some S is P*". By instantiating S, M, P for A, B, C in Theorem 2 with $x > 0$, $y = 1$, and $t > 0$, as $t(x + y - 1) > 0$, it follows that $z' > 0$ (see also Remark 3). Then,

$$p(P|M) > 0, \ p(S|M) = 1, \ \text{and} \ p(M|S \vee M) > 0 \Longrightarrow p(P|S) > 0. \qquad (11)$$

Inference (11) is a probabilistically informative version of Disamis under the existential import assumption.

Bocardo. By instantiating S, M, P for A, B, C in Theorem 2 with $x < 1, y = 1$ and $t > 0$, as $t(y - x) > 0$, it follows that $z'' < 1$. Then, $p(P|M) < 1$, $p(S|M) = 1$, and $p(M|S \vee M) > 0 \Longrightarrow p(P|S) < 1$, which can be rewritten as

$$p(\bar{P}|M) > 0, \ p(S|M) = 1, \ \text{and} \ p(M|S \vee M) > 0 \Longrightarrow p(\bar{P}|S) > 0. \qquad (12)$$

Inference (12) is a probabilistically informative version of Bocardo (*Some M is not P, Every M is S, therefore Some S is not P*) under the existential import. Notice that Bocardo implies Felapton by strengthening the first premise (from $p(\bar{P}|M) > 0$ to $p(\bar{P}|M) = 1$).

Remark 4. We recall that $p(M) = p(M \wedge (S \vee M)) = p(M|S \vee M)p(S \vee M)$. Hence, if we assume that $p(M)$ is positive, then $p(M|S \vee M)$ must be positive too (the converse, however, does not hold). Therefore, the inferences (7)–(12) hold if $p(M|S \vee M) > 0$ is replaced by $p(M) > 0$. The constraint $p(M) > 0$ can be seen as a stronger version of an existential import assumption compared to the conditional event existential import.

Remark 5. We observe that, traditionally, the conclusions of logically valid categorical syllogisms of Figure III are neither in the form of sentence type A (*every*) nor of E (*no*). In terms of our probabilistic semantics, we study which assessments (x, y, t) on $(P|M, S|M, S|S \vee M)$ propagate to $z' = z'' = p(P|S) = 1$ in order to validate A in the conclusion. According to Theorem 2, the following conditions should be satisfied

$$
\begin{cases} (x,y,t) \in [0,1]^3, t(x+y-1) > 0, \\ t(x+y-1) = 1 - t(1-y), \\ t(y-x) \leqslant 0, \end{cases} \iff \begin{cases} (x,y,t) \in [0,1]^3, \\ 1 + yt - t > 0, \\ tx = 1, ty \leqslant 1, \end{cases} \iff \begin{cases} x = 1, \\ 0 < y \leqslant 1, \\ t = 1. \end{cases}
$$

Then, $z' = z'' = 1$ if and only if $(x,y,t) = (1,y,1)$, with $y > 0$. However, for the syllogisms it would be too strong to require $t = 1$ as an existential import assumption, we only require that $t > 0$. Similarly, in order to validate E in the conclusion, it can be shown that assessments (x,y,t) on $(P|M, S|M, S|S \vee M)$ propagate to the conclusion $z' = z'' = p(P|S) = 0$ if and only if $(x,y,t) = (0,y,1)$, with $y > 0$. Therefore, if t is just positive neither A nor E can be validate within in our probabilistic semantics of Figure III.

Application to Default Reasoning. We recall that the *default* $H \mathrel{|\!\sim} E$ denotes the sentence "*E is a plausible consequence of H*" (see, e.g., [27]). Moreover, the *negated default* $H \mathrel{|\!\not\sim} E$ denotes the sentence "*it is not the case, that: E is a plausible consequence of H*". Based on [16, Definition 8], we interpret the default $H \mathrel{|\!\sim} E$ by the probability assessment $p(E|H) = 1$; while the negated default $H \mathrel{|\!\not\sim} E$ is interpreted by the imprecise probability assessment $p(E|H) < 1$. Then, as the probability assessment $p(E|H) > 0$ is equivalent to $p(\bar{E}|H) < 1$, the negated default $H \mathrel{|\!\not\sim} \bar{E}$ is also interpreted by $p(E|H) > 0$. Table 2 presents the syllogisms (7)–(12) of Figure III in terms of inference rules which involve defaults and negated defaults.

Table 2. Syllogisms of Figure III (see Table 1) in terms of defaults and negated defaults.

AII	Datisi	*from* $M \mathrel{|\!\sim} P, M \mathrel{|\!\not\sim} \bar{S}$, *and* $(S \vee M) \mathrel{|\!\not\sim} \bar{M}$ *infer* $S \mathrel{|\!\not\sim} \bar{P}$
AAI	Darapti	*from* $M \mathrel{|\!\sim} P, M \mathrel{|\!\sim} S$, *and* $(S \vee M) \mathrel{|\!\not\sim} \bar{M}$ *infer* $S \mathrel{|\!\not\sim} \bar{P}$
EIO	Ferison	*from* $M \mathrel{|\!\sim} \bar{P}, M \mathrel{|\!\not\sim} \bar{S}$, *and* $(S \vee M) \mathrel{|\!\not\sim} \bar{M}$ *infer* $S \mathrel{|\!\not\sim} P$
EAO	Felapton	*from* $M \mathrel{|\!\sim} \bar{P}, M \mathrel{|\!\sim} S$, *and* $(S \vee M) \mathrel{|\!\not\sim} \bar{M}$ *infer* $S \mathrel{|\!\not\sim} P$
IAI	Disamis	*from* $M \mathrel{|\!\not\sim} \bar{P}, M \mathrel{|\!\sim} S$, *and* $(S \vee M) \mathrel{|\!\not\sim} \bar{M}$ *infer* $S \mathrel{|\!\not\sim} \bar{P}$
OAO	Bocardo	*from* $M \mathrel{|\!\not\sim} P, M \mathrel{|\!\sim} S$, *and* $(S \vee M) \mathrel{|\!\not\sim} \bar{M}$ *infer* $S \mathrel{|\!\not\sim} P$

5　Concluding Remarks

In this paper we proved probability propagation rules for Aristotelian syllogisms of Figure III by using an existential import assumption which we expressed in terms of a probability constraint. Although Aristotelian syllogistics is an ancient reasoning system, our probabilistic semantics allows for various applications including applications to (i) rational nonmonotonic reasoning (we showed how to express basic syllogistic sentence types in terms of defaults and negated defaults; see also [15,16] for connections between syllogisms and default reasoning), (ii) the psychology of reasoning as a new rationality framework (see, e.g., [26,35–38,42]), (iii) the square of opposition [39,40], and to (iv) formal semantics: by setting appropriate thresholds in Theorem 3, we can interpret generalised quantifiers (see, e.g., [33]) probabilistically (like interpreting *Almost all S are P* by $p(P|S) \geq t$, where t is a given—usually context dependent—threshold like $> .9$). Resulting probabilistic syllogisms are a much more plausible rationality framework for studying commonsense contexts compared to traditional Aristotelian syllogisms. We observe that our interpretation of syllogisms relies on conditionals. Thus, future work will be devoted to further generalise Aristotelian syllogisms by iterated conditionals where the S, M, or P terms are replaced by conditional events. We have shown in the context of conditional syllogisms [14,44,45], that the theory of conditional random quantities (see, e.g. [17,18,21–23]) is able manage nested conditionals without running into the notorious Lewis' triviality. Applying these results will yield further generalisations of Aristotelian syllogisms.

Acknowledgments. We thank three anonymous referees.

References

1. Adams, E.W.: A Primer of Probability Logic. CSLI, Stanford (1998)
2. Amarger, S., Dubois, D., Prade, H.: Constraint propagation with imprecise conditional probabilities. In: Proceedings of the UAI-91, pp. 26–34. Morgan Kaufmann (1991)
3. Baioletti, M., Capotorti, A., Galli, L., Tognoloni, S., Rossi, F., Vantaggi, B.: CkC-Check Coherence package (version e6, 2016). http://www.dmi.unipg.it/upkd/paid/software.html
4. Biazzo, V., Gilio, A.: A generalization of the fundamental theorem of de Finetti for imprecise conditional probability assessments. Int. J. Approx. Reasoning **24**(2–3), 251–272 (2000)
5. Biazzo, V., Gilio, A., Lukasiewicz, T., Sanfilippo, G.: Probabilistic logic under coherence: complexity and algorithms. Ann. Math. Artif. Intell. **45**(1–2), 35–81 (2005)
6. Biazzo, V., Gilio, A., Sanfilippo, G.: Coherent conditional previsions and proper scoring rules. In: Greco, S., Bouchon-Meunier, B., Coletti, G., Fedrizzi, M., Matarazzo, B., Yager, R.R. (eds.) IPMU 2012. CCIS, vol. 300, pp. 146–156. Springer, Heidelberg (2012). https://doi.org/10.1007/978-3-642-31724-8_16
7. Chater, N., Oaksford, M.: The probability heuristics model of syllogistic reasoning. Cognit. Psychol. **38**, 191–258 (1999)
8. Cohen, A.: Generics, frequency adverbs, and probability. Linguist. Philos. **22**, 221–253 (1999)
9. Coletti, G., Scozzafava, R.: Characterization of coherent conditional probabilities as a tool for their assessment and extension. Int. J. Uncertainty Fuzziness Knowl. Based Syst. **04**(02), 103–127 (1996)
10. Coletti, G., Scozzafava, R., Vantaggi, B.: Possibilistic and probabilistic logic under coherence: default reasoning and System P. Math. Slovaca **65**(4), 863–890 (2015)
11. Dubois, D., Godo, L., López De Màntaras, R., Prade, H.: Qualitative reasoning with imprecise probabilities. J. Intell. Inf. Syst. **2**(4), 319–363 (1993)

12. Gilio, A.: Probabilistic reasoning under coherence in System P. Ann. Math. Artif. Intell. **34**, 5–34 (2002)
13. Gilio, A., Ingrassia, S.: Totally coherent set-valued probability assessments. Kybernetika **34**(1), 3–15 (1998)
14. Gilio, A., Over, D.E., Pfeifer, N., Sanfilippo, G.: Centering and compound conditionals under coherence. In: Ferraro, M.B., et al. (eds.) Soft Methods for Data Science. AISC, vol. 456, pp. 253–260. Springer, Cham (2017). https://doi.org/10.1007/978-3-319-42972-4_32
15. Gilio, A., Pfeifer, N., Sanfilippo, G.: Transitive reasoning with imprecise probabilities. In: Destercke, S., Denoeux, T. (eds.) ECSQARU 2015. LNCS (LNAI), vol. 9161, pp. 95–105. Springer, Cham (2015). https://doi.org/10.1007/978-3-319-20807-7_9
16. Gilio, A., Pfeifer, N., Sanfilippo, G.: Transitivity in coherence-based probability logic. J. Appl. Logic **14**, 46–64 (2016)
17. Gilio, A., Sanfilippo, G.: Conditional random quantities and iterated conditioning in the setting of coherence. In: van der Gaag, L.C. (ed.) ECSQARU 2013. LNCS (LNAI), vol. 7958, pp. 218–229. Springer, Heidelberg (2013). https://doi.org/10.1007/978-3-642-39091-3_19
18. Gilio, A., Sanfilippo, G.: Conjunction, disjunction and iterated conditioning of conditional events. In: Kruse, R., Berthold, M., Moewes, C., Gil, M., Grzegorzewski, P., Hryniewicz, O. (eds.) Synergies of Soft Computing and Statistics for Intelligent Data Analysis. AISC, vol. 190, pp. 399–407. Springer, Heidelberg (2013). https://doi.org/10.1007/978-3-642-33042-1_43
19. Gilio, A., Sanfilippo, G.: Probabilistic entailment in the setting of coherence: the role of quasi conjunction and inclusion relation. Int. J. Approx. Reasoning **54**(4), 513–525 (2013)
20. Gilio, A., Sanfilippo, G.: Quasi conjunction, quasi disjunction, t-norms and t-conorms: probabilistic aspects. Inf. Sci. **245**, 146–167 (2013)
21. Gilio, A., Sanfilippo, G.: Conditional random quantities and compounds of conditionals. Studia Logica **102**(4), 709–729 (2014)
22. Gilio, A., Sanfilippo, G.: Conjunction of conditional events and t-norms. In: Kern-Isberner, G., Ognjanović, Z. (eds.) ECSQARU 2019. LNAI, vol. 11726, pp. 199–211. Springer, Cham (2019)
23. Gilio, A., Sanfilippo, G.: Generalized logical operations among conditional events. Appl. Intell. **49**(1), 79–102 (2019)
24. Hailperin, T.: Sentential Probability Logic. Lehigh, Bethlehem (1996)
25. Holzer, S.: On coherence and conditional prevision. Bollettino dell'Unione Matematica Italiana **4**(6), 441–460 (1985)
26. Kleiter, G.D., Fugard, A.J.B., Pfeifer, N.: A process model of the understanding of uncertain conditionals. Thinking Reasoning **24**(3), 386–422 (2018)
27. Kraus, S., Lehmann, D., Magidor, M.: Nonmonotonic reasoning, preferential models and cumulative logics. Artif. Intell. **44**, 167–207 (1990)
28. Lad, F., Sanfilippo, G., Agró, G.: Completing the logarithmic scoring rule for assessing probability distributions. In: AIP Conference Proceedings, vol. 1490, no. 1, pp. 13–30 (2012)
29. Lad, F., Sanfilippo, G., Agró, G.: Extropy: complementary dual of entropy. Stat. Sci. **30**(1), 40–58 (2015)
30. Lambert, J.H.: Neues Organon. Wendler, Leipzig (1764)
31. Lehman, R.S.: On confirmation and rational betting. J. Symbolic Logic **20**, 251–261 (1955)

32. Ognjanović, Z., Rašković, M., Marković, Z.: Probability Logics: Probability-Based Formalization of Uncertain Reasoning, 1st edn. Springer, Cham (2016). https://doi.org/10.1007/978-3-319-47012-2
33. Peters, S., Westerståhl, D.: Quantifiers in Language and Logic. Oxford University Press, Oxford (2006)
34. Petturiti, D., Vantaggi, B.: Envelopes of conditional probabilities extending a strategy and a prior probability. Int. J. Approx. Reasoning **81**, 160–182 (2017)
35. Pfeifer, N.: The new psychology of reasoning: a mental probability logical perspective. Thinking Reasoning **19**(3–4), 329–345 (2013)
36. Pfeifer, N.: Reasoning about uncertain conditionals. Studia Logica **102**(4), 849–866 (2014)
37. Pfeifer, N., Douven, I.: Formal epistemology and the new paradigm psychology of reasoning. Rev. Philos. Psychol. **5**(2), 199–221 (2014)
38. Pfeifer, N., Kleiter, G.D.: Towards a mental probability logic. Psychologica Belgica **45**(1), 71–99 (2005)
39. Pfeifer, N., Sanfilippo, G.: Probabilistic squares and hexagons of opposition under coherence. Int. J. Approx. Reasoning **88**, 282–294 (2017)
40. Pfeifer, N., Sanfilippo, G.: Square of opposition under coherence. In: Ferraro, M.B., et al. (eds.) Soft Methods for Data Science. AISC, vol. 456, pp. 407–414. Springer, Cham (2017). https://doi.org/10.1007/978-3-319-42972-4_50
41. Pfeifer, N., Sanfilippo, G.: Probabilistic semantics for categorical syllogisms of Figure II. In: Ciucci, D., Pasi, G., Vantaggi, B. (eds.) SUM 2018. LNCS (LNAI), vol. 11142, pp. 196–211. Springer, Cham (2018). https://doi.org/10.1007/978-3-030-00461-3_14
42. Pfeifer, N., Tulkki, L.: Conditionals, counterfactuals, and rational reasoning. An experimental study on basic principles. Minds Mach. **27**(1), 119–165 (2017)
43. Regazzini, E.: Finitely additive conditional probabilities. Rendiconti del Seminario Matematico e Fisico di Milano **55**, 69–89 (1985)
44. Sanfilippo, G., Pfeifer, N., Gilio, A.: Generalized probabilistic modus ponens. In: Antonucci, A., Cholvy, L., Papini, O. (eds.) ECSQARU 2017. LNCS (LNAI), vol. 10369, pp. 480–490. Springer, Cham (2017). https://doi.org/10.1007/978-3-319-61581-3_43
45. Sanfilippo, G., Pfeifer, N., Over, D., Gilio, A.: Probabilistic inferences from conjoined to iterated conditionals. Int. J. Approx. Reasoning **93**, 103–118 (2018)
46. Williams, P.M.: Notes on conditional previsions. Technical report, School of Mathematical and Physical Sciences, University of Sussex (1975). Reprint in IJAR, 2007, 44(3)

Structural Fusion/Aggregation of Bayesian Networks via Greedy Equivalence Search Learning Algorithm

Jose M. Puerta[1](✉), Juan Ángel Aledo[2](✉), José Antonio Gámez[1](✉), and Jorge D. Laborda[1](✉)

[1] Departamento de Sistemas Informáticos, Universidad de Castilla-La Mancha, 02071 Albacete, Spain
[2] Departamento de Matemáticas, Universidad de Castilla-La Mancha, 02071 Albacete, Spain
{jose.puerta,juanangel.aledo,jose.gamez,jorgedaniel.laborda}@uclm.es

Abstract. Aggregating a set of Bayesian Networks (BNs), also known as BN fusion, has been studied in the literature, providing a precise theoretical framework for the structural phase. This phase depends on a total ordering of the variables, but both the problem of searching for the optimal consensus structure (according to standard problem definition), as well as the one of looking for the optimal ordering are NP-hard.

In this paper we start from this theoretical framework and extend it from a practical point of view. We propose a heuristic method to identify a suitable order of the variables, which allows us to obtain consensus BNs having (by far) less edges than those obtained by using random orderings. Furthermore, we apply an optimization method based on the GES algorithm to remove the extra edges. As GES is a data-driven method and we have not data but a set of incoming networks, we propose to use the independences codified in the incoming networks to determine a score in order to evaluate the goodness of removing a given edge. From the experiments carried out, we observe that our heuristic is very competitive, driving the fusion process to solutions close to the optimal one.

Keywords: Bayesian Networks · Aggregation · Fusion · Consensus · Heuristic orders

1 Introduction

A Bayesian network (BN) [12] is a knowledge representation technique frequently used to design Intelligent Systems in domains where uncertainty is predominant.

A BN $\mathcal{B} = (G, P)$ consists of:

- a directed acyclic graph (DAG) $G = (\mathbf{V}, \mathbf{E})$ which codifies the (in)dependence relationships between a set of variables $\mathbf{V}$ by means of the *d-separation* criterion [12], $\mathbf{E}$ being the set of directed relations (arcs) between such variables, and

This work has been partially funded by FEDER funds, the Spanish Goverment (AEI/MINECO) through the project TIN2016-77902-C3-1-P and the Regional Government (JCCM) by SBPLY/17/180501/000493.

© Springer Nature Switzerland AG 2019
G. Kern-Isberner and Z. Ognjanović (Eds.): ECSQARU 2019, LNAI 11726, pp. 432–443, 2019.
https://doi.org/10.1007/978-3-030-29765-7_36

- a set **P** of conditional probability distributions $\{p(X_i \mid pa_G(X_i)\}_1^n$, which factorizes the joint probability distribution over the domain of the variables[1]

$$P(\mathbf{V}) = P(X_1, \ldots, X_n) = \prod_{i=1}^{n} p(X_i | pa_G(X_i))$$

A BN can be built by domain experts [14], learnt from data [1,3] or using a hybrid approach [7]. Once defined, it can be used for both qualitative reasoning (visualization, relevance analysis, etc.) and quantitative reasoning (predictive or abductive inference). When the BN is built by a single expert, the result can be biased because of the particular expert criteria. A possible solution to this problem is to deal with several experts, each one providing a BN and then to combine/fuse them into a single *consensus* one.

Many real-world problems require learning a BN from data coming from different sources [4,6]. Usually, this is performed by using any standard algorithm which considers jointly all the data as a single training dataset. However, nowadays it is usual that the data are distributed over several datacenters, and because of privacy reasons and/or because of taking advantage of local computing power, local models are learnt at each datacenter and then these models (instead of the data) are sent to a single station. As an example, consider a chain of national supermarkets which have several regional datacenters. They all sell the same catalogue of products, but the model learnt at each datacenter reflects the specific behavior of their customer according to particular demographic factors. Summing up, our goal is to aggregate a set of BNs, possibly learnt in different locations (machines) and from different (but usually similar) datasets.

This problem, also known as BN *fusion*, has been previously studied in the literature both from the structural and parametric viewpoints:

- *Structure.* The goal of structurally combining a given set of BNs is to obtain a new DAG containing only those (conditional) independencies which are satisfied in all the networks to be aggregated. In [8,9], two methods are described to cope with this problem based on the use of a total order of the variables. In [5], the idea of computing a consensus DAG compatible with a given order is also studied. Specifically, their theoretical analysis concludes that, once the input BNs have been arranged to be compatible with respect to such an order of the variables, the union applied over the arcs of these (arranged) DAGs is the operation which ensures that the consensus DAG maintains the independencies codified in all the DAGs of the input set.
  Later, in [11], the author amend some aspects regarding the methods in [8,9] and proves that (1) the consensus DAG satisfying the required independence pattern is not unique; (2) the problem is NP-hard; and (3) the amended methods obtain a proper consensus DAG.
- *Parameters.* When the parameters (probabilistic distributions) are also considered, the whole process is much more complex. To illustrate this fact,

---

[1] We denote by $pa(X_i)$ $(pa_G(X_i))$ the parent set of X_i in G. Analogously, we denote by $ch(X_i)$ $(ch_G(X_i))$ the children set of X_i. We take $|\mathbf{V}| = n$.

in [13] it is shown under the (quite mild) assumption of all the input BNs agreeing on the structure, that no method of combining the parameters produces a consensus parameterization compatible with all the given parametric structures.

To obtain the consensus BN, we follow a two steps approach as the one in [11]: first, we identify the consensus BN structure and, then, we look for the consensus parameters given such structure. In this paper we only deal with the structural part of the problem, as in [5,8,9,11].

In particular, we work on the basis of [5,11]. It should be pointed out that these papers focus on the theoretical properties of the proposed algorithms, which heavily depend on the chosen total order defined between the variables. Although the use of whichever order guarantees to obtain a minimal I-map of the common independencies in the input DAGs (see Sect. 2), it usually leads to a very dense DAG, which is of few utility for posterior qualitative/quantitative reasoning. It is important to remark that, during the transformation of the input DAGs into DAGs compatible with the given order, their number of arcs increases, depending such increment on the election of the order.

The main contribution of this study is the proposal of a heuristic method to identify a suitable order of the variables, which allows us to obtain consensus DAGs having (by far) less arcs than those obtained by using random orders. Furthermore, as the use of a heuristic order does not guarantee to obtain the network with the minimum number of arcs, we apply an optimization phase based on the GES algorithm [3] to remove the extra edges/arcs. Recall that the GES algorithm asymptotically converges to the optimum, and so it can be used as a measure of the quality of the chosen order (the more edges are removed, the farther the consensus DAG is from the optimum). The novelty in this phase is that, instead of using data to measure the goodness of a removal operator, we directly use the independencies codified in the input DAGs. The experiments carried out show that this optimization step removes *few* arcs, which supports our heuristic proposal.

2   Fusion of Bayesian Networks

In this Section we briefly review the structural fusion method described in [11]. Let us start by revising some important notions. Let $\mathbf{X}$, $\mathbf{Y}$ and $\mathbf{Z}$ be three disjoint subsets of variables. An *independence model I* is a collection of conditional independence constraints $\mathbf{X} \perp\!\!\!\perp_I \mathbf{Y}|\mathbf{Z}$, saying that $\mathbf{X}$ is conditionally independent of $\mathbf{Y}$ given $\mathbf{Z}$.

We will denote by $I(G)$ the set of independencies codified in a DAG G by using the d-separation criterion. Given an independence model I between the variables of the DAG G, we say that G is an I-map of I if $I(G) \subseteq I$. Then, G is a *minimal* I-map of I if: (i) G is an I-map of I, and (ii) if we remove any arc from G, the resulting DAGs is not an I-map of I anymore. In particular, given two DAGs G an H, we say that G is a (minimal) I-map of H, if G is a (minimal) I-map of $I(H)$.

Algorithm 1: Fusion

Data: Incoming DAGs: $\{G_1, \ldots, G_g\}$ defined over $\mathbf{V} = \{X_1, \ldots, X_n\}$
Data: Set of orders: Σ
Result: Consensus DAG G^+

1 **for** $i = 1, \ldots, |\Sigma|$ **do**
2   σ_i = i-th element of Σ
3   **for** $j = 1, \ldots, g$ **do**
4          $G_j^{\sigma_i} = A(G_j, \sigma_i)$
5      $E_i = \bigcup_1^g E(G_j^{\sigma_i})$ // $E(G)$ are the edges of G
6 **return** $G^+ = (\mathbf{V}, \arg\min_i |E_i|)$

We say that a total order σ of the nodes (variables) of G is an *ancestral* order for G, if the variable A precedes to the variable B in σ for all arc $A \to B$ in G.

The consensus DAG G^+ for a given set of DAGs $\mathbf{G} = \{G_1, \ldots, G_g\}$ is defined as the DAG having the minimum number of arcs which is an I-map of $\cap_{i=1}^g I(G_i)$ [5,11]. That is, if a conditional independence is in G^+, then it is also present in all the input DAGs.

The following theorem [5] gives the clue to obtain the consensus DAG G^+:

Theorem 1 [5]. *Let $G = (\mathbf{V}, \mathbf{E_1})$ and $H = (\mathbf{V}, \mathbf{E_2})$ be two DAGs. If there exists a common ancestral order σ for both G and H, then the union $G^+ = (\mathbf{V}, \mathbf{E_1} \cup \mathbf{E_2})$ is a minimal I-map of $I(G) \cap I(H)$ and the intersection independence model follows the grafoid axioms[2].*

However, given $\mathbf{G} = \{G_1, \ldots, G_g\}$, there may not exist a common ancestral order σ for all the DAGs G_i, and so the result above cannot be directly applied. From the A method introduced in [8], Peña [11] designed a modified version and proved the following theorem:

Theorem 2 [11]. *Let G^σ be the minimal I-map of a DAG G compatible with a node order σ. Then, the method $A(G, \sigma)$ returns G^σ.*

Taking these two results into account, in Algorithm 1 we show the fusion algorithm. Note that this algorithm can iterate over a number of orders. Ideally, all the possible orders should be used to ensure that the optimum is obtained, but due to the enormous number of orders ($n!$), only a few may be tested. In Sect. 3 we cope with this problem by proposing a greedy heuristic to obtain a suitable order σ to guide the fusion process.

3   Greedy Heuristic Search of a Fusion Oriented Order

In this section we propose a heuristic greedy constructive search algorithm to look for an ancestral order σ suitable to guide the fusion process. Obtaining an ancestral order for a DAG is an simple task: (1) select a *sink*[3] node and place

---

[2] Symmetry, decomposition, weak union, contraction and intersection.
[3] A node with no children.

Algorithm 2: Compute $cost_{sink}$

Data: $G = (\mathbf{V}, \mathbf{E})$, $Y \in \mathbf{V}$
Result: $cost_{sink}(Y, G)$

1 $cost = 0$
2 **for** $Z \in ch(Y)$ **do**
3 $\quad$ **for** $X \in pa(Z)$ **do**
4 $\quad\quad$ **if** $X \to Y \notin \mathbf{E}$ **then**
5 $\quad\quad\quad$ cost++
6 $\quad$ **for** $X \in pa(Y)$ **do**
7 $\quad\quad$ **if** $X \to Z \notin \mathbf{E}$ **then**
8 $\quad\quad\quad$ cost++

9 **return** $cost$

---

it as the first node of the partially constructed order; (2) delete that node and all its incident arcs from the DAG; (3) iterate the previous steps until a total order of the variables is obtained. However, when dealing with a set of DAGs it is quite improbable that a node X_i is a sink in all of them. Therefore, our idea is to identify the node that is *closest* to be a sink in all the DAGs. To do this, we propose to use a score or *cost* which is computed for each node over all the DAGs.

Before defining properly such score, let us introduce some notation. An arc $A \to B$ is *covered* if $pa(A) = pa(B) \setminus \{A\}$. If $A \to B$ is covered, we can *reverse* it, that is, change its direction to $B \to A$ without introducing any additional independence. To transform an arc $A \to B$ into a covered one, we add the necessary arcs from the nodes in $pa(A)$ to B and from the nodes in $pa(B)$ to A to make $pa(A) = pa(B) \setminus \{A\}$.

Definition 1 ($cost_{sink}$). *Given a DAG $G = (\mathbf{V}, \mathbf{E})$ and a node $Y \in \mathbf{V}$, we define the cost of making Y a sink in G, $cost_{sink}(Y, G)$, as the number of arcs to be added to cover each arc $Y \to Z$, $\forall Z \in ch_G(Y)$.*

Algorithm 2 shows the pseudocode to compute $cost_{sink}$, where we count the number of added arcs in a symbolic way, that is, without actually modifying the DAG.

Finally, Algorithm 3 shows the pseudocode to compute the ancestral order by making use of the $cost_{sink}$ heuristic. Lines 5 to 12 use the $cost_{sink}$ heuristic to identify the next sink, that is, the node that induces the minimum number of inserted arcs across all the DAGs. On the other hand, lines 13 to 19 actually apply the arc-reversal operation, that is, covering followed by reversal, over the DAGs for all the arcs $bestY \to Z, \forall Z \in ch(bestY)$. Finally the sink is removed from all the DAGs to continue with the process.

Algorithm 3: Greedy Heuristic Order

    Data: $\mathbf{G} = \{G_1 = (\mathbf{V}, \mathbf{E}_1), \ldots, G_g = (\mathbf{V}, \mathbf{E}_g)\}$

    Result: σ // A complete order for $\mathbf{V}$

1 $\mathbf{G}' = \mathbf{G}$.copy

2 $\mathbf{V}' = \mathbf{V}$.copy

3 $\sigma = \{\}$

4 **while** $\mathbf{V}' \neq \emptyset$ **do**

5 $\langle bestY, bestCost \rangle = \langle null, +\infty \rangle$

6   **for** $Y \in \mathbf{V}'$ **do**

7   cost=0

8   **for** $G \in \mathbf{G}'$ **do**

9 $cost \mathrel{+}= cost_{sink}(Y, G)$

10   **if** $cost < bestCost$ **then**

11 $\langle bestY, bestCost \rangle = \langle Y, cost \rangle$

12 $\sigma = bestY \cdot \sigma$ // concatenate $bestY$ at the beginning of σ

13   **for** $G \in \mathbf{G}'$ **do**

14 $\mathbf{E} = \{bestY \to Z \mid Z \in ch(bestY)\}$

15   **while** $\mathbf{E} \neq \emptyset$ **do**

16   Select $bestY \to Z \in \mathbf{E}$, s.t. $bestY \leftarrow Z$ does not introduce a directed cycle

17   Make $bestY \to Z$ covered and then reverse it

18 $\mathbf{E} = \mathbf{E} \setminus \{bestY \to Z\}$

19   Remove $bestY$ and all its incident arcs from G

20 $\mathbf{V}' = \mathbf{V}' \setminus \{bestY\}$

21   **return** σ

---

4   Greedy Search for an Optimal Fusion DAG

GES (*Greedy Equivalence Search*) [3] is a data-driven score+search structural BN learning algorithm that carries out the search in the space of DAG's equivalence classes[4]. GES consists of two greedy phases:

- FES, a forward phase which starts from the empty graph and incrementally adds an edge/arc, the one which maximizes the used scoring metric. By $score(X; \mathbf{S})$ we denote the score of having $\mathbf{S}$ as set of parents of X in the graph. Due to the decomposability property of this score function, it is enough to evaluate local changes in the DAG to update the scores. The increment $score(X; \mathbf{S}) - score(X; \mathbf{S} \cup \{Y\})$ is used to evaluate the potential inclusion of $Y \to X$ or $Y - X$ in the current DAG. This phase stops when all the increments are negative.
- BES, a backward phase which starts from the solution obtained by FES, and iteratively removes the best arc/edge according to the score increment.

---

[4] Equivalence classes are represented by using a mixed graph structure which contains directed and undirected arcs/edges.

As FES, this phase stops when no arc/edge in the current structure has a positive score increment.

Under certain assumptions, asymptotically the output of FES is an I-map, and, after BES, the output graph is the correct graph with respect to the conditional independence statements codified in the input data. These required assumptions are: (1) there are enough data and the underlying distribution p, from which the data was drawn, is faithful a DAG; (2) the score used is globally and locally consistent; and (3) the delete and add operations assure that the search space properly traverses the I-map representation of the graphs [3].

It should be pointed out that the assumption (1) is rarely satisfied in real data, but, in spite of this, GES performs well in practice. On the other hand, most of the used scoring metrics satisfy the assumption (2), which means that

$$score(X; \mathbf{S}) - score(X; \mathbf{S} \cup \{Y\}) < 0 \Longleftrightarrow X \perp\!\!\!\perp_p Y | \mathbf{S} \tag{1}$$

In this study we use an adaptation of BES, BES_d, which takes as starting point the DAG returned by the fusion process (Algorithm 1), which is an I-map of $I^+ = \cap_{i=1}^g I(G_i)$ (Theorem 1). In that way, the output of $BES_d(Fusion(G_1, \ldots, G_g))$ is the *optimal* consensus graph. Note that we cannot directly apply BES, since we have a set of input DAGs instead of data.

Assume that instead of data as source we can consult an *oracle* to obtain answers about d-separation sentences. Assume also that the model used by the *oracle* is a DAG, gold-standard which is unknown to us, where we could read the d-separation sentences. According to, Eq. (1) a positive answer for a sentence $X \perp\!\!\!\perp_{G^*} Y | \mathbf{S}$ is equivalent to a negative score for having Y as parent of X in our model, and so BES_d can delete this arc/edge. Therefore, by using this *oracle* instead of a score over the data, we can use as initial solution a complete DAG (a trivial I-map) and then use BES_d to iteratively remove the arcs/edges that find conditionally independent via d-separation. The method stops when no more independencies are found[5].

In our framework, we know that the model $I^+ = \cap_{i=1}^g I(G_i)$ follows the graphoid axioms (Theorem 1), which means that we can found a minimal I-map DAG for this model, so there exists a DAG G^+ to represent I^+. Therefore, we can use I^+ as our *oracle*. On the other hand, we also know that the output provided by Algorithm 1 is an I-map of I^+, and so we can use it as seed for BES_d. With all of this, we can state the following theorem:

Theorem 3. *Given a set of DAGs* $\mathbf{G} = \{G_1, \ldots, G_g\}$, $BES_d(Fusion(\mathbf{G}))$ *returns the consensus DAG G^+ for $\mathbf{G}$ with the minimum number of edges that is a minimal I-map of* $I^+ = \cap_{i=1}^g I(G_i)$.

5   Experimental Evaluation

In this section we describe the experiments carried out to evaluate our proposals.

---

[5] It would be easy to show that GES would get the correct gold-standard DAG.

Experimental setup: In this preliminary study we have only used synthetically generated DAGs in order to control the complexity of the process and the different parameters to be analyzed. Furthermore, although the proposed algorithms can deal with any subset of DAGs, we assume that the DAGs to be combined should be similar, because otherwise the complexity (edge density) of the output DAG increases significantly.

We have generated the set of input DAGs by using the following method: First, we randomly generate a DAG G_1 according to a set of parameters (number of nodes, number of parents, etc.) by using the method described in [10]. Then we generate the remaining DAGs $\{G_2, \ldots, G_g\}$ from G_1 by applying some perturbations. Specifically, a perturbation consists in randomly selecting two nodes, and then: if they are linked by an arc, the arc is removed; otherwise and arc between them is added taking care of not introducing a (directed) cycle. During this generation process, the complexity (maximum number of parents and children for each node) of the network is controlled in order to avoid generating so dense DAGs.

The parameters we have used to generate the DAGs are: number of nodes, $n = \{10, 25, 50\}$; number of DAGs in the input set, $g = \{10, 20, 30\}$; maximum number of parents and children, maxP $= 3$ and maxCh $= 4$; maximum number of edges per DAG, maxEdges $= n \times 2.5$; and maximum number of perturbations, maxNoP $= n \times 0.75$. For each combination of n and g (9 configurations), we have generated 10 different sets of DAGs.

Table 1 shows some descriptive features about the complexity of the generated DAGs, averaged over the 10 different sets generated for each configuration. Besides the number of nodes and DAGs, we also show the averaged number of parents (AvPar) and edges (AvEdges). The configurations are numbered from 1 to 9 for their identification in the next sections.

Table 1. Descriptive statistics of the generated (set of) DAGs.

Conf.	#Nodes	#DAGs	AvPar	AvEdges
1	10	10	1,82	18,16
2	10	20	1,82	18,19
3	10	30	1,82	18,18
4	25	10	2,26	56,38
5	25	20	2,26	56,52
6	25	30	2,26	56,57
7	50	10	2,43	121,38
8	50	20	2,43	121,37
9	50	30	2,43	121,37

Results: As the fusion algorithms heavily depend on the ordering σ used to guide the search, we consider two different scenarios:

- Using an ancestral ordering for G_1 to guide the search. As G_1 is the DAG from which the remaining DAGs in each set are obtained, this ordering may be a good election, by far much better than a random one which usually leads to very dense DAGs. The results obtained are shown in Table 2(a).
- Using our proposal heuristic to identify a good ancestral ordering for a given input set of DAGs (Algorithm 3). The results obtained are shown in Table 2(b).

Furthermore, in Table 2(c) we show the results of applying the GES-based algorithm designed in Sect. 4 to the consensus DAG obtained in the second scenario above.

More specifically, in Table 2 we show the following statistics (averaged over the 10 generated sets):

- *AddEdges*: average number of additional edges in the transformed DAGs, that is, after making them coherent with respect to the used ancestral ordering
- *RatioPar*: average number of parents in the consensus DAG divided by the average number of parents in the input DAGs
- *MaxPar* and *MaxCh*: maximum number of parents and children, respectively, observed in any of the 10 runs for the configuration
- *TFusion*: CPU time (ms) spent by the fusion process. In Table 2(c) the TFusion is computed considering both the fusion and the GES steps.

Analysis: From the results obtained (Table 2) we can draw several conclusions:

- Not surprisingly, as the number of DAGs to be aggregated and the number of nodes in these DAGs increase, the resulting fusion graph is much more complex (dense). Actually, the complexity grows when the number of DAGs with a same number of nodes increases (configurations 1-2-3, 4-5-6 and 7-8-9), and when the number of DAGs is fixed and the number of nodes of the DAGs increases (configurations 1-4-7, 2-5-8 and 3-6-9). This pattern is independent of the fusion method used.
- The election of the order to guide the fusion strongly influences the complexity of the obtained consensus DAG. It becomes clear when comparing the statistics in Table 2(a) with the corresponding ones in Table 2(b) and (c). It worths pointing out the decisive influence of that election on the fusion time.
- The heuristic to get the initial order leads to good solutions (in terms of both complexity and fusion time). In fact, note that Table 2(c) shows the statistics corresponding to optimal solutions to the fusion problem, which are very close to their counterpart in Table 2(b). Hence, at sight of the TFusion values, it becomes clear the outstanding trade-off between accuracy and time execution of our proposal.

Table 2. Results for different fusion processes.

Conf.	AddEdges	RatioPar	MaxPar	MaxCh	TFusion
1	1,70	1,72	6,10	6,50	14,30
2	1,46	1,96	6,90	7,20	22,00
3	1,35	2,04	7,40	7,60	26,50
4	2,96	1,59	7,40	8,60	80,00
5	3,44	1,94	9,10	10,50	175,20
6	3,56	2,29	11,50	12,50	243,20
7	18,33	2,21	13,90	18,30	1551,40
8	21,68	2,95	18,00	22,50	5317,50
9	23,20	3,35	19,80	24,50	17773,00

(a) Using an ancestral ordering for G_1

Conf.	AddEdges	RatioPar	MaxPar	MaxCh	TFusion
1	0,67	1,64	5,50	5,90	12,00
2	0,86	1,93	7,10	6,70	15,80
3	0,81	2,00	7,40	7,10	23,70
4	0,41	1,28	5,70	5,80	78,20
5	1,08	1,65	8,20	8,70	142,40
6	1,17	1,91	10,20	10,20	209,90
7	0,98	1,24	7,60	6,30	302,20
8	3,10	1,71	13,50	10,80	615,40
9	4,27	2,10	16,60	13,90	933,40

(b) Using an heuristically generated ordering (Algorithm 3)

Conf.	AddEdges	RatioPar	MaxPar	MaxCh	TFusion
1	0,67	1,57	5,40	6,00	80,90
2	0,86	1,66	7,00	6,00	223,20
3	0,81	1,65	7,20	5,60	254,60
4	0,41	1,27	5,70	5,70	123,10
5	1,08	1,62	8,10	8,10	845,60
6	1,17	1,86	10,20	9,60	1497,30
7	0,98	1,23	7,60	6,60	1043,10
8	3,10	1,70	13,50	10,60	7095,70
9	4,27	2,09	16,60	14,00	93607,40

(c) Applying adapted GES

6   Conclusions

In this paper we deal with the structural fusion of BNs. This problem was previously studied from a theoretical point of view. We go further and provide two main contributions: (1) an algorithm to obtain a heuristic order to guide the fusion process; and (2) an algorithm that, after the fusion process, applies an adapted version of GES (BES), which uses d-separation to score arc deletion instead of doing that from the data.

The experiments show that using our heuristic to select the order produces high-quality consensus DAGs in comparison to those obtained when taking other (arbitrary but also informed) orders. It becomes clear when applying our adapted BES to the output provided by this heuristic, which allows us to check that our solutions are significantly close (regarding complexity) to be optimal.

We should point out that, even being rather positive results, the output consensus network could be dense due to the complexity of the BN fusion process, which may obstruct its use for posterior reasoning (inference). Therefore, as future a work, we aim to investigate the problem of relaxing the conditions required in the BN fusion process in order to get more *usable* networks [2].

References

1. Arias, J., Gámez, J.A., Puerta, J.M.: Structural learning of Bayesian networks via constrained hill climbing algorithms: adjusting trade-off between efficiency and accuracy. Int. J. Intell. Syst. **30**(3), 292–325 (2015)
2. Benjumeda, M., Larrañaga, P., Bielza, C.: Learning Bayesian networks with low inference complexity. Prog. Artif. Intell. **5**(1), 15–26 (2016)
3. Chickering, D.M.: Optimal structure identification with greedy search. J. Mach. Learn. Res. **3**, 507–554 (2003)
4. Crammer, K., Kearns, M., Wortman, J.: Learning from multiple sources. J. Mach. Learn. Res. **9**, 1757–1774 (2008)
5. Del Sagrado, J., Moral, S.: Qualitative combination of Bayesian networks. Int. J. Intell. Syst. **18**(2), 237–249 (2003)
6. Del Sagrado, J.: Learning Bayesian networks from distributed data: an approach based on the MDL principle. In: Proceedings of The 13th Conference of the Spanish Association for Artificial Intelligence (CAEPIA-2009), p. 9 (2009)
7. Julia Flores, M., Nicholson, A.E., Brunskill, A., Korb, K.B., Mascaro, S.: Incorporating expert knowledge when learning Bayesian network structure: a medical case study. Artif. Intell. Med. **53**(3), 181–204 (2011)
8. Matzkevich, I., Abramson, B.: The topological fusion of Bayes nets. In: Proceedings of the Eight Conference on Uncertainty in Artificial Intelligence (UAI-92), pp. 191–198 (1992)
9. Matzkevich, I., Abramson, B.: Deriving a minimal I-map of a belief network relative to a target ordering of its nodes. In: Proceedings of the Ninth Conference on Uncertainty in Artificial Intelligence (UAI-93), pp. 159–165 (1993)
10. Melançon, G., Philippe, F.: Generating connected acyclic digraphs uniformly at random. CoRR cs.DM/0403040 (2004)
11. Peña, J.M.: Finding consensus Bayesian network structures. J. Artif. Intell. Res. **42**(1), 661–687 (2011)

12. Pearl, J.: Probabilistic Reasoning in Intelligent Systems: Networks of Plausible Inference. Morgan Kaufmann Publishers Inc., San Francisco (1988)
13. Pennock, D.M., Wellman, M.P.: Graphical representations of consensus belief. CoRR abs/1301.6732 (2013). http://arxiv.org/abs/1301.6732
14. Zagorecki, A., Druzdzel, M.J.: Knowledge engineering for Bayesian networks: how common are noisy-max distributions in practice? IEEE Trans. Syst. Man Cybern. Syst. **43**(1), 186–195 (2013)

On Irreducible Min-Balanced Set Systems

Milan Studený$^{(\boxtimes)}$ [ID], Václav Kratochvíl [ID], and Jiří Vomlel [ID]

Czech Academy of Sciences, Institute of Information Theory and Automation,
Pod vodárenskou věží 4, 18200 Prague 8, Czech Republic
{studeny,velorex,vomlel}@utia.cas.cz

Abstract. Non-trivial minimal balanced systems (= collections) of sets are known to characterize through their induced linear inequalities the class of the so-called balanced (coalitional) games. In a recent paper a concept of an *irreducible* min-balanced (= minimal balanced) system of sets has been introduced and the irreducible systems have been shown to characterize through their induced inequalities the class of totally balanced games. In this paper we recall the relevant concepts and results, relate them to various contexts and offer a catalogue of permutational types of non-trivial min-balanced systems in which the irreducible ones are indicated. The present catalogue involves all types of such systems on sets with at most 5 elements; it has been obtained as a result of an alternative characterization of min-balanced systems.

Keywords: Balanced set system · Irreducible min-balanced system · Totally balanced games · Exact games

1 Introduction

A central notion of this note, namely the concept of a minimal balanced set system, shortened as a *min-balanced* (set) *system*, is basically a combinatorial concept. Nonetheless, the concept itself has been introduced in the context of cooperative game theory, where it plays quite an important role. Specifically, the well-known Shapley-Bondareva theorem [2,12] says that *balanced* systems of subsets of a non-empty finite basic set N covering N induce linear inequalities characterizing coalitional games over the set of players N with a non-empty *core*. Note that the concept of a core (polytope) is a substantial concept in cooperative game theory [12]. The least class of inequalities characterizing the non-emptiness of the core consists of those inequalities, which are induced by non-trivial *minimal balanced systems*, where the minimality is understood with respect to inclusion of set systems covering N.

For analogous reasons the inequalities induced by min-balanced systems are important in the context of the theory of imprecise probabilities [15]. In that context the basic set N can be interpreted as the sample space for probabilities,

Supported by GAČR project n. 19-04579S.

© Springer Nature Switzerland AG 2019

G. Kern-Isberner and Z. Ognjanović (Eds.): ECSQARU 2019, LNAI 11726, pp. 444–454, 2019.
https://doi.org/10.1007/978-3-030-29765-7_37

(normalized non-negative) games over N correspond to *lower probabilities* and the cores to *credal sets* of probabilities. Thus, (minimal) balanced set systems on N correspond to the inequalities characterizing lower probabilities *avoiding sure loss*, which are the lower probabilities with non-empty credal sets [15, Sect. 3.3.4]. We refer the reader to [7] for further details about the correspondence between game-theoretical concepts and those in imprecise probabilities.

Here is a formal definition: we say that a system $\mathcal{B} = \{S_1, \ldots, S_\ell\}$, $\ell \geq 1$, of non-empty subsets of N is a *balanced* system *on* a non-empty subset $M \subseteq N$ if there exist strictly positive real coefficients $\lambda_i > 0$, $i = 1, \ldots, \ell$, such that

$$\chi_M = \sum_{i=1}^{\ell} \lambda_i \cdot \chi_{S_i}, \text{ where } \chi_S \in \mathbb{R}^N \text{ denotes the incidence vector of } S \subseteq N. \quad (1)$$

In particular, the sets S_i must be subsets of M and $\lambda_i \leq 1$ for $i = 1, \ldots, \ell$. Thus, the concept of a balanced set system on M generalizes a classic combinatorial concept of a *partition* of M (consider $\lambda_i = 1$ for all $i = 1, \ldots, \ell$). For example, in case $N = \{a, b, c, d\}$ the partition $\mathcal{B} = \{\{a\}, \{b, c\}\}$ of $M = \{a, b, c\}$ is an example of a balanced system on M. On the other hand, one can also find links to fuzzy set theory with a little bit of imagination: balanced systems can perhaps be regarded as *fuzzy partitions* [1] of a crisp set M with fuzzy subsets having allowed only two grades, namely 0 and $\lambda_i \in (0, 1]$.

Note that balanced systems, called *balanced collections* in game-theoretical literature, do have some applications in combinatorics and topology. Shapley [13] generalized Sperner's celebrated topological lemma concerning triangulations of a simplex and balanced collections of sets play a crucial role in his generalization [4]. On the other hand, we would like to warn the reader that a combinatorial concept of a *balanced hypergraph* from [11, Sect. 83.1] has apparently nothing common with the concept of a balanced set system; these are different notions.

As explained below, in case of a min-balanced system $\mathcal{B}$ the coefficients in (1) are uniquely determined and the class of *min-balanced systems* on a given basic set is finite. Every permutational type of non-trivial min-balanced systems can be viewed as a combinatorial object: it represents a particular way in which a finite set M can be composed from its proper subsets. Thus, questions of natural interest are what are the permutational types of such systems, whether one can classify/categorize them or even whether an enumeration method generating all these types exists. Note in this context that Peleg [8] proposed an algorithm for inductive generating min-balanced systems on a given basic set. Nonetheless, as far as we know, no public available catalogue of their permutational types has been generated as an output of that algorithm.

1.1 Totally Balanced and Exact Games

Because of the above mentioned Shapley-Bondareva theorem, games with non-empty cores are named *balanced games*. There are two important subclasses of the class of balanced games over a player set N. One of them is the class of *totally balanced games*: these are such games m over N that, for every non-empty subset

$M \subseteq N$, the subgame of m for M is a balanced game over M. An even smaller class is the class of *exact games*: these are such games that each bound defining the core polyhedron is tight (a precise definition can be found below). Since these classes of games play an important role in cooperative game theory [9,10], some effort has been exerted to characterize them in terms of linear inequalities.

Special set systems play an important role in this context, too. Csóka et al. [3] characterized exact games over N by means of an infinite set of linear inequalities which could be associated with the so-called *exactly balanced* set systems on N. Lohman et al. [6] then refined that result and showed that the exact games over N can be characterized by means of a finite set of linear inequalities. Specifically, these are inequalities assigned to *min-balanced* set systems on non-empty subsets $M \subseteq N$ and to the so-called minimal *negatively balanced* set systems on N. Nonetheless, the reader was warned in [6] that this set of inequalities is not the the least possible set of inequalities characterizing the exact games.

In a recent paper [5] the least possible set of inequalities (up to a positive multiple) that characterizes the totally balanced games over N was found. These inequalities are induced by special *irreducible min-balanced* set systems on non-empty subsets $M \subseteq N$ (a formal definition is placed below). Another interesting observation from [5] is as follows: if $\mathcal{B}$ is a non-trivial min-balanced system on $M \subseteq N$ then its *complementary system* relative to M, that is,

$$\mathcal{B}^* := \{M \setminus S : S \in \mathcal{B}\}$$

is also a non-trivial min-balanced system on M. In particular, the non-trivial min-balanced systems on a fixed non-empty set $M \subseteq N$ come in pairs of mutually complementary systems. Moreover, the inequality induced by $\mathcal{B}^*$ is a *conjugate* inequality *with respect to* M to the one induced by $\mathcal{B}$ (also to be defined below).

Finally, a *conjecture* about the least possible set of inequalities characterizing the exact games was formulated in [5]. It says that a game over N is exact if and only if it satisfies the inequalities induced by non-trivial *irreducible min-balanced* systems on non-empty strict subsets $M \subset N$ and their conjugate inequalities with respect to N. The conjecture is known to be true in case $|N| \leq 5$.

Therefore, the question of classifying permutational types of non-trivial irreducible min-balanced systems over a fixed basic set is of great importance for the study of totally balanced and exact games. This is a topic of this note.

1.2   Structure of the Rest of the Paper

We provide a catalogue [14] of permutational types of non-trivial min-balanced systems on small sets in which the irreducible types are indicated. Its initial version describes all such types on sets with at most five elements. Nonetheless, we intend to upgrade it later into an interactive web platform and possibly extend it to involve all types of min-balanced systems on a six-element set. Now, we describe the structure of the rest of the paper.

In Sect. 2 we recall basic concepts and facts. In particular, we describe the way linear inequalities are induced by min-balanced systems and introduce the concept of an irreducible min-balanced system. Section 3 deals with conjugate inequalities and complementary set systems. Section 4 then describes the way our catalogue [14] was obtained. Note that our computations were not based on Peleg's iterative algorithm [8] but on an alternative characterization of the min-balanced systems in terms of linear independence of certain vectors. In Sect. 5 our tools to classify the permutational types are discussed. In last Sect. 6 we mention possible future research directions and open tasks.

2   Basic Concepts and Facts

Let N be a non-empty finite *basic set*. The symbol $\mathcal{P}(N)$ will denote its power set, that is, the collection $\{S : S \subseteq N\}$ of all its subsets. The symbol $\mathbb{R}^N$ will be used to denote the Euclidean space of real vectors $[x_i]_{i \in N}$ whose components are indexed by elements of N. Given $S \subseteq N$, the symbol χ_S will denote the *incidence vector* of S in $\mathbb{R}^N$, that is, its zero-one identifier in $\mathbb{R}^N$ defined by

$$(\chi_S)_i := \begin{cases} 1 & \text{for } i \in S, \\ 0 & \text{for } i \in N \setminus S, \end{cases} \quad \text{whenever } i \in N.$$

2.1   Game-Theoretical Notions

In this context, elements of the basic set N correspond to players and subsets of N to coalitions. A *(transferable-utility coalitional) game over* N is modeled by a real function $m \colon \mathcal{P}(N) \to \mathbb{R}$ such that $m(\emptyset) = 0$. If $\emptyset \neq M \subseteq N$ then the restriction of m to $\mathcal{P}(M)$ is called a *subgame* of m *for* M.

The *core* $C(m)$ of a game m over N is a polyhedron in $\mathbb{R}^N$ defined by

$$C(m) := \big\{\, [x_i]_{i \in N} \in \mathbb{R}^N : \sum_{i \in N} x_i = m(N) \ \& \ \sum_{i \in S} x_i \geq m(S) \text{ for any } S \subseteq N \big\}.$$

We say that a game m over N is *balanced* if $C(m) \neq \emptyset$. It is called *totally balanced* if every its subgame is balanced. Finally, a game over N is *exact* if, for each coalition $S \subseteq N$, a vector $[x_i]_{i \in N} \in C(m)$ exists such that $\sum_{i \in S} x_i = m(S)$. This basically means that every inequality defining the core of m is tight.

A well-known fact is that every exact game is totally balanced (see Remark 1.19 in [10, Sect. V.1]); by definition, every totally balanced game is balanced.

2.2   Min-Balanced Set Systems

Any subset $\mathcal{B}$ of $\mathcal{P}(N)$ is called a *set system*; the union of sets in $\mathcal{B}$ will be denoted by $\bigcup \mathcal{B}$. A set system having at most one set is considered to be trivial; thus, set systems $\mathcal{B} \subseteq \mathcal{P}(N)$ with $|\mathcal{B}| \geq 2$ will be named *non-trivial*.

We say that $\mathcal{B}$ *composes to* a non-empty set $M \subseteq N$ if $M = \bigcup \mathcal{B}$ and the vector χ_M belongs to the conic hull of $\{\chi_S \in \mathbb{R}^N : S \in \mathcal{B}\}$, that is, there exist *non-negative* coefficients $\lambda_S \geq 0$, $S \in \mathcal{B}$, such that $\chi_M = \sum_{S \in \mathcal{B}} \lambda_S \cdot \chi_S$.

Given $\mathcal{B} \subseteq \mathcal{P}(N)$ and $\emptyset \neq M \subseteq N$ we say $\mathcal{B}$ is *min-balanced on M* if it is a minimal set system in $\mathcal{P}(N)$ which composes to M. That means, $\mathcal{B}$ composes to M and, moreover, there is no $\mathcal{C} \subset \mathcal{B}$ such that $\mathcal{C}$ composes to M. The following basic observation was done in [5, Lemma 2.1].

Lemma 1. *A non-empty set system $\mathcal{B} \subseteq \mathcal{P}(N)$ is min-balanced on a non-empty set $M \subseteq N$ iff the following two conditions hold:*

(i) *there exist strictly positive $\lambda_S > 0$, $S \in \mathcal{B}$, such that $\chi_M = \sum_{S \in \mathcal{B}} \lambda_S \cdot \chi_S$,*
(ii) *the incidence vectors $\{\chi_S \in \mathbb{R}^N : S \in \mathcal{B}\}$ are linearly independent.*

The condition (i), mentioned already with (1), means that $\mathcal{B}$ is *balanced*, which is usual terminology in game-theoretical literature. The condition (ii), equivalent to minimality, then implies the uniqueness of the so-called *balancing coefficients* λ_S in (i). One can observe using Lemma 1 that a non-empty $\mathcal{B} \subseteq \mathcal{P}(N)$ is min-balanced on M iff it is a minimal set system satisfying (i), which is a standard definition of a *minimal balanced collection* in game-theoretical literature.

Note that it follows from [5, Lemma 2.2] that any *non-trivial* min-balanced system $\mathcal{B}$ on $M \subseteq N$ consists of least two proper subsets of M. Moreover, the intersection of all sets in $\mathcal{B}$ must be empty and one has at most $|M|$ sets in $\mathcal{B}$.

As mentioned earlier, every non-trivial min-balanced system on $M \subseteq N$ induces a unique non-trivial inequality (up to a positive multiple) for games m over N. More specifically, we know by Lemma 1 that *unique* balancing coefficients $\lambda_S > 0$, $S \in \mathcal{B}$, exist such that $\chi_M = \sum_{S \in \mathcal{B}} \lambda_S \cdot \chi_S$. The *induced inequality* for games m over N has then the form

$$m(M) \geq \sum_{S \in \mathcal{B}} \lambda_S \cdot m(S). \tag{2}$$

One can show that the balancing coefficients λ_S must be rational [5, Sect. 3.3], which allows one to multiply (2) by a positive factor so that the (balancing) coefficients become integers with no common prime divisor. Moreover, it is convenient to introduce a conventional coefficient with the empty set which plays no role in (2) because $m(\emptyset) = 0$ for any game m. The convention is such that one gets, after a re-arrangement, a unique *standardized form* of the inequality

$$\alpha(M) \cdot m(M) + \sum_{S \in \mathcal{B}} \alpha(S) \cdot m(S) + \alpha(\emptyset) \cdot m(\emptyset) \geq 0, \tag{3}$$

where $\alpha(S)$, $S \in \mathcal{B}$, are negative integers with no common prime divisor and $\alpha(M), \alpha(\emptyset)$ are positive integers determined by the standardization conditions:

$$\sum_{S \subseteq N} \alpha(S) = 0 \quad \text{and} \quad \forall i \in N \sum_{S \subseteq N: i \in S} \alpha(S) = 0. \tag{4}$$

The point of this particular convention will be revealed in Sect. 3.

Example 1. Consider $N = \{a, b, c, d\}$ and a set system $\mathcal{B} = \{a, bc, bd, cd\}$, where abbreviations like ab stand for sets like $\{a, b\}$. One has

$$\chi_N = 1 \cdot \chi_a + \frac{1}{2} \cdot \chi_{bc} + \frac{1}{2} \cdot \chi_{bd} + \frac{1}{2} \cdot \chi_{cd},$$

which allows one to observe using Lemma 1 that $\mathcal{B}$ is min-balanced on N. The respective inequality (2) is multiplied by factor 2, which gives $\alpha(N) = 2$, $\alpha(a) = -2$ and $\alpha(S) = -1$ for remaining $S \in \mathcal{B}$. The convention in the first formula of (4) gives $\alpha(\emptyset) = +3$, which leads to the standardized inequality (3)

$$2 \cdot m(abcd) - 2 \cdot m(a) - m(bc) - m(bd) - m(cd) + 3 \cdot m(\emptyset) \geq 0 \qquad (5)$$

for games m over N.

Shapley-Bondareva theorem can be re-formulated as follows [5, Lemma 3.5]:

Proposition 1. *If $|N| \geq 2$ then the least possible set of standardized inequalities characterizing the balanced games m over N is the set of inequalities (3) induced by non-trivial min-balanced systems $\mathcal{B}$ on N.*

2.3   Irreducible Min-Balanced Systems

Let $\mathcal{B}$ be a min-balanced system on $\emptyset \neq M \subseteq N$. We say that $\mathcal{B}$ is *reducible* if there exist a set $\emptyset \neq A \subset M$ such that $\mathcal{B}_A := \{S \in \mathcal{B} : S \subset A\}$ composes to A. Note that one can assume without loss of generality that both $|A| \geq 2$ and $A \notin \mathcal{B}$ because otherwise $\mathcal{B}$ cannot be min-balanced. A min-balanced system $\mathcal{B} \subseteq \mathcal{P}(N)$ that is not reducible is called *irreducible*.

The meaning of the reducibility condition is that the induced inequality (for games over N) is a conic combination of inequalities induced by other min-balanced systems, in particular by the irreducible ones.

Lemma 2. *Given a min-balanced system $\mathcal{B}$ on $\emptyset \neq M \subseteq N$, the reducibility condition with a set $\emptyset \neq A \subset M$ is equivalent to the existence of min-balanced systems $\mathcal{C}$ on A and $\mathcal{D}$ on M such that $A \in \mathcal{D}$, $\mathcal{C} \setminus \mathcal{D} \neq \emptyset$ and $\mathcal{B} = \mathcal{C} \cup \mathcal{D} \setminus \{A\}$.*

Proof. The sufficiency of the condition is easy as $\mathcal{C} \subseteq \mathcal{B}_A$. For the necessity realize using Lemma 1 that $\{\chi_S : S \in \mathcal{B}_A\}$ are linearly independent. Hence, uniquely determined coefficients $\mu_S \geq 0$, $S \in \mathcal{B}_A$, exist such that $\chi_A = \sum_{S \in \mathcal{B}_A} \mu_S \cdot \chi_S$. Put $\mathcal{C} := \{S \in \mathcal{B}_A : \mu_S > 0\}$ and $\mu_T := 0$ for $T \in \mathcal{B} \setminus \mathcal{B}_A$. Again by Lemma 1, unique coefficients $\lambda_S > 0$, $S \in \mathcal{B}$, exist such that $\chi_M = \sum_{S \in \mathcal{B}} \lambda_S \cdot \chi_S$. Let us put $\varepsilon := \min_{C \in \mathcal{C}} \frac{\lambda_C}{\mu_C}$ and introduce $\kappa_A := \varepsilon$, $\kappa_S := \lambda_S - \varepsilon \cdot \mu_S$ for $S \in \mathcal{B}$. Then one has $\chi_M = \sum_{S \in \mathcal{B}} \lambda_S \cdot \chi_S + \varepsilon \cdot (\chi_A - \sum_{S \in \mathcal{B}} \mu_S \cdot \chi_S) = \sum_{S \in \mathcal{B} \cup \{A\}} \kappa_S \cdot \chi_S$ with $\kappa_S \geq 0$ for $S \in \mathcal{B} \cup \{A\}$. One can put $\mathcal{D} := \{S \in \mathcal{B} \cup \{A\} : \kappa_S > 0\}$ and verify the conditions from Lemma 2.

One can extend the arguments used in the above proof to show that the min-balanced systems $\mathcal{C}$ and $\mathcal{D}$ mentioned above are uniquely determined by the set A. Lemma 2 also allows one to observe that the reducibility condition for a min-balanced system $\mathcal{B}$ is equivalent to the original one from [5, Definition 4.1]. The next example illustrates the fact that the "decomposition" of $\mathcal{B}$ into systems $\mathcal{C}$ and $\mathcal{D}$ leads to conic combination of the induced inequalities.

Example 2. The set system $\mathcal{B} = \{a, bc, bd, cd\}$ from Example 1 is reducible. Put $A := \{b, c, d\}$; then $\mathcal{B}_A = \{bc, bd, cd\}$ composes to A because of

$$\chi_A \equiv \chi_{bcd} = \frac{1}{2} \cdot \chi_{bc} + \frac{1}{2} \cdot \chi_{bd} + \frac{1}{2} \cdot \chi_{cd} \,.$$

One gets $\mathcal{C} = \{bc, bd, cd\}$ and $\mathcal{D} = \{a, bcd\}$ in this particular case. The inequality (5) induced by $\mathcal{B}$ is then a conic combination of inequalities induced by $\mathcal{C}$ and $\mathcal{D}$:

$$1 \times \{ 2 \cdot m(bcd) - m(bc) - m(bd) - m(cd) + m(\emptyset) \} \geq 0$$
$$2 \times \{ m(abcd) - m(a) - m(bcd) + m(\emptyset) \} \geq 0 \,,$$

where the (conic) coefficient for $\mathcal{C}$ is 1 and the coefficient for $\mathcal{D}$ is 2.

Thus, reducible systems are superfluous for describing totally balanced games. Nonetheless, the irreducible ones are substantial as shown in [5, Theorem 5.1]:

Proposition 2. *Assume $|N| \geq 2$. The least set of standardized inequalities that characterizes totally balanced games m over N is the set of inequalities (3) induced by non-trivial irreducible min-balanced systems $\mathcal{B}$ on non-empty subsets $M \subseteq N$.*

3   Conjugate Inequalities and Complementary Systems

Every (standardized) inequality (3) for games m (over N) can be viewed as

$$\sum_{S \subseteq N} \alpha(S) \cdot m(S) \geq 0 \,, \quad \text{where the coefficients outside } \mathcal{B} \cup \{\emptyset, M\} \text{ are zeros,}$$

and assigned its *conjugate inequality* for games m (over N) *with respect to* N:

$$\sum_{T \subseteq N} \alpha^*(T) \cdot m(T) \geq 0 \,, \quad \text{where } \alpha^*(T) := \alpha(N \setminus T) \quad \text{for any } T \subseteq N.$$

The importance of this concept for balanced and exact games is apparent from [5, Lemma 3.4], which can be re-phrased as follows.

Proposition 3. *The least set $\mathcal{S}$ of standardized inequalities that characterizes balanced games over N is closed under conjugacy: whenever (3) is in $\mathcal{S}$ then its conjugate inequality is in $\mathcal{S}$. The same holds for the least set of standardized inequalities characterizing exact games over N.*

We know from Proposition 1 that the inequalities in the set $\mathcal{S}$ characterizing balanced games over N correspond to non-trivial min-balanced systems on N. Thus, the conjugate inequality to (3) for a system $\mathcal{B}$ on N also corresponds to a non-trivial min-balanced system on N, which is nothing but the complementary system to $\mathcal{B}$ relative to N. The following is a re-formulation of [5, Corollary 3.1].

Proposition 4. *Let $\mathcal{B}$ be a non-trivial min-balanced system on N inducing (3). Then its complementary system $\mathcal{B}^* := \{N \setminus S : S \in \mathcal{B}\}$ relative to N is also a non-trivial min-balanced system on N, inducing the conjugate inequality to (3).*

The reader can now comprehend the convention from Sect. 2.2, where the coefficient $\alpha(\emptyset)$ with the empty set was introduced. It is just the coefficient $\alpha^*(N)$ in the conjugate inequality, that is, the coefficient with N for the complementary system $\mathcal{B}^*$. This phenomenon is illustrated by the next example.

Example 3. Consider again the system $\mathcal{B} = \{a, bc, bd, cd\}$ on $N = \{a, b, c, d\}$ from Example 1. Its complementary system relative to N is $\mathcal{B}^* = \{ab, ac, ad, bcd\}$. The standardized inequality induced by $\mathcal{B}^*$ is then

$$3 \cdot m(abcd) - m(ab) - m(ac) - m(ad) - 2 \cdot m(bcd) + 2 \cdot m(\emptyset) \geq 0 \,,$$

which is, as the reader can check, the conjugate inequality to the inequality (5) induced by $\mathcal{B}$ (see Example 1). Note that $\mathcal{B}^*$ is an irreducible min-balanced system unlike its complementary system $\mathcal{B}^{**} = \mathcal{B}$ (see Example 2).

4 Catalogue: Procedure

Here we describe the method our catalogue [14] has been obtained. As explained in Sect. 2.2, a min-balanced system on a basic set N contains at most $n := |N|$ sets. Thus, a non-trivial such system $\mathcal{B} \subseteq \mathcal{P}(N)$ can be represented by a zero-one $n \times r$-matrix, where $2 \leq r \leq n$, namely by a matrix whose (distinct) columns are incidence vectors of (all) sets $S \in \mathcal{B}$. To get one-to-one correspondence between such matrices and non-trivial set systems one can choose and fix an order of elements in N and also choose and fix an order of elements in $\mathcal{P}(N)$.

Moreover, by Lemma 1(ii), the columns of a matrix which represents a min-balanced system must be linearly independent, which means that the rank of the matrix is r, the number of its columns. This is something one can easily test using linear algebra computational tools. Thus, the first step of our procedure was computing a list of representatives of permutational types of non-trivial set systems $\mathcal{B} \subseteq \mathcal{P}(N)$ such that $\{\chi_S : S \in \mathcal{B}\}$ are linearly independent in $\mathbb{R}^N$. In our case $n = 5$ we have obtained 1649 such (non-trivial) type representatives.

The second necessary condition for a min-balanced system $\mathcal{B} \subseteq \mathcal{P}(N)$ is that $\mathcal{B}$ composes to $M := \bigcup \mathcal{B}$, that is, non-negative coefficients $\lambda_S \geq 0$, $S \in \mathcal{B}$, exist such that $\chi_M = \sum_{B \in \mathcal{B}} \lambda_S \cdot \chi_S$. If $\mathbb{A}$ is the $n \times r$-matrix representing $\mathcal{B}$ then this condition is equivalent to the existence of a non-negative column vector $\lambda \in \mathbb{R}^r$ such that $\mathbb{A} \cdot \lambda = \chi_M$, which is a standard feasibility task in linear programming. Again, this can be tested computationally by means linear programming software packages. In our case $n = 5$ we found that 934 representatives of those 1649 ones mentioned above describe systems $\mathcal{B}$ composing to $\bigcup \mathcal{B}$.

In case of an $n \times r$-matrix $\mathbb{A}$ of the rank r the solution λ of a linear system $\mathbb{A} \cdot \lambda = \chi_M$ is uniquely determined. Hence, the criterion to decide whether the corresponding system $\mathcal{B}$ is min-balanced is immediate: all the components of the unique solution λ must be strictly positive. This gave massive reduction: only 57 representatives of above mentioned 934 ones describe min-balanced systems.

Testing irreducibility of min-balanced systems using their matrix computer representations comes from the definition in Sect. 2.3: testing whether a set subsystem composes to its union is a linear programming feasibility task. In our case $n = 5$ we recognized 23 irreducible types of 57 of min-balanced ones. The particular numbers are given in Table 1, the details can be found in [14].

Table 1. The numbers of non-trivial min-balanced systems in case $|N| \leq 5$.

Variables	Systems	Permutational types	Irreducible systems	Irreducible types
$|N| = 2$	1	1	1	1
$|N| = 3$	5	3	4	2
$|N| = 4$	41	9	18	5
$|N| = 5$	1291	44	288	15
$|N| \leq 5$	1338	57	311	23

5   Towards Classification of Permutation Types

Here we mention some characteristics which can be used to classify permutational types of non-trivial (irreducible) min-balanced set systems.

5.1   Numerical Characteristics of Permutational Types

Let $\mathcal{B}$ be a non-trivial min-balanced set system on N with $n := |N| \geq 2$. Introduce (set) *cardinality characteristics* of $\mathcal{B}$ by $c_k := |\{S \in \mathcal{B} : |S| = k\}|$ for $k = 1, \ldots, n - 1$. Note that $\sum_{k=1}^{n-1} c_k = |\mathcal{B}|$ is the number of sets in $\mathcal{B}$. The *cardinality vector* $[c_1, \ldots, c_{n-1}]$ can then serve as a characteristic of any permutational type of non-trivial min-balanced systems. Cardinality vectors cannot, however, distinguish between some different permutational types.

An alternative idea comes from *multiplicity characteristics* which are defined by $m_i := |\{S \in \mathcal{B} : i \in S\}|$ for elements $i \in N$. One can order the numbers m_i in an increasing way, say, and get a *multiplicity vector* of the length $|N|$, which can serve as a characteristic of the permutational type of $\mathcal{B}$. The sum of its components $\sum_{i \in N} m_i$ can be viewed as a kind of *multiplicity index* for $\mathcal{B}$. Multiplicity vectors cannot, however, distinguish between different partitions.

5.2   Archetypes

Let $\mathcal{B}$ be a set system on a basic set N. It defines an equivalence relation on N: given $i, j \in N$, $i \sim j$ will mean that, for every $S \in \mathcal{B}$, one has $i \in S \Leftrightarrow j \in S$. For any $i \in N$ put $[i] := \{j \in N : i \sim j\}$ and denote by $\widetilde{N} := \{[i] : i \in N\}$ the *factor set* of $\sim$, that is, the set of equivalence classes of $\sim$. Analogously, any $S \in \mathcal{B}$ can be identified with a subset of $\widetilde{N}$, namely with $\widetilde{S} := \{[i] : i \in S\}$; note that the inverse relation is $S = \bigcup \{[i] : [i] \in \widetilde{S}\}$. The system $\mathcal{B}$ itself can be identified with $\widetilde{\mathcal{B}} := \{\widetilde{S} : S \in \mathcal{B}\}$, which is a set system on $\widetilde{N}$.

Given a set system $\mathcal{B}$ on $N \neq \emptyset$ and a set system $\mathcal{C}$ on $L \neq \emptyset$ we will say that they belong to the *same archetype* if there exists a one-to-one mapping $\psi : \widetilde{N} \rightarrow \widetilde{L}$ from $\widetilde{N}$ onto $\widetilde{L}$ which maps $\widetilde{\mathcal{B}}$ to $\widetilde{\mathcal{C}}$, that is, $\widetilde{\mathcal{C}} = \{ \psi(\widetilde{S}) : \widetilde{S} \in \widetilde{\mathcal{B}} \}$.

It is easy to see that this is an equivalence relation on set systems coarsening their permutational equivalence. Trivial set systems form one equivalence class of this archetypal equivalence; however, such systems are not of our interest.

Lemma 3. *Let $\mathcal{B}$ be a set system on N and $\mathcal{C}$ a set system on L which belong to the same archetype. Then $\mathcal{B}$ is min-balanced iff $\mathcal{C}$ is min-balanced. Moreover, $\mathcal{B}$ is a non-trivial irreducible min-balanced system iff $\mathcal{C}$ is so.*

Proof. It is enough to verify the claims for a set system $\mathcal{B}$ on N and the system $\widetilde{\mathcal{B}}$ on $\widetilde{N}$ in place of the system $\mathcal{C}$ on L. The claim about min-balanced systems follows easily from Lemma 1: realize that one has $\chi_N = \sum_{S \in \mathcal{B}} \lambda_S \cdot \chi_S$ if and only if $\chi_{\widetilde{N}} = \sum_{S \in \mathcal{B}} \lambda_S \cdot \chi_{\widetilde{S}}$ for arbitrary real coefficients λ_S and similar consideration works with zero vectors in place of χ_N and $\chi_{\widetilde{N}}$.

As concerns the irreducible systems it is more convenient to show that $\mathcal{B}$ is reducible iff $\widetilde{\mathcal{B}}$ is reducible. As mentioned in Sect. 2.3 the set A in the definition of reducibility of $\mathcal{B}$ has the form $A = \bigcup \mathcal{B}_A$ with $\mathcal{B}_A = \{ S \in \mathcal{B} ; S \cap A \}$. Such a set A is composed of equivalence classes of $\sim$ and can be identified with a subset of $\widetilde{N}$: one has $\widetilde{A} := \{ [i] : i \in A \}$ and $A = \bigcup \{ [i] : [i] \in \widetilde{A} \}$. Hence, one has $\widetilde{A} \subset \widetilde{N}$ and $\chi_A = \sum_{S \in \mathcal{B}_A} \lambda_S \cdot \chi_S$ iff $\chi_{\widetilde{A}} = \sum_{S \in \mathcal{B}_A} \lambda_S \cdot \chi_{\widetilde{S}}$ for arbitrary real coefficients λ_S. This implies the claim about reducible systems. The claim about trivial/non-trivial systems is evident.

Lemma 3 implies that permutational types can be classified by their archetypes. Any archetype can be canonically represented by an *archetypal* set system, which is such a system $\mathcal{B}$ on N that, for any $i, j \in N$, one has $i \sim j$ iff $i = j$.

Example 4. Consider an irreducible min-balanced system $\mathcal{B} = \{ab, acd, bcd\}$ on $N = \{a, b, c, d\}$. One has $c \sim d$ in this case and the system $\mathcal{B}$ belongs to the same archetype as $\mathcal{C} = \{ab, ac, bc\}$ on $M = \{a, b, c\}$. Clearly, $\mathcal{C}$ is an archetypal system.

6   Conclusions

We would like to find out whether our method of generating (all) types of min-balanced systems based on Lemma 1 can be modified and can lead to some iterative algorithm, which would be able to produce catalogues for $|N| \geq 6$.

One of our open tasks is whether the numerical characteristics from Sect. 5.1 are able to distinguish between any distinct types of min-balanced systems. If this is so then an alternative method of generating types could possibly be designed.

This is also related to the question of finding lower and upper estimates for the numbers (of types) of min-balanced systems and irreducible min-balanced system in terms of $|N|$. The asymptotic behavior of these numbers with increasing $|N|$ would be of our interest, too.

References

1. Bodjanova, S., Kalina, M.: Coarsening of fuzzy partitions. In: The 13th IEEE International Symposium on Intelligent Systems and Informatics, September 17–19, 2015, Subotica, Serbia, pp. 127–132 (2015)
2. Bondareva, O.: Some applications of linear programming methods to the theory of cooperative games (in Russian). Problemy Kibern. **10**, 119–139 (1963)
3. Csóka, P., Herings, P.J.-J., Kóczy, L.Á.: Balancedness conditions for exact games. Math. Methods Oper. Res. **74**(1), 41–52 (2011)
4. Ichiishi, T.: On the Knaster-Kuratowski-Mazurkiewicz-Shapley theorem. J. Math. Anal. Appl. **81**, 297–299 (1981)
5. Kroupa, T., Studený, M.: Facets of the cone of totally balanced games. Math. Methods Oper. Res. (2019, to appear). https://link.springer.com/article/10.1007/s00186-019-00672-y
6. Lohmann, E., Borm, P., Herings, P.J.-J.: Minimal exact balancedness. Math. Soc. Sci. **64**, 127–135 (2012)
7. Miranda, E., Montes, I.: Games solutions, probability transformations and the core. In: JMLR Workshops and Conference Proceedings 62: ISIPTA 2017, pp. 217–228 (2017)
8. Peleg, B.: An inductive method for constructing minimal balanced collections of finite sets. Nav. Res. Logist. Q. **12**, 155–162 (1965)
9. Peleg, B., Sudhölter, P.: Introduction to the Theory of Cooperative Games. Theory and Decision Library, series C: Game Theory, Mathematical Programming and Operations Research. Springer, Heidelberg (2007)
10. Rosenmüller, J.: Game Theory: Stochastics, Information, Strategies and Cooperation. Kluwer, Boston (2000)
11. Schrijver, A.: Combinatorial Optimization: Polyhedra and Efficiency. Springer, Berlin (2003)
12. Shapley, L.S.: On balanced sets and cores. Nav. Res. Logist. Q. **14**, 453–460 (1967)
13. Shapley, L.S.: On balanced games without side payments. In: Hu, T.C., Robinson, S.M. (eds.) Mathematical Programming, pp. 261–290. Academic Press, New York (1973)
14. Studený, M., Kratochvíl, V., Vomlel, J.: Catalogue of min-balanced systems, June 2019. http://gogo.utia.cas.cz/min-balanced-catalogue/
15. Walley, P.: Statistical Reasoning with Imprecise Probabilities. Chapman and Hall, London (1991)

A Recursive Algorithm for Computing Inferences in Imprecise Markov Chains

Natan T'Joens[(✉)], Thomas Krak, Jasper De Bock, and Gert de Cooman

ELIS – FLip, Ghent University, Ghent, Belgium
{natan.tjoens,thomas.krak,jasper.debock,gert.decooman}@ugent.be

Abstract. We present an algorithm that can efficiently compute a broad class of inferences for discrete-time imprecise Markov chains, a generalised type of Markov chains that allows one to take into account partially specified probabilities and other types of model uncertainty. The class of inferences that we consider contains, as special cases, tight lower and upper bounds on expected hitting times, on hitting probabilities and on expectations of functions that are a sum or product of simpler ones. Our algorithm exploits the specific structure that is inherent in all these inferences; they admit a general recursive decomposition. This allows us to achieve a computational complexity that scales linearly in the number of time points on which the inference depends, instead of the exponential scaling that is typical for a naive approach.

Keywords: Imprecise Markov chains · Upper and lower expectations · Recursively decomposable inferences

1 Introduction

Markov chains are popular probabilistic models for describing the behaviour of dynamical systems under uncertainty. The crucial simplifying assumption in these models is that the probabilities describing the system's future behaviour are conditionally independent of its past behaviour, given that we know the current state of the system; this is the canonical *Markov property.*

It is this Markov assumption that makes the parametrisation of a Markov chain relatively straightforward—indeed, as we will discuss in Sect. 2, the uncertain dynamic behaviour is then completely characterised by a transition matrix T, whose elements $T(x_n, x_{n+1}) = P(X_{n+1} = x_{n+1} | X_n = x_n)$ describe the probabilities that the system will transition from any state x_n at time n, to any state x_{n+1} at time $n + 1$. Note that T itself is independent of the time n; this is the additional assumption of *time homogeneity* that is often imposed implicitly in this context. An important advantage of these assumptions is that the resulting matrix T can be used to solve various important inference problems, using one of the many available efficient algorithms.

© Springer Nature Switzerland AG 2019
G. Kern-Isberner and Z. Ognjanović (Eds.): ECSQARU 2019, LNAI 11726, pp. 455–465, 2019.
https://doi.org/10.1007/978-3-030-29765-7_38

In many cases however, the numerical value of the transition matrix T may not be known exactly; that is, there may be additional (higher-order) uncertainty about the model itself. Moreover, it can be argued that simplifying assumptions like the Markov property and time homogeneity are often unrealistic in practice. It is of interest, then, to compute inferences in a manner that is *robust*; both to violations of such simplifying assumptions, and to variations in the numerical values of the transition probabilities.

The theory of *imprecise probabilities* allows us to describe such additional uncertainties by using, essentially, *sets* of traditional ("precise") models. In particular, such a set is comprised of all the models that we deem "plausible"; for instance, we may include all Markov chains whose characterising transition matrix T is included in some given *set* $\mathscr{T}$ of transition matrices. In this way we can also include non-homogeneous Markov chains, by simply requiring that their (now time-dependent) transition matrices remain in $\mathscr{T}$. Moreover, we can even include non-Markovian models in such a set. This leads to the notion of an *imprecise Markov chain*. The robust inferences that we are after, are then the tightest possible lower and upper bounds on the inferences computed for each of the included precise models.

In this work, we present an efficient algorithm for solving a large class of these inferences within imprecise Markov chains. Broadly speaking, this class consists of inferences that depend on the uncertain state of the system at a finite number of time instances, and which can be decomposed in a particular recursive form. As we will discuss, it contains as special cases the *(joint) probabilities* of sequences of states; the *hitting probabilities* and *expected hitting times* of subsets of the possible states; and *time averages* of functions of the state of the system.

Interestingly, existing algorithms for some of these inferences turn out to correspond to special cases of our algorithm, giving our algorithm a unifying character. Time averages, for example, were already considered in [9], and some of the results in [8]—a theoretical study of lower and upper expected hitting times and probabilities—can be interpreted as a special cases of the algorithm presented here. Readers that are familiar with recursive algorithms for credal networks under epistemic irrelevance [1,2,4] might also recognise some of what we do; in fact, many of the ideas behind our algorithm have previously been discussed in this more general context [2, Chapter 7].

In order to adhere to the page limit, all proofs have been relegated to the appendix of an online extended version [13].

2   Preliminaries

We denote the natural numbers, without 0, by $\mathbb{N}$, and let $\mathbb{N}_0 := \mathbb{N} \cup \{0\}$. The set of positive real numbers is denoted by $\mathbb{R}_{>0}$ and the set of non-negative real numbers by $\mathbb{R}_{\geq 0}$. Throughout, we let $\mathbb{I}_A$ denote the indicator of any subset $A \subseteq \mathscr{Y}$ of a set $\mathscr{Y}$; so, for any $y \in \mathscr{Y}$, $\mathbb{I}_A(y) := 1$ if $y \in A$ and $\mathbb{I}_A(y) := 0$ otherwise.

Before we can introduce the notion of an imprecise Markov chain, we first need to discuss general (non-Markovian) stochastic processes. These are arguably most commonly formalised using a measure-theoretic approach; however, the majority of our results do not require this level of generality, and so we will keep the ensuing introduction largely intuitive and informal.

Let us start by considering the *realisations* of a stochastic process. At each point in time $n \in \mathbb{N}$, such a process is in a certain *state* x_n, which is an element of a *finite* non-empty state space $\mathscr{X}$. A realisation of the process is called a *path*, and is an infinite sequence $\omega = x_1 x_2 x_3 \cdots$ where, at each discrete time point $n \in \mathbb{N}$, $\omega_n := x_n \in \mathscr{X}$ is the state obtained by the process at time n, on the path ω. So, we can interpret any path as a map $\omega : \mathbb{N} \to \mathscr{X}$, allowing us to collect all paths in the set $\Omega := \mathscr{X}^{\mathbb{N}}$. Moreover, for any $\omega \in \Omega$ and any $m, n \in \mathbb{N}$ with $m \leq n$, we use the notation $\omega_{m:n}$ to denote the finite sequence of states $\omega_m \cdots \omega_n \in \mathscr{X}^{n-m+1}$.

A stochastic process is now an infinite sequence $X_1 X_2 X_3 \cdots$ of uncertain states where, for all $n \in \mathbb{N}$, the uncertain state at time n is a function of the form $X_n : \Omega \to \mathscr{X} : \omega \mapsto \omega_n$. Similarly, we can consider finite sequences of such states where, for all $m, n \in \mathbb{N}$ with $m \leq n$, $X_{m:n} : \Omega \to \mathscr{X}^{n-m+1} : \omega \mapsto \omega_{m:n}$. These states are uncertain in the sense that we do not know which realisation $\omega \in \Omega$ will obtain in reality; rather, we assume that we have assessments of the probabilities $\mathrm{P}(X_{n+1} = x_{n+1} | X_{1:n} = x_{1:n})$, for any $n \in \mathbb{N}$ and any $x_{1:n} \in \mathscr{X}^n$. Probabilities of this form tell us something about which state the process might be in at time $n + 1$, given that we know that at time points 1 through n, it followed the sequence $x_{1:n}$. Moreover, we can consider probabilities of the form $\mathrm{P}(X_1 = x_1)$ for any $x_1 \in \mathscr{X}$; this tells us something about the state that the process might start in. It is well known that, taken together, these probabilities suffice to construct a global probability model for the *entire* process $X_1 X_2 X_3 \cdots$, despite each assessment only being about a finite subsequence of the states; see e.g. the discussion surrounding [7, Theorem 5.16] for further details on formalising this in a proper measure-theoretic setting. We simply use P to denote this global model.

Once we have such a global model P, we can talk about *inferences* in which we are interested. In general, these are typically encoded by functions $f : \Omega \to \mathbb{R}$ of the unknown realisation $\omega \in \Omega$, and we collect all functions of this form in the set $\mathscr{L}(\Omega)$. To compute such an inference consists in evaluating the (conditional) expected value $\mathrm{E}_{\mathrm{P}}(f | C)$ of f with respect to the model P, where $C \subseteq \Omega$ is an event of the form $X_{m:n} = x_{m:n}$ with $m, n \in \mathbb{N}$ such that $m \leq n$. In particular, if P is a global model in the measure-theoretic sense, then under some regularity conditions like the measurability of f, we would be interested in computing the quantity $\mathrm{E}_{\mathrm{P}}(f | C) := \int_\Omega f(\omega) \, \mathrm{dP}(\omega | C)$. For notational convenience, we will also use $X_{1:0} := \Omega$ as a trivial conditioning event, allowing us to regard unconditional expectations as a special case of conditional ones.

A special type of inferences that will play an important role in the remainder of this work are those for which the function f only depends on a *finite* subsequence of the path ω, thereby vastly simplifying the definition of its expectation.

In particular, if an inference only depends on the states at time points m through n, say, then it can always be represented by a function $f : \mathscr{X}^{n-m+1} \to \mathbb{R}$ evaluated in the uncertain states $X_{m:n}$; specifically, the inference is represented by the composition $f \circ X_{m:n}$, which we will denote by $f(X_{m:n})$. In the sequel, we will call a composite function of this form *finitary*. Moreover, for any $n \in \mathbb{N}$, we denote by $\mathscr{L}(\mathscr{X}^n)$ the set of all functions of the form $f : \mathscr{X}^n \to \mathbb{R}$, and we write $\mathscr{L}_{\mathrm{fin}}(\Omega) \subset \mathscr{L}(\Omega)$ for the set of all finitary functions. For a finitary function $f(X_{1:n})$, the computation of its expected value reduces to evaluating the finite sum

$$\mathrm{E}_{\mathrm{P}}(f(X_{1:n}) | C) = \sum_{x_{1:n} \in \mathscr{X}^n} f(x_{1:n}) \mathrm{P}(X_{1:n} = x_{1:n} | C).$$

Let us now move on from the discussion about general uncertain processes, to the special case of Markov chains. An uncertain process P is said to satisfy the *Markov property* if, for all $n \in \mathbb{N}$ and all $x_{1:n+1} \in \mathscr{X}^{n+1}$, the aforementioned probability assessments simplify in the sense that

$$\mathrm{P}(X_{n+1} = x_{n+1} | X_{1:n} = x_{1:n}) = \mathrm{P}(X_{n+1} = x_{n+1} | X_n = x_n).$$

A process that satisfies this Markov property is called a *Markov chain*. Thus, for a Markov chain, the probability that it will visit state x_{n+1} at time $n+1$ is independent of the states $X_{1:n-1}$, given that we know the state X_n at time n. If the process is moreover *homogeneous*, meaning that $\mathrm{P}(X_{n+1} = y | X_n = x) = \mathrm{P}(X_2 = y | X_1 = x)$ for all $x, y \in \mathscr{X}$ and all $n \in \mathbb{N}$, then the parameterisation of the process becomes exceedingly simple. Indeed, up to the initial distribution $\mathrm{P}(X_1)$—a probability mass function on $\mathscr{X}$—the process' behaviour is then fully characterised by a single $|\mathscr{X}| \times |\mathscr{X}|$ matrix T that is called the transition matrix. It is row-stochastic (meaning that, for all $x \in \mathscr{X}$, the x-th row $T(x, \cdot)$ of T is a probability mass function on $\mathscr{X}$) and its entries satisfy $T(x, y) = \mathrm{P}(X_{n+1} = y | X_n = x)$ for all $x, y \in \mathscr{X}$ and $n \in \mathbb{N}$. The usefulness of this representation comes from the fact that we can interpret T as a linear operator on the vector space $\mathscr{L}(\mathscr{X}) \simeq \mathbb{R}^{|\mathscr{X}|}$, due to the assumption that $\mathscr{X}$ is finite. For $f \in \mathscr{L}(\mathscr{X})$, this allows us to write the conditional expectation of $f(X_{n+1})$ given X_n as a matrix-vector product: for any $x \in \mathscr{X}$, $\mathrm{E}_{\mathrm{P}}(f(X_{n+1}) | X_n = x)$ equals

$$\sum_{y \in \mathscr{X}} f(y) \mathrm{P}(X_{n+1} = y | X_n = x) = \sum_{y \in \mathscr{X}} f(y) T(x, y) = [Tf](x).$$

3   Imprecise Markov Chains

Let us now move on to the discussion about *imprecise* Markov chains. Here, we additionally include uncertainty about the model specification, such as uncertainty about the numerical values of the probabilities $\mathrm{P}(X_{n+1} | X_{1:n})$, and about the validity of structural assessments like the Markov property.

We will start this discussion by regarding the parameterisation of such an imprecise Markov chain. We first consider the (imprecise) *initial model* $\mathcal{M}$;

this is simply a non-empty set of probability mass functions on $\mathscr{X}$ that we will interpret as containing those probabilities that we deem to plausibly describe the process starting in a certain state. Next, instead of being described by a single transition matrix T, an imprecise Markov chain's dynamic behaviour is characterised by an entire *set* $\mathscr{T}$ of transition matrices. So, each element $T \in \mathscr{T}$ is an $|\mathscr{X}| \times |\mathscr{X}|$ matrix that is row-stochastic. In the sequel, we will take $\mathscr{T}$ to be fixed, and assume that it is non-empty and that it has separately specified rows. This last property is instrumental in ensuring that computations can be performed efficiently, and is therefore often adopted in the literature; see e.g. [6] for further discussion. For our present purposes, it suffices to know that it means that $\mathscr{T}$ can be completely characterised by providing, for any $x \in \mathscr{X}$, a non-empty set $\mathscr{T}_x$ of probability mass functions on $\mathscr{X}$. In particular, it means that $\mathscr{T}$ is the set of all row-stochastic $|\mathscr{X}| \times |\mathscr{X}|$ matrices T such that, for all $x \in \mathscr{X}$, the x-row $T(x, \cdot)$ is an element of $\mathscr{T}_x$.

Given the sets $\mathcal{M}$ and $\mathscr{T}$, the corresponding imprecise Markov chain is defined as the largest *set* $\mathscr{P}_{\mathcal{M}, \mathscr{T}}$ of stochastic processes that are in a specific sense *compatible* with both $\mathcal{M}$ and $\mathscr{T}$. In particular, a model P is said to be compatible with $\mathcal{M}$ if $P(X_1) \in \mathcal{M}$, and it is said to be compatible with $\mathscr{T}$ if, for all $n \in \mathbb{N}$ and all $x_{1:n} \in \mathscr{X}^n$, there is some $T \subset \mathscr{T}$ such that

$$P(X_{n+1} = x_{n+1} | X_{1:n} = x_{1:n}) = T(x_n, x_{n+1}) \text{ for all } x_{n+1} \in \mathscr{X}.$$

Notably, therefore, $\mathscr{P}_{\mathcal{M}, \mathscr{T}}$ contains all the (precise) homogeneous Markov chains whose characterising transition matrix T is included in $\mathscr{T}$, and whose initial distribution $P(X_1)$ is included in $\mathcal{M}$. However, in general, $\mathscr{P}_{\mathcal{M}, \mathscr{T}}$ clearly also contains models that do not satisfy the Markov property, as well as Markov chains that are not homogeneous.[1]

For such an imprecise Markov chain, we are interested in computing inferences that are in a specific sense robust with respect to variations in the set $\mathscr{P}_{\mathcal{M}, \mathscr{T}}$. Specifically, for any function of interest $f : \Omega \to \mathbb{R}$, we consider its (conditional) *lower* and *upper* expectations, which are respectively defined by

$$\underline{E}_{\mathcal{M}, \mathscr{T}}(f | C) := \inf_{P \in \mathscr{P}_{\mathcal{M}, \mathscr{T}}} E_P(f | C) \quad \text{and} \quad \overline{E}_{\mathcal{M}, \mathscr{T}}(f | C) := \sup_{P \in \mathscr{P}_{\mathcal{M}, \mathscr{T}}} E_P(f | C).$$

In words, we are interested in computing the tightest possible bounds on the inferences computed for each $P \in \mathscr{P}_{\mathcal{M}, \mathscr{T}}$. These lower and upper expectations are related through conjugacy, meaning that $\underline{E}_{\mathcal{M}, \mathscr{T}}(f | C) = -\overline{E}_{\mathcal{M}, \mathscr{T}}(-f | C)$, so it suffices to consider only the upper expectations in the remaining discussion; any results for lower expectations follow analogously through this relation.

From a computational point of view, it is also useful to consider the dual representation of the set $\mathscr{T}$, given by the *upper transition operator* $\overline{T}$ with respect to this set [5,6]. This is a (non-linear) operator that maps $\mathscr{L}(\mathscr{X})$ into $\mathscr{L}(\mathscr{X})$; it is defined for any $f \in \mathscr{L}(\mathscr{X})$ and any $x \in \mathscr{X}$ as

$$[\overline{T}f](x) := \sup_{T(x, \cdot) \in \mathscr{T}_x} \sum_{y \in \mathscr{X}} T(x, y)f(y).$$

---

[1] Within the field of imprecise probability theory, this model is called an *imprecise Markov chain under epistemic irrelevance* [5,6,9].

So, in order to evaluate $[\overline{T}f](x)$, one must solve an optimisation problem over the set $\mathscr{T}_x$ containing the x-rows of the elements of $\mathscr{T}$. In many practical cases, the set $\mathscr{T}_x$ is closed and convex and therefore, evaluating $[\overline{T}f](x)$ is relatively straightforward: for instance, if $\mathscr{T}_x$ is described by a finite number of (in)equality constraints, then this problem reduces to a simple linear programming task, which can be solved by standard techniques. We will also make use of the conjugate *lower transition operator* $\underline{T}$, defined by $[\underline{T}f](x) := -[\overline{T}(-f)](x)$ for all $x \in \mathscr{X}$ and all $f \in \mathscr{L}(\mathscr{X})$. Results about upper transition operators translate to results about lower transition operators through this relation; we will focus on the former in the following discussion.

Now, the operator $\overline{T}$ can be used for computing upper expectations in much the same way as transition matrices are used for computing expectations with respect to precise Markov chains: for any $n \in \mathbb{N}$, any finitary function $f(X_{n+1})$ and any $x_{1:n} \in \mathscr{X}^n$ it holds that

$$\overline{\mathbb{E}}_{\mathcal{M},\mathscr{T}}(f(X_{n+1})|X_{1:n} = x_{1:n}) = [\overline{T}f](x_n). \tag{1}$$

Observe that the right-hand side in this expression does not depend on the history $x_{1:n-1}$; this can be interpreted as saying that the model satisfies an *imprecise Markov property*, which explains why we call our model an "imprecise Markov chain". Moreover, a slightly more general property holds that will be useful later on:

Proposition 1. *Consider the imprecise Markov chain $\mathscr{P}_{\mathcal{M},\mathscr{T}}$. For any $m, n \in \mathbb{N}$ such that $m \leq n$, any function $f \in \mathscr{L}(\mathscr{X}^{n-m+1})$ and any $x_{1:m-1} \in \mathscr{X}^{m-1}$ and $y \in \mathscr{X}$, we have that*

$$\overline{\mathbb{E}}_{\mathcal{M},\mathscr{T}}\big(f(X_{m:n})\big|X_{1:m-1} = x_{1:m-1}, X_m = y\big) = \overline{\mathbb{E}}_{\mathcal{M},\mathscr{T}}\big(f(X_{1:n-m+1})\big|X_1 = y\big).$$

Finally, we remark that, for any $m, n \in \mathbb{N}$ such that $m \leq n$, a conditional upper expectation $\overline{\mathbb{E}}_{\mathcal{M},\mathscr{T}}\big(f|X_{m:n}\big)$ is itself a (finitary) function depending on the states $X_{m:n}$. Using this observation, we can now introduce the *law of iterated upper expectations*, which will form the basis of the algorithms developed in the following sections:

Theorem 1. *Consider the imprecise Markov chain $\mathscr{P}_{\mathcal{M},\mathscr{T}}$. For all $m \in \mathbb{N}_0$, all $k \in \mathbb{N}$ and all $f \in \mathscr{L}_{\text{fin}}(\Omega)$, we have that*

$$\overline{\mathbb{E}}_{\mathcal{M},\mathscr{T}}\big(f|X_{1:m}\big) = \overline{\mathbb{E}}_{\mathcal{M},\mathscr{T}}\Big(\overline{\mathbb{E}}_{\mathcal{M},\mathscr{T}}\big(f|X_{1:m+k}\big)\Big|X_{1:m}\Big).$$

4   A Recursive Inference Algorithm

In principle, for any function $f \in \mathscr{L}(\mathscr{X}^n)$ with $n \in \mathbb{N}$, the upper expectations of $f(X_{1:n})$ can be obtained by maximising $\mathbb{E}_P(f(X_{1:n}))$ over the set $\mathscr{P}_{\mathcal{M},\mathscr{T}}$ of all precise models P that are compatible with $\mathcal{M}$ and $\mathscr{T}$. Since this will almost always be infeasible if n is large, we usually apply the law of iterated upper expectations in combination with the Markov property in order to divide the optimisation problem into multiple smaller ones. Indeed, because of Theorem 1, we have that

$$\overline{E}_{\mathcal{M},\mathscr{T}}(f(X_{1:n})) = \overline{E}_{\mathcal{M},\mathscr{T}}\left(\overline{E}_{\mathcal{M},\mathscr{T}}(f(X_{1:n})|X_{1:n-1})\right).$$

Using Eq. (1), one can easily show that $\overline{E}_{\mathcal{M},\mathscr{T}}(f(X_{1:n})|X_{1:n-1})$ can be computed by evaluating $[\overline{T}f(x_{1:n-1}\cdot)](x_{n-1})$ for all $x_{1:n-1} \in \mathscr{X}^{n-1}$. Here, $f(x_{1:n-1}\cdot)$ is the function in $\mathscr{L}(\mathscr{X})$ that takes the value $f(x_{1:n})$ on $x_n \in \mathscr{X}$. This accounts for $|\mathscr{X}|^{n-1}$ optimisation problems to be solved. With the acquired function $f'(X_{1:n-1}) := \overline{E}_{\mathcal{M},\mathscr{T}}(f(X_{1:n})|X_{1:n-1})$, we can then compute the upper expectation $\overline{E}_{\mathcal{M},\mathscr{T}}(f'(X_{1:n-1})|X_{1:n-2})$ in a similar way, by solving $|\mathscr{X}|^{n-2}$ optimisation problems. Continuing in this way, we end up with a function that only depends on X_1 and for which the expectation needs to be maximised over the initial models in $\mathcal{M}$. Hence, in total, $\sum_{i=0}^{n-1} |\mathscr{X}|^i$ optimisation problems need to be solved in order to obtain $\overline{E}_{\mathcal{M},\mathscr{T}}(f(X_{1:n}))$. Although these optimisation problems are relatively simple and therefore feasible to solve individually, the total number of required iterations is still exponential in n, therefore making the computation of $\overline{E}_{\mathcal{M},\mathscr{T}}(f(X_{1:n}))$ intractable when n is large.

In many cases, however, $f(X_{1:n})$ can be recursively decomposed in a specific way allowing for a much more efficient computational scheme to be employed; see Theorem 2 further on. Before we present this scheme in full generality, let us first provide some intuition about its basic working principle.

So assume we are interested in $\overline{E}_{\mathcal{M},\mathscr{T}}(f(X_{1:n}))$, which, according to Theorem 1, can be obtained by maximising $E_P(\overline{E}_{\mathcal{M},\mathscr{T}}(f(X_{1:n})|X_1))$ over $P(X_1) \in \mathcal{M}$. The problem then reduces to the question of how to compute $\overline{E}_{\mathcal{M},\mathscr{T}}(f(X_{1:n})|X_1)$ efficiently. Suppose now that $f(X_{1:n})$ takes the following form:

$$f(X_{1:n}) = g(X_1) + h(X_1)\tau(X_{2:n}), \tag{2}$$

for some $g, h \in \mathscr{L}(\mathscr{X})$ and some $\tau \in \mathscr{L}(\mathscr{X}^{n-1})$. Then, because $\overline{E}_{\mathcal{M},\mathscr{T}}$ is a supremum over linear expectations, we find that

$$\overline{E}_{\mathcal{M},\mathscr{T}}(f(X_{1:n})|X_1) = g(X_1) + h(X_1)\overline{E}_{\mathcal{M},\mathscr{T}}(\tau(X_{2:n})|X_1),$$

where, for the sake of simplicity, we assumed that h does not take negative values. Then, by appropriately combining Proposition 1 with Theorem 1, one can express $\overline{E}_{\mathcal{M},\mathscr{T}}(\tau(X_{2:n})|X_1)$ in terms of $\overline{\Upsilon}: \mathscr{X} \to \mathbb{R}$, defined by

$$\overline{\Upsilon}(x) := \overline{E}_{\mathcal{M},\mathscr{T}}(\tau(X_{1:n-1})|X_1 = x) \text{ for all } x \in \mathscr{X}.$$

In particular, we find that

$$\overline{E}_{\mathcal{M},\mathscr{T}}(\tau(X_{2:n})|X_1) = \overline{E}_{\mathcal{M},\mathscr{T}}\left(\overline{E}_{\mathcal{M},\mathscr{T}}(\tau(X_{2:n})|X_{1:2})|X_1\right) = \overline{E}_{\mathcal{M},\mathscr{T}}\left(\overline{\Upsilon}(X_2)|X_1\right)$$
$$= [\overline{T}\,\overline{\Upsilon}](X_1),$$

where the equalities follow from Theorem 1, Proposition 1 and Eq. (1), respectively. So $\overline{E}_{\mathcal{M},\mathscr{T}}(f(X_{1:n})|X_1)$ can be obtained from $\overline{\Upsilon}$ by solving a single optimisation problem, followed by a pointwise multiplication and summation.

Now, by repeating the structural assessment (2) in a recursive way, we can generate a whole class of functions for which the upper expectations can be computed using the principle illustrated above. We start with a function $\tau_1(X_1)$, with $\tau_1 \in \mathscr{L}(\mathscr{X})$, that only depends on the initial state. The upper expectation $\overline{E}_{\mathcal{M},\mathscr{T}}(\tau_1(X_1)|X_1)$ is then trivially equal to $\tau_1(X_1)$. Next, consider $\tau_2(X_{1:2}) = g_1(X_1) + h_1(X_1)\tau_1(X_2)$ for some g_1, h_1 in $\mathscr{L}(\mathscr{X})$. $\overline{E}_{\mathcal{M},\mathscr{T}}(\tau_2(X_{1:2})|X_1)$ is then given by $g_1(X_1) + h_1(X_1)[\overline{T}\,\overline{T}_1](X_1)$, where we let $\overline{T}_1(x) := \overline{E}_{\mathcal{M},\mathscr{T}}(\tau_1(X_1)|X_1 = x) = \tau_1(x)$ for all $x \in \mathscr{X}$ and where we (again) neglect the subtlety that h_1 can take negative values. Continuing in this way, step by step considering new functions constructed by multiplication and summation with functions that depend on an additional time instance, and no longer ignoring the fact that the functions involved can take negative values, we end up with the following result.

Theorem 2. *Consider any imprecise Markov chain $\mathscr{P}_{\mathcal{M},\mathscr{T}}$ and two sequences of functions $\{g_n\}_{n\in\mathbb{N}_0}$ and $\{h_n\}_{n\in\mathbb{N}}$ in $\mathscr{L}(\mathscr{X})$. Define $\tau_1(x_1) := g_0(x_1)$ for all $x_1 \in \mathscr{X}$, and for all $n \in \mathbb{N}$, let*

$$\tau_{n+1}(x_{1:n+1}) := h_n(x_1)\tau_n(x_{2:n+1}) + g_n(x_1) \text{ for all } x_{1:n+1} \in \mathscr{X}^{n+1}.$$

If we write $\{\overline{T}_n\}_{n\in\mathbb{N}}$ and $\{\underline{T}_n\}_{n\in\mathbb{N}}$ to denote the sequences of functions in $\mathscr{L}(\mathscr{X})$ that are respectively defined by $\overline{T}_n(x) := \overline{E}_{\mathcal{M},\mathscr{T}}(\tau_n(X_{1:n})|X_1 = x)$ and $\underline{T}_n(x) := \underline{E}_{\mathcal{M},\mathscr{T}}(\tau_n(X_{1:n})|X_1 = x)$ for all $x \in \mathscr{X}$ and all $n \in \mathbb{N}$, then $\{\overline{T}_n\}_{n\in\mathbb{N}}$ and $\{\underline{T}_n\}_{n\in\mathbb{N}}$ satisfy the following recursive expressions:

$$\begin{cases} \overline{T}_1 = \underline{T}_1 = g_0; \\ \overline{T}_{n+1} = h_n\mathbb{I}_{h_n \geq 0}[\overline{T}\,\overline{T}_n] + h_n\mathbb{I}_{h_n<0}[\underline{T}\,\underline{T}_n] + g_n \text{ for all } n \in \mathbb{N}; \\ \underline{T}_{n+1} = h_n\mathbb{I}_{h_n \geq 0}[\underline{T}\,\underline{T}_n] + h_n\mathbb{I}_{h_n<0}[\overline{T}\,\overline{T}_n] + g_n \text{ for all } n \in \mathbb{N}. \end{cases}$$

Here, we used $\mathbb{I}_{h_n \geq 0} \in \mathscr{L}(\mathscr{X})$ to denote the indicator of $\{x \in \mathscr{X} : h_n(x) \geq 0\}$, and similarly for $\mathbb{I}_{h_n<0} \in \mathscr{L}(\mathscr{X})$. Note that, because we now need to evaluate both $\overline{T}$ and $\underline{T}$ for every iteration, we will in general need to solve $2(n-1)|\mathscr{X}|$ optimisation problems to obtain $\overline{E}_{\mathcal{M},\mathscr{T}}(\tau_n(X_{1:n})|X_1)$ and $\underline{E}_{\mathcal{M},\mathscr{T}}(\tau_n(X_{1:n})|X_1)$ for some $n \in \mathbb{N}$. In order to obtain the unconditional inferences $\overline{E}_{\mathcal{M},\mathscr{T}}(\tau_n(X_{1:n}))$ and $\underline{E}_{\mathcal{M},\mathscr{T}}(\tau_n(X_{1:n}))$, it then suffices to respectively maximise and minimise the expectations of $\overline{E}_{\mathcal{M},\mathscr{T}}(\tau_n(X_{1:n})|X_1)$ and $\underline{E}_{\mathcal{M},\mathscr{T}}(\tau_n(X_{1:n})|X_1)$ over all initial models in $\mathcal{M}$.

5   Special Cases

To illustrate the practical relevance of our method, we now discuss a number of important inferences that fall within its scope. As already mentioned in the introduction section, in some of these cases, our method simplifies to a computational scheme that was already developed earlier in a more specific context. The strength of our present contribution, therefore, lays in its unifying character and the level of generality to which it extends.

Functions that depend on a single time instant. As a first, very simple inference we can consider the upper and lower expectation of a function $f(X_n)$, for some $f \in \mathscr{L}(\mathscr{X})$ and $n \in \mathbb{N}$, conditional on the initial state. The expressions for these inferences are given by $\overline{T}^{n-1} f$ and $\underline{T}^{n-1} f$, respectively [5]. For instance, for any $x \in \mathscr{X}$, $\underline{\mathbb{E}}_{\mathcal{M},\mathscr{T}}(f(X_5))|X_1 = x) = [\underline{T}^4 f](x)$. These expressions can also easily be obtained from Theorem 2, by setting $g_0 := f$ and, for all $k \in \{1, \cdots, n-1\}$, $g_k := 0$ and $h_k := 1$.

Sums of functions. One can also use our method to compute upper and lower expectations of sums $\sum_{k=1}^{n} f_k(X_k)$ of functions $f_k \in \mathscr{L}(\mathscr{X})$. Then we would have to set $g_0 := f_n$ and, for all $k \in \{1, \cdots, n-1\}$, $g_k := f_{n-k}$ and $h_k := 1$. Although we allow the functions f_k to depend on k, it is worth noting that, if we set them all equal to the same function f, our method can also be employed to compute the upper and lower expectation of the time average $1/n \sum_{k=1}^{n} f(X_k)$ of f over the time interval n. The subtlety of the constant factor $1/n$ does not raise a problem here, because upper and lower expectations are homogeneous with respect to non-negative scaling.

Product of functions. Another interesting class of inferences are those that can be represented by a product $\prod_{k=1}^{n} f_k(X_k)$ of functions $f_k \subset \mathscr{L}(\mathscr{X})$. To compute upper and lower expectations of such functions, it suffices to set $g_0 := f_n$ and, for all $k \in \{1, \cdots, n-1\}$, $g_k := 0$ and $h_k := f_{n-k}$. A typical example of an inference than can be described in this way is the probability that the state will be in a set $A \subseteq \mathscr{X}$ during a certain time interval. For instance, the upper expectation of the function $\mathbb{I}_A(X_1)\mathbb{I}_A(X_2)$ gives us a tight upper bound on the probability that the state will be in A during the first two time instances.

Hitting probabilities. The hitting probability of some set $A \subseteq \mathscr{X}$ over a finite time interval n is the probability that the state X_k will be in A somewhere within the first n time instances. The upper and lower bounds on such a hitting probability are equal to the upper and lower expectation of the function $f(X_{1:n}) := \mathbb{I}_{A'_n} \in \mathscr{L}(\Omega)$, where $A'_n := \{\omega \in \Omega : (\exists k \leq n)\,\omega_k \in A\}$. Note that $f(X_{1:n})$ can be decomposed in the following way:

$$f(X_{1:n}) = \mathbb{I}_A(X_1) + \mathbb{I}_A(X_2)\mathbb{I}_{A^c}(X_1) + \cdots + \mathbb{I}_A(X_n) \prod_{k=1}^{n-1} \mathbb{I}_{A^c}(X_k)$$

Hence, these inferences can be obtained using Theorem 2 if we let $g_0 := \mathbb{I}_A$ and, for all $k \in \{1, \cdots, n-1\}$, $g_k := \mathbb{I}_A$ and $f_k := \mathbb{I}_{A^c}$. Additionally, one could also be interested in the probability that the state X_k will *ever* be in A. Upper and lower bounds on this probability are given by the upper and lower expectation of the function $f := \mathbb{I}_{A'} \in \mathscr{L}(\Omega)$ where $A' := \{\omega \in \Omega : (\exists k \in \mathbb{N})\,\omega_k \in A\}$. Since the function f is non-finitary, we are unable to apply our method in a direct way. However, it is shown in [8, Proposition 16] that, if the set $\mathscr{T}$ is convex and closed, the upper and lower bounds on the hitting probability over a finite time interval converge to the upper and lower bounds on the hitting probability over an infinite time interval, therefore allowing us to approximate these inferences by choosing n sufficiently large.

Hitting times. The *hitting time* of some set $A \subseteq \mathcal{X}$ is defined as the time τ until the state is in A for the first time; so $\tau(\omega) := \inf\{k \in \mathbb{N}_0 \colon \omega_k \in A\}$ for all $\omega \in \Omega$. Once more, the function τ is non-finitary, necessitating an indirect approach to the computation of its upper and lower expectation. This can be done in a similar way as we did for the case of hitting probabilities, now considering the finitary functions $\tau_n(X_{1:n})$, where $\tau_n \in \mathcal{L}(\mathcal{X}^n)$ is defined by $\tau_n(x_{1:n}) := \inf\{k \in \mathbb{N} \colon x_k \in A\}$ if $\{k \in \mathbb{N} \colon x_k \in A\}$ is non-empty, and $\tau_n(x_{1:n}) := n+1$ otherwise, for all $n \in \mathbb{N}$ and all $x_{1:n} \in \mathcal{X}^n$. These functions correspond to choosing $g_0 := \mathbb{I}_{A^c}$ and, for all $k \in \{1, \cdots, n-1\}$, $g_k := \mathbb{I}_{A^c}$ and $f_k := \mathbb{I}_{A^c}$. If the set $\mathcal{T}$ is convex and closed, the upper and lower expectations of these functions for large n will then approximate those of the non-finitary hitting time [8, Proposition 10].

6 Discussion

The main contribution of this paper is a single, unified method to efficiently compute a wide variety of inferences for imprecise Markov chains; see Theorem 2. The set of functions describing these inferences is however restricted to the finitary type, and therefore a general approach for inferences characterised by non-finitary functions is still lacking. In some cases, however, as we already mentioned in our discussion of hitting probabilities and hitting times, this issue can be addressed by relying on a continuity argument.

Indeed, consider any function $f = \lim_{n \to +\infty} \tau_n(X_{1:n})$ that is the pointwise limit of a sequence $\{\tau_n(X_{1:n})\}_{n \in \mathbb{N}}$ of finitary functions, defined recursively as in Theorem 2. If $\overline{\mathbb{E}}_{\mathcal{M}, \mathcal{T}}$ is continuous with respect to $\{\tau_n(X_{1:n})\}_{n \in \mathbb{N}}$, meaning that $\lim_{n \to +\infty} \overline{\mathbb{E}}_{\mathcal{M}, \mathcal{T}}(\tau_n(X_{1:n})) = \overline{\mathbb{E}}_{\mathcal{M}, \mathcal{T}}(f)$, the inference $\overline{\mathbb{E}}_{\mathcal{M}, \mathcal{T}}(f)$ can then be approximated by $\overline{\mathbb{E}}_{\mathcal{M}, \mathcal{T}}(\tau_n(X_{1:n}))$ for sufficiently large n. Since we can recursively compute $\overline{\mathbb{E}}_{\mathcal{M}, \mathcal{T}}(\tau_n(X_{1:n}))$ for any $n \in \mathbb{N}$ using the methods discussed at the end of Sect. 4, this yields an efficient way of approximating $\overline{\mathbb{E}}_{\mathcal{M}, \mathcal{T}}(f)$. A completely analogous argument can be used for the lower expectation $\underline{\mathbb{E}}_{\mathcal{M}, \mathcal{T}}(f)$. This begs the question whether the upper and lower expectations $\overline{\mathbb{E}}_{\mathcal{M}, \mathcal{T}}$ and $\underline{\mathbb{E}}_{\mathcal{M}, \mathcal{T}}$ satisfy the appropriate continuity properties for this to work.

Unfortunately, results about the continuity properties of these operators are rather scarce —especially compared to their precise counterparts—and depend on the formalism that is adopted. In this paper, for didactical reasons, we have considered one formalism: we have introduced imprecise Markov chains as being *sets* of "precise" models that are in a specific sense compatible with the given set $\mathcal{T}$. It is however important to realise that there is also an entirely different formalisation of imprecise Markov chains that is instead based on the game-theoretic probability framework that was popularised by Shafer and Vovk; we refer to [10, 11] for details. It is well known that the inferences produced under these two different frameworks agree for finitary functions [3, 9], so the method described by Theorem 2 is also applicable when working in a game-theoretic framework. The continuity properties of the game-theoretic upper and lower expectations, however, are not necessarily the same as those of the measure-theoretic operators that we considered here. So far, the continuity properties of game-theoretic upper and lower expectations are better understood [10–12], making these operators more suitable if we plan to employ the continuity argument above.

Acknowledgments. The work in this paper was partially supported by H2020-MSCA-ITN-2016 UTOPIAE, grant agreement 722734.

References

1. De Bock, J.: Credal networks under epistemic irrelevance. Int. J. Approximate Reasoning **85**, 107–138 (2017)
2. De Bock, J.: Credal networks under epistemic irrelevance: theory and algorithms. Ph.D. thesis, Ghent University (2015)
3. de Cooman, G., Hermans, F.: Imprecise probability trees: bridging two theories of imprecise probability. Artif. Intell. **172**(11), 1400–1427 (2008)
4. de Cooman, G., Hermans, F., Antonucci, A., Zaffalon, M.: Epistemic irrelevance in credal nets: the case of imprecise Markov trees. Int. J. Approximate Reasoning **51**(9), 1029–1052 (2010)
5. de Cooman, G., Hermans, F., Quaeghebeur, E.: Imprecise Markov chains and their limit behaviour. Probab. Eng. Inf. Sci. **23**(4), 597–635 (2009)
6. Hermans, F., Škulj, D.: Stochastic processes. In: Augustin, T., Coolen, F.P.A., De Cooman, G., Troffaes, M.C.M. (eds.) Introduction to Imprecise Probabilities. Wiley (2014). Chapter 11
7. Kallenberg, O.: Foundations of Modern Probability. Springer Science & Business Media, New York (2006)
8. Krak, T'Joens, N., De Bock, J.: Hitting times and probabilities for imprecise Markov chains (2019). Accepted for publication in the conference proceedings of ISIPTA 2019. A preprint can be found at arXiv:1905.08781
9. Lopatatzidis, S.: Robust Modelling and Optimisation in Stochastic Processes using Imprecise Probabilities, with an Application to Queueing Theory. Ph.D. thesis, Ghent University (2017)
10. Shafer, G., Vovk, V.: Probability and Finance: It's Only a Game!. Wiley, New York (2001)
11. T'Joens, N., De Bock, J., de Cooman, G.: In search of a global belief model for discrete-time uncertain processes (2019). Accepted for publication in the conference proceedings of ISIPTA 2019
12. T'Joens, N., De Bock, J., de Cooman, G.: Continuity properties of game-theoretic upper expectations. arXiv:1902.09406 (2019)
13. T'Joens, N., Krak, T., De Bock, J., de Cooman, G.: A Recursive Algorithm for Computing Inferences in Imprecise Markov Chains. arXiv:1905.12968 (2019)

Uncertain Reasoning for Applications

Probabilistic Consensus of the Blockchain Protocol

Bojan Marinković[1], Paola Glavan[2], Zoran Ognjanović[1(✉)], Dragan Doder[3], and Thomas Studer[4]

[1] Mathematical Institute of the Serbian Academy of Sciences and Arts, Belgrade, Serbia
{bojanm,zorano}@mi.sanu.ac.rs
[2] Faculty of Mechanical Engineering and Naval Architecture, Zagreb, Croatia
pglavan@fsb.hr
[3] Université Paul Sabatier – CNRS, IRIT, Toulouse, France
dragan.doder@irit.fr
[4] University of Bern, Bern, Switzerland
thomas.studer@inf.unibe.ch

Abstract. We introduce a temporal epistemic logic with probabilities as an extension of temporal epistemic logic. This extension enables us to reason about properties that characterize the uncertain nature of knowledge, like "agent a will with high probability know after time s same fact". To define semantics for the logic we enrich temporal epistemic Kripke models with probability functions defined on sets of possible worlds. We use this framework to model and reason about probabilistic properties of the blockchain protocol, which is in essence probabilistic since ledgers are immutable with high probabilities. We prove the probabilistic convergence for reaching the consensus of the protocol.

Keywords: Multi-agent systems · Blockchain ·
Temporal epistemic logic with probabilities · Formal model ·
Specification/verification

1 Introduction

Time, knowledge and uncertainty are fundamental properties of distributed systems. In order to be able to deal with these properties we need to represent them and reason about them. Reasoning about time and knowledge started, if not earlier, in the 1950s, 1960s with [9,15]. Since then, epistemic temporal logic has been applied in many fields. Particularly, it has been proven useful in analyzing message-passing based protocols in distributed computer networks [4,6,7], where a suitable semantics was proposed, and modal operators are used to express both agents' knowledge and temporal properties of actions in distributed systems. The idea of extending the epistemic logic with probability operators which enable reasoning about uncertainty seems natural and it is not new, see for example [5,16].

© Springer Nature Switzerland AG 2019
G. Kern-Isberner and Z. Ognjanović (Eds.): ECSQARU 2019, LNAI 11726, pp. 469–480, 2019.
https://doi.org/10.1007/978-3-030-29765-7_39

In this paper, we extend reasoning about temporal and epistemic properties of agents [10] with probability properties. Agents are not rigid, i.e. one agent participate as active or passive in the system. This property of agents implies that knowledge does not satisfy that everything which is known is true (and in that sense it might be also called belief [4]). Knowledge of an agent a is represented using the modal operator K_a, that is interpreted with an accessibility relation in Kripke models. The temporal part of the logic is discrete linear time (future) LTL logic, where the flow of time is isomorphic to the natural numbers, and the corresponding part of the formal language contains the operators Next ($\bigcirc$) and Until (U). Probabilistic part is modeled by introducing probability operators of the form $P_{a,\geqslant s}$, with the meaning that according to agent a some fact holds with the probability greater then or equal to s. Then both K_a and $P_{a,\geqslant s}$, in $K_a(P_{a,\geqslant s}\phi)$, express together probabilistic knowledge i.e. that agent a will with probability at most s know some fact ϕ. We also introduce probabilistic common knowledge operators of the form C_s, with the meaning that common knowledge of the probability of formula holds is at least s.

Nowadays, one of the most popular distributed protocols is the blockchain protocol [13], which is used, for example, to synchronize copies of the public ledger in the bitcoin cryptocurrency. By its nature blockchain is the probabilistic protocol [11] and every agent has its own knowledge which evaluates during the time [10]. In the formal language of our logic we formulate a theory which describes the blockchain. We illustrate expressiveness of the logic by reasoning about probabilistic consensus of agents participating in an execution of the blockchain protocol.

A blockchain is a decentralized, distributed and public digital ledger. The ledger is also immutable and ordered. It is used to record transactions across many computers with the property that transactions can be added only at the end of the ledger and a record cannot be altered retroactively, without the alteration of all subsequent blocks and the consensus of the network. All participants have a large common prefix of the ledger. A blockchain database is managed autonomously, without third authority, using a peer-to-peer network and a distributed time-stamping server. At any point of the protocol execution (round), each participant attempts to increase the length of its own chain by mining for a new block: upon receiving some record m, it picks a random string and checks whether string is a valid proof-of-work (PoW) with respect to m and a pointer to the last block of its current chain. If so, it extends its own local chain and broadcasts it to the all the other participants. Whenever a participant receives a chain that is longer than its own local chain, it replaces its own chain with the longer one [3,14]. It is possible, in the run of the protocol, that two transactions arrive approximately simultaneously. In that case, each participant chooses one transaction and works on it (approximately half choose the first one, and the other half the second one), keeping the other transaction. This situation is called fork. Fork is resolved in some of the next round when the next unique PoW is found and one branch becomes longer; the nodes that were working on the other branch will then switch to the longer one.

In essence the blockchain protocol is probabilistic since the ledgers change with high probabilities. Probabilistic temporal epistemic logic enables us to model and reason about probabilistic characteristics of the blockchain protocol: we are able to prove existence of the probabilistic common knowledge of agents about consensus on the common prefix of the ledger.

The paper [8] analyzes probabilistic conditions to achieve consensus on a public ledger, and presents a model theoretic approach with probabilistic constraints on runs that guarantee that (so called $\Delta - \square$) common knowledge about the ledger is obtained. On the other hand, in this paper a theory which describes the blockchain is used as the starting point to prove existence of probabilistic common knowledge about the ledger. Other related papers are: [2] describes how an agent's knowledge is changed when a new block that might be added to the blockchain arrives; [10] develops a logic to analyze properties of the protocol in terms of knowledge of agents; [3,14] using cryptographic techniques prove that with the high probability honest agents have the same common prefix of the ledger.

The rest of the paper is organized as follows. In Sect. 2 we describe syntax and semantics for the considered temporal epistemic logic with probabilities. Section 3 describes the blockchain protocol as a theory (a set of proper axioms) of the presented logic. We prove important properties of the protocol in this section. Section 4 contains concluding remarks and directions for further work.

2 Temporal Epistemic Logic with Probabilities

2.1 Syntax

Let $\mathbb{N}$ be the set of nonnegative integers, Var a nonempty at most countable set of propositional letters, and $\mathbf{A} = \{a_1, \ldots, a_m\}$, where $m \in \mathbb{N}$, a set of agents. Also, we introduce the set of propositional letters $\mathbf{A}_a = \{A_a | a \in \mathbf{A}\}$. The intuitive meaning of propositional letter A_a is that "agent a is active". The set For of all formulas is the smallest superset of $Var \cup \mathbf{A}_a$ which is closed under the following formation rules:

- $\psi \mapsto *\psi$ where $* \in \{\neg, \bigcirc, \mathsf{K}_a, \mathsf{C}, \mathsf{P}_{a,\geqslant s}, \mathsf{C}_s\}$, where $a \in \mathbf{A}$, $s \in [0,1]_{\mathbb{Q}}$
- $\langle \phi, \psi \rangle \mapsto \phi * \psi$ where $* \in \{\wedge, \mathsf{U}\}$.

The operators $\bigcirc$ and U are standard temporal operators Next and Until. We read the formula $\bigcirc\psi$ "ψ will hold in the next moment", and the formula $\varphi\mathsf{U}\psi$ "φ will hold until ψ becomes true. The remaining Boolean and temporal connectives $\vee$, $\underline{\vee}$, $\rightarrow$, $\leftrightarrow$, F ("sometimes"), and G ("always") are defined in the usual way. The formula $\mathsf{K}_a\psi$ denotes that the agent a knows ψ. The knowledge operator E, which we read "everybody knows", is introduced as $\mathsf{E}\psi =_{def} \bigwedge_{a \in \mathbf{A}} \mathsf{K}_a\psi$. The operator C expresses common knowledge, i.e., the meaning of the formula $\mathsf{C}\psi$ is "that everyone knows that everyone knows that everyone knows... that ψ is true". The formula $\mathsf{P}_{a,\geqslant s}\psi$ represents that the probability of the formula ψ, according to agent a, is at least s. The probabilistic variants of the operators K_a and E are defined as abbreviations, in the following way:

- $K_a^s \psi =_{def} K_a(P_{a, \geqslant s} \psi)$, and
- $E_s \psi =_{def} \bigwedge_{a \in \mathbf{A}} K_a^s \psi$,

while C_s is the operator for probabilistic common knowledge (i.e., common knowledge is that the probability of a formula is at least s). Theories are sets of formulas.

2.2 Semantics

In this paper we will consider time flow which is isomorphic to the set $\mathbb{N}$. Our models are propositional Kripke structures with possible worlds, similar as the interpreted systems from [4, 16].

Definition 1. *A model $\mathcal{M}$ is any tuple $\langle W, R, \pi, \mathcal{A}, \mathcal{K}, \mathcal{P} \rangle$ such that*

- W *is the set of possible worlds,*
- R *is the set of runs, where:*
  - *every* run r *is a countably infinite sequence of possible worlds r_0, r_1, r_2, ..., and*
  - *every possible world belongs to only one run.*
- $\pi = \{\pi_i^r : r \in R, \ i \in \mathbb{N}\}$ *is the set of* valuations:
  - $\pi_i^r(q) \in \{\top, \bot\}$, *for $q \in Var$, associates truth values to propositional letters of the possible world r_i,*
- $\mathcal{A}$ *associates a set of active agents to each possible worlds,*
- $\mathcal{K} = \{\mathcal{K}_i^a : a \in \mathbf{A}\}$ *is the set of binary accessibility relations on R, such that*
  - *if $a \notin \mathcal{A}(r_i)$, then $r\mathcal{K}_i^a r'$ is false for all $r' \in R$.*
- $\mathcal{P}$ *associates a probability space $\mathcal{P}(r_i, a) = (R_{r_i}^a, \xi_{r_i}^a, \mu_{r_i}^a)$ to every possible world r_i and every agent a, such that*
  - $R_{r_i}^a$ *is a non-empty subset of R,*
  - $\xi_{r_i}^a$ *is an algebra of subsets of $R_{r_i}^a$ whose elements are called measurable sets, and*
  - $\mu_{r_i}^a : \xi_{r_i}^a \to [0,1]$ *is a finitely-additive probability measure.*

We denote the class of all models by Mod.

Note that in Definition 1 we consider the general case and did not introduce any restrictions on $\mathcal{K}_i^a$, except introduction of "dead end worlds" in the situations when agents are not active. In order to reason about agents' knowledge, we will consider the case when $\mathcal{K}_i^a$ are equivalence relations for the active agents.

2.3 Satisfiability Relation

The satisfiability relation $\models$ is recursively defined as follows:

Definition 2. *Let $\mathcal{M} = \langle R, \pi, \mathcal{A}, \mathcal{K}, \mathcal{P} \rangle$ be an* Mod *model. The satisfiability relation $\models$ satisfies:*

1. $r_i \models q$ iff $\pi_i^r(q) = \top$, for $q \in Var \cup \mathbf{A}_a$,

2. $r_i \models A_a$ *iff* $a \in \mathcal{A}(r_i)$,
3. $r_i \models \beta_1 \wedge \beta_2$ *iff* $r_i \models \beta_1$ *and* $r_i \models \beta_2$,
4. $r_i \models \neg\beta$ *iff not* $r_i \models \beta$ *(*$r_i \not\models \beta$*)*,
5. $r_i \models \bigcirc\beta$ *iff* $r_{i+1} \models \beta$,
6. $r_i \models \beta_1 U \beta_2$ *iff there is an* $j \geqslant 0$ *such that* $r_{i+j} \models \beta_2$, *and for every* k, *such that* $0 \leqslant k < j$, $r_{i+k} \models \beta_1$,
7. $r_i \models K_a\beta$ *iff* $r_i' \models \beta$ *for all* r' *such that* $r \mathcal{K}_i^a r'$,
8. $r_i \models C\psi$ *iff for every* $n \geqslant 0$, $r_i \models E^n\psi$,
9. $r_i \models P_{a,\geqslant s}\beta$ *iff* $\mu_{r_i}^a(\{r' \in R \mid r_i' \models \beta\}) \geqslant s$, *and*
10. $r_i \models C_r\beta$ *iff for every* $n \geqslant 0$, $r_i \models E_r^n\beta$.

$\square$

Our semantic definition of probabilistic common knowledge is taken from the paper [12], where the operator C_r is introduced for a the first time, as reflexive and transitive closure of E_r.[1]

A set of formulas is *satisfiable* if there is a possible world r_i of a run r in a model $\mathcal{M}$ such that every formula from the set holds in r_i. A formula α is satisfiable if the set $\{\alpha\}$ is satisfiable. A formula is *valid in a model*, if it holds in every world of the model. α is *valid* ($\models \alpha$), if it is valid in each model. A formula α is a *semantic consequence* of a set of formulas F ($F \models \alpha$) if for every model $\mathcal{M}$ in which all formulas from the set F are valid, $\mathcal{M} \models \alpha$.

In order to keep the satisfiability relation well-defined, in this work we consider only so-called *measurable* models. A measurable model is a model in which each set $\{r' \in R \mid r_i' \models \beta\}$ belongs to $\xi_{r_i}^a$ for every possible world r_i and every agent a. The class of all measurable models is denoted by Mod_{Meas}.

We consider models with non rigid sets of active atoms. We can assume that non-active agent (i.e. $a \notin \mathcal{A}(r_i)$) knows everything (i.e. $r_i \models K_a\beta$, for every formula β). However, since satisfiability of knowledge of a group is represented as a conjunction of knowledge of agents from the group, knowledge of a non-active agents do not affect the knowledge of the group of active agents.

3 Blockchain Protocol

The blockchain protocol is used in the process of obtaining the consensus among the agents in distributed environment. The most known used version of the blockchain protocol is the bitcoin and some other cryptocurrencies. In this section we will describe the blockchain protocol [1,3,13,14] and provide the proof that the consensus is achieved with the high probability.

The following properties particularly contribute to the popularity of the blockchain protocol:

- It is managed autonomously, without third authority.
- It solves the long-standing problem of double spending.
- It provides a record that compels offer and acceptance, since the fact that all the transactions are kept in the public ledger.

---

[1] In [5], a slightly different definition is presented, and it is pointed out that both definitions are valid probabilistic generalizations of common knowledge. For the details we refer the reader to [5].

3.1   Overview of the Blockchain Protocol

In the blockchain environment transactions, that record chaining the ownership of goods between the agents in the distributed network, are kept permanently and publicly available. A transaction with the corresponding data (time stamp, identifiers of agents and value of property) are recorded in the blocks that are parts of a digital public ledger. The agents follow certain set of rules how to add and accept new blocks, add them to the ledger and achieve consensus among them. The most known version of such set of rules is so called PoW, i.e. all the agents try to solve unique cryptographic puzzle and all the agents have to accept the first valid solution. Since the fact that the hash function is used, one block in the ledger cannot be replace without the replacement of the whole section of all subsequent blocks.

The blockchain protocol was introduced in the following way (quotation from [13]):

1. New transactions are broadcast to all nodes.
2. Each node collects new transactions into a block.
3. Each node works on finding a difficult PoW for its block.
4. When a node finds a PoW, it broadcasts the block to all nodes.
5. Nodes accept the block only if all transactions in it are valid and not already spent.
6. Nodes express their acceptance of the block by working on creating the next block in the chain, using the hash of the accepted block as the previous hash.

Nodes always consider the longest chain, i.e., the one containing the most proofs-of-work, to be the correct one and will keep working on extending it. If two nodes broadcast different versions of the next block simultaneously, some nodes may receive one or the other first. In that case, they work on the first one they received, but save the other branch in case it becomes longer. The tie will be broken when the next PoW is found and one branch becomes longer; the nodes that were working on the other branch will then switch to the longer one.

A round is described with the above described steps $(1-6)$. Each node tries to increase the length of its own chain by "mining" the new block: find the string that will produce the hash value of the whole block that satisfies the certain property. If several blocks are produced approximately simultaneously every node can choose which branch will try to extend. This situation is called fork. Forks are resolved in later rounds, when all the nodes will accept the longest branch.

We consider the blockchain protocol that runs in a synchronous setting (the time needed to solve a puzzle for one round is much greater than the time to exchange that information among the agents). We do not consider cryptographic properties of the protocol, and we assume that all nodes in the network are perfectly honest and reasonable, and that there are no dishonest nodes trying to exploit cryptographic vulnerabilities of the protocol to gain benefits.

3.2 Modeling of the Blockchain Protocol

The logic presented in this paper extends the temporal epistemic logic with a non-rigid set of agents from [10] to allow probabilistic reasoning. In [10] a theory (set of formulas) in the corresponding language is formulated to describe a simplified version of the blockchain protocol. The simplification concerns avoiding probabilistic behavior which characterizes the blockchain, and there is an axiom which says that forks will be resolved after a fixed number of rounds. Here we overcome this constraint since we can express explicit probabilities. We replace the mentioned axiom of the temporal epistemic logic with a new one ([AB11]) which determines the probability that after z rounds all agents have the same prefix of the ledger. As a consequence, we can consider more realistic (probabilistic) executions of the blockchain and formulate and prove a statement about probabilistic common knowledge among agents.

We define Var as $Var \supseteq \mathbf{POW} \cup \mathbf{ACC}$, where:

- $\mathbf{POW} = \{\mathsf{pow}_{a,i} | a \in \mathbf{A}, i \in \mathbb{N}\}$ is a set of atomic propositions, with the intended meaning of $\mathsf{pow}_{a,i}$ that the agent a produces a PoW for round i, and
- $\mathbf{ACC} = \{\mathsf{acc}_{a,b,i} | a, b \in \mathbf{A}, i \in \mathbb{N}\}$ is a set of atomic propositions, with the intended meaning of $acc_{a,b,i}$ that the agent a accepts the PoW produced for round i by the agent b.

We set

$$e_{a,i} := \bigwedge_{b \in \mathbf{A}} (A_b \rightarrow \mathsf{acc}_{b,a,i})$$

The formulas $e_{a,i}$ mean that every active agent accepts the PoW produced for round i by agent a.

Further we set

$$\mathsf{ech}_{b,i} := \bigvee_{a \in \mathbf{A}} \mathsf{acc}_{b,a,i}$$

The formula $\mathsf{ech}_{b,i}$ means that agent b accepts some PoW produced for round i.

We will use $\mathsf{pf} \in (0,1)$ to denote the probability that a fork occurs in a particular round.

Our theory of blockchain, denoted with **BCT**, consists of the following proper axioms (let a, b and c denote agents from **A**):

AB1 $\bigvee_a A_a$

AB2 $acc_{b,a,i} \to pow_{a,i}$

AB3 $acc_{b,a,i} \to K_b acc_{b,a,i}$

AB4 $acc_{b,a,i} \to \neg acc_{b,c,i}$, for each
 $c \neq a$

AB5 $acc_{a,c,j} \wedge \bigcirc acc_{b,a,i} \to \bigcirc acc_{b,c,j}$,
 for $j < i$

AB6 $A_b \wedge \bigvee_a pow_{a,i} \to ech_{b,i}$

AB7 $ech_{a,i} \to A_a$

AB8 $ech_{a,i+1} \to ech_{a,i}$

AB9 $ech_{b,i} \to \bigcirc \bigvee_a pow_{a,i+1}$

AB10 $\neg ech_{a,i} \to \neg \bigcirc pow_{a,i+1}$

AB11 $ech_{a,i+z} \wedge acc_{a,b,i} \to P_{a,\geqslant(1-pf^z)} e_{b,i}$

Let us briefly discuss the meaning of the above axioms.

AB1 There is always at least one agent active.

AB2 One can only accept PoW that has been produced.

AB3 The agents know if they accept some PoW.

AB4 An agent accepts at most one PoW for a given round.

AB5 If a accepts c's proof of work for round j and (in the next step) b accepts a's PoW for a later round, then b must also accept c's PoW for round j. This essentially means that if b accepts a's PoW, then b accepts the whole history of a.

AB6 If proofs-of-work for some round are produced, then each active agent must accept one of them. Note that we do not have any assumption on how an agent accepts a proof.

AB7 Only active agents can accept proofs-of-work.

AB8 If an agent accepts some PoW for round $i+1$, then the agent also accepts some PoW for round i.

AB9 If an agent accepts some PoW for round i, then in the next round a PoW for round $i+1$ must be available.

AB10 Only an agent that has accepted a PoW for round i can create (in the next step) a PoW for round $i+1$. This models the fact that a PoW depends on the previously accepted history.

AB11 This states how the probability that PoW remains in the common history depends on how deep it is in the ledger. Note that we do not have any assumption on how this consensus is achieved. This formalizes the *common prefix* property from [3].

Let us now briefly discuss the relationship between time instants (from the linear time logic part) and rounds (referenced in the atomic propositions in **POW** and **ACC**).

We start at time instant t and assume that agent b accepts some proof of work for round i, that means agent b accepts a blockchain of length i. Because of [AB9], at time instant $t+1$ some agent a will produce a PoW for round $i+1$. By [AB1] at least one agent, say agent c, will be active at time instant $t+1$. By [AB6] agent c at time instant $t+1$ accepts some proof of work for round $i+1$, that means a blockchain of length $i+1$. Hence with every time instant, the accepted blockchain grows by one block.

However, we do not require that all PoW for round $i+1$ is generated at time instant $t+1$. It is possible that some PoW for round $i+1$ is produced at a later time instant.

The following lemmas will be used to prove the statement in Theorem 1.

Lemma 1. *The set of Blockchain Axioms is satisfiable.*

Lemma 2. *The following holds:*

RPN: if $\mathbf{BCT} \models \beta$ *then* $\mathbf{BCT} \models P_{a,\geqslant 1}\beta$
RICP: if $\mathbf{BCT} \models E_s^i\beta$, *for all* $i \geqslant 0$, *then* $\mathbf{BCT} \models C_s\beta$

A trivial consequence of [AB4] is that there cannot be an agreement of acceptance of two different proofs-of-work.

Lemma 3. *The following holds* $\mathbf{BCT} \models e_{a,i} \to \neg e_{b,i}$ *for* $b \neq a$.

Now we show that the common history persists, i.e., agreements cannot be undone.

Lemma 4. *We have* $\mathbf{BCT} \models e_{a,i} \to \bigcirc e_{a,i}$.

The following lemma says that if an agent accepts the choice of an another agent for a round then it accepts the whole history of that other agent.

Lemma 5. *We have* $\mathbf{BCT} \models A_b \wedge \mathsf{ech}_{a,i} \to \mathsf{ech}_{b,i}$.

For the proof of Theorem 1 we need to prove the following lemma.

Lemma 6. *If* $\mathbf{BCT} \models \alpha \to E_s\alpha$, *then* $\mathbf{BCT} \models \alpha \to (E_s)^k\alpha$ *for any* $k \subset \mathbb{N}$.

Proof
Suppose

$$\mathbf{BCT} \models \alpha \to E_s\alpha \tag{1}$$

By [RPN] $\mathbf{BCT} \models P_{a,=1}(\alpha \to E_s\alpha)$ for some agent $a \in \mathbf{A}$
And also

$$\mathbf{BCT} \models E_1(\alpha \to E_s\alpha) \tag{2}$$

Further from probabilistic logic we have

$$\mathbf{BCT} \models P_{a,=1}(\beta \to \gamma) \to (P_{a,\geqslant s}\beta \to P_{a,\geqslant s}\gamma)$$

for some agent $a \in \mathbf{A}$
Thus we get:

$$\mathbf{BCT} \models E_1(\beta \to \gamma) \to (E_s\beta \to E_s\gamma) \tag{3}$$

Thus 2 and 3 together (with $\beta = \alpha$ and $\gamma = E_s\alpha$) yield $E_s\alpha \to E_sE_s\alpha$ together with 1 we obtain $\mathbf{BCT} \models \alpha \to E_sE_s\alpha$
We can iterate this to obtain $\mathbf{BCT} \models \alpha \to (E_s)^k\alpha$ for any natural number k. $\square$

As a result of Theorem 1 we get the estimation of the probability of the consensus of an agent:

Theorem 1. *We have* $\mathbf{BCT} \models \mathrm{ech}_{a,i+z} \wedge \mathrm{acc}_{a,b,i} \to C_{1-\mathrm{pf}^z} e_{b,i}.$

Proof
For an arbitrary agent c by [AB3]:

$$\mathbf{BCT} \models \mathrm{acc}_{c,b,i} \to K_c \mathrm{acc}_{c,b,i}. \tag{4}$$

Also, $\mathbf{BCT} \models \mathrm{acc}_{c,d,i+z} \to K_c \mathrm{acc}_{c,d,i+z}$, and: $\mathbf{BCT} \models \bigvee_{d \in A} \mathrm{acc}_{c,d,i+z} \to \bigvee_{d \in A} K_c \mathrm{acc}_{c,d,i+z},$

$$\mathbf{BCT} \models \bigvee_{d \in A} \mathrm{acc}_{c,d,i+z} \to K_c \bigvee_{d \in A} \mathrm{acc}_{c,d,i+z},$$

which give us:

$$\mathbf{BCT} \models \mathrm{ech}_{c,i+z} \to K_c \mathrm{ech}_{c,i+z}. \tag{5}$$

By 4 and 5:

$$\mathbf{BCT} \models \mathrm{ech}_{c,i+z} \wedge \mathrm{acc}_{c,b,i} \to K_c \mathrm{ech}_{c,i+z} \wedge K_c \mathrm{acc}_{c,b,i}$$

and

$$\mathbf{BCT} \models \mathrm{ech}_{c,i+z} \wedge \mathrm{acc}_{c,b,i} \to K_c (\mathrm{ech}_{c,i+z} \wedge \mathrm{acc}_{c,b,i}). \tag{6}$$

By [AB11]:
$$\mathbf{BCT} \models \mathrm{ech}_{c,i+z} \wedge \mathrm{acc}_{c,b,i} \to P_{a,\geqslant(1-\mathrm{pf}^z)} e_{b,i},$$

so by 6:
$$\mathbf{BCT} \models \mathrm{ech}_{c,i+z} \wedge \mathrm{acc}_{c,b,i} \to K_c^{1-\mathrm{pf}^z} e_{b,i}.$$

Using Lemma 5 we get

$$\mathbf{BCT} \models A_c \wedge \mathrm{ech}_{a,i+z} \wedge \mathrm{acc}_{c,b,i} \to K_c^{1-\mathrm{pf}^z} e_{b,i}$$

. We have that

$$\mathbf{BCT} \models A_c \wedge e_{b,s} \to \mathrm{acc}_{c,b,s}.$$

Thus we obtain

$$\mathbf{BCT} \models A_c \wedge \mathrm{ech}_{a,i+z} \wedge e_{b,i} \to K_c^{1-\mathrm{pf}^z} e_{b,i}$$

. We have that

$$\mathbf{BCT} \models \neg A_c \to K_c \bot.$$

Hence we have

$$\mathbf{BCT} \models \mathrm{ech}_{a,i+z} \wedge e_{b,i} \to K_c^{1-\mathrm{pf}^z} e_{b,i}.$$

Since c was arbitrary, this gives us

$$\mathbf{BCT} \models \mathrm{ech}_{a,i+z} \wedge e_{b,i} \to E_{1-\mathrm{pf}^z} e_{b,i}.$$

Using Lemma 6 and [RICP] we finally conclude

$$\mathbf{BCT} \models \mathsf{ech}_{a,i+z} \wedge \mathsf{acc}_{a,b,i} \to \mathsf{C}_{1-\mathsf{pf}^z} e_{b,i}.$$

$\square$

As a corollary we get the following result.

Corollary 1 *With high probability the active agents have unique common history:* $\mathbf{BCT} \models \mathsf{ech}_{a,i+z} \to \bigwedge_{k=0}^{i} (\mathsf{acc}_{a,b,k} \to \mathsf{C}_{1-\mathsf{pf}^z} e_{b,k}).$

Proof Let $0 \le k \le i$. $\mathbf{BCT} \models \mathsf{ech}_{a,i+z}$ and [AB8] yields $\mathbf{BCT} \models \mathsf{ech}_{a,k+z}$. Theorem 1 gives $\mathbf{BCT} \models \mathsf{acc}_{a,b,k} \to \mathsf{C}_{1-\mathsf{pf}^z} e_{b,k}$, which implies the statement. $\square$

Corollary 1 corresponds to [3, Theorem 15], [8, Theorem 5.2] and [14, Claim 6.2]. This is, so called, *persistence* property [3]: "a transaction that goes more than k blocks "deep" into the blockchain of one honest player will be included in every honest player's blockchain with overwhelming probability, and it will be assigned a permanent position in the ledger." The main difference with the results given in [8] is that we can express how ledgers are evolving during the execution of the blockchain protocol, while [8] shows how a consensus between all agents can be achieved. Also, in [8] they reason about the probabilities to reach common knowledge, while here we used probabilistic common knowledge.

4   Conclusion

In this paper, we define the semantics of temporal epistemic probabilistic logic. We employ this framework to study the blockchain protocol. We prove that the blockchain protocol has the property of achieving probabilistic common knowledge among a set of agents. i.e. of reaching the consensus of the system with the high probability.

Presented description assumes that the messages are transferred between the agents much faster then the length of the period for the generation of a new PoW. We plan to develop an axiomatic system for our logic and to study proof theoretic properties of the framework. Also, another task could be to use this approach as a base ground for formal generated proof using the proof assistants like, e.g., Coq or Isabelle/HOL.

Acknowledgment. This work was supported by the Serbian Ministry of Education and Science through projects ON174026, OI174010 and III44006, the Swiss National Science Foundation 200021 165549, and by ANR-11-LABX-0040-CIMI.

References

1. Brünnler, K.: Blockchain kurz & gut. O'Reilly, Sebastopol (2018)
2. Brünnler, K., Flumini, D., Studer, T.: A logic of blockchain updates. In: Artemov, S., Nerode, A. (eds.) LFCS 2018. LNCS, vol. 10703, pp. 107–119. Springer, Cham (2018). https://doi.org/10.1007/978-3-319-72056-2_7

3. Garay, J., Kiayias, A., Leonardos, N.: The Bitcoin Backbone Protocol: Analysis and Applications. Cryptology ePrint Archive (2017). https://eprint.iacr.org/2014/765.pdf

4. Fagin, R., Halpern, J., Moses, Y., Vardi, M.Y.: Reasoning About Knowledge. The MIT Press, Cambridge (1995)

5. Fagin, R., Halpern, J.: Reasoning about knowledge and probability. J. ACM (JACM) **41**(2), 340–367 (1994)

6. Halpern, J., Fagin, R.: Modelling knowledge and action in distributed systems. Distrib. Comput. **3**, 159–177 (1989)

7. Halpern, J., Moses, Y.: Knowledge and common knowledge in a distributed environment. J. ACM **37**(3), 549–587 (1990)

8. Halpern, J., Pass, R.: A knowledge-based analysis of the blockchain protocol. In: Lang, J., (eds.) Proceedings of the Sixteenth Conference on Theoretical Aspects of Rationality and Knowledge, TARK 2017, Electronic Proceedings in Theoretical Computer Science, Liverpool, UK, vol. 251, pp. 324–335, 24–26 July 2017. https://arxiv.org/pdf/1707.08751v1.pdf

9. Hintikka, J.: Knowledge and Belief: An Introduction to the Logic of the Two Notions. Cornell University Press, Ithaca (1962)

10. Marinković, B., Glavan, P., Ognjanović, Z., Struder, T.: A temporal epistemic logic with a non-rigid set of agents for analyzing the blockchain protocol. J. Logic Comput. (2019). https://doi.org/10.1093/logcom/exz007

11. Mirto, C., Yu, J., Rahli, V., Esteves-Verissimo, P.: Probabilistic formal methods applied to blockchain's consensus protocol. In: BCRB 2018 DSN Workshop on Byzantine Consensus and Resilient Blockchains, Luxembourg (2018)

12. Monderer, D., Samet, D.: Approximating common knowledge with common beliefs. Games Econ. Behav. **1**(2), 170–190 (1989)

13. Nakamoto, S.: Bitcoin: A Peer-to-Peer Electronic Cash System (2009). https://bitcoin.org/bitcoin.pdf

14. Pass, R., Seeman, L., Shelat, A.: Analysis of the Blockchain Protocol in Asynchronous Networks. Cryptology ePrint Archive. https://eprint.iacr.org/2016/454.pdf (2016)

15. Prior, A.N.: Time and Modality. Clarendon Press, Oxford (1957)

16. Tomović, S., Ognjanović, Z., Doder, D.: Probabilistic common knowledge among infinite number of agents. In: Destercke, S., Denoeux, T. (eds.) ECSQARU 2015. LNCS (LNAI), vol. 9161, pp. 496–505. Springer, Cham (2015). https://doi.org/10.1007/978-3-319-20807-7_45

Measuring Investment Opportunities Under Uncertainty

Jorge Castro, Joaquim Gabarro[✉], and Maria Serna

CS Dept., Universitat Politècnica de Catalunya, Barcelona, Spain
{castro,gabarro,mjserna}@cs.upc.edu

Abstract. In order to make sound economic decisions it is important to measure the possibilities offered by a market in relation to investments. Provided an investment scheme $S = \langle r; R_1, \ldots, R_n \rangle$, where r is a lower bound on the desired investment return and the R_i's are the asset yields, the *power to invest* measures the capability of the scheme to fulfill requirement r. The power to invest is inspired in the Coleman's power of a collectivity to act. We exemplify this approach considering subsets of companies from stock indexes IBEX35 and DAX. We extend the power to invest to investment schemata with imprecise yields. We prove basic relations with the precise yields case and we also show that good monotonicity properties hold. Finally, we propose an analysis, through integrals of the power function, for the case of an unspecific desired return r.

Keywords: Power to act · Investment opportunities · Power to invest · Uncertainty

1 Introduction

The assessment of the economic opportunities offered by a market is an important an unavoidable topic [1,9]. For instance, a start-up will prefer to implement its innovative business idea in a dynamic market rather than in one less promising. In this paper, we consider stock markets and we attempt to provide a quantitative assessment of the intuitive sentence "plentiful of investment opportunities". This issue is a special case of "how to measure the opportunities to do something".

In the framework of cooperative simple games this question was answered precisely by J. Coleman who introduced the *power to act* [3]. Given a collectivity with n players, the power of such a collectivity to act is:

$$\text{Power to Act} = \frac{1}{2^n}\text{Number of Winning Coalitions}$$

J. Castro was partially supported by the Spanish Ministry for Economy and Competitiveness (MINECO) under grant (TIN2017- 89244-R) and the recognition 2017 SGR-856 (MACDA) from AGAUR (Generalitat de Catalunya). J. Gabarro and M. Serna were partially supported by MINECO and FEDER funds under grant GRAMM (TIN2017-86727-C2-1-R) and Generalitat de Catalunya, Agència de Gestió d'Ajuts Universitaris i de Recerca, under project 2017 SGR 786 (ALniBCOM). M. Serna was also supported by MINECO under grant BGSMath (MDM-2014-044).

© Springer Nature Switzerland AG 2019
G. Kern-Isberner and Z. Ognjanović (Eds.): ECSQARU 2019, LNAI 11726, pp. 481–491, 2019.
https://doi.org/10.1007/978-3-030-29765-7_40

We adapt this idea to provide information to investors about the degree of freedom of choice. For a portfolio with n assets, we define the *power to invest* as:

$$\text{Power to Invest} = \frac{1}{2^n - 1} \text{Number of Investments with Adequate Expected Return.}$$

Along the paper, we use the word return of an asset as a synonymous of the yield of this asset. The yield can be defined in different ways depending on the available information or on the investment strategy. The exact meaning of "asset return" in the examples below will be clear from the context. We first introduce the power to invest on portfolios whose assets have precise yield values. We extend later this definition to encompass more realistic scenarios where returns of some assets are uncertain. The power to invest provides a numerical assessment that can be used to compare the investment opportunities in different markets, even in the case that the minimum acceptable return is not precisely defined.

The paper is organized as follows. In Sect. 2 we introduce the power to invest concept for portfolios whose assets have precise returns and the power function is computed for a couple of stock indexes. Section 3 shows basic properties of the power function. A couple of stylized investment schemata are analyzed. Section 4 extends the power function to assets with imprecise returns. Theorem 1 shows good monotonicity properties for the power even in the case of imprecise returns. Section 5 considers the case where the minimum acceptable return for the investments is unspecific. We show how the integral of the power function can be used to measure investment opportunities in this case. Theorem 2 shows that, under an equiprobable investment schemata and having no information about the minimum acceptable return, the integral of the power function is the average of the asset yields. Finally in Sect. 6 we summarize the paper and suggest some possible future developments.

2   Power to Invest

We consider a set $[n] = \{1, \ldots, n\}$ of assets providing returns $A = (R_1, \ldots, R_n)$. We assume that, for each i in $[n]$, $R_i > 0$. An *investment scheme* is a pair $S = \langle r; A \rangle$, where A is the asset return tuple and $r > 0$ is the minimum acceptable return for investments. We roughly identify possible investments with non-empty subsets on $[n]$. In order to associate a return to a non-empty subset I, it is customary to choose a probability distribution on the assets in I. To exemplify our approach, we consider along the paper the uniform distribution which grants a maximal variety and diversification among investments. For an investment $I \subseteq [n]$, the *expected return* is $E(I) = \frac{1}{|I|} \sum_{i \in I} R_i$, where $|I|$ denotes the cardinality of I. An investment I is *feasible* (or acceptable) for S iff $E(I) \geq r$. The set of all feasible investments for scheme S is $F(S) = \{I \mid E(I) \geq r\}$. Therefore, $|F(S)| \leq 2^n - 1$. The following definition is taken from [2] where the power to invest is briefly analyzed in an strategic setting under the angel-daemon approach [7].

Definition 1. *Given an investment scheme $S = \langle r; A \rangle$ the power to invest is the ratio between the number of feasible investments and the total number of investments. Formally, Power$(S) =$ Power$(r; A) = |F(S)|/(2^n - 1)$.*

Name	Price	PER	Dividend Yield
ACERINOX	12.10	14.38	3.73%
DIA	3.66	11,78	4.72%
FERROVIAL	17.23	33,90	3.17%
IAG (IBERIA)	7.15	6.23	4.14%
INDITEX	25.52	20.93	3.24%
REPSOL	13.99	9.73	6.38%
SANTANDER	5.34	10.64	4.03%
SIEMENS GAMESA	12.90	1.,82	1.03%
TECNICAS REUNIDAS	26.34	22.84	5.52%
TELEFONICA	8.05	11.03	4.90%

(a) Information for companies in ECO10 index

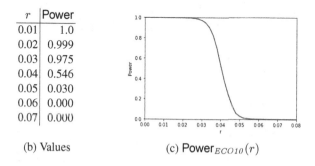

r	Power
0.01	1.0
0.02	0.999
0.03	0.975
0.04	0.546
0.05	0.030
0.06	0.000
0.07	0.000

(b) Values   (c) $\text{Power}_{ECO10}(r)$

Fig. 1. Data corresponding to the Companies in the ECO10 Spanish Index. Returns are considered to be the dividend yields.

Sometimes we fix the asset return tuple A taking the minimum acceptable return r as a variable. In such a case the function $F(r; A)$, as A is fixed, depends only on r and we write $F_A(r)$. Similarly, we write in this case $\text{Power}_A(r) = \frac{1}{2^n-1}|F_A(r)|$.

To get the power to invest for a given market, we need to choose a set of assets. A way to do that is through stock indexes. The IBEX35 [12] is the stock market index of the Bolsa de Madrid and contains the 35 major Spanish stocks. The DAX [10] is a market index consisting of the 25 major German companies trading on the Frankfurt Stock Exchange. The data of the following examples is drawn from the electronic version of the *El Economista* newspaper [6]. In both cases, we consider the *dividend yield* or *dividend-price ratio* i.e., the dividend of a share divided by the price of a share [11], as a proxy of the R_i values.

Example 1. Consider the subset of the Spanish IBEX35 companies chosen by the newspaper elEconomista.es [5] used to build the ECO10 index. Data is shown in Fig. 1a. The tuple of dividend yields is:

$$ECO10 = (0.037, 0.047, 0.031, 0.041, 0.032, 0.063, 0.040, 0.010, 0.055, 0.049).$$

Fig. 1b shows some power values and Fig. 1c provides a complete plot.   □

Name	Dividend Yield	Name	Dividend Yield
ADIDAS	0.00%	E.ON	5.21%
ALLIANZ	4.58%	FRESENIUS SE	1.20%
BASF SE	3.77%	HENKEL	0.00%
BAYER	2.96%	MERCK KGAA	0.00%
BMW	4.52%	MUNICH RE	4.87%
COMMERZBANK	0.00%	PROSIEBEN SAT.1 N	0.00%
CONTINENTAL	0.00%	RWE	6.96%
DAIMLER	5.44%	SAP	1.61%
DEUTSCHE	1.92%	SIEMENS	3.60%
DEUTSCHE BOERSE	6.04%	THYSSENKRUPP	0.00%
DEUTSCHE LUFTHANSA	0.00%	VOLKSWAGEN VORZ	0.00%
DEUTSCHE POST	3.23%	VONOVIA N	0.00%
DEUTSCHE TELEKOM	5.29%		

(a) Information for companies in DAX

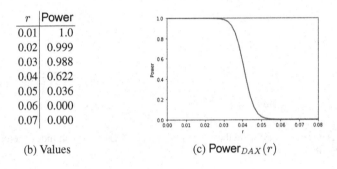

r	Power
0.01	1.0
0.02	0.999
0.03	0.988
0.04	0.622
0.05	0.036
0.06	0.000
0.07	0.000

(b) Values

(c) $\text{Power}_{DAX}(r)$

Fig. 2. Data corresponding to the companies belonging the DAX German Index. We consider as returns the dividend yields

Example 2. Data of the German DAX is drawn from [4]. Yields are given in Fig. 2a. To compute the power we only consider the non-zero yield assets. So, here the number of assets reduces to $n = 15$.

$$DAX = (0.0458, 0.0377, 0.0296, 0.0452, 0.0544, 0.0192, 0.0604,$$
$$0.0323, 0.0529, 0.0521, 0.0120, 0.0487, 0.0696, 0.0161, 0.0360).$$

Figure 2b gives some power values and Fig. 2c shows the function plot.   □

Given a market, represented by A, the power to invest provides a measure of the investment opportunities. We propose to use $\text{Power}_A(r)$ to compare markets. We prefer the market with the highest power to invest. The preference may depend on the specific value of r, as we illustrate in the following example.

Example 3. The power to invest, for ECO10 and DAX, seems quite similar (see Fig. 3a). However, looking in more detail, we get additional insight. For $r \in (0, 0.0426]$, the DAX is better. For $r \geq 0.0427$, $ECO10$ is the best option (see Fig. 3b).   □

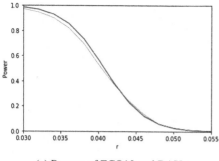

r	Power$_{ECO10}$		Power$_{DAX}$
0.042	0.3831	<	0.3914
0.0425	0.3431	<	0.3481
0.0426	0.3401	<	0.3402
0.0427	0.3294	>	0.3315
0.0428	0.3245	>	0.3239
0.043	0.3167	>	0.3078
0.05	0.0303	>	0.0263

(a) Powers of ECO10 and DAX   (b) Numerical comparison on powers

Fig. 3. Powers corresponding to ECO10 and DAX. Initially DAX outperforms ECO10 in relation to the power. At some point in $[0.0426, 0.0427]$ the relation is reversed.

3   Stylized Properties

We start showing several basic properties of the power function including some elementary monotonicity relationships. Let $A = (R_1, \ldots, R_n)$ and $\hat{A} = (\hat{R}_1, \ldots, \hat{R}_n)$ be two return tuples. As usual we say that $A \leq \hat{A}$ iff for each i in $[n]$ it is $R_i \leq \hat{R}_i$.

Lemma 1. *Let $A = (R_1, \ldots, R_n)$ and $\hat{A} = (\hat{R}_1, \ldots, \hat{R}_n)$ be two return tuples. The following holds*

- *$0 \leq Power_A(r) \leq 1$, for $r > 0$.*
- *$Power_A(r) = 1$, for $r \leq \min_{i \in [n]} R_i$ and $Power_A(r) = 0$, for $r > \max_{i \in [n]} R_i$.*
- *$Power_A(r)$ is decreasing, i.e., for $r \leq r'$, $Power_A(r) \geq Power_A(r')$.*
- *$Power(r; A)$ is increasing with respect to A i.e., if $A \leq \hat{A}$, $Power(r; A) \leq Power(r; \hat{A})$.*

Let us consider some stylized investment schemata. First, we consider a case where all the assets have the same return. We denote n identical returns R by $n : R$. So, we write the scheme $S = \langle r; R, \ldots, R \rangle$ in a shorter way as $S = \langle r; n : R \rangle$.

Lemma 2. *Let $S = \langle r; n : R \rangle$. It holds $Power(S) = 1$ when $r \leq R$ and 0 otherwise.*

For instance, Fig. 4a plots function $Power_A(r)$ for $S = \langle r; A \rangle = \langle r; 0.05 \rangle$.

Lemma 3. *Let S be a scheme with two different returns, $S = \langle r, n_1 : R_1, n_2 : R_2 \rangle$. Assume w.l.o.g that $R_1 \leq R_2$. When $r \leq R_1$ the power is 1 and when $r > R_2$ the power is 0. When $R_1 \leq r \leq R_2$, we have*

$$Power(S) = \frac{1}{2^{n_1+n_2} - 1} \sum_{(k_1,k_2) \in P_S} \binom{n_1}{k_1}\binom{n_2}{k_2}$$

where $P_S = \{(k_1, k_2) \mid (k_1 + k_2)r \leq k_1 R_1 + k_2 R_2, k_1 \leq n_1, k_2 \leq n_2, k_1 + k_2 > 0\}$.

Consider the return tuples $A_1 = (1 : 0.05, 1 : 0.15)$, $A_2 = (2 : 0.05, 1 : 0.15)$ and $A_3 = (4 : 0.05, 7 : 0.15)$. The respective power plots are given in Figs. 4b, c and d. All plots have stepwise shapes. This type of shape also holds in the general case.

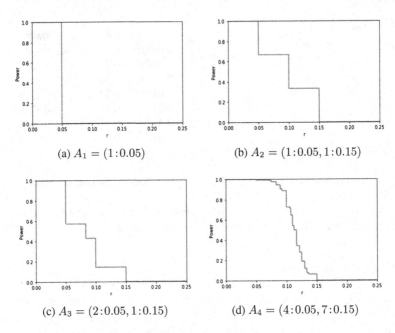

(a) $A_1 = (1\!:\!0.05)$

(b) $A_2 = (1\!:\!0.05, 1\!:\!0.15)$

(c) $A_3 = (2\!:\!0.05, 1\!:\!0.15)$

(d) $A_4 = (4\!:\!0.05, 7\!:\!0.15)$

Fig. 4. Function $\mathsf{Power}(r, n_1\!:\!0.05, n_2\!:\!0.15)$ for several values of n_1 and n_2.

Lemma 4. *Let $A = (R_1, \ldots, R_n)$ be an asset return tuple. The function $\mathsf{Power}_A(r)$ is a non-increasing stepwise function having a finite number of discontinuity points in the interval $[x, y]$ where $x = \min_{i \in [n]} R_i$ and $y = \max_{i \in [n]} R_i$. Furthermore, $\mathsf{Power}_A(r) = 1$, for $r \in [0, x]$, and $\mathsf{Power}_A(r) = 0$, for $r \in (y, +\infty)$.*

Proof. Given $A = (R_1, \ldots, R_n)$, we consider the function $f_I(r) : \mathcal{P}([n]) \times \mathbb{R}^+ \to \{0, 1\}$ defined to be 1 when $\frac{1}{|I|} \sum_{i \in I} R_i \geq r$ and 0 otherwise. Observe that, for each $I \neq \emptyset$, $f_I(r)$ is stepwise and has an unique discontinuity point. The result follows from the equality $\mathsf{Power}_A(r) = (\sum_{I \subseteq [n]} f_I(r))/(2^n - 1)$. $\qquad\square$

4   Imprecise Returns

We consider a simple model of imprecision where each asset yield is an unspecific value in a closed interval.

Definition 2. *A tuple of imprecise returns A is formed by $(R_1, \ldots, R_n)$ where each R_i is a closed interval $[R_i^-, R_i^+]$. Return of asset i is some unspecific value in R_i. An investment scheme on imprecise returns is $S = \langle r; A \rangle$ where tuple A is imprecise.*

Observe that any precise return tuple $(R_1, \ldots, R_n)$ can be understood as an imprecise one by defining $R_i = [R_i, R_i]$. On the other hand, an investment scheme on imprecise returns $S = \langle r; R_1, \ldots, R_n \rangle$ defines two natural extreme investment schemata on precise ones: $S^- = \langle r; R_1^-, \ldots, R_n^- \rangle$ and $S^+ = \langle r; R_1^+, \ldots, R_n^+ \rangle$. For imprecise return tuple $A = (R_1, \ldots, R_n)$ we also consider the two extreme precise return tuples A^- and A^+. Under equiprobable weights, for $I \subseteq [n]$, the expected returns under those extreme cases are $E^-(I) = \frac{1}{|I|} \sum_{i \in I} R_i^-$ and $E^+(I) = \frac{1}{|I|} \sum_{i \in I} R_i^+$. We also consider $F(S^+) = \{I \mid E^+(I) \geq r\}$ and $F(S^-) = \{I \mid E^-(I) \geq r\}$. The following monotonicity properties hold.

Lemma 5. *Given a tuple of imprecise returns $A = (R_1, \ldots, R_n)$ where, for $i \in [n]$, $R_i = [R_i^-, R_i^+]$, it holds $E^-(I) \leq E^+(I)$ and $F(S^-) \subseteq F(S^+)$.*

We extend the power function to the case of imprecise returns by considering the following intuition. Given scheme S on imprecise returns in the previous paragraph, investment $I \in F(S^-)$ contributes $1/(2^n - 1)$ to the power sum. This is because for any returns choice in the imprecise intervals, investment I will be feasible. On the contrary, when $I \notin F(S^+)$, whatever the yield values are the investment is not feasible and the contribution is zero. Finally, for the intermediate case where $I \in F(S^+) \setminus F(S^-)$ the contribution of I can be considered to be $(E^+(I) - r)/(E^+(I) - E^-(I))$ times $1/(2^n - 1)$. Summing up, we provide the following extension of the power definition.

Definition 3. *Let $S = \langle r; R_1, \ldots, R_n \rangle$ be an investment scheme on imprecise returns where for each $i \in [n]$ interval R_i is $[R_i^-, R_i^+]$. The power to invest is defined as*

$$\text{Power}(S) = \frac{1}{2^n - 1} \sum_{I \in F(S^+)} \frac{E^+(I) - \max\{r, E^-(I)\}}{E^+(I) - E^-(I)}$$

Here the expression $(E^+(I) - E^-(I))/(E^+(I) - E^-(I))$ is considered to be 1 when $E^+(I) = E^-(I)$ so that Power for imprecise returns extends Power for precise ones.

We first analyze some basic properties of Power. As one can expect the power of an investment scheme on imprecise returns lies between the power of its two extreme investment schemata.

Lemma 6. *Given an imprecise scheme S, it holds $\text{Power}(S^-) \leq \text{Power}(S) \leq \text{Power}(S^+)$.*

Example 4. Consider the scheme $S = \langle r; n:R \rangle$ given in Lemma 2. We assume a volatility σ on R such that $R > \sigma$. We model this volatility into imprecision by considering $R' = [R - \sigma, R + \sigma]$ and defining the imprecise return tuple $A = (n:R')$. In this case we have $A^- = (n:R - \sigma)$ and $A^+ = (n:R + \sigma)$. From Lemma 2, we have,

$$\text{Power}_{A^-}(r) = \begin{cases} 1 & \text{if } r \leq R - \sigma \\ 0 & \text{otherwise} \end{cases} \quad \text{Power}_{A^+}(r) = \begin{cases} 1 & \text{if } r \leq R + \sigma \\ 0 & \text{otherwise} \end{cases}$$

For an investment $I \subseteq [n]$, $E^-(I) = R - \sigma$ and $E^+(I) = R + \sigma$. Therefore,

$$\text{Power}_A(r) = \begin{cases} 1 & \text{if } r \leq R - \sigma, \\ \frac{R + \sigma - r}{2\sigma} & \text{if } R - \sigma < r \leq R + \sigma, \\ 0 & \text{if } R + \sigma < r. \end{cases}$$

Power_A is a "smooth average" between Power_{A^-} and Power_{A^+}. See Fig. 5a.   □

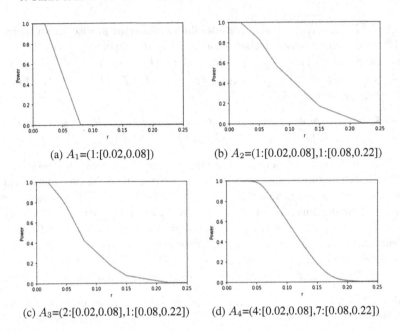

Fig. 5. Power($r, n_1 : [0.02, 0.08], n_2 : [0.08, 0.22]$) for several n_1 and n_2 values.

Example 5. Consider the investment scheme $S = \langle r; n_1 : R_1, n_2 : R_2 \rangle$ given in Lemma 3, where $R_1 < R_2$. We assume volatilities σ_1 and σ_2 on, respectively, returns R_1 and R_2 such that $R_1 > \sigma_1$ and $R_2 > \sigma_2$. We model this volatility into imprecision by considering $R_1' = [R_1 - \sigma_1, R_1 + \sigma_1]$ and $R_2' = [R_2 - \sigma_2, R_2 + \sigma_2]$, and defining the imprecise return tuple $A = (n_1 : R_1', n_2 : R_2')$. Figure 5 shows the plot for $R_1 = 0.05$, $R_2 = 0.15$, $\sigma_1 = 0.03$ and $\sigma_2 = 0.07$ (so, $R_1' = [0.02, 0.08]$ and $R_2' = [0.08, 0.22]$) and several values of n_1 and n_2.   □

We introduce now a partial order among imprecise returns

Definition 4. *We say that the imprecise return tuple* $\hat{A} = (\hat{R}_1, \ldots, \hat{R}_n)$ *improves imprecise tuple* $A = (R_1, \ldots, R_n)$, *denoted as* $A \sqsubseteq \hat{A}$, *when for each* $i \in [n]$ *it is* $R_i^- \leq \hat{R}_i^-$ *and* $R_i^+ \leq \hat{R}_i^+$. *When* $S = \langle r; A \rangle$ *and* $\hat{S} = \langle r; \hat{A} \rangle$ *and* $A \sqsubseteq \hat{A}$ *we say that* $\hat{S}$ *improves the investment scheme* S. *Formally we write* $S \sqsubseteq \hat{S}$.

Our next results shows that **Power** is monotonic with respect to this partial order.

Theorem 1. *Let* S *and* $\hat{S}$ *be two investment schemata on imprecise returns. If* $S \sqsubseteq \hat{S}$ *then* Power(S) $\leq$ Power($\hat{S}$).

Proof. Assume that $S = \langle r; A \rangle$ and $\hat{S} = \langle r; \hat{A} \rangle$. As $A \sqsubseteq \hat{A}$, we have that $F(S^+) \subseteq F(\hat{S}^+)$ and $F(S^-) \subseteq F(\hat{S}^-)$. For an investment S and a subset $I \subseteq [n]$ we consider the function

$$f_{S,I}(r) = \begin{cases} 1 & \text{if } r \leq E^-(I) \\ \frac{E^+(I) - r}{E^+(I) - E^-(I)} & \text{if } E^-(I) < r \leq E^+(I) \\ 0 & \text{if } E^+(I) < r. \end{cases}$$

In the following, with an abuse of notation, we use $\hat{E}$ to denote the expected returns under returns $\hat{A}$. As $A \sqsubseteq \hat{A}$, we have that $E^-(I) \leq \hat{E}^-(I)$ and $E^+(I) \leq \hat{E}^+(I)$. We have to study the relationship among $f_{S,I}$ and $f_{\hat{S},I}$ depending on the value of r. When $r \leq E^-(I)$, $f_{S,I}(r) = f_{\hat{S},I}(r) = 1$. When $r > \hat{E}^+(I)$, $f_{S,I}(r) = f_{\hat{S},I}(r) = 0$. Now we have to consider two cases.

Case 1: $E^+(I) < \hat{E}^-(I)$. In this case we have three relevant intervals to consider for r. When $r \in [E^-(I), E^+(I)]$, $f_{S,I}(r) \leq 1$ and $f_{\hat{S},I}(r) = 1$. When $r \in (E^+(I), \hat{E}^-(I)]$, $f_{S,I}(r) = 0$ and $f_{\hat{S},I}(r) = 1$. Finally, when $r \in (\hat{E}^-(I), \hat{E}^+(I)]$, $f_{S,I}(r) = 0$ and $0 \leq f_{\hat{S},I}(r) \leq 1$. We conclude that $f_{S,I} \leq f_{\hat{S},I}$.

Case 2: $\hat{E}^-(I) \leq E^+(I)$. Again we have to consider three intervals for r. When $r \in [E^-(I), \hat{E}^-(I)]$, $f_{S,I}(r) \leq 1$ and $f_{\hat{S},I}(r) = 1$. When $r \in (E^+(I), \hat{E}^+(I)]$, $f_{S,I}(r) = 0$ and $0 \leq f_{\hat{S},I}(r) \leq 1$. Finally, when $r \in [\hat{E}^-(I), E^+(I)]$, $f_{S,I}(r) = (E^+(I) - r)/(E^+(I) - E^-(I))$ and $f_{\hat{S},I}(r) = (\hat{E}^+(I) - r)/(\hat{E}^+(I) - \hat{E}^-(I))$. Introducing the non-negative values $a = \hat{E}^-(I) - E^-(I)$, $b = r - \hat{E}^-(I)$, $c = E^+(I) - r$ and $d = \hat{E}^+(I) - E^+(I)$, we can write $f_{S,I}(r) = c/(c+a+b)$ and $f_{\hat{S},I}(r) = (c+d)/(c+d+b)$. From this expression, after some algebraic calculation, it follows that $f_{S,I} \leq f_{\hat{S},I}$. As it holds that $\mathsf{Power}(S) = (\sum_{I \subseteq [n]} f(S,I)))/(2^n - 1)$ the claim follows. $\square$

For instance, given imprecise return tuples $A = (2 : [0.02, 0.08], 1 : [0.08, 0.22])$ and $\hat{A} = (2 : [0.02, 0.09], 1 : [0.085, 0.23])$ it holds that $A \sqsubseteq \hat{A}$ and thus, by Theorem 1 $\mathsf{Power}_{\hat{A}}(r)$ is greater or equal than $\mathsf{Power}_A(r)$ for any value of r. Independently of the minimum acceptable return, we can conclude that the investment scheme defined on $\hat{A}$ will offer more investment opportunities.

5   Power Under Uncertain Minimum Return

In this section we offer a quantitative assessment tool of opportunities for the case of an investor attempting to ensure an unspecific acceptable return in a predefined interval.

A natural way to provide a valuation on an interval of acceptable returns is to consider the integral of the power function on such interval. The larger the integral value is, more business opportunities the market will offer on average, although it may have a worse performance for some return values in the uncertainty interval.

We introduce a bit of notation. For a tuple of imprecise yields A and two real numbers $0 \leq x \leq y$ defining the interval of acceptable investment returns, let us define:

$$\mathsf{IntegralPower}_A(x, y) = \int_x^y \mathsf{Power}_A(r)\,dr.$$

We analyze the case of having no information at all about the acceptable return. We do the valuation of this case with the $\mathsf{IntegralPower}(0, +\infty)$ value, that will be denoted in short by $\mathsf{IntegralPower}(0)$.

Lemma 7. *Given a tuple of imprecise returns* $A = (R_1, \ldots, R_n)$

$$\text{IntegralPower}_A(0) = \frac{\text{IntegralPower}_{A^-}(0) + \text{IntegralPower}_{A^+}(0)}{2}.$$

Using the fact that inside of each investment we are using the uniform distribution we can derive another expression for the total integral value.

Theorem 2. *For a tuple of imprecise returns* $A = (R_1, \ldots, R_n)$,
$$\text{IntegralPower}_A(0) = \frac{1}{n}\sum_{i=1}^{n} \frac{R_i^- + R_i^+}{2}.$$

For the case of precise returns the following holds.

Corollary 1. *Given a tuple of precise returns* $A = (R_1, \ldots, R_n)$,
$$\text{IntegralPower}_A(0) = \left(\sum_{i=1}^{n} R_i\right)/n.$$

Example 6. We consider the tuple of asset returns corresponding to *ECO10* and *DAX* given in Examples 1 and 2. From Corollary 1, $\text{IntegralPower}_{ECO10}(0) = 0.0405$ and $\text{IntegralPower}_{DAX}(0) = 0.0408$. We can interpret these values as follows. For the *ECO10* case, choosing an investment uniformly at random is expected —as average— a return of value 4.05%. In contrast, we expect a bit greater return (4.08%) for a random investment on the *DAX* index.   □

6   Conclusion and Open Problems

We have developed different facets of the power to invest. We have seen that this notion has both interesting mathematical properties and potential applicability. The power to invest has the flexibility to be adapted to deal with uncertainty in two dimensions. Uncertainty can arise on the asset yields or on the minimum acceptable investment return.

We have developed the equiprobable approach, but there are other possibilities. Harry Markowitz introduced the mean-variance approach [8]. Given $R_1, \ldots R_n$ as a part of and investment scheme $S = \langle r, R_1, \ldots R_n \rangle$, a *portfolio* $w = (w_1, \ldots, w_n) \in \Delta_n$ provides a probability distribution on $[n]$, i.e., positive weights with $\sum_{i \in [w]} w_i = 1$. The expected return (the mean) is $E(S, w) = \sum_{i \in [n]} w_i R_i$. For investment $I \subseteq [n]$ the expected return equals to $E(I) = \sum_{i \in I} w_i R_i / \sum_{i \in I} w_i$. From these values a power function $\text{Power}(S, w)$ can be also defined. We are working towards analyzing properties of this function for generic portfolios.

In [2] we consider the angel-daemon approach [7] to investment schemata. The approach tries to tune cases in-between the worst and the best scenarios and analyses them through game theory. The relation between angel-daemon games and the current approach requires further analysis.

References

1. Adam, T., Goyal, V.: The investment opportunity set and its proxy variables. J. Financ. Res. **31**, 41–63 (2008). https://doi.org/10.1111/j.1475-6803.2008.00231.x
2. Castro, J., Gabarro, J., Serna, M.: Power to invest. In: Sarabia, J., Prieto, F., Guillén, M. (eds.) Contributions to Risk Analysis: RISK 2018. Cuadernos de la Fundación, vol. 223, pp. 135–142. Insurance Sciences Institute of Fundación MAPFRE (2018)
3. Coleman, J.: Control of collectivities and the power of a collectivity to act. In: Lieberman, B. (ed.) Social Choice, pp. 269–300. Gordon and Breach (1971). Reedited in Routledge Revivals, 2011
4. elEconomista.es: Dax30, rentabilidad/dividendo. http://www.eleconomista.es/indice/DAX-30/resumen/Rentabilidad-Dividendo
5. elEconomista.es: Eco10, rentabilidad/dividendo. http://www.eleconomista.es/indice/ECO10/resumen/Rentabilidad-Dividendo
6. elEconomista.es: Portada. http://www.eleconomista.es
7. Gabarro, J., Serna, M., Stewart, A.: Analysing web-orchestrations under stress using uncertainty profiles. Comput. J. **57**(11), 1591–1615 (2014). https://doi.org/10.1093/comjnl/bxt063
8. Markowitz, H.: Portafolio Selection. John Wiley, London (1959)
9. Skinner, D.: The investment opportunity set and accounting procedure choice: preliminary evidence. J. Account. Econ. **16**, 407–445 (1993) https://doi.org/10.1016/0165-4101(93)90034-D
10. WikipediA: Dax. https://en.wikipedia.org/wiki/DAX
11. WikipediA: Divident yield. https://en.wikipedia.org/wiki/Dividend_yield
12. WikipediA: Ibex35. https://en.wikipedia.org/wiki/IBEX_35

Balancing Schedules
Using Maximum Leximin

Federico Toffano and Nic Wilson[✉]

Insight Centre for Data Analytics, School of CS and IT,
University College Cork, Cork, Ireland
{federico.toffano,nic.wilson}@insight-centre.org

Abstract. We consider the problem of assigning, in a fair way, time limits for processes in manufacturing a product, subject to a deadline where the duration of each activity can be uncertain. We focus on an approach based on choosing the maximum element according to a leximin ordering, and we prove the correctness of a simple iterative procedure for generating this maximally preferred element. Our experimental testing illustrates the efficiency of our approach.

Keywords: Fair division · Preferences · Scheduling under uncertainty

1 Introduction

We consider a network representing the manufacturing processes required to make a particular product. Each of the n edges represents one of the activities (i.e., processes) involved, and the network structure implies precedence constraints between the activities, allowing activities to be in series or in parallel. We assume an overall time limit D, and we wish to assign a time limit (i.e., maximum duration) $\mathrm{dur}_j \geq 0$ to each activity j in a way that is consistent with the overall time limit, i.e., so that the makespan (the length of the longest path) is at most D, when the length of edge j is equal to dur_j.

Computing such time limits can be useful in a number of ways: given a delayed order, we can understand which are the activities most to blame by considering their *lateness* defined as $C_j - \mathrm{dur}_j$, where C_j is the completion time of activity j. With repeated data and considering the probability distribution of the duration of each activity, we can see which activities are most often to blame, which may motivate more detailed exploration of why this is the case. This analysis can also help to identify the most critical activities, so that one can assign more or less resources to a specific job. It can also give us information about how reasonable the overall deadline D is.

We aim to assign the time limits dur_j to be *slack-free* and *balanced*, given the overall time limit D (and potentially other constraints on the individual

This material is based upon works supported by the Science Foundation Ireland under Grant No. 12/RC/2289 which is co-funded under the European Regional Development Fund, and by United Technologies Corporation under the Insight-UCC Collaboration Project.

© Springer Nature Switzerland AG 2019
G. Kern-Isberner and Z. Ognjanović (Eds.): ECSQARU 2019, LNAI 11726, pp. 492–503, 2019.
https://doi.org/10.1007/978-3-030-29765-7_41

time limits). Slack-free means that it is impossible to increase the limits whilst still satisfying the constraints. If the time limits vector is not slack-free then the durations of activities could exceed these time limits whilst still being consistent with the constraints.

Being balanced is a more complex notion, but the fundamental idea is being fair to the different activities. We introduce parameterised forms of each time limit $dur_j = f_j(\alpha)$, chosen so that for any given value of α, and for any activities i and j, a time limit of $f_i(\alpha)$ for the duration of activity i is an equally reasonable requirement as a time limit $f_j(\alpha)$ for the duration of activity j. These functions f_j, which we call *commensuracy functions*, are assumed to be continuous and strictly increasing. Given these functions f_j, a collection of parameters $r(j) : j \in \{1, \dots, n\}$, generates a collection of time limits $dur_j = f_j(r(j)) : j \in \{1, \dots, n\}$, so that it is sufficient to choose the vector r of parameters. We sometimes abbreviate $r(j)$ to r_j.

Linear Case: We first consider the simple case where all the activities are in series. In this case, we must have $\sum_{j=1}^{n} dur_j \leq D$, or in terms of a parameters vector r, $\sum_{j=1}^{n} f_j(r_j) \leq D$. Then r is slack-free if and only if $\sum_{j=1}^{n} f_j(r_j) = D$. Since there is only one complete path, all the n activities are similar in the sense described above, so to also satisfy the basic balance property we need that for all $j = 1, \dots, n$, $r_j = \alpha$, for the unique value of α such that $\sum_{j=1}^{n} f_j(\alpha) = D$.

For more general networks, the situation is more complicated. It will typically not be possible to set the values of r to be all equal, without breaking the slack-free property. It would mean potentially penalising an activity j whose duration is greater than $f_j(r_j)$, even though the overall time limit D (the makespan limit) is still maintained. However, we do not want to penalise any activities unnecessarily.

Thus, the input of the problem is a graph $\mathcal{G}$, where with each edge j, representing an activity, is associated a commensuracy function f_j, and an overall time limit D. The output is a balanced and slack-free deadline dur_j for each activity j.

In this paper we focus on balancing schedules by using a standard notion of fairness, based on maximising leximin, which is a refinement of maximising the minimum value (see e.g., [13,24] for a deep investigation of fairness in many different contexts). The final output is fair in the sense that there is no way to increase the parameter r_j of an activity j (and therefore the corresponding time limit $f_j(r_j)$) without decreasing another parameter r_i which is lower or equal.

A standard form of algorithm for obtaining the max leximin element for many problems is sometimes referred to as the *water filling* algorithm[1]; the idea is to increase the levels of each component together until one of the constraints becomes tight; this gives a maximin solution; the components in tight constraints are then fixed (since reducing any such component will give a solution with worse min value, and increasing any such component will give a vector that fails to satisfy the tight constraint). The non-fixed components are increased again until

---

[1] This is different from the classic water pouring/filling algorithms for allocating power [17].

a new constraint involving one of them becomes tight, and so on. We prove the correctness of this algorithm in a rather general setting, which makes no assumption of convexity of the spaces (in contrast with the unifying framework in [18]).

Different Forms of Commensuracy Function: There are different approaches for generating the commensuracy functions f_j. One kind of method involves making use of a probability distribution over the durations of activity j (or an approximation of this based on past data). For instance, one might define $f_j(\alpha)$ to be equal to $\mu_j + \alpha\sigma_j$, where μ_j is the mean of the distribution of the jth duration, and σ_j is its standard deviation. Alternatives include $f_j(\alpha) = \mu_j + \alpha\sigma_j^2$; or the quantile function: $f_j(\alpha)$ is the value d such that the probability that the duration is less than or equal to d is equal to α.

We first, in Sect. 2, give an example to illustrate the method and our notation. In Sect. 3 we define some notation that we use throughout the paper. Section 4 describes the maximum leximin method for balancing the schedules, and defines a simple iterative method that we prove generates the unique most balanced schedule. Other balancing methods are also possible, but may lack some natural properties. Section 5, describes the experimental testing, with the related work being discussed in Sect. 6, and Sect. 7 concluding.

2   Running Example

Table 1. Parameters

Fig. 1. Activities graph

	e_1	e_2	e_3	e_4	e_5
μ_j	5	5	5	4	4
σ_j	1	1	3	2	2

Consider the graph in Fig. 1, where each edge e_j is associated with the commensuracy function $f_j(r_j) = \mu_j + r_j\sigma_j$. Assuming a global deadline $D = 20$, and the parameters of the activities shown in Table 1, we want to compute a slack-free and balanced vector r. Starting with $r_j = 0$ for each activity j, we increase the values of all the parameters r_j at the same rate until we find the first complete path π_1 with length D. In this example, given the assignment $r = (1,1,1,1,1)$, the path $\pi_1 = \{e_1, e_2, e_3\}$ has length $D(= 20)$; we can then fix the values r_j of the activities in π_1 (i.e. $r(1) = r(2) = r(3) = 1$) and keep increasing the remaining non-fixed r_j. Repeating this procedure until all the parameters are fixed, we obtain a slack-free assignment r that is balanced w.r.t. (with respect to) the commensuracy functions f_j. In the current example this requires three iterations: the first to fix the parameters of π_1, the second to fix $r(4) = 4$ ($\pi_2 = \{e_4, e_3\}$), and the third to fix $r(5) = 5$ ($\pi_3 = \{e_1, e_5\}$). The final assignment is therefore $r = (1,1,1,4,5)$ with associated time limits vector $dur = (6,6,8,12,14)$.

3   Formal Definitions

Graphical Structure: We assume a finite directed acyclic graph $\mathcal{G}$ containing a source node and a sink node and n edges, each of which we label with a different value in $\{1, \ldots, n\}$. Apart from the source (with only out-edges) and sink (with only in-edges), every node has at least one in-edge and at least one out-edge. A *complete path* is defined to be a path from source to sink; we identify this with its set of edges π. The assumptions above imply that every edge is in at least one complete path. We write the set of complete paths as $\mathcal{C}_{\mathcal{G}}$. As discussed in Sect. 1, each edge is intended to represent an activity required for making a product, with the topology encoding precedence constraints.

Commensuracy Functions f_j: For each $j \in \{1, \ldots, n\}$, we assume a strictly monotonic continuous function f_j from closed interval I to the non-negative reals. We write $I = [L_I, U_I]$, where we make some assumptions on I below.

Assignments: A *complete assignment* is a function from $\{1, \ldots, n\}$ to I. The set of all complete assignments is written as CA. Given a complete assignment r, we consider $f_j(r_j)$ as the length of edge j in the graph $\mathcal{G}$, which we also consider as the time limit (maximum duration) of the jth activity.

   An *assignment* is a function b from some subset B of $\{1, \ldots, n\}$ to I. We write $Dom(b) = B$, and write AS as the set of all assignments. A *partial assignment* is an element of $PA = AS \setminus CA$, i.e., a function from a proper subset of $\{1, \ldots, n\}$ to I. For convenience we define $\diamond$ as the empty assignment, i.e., the (trivial) function from $\emptyset$ to I.

   Consider two assignments $a : A \to I$ and $b : B \to I$, where $A \subseteq B$. We say that a is the projection of b to A if for all $j \in A$, $a(j) = b(j)$. We write $a = b^{\downarrow A}$. We also then say that b *extends* a. If, in addition, $a \neq b$ then b *strictly extends* a. For instance (see the running example in Sect. 2), complete assignment $(1, 1, 1, 4, 5)$ strictly extends partial assignment $(1, 1, 1, \_, \_)$.

Pareto Dominance: Consider two complete assignments r and s. We write $r \geqq s$ if and only if for all $j \in \{1, \ldots, n\}$, $r(j) \geq s(j)$. We say that r *Pareto-dominates* s if $r \geqq s$ and $r \neq s$, and write $r \gneqq s$. We say that r is Pareto-undominated in a set $T \subseteq CA$ if $r \in T$ and there exists no element of T that Pareto-dominates r.

Length of Path: With the graphical case of $\mathcal{C}_{\mathcal{G}}$, for assignment s and $\pi \in \mathcal{C}_{\mathcal{G}}$ such that $\pi \subseteq Dom(s)$, we define $Len_s(\pi) = \sum_{j \in \pi} f_j(s_j)$. This is the length of π, given s. We also define $Makespan(r) = \max_{\pi \in \mathcal{C}_{\mathcal{G}}} Len_r(\pi)$, which is the length of the longest complete path, given complete assignment r.

Consistent and Slack-Free Assignments

We will consider a somewhat more general setting than that purely based on constraints on the maximum lengths of paths, enabling our approach to be more generally applicable. We can add, for instance, upper bounds on time limits and on the lengths of incomplete paths, representing e.g., completion of a part of the product (implemented using a modified makespan relative to an internal node in the graph) as well as more complex constraints, which may cause the set of consistent complete assignments (and also the set of feasible duration vectors) to be non-convex. This can include upper bounds on more complex sums and averages, such as OWAs (ordered weighted averages) [7,21], and use of soft minimums to represent such constraints as *either Part-1 is completed early or Part-2 is*. (In addition, direct constraints on the complete assignments in the preference space may well lead to a non-convex space of feasible duration vectors, via non-linear commensuracy functions.)

We are especially interested in an upper bound constraint on the makespan, leading to a constraint for each complete path π that can be written as $Len_r(\pi) - D \leq 0$. The left-hand-side is a continuous function of r that is strictly increasing in each component of π. We consider more general constraints of this form.

We assume a set $\mathcal{C}$ of non-empty subsets of $\{1, \ldots, n\}$, such that every $j \in \{1, \ldots, n\}$ is in some element of $\mathcal{C}$, and associated with each $\pi \in \mathcal{C}$ is a continuous function H_π that maps assignments with domain π to real numbers, and that is strictly increasing w.r.t. each argument in π. We use H_π to express constraints, for example, a time limit on a sub-path. For complete assignment r we also write $H_\pi(r)$ as an abbreviation for $H_\pi(r^{\downarrow \pi})$.

We also have a value L_j associated with each $j \in \{1, \ldots, n\}$ (the *lower bound* for component j), and we assume, without loss of generality, that $L_j \geq L_I$.

Let $r \in \text{CA}$ be a complete assignment and let $\pi \in \mathcal{C}$. We say that r *satisfies the lower bound constraints* if for all $j \in \{1, \ldots, n\}$, $r(j) \geq L_j$. We say that r *satisfies [the constraint for]* π if $H_\pi(r) \leq 0$.

We also say that π is *tight w.r.t. r* if $H_\pi(r) = 0$. We define $UT(r)$ to be the union of all $\pi \in \mathcal{C}$ such that π is tight w.r.t. r.

In the running example, for each complete path π, $H_\pi(r) = Len_r(\pi) - D \leq 0$ is the constraint representing the upper bound limit D for the sum of the durations of the activities in the path π under the assignment r. $H_{\pi_3}(1, 1, 1, 4, 4) = H_{\pi_3}(1, \_, \_, \_, 4) = f_1(1) + f_5(4) - 20 = -2$, with $\pi_3 = \{e_1, e_5\}$. Thus, π_3 is not tight w.r.t. $(1, 1, 1, 4, 4)$. We have $UT(1, 1, 1, 4, 4) = \{e_1, e_2, e_3, e_4\}$.

Defining Consistency, $\mathcal{R}$ and $\mathcal{S}$: We say that complete assignment r is *consistent* if r satisfies the lower bound constraints and each $\pi \in \mathcal{C}$. We write $\mathcal{S}$ for the set of consistent complete assignments. Partial assignment b is said to be *consistent* if there exists an element of $\mathcal{S}$ extending b. We say that consistent complete assignment r is *slack-free* if for all $j \in \{1, \ldots, n\}$ there exists some $\pi \in \mathcal{C}$ containing j that is tight with respect to r; in other words, if $UT(r) = \{1, \ldots, n\}$. It can be shown that r in $\mathcal{S}$ is slack-free if and only if r is Pareto-undominated in $\mathcal{S}$. We write $\mathcal{R}$ for the set of slack-free consistent complete assignments.

Assumptions on L_I, U_I and on the Lower Bounds: We assume that the empty assignment $\Diamond$ is consistent, i.e., that $\mathcal{S}$ is non-empty. Because of the monotonicity of each H_π, this is equivalent to the assumption that $r^L \in \mathcal{S}$, where, for all $j \in \{1, \ldots, n\}$, $r^L(j)$ is defined to be L_j. We also assume that every partial assignment can be extended to a complete assignment not in $\mathcal{S}$. This is equivalent to the assumption that for each $i \in \{1, \ldots, n\}$, $r_L^i \notin \mathcal{S}$, for r_L^i defined by $r_L^i(i) = U_I$ and $r_L^i(j) = L_j$ for $j \in \{1, \ldots, n\} \setminus \{i\}$.

The Case of $\mathcal{C}_\mathcal{G}$: When the set of constraints is $H_\pi(r) = Len_r(\pi) - D \leq 0$ for each complete path $\pi \in \mathcal{C}_\mathcal{G}$, then, the definition of $\mathcal{S}$ simplifies to: $r \in \mathcal{S} \iff r$ satisfies the lower bound constraints and $Makespan(r) \leq D$. This helps computationally since it enables us to deal with the exponential number of constraints in a compact way. For this case, we can show a further characterisation[2] of the set $\mathcal{R}$ of consistent slack-free complete assignments: for complete assignment r satisfying the lower bound constraints, $r \in \mathcal{R}$ if and only if every element of $\mathcal{C}_\mathcal{G}$ is tight w.r.t. r.

Proposition 1. *Suppose $\pi \in \mathcal{C}_\mathcal{G}$ is a complete path and $\pi \subseteq UT(r)$ for some complete assignment $r \in \mathcal{S}$. Then, π is tight w.r.t. r, i.e., $Len_r(\pi) = D$. For $r \in \mathcal{S}$, we have $r \in \mathcal{R}$ if and only if every element of $\mathcal{C}_\mathcal{G}$ is tight w.r.t. r.*

4   Leximin Maximising: Iterative Method

We aim to find a most balanced consistent slack-free complete assignment r. For the process graph $\mathcal{G}$, we then can define the time limit of activity j to be $f_j(r_j)$. The basic idea is to maximise the minimum value (over all the n co-ordinates of r). However, there are many such vectors; it is thus natural to iterate this process, which leads to leximin maximising.

4.1   Most Balanced Schedule

Given complete assignment r, define $r^\uparrow$ to be the vector in $\mathbb{R}^n$ formed by permuting the co-ordinates of r in such a way that $r^\uparrow(1) \leq r^\uparrow(2) \leq \cdots \leq r^\uparrow(n)$.

We define the leximin order relation $\leq_{lexm}$ by $s \leq_{lexm} r$ if and only if either $s^\uparrow = r^\uparrow$, or there exists some $i \in \{1, \ldots, n\}$ such that $s^\uparrow(i) < r^\uparrow(i)$, and for all $j < i$, $s^\uparrow(j) = r^\uparrow(j)$. It follows easily that $\leq_{lexm}$ is a total pre-order (a transitive and complete relation); also, if r is a complete assignment, and r' is generated from r by permuting the n co-ordinates in some way, then r and r' are equivalent in the order. We always have $r \geq s \Rightarrow r \geq_{lexm} s$. The maximal leximin r in $\mathcal{S}$ are all those vectors $r \in \mathcal{S}$ such that for all $s \in \mathcal{S}$, $r \geq_{lexm} s$.

The main result of this section, Theorem 1, implies that there exists a unique leximin-maximal element in $\mathcal{R}$ (and also in $\mathcal{S}$), and this can be obtained through a sequence of maximisations over one-dimensional sets. This allows efficient implementation, as discussed in Sect. 5.

4.2   The Basic Iteration Operation

We will define an operation that takes a consistent partial assignment b and generates an assignment b_* that strictly extends b. Iterating this operation will

---

[2] For space reasons, almost all the proofs have been omitted; they can be found in the longer version [20], which also contains auxiliary results and many more details about the implementation and experimental testing.

lead to a complete assignment in $\mathcal{S}$, which we prove later (in Theorem 1) to be the unique maximum leximin complete assignment extending b.

Notation b^α, $\widetilde{b}$, Z_b, $\gamma(b)$, Fix(b): We will define a method for extending a consistent partial assignment b to a complete assignment $\widetilde{b}$ in $\mathcal{S}$. For partial assignment $b \in$ PA and $\alpha \in I$ we first define b^α to be the complete assignment extending b given by, for $j \in \{1, \ldots, n\} \setminus Dom(b)$, $b^\alpha(j) = \max(\alpha, L_j)$. For partial assignment $b \in$ PA we define Z_b to be the set of all $\alpha \in I$ such that $b^\alpha \in \mathcal{S}$. This is non-empty if and only if b is consistent. Our assumption about U_I implies that $U_I \notin Z_b$.

For consistent partial assignment b, we define $\gamma(b)$ to be the supremum of Z_b. We also define $\widetilde{b}$ to be $b^{\gamma(b)}$. We write $UT(\widetilde{b})$ as Fix(b); these are the variables that are in some π that is tight w.r.t. $\widetilde{b}$ (and they are fixed given b at the end of each stage of the iterative sequence described in Sect. 4.3).

The lemma below, giving some basic properties, can be proved making use of the fact that $H_\pi(b^\alpha)$ is an increasing continuous function of α.

Lemma 1. *Suppose that $b \in$ PA is a consistent partial assignment. Then, $\gamma(b) \in Z_b$ and $\widetilde{b} \in \mathcal{S}$. Also, there exists $\pi \in \mathcal{C}$ such that $\pi \not\subseteq Dom(b)$ and π is tight w.r.t. $\widetilde{b}$. Thus, Fix(b) $\not\subseteq Dom(b)$.*

Definition of b_*: For consistent partial assignment $b \in$ PA, we define b_* to be the projection of $\widetilde{b}$ to $Dom(b) \cup$ Fix(b). Thus, $Dom(b_*) = Dom(b) \cup$ Fix(b).

In the running example with $I = [0, 10]$ and $b_2 = (1, 1, 1, \_, \_)$, we have $Dom(b_2) = \{e_1, e_2, e_3\}$, $b_2^\alpha = (1, 1, 1, \alpha, \alpha)$, $\gamma(b_2) = 4$, $\widetilde{b}_2 = (1, 1, 1, 4, 4)$, Fix($b_2$) $= \{e_1, e_2, e_3, e_4\}$, and $(b_2)_* = (1, 1, 1, 4, \_)$.

Proposition 2. *Assume that partial assignment b is consistent. Then b_* is consistent and strictly extends b. If $Dom(b_*) \neq \{1, \ldots, n\}$ then $\widetilde{b}_* \gneqq \widetilde{b}$ and Fix(b_*) $\gneqq$ Fix(b).*

The following proposition gives key properties related to leximin dominance. Regarding (i), the point is that for all $j \in$ Fix(b) there exists a $\pi \in \mathcal{C}$ that is tight w.r.t. $\widetilde{b}$, and also with respect to any extension of b_*. Strict monotonicity of H_π implies that we cannot increase the value of any such j from its value in b_* without violating the constraint π. Thus, r is equal to $\widetilde{b}$ on Fix(b), and so, r extends b_* (since it extends b).

(ii) includes an interesting (very partial) converse of the property $r \geqq s \Rightarrow r \geq_{lexm} s$. The rough idea is that if $r \in \mathcal{S}$ and r extends b and $r \not\geqq b^\alpha$ then there exists $j \in \{1, \ldots, n\} \setminus Dom(b)$ such that $r(j) < \alpha$, which leads to $b^\alpha >_{lexm} r$. Then, using $\alpha = \gamma(b)$ and so, $b^\alpha = \widetilde{b}$, we can chain (ii) and (i) to obtain (iii).

Proposition 3. *Let r be an element of $\mathcal{S}$ that extends assignment b.*

(i) If for all $j \in Dom(b_*)$, $r(j) \geq b_*(j)$ then r extends b_*.
(ii) For any α such that $b^\alpha \in \mathcal{S}$, $r \geq_{lexm} b^\alpha \iff r \geqq b^\alpha$.
(iii) If $r \geq_{lexm} \widetilde{b}$ then r extends b_*.

4.3   The Iterative Sequence Generated from b

Given consistent partial assignment b, we define a sequence of assignments, $b_1, b_2, \ldots, b_m$, in an iterative fashion, as follows:

Define $b_1 = b$. Assume now that b_i has been defined, for some $i \geq 1$. If b_i is consistent and $Dom(b_i) \neq \{1, \ldots, n\}$, we let b_{i+1} equal $(b_i)_*$; otherwise, we end the sequence with i, and let $m = i$. We call $b_1, b_2, \ldots, b_m$ the *iterative sequence of assignments generated from b_1*, and we say that b_m is its *result*.

We are mainly interested in the case in which the initial partial assignment b_1 is the empty assignment $\Diamond$. However, allowing other b_1 enables a simple representation of a situation in which certain of the components are fixed in advance, i.e., some of the durations are fixed. Table 2 shows the iterative sequence of assignments generated from $b_1 = \Diamond$ in the running example, where $b_m = b_4 = (b_3)_*$.

Table 2. Progress of the algorithm

i	$\gamma(b_i)$	$\widetilde{b}_i$	$(b_i)_*$
1	1	$(1,1,1,1,1)$	$(1,1,1,\_,\_)$
2	4	$(1,1,1,4,4)$	$(1,1,1,4,\_)$
3	5	$(1,1,1,4,5)$	$(1,1,1,4,5)$

Proposition 2 implies that each b_i in the sequence is consistent, and strictly extends the previous element. This implies that the sequence terminates with some complete assignment $t = b_m$, with each earlier b_i being a partial assignment. The other parts of Proposition 2 can be used to show that $t \geq \widetilde{b}_i$ and $UT(t) \cup Dom(b_1) = \{1, \ldots, n\}$.

Proposition 4. *Consider the iterative sequence of assignments $b_1, b_2, \ldots, b_m$, generated by consistent assignment b_1, and let $t = b_m$ be its result. Then, t is a complete assignment in S that extends each b_i, and for all $i = 1, \ldots, m-1$, $t \geq \widetilde{b}_i$, and b_i is a consistent partial assignment. Also, $UT(t) \cup Dom(b_1) = \{1, \ldots, n\}$.*

Propositions 3 and 4 lead to the following theorem, which shows that there is a uniquely maximally leximin element in S (and in $\mathcal{R}$), and the iterative sequence can be used as the basis of an algorithmic procedure for finding it.

Theorem 1. *The result of the iterative sequence of assignments generated from b is the unique maximal leximin element in S_b, where S_b is the set of elements of S that extend b. If b is the empty assignment then the result is in $\mathcal{R}$ and is thus the unique maximal leximin element in S and the unique maximal leximin element in $\mathcal{R}$.*

Proof: Let $b_1, b_2, \ldots, b_m$ be the iterative sequence of assignments generated by $b = b_1$ and with result $t = b_m$. By Proposition 4, t is in S_b, and for all $i = 1, \ldots, m - 1$, $t \geq \widetilde{b}_i$ and b_i is a consistent partial assignment.

Consider any element r of S_b such that $r \geq_{lexm} t$. This implies that, for all $i = 1, \ldots, m - 1$, $r \geq_{lexm} \widetilde{b}_i$, because $t \geq \widetilde{b}_i$ (and thus, $t \geq_{lexm} \widetilde{b}_i$). Since

r extends $b = b_1$, Proposition 3(iii) applied iteratively shows that r extends each b_i, for $i = 1, \ldots, m$, and, in particular, r extends t. Since, t is a complete assignment, we have $r = t$.

We have shown that for any $r \in \mathcal{S}_b$ if $r \geq_{lexm} t$ then $r = t$. Thus, if $r \neq t$ then $r \ngeq_{lexm} t$, i.e., $t >_{lexm} r$. Thus, t is the unique maximally leximin element in $\mathcal{S}_b$.

Now, consider the case when b is the empty assignment $\Diamond$. Since $\mathcal{S}_\Diamond = \mathcal{S}$, assignment t is then the unique maximal leximin element in $\mathcal{S}$. Proposition 4 implies that $UT(t) = \{1, \ldots, n\}$, and so $r \in \mathcal{R}$. Therefore, because $\mathcal{R} \subseteq \mathcal{S}$, assignment t is the unique maximal leximin element in $\mathcal{R}$.   □

A vector t is said to be *max-min fair* on $\mathcal{X} \subseteq \mathbb{R}^n$ if and only if for all $s \in \mathcal{X}$ and $j \in \{1, \ldots, n\}$ if $s(j) > t(j)$ then there exists $i \in \{1, \ldots, n\}$ such that $s(i) < t(i) \leq t(j)$. Thus, increasing some component $t(j)$ must be at the expense of decreasing some already smaller component $t(i)$. If there exists a max-min fair element, then it is unique and equals the unique maximum leximin element [18]. Theorem 1 can be used to show that there is a max-min fair element in $\mathcal{S}$, i.e., the max leximin element t. The idea behind the proof is that if max-min fairness were to fail for t, then one can show that there would exist s and j such that $s(j) > t(j)$ and for all $i \in Dom(b_k)$, $s(i) \geq t(i)$, where k is minimal in the iterative sequence such that $Dom(b_k) \ni j$. Applying Proposition 3(i) iteratively would imply that s extends b_k, contradicting $s(j) > t(j) = b_k(j)$.

Corollary 1. *For any consistent partial assignment b, the maximum leximin element of $\mathcal{S}_b$ is max-min fair on $\mathcal{S}_b$.*

Corollary 2. *For the maximum leximin element r in $\mathcal{S}$, every component is maximal w.r.t. some constraint, i.e., for each $j \in \{1, \ldots, n\}$ there exists $\pi \in \mathcal{C}$ such that j is maximal w.r.t. π. For the graph case when $\mathcal{C} = \mathcal{C}_\mathcal{G}$, for r in $\mathcal{S}$, we have r is the maximum leximin element in $\mathcal{S}$ if and only if r is slack-free and every component is maximal w.r.t. some constraint.*

5   Implementation

We have implemented a version of the water filling algorithm for the graph-based case using $\mathcal{C}_\mathcal{G}$, and with both linear and non-linear commensuracy functions f_j (see Sect. 3). Our algorithm constructs the iterative sequence generated from the empty assignment $\Diamond$ (see Sect. 4.3). To implement this, we need, for partial assignment b, to compute b_* (see Sect. 4.2), by first computing $\gamma(b)$, and setting b_* to be the projection of $b^{\gamma(b)}$ to Fix(b), where Fix(b) is determined using a simple forward and backward dynamic programming algorithm. We use an obvious binary/logarithmic search algorithm to approximate $\gamma(b)$ within a chosen number $\epsilon > 0$, using the fact that $\gamma(b)$ is the maximal real β such that $Makespan(b^\beta) \leq D$. We also implemented a variation of this iterative approach for computing $\gamma(b)$, based on iterating over paths: given an upper bound β for $\gamma(b)$, we generate the longest path π w.r.t. b^β and then update β to a better upper bound β', defined to be the unique solution of the equation $Len_{b^{\beta'}}(\pi) = D$; we iterate until $Makespan(b^\beta) = D$. In our experimental testing we were able to solve problems with hundreds of activities in few seconds [20].

6   Related Work

A branch of research, that is somewhat related to our problem of computing time limits within a network of activities, is that concerned with the minimization of the *tardiness* of a set of jobs [3,4,9,22]. The tardiness is defined as $max(C_i - d_i, 0)$, where C_i is the completion time for a job and d_i is the due time. In our scenario, dur_i is equivalent to d_i, and we want to evaluate the *tardiness* as well; but in this case d_i is given as input, and the goal is finding the best scheduling of jobs in an assembly line where a machine is able to treat only one job at a time. Flexible constraints for scheduling (under uncertainty) have also been considered in [5].

Max leximin (and max-min) fairness has been widely studied and applied. For example, there are applications for balancing social welfare in markets [8], for a price-based resource allocation scheme [12], for kidney exchange [23], and for allocating unused classrooms [11]. Optimising leximin on constraint networks is studied in [2], and for systems of fuzzy constraints in [6]. Regarding the work which, like our framework, uses continuous variables, there is a substantial literature related to bandwidth allocation problems, e.g., [1,10,14,15,18,19]; see [16] for a survey of fair optimisation for networks. The most general framework of this kind seems to be that in [18], showing, under relatively general conditions, that the water filling algorithm generates the unique max leximin, which equals the max-min fair solution. Although the applications they focus on are very different from our form of scheduling problem, their theoretical framework and results still apply if the constraints on durations are linear inequalities, since their framework assumes convexity (and compactness) of the space of durations. In contrast, our framework makes no assumption of convexity.

7   Summary and Discussion

We have explored the problem of fairly assigning time limits for processes in manufacturing a product whose duration can be uncertain, subject to a deadline. We proved that a simple iterative procedure (a version of the water filling algorithm) can be used to generate the unique most balanced solution, w.r.t. max leximin, in a very general setting, not making any assumptions of convexity. This allows a wide range of side constraints to be added to the problem, whilst maintaining the same structure of the algorithm. We go on to prove, in this very general setting, that the maximum leximin element is still max-min fair, and discuss further properties. The experimental results of our implementation indicate that it is scalable to large problems. In the future, we plan to apply the method for real industrial problems, and to develop automatic methods for suggesting remedial actions for problematic schedules.

References

1. Bonald, T., Massoulié, L., Proutiere, A., Virtamo, J.: A queueing analysis of max-min fairness, proportional fairness and balanced fairness. Queueing Syst. **53**(1–2), 65–84 (2006)
2. Bouveret, S., Lemaître, M.: Computing leximin-optimal solutions in constraint networks. Artif. Intell. **173**(2), 343–364 (2009)
3. Chen, J.F.: Minimization of maximum tardiness on unrelated parallel machines with process restrictions and setups. Int. J. Adv. Manufact. Technol. **29**(5), 557–563 (2006)
4. Du, J., Leung, J.Y.T.: Minimizing total tardiness on one machine is NP-hard. Math. Oper. Res. **15**(3), 483–495 (1990)
5. Dubois, D., Fargier, H., Fortemps, P.: Fuzzy scheduling: modelling flexible constraints vs. coping with incomplete knowledge. Eur. J. Oper. Res. **147**(2), 231–252 (2003)
6. Dubois, D., Fortemps, P.: Computing improved optimal solutions to max-min flexible constraint satisfaction problems. Eur. J. Oper. Res. **118**(1), 95–126 (1999)
7. Fodor, J., Marichal, J.L., Roubens, M.: Characterization of the ordered weighted averaging operators. IEEE Trans. Fuzzy Syst. **3**(2), 236–240 (1995)
8. Gopinathan, A., Li, Z.: Strategyproof auctions for balancing social welfare and fairness in secondary spectrum markets. In: INFOCOM, 2011 Proceedings IEEE, pp. 3020–3028. IEEE (2011)
9. Ho, J.C., Chang, Y.L.: Heuristics for minimizing mean tardiness for m parallel machines. Nav. Res. Logist. (NRL) **38**(3), 367–381 (1991)
10. Huang, X.L., Bensaou, B.: On max-min fairness and scheduling in wireless ad-hoc networks: analytical framework and implementation. In: Proceedings of the 2nd ACM International Symposium on Mobile Ad Hoc Networking & Computing, pp. 221–231. ACM (2001)
11. Kurokawa, D., Procaccia, A.D., Shah, N.: Leximin allocations in the real world. In: Roughgarden, T., Feldman, M., Schwarz, M. (eds.) Proceedings of the Sixteenth ACM Conference on Economics and Computation, EC 2015, Portland, OR, USA, June 15–19, 2015, pp. 345–362. ACM (2015)
12. Marbach, P.: Priority service and max-min fairness. In: INFOCOM 2002, Twenty-First Annual Joint Conference of the IEEE Computer and Communications Societies, Proceedings, IEEE, vol. 1, pp. 266–275. IEEE (2002)
13. Moulin, H.: Fair Division and Collective Welfare. MIT Press, London (2004)
14. Nace, D., Orlin, J.B.: Lexicographically minimum and maximum load linear programming problems. Oper. Res. **55**(1), 182–187 (2007)
15. Nace, D., Pioro, M., Doan, L.: A tutorial on max-min fairness and its applications to routing, load-balancing and network design. In: 4th IEEE International Conference on Computer Sciences Research, Innovation and Vision for the Future (RIVF 2006) (2006)
16. Ogryczak, W., Luss, H., Pióro, M., Nace, D., Tomaszewski, A.: Fairoptimization and networks: a survey. J. Appl. Math. **2014**, 25 (2014)
17. Palomar, D.P., Fonollosa, J.R.: Practical algorithms for a family of waterfilling solutions. IEEE Trans. Signal Process. **53**(2), 686–695 (2005)
18. Radunović, B., Boudec, J.Y.L.: A unified framework for max-min and min-max fairness with applications. IEEE/ACM Trans. Netw. (TON) **15**(5), 1073–1083 (2007)

19. Tassiulas, L., Sarkar, S.: Maxmin fair scheduling in wireless networks. In: INFO-COM 2002, Twenty-First Annual Joint Conference of the IEEE Computer and Communications Societies, Proceedings, IEEE, vol. 2, pp. 763–772. IEEE (2002)
20. Toffano, F., Wilson, N.: Balancing Schedules Using Maximum Leximin (Extended version including proofs). Unpublished Document (2019)
21. Yager, R.R.: On ordered weighted averaging aggregation operators in multicriteria decisionmaking. IEEE Trans. Syst. Man Cybern. **18**(1), 183–190 (1988). https://doi.org/10.1109/21.87068
22. Yalaoui, F., Chu, C.: Parallel machine scheduling to minimize total tardiness. Int. J. Prod. Econ. **76**(3), 265–279 (2002)
23. Yilmaz, Ö.: Kidney exchange: an egalitarian mechanism. J. Econ. Theory **146**(2), 592–618 (2011)
24. Young, H.P.: Equity: In Theory and Practice. Princeton University Press, Princeton (1995)

Author Index

Printed in the United States
By Bookmasters